OPTIMIZATION THEORY WITH APPLICATIONS

Donald A. Pierre
Department of Electrical Engineering
Montana State University
Bozeman, Montana

DOVER PUBLICATIONS, INC., NEW YORK

Copyright © 1969, 1986 by Donald A. Pierre.
All rights reserved under Pan American and International Copyright Conventions.

Published in Canada by General Publishing Company, Ltd., 30 Lesmill Road, Don Mills, Toronto, Ontario.
Published in the United Kingdom by Constable and Company, Ltd.

This Dover edition, first published in 1986, is an unabridged and corrected republication of the work first published by John Wiley & Sons, Inc., New York, 1969. The author has written a new Preface for this edition.

Manufactured in the United States of America
Dover Publications, Inc., 31 East 2nd Street, Mineola, N.Y. 11501

Library of Congress Cataloging-in-Publication Data

Pierre, Donald A.
  Optimization theory with applications.

  Reprint. Originally published: New York : Wiley, c1969. With new pref.
  Includes bibliographies and index.
  1. Mathematical optimization. 2. Programming (Mathematics)
I. Title.
QA402.5.P5 1986   519   86-6278
ISBN 0-486-65205-X

PREFACE TO THE DOVER EDITION

The advances in computing capabilities during the past two decades have been impressive. Many types of optimization algorithms are routinely available to the user of a computer center, and small software packages for personal computers are widespread. The understanding of the theory behind these computer routines is important for the user who wishes to gain the most from them or who wishes to embed them as part of a larger system. The strength of this book lies in the breadth of optimization concepts that are presented.

In looking over the contents of the book, my main regret is that I had not included enough on the first- and second-order Kuhn-Tucker conditions which have come to play a major role in many nonlinear programming algorithms. To compensate for this, the major features of the Kuhn-Tucker conditions are presented near the end of this preface; the presentation is addressed to the reader who is familiar with the material in Chapters 2, 5, and 6.

Enhanced computational algorithms continue to be developed for optimization problems. For linear programming (LP) problems, the simplex theory remains as the cornerstone for understanding the nature of the LP problem structure. There have been challenges to Dantzig's basic simplex algorithm as to its efficiency. One of these by L. G. Khachian in 1979 was shown to be more efficient in principle, but in practice proved to be far less efficient. In 1984 another challenge to the simplex algorithm was announced by Narendra Karmarkar. It is claimed that Karmarkar's algorithm requires on the order of log (N) iterations to solve those problems which require N iterations with the simplex algorithm; but one iteration of Karmarkar's algorithm requires more computation than is required for one iteration of the simplex algorithm. Testing is currently being done to determine the range of LP problems for which Karmarkar's algorithm may indeed be superior to variations of the simplex algorithm.

In unconstrained optimization of continuous nonlinear functions, the Davidon-Fletcher-Powell (DFP) algorithm and self-scale variations of the DFP algorithm were viewed as the best available in the early 1970's. These methods are quasi-Newton methods that generate approximations to the inverse of the Hessian matrix, and have been called inverse positive-definite secant methods. In the standard implementation of these methods, numerical roundoff and finite arithmetic can result in the inverse Hessian approximation becoming indefinite in some cases. It is not easy to detect automatically when this is occurring, and the solution algorithm may "spin its wheels" if the algorithm is not reinitialized. When reinitilization is imposed periodically, however, valuable Hessian information is lost for the majority of applications.

One solution to the above problem is embedded in the now popular direction-update method known as the Broyden-Fletcher-Goldfarb-Shanno (BFGS) update. In the BFGS update, an approximation to the Hessian is constructed recursively (rather than an approximation to the inverse of the Hessian as in the DFP method). The BFGS update has proved to be the best of a class of positive-definite secant update formulas. The fact that the Hessian is approximated, rather than the inverse of the Hessian, would appear at first glance to be counterproductive because the search-direction generation makes direct use of an inverse Hessian approximation—ordinary matrix inversion requires on the order of n-cubed multiplications whereas the DFP update requires on the order of n-squared multiplications per iteration. The answer to this is that the BFGS update can be restructured using matrix factorization (in particular, Cholesky factorization) to reduce the total effort to the order of n-squared multiplications per iteration. A disadvantage of the approach is that considerably more code is required to implement it. A key advantage, however, is that in the process of these matrix manipulations, the condition number of the Hessian approximation is obtained. The condition number of a matrix is an accurate measure of the degree of ill-conditioning of the matrix. When the condition number of the Hessian approximation in the BFGS method exceeds a threshold, one dependent on the numerical accuracy of the computer being used, the search direction can be reinitialized.

Many computer algorithms have been advanced to solve the nonlinear programming problem (NLP). One version of the NLP problem is as follows: Find $\mathbf{x}^*$ and $f(\mathbf{x}^*)$ such that

$$f(\mathbf{x}^*) = \max_{\mathbf{x}} f(\mathbf{x}) \qquad \text{(P-1)}$$

subject to the constraints

$$p_i(\mathbf{x}) = a_i, i = 1, 2, \ldots, m_1 < n \qquad \text{(P-2)}$$

and

$$q_j(\mathbf{x}) \leq b_j, j = 1, 2, \ldots, m_2 \qquad \text{(P-3)}$$

where $f(\mathbf{x})$, $p_i(\mathbf{x})$'s, and $q_j(\mathbf{x})$'s are functions of the n real variables contained in the vector $\mathbf{x}$.

PREFACE TO THE DOVER EDITION

The Kuhn-Tucker optimality conditions are concerned with first-order and second-order properties and help in the identification of $\mathbf{x}^*$ values which are constrained local maximum points for the above NLP problem. The following development is restricted to the *normal case*. A sufficient condition for the normal case to apply is that the gradient vectors of the active constraints at $\mathbf{x}^*$ be linearly independent. All equality constraints (P-2) are active at any $\mathbf{x}$; whereas the jth inequality constraint is active at $\mathbf{x}^*$, only if $q_j(\mathbf{x}^*) \geq b_j$. If the gradient vectors of the active constraints are linearly independent at $\mathbf{x}^*$, $\mathbf{x}^*$ is called a *regular point*.

Of central interest in the Kuhn-Tucker conditions is the concept of a Lagrangian function L for the NLP problem. Let L be defined by

$$L \stackrel{\Delta}{=} f + \boldsymbol{\alpha}'(\mathbf{a} - \mathbf{p}) + \boldsymbol{\beta}'(\mathbf{b} - \mathbf{q}) \qquad \text{(P-4)}$$

where $\boldsymbol{\alpha}$ is a vector of Lagrange multipliers for the equality constraints, and $\boldsymbol{\beta}$ is a vector of Lagrange multipliers for the inequality constraints. The vectors $\mathbf{a}$, $\mathbf{p}$, $\mathbf{b}$, and $\mathbf{q}$ contain the associated scalar terms from (P-2) and (P-3).

For first-order Kuhn-Tucker conditions, the functions in (P-1), (P-2), and (P-3) are assumed to have continuous first derivatives at $\mathbf{x}^*$. In the normal case, a first-order necessary condition for $f^* = f(\mathbf{x}^*)$ to be a constrained local maximum of f is given by the following statement:

There exist Lagrange multipliers $\boldsymbol{\alpha}^*$ and $\boldsymbol{\beta}^*$ such that

$$\nabla L(\mathbf{x}^*) = \mathbf{0} \qquad \text{(P-5a)}$$

$$\mathbf{p}(\mathbf{x}^*) = \mathbf{a} \qquad \text{(P-5b)}$$

$$\mathbf{q}(\mathbf{x}^*) \leq \mathbf{b} \qquad \text{(P-5c)}$$

$$(b_j - q_j(\mathbf{x}^*))\beta_j^* = 0, j = 1, 2, \ldots, m_2 \qquad \text{(P-5d)}$$

and

$$\boldsymbol{\beta}^* \geq 0 \qquad \text{(P-5e)}$$

The $\nabla L(\mathbf{x}^*)$ of (P-5a) is the gradient vector of the Lagrangian function evaluated at $\mathbf{x}^*$. Relation (P-5a) indicates that the gradient of the performance measure is a linear combination of the gradients of the active constraints; the multipliers serve as the weighting factors of the constraint gradients to form ∇f^*, where the components of $\boldsymbol{\beta}^*$ are required to be greater than or equal to zero, as in (P-5e). Relations (P-5b) and (P-5c) simply indicate that the constraints are satisfied at $\mathbf{x}^*$. And relation (P-5d) indicates that either a given inequality constraint is active or its associated multiplier is zero, or both. For the general nonlinear programming problem, subject to the assumed normal case, this first-order condition is a necessary condition for f^* to be a constrained local maximum of f. In the special case that f, $\mathbf{p}$, and $\mathbf{q}$ are linear in $\mathbf{x}$, however, the above first-order condition for the normal case is both necessary and sufficient for the global constrained maximum of f.

Second-order Kuhn-Tucker conditions for a regular point x^* to be a constrained local maximum point are considered next. Let I1 denote the set of integer j's for which the corresponding β_j^* multipliers are strictly greater than zero, and let I2 be the set of integer k's for which both $\beta_k^* = 0$ and $q_k(x^*) = b_k$. We restrict this development to NLP cases where: (1) x^* is a regular point of the constraints; (2) first-order conditions (P-5) are satisfied at x^*; and (3) all problem functions are twice continuously differentiable in an open neighborhood of x^*. Under these assumptions, a necessary condition for x^* to be a constrained local maximum point is that

$$\Delta x' \nabla^2 L^* \Delta x \leq 0 \quad \text{(P-6)}$$

for every nonzero Δx vector that satisfied $\Delta x' \nabla p_i^* = 0$, $i = 1, 2, \ldots, m_1$, $\Delta x' \nabla q_j^* = 0$ for all j in I1, and $\Delta x' \nabla q_k^* \leq 0$ for all k in I2. The $\nabla^2 L^*$ in (P-6) is the Hessian matrix of L, the matrix of second partial derivatives of L, evaluated at x^*.

If the inequality in (P-6) is satisfied in the strict inequality sense, the second-order necessary condition of the preceding paragraph is converted to a sufficient condition for x^* to be a constrained local maximum point for the NLP.

Note that because of the restrictions on the Δx's which need to satisfy (P-6), x^* can be a constrained local maximizing point for f without being an unconstrained local maximizing point for the Lagrangian L. One of the computational approaches to solving the NLP is to subtract penalty terms from L to form an augmented Lagrangian, one which has the property of being maximized with respect to x at x^* whenever x^* is a constrained maximizing point for f. This assumes that the correct multipliers are used in both cases. Computationally, the above approach can be separated into two iterative stages. In one stage the multipliers can be held fixed while an unconstrained search is conducted for the maximum of the augmented Lagrangian with respect to x; while in the other stage, the multipliers can be updated in an attempt to satisfy Kuhn-Tucker conditons. It can be shown that augmented Lagrangian approaches of the above type do not require excessively large penalty weights in the normal case; this is in contrast to the large penalty weights required for accuracy in conventional penalty-function approaches (see Chapter 6). Thus, augmented Lagrangian approaches generally have a significant computational advantage over conventional penalty-function approaches.

The above development is given as an introduction to some of the significant advances in computational optimization. Certainly other advances await the interested reader who pursues the current literature. As an aid in doing this, the present book is designed to help the reader develop an understanding of many important optimization concepts.

PREFACE TO THE DOVER EDITION

Methods of optimization are utilized in all fields of endeavor to enhance the design or the performance of systems. When the term "optimum" is used, one must always ask the question: "Optimum with respect to what, and subject to what constraints?" In many cases, problems can be formulated so that well-known algorithms can be used to obtain both optimal solutions and sensitivity information. In other cases, either because of the complexity of the problems encountered or because of the lack of adequate problem structure, heuristic approaches are used, with more formal methods being relegated to work on relatively small parts of the overall problem. One of the challenges of the coming decades will be to see to what extent optimization can be incorporated in computer systems that are designed to exhibit artificial intelligence.

DONALD A. PIERRE

Bozeman, Montana
November 1985

PREFACE TO THE FIRST EDITION

Optimization principles are of undisputed importance in modern design and system operation. Over the past few decades there has been a steady shift in applied optimization from the status of an art to that of a scientific discipline. To a large degree, this trend has been fostered by the development of high-speed computers with which large-scale problems can be solved with an exactness that previously was unapproachable. Computer availability has given rise to new optimization techniques and has enhanced previously developed ones.

In this book a balanced view of optimization is given with emphasis on basic techniques from both classical and modern work. Relations between these techniques are appropriately noted, and relative merits are examined. The material is developed in a logical sequence: Classical necessary and sufficient conditions for relative extrema are examined first; these are followed by modern programming approaches, the structures of which are often influenced by the results of classical theory; the final chapter illustrates the converse—dynamic programming gives valuable insight into the maximum principle extension of classical calculus of variations. While the specifics of many techniques are examined, others are referenced for additional reading, and more specialized applications are necessarily omitted for lack of space.

The book is designed for use in a first course in optimization for advanced undergraduates, graduate students, and practicing engineers and system designers. The book is based on notes developed and used over the past five years in a three-quarter graduate course sequence. Basic differential equation theory and matrix operations are required background; an appendix on matrix identities and operations is included, as is an appendix on the two-sided Laplace transform which is used in Chapter 4.

In keeping with the purpose of the book, those mathematical derivations which add to understanding of the theory are given, but certain other detailed proofs are by-passed, with references cited, in favor of examples that clarify the theory. Chapters 2 through 8 are accompanied by ample problems, some of which are intended to convey a deeper understanding of the theory, whereas others have significant practical merit. In Chapter 1, unifying principles of optimization are presented, notation is introduced, and the scope of the material is examined with reference to a table of problem types. In Chapter 2, the classical theory of minima and maxima is considered: Necessary and sufficient conditions for relative extrema are developed; Lagrange multipliers are used to account for constraints; and the sensitivity of solutions to parameter changes is examined. Sensitivity considerations are stressed throughout the book. In Chapter 3, necessary and sufficient conditions for relative extrema of functionals are developed from the viewpoint of the Euler-Lagrange formalism of the calculus of variations; properties of functionals, in contrast to functions, are studied; differential and integral equation constraints are incorporated in the theory; and direct methods of solution are presented. Whereas the material of Chapter 4 is of the calculus-of-variations nature, it is restricted to linear time-invariant systems for which significant results can be obtained via transform methods with a minimum of computational difficulty. In Chapter 5, emphasis is placed on applied problems which can be converted to a standard problem form for linear programming solution—many seemingly nonlinear problems can be so converted; the fundamentals of convex sets and of the simplex technique for solution are given detailed attention, but with an appreciation of the fact that standard linear programming routines are readily available at modern digital computer centers. In Chapter 6, global aspects of functions are examined; a variety of search techniques with variable merits in different applications are presented; and penalty-function approaches to satisfy constraints are detailed. The dynamic programming of Chapter 7 has as its cornerstone Bellman's principle of optimality—a multitude of problems are encompassed by this principle. Finally, the maximum principle of Chapter 8 is an extension of classical calculus of variations in which inequality constraints on control or policy variables are readily treated.

When the book is used in a three-quarter or two-semester sequence, essentially all of the material can be covered in a comprehensive manner. For a two-quarter sequence, the topics covered will depend on student background and instructor emphasis; a logical choice of material would include all of Chapters 1 and 2, the first halves of Chapters 3 and 5, and major portions of Chapters 6 and 7.

Montana State University supported the typing of preliminary notes and allowed me ample use of computer facilities for which I am grateful. Present and former graduate students who studied my notes between 1964 and 1969 gave constructive criticism and comment which resulted in a marked increase in the clarity of the manuscript. I gratefully acknowledge their assistance, and I acknowledge the thoughtful comments of Professors T. J. Higgins, S. J. Kahne, E. B. Lee, and S. Shapiro who read the next-to-final version of the manuscript. A welcomed suggestion by the publisher in February, 1969, resulted in the apt placement of the book in Wiley's series of books on Decision and Control.

Finally, I am thankful for the understanding of my wife, Mary, and family while my attendance to work left many personal items wanting.

DONALD A. PIERRE

Bozeman, Montana
March 1969

CONTENTS

1	**INTRODUCTION**	**1**
	1-1 Optimization in Perspective	1
	1-2 The Concepts of System and State	3
	1-3 Performance Measures	5
	1-4 Constraints	7
	1-5 Optimization Problems	8
	1-6 Conditions for Optimality	18
	1-7 Approaches to Solution	19
	1-8 Forms of Solutions	22
	1-9 Sensitivity and Identification	24
	1-10 Discussion	26
2	**CLASSICAL THEORY OF MINIMA AND MAXIMA**	**29**
	2-1 Introduction	29
	2-2 Basic Concepts and Notation	30
	2-3 Functions of One Variable	31
	2-4 Functions of Several Variables	35
	2-5 Equality Constraints and a Lagrange Multiplier	36
	2-6 General Case of Equality Constraints	42
	2-7 Inequality Constraints	43
	2-8 Extremization of Integrals	45
	2-9 Sensitivity Analysis	47
	2-10 Conclusion	51
3	**CLASSICAL CALCULUS OF VARIATIONS**	**62**
	3-1 Introduction	62
	3-2 Preliminary Concepts	64

CONTENTS

	a. Continuity, Extrema, and Variations	64
	b. Classes of Problems and Equivalence Relations	66
3-3	The Problem of Lagrange: Scalar Case	68
	a. Problem Statement and the First Variation	68
	b. Fundamental Lemma	70
	c. First Necessary Condition and First-Variational Curves	71
	d. A Corner Condition	72
	e. The Euler-Lagrange Equation	73
3-4	Isoperimetric Constraints	75
3-5	Variable End-Point Conditions	82
3-6	Corner Conditions	87
3-7	The Problem of Lagrange: State-Vector Case	90
3-8	Constraints	91
	a. Isoperimetric Constraints	91
	b. Constraints of the Form $g_i(\mathbf{x},\dot{\mathbf{x}},t) = 0$	92
	c. Constraints of the Form $z_i(\mathbf{x},t) = 0$	94
	d. Inequality Constraints	95
3-9	A General Control Problem	100
3-10	Sufficient Conditions and Additional Necessary Conditions	104
	a. Discussion	104
	b. The Second Variation and the Legendre Condition	105
	c. Fields of Solutions	107
	d. The Jacobi Condition	109
	e. Sufficient Condition for Weak Extrema	112
	f. Green's Theorem	114
	g. The Weierstrass Condition and Strong Extrema	115
3-11	Direct Methods	118
	a. Discussion	118
	b. Series Approximations	119
	c. Finite Differences	121
3-12	Sensitivity Considerations	123
3-13	Conclusion	124

4 WIENER-HOPF SPECTRUM FACTORIZATION AND FREQUENCY-DOMAIN OPTIMIZATION   **135**

4-1	Introduction	135
4-2	Filter, Control, and Predictor Problems	136
	a. Description of the System	136
	b. Integral-Square-Error Problems	137
	c. Mean-Square-Error Problems	140
4-3	A General Optimal Pulse-Shape Problem	141
	a. Description of the System	141

CONTENTS

	b. Maximum Peak Output	143
	c. Maximum of the Average Output	144
4-4	A General Problem and Solution	144
	a. The Problem	144
	b. Initial Steps to Solution	144
	c. Notation for Spectrum Factorization	147
	d. Solution Using Wiener-Hopf Spectrum Factorization	149
4-5	Solutions to Filter, Control, and Predictor Problems	150
4-6	Solutions to Optimal Pulse-Shape Problems	162
4-7	Non-Wiener-Hopf-Type Frequency-Domain Problems	170
	a. General Comments	170
	b. Pulse Shape for Maximum Output Energy	170
4-8	Sensitivity Considerations in Design	173
4-9	Conclusion	182

5 THE SIMPLEX TECHNIQUE AND LINEAR PROGRAMMING 193

5-1	Introduction	193
5-2	The General Problem and Its Standard Form	194
5-3	Conversion to the Standard Form	195
5-4	Analytical Basis	200
	a. Prelude	200
	b. Convexity	201
	c. Extreme Point and Verticy Properties	203
	d. Optimal P at a Vertex	204
5-5	Simplex Algorithm Theory	204
5-6	Simplex Algorithm Mechanics: The Simplex Tableau	209
5-7	Initializing and Scaling	215
	a. Avoiding Initial Degeneracy	215
	b. Generating an Initial Basic Feasible Solution	217
	c. Scaling	218
5-8	Upper-Bounding Algorithm	224
5-9	Dual Problems	227
	a. Duals in General	227
	b. Symmetric Duals	227
	c. Other Duals	229
5-10	Sensitivity Analysis	232
5-11	Analog Solutions	234
	a. Analogies	234
	b. Linear Programming on the General-Purpose Analog Computer	235
5-12	Applications	239
	a. Problems of Economics	239
	b. Control Problems	240

c. Communications Problems		244
d. Circuit Design Applications		248
e. Field Problems		251
f. Other Applications		253
5-13	Conclusion	254

6 SEARCH TECHNIQUES AND NONLINEAR PROGRAMMING — 264

6-1	Introduction	264
6-2	Geometrical Interpretation and Scaling	266
	a. Local Properties	266
	b. Regional Properties	267
	c. Scaling and Change of Variables	270
	d. Noise Considerations	271
	e. Constraint Geometry	272
6-3	One-Dimensional Search	272
	a. Newton-Raphson Search	272
	b. Cubic-Convergent Search without Second Derivatives	274
	c. Quadratic-Convergent Search without Derivatives	277
	d. Fibonacci Search	280
	e. Search by Golden Section	284
	f. One-Dimensional Search in n-Dimensional Space	286
6-4	Nonsequential Methods	287
	a. Nonsequential Random Search	288
	b. Nonsequential Factorial Search	288
6-5	Univariate and Relaxation Search	292
	a. Univariate Search	292
	b. Southwell's Relaxation Search	292
	c. Southwell-Synge Search	293
6-6	Basic Gradient Methods	296
	a. Common Features	296
	b. Continuous Steepest Ascent (Descent)	297
	c. Discrete Steepest Ascent (Descent)	299
	d. Newton Search	307
6-7	Acceleration-Step Search	309
	a. Two-Dimensional Case	309
	b. n-Dimensional Case: PARTAN	312
6-8	Conjugate-Direction Methods	314
	a. Conjugate Directions	314
	b. Method of Fletcher and Reeves	319
	c. Davidon's Method via Fletcher and Powell (The DFP Method)	320
6-9	Other Search Methods	322

CONTENTS

	a. Discussion	322
	b. Pattern Search	322
	c. Search by Directed Array	324
	d. Creeping Random Methods	329
	e. Centroid Methods	329
6-10	Combined Use of Indirect and Direct Methods	331
	a. Equation Solution by Search	331
	b. Reduction of Dimensionality	332
6-11	Constraints	333
	a. The Nonlinear Programming Problem	333
	b. Outside Penalty Functions for Inequality Constraints	334
	c. Penalty Functions for Equality Constraints	335
	d. Minimization of the Penalized Performance Measure	335
	e. Inside Penalty Functions	339
	f. Equality Constraints and Classical Lagrange Multipliers	341
	g. General Constraints and Lagrange Multipliers	342
6-12	Comparison of Techniques	345
6-13	Conclusion	350

7 A PRINCIPLE OF OPTIMALITY AND DYNAMIC PROGRAMMING   **367**

7-1	Introduction	367
7-2	Allocation Problems	369
	a. Problem Statement and Applications	369
	b. Dynamic Programming Approach to Solution	371
7-3	Efficiency Comparison	378
7-4	Redundancy to Improve Reliability	378
7-5	Minimal Chain Problems	380
	a. Chain Networks	380
	b. Forward Solution I	381
	c. Backward Solution I	383
	d. Backward Solution II	383
	e. Comparison of Forward and Backward Solutions	385
7-6	A Control Problem	386
	a. Statement of the Problem	386
	b. Backward Solutions	387
	c. Forward Solutions	390
7-7	Numerical Considerations	394
7-8	A Principle of Optimality	402
7-9	Placement of Transmission-Line Towers	404
7-10	n State Variables: Discrete Processes	408
	a. Problems and Difficulties	408

CONTENTS

- b. Series Approximations — 411
- c. Lagrange Multipliers — 412
- d. Region-Limiting Strategies and Iterated Dynamic Programming — 414
- 7-11 Approximations in Function and Policy Space — 418
  - a. A Control Problem — 418
  - b. An Approximation in Function Space — 419
  - c. An Approximation in Policy Space — 420
  - d. Nonoriented Minimal Chain Problems — 422
- 7-12 Continuous Decision Processes: Discrete Approximations with n State Variables — 423
  - a. A General Control Problem — 423
  - b. Recurrence Relations with Prespecified Time Increments — 424
  - c. A Continuous Recurrence Relation — 426
  - d. Recurrence Relations with Controlled Time Increments — 428
- 7-13 Continuous Decision Problems: Calculus of Variations and Extensions — 431
  - a. The Problem and Its Forward Recurrence Relation — 431
  - b. Hamilton-Jacobi Equations — 433
  - c. Costate Equations — 434
  - d. Hamiltonian Functions — 436
  - e. Necessary Conditions: A Maximum Principle — 436
  - f. Necessary Conditions: Classical Calculus of Variations — 439
- 7-14 Quadratic Minimum-Cost Function and Closed-Loop Control — 440
  - a. A General Case — 440
  - b. Steady-State Riccati Equations — 447
- 7-15 A Stochastic Control Problem — 450
- 7-16 Estimation of State Variables in the Presence of Noise — 452
  - a. Modal Trajectory Estimation — 452
  - b. Discrete Kalman-Bucy Filter — 457
- 7-17 Conclusion — 462

8 A MAXIMUM PRINCIPLE — 478

- 8-1 Introduction — 478
- 8-2 Preliminary Concepts — 479
- 8-3 A Canonical Problem Form and Equivalent Problems — 481
- 8-4 A Maximum Principle — 485
- 8-5 The Constancy of $\mathscr{H}^*$ — 486
- 8-6 The General Transversality Condition — 493
- 8-7 Time Optimal Control — 500
  - a. Comments — 500
  - b. A Second-Order System — 502

CONTENTS

	c. Optimal Switch-Time Evaluation	510
8-8	Search Techniques for Solution of Boundary-Value Problems	513
	a. Comments	513
	b. Utilization of $\mathscr{H}$ in a Search Solution	515
	c. A Newton-Raphson Algorithm for Linearization of Differential Equations and Solution of Two-Point Boundary-Value Problems	517
	d. Iterative Solutions with Stabilization via Riccati Equations	519
	e. A Riccati Transformation	521
8-9	Non-Normal Solutions	523
8-10	Singular Solutions	525
8-11	Equivalent Principles	536
	a. An Equivalent Minimum Principle	536
	b. Necessary Conditions for End-Point Functionals	537
8-12	Conclusion	539

APPENDICES

A. MATRIX IDENTITIES AND OPERATIONS — 555

B. TWO-SIDED LAPLACE TRANSFORM THEORY — 566

C. CORRELATION FUNCTIONS AND POWER-DENSITY SPECTRA — 572

D. INEQUALITIES AND ABSTRACT SPACES — 580

AUTHOR INDEX — 593

SUBJECT INDEX — 599

INTRODUCTION

1

1-1. OPTIMIZATION IN PERSPECTIVE

When an individual is confronted with a problem, he must progress through an alternating sequence of evaluations and decisions. Greber [1.3][1] lists six cardinal steps on which evaluations and decisions are made in the solution of engineering problems, namely,

1. Recognition of need.
2. Formulation of the problem.
3. Resolving the problem into concepts that suggest a solution.
4. Finding elements for the solution.
5. Synthesizing the solution.
6. Simplifying and optimizing the solution.

The order in which these steps are followed can differ considerably from one problem to another. Insight gained at any given step may be employed to modify conclusions of other steps: we should visualize a set of feedback paths which allow transition from any step to any preceding step in accordance with the dictates of a given problem. For example, the step of "synthesizing the solution" or the "formulation of the problem" may be modified or augmented by considerations associated with "simplifying and optimizing the solution."

Problems are generally associated with physical things; without a thorough understanding of the physical principles upon which a given problem solution depends, the application of optimization principles is of dubious value. There is no substitute for knowledge of physical principles and

---
[1] References are associated with numbers in brackets and are listed at the end of each chapter.

devices, nor is there any substitute for an inventive idea. The ideal role that optimization plays in the solution of problems is evidenced in the following statement: After constraints that must be satisfied by the problem solution are defined, either directly or indirectly, all significant forms of solution which satisfy the constraints should be conceived; and from the generally infinite number of such solutions, the one or ones which are best under some criteria of goodness should be extracted by using optimization principles. As with most ideals, the ideals of optimization are not easily achieved: the identification of *all* significant forms of solution to a given problem can be accomplished in special cases only, and limitations on time available to produce *the* solution to a given problem are always present. Thus, the good designer or manager does the best that he can, all factors considered.

Problems which involve the operation or the design of systems are generally of the type to which optimization principles can be beneficially applied. Moreover, problems of analysis can be viewed as optimization problems, albeit trivial ones; for example, if a linear circuit is given with specified voltage and current sources and specified initial conditions, the problem of finding the current distribution in the circuit as a function of time admits to a unique solution, and we could say in such cases that the unique solution is the optimal solution.

Whenever we use "best" or "optimum" to describe a system, the immediate question to be asked is, "Best with respect to what criteria and subject to what limitations?" Given a specific measure of performance and a specific set of constraints, we can designate a system as optimum (with respect to the performance measure and the constraints) if it "performs" as well as, if not better than, any other system which satisfies the constraints. The term *suboptimum* is used to describe any system which is not optimum (with respect to the given performance measure and constraints). Specific uses of the term suboptimum vary. It can be used in reference to systems which are not optimum because of parameter variations, or in reference to systems which are not optimum because they are designed to satisfy additional constraints, or in reference to any system which is to be compared to a reference optimum one.

Great advances have been made in optimization theory since 1940. For example, almost all of the material in the last five chapters of this book has been developed since that time. In the words of Athans [1.1], "At the present time, the field of optimization has reached a certain state of maturity, and it is regarded as one of the areas of most fervent research." The one factor that has influenced this rapid growth of optimization theory more than any other has been the parallel development of computer equipment with which optimization theory can be applied to broad classes of problems. In the remainder of this chapter, we examine the general nature of problems that

1-2. THE CONCEPTS OF SYSTEM AND STATE

The concept of system is fundamental to optimization theory. We can speak of systems of equations or of physical systems, both of which fall under the broad meaning of the term *system*. An important attribute of any system is that it be describable, perhaps only approximately and perhaps with probabilistic data included in the description. Systems are governed by rules of operation; systems generally receive inputs; and systems exhibit outputs which are influenced by inputs and by the rules of operation. Concisely put [1.5], "A system is a collection of entities or things (animate or inanimate) which receive certain inputs and is constrained to act concertedly upon them to produce certain outputs with the objective of maximizing some function of the inputs and outputs." Although the latter part of this definition tends to be restrictive, the flavor which it adds is in keeping with the objectives of this book.

In addition to inputs, outputs, and rules of operation, systems generally require the concept of state for complete description. To bring the concept of state into focus, suppose that the characterizing equations of a system are known and that outputs are to be determined which result from a set of deterministic inputs and/or changes in the system after a certain time τ. To accurately predict the response (outputs) of the system after time τ, knowledge of the state of the system at time τ is the additional information required. The following examples will serve to clarify the preceding statements.

Consider the output represented by $x \equiv x(t)$ which is associated with the system characterized by

$$\ddot{x} + a\dot{x} + bx = m(t) \tag{1-1}$$

where a and b are constants, $m(t)$ represents an input, and $\dot{x}$ equals dx/dt. To compute $x(t)$ for t greater than τ, knowledge of $m(t)$ for t greater than τ and knowledge of values of $x(\tau)$ and $\dot{x}(\tau)$ are sufficient. In this case, x and $\dot{x}$ are a set of *state variables* for the system.[2] That the state variables of systems can be specified in an unlimited number of ways is illustrated as follows:

Let x_1 and x_2 be defined by

$$x_1 \stackrel{\Delta}{=} a_{11}x + a_{12}\dot{x} \tag{1-2}$$

---

[2] These particular state variables are often referred to as phase variables. Corresponding to an nth-order ordinary differential equation, the dependent variable and its first $n - 1$ derivatives form the set of n *phase variables*.

and
$$x_2 \stackrel{\Delta}{=} a_{21}x + a_{22}\dot{x} \qquad (1\text{-}3)$$
where the a_{ij}'s are arbitrary constants that satisfy
$$\begin{vmatrix} a_{11} & a_{12} \\ a_{21} & a_{22} \end{vmatrix} \neq 0 \qquad (1\text{-}4)$$

The variables x_1 and x_2 could be defined as state variables for the system of Equation 1-1 because they contain the same information about the system as do x and $\dot{x}$—in fact, x and $\dot{x}$ can be found in terms of x_1 and x_2.

In the preceding example, the state of the system at an arbitrary instant $t = \tau$ is given by two numbers. For many systems, a finite set of numbers suffices for the state description at any instant τ, but not all systems are so characterized. For example, consider the differential-difference equation

$$\dot{x}(t) + x(t - 1) = m(t) \qquad (1\text{-}5)$$

and suppose that $m(t)$ is known for $t \geq \tau$. To compute $x(t)$ for $t > \tau$, knowledge of $x(t)$ is required in the closed interval defined by $\tau - 1 \leq t \leq \tau$. Thus, a continuum of values of x is the state of this system at any time τ.

There are many classes of systems, some of which are of such fundamental importance that, to certain people who work with a particular class, the word "system" means essentially *their* class of system. Some of the more important categorizations of systems are summarized as follows:

There are *zero-memory* systems wherein inputs determine outputs directly without any need for the concept of state. There is the classification of *linear* system versus *nonlinear* system, the linear system being described in terms of strictly linear equations, such as linear algebraic or linear differential equations. There is the categorization of *discrete* system versus *continuous* system, the former being generally describable in terms of difference equations, whereas the latter is typically describable in terms of either ordinary or partial differential equations. There is the concept of *feedback* in systems wherein control actions are influenced by state variables. There is the concept of *adaptation* in systems wherein the structure of one part of a system is intentionally changed as a consequence of changes in other parts of the system structure or as a consequence of differences in inputs. There is the concept of *learning* in systems wherein effects of past control actions are weighted in decisions pertaining to future control actions. There is the classification of *multivariable* systems with multiple inputs and/or interdependent outputs. There is the classification of *stochastic* systems in which system inputs and/or parameters assume random properties. There are so-called *logic* systems in which inputs and outputs are stratified at discrete levels (the most common being two-level logic)—and logic systems are subdivided into either *combinational* logic systems (of the zero-memory type) or *sequential* logic systems (in which state variables exist).

For zero-memory systems, the concept of state has no obvious meaning; but it will be observed in the programming chapters (5, 6, and 7) that certain methods of solving problems associated with zero-memory systems lead to an analogous state-type concept where, because of the sequential nature of a given solution, the *state of the solution* is important. In fact, we can view these solution *algorithms*[3] as systems in themselves.

1-3. PERFORMANCE MEASURES

To design or plan something so that it is best in some sense, is to use optimization. The sense in which the something (e.g., a system) is best, or is to be best, is a very pertinent factor. The term *performance measure* is used here to denote that which is to be maximized or minimized (to be extremized). Other terms are used in this regard, e.g., performance index, performance criterion, cost function, return function, and figure of merit.

If a performance measure and system constraints are clearly evident from the nature of a given problem, the sense in which the corresponding system is to be optimum is objective, rather than subjective. On the other hand, if the performance measure and/or system constraints are partially subjective, a matter of personal taste, then the attainment of an "optimal" design is also subjective. In many cases, subjective criteria are more prevalent than are the objective. In the words of Churchman [1.2], "Probably the most startling feature of twentieth-century culture is the fact that we have developed such elaborate ways of doing things and at the same time have developed no way of justifying any of the things we do." The preceding comment should not be interpreted in its strictest sense, for as noted by Hall [1.4], "If an objective (a goal) cannot be well defined it probably is not worth much, and the problem may as well be simplified by eliminating it." The process through which we proceed to set good goals can be quite challenging [1.7, and Chapter 13 of 1.4].

Examples of partially subjective conditions serve to clarify the preceding statements. In the first place, suppose that we wish to minimize the magnitude of a system error $e \equiv e(t)$ which is a function of time and is characterized by

$$\dot{e} = q(e,m,t), \qquad t_a \leq t \leq t_b \tag{1-6}$$

where t_a and t_b are specified, $e(t_a)$ is given, $q \equiv q(e,m,t)$ is a given real-valued function of its arguments, and $m \equiv m(t)$ is a bounded control action

---

[3] An algorithm is a detailed process, often iterative in nature, for obtaining a solution to a problem. Algorithms generally can be implemented by use of computer programs.

which is to be selected to attain minimum error over the time period defined by $t_a \leq t \leq t_b$. Because of the differential equation characterization of $e(t)$, $e(t)$ at one instant of time is dependent on $e(t)$ at other instants of time so that the sense in which the magnitude of $e(t)$ is to be minimized is not obvious. We may choose to minimize the integral of the absolute value of the error over the closed interval $[t_a, t_b]$; we may choose to minimize the integral of the squared error; or we may choose to minimize some other integral of a positive-definite function[4] of the system error. Commonly, the form of the "optimum" system depends on the particular performance measure employed. Note that any one of the just-noted performance measures is a function of the error $e(t)$; that is, these performance measures are functions of a function. The term *functional* is used to denote a function which maps a function of independent variables into scalar values.

As a second example of partially subjective conditions, consider two companies which design, manufacture, and sell similar products. Company A has a reputation for producing quality products which withstand the harshest of treatment. Company B on the other hand has a reputation for producing good but not quite so durable products with a lower price than similar products of company A. If an employee changes jobs and shifts from company B to company A, his subjective criteria for "goodness" of a product must also change if he is to be successful in working for company A.

A performance measure is not necessarily a single entity, such as cost, but may represent a weighted sum of various factors of interest: e.g., cost, reliability, safety, economy of operation and repair, accuracy of operation, size and weight, user comfort and convenience, etc. To illustrate this point, consider a case where cost C and reliability R are the factors to be included in the performance measure (other factors of interest are generally required to satisfy constraint relationships; constraint considerations are examined in the following section). It is desired that C be small but R be large. Suppose the performance measure P is

$$P = w_1 C + w_2 R \qquad (1\text{-}7)$$

where w_1 is a dimensionless, positive weighting factor; w_2 is a negative weighting factor with an appropriate monetary dimension; and P is to be minimized with respect to *allowable* functions or parameters (those which satisfy all constraints imposed on the functions or parameters) upon which C and R depend. Note that w_2 is assigned a negative value; the reason for this can be gleaned from the limiting case where w_1 is assigned the value of zero, in which case the minimum of $w_2 R$ corresponds to the maximum of R, which is desired, because w_2 is negative. Analytically,

$$\text{maximum } R = -\text{minimum }(-R) \qquad (1\text{-}8)$$

---

[4] A function $f(e)$ is positive-definite if $f(e) > 0$ for $e \neq 0$, and $f(e) = 0$ for $e = 0$.

In general, the factors that are deemed more important in a given case should be weighted more heavily in the associated performance measure. This weighting process is partially subjective if strictly objective criteria are not evident—because of this, results obtained by using optimization theory should be carefully examined from the standpoint of overall acceptability.

1-4. CONSTRAINTS

Any relationship that must be satisfied is a constraint. Constraints are classified either as equality constraints or as inequality constraints. Arguments of constraint relationships are related in some well-defined fashion to arguments of corresponding performance measures. Thus, if a particular performance measure depends on parameters and functions to be selected for the optimum, the associated constraints depend, either directly or indirectly, on at least some of the same parameters and functions. Constraints limit the set of solutions from which an optimal solution is to be found.

If certain parts of a system are fixed, the equations which characterize the interactions of these fixed parts are constraint equations. For example, a control system is usually required to control some process; the dynamics of the process to be controlled are seldom at the discretion of the control system designer and, therefore, are constraints with which he must contend. Likewise, if a system is but a part of a larger system, in which case the former may be referred to as a *subsystem*, the larger system may impose constraints on the subsystem; e.g., only certain specific power sources may be available to the subsystem, or the subsystem may be required to fit into a limited space and to weigh no more than a specified amount, or we may be required to design the subsystem using only a limited set of devices because of some *a priori* decisions made in regard to the larger system.

Of course, physical systems are designed to satisfy some need. In any particular case, conditions exist which must be satisfied if the system is to fulfill its intended purpose. If the gain of some amplifier is below a predetermined acceptable level, it must be rejected; if the percentage of step-response overshoot of a control system exceeds a predetermined value, it is not acceptable; if the reliability of a system is below a predetermined level, the system will probably not fulfill its task; if a moon-bound rocket from earth misses the moon because of faulty control or design, it has not satisfied an obvious goal; and so forth.

Constraints also arise from the operating environment of physical systems; for example, a physical system must operate satisfactorily over some specified range of temperatures and must be able to withstand some degree of vibrational stress. An important aspect of optimization is the *sensitivity* of system performance with respect to environmental changes or uncertainties in the parameters and factors that characterize the system. Thus, we may wish to include certain constraints in a design for the sole purpose of obtaining an assured degree of insensitivity in the system. Sensitivity concepts are examined further in Section 1-9.

Finally, constraints for physical systems invariably arise from the physical nature of things. Passive resistors, capacitors, and inductors cannot assume negative values; negative amounts of physical resources such as space or weight are not acceptable; a physical system must be *realizable* in the sense that it cannot respond to an input before the input is applied; infinite amounts of power and energy are not available; and so forth. If a particular parameter or function satisfies all conditions imposed upon it, it is said to be *admissible*. If a set of admissible parameters and functions which constitute a design satisfy all constraints imposed upon the corresponding system, the design is said to be a *feasible* design for the system. Obviously, the set of feasible designs for a practical system must be nonempty; and only if more than one feasible design exists will optimization be nontrivial.

All that is said of performance measures (Section 1-3) in regard to objectivity and subjectivity can be said of constraints. In fact, it is often difficult to decide whether a particular aspect of a given problem should be treated as a constraint or as part of the performance measure. We may decide, for example, to hold cost below some value (to constrain cost) and to maximize reliability; or under quite similar circumstances, we may decide to hold reliability above a certain level (to constrain reliability) and to minimize cost.

1-5. OPTIMIZATION PROBLEMS

It is easy to categorize optimization problems according to mathematical characteristics, as is done in this section, but it should be clearly understood that any problem associated with a physical system generally fits into one of several classes, depending on the assumptions and approximations that are made in mathematical characterizations of the system and its associated performance measure. We often form a given system model so that a convenient type of analysis is particularly appropriate; we should always go back to the actual system (or to a more realistic model) to check results obtained. Moreover, even if the mathematical structure of a problem is precisely defined, the solution of the problem is generally approachable by use of several different optimization techniques, each of which has relative advantages. As graphically illustrated by Mulligan [1.6], therefore, we should not tie a given problem form too rigidly to a single optimization technique. Similarly, we should not artificially limit a given approach to solution to a particular problem type—the concepts presented in this book have a broad range of applicability so that even if a given problem does not fall exactly under one of the problem types treated herein, a solution of the problem may yet be obtainable with the aid of concepts presented.

Table 1-1 contains the major categories of problems that are considered in this book. Numerous applications and particular cases are cited in the chapters that follow. Only the barest outlines of the problems are depicted in Table 1-1; for certain approaches to solution to be applicable, additional conditions in regard to continuity must be satisfied by functions in the table, and these conditions are given in appropriate sections of the book. Any one or all of the constraint forms associated with a given performance measure in Table 1-1 can be present in a given problem. Also, several problems which are treated in the chapters that follow are not easily categorized under any one of the problem types shown here. Thus, even though the list of problems in Table 1-1 is rather sweeping, it is by no means complete.

For the problems listed under I of Table 1-1, adjustable parameters (variables) are to be assigned values which yield the maximum or the minimum of a performance measure P that is a real-valued function of the variables; but the variables are restricted to satisfy prescribed constraints. The constraints are given in terms of conditions that the variables to be selected must satisfy and in terms of functions of the variables to be specified. The constraints limit the domain from which the optimal solution can be obtained.

For the problems listed under II, III, and IV of Table 1-1, a set of dependent variables (i.e., functions of independent variables) are to be specified over ranges of the independent variables so as to obtain the maximum or the minimum of a performance measure which is a real-valued functional of the functions to be specified. As previously noted, a functional is a function of functions; corresponding to specific assignments of the functions of the independent variables, the functionals of Table 1-1 assume scalar values. Constraints associated with functional-type performance measures can assume varied forms, including the following: differential equation relationships between the functions to be specified; functional-type constraints; magnitude constraints on the dependent variables; and constraints on the end values of the independent variables and on the values of the dependent variables at the end values of the independent variables.

No conceptual difficulty is encountered in the treatment of problems that are mixtures of the types described in the two preceding paragraphs. In fact, the former type generally can be viewed as a special case of the latter (see Section 3-13).

Despite the foibles linked to the assignment of names to problems and solution methods, the problems in Table 1-1 are often associated with

TABLE 1-1. MATHEMATICAL STRUCTURE OF REPRESENTATIVE OPTIMIZATION PROBLEMS

A	B	C	D
FORM OF PERFORMANCE MEASURE	TYPICAL SPECIAL CASES OF PERFORMANCE MEASURE	TYPICAL CONSTRAINT FORMS	CHAPTERS AND SECTIONS WITH DIRECTLY RELATED MATERIAL
I $P = f_0(\mathbf{x})$ where $\mathbf{x} = [x_1 \ x_2 \ \cdots \ x_n]'$ is to be selected for the optimum and $f_0(\mathbf{x})$ is a given function of $\mathbf{x}$.	I_a General	—	2, 6
		$f_i(\mathbf{x}) = c_i$ where c_i is a constant	2-5, 2-6, 6-2e, 6-11
		$f_i(\mathbf{x}) \leq c_i$ or $f_i(\mathbf{x}) \geq c_i$	2-7, 6-2e, 6-11
	I_b $\int_{t_a(\mathbf{x})}^{t_b(\mathbf{x})} f_0(\mathbf{x},t)\, dt$ where $t_a(\mathbf{x})$ and $t_b(\mathbf{x})$ are given functions of $\mathbf{x}$	Same as those above with $\mathbf{x}$ independent of t	2-8, 3-13
		$\mathbf{x}$ independent of t and $c_i = \int_{t_{ai}(\mathbf{x})}^{t_{bi}(\mathbf{x})} f_i(\mathbf{x},t)\, dt$	2-8
	I_c $\mathbf{c}'\mathbf{x} = \sum_{j=1}^{n} c_j x_j$	$0 \leq x_i$ $\sum_{j=1}^{n} a_{ij} x_j = b_i$ where a_{ij} and b_i are constants	5, 5-2
		$d_{j1} \leq x_j \leq d_{j2}$ $\sum_{j=1}^{n} a_{ij} x_j = b_i$	5-3, 5-8

TABLE 1-1 (continued)

A	B	C	D
I (continued)	I_c (continued)	$\sum_{j=1}^{k} a_{ij}x_j \leq b_i$ or $\sum_{j=1}^{k} a_{ij}x_j \geq b_i$	5-3, 5-11
	I_d $\|\mathbf{c'x}\|$	Same constraint forms as associated with I_c	5-3
	I_e $\sum_{j=1}^{n} c_j\|x_j\|$	Same constraint forms as associated with I_c	5-3, 5-12b
	I_f $\sum_{j=1}^{K} f_j(x_j)$	$\sum_{j=1}^{K} g_{ij}(x_j) \leq b_i$ where $g_{ij}(x_j)$ is a given function of x_j	6, 7, 7-2, 7-7, 7-8, 7-10
		x_j restricted to integer values	7-2, 7-10
		$d_{j1} \leq x_j \leq d_{j2}$ where d_{j1} and d_{j2} are constants	7-2, 7-7, 7-10
	I_g $\prod_{j=1}^{K} f_j(x_j)$	Same constraint forms as associated with I_f	7-4

TABLE 1-1 (*continued*)

A	B	C	D
I (*continued*)	I_h $\sum_{j=1}^{K} f_j(\mathbf{x}_j, \mathbf{m}_{j-1})$	$x_{i,j+1} = q_i(\mathbf{x}_j, \mathbf{m}_j, j)$ which is a given function of its arguments, and where $\mathbf{x}_j = [x_{1j} \ x_{2j} \ \cdots \ x_{nj}]'$ and $\mathbf{m}_j = [m_{1j} \ m_{2j} \ \cdots \ m_{rj}]'$	5-12b, 7-6, 7-7, 7-8, 7-10, 7-11, 7-12
		$c_{ij1} \leq x_{ij} \leq c_{ij2}$ $d_{ij1} \leq m_{ij} \leq d_{ij2}$	5-12b, 7-6, 7-10, 7-11
		$\mathbf{x}_0$ and $\mathbf{x}_k$ free or fixed	7-6
		$\mathbf{m}_j \in U_j$ where U_j is a specified closed and bounded set	5-12b, 7-6, 7-10, 7-11
II $J = \int_{t_a}^{t_b} f_0(\mathbf{x}, \dot{\mathbf{x}}, t) \, dt$ where $\mathbf{x} = \mathbf{x}(t) = [x_1(t) \ x_2(t) \ \cdots \ x_n(t)]'$ is to be selected for the optimum	II_a General	—	3, 3-3, 3-7, 7-13f, 8
		$\int_{t_a}^{t_b} f_i(\mathbf{x}, \dot{\mathbf{x}}, t) \, dt = K_i$ where K_i is a specified constant	3-4, 3-8a, 8-3
		$t_a, t_b, \mathbf{x}(t_a),$ and $\mathbf{x}(t_b)$ fixed	3-5

Sect. 1-5 OPTIMIZATION PROBLEMS 13

TABLE 1-1 *(continued)*

A	B	C	D
II *(continued)*	II$_a$ *(continued)*	t_a, t_b, $\mathbf{x}(t_a)$, and $\mathbf{x}(t_b)$ related by specified equations	3-5, 3-7, 8-6
		$g_i(\mathbf{x},\dot{\mathbf{x}},t) = 0$	3-8b
		$\dot{x}_i = q_i(\mathbf{x},t)$	3-8b, 7-13f
		$z_i(\mathbf{x},t) = 0$	3-8c
		$c_{i1} \leq g_i(\mathbf{x},\dot{\mathbf{x}},t) \leq c_{i2}$	3-8d
	II$_b$ $\int_{t_a}^{t_b} f_0(\mathbf{x},\mathbf{m},t)\, dt$ where $\mathbf{m} = [m_1(t)\ \ m_2(t)\ \cdots\ m_r(t)]'$ is to be selected for the optimum	$\dot{x}_i = q_i(\mathbf{x},\mathbf{m},t)$	3-9, 7-12, 7-13, 7-14, 8, 8-8
		$c_{i1} \leq m_i \leq c_{i2}$	7-12, 7-13, 8, 8-7
		$\mathbf{m}(t) \in U$ where U is a specified closed and bounded set in an r-dimensional space of real numbers	7-12, 7-13, 8, 8-4

TABLE 1-1 (*continued*)

A	B	C	D
II (*continued*)	II$_b$ (*continued*)	Same comments in regard to t_a, t_b, $\mathbf{x}(t_a)$, and $\mathbf{x}(t_b)$ as given under II$_a$	8-6
		$K_i = \int_{t_a}^{t_b} f_i(\mathbf{x},\mathbf{m},t)\, dt$	8-3
	II$_c$ $f_0(\mathbf{x},t)\vert_{t_a}^{t_b}$	Same constraint forms as associated with II$_a$ and II$_b$	3-2b, 8-3, 8-11b
	II$_d$ $\sum_{i=1}^{n} c_i x_i(t_b)$	Same constraint forms as associated with II$_a$ and II$_b$	3-2b, 8-3, 8-11b
	II$_e$ $\int_{t_a}^{t_b} f(\mathbf{x},t)\, dt$	$\dot{\mathbf{x}} = \mathbf{a}(\mathbf{x},t) + \mathbf{B}(\mathbf{x},t)\mathbf{m}$ and $c_{i1} \leq m_i(t) \leq c_{i2}$	7-13, 8
	II$_f$ $\int_{t_a}^{t_b} (\mathbf{e'We} + \mathbf{m'Vm})\, dt$	where $\mathbf{e} = \mathbf{x} - \mathbf{x}_d$, $\mathbf{x}_d \triangleq$ desired state vector as a function of time, $\mathbf{V}$ = positive-definite $r \times r$ matrix, $\dot{\mathbf{x}} = \mathbf{A}(t)\mathbf{x} + \mathbf{B}(t)\mathbf{m}$	7-14, 8-8d, 8-8e

Sect. 1-5 OPTIMIZATION PROBLEMS 15

TABLE 1-1 (*continued*)

A	B	C	D
III $\dfrac{1}{2\pi j}\int_{-j\infty}^{j\infty}[A_1(s)F(s)F(-s) + A_2(s)F(s) + A_3(s)F(-s) + A_4(s)]\,ds$ $= \int_{t_a}^{t_b}\int_{t_a}^{t_b} f(t_1)f(t_2)a_1(t_1 - t_2)\,dt_1\,dt_2 + \int_{t_a}^{t_b} f(t_1)[a_2(-t_1) + a_3(t_1)]\,dt_1 + a_4(0)$	The function $f(t)$ is to be selected for the optimum subject to the constraint that $f(t)$ equals identically zero for t outside the closed interval $[t_a,t_b]$; the $A_i(s)$'s are given two-sided Laplace transforms of corresponding $a_i(t)$'s, with a strip of convergence which includes the imaginary axis of the s plane.	—	4, 3-2a
		$t_a = 0$ $t_b \to \infty$	4-2, 4-4, 4-5
		$t_a = 0$ $t_b = T_f$	4-3, 4-4, 4-6
		$t_a \to -\infty$ $t_b = 0$	4-6
		$K = \dfrac{1}{2\pi j}\int_{-j\infty}^{j\infty} F(s)F(-s)G(s)\,ds$ or $K = \dfrac{1}{2\pi j}\int_{-j\infty}^{j\infty} F(s)G(s)\,ds$ where $G(s)$ is a given function of s	4

TABLE 1-1 (*continued*)

A	B	C	D
IV The function $f(t)$ equals identically zero for all t outside the closed interval $[t_a, t_b]$. $\int_{t_a}^{t_b}\int_{t_a}^{t_b} f(t_1)f(t_2)\alpha(t_1,t_2)\,dt_1\,dt_2$	IV$_a$ —	$K = \int_{t_a}^{t_b} [f(t_1)]^2\,dt_1$ The kernel $\alpha(t_1,t_2)$ is a given function of t_1 and t_2.	4-7
	IV$_b$ $\int_0^{T_o} [v_o(t)]^2\,dt$	$K = \int_0^{T_i} [v_i(t)]^2\,dt$ $t_a = 0$ $t_b = T_i$ $V_o(s) = V_i(s)G(s)$ where $V_o(s) = \mathscr{L}[v_o(t)]$ and $V_i(s) = \mathscr{L}[v_i(t)]$	4-7

Sect. 1-5   OPTIMIZATION PROBLEMS   17

particular names. Methods that can be applied to solve problems of the form I_a or I_b in Table 1-1 are grouped under the heading of nonlinear programming—nonlinear because nonlinear functions are involved; and programming, originally in the sense of scheduling or selecting (programming) the various variables for the optimum, but more and more in the sense of computer programming which plays a leading role in many methods of solution. Similarly, methods that are particularly suited for use in the solution of problems of the form I_c, I_d, or I_e are grouped under the heading of *linear programming*. To the contrary, however, whereas *dynamic programming* theory may be applied to obtain numerical solutions to problems of the form I_f, I_g, and I_h, other nonlinear programming methods are also valuable in the solution of such problems. This is because dynamic programming is based on a specific principle, Bellman's principle of optimality, and its scope is such that it can also be applied to obtain analytical results associated with problems of category II in Table 1-1. Yet other "programming" names are common: methods of *integer programming* are used to solve problems in which the variables to be specified are restricted to integer values; and methods of *quadratic programming* are used to solve problems that are characterized by quadratic algebraic equations, in addition (perhaps) to linear algebraic equations. As for categories II, III, and IV in Table 1-1, these are classified under the general heading of *calculus of variations*, but special names are given to special problem forms; and, as previously noted, dynamic programming theory can be used to derive classical results associated with these problems.

Practical problems frequently contain probabilistic data. The incorporation of probabilistic data in the solution of problems usually leads to problems of the form given in Table 1-1. To illustrate this point, suppose that we desire to maximize the *expected value* of a given function $f(x)$ where the argument x of $f(x)$ is known to equal $\mu + \xi$—μ is a design-center value to be selected and ξ is a random variable with associated probability density function $p(\xi)$. To maximize the expected value $P = E\{f(\mu + \xi)\}$ with respect to μ, we must maximize

$$P = E\{f(\mu + \xi)\} = \int_{-\infty}^{\infty} f(\mu + \xi) p(\xi) \, d\xi \qquad (1\text{-}9)$$

and this, of course, is a special case of I_b of Table 1-1.

More detailed examples of the incorporation of probabilistic data in problem solutions are considered in appropriate chapters: a special case of III in Table 1-1 is shown in Section 4-2 to be the problem of minimizing a mean-square error which exists in a linear system that is subjected to random disturbances and signals with known correlation functions; and a special case of I_f in Table 1-1 is shown in Section 7-2 to be the problem of allocating the tolerances of components about known design-center values so as to obtain a satisfactory system at the minimum-possible cost (this type of problem is of especial interest to designers of electronic circuits).

1-6. CONDITIONS FOR OPTIMALITY

Of all possible solutions that satisfy the constraints of a given problem, an optimal solution is one which yields the maximum (minimum) of the given performance measure. Only in the most trivial cases is it feasible to test all possible solutions that satisfy the constraints of a problem to determine by comparison a solution which yields the maximum (minimum). Thus, conditions are desired that can be used both to generate candidates for the optimum and to test for optimal properties.

Optimal solutions need not be unique; more than one solution may result in the same maximum (minimum). It is generally possible to find conditions that an optimal solution *must* satisfy, but which other, nonoptimal solutions *may* also satisfy. Such conditions are called *necessary conditions* for optimality. In Chapters 2, 3, 4, and 8 the necessary conditions derived are based primarily on differential properties associated with optimal solutions. In Chapter 5, necessary conditions for optimality are based on analytic properties of linear algebraic equations. In Chapter 7, necessary conditions are based on Bellman's principle of optimality. As an elementary example of necessary conditions that stem from differential properties, consider a function $f(x)$ that is continuous and that has a continuous first derivative (such a function is said to be of class C^1). Assuming that the maximum (minimum) of $f(x)$ does not occur at $x = \pm\infty$, the maximum (minimum) of $f(x)$ can occur only at points where the slope $df(x)/dx$ equals zero. Thus, a necessary condition for the optimum is that $df(x)/dx$ equals zero at the optimal value of x.

A *sufficient condition* for optimality is one which, if satisfied by a given solution, guarantees that the given solution is optimum; but if a given solution does not pass a sufficient-condition test, it does not necessarily mean that the given solution is nonoptimum. Only if a given optimality condition is both *necessary and sufficient* are we guaranteed that optimal solutions, and only optimal solutions, satisfy it.

For linear programming (Chapter 5), there is a straightforward necessary-and-sufficient-condition test for optimality; in many dynamic programming applications (Chapter 7), we are guaranteed to obtain the absolute maximum (minimum) by the nature of the computational process; and in the spectrum-factorization solution process (Chapter 4), optimal solutions are assured for the applications cited. For many other general problem forms, however, necessary-and-sufficient conditions for optimality are not available. Often, the best that can be done from the standpoint of a necessary-and-sufficient-condition test is to show that a given solution is a *relative maximum (minimum)*. A solution corresponds to a relative maximum (minimum) if sufficiently small and permissible variations away from the solution always result in a smaller (larger) value of the performance measure.[5]

1-7. APPROACHES TO SOLUTION

A classical approach to solution of optimization problems is the following: (1) find necessary conditions that the optimum must satisfy by using differential properties of certain optimal solutions; (2) solve the equations that constitute the necessary conditions to obtain candidates for the optimum; and (3) test the candidates for the optimum by using necessary-and-sufficient-condition tests. Optimization procedures that parallel the preceding approach are generally referred to as *indirect methods* of solution—indirect only in the sense that optimal solutions are determined primarily on the basis of differential properties of the functions or functionals involved. In contrast, *direct methods* of solution require the direct use of the performance measure and the constraint equations of a given problem, and systematic recursive methods are employed to obtain an optimal solution. It is not always possible, nor is it necessary, to clearly distinguish between direct and indirect methods; a comprehensive optimization procedure may beneficially employ both.

Chapters 2, 3, 4, and 8 are primarily devoted to the so-called indirect methods of solution. Necessary conditions based on differential properties of the optimum are derived in Chapter 2 for some problems of type I in Table 1-1; for problems of type II, Chapters 3 and 8 are pertinent; and for those of types III and IV, Chapter 4 is pertinent.

Geometrical interpretations of problems often afford insight into methods of solution. Solutions and solution techniques may be considered in terms of a multidimensional Euclidean space (E^n or E_n are sometimes used to denote an n-dimensional Euclidean space). For example, solutions of type I problems in Table 1-1 often assume the form of a sequence of trials in a multidimensional space; at any given stage of the solution, the current trial point and pertinent information concerning the current and preceding trial points represent the *state* of the solution in a multidimensional space, and this state information in conjunction with the equations that govern the solution scheme

---

[5] More rigorous definitions of relative extrema are given in Chapters 2 and 3. The phrase "a maximum" is commonly used in place of "a relative maximum" in the literature, whereas "the maximum" is commonly used in reference to "the absolute maximum."

is used to determine the next trial point—the sequential search techniques of Chapter 6 are of this type. Geometrically, most constraint relationships define allowable regions in Euclidean space. The performance measure and the constraints associated with linear programming theory (Chapter 5) are particularly well suited for geometrical interpretations; the region in which a general linear programming solution is constrained to lie is a convex *hyperpolyhedron* (a hyperpolyhedron is a polyhedron in a Euclidean space of more than three dimensions). It is shown in Chapter 5 that an optimal solution to a general linear programming problem is always associated with one (or more) of the vertices of a convex hyperpolyhedron.

Geometrical insight is also of value in the solution of type II problems in Table 1-1. We may view the column matrix $\mathbf{x}(t)$ as a state vector in an n-dimensional Euclidean space called state space, and the control matrix $\mathbf{m}(t)$ as a vector in an r-dimensional Euclidean space. Each *trajectory* in control space gives rise to a trajectory in state space. With automatic feedback control, the opposite is partially true: State information at any given instant of time influences the control applied at that instant of time or at a slightly later instant. These geometrical interpretations facilitate the development of necessary conditions for the optimum.

Constraint relationships are taken into account in optimization procedures in many ways, but one way that prevails through major classes of problems, whether functions or functionals are involved, is that of *performance weighting* which is closely associated with *Lagrange multipliers* on the one hand and with *penalty coefficients* on the other. Reasons that we might wish to weight several factors of interest in a performance measure are considered in Section 1-3. As a very limited example of the relationship of this weighting process to the use of Lagrange multipliers and penalty coefficients, consider the minimization of the performance measure P_a:

$$P_a = f_0(x_1,x_2) + hf_1(x_1,x_2) \tag{1-10}$$

where h is a positive weighting factor, and both $f_0 \equiv f_0(x_1,x_2)$ and $f_1 \equiv f_1(x_1,x_2)$ are bounded real-valued functions. It is assumed that a minimum of both f_0 and f_1 is desired, but because of their mutual dependence on x_1 and x_2, a trade-off must be effected. If h approaches zero, the minimum of P_a approaches the minimum of f_0; but if h is allowed to be arbitrarily large, the minimum of P_a/h approaches the minimum of f_1.

As a modification of the above conditions, suppose that f_1 is required to be a constant c_1. But suppose that we proceed to minimize P_a of Equation 1-10, with h not specified in advance, and find the minimum $P_a^*(h)$ of P_a in terms of h, and also find the corresponding values $x_1^*(h)$ and $x_2^*(h)$ of x_1 and x_2. If h can then be evaluated so that $f_1[x_1^*(h),x_2^*(h)]$ equals c_1, the desired result is obtained (a rigorous development of this fact is given in

Chapters 2 and 6). In this case, h is called a Lagrange multiplier, after the famous mathematician Joseph Louis Lagrange (1736–1813) who introduced this approach.

Alternatively, suppose f_1 is required to equal c_1, as before, but suppose that P_p is minimized, rather than P_a, where P_p is a penalized performance measure and is expressed by

$$P_p = f_0(x_1,x_2) + h[f_1(x_1,x_2) - c_1]^2 \qquad (1\text{-}11)$$

If h is assigned a very large value, values of x_1 and x_2 that give $f_1 \neq c_1$ generally result in an inordinately large value of P_p; that is, the performance measure is harshly penalized when the constraint is violated (much), and therefore those values of x_1 and x_2 which yield the minimum of P_p also yield (within a controllable degree of accuracy) the constrained minimum of $f_0(x_1,x_2)$.

One of the advantages of the so-called indirect methods is that closed-form solutions are obtainable for certain forms of problems. But for sufficiently complex problems, all feasible approaches to solution, including the indirect approaches, require the use of high-speed, general-purpose computers during some phase of the solution. The direct methods are specifically suited to computer approaches to solution, and, as with any numerical solution scheme, scaling of variables and the appropriate introduction of new variables (for example, a linear transformation of variables) may significantly influence the accuracy of the solution and the time required to obtain the solution.

Another approach to solution of optimization problems involves the concept of dual problems. For each problem of linear programming, for example, there exists a well-defined dual problem. The solutions of the dual and primal (original) problems are related in a definite way, and the problem that is easier to solve in a given case should be solved to obtain solutions to both problems. Although references are cited for duals in general, only dual problems of linear programming are given detailed attention in this book.

The dynamic programming approach to solution is based on Bellman's principle of optimality. This principle applies to systems for which a concept of state (Section 1-2) can be inferred and for which an ordered sequence of *stages of solution* exists; at each stage of the solution, a decision is made which affects the present and subsequent stages of solution. For such problems, the principle of optimality is embodied in the following statement: whatever the initial state and initial decision are, the decisions applied at remaining stages of solution must be optimum, with respect to the state resulting from the initial decision, if the overall decision process is to be potentially optimum.

Finally, for special classes of problems, inequality relationships (Appendix D) can be used to deduce optimal solutions. This is especially true of those problems associated with linear dynamic systems for which performance is measured in terms of an appropriate *norm* on abstract Hilbert or Banach spaces. Concise statements of the requisite properties of these spaces are given in Appendix D, and illustrative problems are solved therein, but thorough development of the underlying theory is left to reference sources.

1-8. FORMS OF SOLUTIONS

As one of the final steps in the principal solution procedures of Chapter 2, sets of equations have to be solved. The number of unknowns in the equations always equals the number of equations; but because of the generally nonlinear characteristics of the functions embodied in the equations, multiple solutions are possible—so comparisons between the solutions must be made to determine absolute maxima (minima).

The principal solution procedures of Chapters 3 and 8 involve, as one of the final steps, the solution of sets of differential equations with which two-point boundary values are generally associated. For example, values of $x_1(t)$, a function to be determined in the optimization process, may be required to equal given values at an initial point t_a and at a terminal point t_b. If the differential equations are linear and time-invariant or are of other special forms for which closed-form solutions are available, the two-point, boundary-value problems are relatively easy to solve; but, in general, the solution of a two-point, boundary-value problem is a difficult task that must be relegated to specially designed computer routines. Fundamental approaches to solution of these problems are considered in Chapter 8. Additional solution forms arise in Chapters 3 and 8. In Section 3-11, direct methods of the calculus of variations are used to approximate problems of type II in Table 1-1 by problems of type I. In special cases, called *singular cases*, the necessary conditions that constitute the maximum principle in Chapter 8 do not give sufficient information to obtain a solution, and additional procedures are required, as considered in Chapter 8.

For the control problems in Chapters 3 and 8, the control vector **m** may be found as an explicit function of time, $\mathbf{m} = \mathbf{m}(t)$, in which case $\mathbf{m}(t)$ is called a *control function*; it is often desirable, however, that **m** should be found as an explicit function of the state vector **x** and external inputs, in which case **m** is called a *control law*. A control function gives essentially open-loop control, whereas a control law results in feedback control. If a system is designed with one set of inputs and conditions in mind, but must respond with satisfactory performance to a wide variety of possible inputs and conditions, the control-law approach is desirable.

FORMS OF SOLUTIONS

The solutions in Chapter 4 are in terms of Laplace transforms. Pertinent information concerning two-sided Laplace transforms is given in Appendix B. Classes of filter problems, control problems, prediction problems, and pulse-shape problems are shown in Chapter 4 to be special cases of type III in Table 1-1. The final solution forms for these problems are in terms of Laplace transforms on which special operations, operations of spectrum factorization, are required. Other problems in Chapter 4 reduce to type IV in Table 1-1, and the final step in the solution of these problems involves the solution of an integral equation.

The theory developed in Chapter 5 forms a base for the simplex technique of linear programming. This technique and its modifications are the best available for the solution of general problems of types I_c, I_d, and I_e in Table 1-1. The theory associated with a standard form of linear programming problem is treated in depth, and methods of converting numerous other problem forms, some of which are nonlinear, to the standard form are developed. For manual computations, the simplex technique consists of the construction of tabular arrays in sequence with elementary operations of multiplication and addition employed in the generation of array upon array; the optimal solution is obtained on the basis of a finite number of the arrays. Manual computations suffice for small linear programming problems but, with the aid of general purpose computers, solutions can be obtained for linear programming problems that involve, literally, thousands of variables and constraints—a most attractive feature.

The use of computers is also requisite to the effective application of most search techniques (Chapter 6). Search techniques can possess sequential and/or nonsequential properties. With sequential types of search, the selection of future search "points" is based on previously generated data. Nonsequential search methods find limited applications; they are of use if a long time is required to evaluate the results of the selection of specific search points (e.g., the yield of a particular type of grain under different moisture, fertilizer, and lighting conditions), and they are of use as initial probes to determine likely starting points for sequential search techniques. The number of different sequential search techniques is unlimited. (You too can develop one!) Some efforts have been made to compare techniques, especially to find good, general-purpose techniques for programming on computers. Poorly designed search techniques have a tendency to converge too slowly to relative extrema or to stop prematurely in a search. With a good sequential search technique, relative extrema are obtained, but even so, the fact that the absolute maximum (minimum) has been obtained during a given search is assured only if the performance measure involved exhibits special properties. For example, a given performance measure may be known to be convex or concave; if convex (concave), any relative minimum (maximum) must be the absolute minimum (maximum). Sequential search techniques can be divided into derivative and nonderivative methods; with the latter, partial derivatives of the performance measure are not used in the search. Of the derivative methods, those that exhibit *quadratic convergence* are desirable because their use results in the attainment of the minimum (maximum) of a quadratic function in a finite number of steps, provided round-off errors are negligible. Because continuously differentiable functions generally exhibit quadratic characteristics in the immediate vicinity of relative minima (maxima), quadratic convergent search methods are very effective when employed in the vicinity of relative extrema.

As with all "programming" approaches to optimization, dynamic programming theory (Chapter 7) leads to numerical procedures that are best handled via digital computers. A problem to be solved is embedded in a class of similar problems the solutions of which are coupled and are of scaled degrees of difficulty. Computer storage limitations tend to restrict the class of problems that can be treated by the straightforward dynamic programming approach; but for special problem forms, or with the aid of certain approximation techniques, dynamic programming concepts often lead to computationally feasible methods.

1-9. SENSITIVITY AND IDENTIFICATION

Physical systems cannot be constructed or operated *exactly* according to completely deterministic specifications because of inherent limitations on the measurement of physical entities. Even if one could be so constructed or operated, its characteristics would change to some extent over a period of time because of environmental factors (e.g., temperature, pressure, aging, etc.). It is fortunate, therefore, that systems operate satisfactorily, provided that certain tolerances are held by elements that influence system operation. The specification of tolerances is an essential part of any complete solution to a practical problem. To specify tolerances realistically, we should know how characteristics of the system under consideration are influenced by changes in system elements. In fact, the tolerances of system elements may be selected so as to minimize the total cost of the system subject to the constraint that assured degrees of tolerances are held by essential system characteristics.

A characteristic of a system is said to be very *sensitive* with respect to an element of the system if the characteristic is greatly influenced by relatively small changes in the element. When a part or subsystem of a system is predetermined (i.e., not at the discretion of the designer), the tolerances on that part may be quite loose. With deference to this, the *sensitivity* of essential

SENSITIVITY AND IDENTIFICATION

system characteristics depends on the way in which the remainder of the system is designed about the predetermined part(s). It is quite appropriate, therefore, that sensitivity considerations be included in design.

For certain types of physical systems, it is feasible to continuously or periodically monitor some of the factors upon which system performance is based. The monitored data constitute useful information if it can be processed and used to improve overall system performance. The problem of monitoring measurable entities, and of estimating therefrom the factors of interest, is known as the *identification problem*. A major use of on-line identification is that of keeping system performance as good as possible under changing conditions. There is often a trade-off between identification and sensitivity; the more insensitive that system performance is made with respect to changes in a given parameter, the less need there is to accurately identify the parameter.

Indicators of sensitivity, *sensitivity measures*, can be based on either macroscopic or microscopic (incremental) concepts. Macroscopic sensitivity of a system characteristic is sensitivity in the large; for example, if a range of values of a particular system parameter is known to give rise to a range of values for a certain characteristic of a system, a design-center value of this system parameter may be selected at a point in the range where the system characteristic is relatively insensitive to changes in the parameter. On the other hand, incremental sensitivity of a system characteristic is applicable to small regions about some design center—incremental sensitivity measures are usually based on truncated Taylor's series.

The sensitivity measures of Chapter 2 (Section 2-9) are of the incremental variety. Given any system characteristic that can be represented as a function of system parameters, the fractional change in this characteristic can be estimated on the basis of fractional changes in the parameters. If the fractional changes in the parameters are known to be bounded by plus or minus certain values, worst-case sensitivities of system characteristics can be estimated.

For a system in which certain characteristics are measured in terms of functionals, sensitivity may be measured (Section 3-12) with respect to changes in system "elements" which are functions (e.g., a given function of time). Deviations in such functions from design-center forms generally result in changes in values of functionals that characterize system performance. The way in which these functions are generated has a direct bearing on sensitivity considerations. An example of this is given in Section 4-8, where three different system configurations are shown to exhibit the same optimal input-output performance, but exhibit different sensitivities with respect to change.

Sensitivity measures in linear programming (Section 5-10) are typically expressed in terms of macroscopic changes in the coefficients of linear algebraic equations. For example, a given c_i of I_c in Table 1-1 has an associated sensitivity range; if, and only if, the value of c_i does not deviate from its sensitivity range (assuming all other c_j's are fixed), the x_j's which are zero in the optimal solution remain at zero in any new solution.

The application of search techniques (Chapter 6) results, in general, in the accumulation of large amounts of macroscopic sensitivity information. Similarly, with dynamic programming (Chapter 7), much macroscopic sensitivity information is obtained as a direct consequence of the nature of dynamic programming solutions.

1-10. DISCUSSION

Principles of optimization can be used for many purposes: optimal design of systems, optimal operation of systems, determination of performance limitations of systems, or perhaps simply the solution of sets of equations. Both classical and modern approaches are considered in this book. Classical min-max theory (Chapter 2) and calculus of variations (Chapter 3) are considered first. Concepts from the calculus of variations are then applied (Chapter 4), in conjunction with Laplace transform theory, to problems involving ordinary linear differential equations. Linear programming (Chapter 5) is also applicable to linear systems, but in this case, systems that can be characterized by linear algebraic equations. Although search techniques (Chapter 6) are applicable to problems of the same form as those considered in Chapter 2, the approaches complement, rather than displace, one another. By using dynamic programming theory (Chapter 7), aspects of min-max theory and the theory of variational calculus are brought together with one principle, Bellman's principle of optimality, which is used in a derivation of the maximum principle of Pontryagin (Chapter 8).

In this brief introductory chapter, the aim has been both to outline the contents of the book and to introduce those system concepts that prevail throughout optimization problems of all types. Many important system concepts, which are of significance in particular classes of optimization problems, have thereby been by-passed at this stage of the development. These are introduced as necessary in subsequent chapters. Thus, "...ability" concepts[6] are considered, but it would be a mistake to suppose that their meanings and uses are displayed to their full extent in this book. Even the concept of stability is one which, for nonlinear systems, has a variety of

---

[6] Stability, controllability, reliability, observability, maintainability, computational feasibility, repeatability, susceptibility, acceptability, reachability, etc.

specific meanings each of which is identified by an appropriate descriptive phrase which ends in the word "stability."

In all the chapters that follow, references are given to much pertinent work which the reader will find to be of value in extending his knowledge; other valuable work is not brought to light herein because of limited space and because of this author's limited resources; and yet other pertinent work remains for the future. In regard to these statements, the following remarks by Lord Rayleigh [1.8] are most appropriate:

"Every year the additions to the common stock of knowledge become more bulky, if not more valuable; and one is impelled to ask, 'where is this to end?' Most students of science, who desire something more than a general knowledge, feel that their powers of acquisition and retention are already severely taxed. It would seem that any considerable addition to the burden would make it almost intolerable.

"It may be answered that the tendency of real science is ever towards simplicity; and that those departments which suffer seriously from masses of undigested material are also those which least deserve the name of science. Happily, there is much truth in this. A new method, or a new mode of conception, easily grasped when once presented to the mind, may supersede at a stroke the results of years of labor, making clear what was before obscure, and binding what was fragmentary into a coherent whole. True progress consists quite as much in the more complete assimilation of the old, as in the accumulation of new facts and inferences which in many cases ought to be regarded rather as the raw materials of science than as science itself. Nevertheless, it would be a mistake to suppose that the present generation can afford to ignore the labors of its predecessors, or to assume that so much of them as is really valuable will be found embodied in recent memoirs and treatises."

REFERENCES

[1.1] Athans, M. "The Status of Optimal Control Theory and Applications for Deterministic Systems." *IEEE Transactions on Automatic Control*, **AC-11**, 580–596, July 1966.

[1.2] Churchman, C. W. *Prediction and Optimal Decision.* Prentice-Hall, Englewood Cliffs, N.J., 1961.

[1.3] Greber, H. "The Philosophy of Engineering." *IEEE Spectrum*, **3**, 112–115, Oct. 1966.

[1.4] Hall, A. D. *A Methodology for Systems Engineering.* Van Nostrand, Princeton, N.J., 1962.

[1.5] Kershner, R. B. "Introduction (to the Proceedings of the Workshop on Systems Engineering)." *IRE Transactions on Education*, **E-5**, 57–60, June 1962.

[1.6] Mulligan, J. E. "Basic Optimization Techniques—A Brief Survey." *Journal of Industrial Engineering*, **16**, 192–197, May–June 1965.

[1.7] Shelly, M. W., II, and G. L. Bryan (Editors). *Human Judgments and Optimality*. Wiley, New York, 1964.

[1.8] Strutt, J. W. (Lord Rayleigh). *Scientific Papers*, Vols. I and II (in one volume, see especially paper no. 29, 1874). Dover Publications, New York, 1964.

CLASSICAL THEORY OF MINIMA AND MAXIMA

2

2-1. INTRODUCTION

The use of optimization techniques is of fundamental importance in system design; it is necessitated by the practical fact that the system design which is best in some specified sense is the one which sells, all other things being equal. Granted, in certain instances the optimal design is obvious, e.g., the greater the value of a certain parameter x, the better the system performance—but this is a trivial case. In many instances, constraints are imposed on the parameters of a system, and the treatment of these constraints requires the application of more sophisticated optimization techniques.

It is interesting to note that nature itself generally takes an optimal course. For example, in a classic work by James Clerk Maxwell [2.19], it is noted that the current distribution in a resistor-source network is the one which, in addition to satisfying Kirchhoff's laws, results in the minimum of energy dissipation (see Problem 2.16).

In this chapter, the classic theory of minima and maxima (min-max theory) is developed within specified limits. In Section 2-2, essential notation is presented; Sections 2-3 and 2-4 contain necessary and sufficient conditions for the optimization of continuously differentiable functions of several variables; equality constraints are taken into account in Sections 2-5 and 2-6; inequality constraints are considered in Section 2-7; optimization of integrals is covered in Section 2-8; and, finally, the sensitivity of optimal solutions to parameter changes is considered in Section 2-9. For a supplement to the theory presented, the reader is referred to the long-established work of Hancock [2.14].

It is to be emphasized that considerable effort is generally required to translate verbal statements about a given physical system into analytical statements which can be treated by the techniques of this chapter. Moreover, only in the exceptional case is there a unique way to specify the desired performance of a system in terms of an analytical performance measure. The examples and exercises given herein illustrate subtle points which must be considered in this regard.

2-2. BASIC CONCEPTS AND NOTATION

Consider the performance measure P:

$$P = f(x_1, x_2, \ldots, x_n) \tag{2-1}$$

where $f(x_1, x_2, \ldots, x_n) \equiv f$ is a known real-valued function of the arguments $x_1, x_2, \ldots, x_n$. The set $\{x_1, x_2, \ldots, x_n\}$ of real arguments of f is denoted by $\mathbf{x}$. For operational purposes, it is often convenient to view $\mathbf{x}$ as an $n \times 1$ column matrix with entries x_i. Unless otherwise restricted, $\mathbf{x}$ may assume any value in n-dimensional Euclidean space E^n. When n is 2, for example, $\mathbf{x}$ assumes values in the ordinary two-dimensional space of analytic geometry.

One of the principal problems of optimal design is to determine the particular values of $\mathbf{x}$ (values of the entries of $\mathbf{x}$) which result in the attainment of *local maxima* and *local minima* of the performance measure P in a subset R_X of n-dimensional Euclidean space. The subset R_X of interest varies from one problem to another. A point $\mathbf{x}_a$ of R_X is said to be a *point of strong local maximum* (or simply, a point of local maximum)[1] in R_X if there exists a real number $\Delta > 0$ such that

$$f(\mathbf{x}_a) - f(\mathbf{x}) > 0 \tag{2-2}$$

for any $\mathbf{x}$ which satisfies the following three conditions:

$$\mathbf{x} \neq \mathbf{x}_a \tag{2-3}$$

$$\mathbf{x} \in R_X \text{ (}\mathbf{x}\text{ is an element of } R_X\text{)} \tag{2-4}$$

and

$$\sum_{i=1}^{n} |x_i - x_{ia}| < \Delta \tag{2-5}$$

It is supposed in the preceding definition that the deleted neighborhood of $\mathbf{x}_a$ in R_X is nonempty, i.e., that the conditions of expressions 2-3, 2-4, and

---

[1] The phrase "local maximum" is used in this book to denote a "strong" local maximum. In contrast to a strong local maximum, a "weak" local maximum is one for which strict inequality in expression 2-2 does not necessarily apply. It is not uncommon to have "local" replaced by "relative" in definitions such as these.

2-5 are satisfied by some **x** for any real $\Delta > 0$. This assumption would not be satisfied if R_X were defined, say, to be the set of all integer n-tuples (each x_i restricted to integer values); the associated optimization problem then would be an integer programming problem (such problems, among others, are considered in Chapters 6 and 7).

If a point $\mathbf{x}_a$ is one of local maximum, the value $f(\mathbf{x}_a)$ is called a *local maximum* of $f(\mathbf{x})$ in R_X. Obviously, a *local minimum* and a *point of local minimum* are defined in an analogous fashion (it is left to the reader to do so).

If a point in R_X is either a point of local maximum or a point of local minimum, it is referred to as a *relative extremum point*.[2] For purposes of optimization, it is most important to determine the relative extremum points in R_X at which the greatest value of P, the *absolute maximum* (the maximum) of P, occurs and at which the least value of P, the *absolute minimum* (the minimum) of P, occurs. That these values exist for every function that is continuous on a closed and bounded set is assured by a classical and almost intuitively obvious theorem of Weierstrass, namely: Such functions possess a largest and a smallest value either in the interior or on the boundary of the set. This basic theorem is consoling when we have the not altogether easy task of determining the point of absolute maximum—at least an answer is known to exist, even though it may be difficult to find.

Two additional definitions are useful in the work that follows. A point $\mathbf{x}_a$ in R_X is called a *stationary point* if all of the first partial derivatives of $f(\mathbf{x})$ with respect to the x_i's are equal to zero at $\mathbf{x} = \mathbf{x}_a$. The use of the term "stationary" in this regard has its origin in mechanics: Consider the case where $n = 2$ and $f(\mathbf{x})$ represents the elevation of a surface above the x_1, x_2 plane in E^3. A constant gravitational force field is assumed perpendicular to the x_1, x_2 plane. The points on the $f(\mathbf{x})$ surface corresponding to $\partial f/\partial x_1 = \partial f/\partial x_2 = 0$ are points of stable or unstable equilibrium, points where a ball under the sole influence of gravity can be placed so as to remain stationary. Stationary points can be points of local maximum, points of local minimum, or *saddle points*. A saddle point is a stationary point which is neither a point of local maximum nor a point of local minimum. Stationary points are also referred to as *singular points*.

2-3. FUNCTIONS OF ONE VARIABLE

The concepts introduced in the preceding section are illustrated most directly by considering a continuous function of one variable:

$$P = f(x) \tag{2-6}$$

---

[2] The literature is not completely consistent in regard to this notation. What is called a "stationary point" here has also been termed an "extremum point"; what is called a "relative extremum point" here has also been called an "extreme point."

which is real valued and which is defined on the set R_x consisting of the closed interval $[a_0, a_f]$.[3] In regard to the extremization of such functions, the literature is voluminous; typical examples are given in references [2.4, 2.6, 2.7, 2.15, 2.17, and 2.28].

An extremum of such a function (see Figure 2-1) can occur only at stationary points or at points at which the derivative of the function is not completely defined. In Figure 2-1, the stationary points are a_1, a_2, and a_3; of these a_1 is a point of local maximum, a_2 is a point of local minimum, and a_3 is a saddle point. At the points a_0, a_4, and a_f, the derivative of $f(x)$ is not uniquely defined; a_4 is the point of absolute maximum in R_x, the left-hand end point a_0 is the point of absolute minimum in R_x, and the right-hand end point a_f is a point of local minimum in R_x.

In place of the $f(x)$ of Figure 2-1, consider any real-valued function $f(x)$ which has a continuous derivative at all points within R_x. In mathematical parlance, a function with this property is said to be of class C^1 in R_x. For such functions, the end points a_0 and a_f of R_x are relative extremum points; and the points which satisfy

$$\frac{df(x)}{dx} = 0 \qquad (2\text{-}7)$$

are either relative extremum points or saddle points. Thus, for an interior point x_a of R_x to be a relative extremum point of a function of class C^1, it is necessary—but not sufficient—that x_a be a stationary point.

Two questions are raised by the above statements: (1) When is a stationary point a relative extremum point? (2) Under what conditions is a relative extremum point a point of local maximum (minimum)? The answers to these questions are developed as follows.

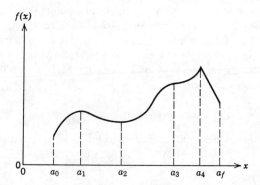

Figure 2-1. Local maxima, local minima, and stationary points.

---

[3] $[a_0, a_f]$ denotes the set of x's which satisfy $a_0 \leq x \leq a_f$.

FUNCTIONS OF ONE VARIABLE

Consider the Taylor's series expansion of $f(x)$:

$$f(x + \Delta x) = f(x) + \frac{\Delta x f^{(1)}(x)}{1!} + \frac{(\Delta x)^2 f^{(2)}(x)}{2!} + \frac{(\Delta x)^3 f^{(3)}(x)}{3!} + \cdots \quad (2\text{-}8)$$

In writing this equation, $f(x)$ is assumed to be a class C^3 or greater—$f(x)$ is of class C^3 if the derivatives of $f(x)$ up to and including the third are continuous in R_x. Suppose $x = x_a$ corresponds to a stationary point; then $f^{(1)}(x_a)$ equals zero, and from Equation 2-8 it follows that

$$f(x_a + \Delta x) - f(x_a) \simeq \frac{(\Delta x)^2 f^{(2)}(x_a)}{2} \quad (2\text{-}9)$$

if Δx is made sufficiently small.

In the case that

$$f^{(2)}(x_a) > 0 \quad (2\text{-}10)$$

the right-hand member of Equation 2-9 is always positive, independent of the sign of Δx; and it follows directly from the definitions given in Section 2-2 that $f(x_a)$ is a local minimum.

On the other hand, if

$$f^{(2)}(x_a) < 0 \quad (2\text{-}11)$$

then the right-hand member of 2-9 is always negative, independent of the sign of Δx; again by application of the appropriate definition of Section 2-2, $f(x_a)$ is a local maximum.

In the instance that $f^{(2)}(x_a) = 0$, Equation 2-8 gives

$$f(x_a + \Delta x) - f(x_a) \simeq \frac{(\Delta x)^3}{3!} f^{(3)}(x_a) \quad (2\text{-}12)$$

when Δx is made sufficiently small. Assuming that $f^{(3)}(x_a) \neq 0$, the sign of the right-hand member of Equation 2-12 changes when the sign of Δx changes, no matter how small in absolute value Δx is taken. Thus, the singular point associated with $f^{(1)}(x_a) = f^{(2)}(x_a) = 0 \neq f^{(3)}(x_a)$ is neither a local maximum nor a local minimum; hence, a saddle point. If it should occur that both $f^{(2)}(x_a) = 0$ and $f^{(3)}(x_a) = 0$, the test must be extended to include consideration of $f^{(4)}(x_a)$—see Problem 2.2.

The one question remaining is: When is a local maximum (minimum) the absolute maximum (minimum)? It must be answered: The only general way to determine the absolute maximum (minimum) is to effect a direct comparison of all local maxima (minima) for the purpose of picking out the greatest (least). Fortunately, in many instances this is a minor problem.

The practical optimal design procedure which is associated with the above theory is as follows: First, Equation 2-7 is solved to determine stationary points—sometimes the solution of 2-7 can be found in closed form, but often an iterative numerical technique, such as the Newton-Raphson method (see Chapter 6), is required; next, the nature of each stationary point is determined by using Equations 2-10 and 2-11; and finally, the absolute maximum (minimum) is obtained by comparison of the local maxima (minima).

Example 2-1. The lead network of Figure 2-2 is an electric circuit which is employed for the purpose of phase-shift compensation, especially in the design of compensators for feedback control systems. R_1 and R_2 are resistors, C is a capacitor, $v_i(t)$ is the applied voltage, and $v_o(t)$ is the output voltage.

In the sinusoidal steady state, the complex-number representation $V_o(j\omega)$ of the output $v_o(t)$ is related to the complex-number representation $V_i(j\omega)$ of the input $v_i(t)$ by the equation

$$\frac{V_o(j\omega)}{V_i(j\omega)} \triangleq G_c(j\omega) = \alpha \frac{1 + j\omega T}{1 + \alpha j\omega T} \tag{2-13}$$

where $\alpha \triangleq R_2/(R_1 + R_2)$, $T \triangleq R_1 C$, and ω (in radians/second) is the frequency of the sinusoidal input.

The phase angle ϕ associated with $G_c(j\omega)$ is

$$\phi = \underline{/G_c(j\omega)} = \tan^{-1} \omega T - \tan^{-1} \alpha\omega T \tag{2-14}$$

The problem: assuming that α is fixed, determine a value x^* of $\omega T \triangleq x$ which results in the absolute maximum ϕ^* of ϕ. Here, R_x is $[0, \infty]$.

Stationary points are obtained by using Equations 2-7 and 2-14:

$$\frac{d\phi}{dx} = \frac{d}{dx} [\tan^{-1} x - \tan^{-1} \alpha x]$$

$$= \frac{1}{1 + x^2} - \frac{\alpha}{1 + (\alpha x)^2} = 0 \quad \text{(for } x = x_a, \text{ a stationary point)} \tag{2-15}$$

which is solved for $x = x_a$,

$$x_a = (1/\alpha)^{1/2} \tag{2-16}$$

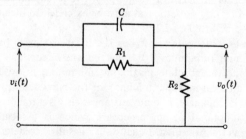

Figure 2-2. Lead compensator.

The nature of this unique stationary point is established by using Equation 2-11:

$$\left.\frac{d^2\phi}{dx^2}\right|_{x=\alpha^{-1/2}} = \frac{-2x}{(1+x^2)^2} + \left.\frac{2\alpha^3 x}{[1+(\alpha x)^2]^2}\right|_{x=\alpha^{-1/2}} = \frac{-2\alpha^{3/2}(1-\alpha)}{(1+\alpha)^2} \quad (2\text{-}17)$$

which is less than zero because α is less than one and greater than zero. Hence, $x = \alpha^{-1/2}$ is a point of local maximum. That it is also the unique point x^* of absolute maximum ϕ^* of ϕ is readily established by noting that

$$\phi^* = \phi|_{\omega T = \alpha^{-1/2}} = \tan^{-1} \alpha^{-1/2} - \tan^{-1} \alpha^{1/2} = \sin^{-1}\left(\frac{1-\alpha}{1+\alpha}\right) \quad (2\text{-}18)$$

is greater than the value of ϕ which is obtained at either $x = 0$ or $x = \infty$:

$$\phi|_{\omega T = 0} = \phi|_{\omega T = \infty} = 0.$$

●

2-4. FUNCTIONS OF SEVERAL VARIABLES

For real-valued functions of several variables, the concepts of the preceding section are readily extended if the functions are of class C^2. A necessary condition for a point to be a point of local maximum (minimum) of a real-valued function $f(x_1, x_2, \ldots, x_n) = f(\mathbf{x})$ of class C^2 is that it be a stationary point; i.e., a point which satisfies

$$\frac{\partial f(\mathbf{x})}{\partial x_i} = 0 \quad \text{for } i = 1, 2, \ldots, n \quad (2\text{-}19)$$

For a given $\mathbf{x} = \mathbf{x}_a$ which satisfies Equation 2-19, the nature of the stationary point $\mathbf{x}_a$ is determined in the following manner.

Initially, determinants A_k, $k = 1, 2, \ldots, n$, are evaluated:

$$A_k = \begin{vmatrix} f_{x_1 x_1} & f_{x_1 x_2} & \cdots & f_{x_1 x_k} \\ f_{x_2 x_1} & & & \vdots \\ \vdots & & \ddots & \\ f_{x_k x_1} & \cdots & & f_{x_k x_k} \end{vmatrix}_{\mathbf{x} = \mathbf{x}_a} \quad (2\text{-}20)$$

where $f_{x_i x_j} = \partial^2 f / \partial x_i \, \partial x_j$. The stationary point $\mathbf{x} = \mathbf{x}_a$ is a point of local minimum if the matrix which corresponds to the determinant A_n is positive-definite;[4] i.e., if

$$A_k > 0 \quad k = 1, 2, \ldots, n \quad (2\text{-}21)$$

[4] An $n \times n$ real symmetric matrix A is positive-definite if $\mathbf{x}'A\mathbf{x}$ is positive for all real, nonzero $n \times 1$ column matrices x. The test for positive-definiteness associated with expression 2-21 is known as Sylvester's theorem (see Ogata [2.20], for example).

On the other hand, the stationary point $\mathbf{x} = \mathbf{x}_a$ is a point of local maximum if both

$$A_k > 0 \quad k = 2, 4, 6, \ldots \tag{2-22}$$

and

$$A_k < 0 \quad k = 1, 3, 5, \ldots \tag{2-23}$$

As in the one-dimensional case, these tests do not exclude the possibility that, when certain $A_k = 0$, a local minimum or a local maximum may yet exist. Proofs of the above tests can be developed on the basis of a Taylor's series expansion in n variables in somewhat the same fashion as the corresponding proof of Section 2-3 (see Problem 2.14).

2-5. EQUALITY CONSTRAINTS AND A LAGRANGE MULTIPLIER

In this section, the performance measure P,

$$P = f_0(x_1, x_2) \equiv f_0 \tag{2-24}$$

is to be extremized, subject to the following equality constraint:

$$c_1 = f_1(x_1, x_2) \equiv f_1 \tag{2-25}$$

where c_1 is a constant and where f_0 and f_1 are assumed to be of class C^2 with respect to x_1 and x_2. A set of real values of x_1 and x_2 is assumed to exist for which the constraint 2-25 is satisfied. Constraints of the form of 2-25 indicate that the parameters x_1 and x_2 are dependent, a common occurrence in many practical problems.

Prior to the presentation of the Lagrange-multiplier method, consider first the straightforward approach which is sometimes used when 2-25 can be solved explicitly for x_1 in terms of x_2, say, $x_1 = g(x_2)$. In this case, $g(x_2)$ is substituted for x_1 in the right-hand member of 2-24, and ordinary (unconstrained) min-max theory of Section 2-3 is applied to test for relative extrema of $f_0[g(x_2), x_2]$. In addition to the possibility of not being able to solve 2-25 directly for x_1 in terms of x_2, the above "straightforward" procedure is often more cumbersome in actual use than is the Lagrange-multiplier method which is developed as follows.

As a first step in the derivation, the derivatives of 2-24 and 2-25 with respect to x_1 are taken by applying the chain rule of differentiation, as follows:

$$\frac{dP}{dx_1} = \frac{\partial f_0}{\partial x_1} + \frac{\partial f_0}{\partial x_2}\frac{dx_2}{dx_1} \tag{2-26}$$

Sect. 2-5 EQUALITY CONSTRAINTS AND A LAGRANGE MULTIPLIER

and

$$\frac{dc_1}{dx_1} = 0 = \frac{\partial f_1}{\partial x_1} + \frac{\partial f_1}{\partial x_2}\frac{dx_2}{dx_1} \qquad (2\text{-}27)$$

In order that P be a relative extremum at a given point, dP/dx_1 in Equation 2-26 must equal zero at that point. Thus, from 2-26, a necessary condition for a relative extremum is that

$$\frac{\partial f_0}{\partial x_1} + \frac{\partial f_0}{\partial x_2}\frac{dx_2}{dx_1} = 0 \qquad (2\text{-}28)$$

A more useful necessary condition for relative extrema would be one that is independent of the term dx_2/dx_1. Such an equation is readily obtained by elimination of dx_2/dx_1 between 2-27 and 2-28. Assuming that the partial derivatives in 2-27 and 2-28 are nonzero, these equations can be manipulated to obtain

$$\frac{dx_2}{dx_1} = \frac{-\partial f_1/\partial x_1}{\partial f_1/\partial x_2} = \frac{-\partial f_0/\partial x_1}{\partial f_0/\partial x_2} \qquad (2\text{-}29)$$

Note that the two equations, 2-25 and 2-29, are sufficient to determine the unknowns x_1 and x_2 which result in stationary points; and, in truth, the analysis of this section could be terminated here but for the fact that the results are in an awkward form and are not amenable to generalization at this stage.

The required modification is obtained by rearranging 2-29 and by introducing a new variable h, a *Lagrange multiplier*, as follows:

$$h \triangleq \frac{-\partial f_0/\partial x_1}{\partial f_1/\partial x_1} = \frac{-\partial f_0/\partial x_2}{\partial f_1/\partial x_2} \qquad (2\text{-}30)$$

which equation is more conveniently expressed by two equations:

$$\frac{\partial f_0}{\partial x_1} + h\frac{\partial f_1}{\partial x_1} = 0 \qquad (2\text{-}31\text{a})$$

and

$$\frac{\partial f_0}{\partial x_2} + h\frac{\partial f_1}{\partial x_2} = 0 \qquad (2\text{-}31\text{b})$$

For notational simplification at this point, an *augmented performance measure* f_a is introduced and is defined to be equal to $f_0(x_1,x_2)+hf_1(x_1,x_2)$. The utility of this augmented performance measure is that Equations 2-31a and 2-31b can be compactly replaced by

$$\frac{\partial f_a}{\partial x_1} = 0 \qquad (2\text{-}32\text{a})$$

and

$$\frac{\partial f_a}{\partial x_2} = 0 \tag{2-32b}$$

in which the Lagrange multiplier h is treated as being independent (as far as the partial differentiation is concerned) of x_1 and x_2. Thus, 2-25, 2-32a, and 2-32b form a set of three equations which can be solved for those values of x_1, x_2, and h which correspond to stationary points (stationary points of the augmented performance measure).

In many practical problems, Equations 2-32a and 2-32b admit to another interpretation of the Lagrange multiplier h. Consider, for example, any case where a maximum of $f_0(x_1,x_2)$ is desired, but where $f_1(x_1,x_2)$ represents a "cost" associated with the variables x_1 and x_2. Under these conditions, the augmented function $f_0(x_1,x_2) + hf_1(x_1,x_2)$—with *cost coefficient* h assigned a negative value—may be viewed as a *penalized* performance measure. For any particular value of the cost coefficient h, the maximum $f_a[x_1^*(h), x_2^*(h)]$ of $f_a(x_1,x_2)$ corresponds to a solution of Equations 2-32a and 2-32b. The values $x_1^*(h)$ and $x_2^*(h)$ of x_1 and x_2, respectively, yield the value $f_0[x_1^*(h), x_2^*(h)]$ of the original performance measure and the value $f_1[x_1^*(h), x_2^*(h)]$ of the cost function.

At this point an analytical test is desired, that can be used to determine the nature of the stationary points resulting from the application of the foregoing theory. The test that is developed in the following paragraphs is more cumbersome than the corresponding tests of the past two sections and is therefore less amenable to generalization.

The initial step in the derivation is the same as that of Section 2-3: if the second (total) derivative of P with respect to x_1 is greater than zero at a given stationary point, the stationary point is a point of local minimum; if the second (total) derivative of P with respect to x_1 is less than zero at a given stationary point, then the stationary point is a point of local maximum. Here, the second derivative of P with respect to x_1 is obtained by differentiating Equation 2-26:

$$\frac{d^2 P}{dx_1^2} = \frac{\partial^2 f_0}{\partial x_1^2} + \frac{\partial^2 f_0}{\partial x_1 \partial x_2}\frac{dx_2}{dx_1} + \frac{\partial f_0}{\partial x_2}\frac{d^2 x_2}{dx_1^2} + \frac{dx_2}{dx_1}\left[\frac{\partial^2 f_0}{\partial x_2 \partial x_1} + \frac{\partial^2 f_0}{\partial x_2^2}\frac{dx_2}{dx_1}\right]$$

$$= \frac{\partial^2 f_0}{\partial x_1^2} + 2\frac{dx_2}{dx_1}\frac{\partial^2 f_0}{\partial x_1 \partial x_2} + \left(\frac{dx_2}{dx_1}\right)^2 \frac{\partial^2 f_0}{\partial x_2^2} + \frac{\partial f_0}{\partial x_2}\frac{d^2 x_2}{dx_1^2} \tag{2-33}$$

From Equation 2-29, $dx_2/dx_1 = -(\partial f_1/\partial x_1)/(\partial f_1/\partial x_2) \triangleq z$ can be evaluated; from Equation 2-27, $d^2x_2/dx_1^2 = -[(\partial^2 f_1/\partial x_1^2) + 2(dx_2/dx_1)(\partial^2 f_1/\partial x_1 \partial x_2) + (dx_2/dx_1)^2(\partial^2 f_1/\partial x_2^2)]/(\partial f_1/\partial x_2)$ is obtained by differentiating and solving for d^2x_2/dx_1^2; the above identities are substituted appropriately into 2-33 which

Sect. 2-5 EQUALITY CONSTRAINTS AND A LAGRANGE MULTIPLIER 39

is then rearranged to obtain

$$\frac{d^2P}{dx_1^2} = \left[\frac{\partial^2}{\partial x_1^2} + 2z\frac{\partial^2}{\partial x_1 \partial x_2} + z^2\frac{\partial^2}{\partial x_2^2}\right] f_a(x_1, x_2) \qquad (2\text{-}34)$$

If d^2P/dx_1^2 is greater (less) than zero at a given stationary point, the stationary point is a point of local minimum (maximum).

Example 2-2. (Reference [2.22])

A common frequency-division circuit exhibits a voltage waveshape depicted in Figure 2-3. This voltage transient could be generated, for example, by a transistor blocking-oscillator circuit; in which case, the input voltage pulses of amplitude $V_s \equiv 2\,\Delta V$ are superimposed on the voltage discharge of a timing capacitor C so that the circuit regenerates—the capacitor is recharged to its initial V_i value—immediately after the composite signal decreases below the threshold voltage V_t.

From preliminary analysis, it is evident that the larger V_i can be made, the greater will be the distinction in voltage levels between the desired frequency-division of k and false frequency-division of $k-1$ or $k+1$. The value of V_i is

$$V_i = V_t + \Delta V + \frac{1}{C}\int_{kt_s^-}^{kt_s^+} \frac{dq}{dt}\,dt \qquad (2\text{-}35)$$

in which dq/dt is the rate at which electric charge enters C; dq/dt is positive

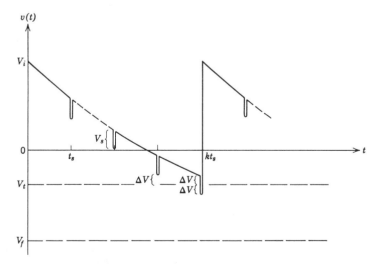

Figure 2-3. Waveform of Example 2-2.

during the recharging transient. To make V_i large, the timing capacitor C should obviously be selected as small as possible; and yet its selection must be consistent with other circuit requirements, such as the width of the output pulse produced during the "instantaneous" recharging of the capacitor to V_i. The fact that the capacitor does not actually recharge instantaneously suggests that some slight modification of the equations to be derived is necessary in practice.

The capacitor C having been selected in accordance with the above considerations, the remaining circuit parameters are to be adjusted for the best-possible circuit operation. It is assumed that the input pulses have a constant period t_s and that the voltage V_f to which the capacitor tends to discharge is fixed. If $-V_f$ were adjustable, it would be made as large as possible, subject to practical limitations, for the same just-mentioned reasons that V_i is made as large as possible.

The problem is now to select the amplitude V_s of the input pulses, the triggering level V_t, and the time constant $1/a$ of the capacitor discharge. The optimization problem to be solved is that of selecting a, V_s, and V_t so that the period of the output of the circuit has the maximal probability of being a kth multiple of the input period t_s. Thus, because circuit elements vary with age and environment and because element tolerances are such that circuits in production seldom operate at design center, the circuit has the most likelihood of proper operation if the design-center value of $\Delta V = V_s/2$ (Figure 2-3) is made as large as possible.[5]

From Figure 2-3 it is seen that ΔV is related to other circuit parameters by two equations: one corresponding to the instant of time $t = (k - 1)t_s$; the other corresponding to the instant of time $t = kt_s$.

When $t = kt_s$,

$$V_f - (V_f - V_i)e^{-akt_s} - \Delta V = V_t \tag{2-36}$$

When $t = (k - 1)t_s$,

$$V_f - (V_f - V_i)e^{-a(k-1)t_s} - 3\Delta V = V_t \tag{2-37}$$

By solving Equation 2-37 for ΔV, the function $f_0(a, V_t)$ to be maximized is found in the form

$$f_0(a, V_t) \triangleq \Delta V = (1/3)[V_f - (V_f - V_i)e^{-a(k-1)t_s} - V_t] \tag{2-38}$$

Next, a constraint equation, which relates V_t to a, is found by substitution of the right-hand member of 2-38 into 2-36; rearranging the result of the

---

[5] The tolerance on an electrical component or element is usually symmetric with respect to a mean value. When a circuit consists solely of such elements, circuit operation is said to be at design center if each element of the circuit assumes its mean value.

Sect. 2-5 EQUALITY CONSTRAINTS AND A LAGRANGE MULTIPLIER

above substitution gives

$$f_1(a, V_t) \triangleq 2V_f = 2V_t + (V_f - V_i)[3e^{-akt_s} - e^{-a(k-1)t_s}] \quad (2\text{-}39)$$

Equations 2-38 and 2-39 correspond to 2-24 and 2-25, respectively; to find the stationary points, therefore, the following partial derivatives of 2-38 and 2-39 are taken initially:

$$\frac{\partial f_0}{\partial a} = (k-1)\left(\frac{t_s}{3}\right)(V_f - V_i)e^{-a(k-1)t_s} \quad (2\text{-}40)$$

$$\frac{\partial f_0}{\partial V_t} = -\tfrac{1}{3} \quad (2\text{-}41)$$

$$\frac{\partial f_1}{\partial a} = (V_f - V_i)t_s[(k-1)e^{-a(k-1)t_s} - 3ke^{-akt_s}] \quad (2\text{-}42)$$

and

$$\frac{\partial f_1}{\partial V_t} = 2 \quad (2\text{-}43)$$

A necessary condition for the maximum ΔV^* of ΔV is obtained by insertion of the right-hand members of 2-40 through 2-43 into 2-31a and 2-31b; when the Lagrange multiplier h is eliminated between the resulting equations, we obtain

$$k - 1 = ke^{-at_s} \quad (2\text{-}44)$$

The optimal value a^* of a, found by solving 2-44 for a, is

$$a^* = \frac{1}{t_s} \ln \frac{k}{k-1} \quad (2\text{-}45)$$

When a^* of 2-45 is substituted into the right-hand member of 2-39, the optimal value V_t^* of V_t is found to be

$$V_t^* = V_f - (V_f - V_i)\left(\frac{k-1}{k}\right)^{k-1}\left(\frac{2k-3}{2k}\right) \quad (2\text{-}46)$$

Finally, the equation which gives the maximal value ΔV^* of ΔV is obtained by substituting a^* (2-45) and V_t^* (2-46) into the right-hand member of 2-38; thus,

$$\Delta V^* = (V_i - V_f)\frac{[(k-1)/k]^k}{2(k-1)} \quad (2\text{-}47)$$

It is left to the reader to show that the stationary point obtained above is indeed the point of absolute maximum.

●

2-6. GENERAL CASE OF EQUALITY CONSTRAINTS

In this section, the necessary conditions of Section 2-5 are extended to the general case of n selectable parameters and m equality constraints. It is assumed that both the constraint equations and the performance measure are of class C^1, and that the constraint equations admit to a set of real solutions. Only the resulting procedure is given—the interested reader is referred to established proofs [2.3, 2.9, and 2.14].

A performance measure P is to be extremized with respect to selection of x_i's:

$$P = f_0(\mathbf{x}) \equiv f_0 \tag{2-48}$$

where the x_i's of $\mathbf{x} = \{x_1, x_2, \ldots, x_n\}$ are restricted to satisfy m independent constraint equations; namely,

$$c_i = f_i(\mathbf{x}) \equiv f_i, \quad i = 1, 2, \ldots, m \tag{2-49}$$

where m is less than n and the c_i's are constants.

Solution of this problem, based on use of Lagrange multipliers, proceeds as follows. First, an augmented function f_a is formed,

$$f_a = f_0 + h_1 f_1 + h_2 f_2 + \cdots + h_m f_m \tag{2-50}$$

where the h_i's are independent of the x_i's and are Lagrange multipliers. Next, it is observed that stationary points of f_a are required to satisfy

$$0 = \frac{\partial f_a}{\partial x_i}$$

$$= \frac{\partial f_0}{\partial x_i} + h_1 \frac{\partial f_1}{\partial x_i} + \cdots + h_m \frac{\partial f_m}{\partial x_i} \tag{2-51}$$

for $i = 1, 2, \ldots, n$. In 2-49 and 2-51, there are $n + m$ equations. The unknowns in these equations are the x_i's, of which there are n, and the h_i's, of which there are m. Thus, the number of unknowns is also $n + m$. In general, Equations 2-49 and 2-51 may have several solutions. The equations are often quite involved, and numerical solution is often required (see Section 6-10).

As in Section 2-5, an analytical test could be developed here to expose the nature of the stationary points (solutions of 2-49 and 2-51). A general test of this sort is cumbersome, however, and is omitted in favor of the more direct method of comparing the values of P obtained at the stationary points. Finally, it must be noted that the Lagrange multiplier approach is more generally applicable to function optimization than is indicated by the classical material of Sections 2-5 and 2-6. The differentiability aspects of the preceding formulation are not requisite for certain generalized Lagrange multiplier concepts considered in Chapter 6.

2-7. INEQUALITY CONSTRAINTS

The general case of inequality constraints is much more difficult to treat by use of the methods of this chapter than is the general case of equality constraints which is presented in Section 2-6. In many cases, gradient and other direct search techniques (Chapter 6) are preferable. Despite this fact, functions of only a few variables are often handled with ease, a point supported by the illustrative case treated in this section.

Consider the performance measure

$$P = f_0(x_1,x_2) \equiv f_0 \tag{2-52}$$

where

$$c_1 \geq f_1(x_1,x_2) \equiv f_1 \tag{2-53}$$

with the additional constraint that

$$a_1 \leq x_1 \leq b_1 \tag{2-54}$$

where f_0 and f_1 are assumed to be of class C^2 over the region R_X defined by Equations 2-53 and 2-54—see Figure 2-4, for example.

The above problem is subdivided into several problems. First, the stationary points of $f_0(x_1,x_2)$ are determined in the ordinary way. Any of these points which fall outside the region R_X (see Figure 2-4) are immediately discarded. All that remain are possible points of relative extrema within the boundaries of the allowable region of the x_1,x_2 plane. In general, the above procedure will not extract all of the relative extremum points which lie on the boundaries of the allowable x_1,x_2 region; to determine these extrema, the equations which apply at the boundaries must be examined, as follows. First, find the relative extrema associated with $f_0(a_1,x_2)$ and keep only those that satisfy $f_1(a_1,x_2) \leq c_1$; second, find the relative extrema associated with $f_0(b_1,x_2)$ and keep only those that satisfy $f_1(b_1,x_2) \leq c_1$; third, find the relative extrema associated with $f_0(x_1,x_2) + hf_1(x_1,x_2)$—$h$ is a Lagrange multiplier and constraint 2-53 is taken in the equality sense—and keep only those that satisfy $a_1 \leq x_1 \leq b_1$; and finally, find any relative extrema which may be associated with the intersection points which are determined by $f_1(a_1,x_2) = c_1$ and $f_1(b_1,x_2) = c_1$. In the above development, it is tacitly assumed that no extrema exist at the infinite limits associated with x_2. By comparing all of the relative extrema remaining after the above selection process, the absolute maximum and the absolute minimum can be determined.

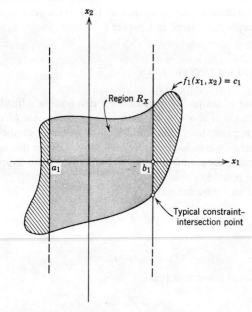

Figure 2-4. An allowable region R_X.

A more formal way of solving the preceding optimization problem is by using slack variables.

A *slack variable* is a real variable which is introduced to take up the slack in an inequality constraint, i.e., to convert an inequality constraint to an equality constraint. For example, let x_3 be a slack variable, and suppose that x_3, as well as x_1 and x_2, is constrained by the relationship

$$f_1(x_1,x_2) + x_3^2 = c_1 \tag{2-55}$$

The value of x_3^2 literally takes up the slack in the inequality constraint 2-53. Whatever real value x_3 assumes, $f(x_1,x_2)$ in 2-55 is guaranteed of being less than or equal to c_1, as required by the original inequality constraint 2-53. Thus, Equation 2-55 can be used in place of 2-53. Similarly, let slack variable x_4 be introduced as follows:

$$x_4^2 - (b_1 - x_1)(x_1 - a_1) = 0 \tag{2-56}$$

As long as x_1 satisfies the original constraint 2-54, x_4^2 in 2-56 is positive with the result that x_4 is real; but if x_1 does not satisfy 2-54, x_4^2 of 2-56 is negative which does not yield a real value of x_4. Thus, 2-56 can be used in place of 2-54. The reader is invited (Problem 2.28) to use the Lagrange multiplier approach of the preceding section on the optimization problem associated with 2-52, 2-55, and 2-56 to obtain necessary conditions for the optimum.

In summary, it has been shown that the solution of a problem with inequality constraints can be obtained through the solution of a set of ordinary min-max problems, ones with equality constraints at most. There is an obvious drawback to the application of the given procedure to problems with many variables and with several inequality constraints: the number of ordinary min-max problems which replace the given problem is exorbitant when the number of variables and the number of inequality constraints is yet relatively small. The programming techniques of Chapters 5, 6, and 7 are better suited for use in solving complicated problems of this type.

2-8. EXTREMIZATION OF INTEGRALS

In this section, the problem considered is that of determining relative extrema of the integral

$$P = \int_{t_a(\mathbf{x})}^{t_b(\mathbf{x})} f(\mathbf{x},t) \, dt \tag{2-57}$$

where t is the independent variable of integration, $\mathbf{x}$ equals $\{x_1, x_2, \ldots, x_n\}$, $f(\mathbf{x},t)$ is a specified real-valued function of class C^2 with respect to its arguments, and the limits of integration $t_a = t_a(\mathbf{x})$ and $t_b = t_b(\mathbf{x})$ are real-valued functions of class C^2 given in terms of $\mathbf{x}$—of course, t_a and t_b may simply be constants in a given problem.

Only the case in which the x_i's are not functions of the variable of integration in 2-57 is considered in this section. The case in which the x_i's in the integrand $f(\mathbf{x},t)$ are allowed to be functions of the variable of integration t is a special class of calculus-of-variation problem, as developed in Chapter 3.

As before, a necessary condition for a stationary point of P is that

$$\frac{\partial P}{\partial x_i} = 0, \quad i = 1, 2, \ldots, n \tag{2-58a}$$

By application of the chain rule of calculus,

$$\begin{aligned}
\frac{\partial P}{\partial x_i} &= \frac{\partial P}{\partial t_b} \frac{\partial t_b}{\partial x_i} + \frac{\partial P}{\partial t_a} \frac{\partial t_a}{\partial x_i} + \int_{t_a}^{t_b} \frac{\partial f(\mathbf{x},t)}{\partial x_i} \, dt \\
&= \frac{\partial t_b}{\partial x_i} \frac{\partial}{\partial t_b} \left[\int_{t_b}^{t_b} f(\mathbf{x},t) \, dt \right] + \frac{\partial t_a}{\partial x_i} \frac{\partial}{\partial t_a} \left[-\int_{t_b}^{t_a} f(\mathbf{x},t) \, dt \right] + \int_{t_a}^{t_b} \frac{\partial f(\mathbf{x},t)}{\partial x_i} \, dt \\
&= f[\mathbf{x}, t_b(\mathbf{x})] \frac{\partial t_b}{\partial x_i} - f[\mathbf{x}, t_a(\mathbf{x})] \frac{\partial t_a}{\partial x_i} + \int_{t_a}^{t_b} \frac{\partial f(\mathbf{x},t)}{\partial x_i} \, dt \tag{2-58b}
\end{aligned}$$

All of the derivatives in this equation exist because of the assumed conditions on $f(\mathbf{x},t)$, $t_a(\mathbf{x})$, and $t_b(\mathbf{x})$.

It is obvious that constraints of the form

$$c_i = \int_{t_{a_i}(\mathbf{x})}^{t_{b_i}(\mathbf{x})} f_i(\mathbf{x},t)\, dt, \qquad i = 1, 2, \ldots, m < n \qquad (2\text{-}59)$$

can be treated by using Lagrange multipliers in classical min-max problems. That this method of Lagrange multipliers is applicable even when the x_i's are allowed to be functions of t is demonstrated in Chapter 3.

The following example illustrates one application, of which there are many (e.g., [2.12 and 2.27]), of the theory of this section.

Example 2-3. Consider solution of the time-variant differential equation

$$\frac{dv}{dt} + tv = 0 \qquad (2\text{-}60)$$

satisfying $v = 1$ when $t = 0$. The closed-form solution is $v = e^{-t^2/2}$ (the reader may verify this); in general, however, such closed-form solutions to time-variant differential equations cannot be obtained. The method of Galerkin can be used to obtain approximate solutions to such equations; Kantorovich and Krylov [2.16] give a general description of the method. In this example, the method of Galerkin is employed to find an approximate solution of Equation 2-60 over the time interval defined by $0 \le t \le 1$. The exact answer affords a check on the approximate solution in this case.

At the outset, it is assumed that v can be adequately approximated by the relation

$$v \cong 1 + x_1 t + x_2 t^2 \qquad (0 \le t \le 1) \qquad (2\text{-}61)$$

where x_1 and x_2 are real constants to be determined such that the approximation is as good as possible (in a sense which will be defined). When $1 + x_1 t + x_2 t^2$ is substituted for v in Equation 2-60, the result is

$$(x_1 + 2x_2 t) + (t + x_1 t^2 + x_2 t^3) = \gamma(x_1, x_2, t) \qquad (2\text{-}62)$$

Note that the function $\gamma(x_1, x_2, t)$, rather than zero, is used as the right-hand member of 2-62 because of the approximation indicated in 2-61. Ideally, the *residue* $\gamma(x_1, x_2, t)$ would be zero for any t in the closed interval [0, 1]. The best that can be done, however, is that some appropriate measure of the residue $\gamma(x_1, x_2, t)$ be made as small as possible—a very tractable measure is the integral-square measure

$$P = \int_0^1 [\gamma(x_1, x_2, t)]^2\, dt \qquad (2\text{-}63)$$

Thus the problem is reduced to one in which two time-invariant parameters,

x_1 and x_2, are to be selected so as to minimize an integral. Equation 2-58 can therefore be employed to determine the stationary point(s); in this case, set $t_a = 0$, $t_b = 1$, and $f(\mathbf{x},t) = [\gamma(x_1,x_2,t)]^2$; thus,

$$\frac{\partial P}{\partial x_1} = \int_0^1 2\gamma(x_1,x_2,t) \frac{\partial \gamma(x_1,x_2,t)}{\partial x_1} dt = 0 \qquad (2\text{-}64)$$

and

$$\frac{\partial P}{\partial x_2} = \int_0^1 2\gamma(x_1,x_2,t) \frac{\partial \gamma(x_1,x_2,t)}{\partial x_2} dt = 0 \qquad (2\text{-}65)$$

where $\gamma(x_1,x_2,t)$ is given by 2-62. It follows from 2-62 that

$$\frac{\partial \gamma(x_1,x_2,t)}{\partial x_1} = 1 + t^2 \qquad (2\text{-}66)$$

and

$$\frac{\partial \gamma(x_1,x_2,t)}{\partial x_2} = 2t + t^3 \qquad (2\text{-}67)$$

The identities given by Equations 2-62, 2-66, and 2-67 are substituted appropriately into Equations 2-64 and 2-65. The indicated integrations are then effected, resulting in the following two algebraic equations in x_1 and x_2:

$$1.867 x_1 + 1.917 x_2 = -0.750 \qquad (2\text{-}68)$$

and

$$1.917 x_1 + 2.276 x_2 = -0.867 \qquad (2\text{-}69)$$

which are solved simultaneously to yield $x_1 = -0.071$ and $x_2 = -0.323$. That these values define the point of minimum for P can be established in the standard way (Section 2-4). Hence, the appropriate solution for $v = v(t)$ is

$$v(t) \cong 1 - 0.071t - 0.323t^2, \qquad 0 \le t \le 1 \qquad (2\text{-}70)$$

which checks quite closely with the exact solution given below 2-60.

•

2-9. SENSITIVITY ANALYSIS [2.23]

From a practical standpoint, the selection of system parameters is a process which requires allowance for component variations. If the tolerance on components is, say, $\pm 5\%$, the resulting system performance may or may not be far from optimum, depending on the sharpness of the peak which is associated with the point of maximum (minimum). In many cases, tolerance limits can be taken into account directly in a performance measure and in constraint equations; for example, a performance measure could be optimized under the condition that system parameters are assumed to be at tolerance extremes which are least favorable to optimal performance; while in a given constraint equation, system parameters are assumed to be at tolerance limits which are most likely to cause the constraint equation to be violated. A general mathematical formulation of this design approach would do little to clarify it—instead, the reader is referred to typical applications [2.2 and 2.21] and to Section 5-12d.

In the remainder of this section, a different approach is taken: definitions of *sensitivity* are given which enable us to estimate likely variations from design-center values of figures of merit. In particular, an examination is made of the variation of the performance measure

$$P = f(\mathbf{x}) \tag{2-71}$$

where certain (perhaps all) of the elements of the set $\mathbf{x} = \{x_1, x_2, \ldots, x_n\}$ are so selected at the design center as to extremize P which is assumed to be twice differentiable with respect to the x_i's.

The classical sensitivity measure of Bode [2.1] assumes the form

$$S_{x_i}^P = \frac{x_i}{P} \frac{\partial P}{\partial x_i}\bigg|_{\mathbf{x}=\mathbf{x}^*} \qquad i = 1, 2, \ldots, n \tag{2-72}$$

where $\mathbf{x}^*$ is the design-center value of $\mathbf{x}$. In the case that the design-center values of the x_i's correspond to a stationary point for P, the $S_{x_i}^P$ values are zero—a result of dubious value in that even small deviations of the x_i's from the design-center values change the value of P in general.

The sensitivity measures presented here are matrix sensitivity measures and are meaningful for optimally selected parameters. As with the sensitivity measure of 2-72, the extensions are obtained from the Taylor's series of $f(\mathbf{x})$ expanded about the design-center value $\mathbf{x}^*$,

$$f(\mathbf{x}) = f(\mathbf{x}^*) + (\mathbf{x} - \mathbf{x}^*)' \nabla f(\mathbf{x}^*) + \tfrac{1}{2}(\mathbf{x} - \mathbf{x}^*)' \mathbf{A} (\mathbf{x} - \mathbf{x}^*) + \cdots \tag{2-73}$$

where $\mathbf{x}$ and $\mathbf{x}^*$ are to be interpreted as $n \times 1$ column vectors; $\nabla f(\mathbf{x}) = \left[\dfrac{\partial f}{\partial x_1} \ \dfrac{\partial f}{\partial x_2} \ \cdots \ \dfrac{\partial f}{\partial x_n}\right]'$ is the *gradient* of $f(\mathbf{x})$ in column-vector form; and $\mathbf{A}$ is the following $n \times n$ matrix of second partial derivatives:

$$\mathbf{A} = \begin{bmatrix} \dfrac{\partial^2 f}{\partial x_1^2} & \dfrac{\partial^2 f}{\partial x_1 \, \partial x_2} & \cdots & \dfrac{\partial^2 f}{\partial x_1 \, \partial x_n} \\ \vdots & & \ddots & \vdots \\ \dfrac{\partial^2 f}{\partial x_n \, \partial x_1} & \cdots & \cdot & \dfrac{\partial^2 f}{\partial x_n^2} \end{bmatrix}_{\mathbf{x}=\mathbf{x}^*} \tag{2-74}$$

SENSITIVITY ANALYSIS

which is sometimes called the *Hessian matrix* of f evaluated at $\mathbf{x} = \mathbf{x}^*$.

Let Ω and Λ be defined as follows:

$$\Omega \triangleq \left[\frac{x_1 - x_1^*}{x_1^*} \quad \frac{x_2 - x_2^*}{x_2^*} \quad \cdots \quad \frac{x_n - x_n^*}{x_n^*}\right]' \tag{2-75}$$

and

$$\Lambda \triangleq \begin{bmatrix} x_1^* & 0 & 0 & \cdots & 0 \\ 0 & x_2^* & & & \vdots \\ \vdots & & \ddots & & \\ 0 & & \cdots & & x_n^* \end{bmatrix} \tag{2-76}$$

Note that $\Lambda\Omega$ equals $\mathbf{x} - \mathbf{x}^*$. With these identities, 2-73 is truncated and rearranged to obtain

$$\frac{f(\mathbf{x}) - f(\mathbf{x}^*)}{f(\mathbf{x}^*)} \cong \Omega'\left[\frac{\Lambda \nabla f(\mathbf{x}^*)}{f(\mathbf{x}^*)}\right] + \Omega'\left[\frac{\Lambda A \Lambda}{2f(\mathbf{x}^*)}\right]\Omega$$

$$\cong \Omega' S_\mathbf{x}^f + \Omega' Sq_\mathbf{x}^f \Omega \tag{2-77}$$

in which the *column sensitivity matrix* $S_\mathbf{x}^f$ is defined by

$$S_\mathbf{x}^f \triangleq \frac{\Lambda \nabla f(\mathbf{x}^*)}{f(\mathbf{x}^*)} \tag{2-78}$$

and the *square sensitivity matrix* $Sq_\mathbf{x}^f$ is defined by

$$Sq_\mathbf{x}^f \triangleq \frac{\Lambda A \Lambda}{2f(\mathbf{x}^*)} \tag{2-79}$$

In application, $S_\mathbf{x}^f$ and $Sq_\mathbf{x}^f$ are evaluated by using 2-78 and 2-79 with $f(\mathbf{x})$ and $\mathbf{x}^*$ known. Equation 2-77 then yields the fractional change in $f(\mathbf{x})$ which results from a specified Ω matrix, i.e., from specified fractional changes in the x_i's. Note that $S_\mathbf{x}^f = \mathbf{0}$ in the important special case that $\mathbf{x}^*$ corresponds to a stationary point of $f(\mathbf{x})$.

In many applications, the value of $(x_i - x_i^*)/x_i^*$ is known to lie within a given closed interval, say,

$$\left|\frac{x_i - x_i^*}{x_i^*}\right| \leq \epsilon_i, \quad i = 1, 2, \ldots, n \tag{2-80}$$

or in matrix form,

$$-\mathbf{e} \leq \Omega \leq \mathbf{e} \tag{2-81}$$

An important practical question to be answered in this case is, "What is the maximum of the absolute value of $[f(\mathbf{x}) - f(\mathbf{x}^*)]/f(\mathbf{x}^*)$ in 2-77 with respect to allowed variations of Ω, as in 2-81?" The answer to this question can be obtained by using nonlinear programming (Chapter 6).

It is important to bear in mind that the sensitivity measures of Equations 2-78 and 2-79 do not have practical merit in all cases; e.g., when $f(\mathbf{x}^*) = 0$ or when any $x_i^* = 0$, the fractional changes in 2-77 are useless—in such cases, we do better to deal simply in terms of magnitude of change.

Example 2-4. In Example 2-1, the function ϕ which is maximized is a function of two variables, $x_1 = \omega T$ and $x_2 = \alpha$. Let $\phi = f(x_1, x_2)$, and from 2-14 it follows that

$$\phi = \tan^{-1} x_1 - \tan^{-1} x_1 x_2 \tag{2-82}$$

At design-center values, let $x_2^* = \alpha^* = k$ (a specified constant), and let $x_1^* = 1/k^{1/2}$; as shown in Example 2-1, $\phi(\mathbf{x})$ is maximized with respect to x_1 under this condition.

Partial derivatives required are

$$(\partial \phi / \partial x_1)|_{\mathbf{x}=\mathbf{x}^*} = 0 \tag{2-83}$$

$$(\partial \phi / \partial x_2)|_{\mathbf{x}=\mathbf{x}^*} = -x_1/(1 + x_1^2 x_2^2)|_{\mathbf{x}=\mathbf{x}^*} = -1/k^{1/2}(1 + k) \tag{2-84}$$

$$(\partial^2 \phi / \partial x_1^2)|_{\mathbf{x}=\mathbf{x}^*} = 2k^{3/2}(k - 1)/(k + 1)^2 \tag{2-85}$$

$$(\partial^2 \phi / \partial x_2^2)|_{\mathbf{x}=\mathbf{x}^*} = 2/k^{1/2}(1 + k)^2 \tag{2-86}$$

and

$$(\partial^2 \phi / \partial x_1 \, \partial x_2)|_{\mathbf{x}=\mathbf{x}^*} = (k - 1)/(k + 1)^2 \tag{2-87}$$

The value of $\phi(\mathbf{x}^*)$ is

$$\phi(\mathbf{x}^*) = \tan^{-1} k^{-1/2} - \tan^{-1} k^{1/2} = \sin^{-1}[(1 - k)/(1 + k)] \tag{2-88}$$

The results of 2-83 through 2-88 are substituted appropriately into 2-78 and 2-79 to obtain

$$S_{\mathbf{x}}^{\phi} = [0 \quad -k^{1/2}/(1 + k)]'/\sin^{-1}[(1 - k)/(1 + k)] \tag{2-89}$$

and

$$Sq_{\mathbf{x}}^{\phi} = \frac{\begin{bmatrix} 2k^{1/2}(k - 1) & k^{1/2}(k - 1) \\ k^{1/2}(k - 1) & 2k^{3/2} \end{bmatrix}}{2(1 + k)^2 \sin^{-1}[(1 - k)/(1 + k)]} \tag{2-90}$$

Numerically, suppose $k = 0.1$, a typical value, and that tolerances on x_1 and x_2 are $\pm 20\%$. Equations 2-89 and 2-90 reduce to

$$S_{\mathbf{x}}^{\phi} = \begin{bmatrix} 0 \\ -0.300 \end{bmatrix} \quad \text{and} \quad Sq_{\mathbf{x}}^{\phi} = \begin{bmatrix} -0.2456 & -0.1228 \\ -0.1228 & 0.0272 \end{bmatrix} \tag{2-91}$$

In the worst case, it is readily verified that

$$\frac{x_1 - x_1^*}{x_1^*} = \frac{x_2 - x_2^*}{x_2^*} = 0.2 \tag{2-92}$$

The results of 2-91 and 2-92 are used in 2-77 to obtain

$$\frac{\phi(\mathbf{x}) - \phi(\mathbf{x}^*)}{\phi(\mathbf{x}^*)} = [0.2 \quad 0.2] \begin{bmatrix} 0.000 \\ -0.300 \end{bmatrix}$$

$$+ [0.2 \quad 0.2] \begin{bmatrix} -0.2456 & -0.1228 \\ -0.1228 & 0.0272 \end{bmatrix} \begin{bmatrix} 0.2 \\ 0.2 \end{bmatrix}$$

$$= -0.0600 - 0.0186 = -0.0786 \qquad (2\text{-}93)$$

If $Sq_\mathbf{x}^\phi$ were ignored in this example, the estimate of $(\phi - \phi^*)/\phi^*$ would be -0.060 rather than the more accurate result of -0.0786—a nontrivial difference.

●

2-10. CONCLUSION

The theory developed in this chapter can be used to effect the optimization of systems that are characterized by performance measures which are differentiable functions of several variables (Sections 2-2, 2-3, and 2-4). When equality constraint equations are present, they are taken into account by using the Lagrange multiplier technique (Sections 2-5 and 2-6). Inequality constraints can also be handled, but with less ease (Section 2-7). The optimization of integrals, the integrands of which contain system parameters, is a straightforward extension of the preceding theory (Section 2-8), and the sensitivity of the optimal solution with respect to parameter variations can be estimated (Section 2-9).

Although no mention is made of optimization problems wherein certain quantities are known only in terms of probability distributions, such problems can often be reduced to a form which corresponds to one of those of this chapter. For example, the expected value of a function of several variables, rather than the function itself, can be maximized.

Much of the theory presented thus far has been known for centuries—since the advent of calculus. It is not surprising, therefore, that for the solution of certain optimization problems, the techniques of this chapter are feeble in comparison with search techniques and nonlinear programming (Chapter 6) and dynamic programming techniques (Chapter 7). Hence, when the application of the techniques of this chapter lead to apparently insurmountable obstacles, the application of those of Chapters 6 or 7 may give more insight to the problem solution, and vice versa. No favoritism should be shown to a restricted approach to a solution when many other avenues are open. Also, the extremization of a linear algebraic equation subject to satisfaction of linear algebraic constraints is best performed by use of the theory of linear programming (Chapter 5).

REFERENCES

[2.1] Bode, H. W. *Network Analysis and Feedback Amplifier Design.* Van Nostrand, Princeton, N.J., 1945.

[2.2] Bongenaar, W., and N. C. de Trove. "Worst-Case Considerations in Designing Logical Circuits." *IEEE Transactions on Electronic Computers,* EC-14, 590–599, Aug. 1965.

[2.3] Brand, L. *Advanced Calculus.* Wiley, New York, 1955 (third printing, revised, November, 1960), pp. 194–195.

[2.4] Brugler, J. S. "Optimum Shunt Voltage Regulator Design." *Proceedings of the IEEE,* 53, 312, March 1965.

[2.5] Chang, S. H. "Optimum Probabilities of Messages with Preassigned Code Words." *Proceedings of the IEEE,* 51, 866–867, May 1963.

[2.6] Davis, S. A. *Outline of Servo-Mechanisms.* Regents, New York, 1966.

[2.7] Drake, F. D. "Optimum Size of Radio Astronomy Antennas." *Proceedings of the IEEE,* 52, 108–109, Jan. 1964.

[2.8] Duda, R. O. "On Resistive Networks with Optimum Circuit Reproducibility." *IRE Transactions on Circuit Theory,* CT-9, 297–298, Sept. 1962.

[2.9] Franklin P. *A Treatise on Advanced Calculus.* Wiley, New York, 1940, 353–354.

[2.10] Gartner, W. W. "Maximum Available Power Gain of Linear Fourpoles." *IRE Transactions on Circuit Theory,* CT-5, 375–376, Dec. 1958.

[2.11] Goodman, L. M. "Optimum Rate Allocation for Encoding Sets of Analog Messages." *Proceedings of the IEEE,* 53, 1776–1777, Nov. 1965.

[2.12] Goodman, L. M. "Optimum Sampling and Quantizing Rates." *Proceedings of the IEEE,* 54, 90–92, Jan. 1966.

[2.13] Greene, J. C., and E. W. Sard. "Optimum Noise and Gain-Bandwidth Performance for a Practical One-Port Parametric Amplifier." *Proceedings of the IRE,* 48, 1583–1590, Sept. 1960.

[2.14] Hancock, H. *Theory of Maxima and Minima.* Dover Publications, New York, 1960. Originally printed by Ginn and Company, New York, 1917.

[2.15] Josephs, H. C. "Conditions for Optimizing FET Operation." *Proceedings of the IEEE,* 53, 199–200, Feb. 1965.

[2.16] Kantorovich, L. V., and V. I. Krylov. *Approximate Methods of Higher Analysis.* Interscience, New York, 1958 (3rd ed.).

[2.17] Kundert, W. R. "The RC Amplifier-Type Active Filter: A Design Method for Optimum Stability." *IEEE International Convention Record,* 12, part 6, 384–392, 1964.

[2.18] Ling, S. T., Y. Tezuka, and Y. Kasahara. "Optimal Allocation of Channels in an Alternate Route Communication Network." *IEEE Transactions on Communication Systems,* CS-12, 185–190, June 1964.

[2.19] Maxwell, J. C. *A Treatise on Electricity and Magnetism.* Clarendon Press, Oxford, England, 1873. (See p. 408 of vol. 1 of the unabridged 3rd ed. of 1891, Dover Publications, New York, 1954.)

[2.20] Ogata, K. *State Space Analysis of Control Systems.* Prentice-Hall, Englewood Cliffs, N.J., 1967.

[2.21] Pierre, D. A. "Maximizing Current Gain in a Transistor Switching Amplifier." *Electronic Design*, **11**, 62–65, March 1962.
[2.22] Pierre, D. A. "Optimization of Pulse and Digital Circuits by Use of the Lagrange Multiplier Technique." *IEEE Transactions on Electronic Computers*, **EC-12**, 488–491, Oct. 1963.
[2.23] Pierre, D. A. "Sensitivity Measures for Optimally Selected Parameters." *Proceedings of the IEEE*, **54**, 321–322, Feb. 1966.
[2.24] Pierre, D. A., V. Lorchirachoonkul, and M. E. Ross. "Deadbeat Response with Minimal Overshoot Compromise." *Proceedings*, 20th Annual Conference of the Instrument Society of America, part III (Advances in Instrumentation), 1965.
[2.25] Scanlan, J. O., and J. S. Singleton. "The Gain and Stability of Linear Two-Port Amplifiers." *IRE Transactions on Circuit Theory*, **CT-9**, 240–246, Sept. 1962.
[2.26] Scanlan, J. O., and J. S. Singleton. "Two-Ports-Maximum Gain for a Given Stability Factor." *IRE Transactions on Circuit Theory*, **CT-9**, 428–429, Dec. 1962.
[2.27] Smith, G. W., Jr. "Dual Requirements for Optimizing Orthonormal Approximations of Impulse Responses." *IRE Transactions on Circuit Theory*, **CT-8**, 486–487, Dec. 1961.
[2.28] Sommer, R. C. "On the Optimization of Random-Access Discrete Address Communications." *Proceedings of the IEEE*, **52**, 1255, Oct. 1964.

PROBLEMS

2.1 A control motor has a stall torque T_s (corresponding to a specified control signal level). The static friction torque of the load at zero speed is a known value F. A gear ratio x is to be specified such that the load acceleration α_L from standstill is maximized. Under standstill conditions, the torque equation is

$$(J_m x^2 + J_L)\alpha_L = T_s x - F, \quad \begin{array}{l} J_m \triangleq \text{motor rotor inertia} \\ J_L \triangleq \text{load inertia} \end{array}$$

Solve for α_L in terms of x and find the value x^* of x which results in a maximum of α_L. What is the expression for max α_L? (A version of this problem is treated by Davis [2.6].)

2.2 Given $f(x)$ and the fact that $(df/dx) = (d^2f/dx^2) = (d^3f/dx^3) = 0$ at $x = x_a$, determine the nature of the stationary point at $x = x_a$ in terms of (d^4f/dx^4).

2.3 Find the stationary point(s) of

$$f(x) = (1 - e^{-ax})^x$$

with respect to x and determine the nature of the stationary point(s). (A problem of this type is treated by Sommer [2.28].)

2.4 Determine the conditions on the parameters a, b, c, and d under which $f(x) = ax^3 + bx^2 + cx + d$ has stationary points with respect to *real* values of x. Find expressions for the stationary points and determine their nature.

2.5 Consider a stable linear system with Laplace-transform transfer function $G(s)$, $G(s) = 1/[(s + \alpha)^2 + \beta^2]$. A unit step input, the Laplace transform of which is $1/s$, is applied to the system with the result that the output $f(t)$ of the system is

$$f(t) = (1/\beta_0^2) + (1/\beta_0\beta)e^{-\alpha t} \sin(\beta t - \psi), \qquad t \geq 0$$

where $\beta_0^2 \triangleq \alpha^2 + \beta^2$ and $\psi \triangleq \tan^{-1}(\beta/-\alpha)$. At what value t^* of t does $f(t)$ attain its maximum value? Find an expression for $f(t)_{\max}$.

2.6 Find the stationary point of

$$f(x) = x^{1/2}(1 - ax^{1/2})$$

where a is a constant and determine the nature of the stationary point. (A problem of this form is considered by Josephs [2.15].)

2.7 Find the value x^* of x which results in a minimum of P:

$$P = \frac{1}{x-1}\left\{\frac{x^2(k^2 + k + 1)}{3} - \frac{x(k + 3)}{2} + 1\right\}$$

where k is a constant. (A problem of this form is considered by Brugler [2.4].)

2.8 Find the stationary point of $f(x) = [x/(x - 1)](1 + ax) - 1$ with respect to x and determine the nature of the stationary point. (A problem of this form is considered by Greene and Sard [2.13].)

2.9 Find the stationary point of $f(x) = ax^{\alpha-1} + (b/x)$ with respect to x and determine the nature of the stationary point. (A problem of this form is considered by Drake [2.7].)

2.10 Find the nature of the stationary point of

$$f(x) = [(2x - a)/a] + (b/ax)(x^2 - ax + 1)$$

with respect to x where a is assumed to be a positive constant less than 2, and x is restricted to be greater than $a/2$. (A problem of this form is considered by Kundert [2.17].)

2.11 Find stationary points of $f(x_1,x_2)$ and determine their nature, where

$$f(x_1,x_2) = (1 + a - bx_1 - bx_2)^2 + (b + x_1 + ax_2 - bx_1x_2)^2$$

and a and b are constants. (Motivation for this problem and a partial solution can be found in Example 6-2 of Chapter 6.)

2.12 Consider the function $f(x_1,x_2)$ of Problem 2.11 under the condition that $a = 10$ and $b = 1$. What is the nature of the stationary point of $f(x_1,x_2)$ at $\mathbf{x} = [10 \ \ 1]'$?

2.13 Consider the function $f(x_1,x_2)$ of Problem 2.11 under the condition that $a = 1$ and $b = 1$. What is the nature of the stationary point of $f(x_1,x_2)$ at $\mathbf{x} = [1 \ \ 1]'$?

2.14 a. Derive Equation 2-73. (*Hint:* Let $\mathbf{x} = \mathbf{x}^* + \epsilon \, \Delta \mathbf{x}$ where ϵ is a real scalar. Expand $f(\mathbf{x}^* + \epsilon \, \Delta \mathbf{x})$ in a Maclaurin's series with respect to the scalar ϵ and take $\epsilon = 1$ in the series so obtained.)

b. Use Equation 2-73 to prove that a sufficient condition for a local minimum of a real-valued function $f(\mathbf{x})$ at a stationary point $\mathbf{x}_a$ is that the $\mathbf{A}$ matrix be positive-definite at $\mathbf{x} = \mathbf{x}_a$.

2.15 It is clear that the points of local maxima of any function $f(\mathbf{x})$ are the points of local minima of $-f(\mathbf{x})$. Based on this fact and on the test for a local minimum which is associated with Equation 2-21, derive the test for a local maximum which is associated with Equations 2-22 and 2-23.

2.16 Consider the circuit diagram of Figure 2-P16. The rate at which energy is delivered from the batteries is

$$P_d = E_1 i_1 + E_2(i_3 - i_2)$$

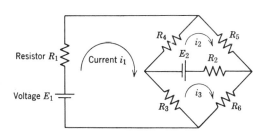

Figure 2-P16.

The rate at which energy is delivered to the resistors is

$$P_r = i_1^2 R_1 + (i_2 - i_3)^2 R_2 + (i_1 - i_3)^2 R_3 + (i_1 - i_2)^2 R_4 + i_2^2 R_5 + i_3^2 R_6$$

A known constraint is

$$0 = P_d - P_r$$

Show that the i_i's which satisfy this constraint and simultaneously minimize P_d are the i_i's which satisfy Kirchhoff's loop equations. (*Hint:* Minimize the augmented function $f_a = P_d + h(P_d - P_r)$ and show that the Lagrange multiplier h must equal 1.)

2.17 Consider the circuit diagram of Figure 2-P17 in which E is the RMS value of a steady-state sinusoidal signal. The average rate at which energy is delivered to the load resistance R_2 is $P = |I|^2 R_2$. Kirchhoff's laws give

$$E = I(R_1 + R_2 + jX_1 + jX_2).$$

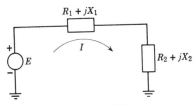

Figure 2-P17.

a. Assuming E, R_1, X_1, and X_2 are fixed, find the value R_2^* of R_2 which results in a maximum of P.
b. Assuming E, R_1, X_1, and R_2 are fixed, find the value X_2^* of X_2 which results in a maximum of P.
c. Assuming E, X_1, R_2, and X_2 are fixed, find the value R_1^* of R_1 which results in a maximum of P.

(More challenging versions of this type of problem are considered in references [2.10, 2.25, and 2.26].)

2.18 Find the minimum of σ^2:

$$\sigma^2 = \sum_{i=1}^{n} T_i^2 x_i^2$$

with respect to the set of x_i's, subject to the constraint that

$$c_1 = \sum_{i=1}^{n} x_i$$

where the T_i's and c_1 are constants. (A problem of this type is considered by Duda [2.8].)

2.19 Consider the performance measure P:

$$P = (x_1 - 1)^2 + x_n^2 + \sum_{k=1}^{n-1} (x_{k+1} - x_k)^2$$

a. What is the stationary point of P?
b. Find the stationary point of P under the condition that

$$ax_1 + a^2 x_2 + \cdots + a^n x_n = c_1$$

where c_1 and a are constants.
(These problems are of the form considered by Pierre et al. [2.24].)

2.20 Suppose that $P = x_1 + x_2$ is to be maximized with respect to the selection of real values of x_1 and x_2, but subject to the constraint that $x_1^2 - 2x_1 + x_2^2 = -2$. What is inconsistent with the problem as stated?

2.21 A minimum of $\overline{e^2(t)}$ is desired:

$$\overline{e^2(t)} = \sum_{i=1}^{n} k_i 2^{-x_i/w_i}$$

subject to the constraint that

$$c_1 = \sum_{i=1}^{n} x_i$$

The terms c_1, k_i, and w_i are constants.

Find expressions for the optimal x_i's, x_i^*'s, which result in the minimum of $\overline{e^2(t)}$. (A problem of this form is considered by Goodman [2.11].)

2.22 Consider the design of a monostable multivibrator circuit which exhibits the voltage waveshape $v(t)$ shown in Figure 2-P22; this waveshape controls the length of the time interval T during which the circuit operates in its unstable state. When $v(t)$ tends to exceed the constant level V_t, an active element of the circuit is triggered causing the circuit to return to its stable state.

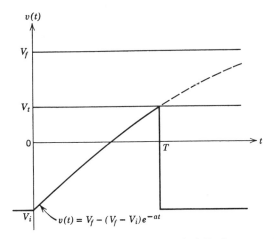

Figure 2-P22. Waveshape which governs the period T of a monostable multivibrator.

The problem is: Given a desired period T, and having both the initial voltage V_i and the would-be final voltage V_f specified, find the time constant $1/a$ and the voltage level V_t which result in the steepest possible slope of $v(t)$ at $t = T$. Assuming that all circuit parameters except possibly V_t can be held within fairly strict limits over operating conditions the above design is optimum in the sense that small variations in V_t have the least effect possible on the "fixed" period T.

Two equations are of importance. The constraint equation associated with $t = T$ is

$$f_1(a, V_t) \equiv V_f = V_t + (V_f - V_i)e^{-aT}$$

and the function $f_0(a, V_t)$ to be maximized is

$$f_0(a, V_t) \equiv \left.\frac{dv(t)}{dt}\right|_{t=T} = a(V_f - V_i)e^{-aT}$$

(This problem is considered by Pierre [2.22].)

2.23 A gating circuit to be designed is represented by the equivalent circuit shown in Figure 2-P23. When switch S_1 is open, diode d_1 is back biased, and

58 CLASSICAL THEORY OF MINIMA AND MAXIMA Ch. 2

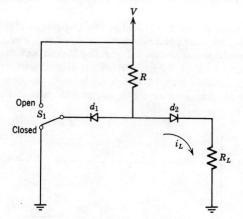

Figure 2-P23. Schematic diagram of a gating circuit.

current i_L flows through the load R_L. It is this current

$$f_0(R,V) \equiv i_L = \frac{V}{R + R_L}$$

that is to be maximized with respect to both R and V. However, when switch S_1 is closed, diode d_1 is forward biased, and $P = V^2/R$ is the rate at which energy is dissipated in the resistor R. In keeping with the trend toward microminiaturization, the resistor R (and possibly many such resistors in the overall system) is to be small in size, having a limit P_m of safe dissipation. Thus, a constraining relationship between V and R is

$$f_1(R,V) \equiv P_m = \frac{V^2}{R}$$

Find optimal values R^* and V^* of R and V in terms of R_L and P_m. (This problem is considered by Pierre [2.22].)

2.24 Consider the clipping circuit (Figure 2-P24a) which has the input-output voltage characteristic shown in Figure 2-P24b. It is assumed that the

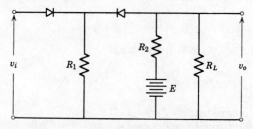

Figure 2-P24a. Schematic diagram of a clipping circuit.

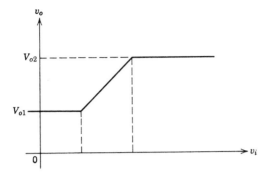

Figure 2-P24b. Transfer characteristic of the clipping circuit of Figure 2-P24a.

load resistor R_L and the voltage levels V_{o1} and V_{o2} are specified, and that the voltage E and the resistors R_1 and R_2 are to be selected on the basis that (a) the desired input-output characteristics are obtained, and (b) the energy drain on the E-voltage supply is as low as possible when v_i is less than V_{o1}.

The describing equations are obtained on the basis of the circuit (Figure 2-P24a) and its characteristics (Figure 2-P24b). The power P to be minimized is

$$P = \frac{(E - V_{o1})^2}{R_2} + \frac{V_{o1}^2}{R_1} + \frac{V_{o1}^2}{R_L}$$

Constraints on the optimization process are

$$V_{o1} = R_1\left[\frac{E - V_{o1}}{R_2} - \frac{V_{o1}}{R_L}\right]$$

and

$$V_{o2} = \frac{(E - V_{o2})R_L}{R_2}$$

Find optimal expressions for R_1, R_2, and E. (This problem is considered by Pierre [2.22].)

2.25 Find the maximum of

$$P = \left(-\sum_{k=1}^{n} w_k x_k \log_2 x_k\right) \Big/ \left(\sum_{k=1}^{n} k w_k x_k\right)$$

subject to the constraint that

$$\sum_{k=1}^{n} w_k x_k = 1$$

where the w_k's are specified positive integers and the x_k's are to be selected for the maximum of P. (A problem of this form is considered by Chang [2.5].)

2.26 A component has a nonlinear characteristic $g(x)$ which is to be approximated by a second-order polynomial $p(x) = ax^2 + bx + c$ over the interval defined by $1 \leq x \leq 5$. Values of $g(x)$ are listed as follows:

x	1	2	3	4	5
$g(x)$	3	5	4	2	1

At the initial value $x = 1$, it is required that $p(1) = g(1)$; and therefore,

$$a + b + c = 3$$

Subject to this constraint, determine optimal values of a, b, and c in the sense that $f(a,b,c)$ is minimized, where

$$f(a,b,c) \triangleq \sum_{k=2}^{5} [p(k) - g(k)]^2$$

2.27 Consider a sampled-data control system in which the state of the system at the kth sampling instant is x_k; and x_k is related to x_{k-1} and to the control action m_{k-1} by

$$x_k = ax_{k-1} + bm_{k-1}$$

for all integer values of k where a and b are real constants. At each sampling instant j, m_j is to be generated such that a minimum of P is obtained, where

$$P = \sum_{k=j}^{j+2} \tfrac{1}{2}[(x_{k+1} - x_{k+1}^d)^2 + hm_k^2]$$

in which x_k^d is a desired value of x_k, and h equals a specified constant. The optimal value m_j^* of m_j is to be determined solely in terms of a measured value of x_j and in terms of known values of h, x_{j+1}^d, x_{j+2}^d, and x_{j+3}^d. In this way,

$$m_j^* = m_j^*(x_j, x_{j+1}^d, x_{j+2}^d, x_{j+3}^d, h)$$

results in closed-loop (feedback) control.
  a. Determine the form of $m_j^*(x_j, x_{j+1}^d, x_{j+2}^d, x_{j+3}^d, h)$ for the case that b, a, and h equal 1.
  b. Outline the general procedure for solution when x_k is replaced by an $n \times 1$ matrix $\mathbf{x}_k$, a is replaced by an $n \times n$ matrix $\mathbf{A}$, b is replaced by an $n \times 1$ matrix $\mathbf{b}$, and $(x_{k+1} - x_{k+1}^d)^2$ is replaced by $(\mathbf{x}_{k+1} - \mathbf{x}_{k+1}^d)'(\mathbf{x}_{k+1} - \mathbf{x}_{k+1}^d)$.

2.28 Use the slack-variable approach suggested in Section 2-7 and obtain the necessary conditions which must be satisfied by the optimal solution of the example in Section 2-7.

2.29 A given function $g(t)$, $0 \leq t \leq \infty$, is to be approximated by $g_a(t)$, where

$$g_a(t) = \sum_{k=1}^{n} x_k \psi_k(t)$$

in which the x_k's are to be selected in the approximation process, and the $\psi_k(t)$'s are orthonormal functions of time over the time interval $0 \leq t \leq \infty$; that is,

$$\int_0^\infty \psi_k(t)\psi_j(t)\,dt = \begin{cases} 1 & \text{when } j = k \\ 0 & \text{when } j \neq k \end{cases}$$

The approximation $g_a(t)$ is required to satisfy

$$\int_0^\infty g_a(t)\,dt = \int_0^\infty g(t)\,dt = c_1 \quad \text{(a constant)}$$

Find expressions for the optimal x_k's, x_k^*'s, which satisfy this constraint and which minimize the integral-square error I:

$$I = \int_0^\infty [g_a(t) - g(t)]^2\,dt$$

(This problem is treated by Smith [2.27].)

2.30 Find an equation which constitutes a necessary condition for the minimum of $\overline{e^2(t)}$,

$$\overline{e^2(t)} = g\left(\frac{R}{2x}\right)\int_0^x S(f)\,df + \int_x^\infty S(f)\,df$$

with respect to x. The parameter R is a constant, and the functions $S(f)$ and $g(R/2x)$ are explicit real-valued functions of their respective arguments. (This problem is considered by Goodman [2.12].)

2.31 Suppose that a parameter x is to be selected where it is known that x may change by $\pm 10\%$ over a time period of interest. It is desired that $f(x)$ be maximized, where $f(x)$ is depicted in Figure 2-P31. Which value of x should be specified and why?

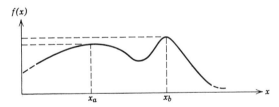

Figure 2-P31.

2.32 Conduct a sensitivity study of the solution of a problem from this chapter.

CLASSICAL CALCULUS OF VARIATIONS

3

3-1. INTRODUCTION

In its most general form, calculus of variations is the body of theory with which a set of functions of several independent variables is selected so as to extremize a given measure of the set of functions, subject to known constraints imposed on the functions. The phrase "calculus of variations" was first used to describe theory associated with the solution of such problems as a result of notation used by Lagrange about the year 1760. The word "variation" is a key word in the classical solution of such problems, as is shown in this chapter.

In Chapter 2, the problem of determining the *points of extrema* of a function was of interest. Here the object is to find the *functions of extrema* of a *functional*. A functional, then, has functions and perhaps derivatives of functions for its arguments. For a specific assignment of the functions, a scalar functional attains a scalar value.

The functionals treated in this chapter are restricted to those which can be represented in the form of an integral of a function of several dependent functions. Only one *independent* variable of integration is considered. The majority of variational problems treated in the literature are or can be put in the above form. For those which cannot, the attainment of necessary conditions for the optimum usually follows by reasoning which is analogous to that presented in the following sections.

In some problems, the functional to be extremized is an obvious facet of the problem, e.g., a functional representing expended energy which is to be minimized. In other problems, however, the functional to be used is a matter of choice. This is especially true for problems in which an error as a continuous function of time is to be minimized; the functional for such a problem could be the integral of the squared error (ISE criterion); it could be the integral of the absolute value of the error; it could be the integral of the time-weighted absolute value of the error (ITAE criterion); and so on, there being an unlimited number of such measures which could be used. For a particular design problem, one choice may be preferable to the others on the basis of practical considerations, and a different choice may be preferable to others because of the mathematical finesse with which it leads to a solution of the problem. For example, the ISE criterion is a tractable one with which emphasis is placed on the reduction of large error magnitudes; whereas the ITAE criterion is a less tractable one, but sometimes a more desirable one, with which emphasis on error magnitude reduction is increased linearly with time.

The theory developed in this chapter is the Euler-Lagrange formalism of the calculus of variations, in contrast to the Hamiltonian approach and the Hamilton-Jacobi approach. These last two approaches are considered further in the chapters which are primarily concerned with Pontryagin's maximum principle and dynamic programming. In Section 3-2, fundamental concepts are presented, and equivalence relations are established between three different forms of functionals. In Sections 3-3 through 3-6, fundamental necessary conditions for the optimum are developed for integral functionals with one dependent function to be selected for the attainment of an optimal solution. In Sections 3-7 and 3-8, these necessary conditions are generalized to the case of n dependent functions, and various constraint conditions are incorporated in the theory by use of Lagrange multipliers. These Lagrange multipliers are often functions of the independent variable of integration. Various constraints can be imposed on the end-point values of independent and dependent variables.

Design examples are given throughout to illustrate applications of the theory; and in Section 3-9, a general Euler-Lagrange formulation for a class of control problems is given with its limitations.

A solution which is obtained on the basis of "necessary conditions" should be tested by "sufficient conditions" to establish with certainty the optimality of the solution. In many problems, sufficiency can be established by considering the physical nature of the process involved. When this is not feasible, mathematical sufficiency conditions may be used to establish the optimality of a solution. Such conditions are considered in Section 3-10.

In Section 3-11, attention is centered on *direct methods* of finding the optimal solutions to variational problems—direct methods are those which do not depend on the application of classical necessary-and-sufficient conditions. And in Section 3-12, functionals are examined from the standpoint of sensitivity with respect to changes in their argument functions.

3-2. PRELIMINARY CONCEPTS

3-2a. Continuity, Extrema, and Variations

Pertinent properties of functionals are introduced by considering the single dependent function case:

$$J = P(x,t) \tag{3-1}$$

where for a specified range of the real independent variable t and for a given real-valued function $x \equiv x(t)$, the performance measure $P(x,t)$ yields a scalar value.

In this chapter, interest is centered primarily on those $x(t)$'s which are continuous and for which derivatives are uniquely defined, except possibly at a finite number of points within the range associated with the independent variable t. Curves (functions) which exhibit these properties are often referred to as *admissible curves* in the classical literature. Of course, we should not conclude that only this type of "admissible" curve results in the extremization of functionals. For example, consider the particular functional

$$\int_{-c}^{c} t^2 \dot{x}^2 \, dt$$

where c is a positive constant, $x(-c) = -1$, and $x(c) = 1$. This functional does not contain a relative minimum within the above-noted class of admissible curves; the minimum of zero is obtained when x equals

$$x^* = \lim_{k \to \infty} (\tan^{-1} kt)/(\tan^{-1} kc),$$

which has a jump discontinuity at $t = 0$.

Consider the case in which a function $x_\alpha \equiv x_\alpha(t)$ results in a relative minimum $J(x_\alpha)$, Equation 3-1, for the range of t between given values t_a and t_b, $t_a < t_b$. The implications of the above statement are not as straightforward as those of the analogous statement for ordinary min-max problems. Here the functional $J(x_\alpha)$ may be a minimum relative to functions x not identically equal to x_α in the sense that $J(x_\alpha) - J(x) < 0$ whenever $|x_\alpha - x| < \Delta_1$ for some real value Δ_1 and all t contained in $[t_a,t_b]$, in which case $J(x_\alpha)$ is said to be a *strong relative minimum*. On the other hand, $J(x_\alpha)$ may be a minimum relative to functions x not identically equal to x_α in the sense that $J(x_\alpha) - J(x) < 0$ whenever $|x_\alpha - x| < \Delta_1$ and $|\dot{x}_\alpha - \dot{x}| < \Delta_2$ for some real values Δ_1 and Δ_2 and all t contained in $[t_a,t_b]$, in which case $J(x_\alpha)$ is said to be a *weak relative minimum*.[1] A strong relative minimum is therefore a special case of a

---

[1] These defining conditions can be expressed more compactly in terms of a norm on the space $C(t_a,t_b)$ of continuous functions (see Appendix D). The definitions of strong and weak relative extrema given here are to be distinguished from those of Chapter 2.

Sect. 3-2   PRELIMINARY CONCEPTS   65

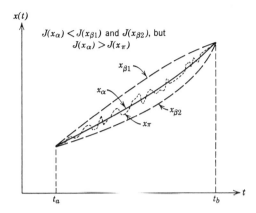

Figure 3-1. Example of a weak relative minimum.

weak relative minimum in which Δ_2 can be assigned an arbitrarily large value. Similar statements apply to strong and weak relative maxima.

The curves in Figure 3-1 illustrate a hypothetical weak relative minimum. The curves $x_{\beta 1}$ and $x_{\beta 2}$ are representative functions of a class of continuous functions for which $|\dot{x}_\beta - \dot{x}_\alpha|$ is less than some number, say 10, for all t in $[t_a, t_b]$; whereas the curve x_π is representative of a class of continuous functions for which $|\dot{x}_\pi - \dot{x}_\alpha| \geq 10$ for some nontrivial subset of $[t_a, t_b]$. Corresponding to the curve $x_\alpha(t)$, the value $J(x_\alpha)$ is less than $J(x_{\beta 1})$ or $J(x_{\beta 2})$. On the other hand, $J(x_\alpha)$ is greater than $J(x_\pi)$, this being the case for some x_π even when $|x_\pi(t) - x_\alpha(t)|$ is arbitrarily small over $[t_a, t_b]$. On the basis of this example, it should be evident that relative extrema associated with variational problems of a physical origin are generally strong relative extrema.

The functional $J(x)$ is said to be *strongly continuous* (or, simply, *continuous*) if a small modification in the function x always results in a small change in the scalar value of the functional $\hat{J}(x)$. Specifically, let $x_\alpha = x_\alpha(t)$ and $x_\beta = x_\beta(t)$ denote typical functions. Then the functional $J(x)$ is termed strongly continuous about $x_\alpha(t)$ if for any real number ϵ_1, $\epsilon_1 > 0$, there exists a real number ϵ_2 such that, for every function x_β which satisfies $|x_\alpha - x_\beta| < \epsilon_2$ throughout the range of t of interest, the value of $|J(x_\alpha) - J(x_\beta)|$ is less than ϵ_1.

The concept of a *variation* is paramount to the calculus of variations. When a function $x = x(t)$ is modified by an amount $\delta x = \delta x(t)$ to $x + \delta x$, the modification δx in the function is called a *variation of the function* x. Any variation δx can be viewed as a special case of a variation $\epsilon \, \delta x$ where ϵ is a real independent variable. Corresponding to a variation $\epsilon \, \delta x$ in the function x, the functional J of Equation 3-1 assumes the value $J(x + \epsilon\, \delta x)$. Let the *increment* ΔJ be defined by

$$\Delta J \triangleq J(x + \epsilon\, \delta x) - J(x) \tag{3-2}$$

and assume that the range of t of interest is fixed. Also, for a given x and δx, suppose that $J(x + \epsilon\, \delta x)$ possesses finite derivatives of all orders with respect to ϵ in an open neighborhood of $\epsilon = 0$. It follows that a Maclaurin's series expansion with respect to ϵ can be made of the first term in the right-hand member of 3-2 with the result that

$$\Delta J = \left.\frac{\partial J(x + \epsilon\, \delta x)}{\partial \epsilon}\right|_{\epsilon=0} \epsilon + \left.\frac{\partial^2 J(x + \epsilon\, \delta x)}{\partial \epsilon^2}\right|_{\epsilon=0} \frac{\epsilon^2}{2} + \cdots \tag{3-3a}$$

By convention (dating back to Lagrange) the first term $[\partial J(x + \epsilon\, \delta x)/\partial \epsilon]_{\epsilon=0}\, \epsilon$ is called the *first variation of the functional* J evaluated at x and is denoted by δJ. As is shown in Section 3-3, a necessary condition for a relative extremum of J evaluated at a particular x, whether it be strong or weak, is that δJ must equal zero for that x when $\epsilon\, \delta x$ is an admissible variation.[2] In like manner, the term $[\partial^2 J(x + \epsilon\, \delta x)/\partial \epsilon^2]_{\epsilon=0}\, \epsilon^2$ of 3-3a is called the *second variation of the functional* J evaluated at x and is denoted by $\delta^2 J$. The second variation plays an important role in the establishment of sufficient conditions for weak relative extrema. In general, the increment ΔJ can be expressed in the form

$$\Delta J = \delta J + \frac{\delta^2 J}{2!} + \frac{\delta^3 J}{3!} + \cdots \tag{3-3b}$$

in which the first variation δJ is linear in ϵ, the second variation $\delta^2 J$ is linear in ϵ^2, etc. The evaluation of the series 3-3 at $\epsilon = 1$ is a special case of particular interest.

3-2b. Classes of Problems and Equivalence Relations

In this subsection, three related variational problems are examined. The distinguishing feature of these problems is the form of the functional employed—the three functionals are J, J_μ, and J_β, as follows:

$$J = \int_{t_a}^{t_b} f(\mathbf{x},\dot{\mathbf{x}},t)\, dt \qquad \text{(the problem of Lagrange)} \tag{3-4}$$

$$J_\mu = f_\mu(\mathbf{x},t)\big|_{t_a}^{t_b} = f_\mu[\mathbf{x}(t_b),t_b] - f_\mu[\mathbf{x}(t_a),t_a] \qquad \text{(the problem of Mayer)} \tag{3-5}$$

and

$$J_\beta = f_{\beta 1}(\mathbf{x},t)\big|_{t_a}^{t_b} + \int_{t_a}^{t_b} f_{\beta 2}(\mathbf{x},\dot{\mathbf{x}},t)\, dt \qquad \text{(the problem of Bolza)} \tag{3-6}$$

---

[2] An admissible variation is one that satisfies the properties of an admissible curve.

where $\mathbf{x} \equiv \mathbf{x}(t)$ represents the set $\{x_1(t), x_2(t), \ldots, x_n(t)\}$ of real-valued functions of time, $\dot{\mathbf{x}}$ represents the set $\{\dot{x}_1(t), \dot{x}_2(t), \ldots, \dot{x}_n(t)\}$ of derivatives of $\mathbf{x}$, and $f(\mathbf{x}, \dot{\mathbf{x}}, t)$, $f_\mu(\mathbf{x}, t)$, $f_{\beta 1}(\mathbf{x}, t)$, and $f_{\beta 2}(\mathbf{x}, \dot{\mathbf{x}}, t)$ are given real-valued functions of class C^2 with respect to their arguments, that is, partial derivatives of these functions up to and including the second are assumed to exist and to be continuous. It is convenient to express $\mathbf{x}$ in column matrix form,

$$\mathbf{x} = [x_1(t) \quad x_2(t) \quad \cdots \quad x_n(t)]' \tag{3-7}$$

and $\mathbf{x}$ is called a state vector.

The functionals of 3-4, 3-5, and 3-6 are typically used as performance measures for the design and/or operation of physical systems. The general problem associated with the extremization of J, Equation 3-4, is known as the problem of Lagrange; that associated with J_μ, Equation 3-5, is known as the problem of Mayer; and that associated with J_β, Equation 3-6, is known as the problem of Bolza [3.7]. In a given problem, the objective is to find a trajectory $\mathbf{x}^*(t)$ which when substituted for $\mathbf{x}(t)$ in a particular performance measure results in the optimum of the performance measure. Of course, certain of the x_i's and $\dot{x}_i$'s may be constrained to satisfy known physical laws; such constraints typically take the form of differential equations—these and other constraint forms are considered later in this chapter. Similarly, the end conditions $\mathbf{x}(t_a)$, $\mathbf{x}(t_b)$, t_a, and t_b may be constrained to satisfy prescribed relationships; these may either partially or completely prescribe $\mathbf{x}(t_a)$, $\mathbf{x}(t_b)$, t_a, and t_b. When not completely prescribed, the optimal end conditions may be found by use of *transversality conditions* (Sections 3-5 and 3-7).

In this section, the objective is simply to show that the problems of Bolza and Mayer can be placed in the form of the problem of Lagrange. Thus, the theory that is developed in this chapter for solution of Lagrange's problem can be applied to the solution of the other two problems as well.

Consider first the problem of Mayer associated with 3-5. Because of the assumed differentiability of $f_\mu(\mathbf{x}, t)$, J_μ of 3-5 can be placed in the form

$$\begin{aligned} J_\mu &= \int_{t_a}^{t_b} \frac{df_\mu(\mathbf{x}, t)}{dt} \, dt \\ &= \int_{t_a}^{t_b} \left\{ \frac{\partial f_\mu(\mathbf{x}, t)}{\partial t} + \left[\frac{\partial f_\mu(\mathbf{x}, t)}{\partial \mathbf{x}}\right]' \dot{\mathbf{x}} \right\} dt \end{aligned} \tag{3-8}$$

in which $\partial f_\mu(\mathbf{x}, t)/\partial \mathbf{x}$ is the *gradient* of $f_\mu(\mathbf{x}, t)$ with respect to $\mathbf{x}$ and is defined, as expressed in column matrix form, by

$$\frac{\partial f_\mu(\mathbf{x}, t)}{\partial \mathbf{x}} \overset{\Delta}{=} \left[\frac{\partial f_\mu(\mathbf{x}, t)}{\partial x_1} \quad \frac{\partial f_\mu(\mathbf{x}, t)}{\partial x_2} \quad \cdots \quad \frac{\partial f_\mu(\mathbf{x}, t)}{\partial x_n}\right]' \tag{3-9}$$

Obviously, J_μ of 3-8 is of the form of J of 3-4, and the desired equivalence is established.

For the problem of Bolza associated with 3-6, the same approach as followed in the preceding paragraph can be applied to the first term of the right-hand member of 3-6, with the result that

$$J_\beta = \int_{t_a}^{t_b} \left\{ f_{\beta 2}(\mathbf{x},\dot{\mathbf{x}},t) + \frac{\partial f_{\beta 1}(\mathbf{x},t)}{\partial t} + \left[\frac{\partial f_{\beta 1}(\mathbf{x},t)}{\partial \mathbf{x}}\right]' \dot{\mathbf{x}} \right\} dt \qquad (3\text{-}10)$$

As before, this expression is in the form of J of 3-4, and the desired relationship between Bolza's problem and Lagrange's problem is thereby established.

3-3. THE PROBLEM OF LAGRANGE: SCALAR CASE

3-3a. Problem Statement and the First Variation

Consider the problem of determining the particular real curve $x^* \equiv x^*(t)$ which, when substituted for x, yields the minimum (maximum) of the functional J:

$$J \equiv J(x) = \int_{t_a}^{t_b} f(x,\dot{x},t)\, dt \qquad (3\text{-}11)$$

where t_a, t_b, $x(t_a) = c_a$, and $x(t_b) = c_b$ are fixed. The necessary and the sufficient conditions that are developed in this chapter are applicable to the extremization of J, Equation 3-11, provided that the real-valued function $f \equiv f(x,\dot{x},t)$ is of class C^2 with respect to all of its arguments.

The preceding statements constitute the most basic form of the problem known as the problem of Lagrange in the calculus of variations. All of the results that are derived for the solution of this problem in this section are extendible to more complex situations.

The optimal $x(t)$, $x^*(t)$, is assumed to belong to a family of functions with certain properties in common. One property of this family is that the end conditions must be satisfied in each case, i.e., for each x of the family, $x(t_a) = c_a$ and $x(t_b) = c_b$. A second property of the family is that $x_\alpha = x + \epsilon(x_\beta - x)$ is a member of the family if x and x_β are members of the family where ϵ is any real number. Notice in particular that if x^* is the optimal curve, then $x_\alpha = x^* + \epsilon\, \delta x$ is a member of the family where $\epsilon\, \delta x = \epsilon(x_\beta - x^*)$ is a *variation* of x^*, as defined in the preceding section.

Consider the nonoptimal value of J given by

$$J(x^* + \epsilon\, \delta x) = \int_{t_a}^{t_b} f(x^* + \epsilon\, \delta x, \dot{x}^* + \epsilon\, \delta \dot{x}, t)\, dt \qquad (3\text{-}12)$$

where $\delta \dot{x}$ is used to denote $d\, \delta x/dt$.

For the sake of clarity, assume that $J(x^*)$ is the minimum, rather than the maximum of $J(x)$. Then for any particular δx, a plot of J versus ϵ similar

Sect. 3-3 THE PROBLEM OF LAGRANGE: SCALAR CASE

to that of Figure 3-2a can be drawn. Corresponding to the definition given in the preceding section, the quantity ϵ multiplied by the slope of the curve at $\epsilon = 0$ in Figure 3-2a is the *first variation of J* for the particular case that x equals x^*. In general, the first variation δJ is given by

$$\delta J = \left.\frac{\partial J(x + \epsilon\,\delta x)}{\partial \epsilon}\right|_{\epsilon=0} \epsilon \qquad (3\text{-}13)$$

Observe that the slope $[\partial J(x^* + \epsilon\,\delta x)/\partial \epsilon]|_{\epsilon=0}$ in Figure 3-2a must equal zero for any allowable δx if $J(x^*)$ is to be truly the minimum; therefore, a *necessary condition* for the minimum of J is that $[\partial J(x + \epsilon\,\delta x)/\partial \epsilon]|_{\epsilon=0}$ must equal zero when x equals x^*, where

$$\left.\frac{\partial J(x + \epsilon\,\delta x)}{\partial \epsilon}\right|_{\epsilon=0} = \left\{\frac{\partial}{\partial \epsilon}\int_{t_a}^{t_b} f(x + \epsilon\,\delta x, \dot x + \epsilon\,\delta \dot x, t)\,dt\right\}_{\epsilon=0} \qquad (3\text{-}14\text{a})$$

The integrand of this equation can be expanded in a Taylor's series about x, $\dot x$, and t; thus,

$$\begin{aligned}\left.\frac{\partial J(x + \epsilon\,\delta x)}{\partial \epsilon}\right|_{\epsilon=0} &= \left\{\frac{\partial}{\partial \epsilon}\int_{t_a}^{t_b}\left[f(x,\dot x,t) + \epsilon\,\delta x\,\frac{\partial f(x,\dot x,t)}{\partial x}\right.\right.\\ &\qquad\left.\left.+ \epsilon\,\delta \dot x\,\frac{\partial f(x,\dot x,t)}{\partial \dot x} + \epsilon o(\epsilon)\right]dt\right\}_{\epsilon=0}\\ &= \left\{\int_{t_a}^{t_b}\left[\frac{\partial f(x,\dot x,t)}{\partial x}\,\delta x + \frac{\partial f(x,\dot x,t)}{\partial \dot x}\,\delta \dot x + o_1(\epsilon)\right]dt\right\}_{\epsilon=0}\\ &= \int_{t_a}^{t_b}(f_x\,\delta x + f_{\dot x}\,\delta \dot x)\,dt \qquad (3\text{-}14\text{b})\end{aligned}$$

in which f_x is the partial derivative of $f(x,\dot x,t)$ with respect to x (meaning that $\dot x$ and t are treated as constants in the differentiation), and $f_{\dot x}$ is the partial derivative of $f(x,\dot x,t)$ with respect to $\dot x$ (meaning that x and t are

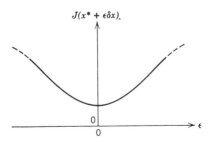

Figure 3-2a. Plot of $J(x^* + \epsilon\,\delta x)$ versus ϵ: $J(x^*)$ minimum, and δx fixed but typical.

treated as constants in the differentiation). Both $o(\epsilon)$ and $o_1(\epsilon)$ in 3-14b are functions of order ϵ, meaning that $\lim_{\epsilon \to 0+} o_1(\epsilon) = 0$. The fact that the right-hand member of 3-14b must equal zero when x equals x^* is of little use in the present form; what is desired is an expression which is independent of δx and which can be used with other facts to find $x^*(t)$. Such an expression is derived as follows.

The first term of the integrand of 3-14b is integrated by parts to obtain

$$\frac{\partial J(x + \epsilon \delta x)}{\partial \epsilon}\bigg|_{\epsilon=0} = \left[\delta x \int f_x \, dt\right]_{t_a}^{t_b} + \int_{t_a}^{t_b} \left(f_{\dot{x}} - \int f_x \, dt\right) \delta \dot{x} \, dt \quad (3\text{-}15)$$

In this section, the end conditions are assumed to be fixed, and therefore $\delta x(t_a) = \delta x(t_b) = 0$. Thus,

$$\frac{\partial J(x + \epsilon \delta x)}{\partial \epsilon}\bigg|_{\epsilon=0} = \int_{t_a}^{t_b} N \, \delta \dot{x} \, dt \quad (3\text{-}16)$$

where

$$N \triangleq f_{\dot{x}} - \int f_x \, dt \quad (3\text{-}17)$$

The condition on N which assures that $\delta J = 0$ is embodied in the following lemma which is known as the *fundamental lemma* of the calculus of variations.

3-3b. Fundamental Lemma

Of all bounded, single-valued functions $N = N(t)$ which are continuous on the interval $[t_a, t_b]$, except possibly at a finite number of jump-type discontinuities, only those $N(t)$ which equal a constant on $[t_a, t_b]$ result in zero values of the integral

$$\int_{t_a}^{t_b} N(t) \, \delta \dot{x}(t) \, dt \quad (3\text{-}18)$$

for every admissible $\delta x(t)$ where $\delta x(t_a) = \delta x(t_b) = 0$.

In other words, of the class of functions considered for $N(t)$, it is necessary and sufficient that $N(t)$ equal a constant c to insure that $\delta J = 0$. That it is sufficient for $N(t)$ to equal c is established by integrating 3-18 by parts,

$$\int_{t_a}^{t_b} N(t) \, \delta \dot{x}(t) \, dt = [N(t) \, \delta x]_{t_a}^{t_b} - \int_{t_a}^{t_b} \frac{dN(t)}{dt} \delta x \, dt \quad (3\text{-}19)$$

The first term of the right-hand member of 3-19 is zero because $\delta x(t_a) = \delta x(t_b) = 0$; and if $N(t) = c$, a constant, the second term of the right-hand member of 3-19 is also zero.

Sect. 3-3   THE PROBLEM OF LAGRANGE: SCALAR CASE

The necessary part of the lemma is established by considering a particular variation which is given by

$$\delta\dot{x}(t) = N(t) - c \tag{3-20}$$

and by noting that

$$\int_{t_a}^{t_b} [N(t) - c]\, \delta\dot{x}(t)\, dt \tag{3-21}$$

must vanish for all admissible $\delta x(t)$ if 3-18 is to vanish also. It follows, upon substitution of the right-hand member of 3-20 for $\delta\dot{x}(t)$ in 3-21, that

$$\int_{t_a}^{t_b} [N(t) - c]^2\, dt = 0 \tag{3-22}$$

only if $N(t) = c$.

3-3c. First Necessary Condition and First-Variational Curves

The important result thus far is that the equation

$$f_{\dot{x}} = \int f_x\, dt + c \tag{3-23}$$

must be satisfied by the optimal $x = x^*(t)$ which results in the minimum (maximum) of J. Unfortunately, more than one function may satisfy 3-23. For example (see Figure 3-2b), the first variation δJ of J about the curve x_s is zero, but x_s is obviously not optimum. However, in the case that the solution of 3-23 is unique and the optimal $x = x^*$ is from the class of continuous functions considered in this chapter, it follows necessarily that the solution is optimal. In short, Equation 3-23 constitutes a necessary condition—often called the *first necessary condition*—for the optimal x, but not a sufficient one.

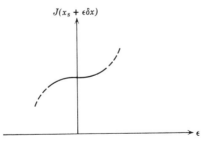

Figure 3-2b. Plot of $J(x_s + \epsilon\, \delta x)$ versus ϵ: $J(x_s)$ nonoptimal, δx fixed but typical, and $[\partial J(x_s + \epsilon\, \partial x)/\partial \epsilon]|_{\epsilon=0}$ equals zero.

Curves $x(t)$ which satisfy the first necessary condition are named *extremal curves* in the classical literature, an unfortunate designation in that all such curves do not result in relative extrema of J. In analogy with the designation of stationary points in Chapter 2, we could conceivably refer to such curves as stationary curves, but this too is nondescriptive, if not altogether misleading. Instead, curves which satisfy 3-23 are designated as *first-variational curves* in this book because they are obtained by operating on the first variation of the functional J.

3-3d. A Corner Condition

Equation 3-23 is now employed in developing two important results which are used in the determination of first-variational curves. The first of these results involves the concept of a *corner point* of a function. In Figure 3-3, the function $x(t)$ has a corner point at $t = t_c$. A corner point of $x(t)$, therefore, is a point at which the derivative dx/dt is not uniquely defined; i.e., a point at which dx/dt possesses a jump discontinuity. Corner points of functions are found in problems which involve reflection or refraction, for example.

Suppose $x(t)$ is a continuous first-variational curve; the question is, "Does $x(t)$ possess corner points, and if so, where are they?" As a first step toward the answer to this question, observe that the integrand of the indefinite integral in 3-23 exhibits, at most, bounded jump discontinuities as a function of t when $x(t)$ is a continuous first-variational curve. The integral of such a function is continuous, so the right-hand member of 3-23 is continuous. But then the left-hand member of 3-23 must be continuous too. Thus, if $x(t)$ has a corner point at $t = t_c$, it must still be true that[3]

$$f_{\dot{x}}|_{t=t_c-0} = f_{\dot{x}}|_{t=t_c+0} \tag{3-24}$$

The above condition is one part of the Erdmann-Weierstrass corner con-

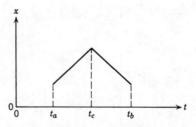

Figure 3-3. Corner point at $t = t_c$.

---
[3] The expressions $t_c - 0$ and $t_c + 0$ denote, respectively, $\lim_{\epsilon \to 0+} t_c - \epsilon$ and $\lim_{\epsilon \to 0+} t_c + \epsilon$.

Sect. 3-3   THE PROBLEM OF LAGRANGE: SCALAR CASE   73

ditions which are treated in more detail in Section 3-6. In conjunction with the results that follow, the Erdmann-Weierstrass corner conditions can be used to determine locations of corner points, if they exist.

3-3e. The Euler-Lagrange Equation

The fact that the left- and right-hand members of 3-23 are continuous enables us to differentiate and obtain

$$\frac{d}{dt}(f_{\dot{x}}) - f_x = 0 \qquad (3\text{-}25)$$

which is known as the *Euler-Lagrange equation*—unquestionably the most famous result of the calculus of variations.

At values of t between corners of $x(t)$, differentiation of the first term in 3-25 can be effected to yield

$$\frac{d^2x}{dt^2}f_{\dot{x}\dot{x}} + \frac{dx}{dt}f_{\dot{x}x} + f_{\dot{x}t} = f_x \qquad (3\text{-}26)$$

where $x(t_a) = c_a$ and $x(t_b) = c_b$ are the given boundary conditions. Because of the specified end conditions, the problem of solving 3-26 is called a *two-point boundary-value problem*. In some cases, e.g., the detailed examples in this chapter, the Euler-Lagrange equations can be solved in closed form, and two-point boundary conditions are easily incorporated in the solution. In other cases, however, special techniques ([3.5, 3.11, 3.15, 3.17, and 3.26] for example) may be required for solution of corresponding two-point boundary-value problems. Dynamic programming (Chapter 7), direct methods (Section 3-11), and search techniques (Chapter 6 and Section 8-8) are of use in this regard. Several important cases of Euler-Lagrange equations are given below.

Example 3-1. ($f(x,\dot{x},t)$ quadratic in x and $\dot{x}$.)
Suppose that

$$f = a_1\dot{x}^2 + a_2\dot{x}x + a_3x^2 + g_1(t)\dot{x} + g_2(t)x \qquad (3\text{-}27)$$

is the integrand of 3-11 where a_1, a_2, and a_3 are real constants, and $g_1(t)$ and $g_2(t)$ are given, differentiable, real-valued functions. The occurrences of this type of problem and related problems in technical fields are many and varied ([3.13, 3.19, and 3.20] for example). In this case, the Euler-Lagrange equation 3-26 between corners reduces to

$$2a_1\ddot{x} - 2a_3x = g_2(t) - \dot{g}_1(t) \qquad (3\text{-}28a)$$

which is a nonhomogeneous, second-order, linear differential equation.

Assume that a_1 and a_3 are greater than zero; and set $\alpha^2 \triangleq (a_3/a_1)$ and $g_0(t) \triangleq [g_2(t) - \dot{g}_1(t)]/2a_1$. The solution of 3-28a for $t \geq t_a$ is

$$x(t) = c_1 e^{-\alpha t} + c_2 e^{\alpha t} + \frac{1}{2\alpha} \int_0^{t-t_a} (e^{\alpha \tau} - e^{-\alpha \tau}) g_0(t - \tau) \, d\tau \quad (3\text{-}28\text{b})$$

where c_1, c_2, and t_a are constants that are determined so as to satisfy boundary conditions in any given case.

●

Example 3-2. (f formally independent of t.)

Assume that f is formally independent of t, meaning that t is not an explicit argument of f, whereas $x(t)$ and $\dot{x}(t)$ may be, and therefore $\partial f/\partial t$ is zero. Consider the identity

$$\frac{d}{dt}(\dot{x}f_{\dot{x}} - f) = \dot{x}\ddot{x}f_{\dot{x}\dot{x}} + \dot{x}^2 f_{x\dot{x}} + \ddot{x}f_{\dot{x}} - \dot{x}f_x - \ddot{x}f_{\dot{x}}$$

$$= \dot{x}(\ddot{x}f_{\dot{x}\dot{x}} + \dot{x}f_{x\dot{x}} - f_x) \quad (3\text{-}29)$$

The term within the parentheses of the right-hand member of 3-29 equals zero for any x which satisfies the Euler-Lagrange equation 3-26 with $f_{\dot{x}t} = 0$. Hence, in this case the Euler-Lagrange equation has a *first integral* given by

$$\dot{x}f_{\dot{x}} - f = c \quad (3\text{-}30)$$

where c is a constant. This first integral may be solved by solving 3-30 for $\dot{x}$ in terms of x and c followed by application of the method of separation of variables.

●

Example 3-3. (f formally independent of $\dot{x}$.)

If $f_{\dot{x}\dot{x}} = f_{\dot{x}x} = f_{\dot{x}t} = 0$, the Euler-Lagrange equation reduces to

$$f_x = 0 \quad (3\text{-}31)$$

The solution of this equation involves no arbitrary constants, and therefore, boundary conditions cannot be specified arbitrarily, unless of course the class of functions of which the optimal x is a member is broadened to include functions with jump-type discontinuities at the ends of the interval.

In many practical problems of the preceding type, end conditions on x are not given explicitly. For example, consider the problem of minimizing $\overline{e^2}$, the mean-square error between a pulse-code-modulated signal and its detected interpretation [3.2]:

$$\overline{e^2} = \frac{1}{2} \int_{-\infty}^{\infty} [(1 - x)^2 p(\eta - a) + x^2 p(\eta + a)] \, d\eta \quad (3\text{-}32)$$

where $p(\eta - a)$ denotes the probability density function for noise associated with a "positive" pulse when received in the presence of additive Gaussian noise at a signal-to-noise ratio of a^2, $p(\eta + a)$ is defined correspondingly but for a "negative" pulse, and $x = x(\eta)$ is the desired detector characteristic. The Euler-Lagrange equation is

$$f_x = -(1-x)p(\eta - a) + xp(\eta + a) = 0 \qquad (3\text{-}33)$$

from which the optimal $x(\eta)$, $x^*(\eta)$, is obtained:

$$x^*(\eta) = \frac{p(\eta - a)}{p(\eta - a) + p(\eta + a)} \qquad (3\text{-}34)$$

●

3-4. ISOPERIMETRIC CONSTRAINTS

As an additive feature to the problem given in Section 3-2a, consider a constraint equation of the form

$$K_1 = \int_{t_a}^{t_b} f_1(x,\dot{x},t)\, dt \qquad (3\text{-}35)$$

where K_1 is a constant and $f_1 \equiv f_1(x,\dot{x},t)$ is a known real-valued function of its arguments with properties equivalent to those assumed for f in 3-11. Numerous examples of such constraints occur in practice because of limited resources (e.g., limited energy, limited fuel, limited surface area, etc.) available. In the classical literature, the first problems of this sort to be considered were problems concerning the determination of the maximum area of an enclosed planar surface where the "perimeter" of the area was constrained; hence, the term "isoperimetric" constraint was used, and this terminology remains in use today for constraints of the form of 3-35.

Isoperimetric constraints are easily incorporated in the problem of Section 3-3 by use of Lagrange multipliers—the Lagrange multiplier is used in the same manner as in Chapter 2.

Isoperimetric Theorem. *Assume that the function $x^*(t)$ is a first-variational curve which results in the maximum of the functional*

$$J_a(x) = \int_{t_a}^{t_b} [f(x,\dot{x},t) + h_1 f_1(x,\dot{x},t)]\, dt \qquad (3\text{-}36)$$

That is,

$$J_a(x) \le J_a(x^*) = \int_{t_a}^{t_b} [f(x^*,\dot{x}^*,t) + h_1 f_1(x^*,\dot{x}^*,t)]\, dt \qquad (3\text{-}37)$$

where h_1 is independent of x and t. Also, assume that constraint 3-35 is satisfied by

$x^*(t)$. Then $x^*(t)$ gives rise to the maximum of J in 3-11 subject to the isoperimetric constraint 3-35.

The proof of the theorem follows by considering the result which would hold if the theorem were false, i.e., assume that there exists a function $r(t)$ which has the property

$$\int_{t_a}^{t_b} f(r,\dot{r},t)\,dt > \int_{t_a}^{t_b} f(x^*,\dot{x}^*,t)\,dt \qquad (3\text{-}38)$$

and which satisfies the constraint equation

$$\int_{t_a}^{t_b} f_1(r,\dot{r},t)\,dt = K_1 \qquad (3\text{-}39)$$

Then

$$\int_{t_a}^{t_b} f(r,\dot{r},t)\,dt + h_1 \int_{t_a}^{t_b} f_1(r,\dot{r},t)\,dt = \int_{t_a}^{t_b} f(r,\dot{r},t)\,dt + h_1 K_1 \qquad (3\text{-}40)$$

But now the right-hand member of 3-40 is greater than that of 3-37, hence, a contradiction! The assumption that the theorem is not true is invalid, and the validity of the theorem follows.

•

With a change in inequality signs, the above argument applies equally well to the case that the minimum, rather than the maximum, of J is desired. Also, if two or more constraints of the form 3-35 exist, they can be treated by assigning a separate Lagrange multiplier to each constraint and by proceeding as before. Thus, the procedure to be used in solving such problems is similar to that used in Chapter 2 to treat equality constraints associated with classical min-max problems.

To utilize the theorem, one obtains the Euler-Lagrange equation corresponding to the *augmented functional* 3-36 and solves for first-variational curves in terms of t and the Lagrange multiplier h_1. These are then substituted into the constraint Equation 3-35 which is integrated to obtain an explicit value of h_1 for each such curve in terms of the constant K_1. In some cases, it is easier to alter this procedure and to treat K_1 as a parameter. When this is true, various values of h_1 are assumed, and corresponding values of K_1 are evaluated. Finally, the maximum (minimum) $J(x^*)$ of J may be found by direct comparison of those values (of J) which correspond to first-variational curves (of J_a) that satisfy 3-35.

Example 3-4. In the circuit shown in Figure 3-4, the input $v_i(t)$ is a voltage pulse of width T. The pulse shape $v_i \equiv v_i(t)$ is to be selected such that the average value J of the output voltage $v_o \equiv v_o(t)$ is maximized over the

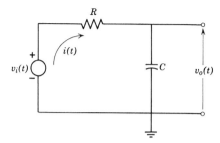

Figure 3-4. Voltage $v_i(t)$ to be selected for the optimum.

interval $0 \leq t \leq T$.

$$J = \int_0^T \frac{v_o}{T} dt \tag{3-41}$$

The maximum value of J is limited, however, by the limited energy K which can be dissipated in the resistor R over the period $0 \leq t \leq T$. Thus, a constraint is

$$K = \int_0^T i^2 R \, dt \tag{3-42}$$

where $i \equiv i(t)$ is the current through the resistor. End-point conditions are given at $t = 0$ and $t = T$: $v_o(0) = 0$, and $v_o(T) = V$ where V is a given constant.

Problem: Find the optimal $v_o(t)$ and the corresponding $v_i(t)$.

As a first step in the solution, the identity $C\dot{v}_o = i$ is used in place of i in 3-42 to reduce the problem to one involving one dependent function. Thus,

$$K = \int_0^T RC^2 \dot{v}_o^2 \, dt \tag{3-43}$$

and the augmented functional J_a is formed,

$$J_a = \frac{1}{T} \int_0^T (v_o + h_1 RC^2 T \dot{v}_o^2) \, dt \tag{3-44}$$

from which the Euler-Lagrange equation is obtained as a special case of 3-28:

$$\frac{\ddot{v}_o}{\beta} - 1 = 0, \quad v_o(0) = 0 \quad \text{and} \quad v_o(T) = V \tag{3-45}$$

where $\beta = 1/(2h_1 RC^2 T)$.

The solution to 3-45 is

$$v_o = c_2 t^2 + c_1 t + c_0 \tag{3-46}$$

where the constants c_0, c_1, and c_2 must be selected to satisfy the prescribed boundary conditions and to satisfy constraint 3-43. (Note that the Lagrange multiplier h_1 is absorbed in c_0, c_1, and c_2.) At $t = 0$, $v_o(0)$ equals 0, and c_0 must therefore be zero. At $t = T$, $v_o = V$ and from 3-46

$$V = c_2 T^2 + c_1 T \tag{3-47}$$

A second relationship between c_1 and c_2 is obtained by using the identity $\dot{v}_o = 2c_2 t + c_1$ in constraint equation 3-43:

$$\frac{K}{RC^2} = \int_0^T (2c_2 t + c_1)^2 \, dt$$
$$= \tfrac{1}{3}(4c_2^2 T^3 + 6c_2 c_1 T^2 + 3c_1^2 T) \tag{3-48}$$

Simultaneous solution of 3-47 and 3-48 results in

$$c_1 = \frac{1}{T}\left\{ V \pm \left[\frac{3KT}{RC^2} - 3V^2\right]^{1/2} \right\} \tag{3-49}$$

and

$$c_2 = \mp \frac{1}{T^2}\left[\frac{3KT}{RC^2} - 3V^2\right]^{1/2} \tag{3-50}$$

In Equations 3-49 and 3-50, the choice of the upper sign in the $\pm$ and $\mp$ terms corresponds to a maximum of J.

It should be observed that a real answer is obtainable if and only if the terms under the square-root signs in 3-49 and 3-50 are non-negative, that is, K must be greater than or equal to $V^2 RC^2/T$ if c_1 and c_2 are to be real. This means, in essence, that if $K < V^2 RC^2/T$, the output voltage cannot be changed from zero volts at $t = 0$ to V volts at $t = T$ because of the lack of allowable dissipated energy.

Note that $\dot{v}_o^2$ appears as an additive term in the integrand of 3-44. If v_o should exhibit any radical changes, such as jump-type discontinuities, $\dot{v}_o$ would exhibit even more pronounced changes, such as impulse-type discontinuities, and the effect of these on the integral of $\dot{v}_o^2$ would be to radically increase K.[4] Thus, the optimal $v_o(t)$, $v_o^*(t)$, must be continuous and a first-variational curve. Because the derived first-variational curve (3-46) satisfies all prescribed conditions, it must be the optimal curve $v_o^*(t)$.

The required optimal $v_i(t)$, $v_i^*(t)$, is determined by using 3-46 in conjunction with the relation

$$v_i^* = RC\dot{v}_o^* + v_o^*$$
$$= c_2 t^2 + (c_1 + 2RCc_2)t + RCc_1 \tag{3-51}$$

where, as before, c_1 and c_2 are given by 3-49 and 3-50, respectively.

---

[4] Over an interval $[t_1, t_1 + \epsilon]$ contained in $[0,T]$, suppose that $v_o(t) = v_o(t_1) + [(t - t_1)/\epsilon][v_o(t_1 + \epsilon) - v_o(t_1)]$, and therefore $\dot{v}_o(t) = [v_o(t_1 + \epsilon) - v_o(t_1)]/\epsilon$ for $t \in (t_1, t_1 + \epsilon)$. The integral of $\dot{v}_o^2$ from t_1 to $t_1 + \epsilon$ is $[v_o(t_1 + \epsilon) - v_o(t_1)]^2/\epsilon$. This integral can be made arbitrarily large through selection of ϵ, $0 < \epsilon \ll 1$, if $v_o(t_1 + \epsilon)$ differs from $v_o(t_1)$ by a fixed amount, say, 1.

Finally, the maximum value of J is found by substituting the right-hand member of 3-46 for $v_o(t)$ in 3-41 as follows:

$$J = \frac{1}{T}\int_0^T (c_2 t^2 + c_1 t)\, dt$$

$$= \frac{c_2 T^2}{3} + \frac{c_1 T}{2}$$

$$= \frac{1}{6}\left\{3V + \left[\frac{3KT}{RC^2} - 3V^2\right]^{1/2}\right\} \tag{3-52}$$

•

Example 3-5. Consider the regulating control system shown in block-diagram form in Figure 3-5. The purpose of the controller is to generate a control signal $v(t)$ (volts) so that the difference between the desired output θ_d (degrees) and the actual output θ (degrees) is minimized in some sense over time. The function $G(s)$ is a transfer function; $G(s)$ is the ratio of the one-sided Laplace transform of $\theta(t)$ to that of $v(t)$ under the condition that both $\theta(0)$ and $\dot{\theta}(0)$ are zero. $G(s)$ characterizes an electromechanical, rotational system. The constant J_m (newton-meter-sec²/degree) in Figure 3-5 is the effective inertia of the motor and mechanical load, as reflected to the output shaft; B_m (newton-meter-sec/degree) is the effective viscous friction; and K_m (newton-meter/volt) is a torque conversion factor.

The following simplifying conditions are assumed: $K_m = 1$; $\theta(0) = 0$; for $-\infty < t < 0$, $\theta_d = 0$; and for $0 \le t \le \infty$, θ_d equals the real constant b. It is desired that the functional J be minimized:

$$J = \int_0^\infty (\theta - \theta_d)^2\, dt \tag{3-53}$$

which is the *integral of the squared error* (ISE).

Problem: Find the transfer function of the controller which minimizes J.

For J of Equation 3-53 to be finite, $\theta(t)$ must approach b as t approaches infinity; and since $\theta(0) = 0$, both end-point conditions are specified.

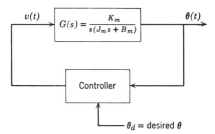

Figure 3-5. A feedback controller.

A limitation to be imposed on the minimization process is that the viscous frictional energy E_B divided by b^2 is not to exceed a specified value K; that is, $K = E_B/b^2$ where E_B is the energy which is dissipated as heat at the output of the controlled system as a result of a step change b of the desired output:

$$K = \int_0^\infty \frac{B_m \dot\theta^2}{b^2} \, dt \tag{3-54}$$

The augmented functional J_a is obtained by combining the right-hand members of 3-53 and 3-54 as follows:

$$J_a = \int_0^\infty \left[(\theta - \theta_d)^2 + \frac{h_1 B_m \dot\theta^2}{b^2} \right] dt \tag{3-55}$$

From the integrand of the above equation, the Euler-Lagrange equation is obtained in the form

$$\ddot\theta - \alpha^2 (\theta - \theta_d) = 0 \tag{3-56}$$

where $\alpha^2 = b^2/h_1 B_m$. The general solution of this equation is of the form

$$\theta - \theta_d = c_1 e^{-\alpha t} + c_2 e^{\alpha t} \tag{3-57}$$

where c_1 and c_2 must be specified to satisfy the boundary conditions imposed on θ; thus, $c_2 = 0$, and $c_1 = -b$ with the result that

$$\theta = b(1 - e^{-\alpha t}) \tag{3-58}$$

In the above solution for θ, the parameter $\alpha^2 = b^2/h_1 B_m$ is yet to be evaluated. This is accomplished by substituting $\alpha b e^{-\alpha t}$ for $\dot\theta$ in the constraint equation (3-54) as follows:

$$K = \int_0^\infty B_m \alpha^2 e^{-2\alpha t} \, dt = \frac{B_m \alpha}{2} \tag{3-59}$$

from which

$$\alpha = \frac{2K}{B_m} \tag{3-60}$$

As in Example 3-4, jump-type discontinuities in the optimal solution are ruled out by the particular way in which the derivative of this solution appears in the integrand of the functional J_a. Thus, the optimal θ, θ^*, must be a first-variational function; and since 3-58 is the only such function which satisfies all prescribed conditions, it must be optimal. Thus,

$$\theta^*(t) = b[1 - e^{-(2K/B_m)t}] \tag{3-61}$$

Next, a specific form for the controller is found by noting that the Laplace transform $V^*(s)$ of the optimal control signal $v^*(t)$ is

Sect. 3-4   ISOPERIMETRIC CONSTRAINTS   81

$$V^*(s) = \frac{\Theta^*(s)}{G(s)} = b\left[\frac{1}{s} - \frac{1}{s+\alpha}\right]s(J_m s + B_m)$$

$$= \frac{b\alpha(J_m s + B_m)}{s+\alpha} \tag{3-62}$$

The above can be realized as a feedback signal by multiplying the error signal $\Theta_d(s) - \Theta^*(s)$ by $\alpha(J_m s + B_m)$. Hence, the block diagram of the optimal system is shown in Figure 3-6.

Several factors limit the practicality of the solution obtained. First, the controller of Figure 3-6 requires a "pure" differentiator which can be obtained only approximately in practice. Second, an instantaneous velocity change of the mechanical load takes place at $t = 0$. This instantaneous change would require an infinite acceleration at $t = 0$, an acceleration which can only be approached in practice and one which is undesirable because of the mechanical stresses which would result. And third, the question arises as to whether or not the frictional power absorbed by the controlled system is always greater than the inertial power released; if not, energy is being supplied over a period of time from the mechanical part of the system back to the electrical part, and this energy is usually wasted in the form of heat. Thus, in order to determine when this third consideration is not important, the condition under which

$$P_B(t) \geq P_J(t) \tag{3-63}$$

must be established: $P_B(t) = B_m \dot{\theta}^2$ is the power supplied to the frictional dissipating element, and $P_J(t) = -J_m \dot{\theta}\ddot{\theta}$ is the inertial power which is released. Thus, condition 3-63 is satisfied if

$$B_m \dot{\theta}^2 \geq -J_m \dot{\theta}\ddot{\theta} \tag{3-64}$$

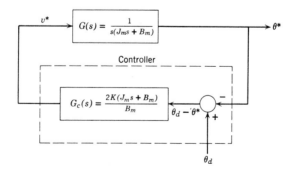

Figure 3-6. The "optimum" system of Example 3-5.

or because $\dot{\theta}^*$ is greater than zero for all $t > 0$,

$$B_m \dot{\theta}^* \geq -J_m \ddot{\theta}^* \tag{3-65}$$

With values of $\dot{\theta}^*$ and $\ddot{\theta}^*$ obtained from 3-58, the above reduces to

$$\frac{B_m^2}{2J_m} \geq K \tag{3-66}$$

If inequality 3-66 is not satisfied, we may wish to include additional constraints in the statement of the problem.

More complex forms of the above problem, and related problems, are of interest [3.13]; and another version of this problem is considered in Example 3-10.

•

3-5. VARIABLE END-POINT CONDITIONS

End-point conditions are conditions imposed on the limits of integration t_a and t_b of the functional J and on the boundary values $x(t_a)$ and $x(t_b)$ of first-variational curves. When some or all of the end-point conditions are not fixed in advance, additional relations are needed to determine the particular end-point conditions, of those allowable, which result in the optimal solution. In the first part of this section, the most general case is treated; the values of t_a, t_b, $x(t_a)$, and $x(t_b)$ are assumed to satisfy arbitrary constraint equations, and conditions which the optimal values of the above must satisfy are found. In the latter part of this section, various special forms of constraint relationships are assumed to exist between the end-point values, and the corresponding end-point conditions which the optimal end-point values must satisfy are obtained.

The key result is given first. The *transversality condition* for the functional 3-11 is

$$[(f - \dot{x} f_{\dot{x}}) \, \delta t]_{t_a}^{t_b} + [f_{\dot{x}} \, \delta x(t)]_{t_a}^{t_b} = 0 \tag{3-67}$$

where the meanings of the end-point variations δt_a, δt_b, $\delta x(t_a)$, and $\delta x(t_b)$ are shown in the following. Optimal values of t_a, t_b, $x(t_a)$, and $x(t_b)$ must satisfy 3-67. In fact, Equation 3-67, in conjunction with other necessary conditions, is used to obtain the optimal values of t_a, t_b, $x(t_a)$, and $x(t_b)$.

Equation 3-67 is sometimes denoted by

$$[(f - \dot{x} f_{\dot{x}}) \, dt + f_{\dot{x}} \, dx]_{t_a}^{t_b} = 0 \tag{3-68}$$

It is to be emphasized, however, that the differentials dx and dt in 3-68 are to be interpreted as small variations which approach differentials if taken

Sect. 3-5   VARIABLE END-POINT CONDITIONS   83

sufficiently small in magnitude (see Case B of the latter part of this section).

If the conditions at one end point are fixed—for example, if t_a and $x(t_a)$ equal specified constants—then the variations δt_a and $\delta x(t_a)$ are zero, and 3-67 reduces to

$$[(f - \dot{x} f_{\dot{x}}) \, \delta t]_{t_b} + [f_{\dot{x}} \, \delta x(t)]_{t_b} = 0 \tag{3-69}$$

The derivation of this equation is given below. It is clear that the proof of 3-67 is a simple extension of the following.

If a solution exists to the stated problem, at least one pair of optimal end points must exist in the x,t plane. If an optimal curve which connects these points has a time derivative that is piecewise continuous, it is necessarily a first-variational curve, as shown in preceding sections. Thus, classical admissible candidates for the optimum in variable end-point cases can be restricted to first-variational curves.

Consider the curves shown in Figure 3-7a. These curves are *assumed to be two of the first-variational curves* of the functional in question. In general, there exists a family of such curves, each member of the family corresponding to one particular value of $x(t_b)$ at $t = t_b$. For convenience in this derivation, it is assumed that the members of the family intersect at $t = t_a$ only. If other points of intersection exist in a given case, they can be detected by using the Jacobi condition (Section 3-10d).

The increment ΔJ in the value of J for the two curves shown in Figure 3-7a

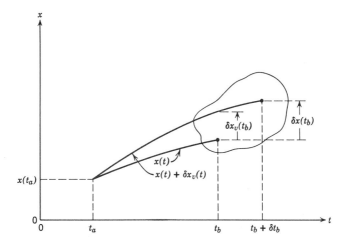

Figure 3-7a. Two first-variational curves; t_b and $x(t_b)$ to be selected for the optimum.

is of interest:

$$\Delta J = \int_{t_a}^{t_b} [f(x + \delta x_v, \dot{x} + \delta \dot{x}_v, t) - f(x, \dot{x}, t)] \, dt$$

$$+ \int_{t_b}^{t_b + \delta t_b} f(x + \delta x_v, \dot{x} + \delta \dot{x}_v, t) \, dt \qquad (3\text{-}70)$$

Excluding terms *of the order* of $(\delta t_b)^2$ or smaller, the value of the second integral in 3-70 is approximated by

$$f(x, \dot{x}, t)|_{t = t_b} \delta t_b \qquad (3\text{-}71)$$

As for the first integral in 3-70,

$$\int_{t_a}^{t_b} [f(x + \delta x_v, \dot{x} + \delta \dot{x}_v, t) - f(x, \dot{x}, t)] \, dt \cong \int_{t_a}^{t_b} [f_x \, \delta x_v + f_{\dot{x}} \, \delta \dot{x}_v] \, dt \qquad (3\text{-}72)$$

is obtained by a Taylor's series expansion of $f(x + \delta x_v, \dot{x} + \delta \dot{x}_v, t)$ about x, $\dot{x}$, t. Furthermore, integrating by parts reduces the right-hand member of 3-72 to

$$[f_{\dot{x}} \, \delta x_v]_{t_a}^{t_b} + \int_{t_a}^{t_b} \left(f_x - \frac{d}{dt} f_{\dot{x}} \right) \delta x_v \, dt \qquad (3\text{-}73)$$

in which the term $f_x - df_{\dot{x}}/dt = 0$ because x is assumed to be a first-variational curve which satisfies 3-25, and $\delta x_v(t_a) = 0$ because $x(t_a)$ is assumed to be fixed. Thus, 3-71 and 3-73 lead to the following value of ΔJ for small (in magnitude) variations δt_b and $\delta x_v(t_b)$.

$$\Delta J \cong \delta J = [f(x, \dot{x}, t) \, \delta t]_{t = t_b} + [f_{\dot{x}} \, \delta x_v]_{t = t_b} \qquad (3\text{-}74)$$

Notice from Figure 3-7b that

$$\delta x_v(t_b) \cong \delta x(t_b) - \dot{x}(t_b) \, \delta t_b \qquad (3\text{-}75)$$

which is used in 3-74 to obtain

$$\delta J = [(f - \dot{x} f_{\dot{x}}) \, \delta t]_{t = t_b} + [f_{\dot{x}} \, \delta x]_{t = t_b} \qquad (3\text{-}76)$$

If the above variation of the functional J is different from zero, some incremental changes in t_b or $x(t_b)$ or both would result in an incremental increase in J, while other incremental changes would result in decreases in J. Thus, a necessary condition for an extremum of J is that δJ of 3-76 equals zero, in which case 3-76 reduces to 3-69 which was to be proved. Important special cases are considered next.

Sect. 3-5   VARIABLE END-POINT CONDITIONS   85

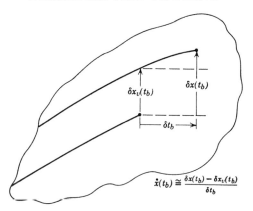

Figure 3-7b. Enlarged view of part of Figure 3-7a.

Case A. (All variations independent.)

When δt_a, δt_b, $\delta x(t_a)$, and $\delta x(t_b)$ are independent of one another, the only way in which 3-67 can be satisfied is that

$$[f - \dot{x}f_{\dot{x}}]_{t=t_a} = [f - \dot{x}f_{\dot{x}}]_{t=t_b} = [f_{\dot{x}}]_{t=t_a} = [f_{\dot{x}}]_{t=t_b} = 0 \quad (3\text{-}77)$$

and these four relations, in conjunction with the fact that x is a first-variational curve, are used in the determination of optimal values of t_a, t_b, $x(t_a)$, and $x(t_b)$.

Case B. (t_a and $x(t_a)$ fixed; $x(t)$ intersects $\phi(t)$ at $t = t_b$.)

In this case the right-hand end point of x is constrained to lie on a given curve $\phi(t)$ where $\dot\phi(t)$ exists and is continuous. Note from Figure 3-8a that

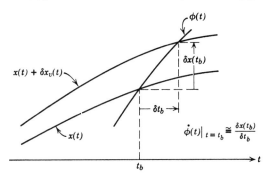

Figure 3-8a. The optimal curve constrained to intersect a given curve $\phi(t)$ at $t = t_b$.

$\delta x(t_b)/\delta t_b$ approaches $\dot{\phi}(t_b)$ as δt_b is decreased toward zero. It follows that $\dot{\phi}(t_b) \, \delta t_b$ can be substituted for $\delta x(t_b)$ in 3-69 with the result that

$$[f - \dot{x}f_{\dot{x}} + f_{\dot{x}}\dot{\phi}]_{t=t_b} \, \delta t_b = 0 \tag{3-78}$$

and since $\delta t_b \neq 0$ in general,

$$[f - \dot{x}f_{\dot{x}} + \dot{\phi}f_{\dot{x}}]_{t=t_b} = 0 \tag{3-79}$$

This equation and the given relation $x(t_b) = \phi(t_b)$ are used in the determination of the best values of t_b and $x(t_b)$ corresponding to the first-variational curves.

Case C. ($x(t_a)$, $x(t_b)$, and t_a fixed; t_b variable.)

With $x(t_a)$, $x(t_b)$, and t_a fixed, the variations $\delta x(t_a)$, δt_a, and $\delta x(t_b)$ are constrained to be zero; but note that $\delta x_v(t_b)$ in addition to δt_b need not be zero (Figure 3-8b). For sufficiently small $|\delta t_b|$, the ratio of $\delta x_v(t_b)$ to $-\delta t_b$ approaches $\dot{x}(t_b)$.[5] With this fact and Equation 3-74,

$$0 = [f - f_{\dot{x}}\dot{x}(t)]_{t=t_b} \, \delta t_b \tag{3-80}$$

and since δt_b need not be zero,

$$\left[f(x,\dot{x},t) - \dot{x}(t)\frac{\partial f(x,\dot{x},t)}{\partial \dot{x}}\right]_{t=t_b} = 0 \tag{3-81}$$

Observe that this result is the special case of 3-79 in which $\dot{\phi}$ is zero and, therefore, ϕ is a constant.

Case D. ($x(t_a)$, $x(t_b)$ fixed; $t_b - t_a = c$.)

With $x(t_a)$ and $x(t_b)$ fixed, $\delta x(t_a)$ and $\delta x(t_b)$ are 0. The end points t_b and t_a are related by the equation $t_b - t_a = c$, where c is a constant. Hence, the

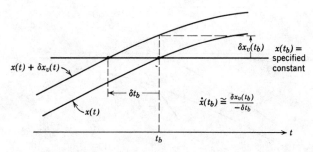

Figure 3-8b. The variations $\delta x(t_a)$, δt_a, and $\delta x(t_b)$ are constrained to be zero.

[5] The counter direction of the arrow associated with δt_b to that associated with the t axis is indicative that δt_b itself is a negative entity in Figure 3-8b.

variations δt_a and δt_b are related by

$$(t_b + \delta t_b) - (t_a + \delta t_a) = c \tag{3-82}$$

which requires that $\delta t_b = \delta t_a$. The transversality condition 3-67 therefore reduces to

$$(f - \dot{x}f_{\dot{x}})_{t_a}^{t_b} \delta t_a = 0 \tag{3-83}$$

and since $\delta t_a \neq 0$ in general,

$$(f - \dot{x}f_{\dot{x}})_{t_a}^{t_b} = 0 \tag{3-84}$$

The fact that $t_b - t_a = c$ and knowledge of Equations 3-25 and 3-84 enable the formal determination of a subclass of the class of first-variational curves and, hopefully, the optimal curve $x^*(t)$.

Example 3-6. In Example 3-4, the final output voltage $v_o(T)$ is assumed to be fixed. In this example, $v_o(T) = V$ is to be selected so as to maximize J_a of 3-44. With T and $v_o(0)$ fixed, the transversality condition 3-67 reduces to

$$\left[\frac{\partial f_a}{\partial \dot{v}_o}\right]_{t=T} \delta v_o(T) = 0 \tag{3-85}$$

which is satisfied when $[\partial f_a/\partial \dot{v}_o]_{t=T} = 0$. From 3-44, $[\partial f_a/\partial \dot{v}_o] = 2h_1 RC^2 \dot{v}_o$. Thus, Equation 3-46 and the above statements give

$$\dot{v}_o(T) = 2c_2 T + c_1 = 0 \tag{3-86}$$

which is solved simultaneously with the energy constraint equation (3-48) to yield

$$c_1 = 2T\left[\frac{3K}{4RC^2T^3}\right]^{1/2} \quad \text{and} \quad c_2 = -\left[\frac{3K}{4RC^2T^3}\right]^{1/2} \tag{3-87}$$

The optimal value V^* of $v_o(T)$, obtained by substituting the above identities appropriately into 3-47, is

$$V^* = \left[\frac{3KT}{4RC^2}\right]^{1/2} \tag{3-88}$$

The reader can check this solution by finding the maximum of J in 3-52 with respect to V.

●

3-6. CORNER CONDITIONS

The concept of a corner point is introduced in Section 3-3d. It is important to note that the occurrence of corner points is not always evident from the form of the functional J. There are certain necessary conditions, however, which are derived here and which can be used to determine the locations of corner points, if any exist. The derivation is given for fixed end-point conditions, but it should be clear that the conditions also apply for variable end-point problems.

The functional J of 3-11 can be expressed in the form

$$J = \int_{t_a}^{t_c} f\, dt + \int_{t_c}^{t_b} f\, dt \tag{3-89}$$

where t_c, as yet unspecified, is to be examined for corner-point properties. If more than one corner point exist, the integral could be subdivided further, but this is not essential for the derivation that follows.

A typical corner condition is shown in Figure 3-3. Once t_c is determined, it is, of course, necessary that both segments of the curve in Figure 3-3 be first-variational curves. Following the line of reasoning of the preceding section, the values t_c and $x(t_c)$ can be viewed as variable end conditions on each of the integrals in 3-89. A transversality condition similar to 3-67 is applicable, therefore, as follows:

$$[(f - \dot{x}f_{\dot{x}})\, \delta t]_{t=t_c+0}^{t=t_c-0} + [f_{\dot{x}}\, \delta x]_{t=t_c+0}^{t=t_c-0} = 0 \tag{3-90}$$

But since the variations δt_c and $\delta x(t_c)$ are independent, it follows that the conditions

$$[f - \dot{x}f_{\dot{x}}]_{t=t_c-0} = [f - \dot{x}f_{\dot{x}}]_{t=t_c+0} \tag{3-91}$$

and

$$f_{\dot{x}}|_{t=t_c-0} = f_{\dot{x}}|_{t=t_c+0} \tag{3-92}$$

must be satisfied at a corner point of any first-variational curve. Equations 3-91 and 3-92 are known as the *Erdmann-Weierstrass corner conditions*.

Example 3-7. Suppose the integrand f of the functional J, Equation 3-11, is given by

$$f(x,\dot{x},t) = \dot{x}^2 f_1(x,t) + f_2(x,t) \tag{3-93}$$

where $f_1 = f_1(x,t)$ and $f_2 = f_2(x,t)$ are of class C^2.

Problem: Determine the locations of corner points if they exist.

To apply condition 3-92, $f_{\dot{x}} = 2\dot{x}f_1(x,t)$ is obtained from 3-93 and the result is used in 3-92 as follows:

$$2\dot{x}f_1|_{t=t_c-0} = 2\dot{x}f_1|_{t=t_c+0} \tag{3-94}$$

Because f_1 is continuous in t and x, and assuming that $f_1(x,t)$ is nonzero over the time interval $[t_a, t_b]$, it follows from 3-94 that $\dot{x}(t_c - 0) = \dot{x}(t_c + 0)$, which fact obviates the possibility that any corners exist.

●

Sect. 3-6   CORNER CONDITIONS

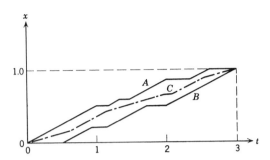

Figure 3-9. An example of first-variational curves with corner points.

Example 3-8. Suppose the integrand f of the functional J, Equation 3-11, is given by

$$f = \dot{x}^2(2\dot{x} - 1)^2 \qquad (3\text{-}95)$$

In this case, f is formally independent of t, and the first integral—Equation 3-30—insures that condition 3-91 is satisfied. The second condition, Equation 3-92, is satisfied if

$$p(2p - 1)(4p - 1) = q(2q - 1)(4q - 1) \qquad (3\text{-}96)$$

where p is identified with $\dot{x}(t_c - 0)$ and q is identified with $\dot{x}(t_c + 0)$. There are six ways in which 3-96 can be satisfied without p being equal to q; namely, $p = 0$ and $q = \frac{1}{2}$, or $p = 0$ and $q = \frac{1}{4}$, or $p = \frac{1}{2}$ and $q = 0$, or $p = \frac{1}{2}$ and $q = \frac{1}{4}$, or $p = \frac{1}{4}$ and $q = 0$, or finally, $p = \frac{1}{4}$ and $q = \frac{1}{2}$. Not all of these pairs result in the optimal value of J since 3-96 is a necessary condition, but not a sufficient one.

It is left to the reader (Problem 3.12) to show that the first-variational curves in this case are of the linear form

$$x = c_1 t + c_0 \qquad (3\text{-}97)$$

between corners, where c_1 and c_0 are constants.

For the first-variational curves drawn in Figure 3-9, the preceding statements are used in verifying that curves labeled A and B (and any other curve which has either the slope $\frac{1}{2}$ or the slope zero on any given interval and which satisfies the indicated boundary values) result in the minimum of the functional in which case $J = 0$; whereas curves such as C, even though it is a first-variational curve between corners, do not minimize J.

●

3-7. THE PROBLEM OF LAGRANGE: STATE-VECTOR CASE

The results of the preceding sections of this chapter are generalized here, using argumentative proofs only, to account for the case in which $f \equiv f(\mathbf{x}, \dot{\mathbf{x}}, t)$ is a given real-valued function of its arguments: $\mathbf{x} \equiv \{x_1, x_2, \ldots, x_n\}$ and $\dot{\mathbf{x}} \equiv \{\dot{x}_1, \dot{x}_2, \ldots, \dot{x}_n\}$ are sets of real-valued functions of the independent variable t. As before, f is assumed to be of class C^2 with respect to its arguments.

The functional to be extremized with respect to $\mathbf{x}$ is

$$J \equiv J(\mathbf{x}) = \int_{t_a}^{t_b} f(\mathbf{x}, \dot{\mathbf{x}}, t) \, dt \qquad (3\text{-}98)$$

The optimal $\mathbf{x}$, $\mathbf{x}^*$, must satisfy certain necessary conditions. The principal one of these is that the first variation of J with respect to a variation in any x_k must be zero. This follows from the fact that if all of the x_i's except x_k have been selected in terms of t in an optimal fashion, then 3-98 reduces to 3-11 in terms of the one remaining $x_k(t)$ to be selected. Thus, the above condition imposed on the first variation leads to a set of Euler-Lagrange equations, each of which is similar to that obtained in Section 3-3, which must be satisfied by the optimal $\mathbf{x}$. The Euler-Lagrange equations have the simple form

$$f_{x_k} - \frac{d}{dt} f_{\dot{x}_k} = 0, \qquad k = 1, 2, \ldots, n \qquad (3\text{-}99)$$

and this set of equations can be solved for sets of first-variational curves.

When some of the end-point conditions on the first-variational curves are unspecified, or when certain end-point conditions are related by given functions, the transversality condition

$$\left[\left(f - \sum_{k=1}^{n} f_{\dot{x}_k} \dot{x}_k \right) dt + \sum_{k=1}^{n} f_{\dot{x}_k} \, dx_k \right]_{t=t_a}^{t=t_b} = 0 \qquad (3\text{-}100)$$

must be satisfied by the optimal solution. Again, as in Section 3-5, it is emphasized that dt and dx_k in 3-100 denote infinitesimal *variations* of the variables t and x_k, respectively. For example, if $x_k(t)$ must equal a given differentiable function $\phi(t)$ at $t = t_b$, then the *ratio* $[dx_k(t_b)/dt_b]$ equals $[d\phi(t)/dt]_{t=t_b}$. Most emphatically, this ratio of $dx_k(t_b)$ to dt_b should not be identified with $\dot{x}_k(t_b)$! Here as in Section 3-5, special forms of 3-100 can be derived in special cases (Problems 3.14 and 3.15, for example). Also note that when Euler-Lagrange equations are formed on the basis of an augmented performance measure, the f in 3-100 must be replaced by the integrand of the augmented performance measure. Additional insight into transversality conditions can be obtained from the deliberations in Sections 3-9 and 8-6.

Similarly, the corner conditions for the n-dependent-function case are listed:

$$\left.\frac{\partial f}{\partial \dot{x}_k}\right|_{t=t_c-0} = \left.\frac{\partial f}{\partial \dot{x}_k}\right|_{t=t_c+0}, \qquad k = 1, 2, \ldots, n \qquad (3\text{-}101)$$

and

$$\left[-f + \sum_{k=1}^{n} f_{\dot{x}_k}\dot{x}_k\right]_{t=t_c-0} = \left[-f + \sum_{k=1}^{n} f_{\dot{x}_k}\dot{x}_k\right]_{t=t_c+0} \qquad (3\text{-}102)$$

Proofs of conditions 3-100, 3-101, and 3-102 can be supplied by straightforward extensions of the corresponding proofs in Section 3-5.

One additional relation is given below without proof (but see Problem 3.16). This relation applies to the case in which one (or more) of the x_k's has derivatives, of order higher than the first, which appear as arguments of f in 3-98. If $x_j^{(m)}$ is the highest-order derivative of x_j which appears in f, the Euler-Lagrange equation associated with x_j is

$$f_{x_j} - \frac{d}{dt}(f_{\dot{x}_j}) + \cdots + (-1)^m \frac{d^m}{dt^m}[f_{x_j^{(m)}}] = 0 \qquad (3\text{-}103)$$

3-8. CONSTRAINTS

The theory developed thus far in this chapter is applicable to a rather limited class of practical problems because methods for treating constraint relationships—which almost always exist in problems with physical origins—between the x_k's have been neglected almost entirely. Fortunately, only minor modifications of the theory are required in order to treat more involved problems, and these are considered in the following four subsections, the first three being devoted to equality constraints of various forms, and the fourth to inequality constraints. Proofs given in this section are indicative in nature; the reader who desires to study rigorous proofs is referred to any of several books ([3.1, 3.6, 3.10, or 3.12] for example). Though the various constraints are treated separately in the following subsections, *problems having several different types of constraints are handled by treating each constraint semi-independently of the others*, as illustrated in Examples 3-9 and 3-10.

3-8a. Isoperimetric Constraints

Consider constraint equations of the form

$$K_i = \int_{t_a}^{t_b} f_i(\mathbf{x},\dot{\mathbf{x}},t)\, dt, \qquad i = 1, 2, \ldots, m \qquad (3\text{-}104)$$

where, as before, $\mathbf{x}$ denotes $\{x_1, x_2, \ldots, x_n\}$, K_i is a constant, and f_i is a given real-valued function of class C^2 with respect to its arguments. The constraints of 3-104 are called isoperimetric constraints, a basic form of which is treated in Section 3-4. As in Section 3-4, an isoperimetric theorem applies.

Isoperimetric Theorem. *Assume that* $\mathbf{x}^*(t)$ *results in the maximum of the functional* $J_a(\mathbf{x})$:

$$J_a(\mathbf{x}) = \int_{t_a}^{t_b} \left[f + \sum_{i=1}^{m} h_i f_i \right] dt$$

$$= \int_{t_a}^{t_b} f_a(\mathbf{x}, \dot{\mathbf{x}}, t, h_1, h_2, \ldots, h_m) \, dt \tag{3-105}$$

That is,

$$J_a(\mathbf{x}) \le J_a(\mathbf{x}^*) \tag{3-106}$$

for any admissible $\mathbf{x}$, *where the* h_i's *are called Lagrange multipliers and are independent of* $\mathbf{x}$ *and* t. *Also, assume that 3-104 is satisfied by* $\mathbf{x}^*$. *Then* $\mathbf{x}^*$ *yields a maximum of 3-98 subject to the isoperimetric constraints* (3-104).

A proof of this theorem can be obtained on the basis of an expanded version of that given in Section 3-4 and is not considered here. An analogous theorem applies for the case that a minimum of J is desired. To apply the theorem, we first obtain the first-variational curves corresponding to the functional J_a in 3-105. These curves are functions of t and of the constant h_i's. The h_i's are evaluated by the requirement that Equations 3-104 must be satisfied. When applicable, corner conditions, transversality conditions, and sufficiency conditions should also be applied in the determination of the optimal first-variational curves.

3-8b. Constraints of the Form $g_i(\mathbf{x}, \dot{\mathbf{x}}, t) = 0$

Consider constraint equations of the form

$$g_i \equiv g_i(\mathbf{x}, \dot{\mathbf{x}}, t) = 0, \qquad i = 1, 2, \ldots, j \tag{3-107}$$

The g_i's are known real-valued functions of class C^2 with respect to the arguments $\mathbf{x}$, $\dot{\mathbf{x}}$, and t; and the Equations 3-107 must be satisfied by any $\mathbf{x} = \{x_1, x_2, \ldots, x_n\}$ which extremizes J of 3-98. If a Lagrange multiplier $\lambda_i = \lambda_i(t)$ is associated with each g_i, an augmented functional J_a can be written in the form

$$J_a = \int_{t_a}^{t_b} \left[f + \sum_{i=1}^{j} \lambda_i(t) g_i \right] dt$$

$$= \int_{t_a}^{t_b} f_a(\mathbf{x}, \dot{\mathbf{x}}, t, \lambda_1, \lambda_2, \ldots, \lambda_j) \, dt \tag{3-108}$$

Because an optimal **x**, **x***, must satisfy Equations 3-107, the above functional is formally equivalent to that of 3-98. To find first-variational curves, therefore, Euler-Lagrange equations are derived by using the integrand $f_a \equiv f_a(\mathbf{x}, \dot{\mathbf{x}}, t, \lambda_1, \lambda_2, \ldots, \lambda_j)$ of 3-108, as follows:

$$\frac{\partial f_a}{\partial x_i} - \frac{d}{dt}\frac{\partial f_a}{\partial \dot{x}_i} = 0, \qquad i = 1, 2, \ldots, n \qquad (3\text{-}109\text{a})$$

and

$$\frac{\partial f_a}{\partial \lambda_i} - \frac{d}{dt}\frac{\partial f_a}{\partial \dot{\lambda}_i} = 0, \qquad i = 1, 2, \ldots, j \qquad (3\text{-}109\text{b})$$

where the latter set of Euler-Lagrange equations (3-109b) reduces to the original constraint equations (3-107) because f_a is formally independent of $\dot{\lambda}_i$, $i = 1, 2, \ldots, j$.

It is emphasized that, in contrast to the Lagrange multipliers of Section 3-8a, the Lagrange multipliers in 3-108 and 3-109a are functions of t, in which case the differentiation with respect to t in 3-109a results in terms which contain $\dot{\lambda}_i$ as a factor. The indicated differentiation in 3-109a is effected to obtain

$$(f_a)_{x_i} = (f_a)_{\dot{x}_i t} + \sum_{k=1}^{n}[(f_a)_{\dot{x}_i \dot{x}_k}\ddot{x}_k + (f_a)_{\dot{x}_i x_k}\dot{x}_k] + \sum_{k=1}^{j}(f_a)_{\dot{x}_i \lambda_k}\dot{\lambda}_k \qquad (3\text{-}110)$$

which is determined for $i = 1, 2, \ldots, n$.

Equations 3-107 and 3-110 constitute a set of $n + j$ differential equations, associated with which there are $n + j$ unknown functions of t, the λ_i's and the x_i's. These equations are solved for the x_i's which are first-variational curves; as before, the first-variational curves are subject to corner conditions, transversality conditions, and sufficiency tests before an optimal solution is established.

An important special case of 3-110 is that in which Equations 3-107 are or can be placed in the form

$$\dot{x}_i - q_i(\mathbf{x}, t) = 0, \qquad i = 1, 2, \ldots, j \qquad (3\text{-}111\text{a})$$

In this case, the Euler-Lagrange equations (3-110) reduce to

$$f_{x_i} - \sum_{k=1}^{j}\lambda_k\frac{\partial q_k}{\partial x_i} = f_{\dot{x}_i t} + \sum_{k=1}^{n}(f_{\dot{x}_i \dot{x}_k}\ddot{x}_k + f_{\dot{x}_i x_k}\dot{x}_k) + \dot{\lambda}_i \qquad (3\text{-}111\text{b})$$

for $i = 1, 2, \ldots, n$.

There are several ways in which we can justify the preceding approach. One of these is given in this paragraph for the case that $t_a = 0$ and $j = 1$ in

3-107. The problem is then to find the minimum (maximum) of

$$J = \int_0^{t_b} f(\mathbf{x},\dot{\mathbf{x}},t)\, dt \tag{3-112a}$$

subject to satisfying the constraint that

$$g_1(\mathbf{x},\dot{\mathbf{x}},t) = 0 \tag{3-112b}$$

An approximate form of 3-112a is

$$J_d = \sum_{k=0}^{N} f\left[\mathbf{x}(k\,\Delta t), \frac{\Delta \mathbf{x}(k\,\Delta t)}{\Delta t}, k\,\Delta t\right] \Delta t \tag{3-113a}$$

where $\Delta \mathbf{x}(k\,\Delta t) = \mathbf{x}[(k+1)\,\Delta t] - \mathbf{x}(k\,\Delta t)$, Δt is a small increment, and $N\,\Delta t = t_b - \Delta t$. Similarly, an approximate form of 3-112b is

$$g_1\left[\mathbf{x}(k\,\Delta t), \frac{\Delta \mathbf{x}(k\,\Delta t)}{\Delta t}, k\,\Delta t\right] = 0, \quad k = 0, 1, \ldots, N \tag{3-113b}$$

To solve the approximate problem associated with 3-113a and 3-113b, classical min-max theory can be applied. Thus, a Lagrange multiplier h_k is introduced for each constraint relationship of 3-113b, and an augmented performance measure J_a is formed,

$$J_a = \sum_{k=0}^{N} \left\{ f\left[\mathbf{x}(k\,\Delta t), \frac{\Delta \mathbf{x}(k\,\Delta t)}{\Delta t}, k\,\Delta t\right] \Delta t + h_k g_1\left[\mathbf{x}(k\,\Delta t), \frac{\Delta \mathbf{x}(k\,\Delta t)}{\Delta t}, k\,\Delta t\right] \right\}$$

Let $\lambda_1(k\,\Delta t) \triangleq h_k/\Delta t$ in the above expression with the result that

$$J_a = \sum_{k=0}^{N} \left\{ f\left[\mathbf{x}(k\,\Delta t), \frac{\Delta \mathbf{x}(k\,\Delta t)}{\Delta t}, k\,\Delta t\right] \right.$$
$$\left. + \lambda_1(k\,\Delta t) g_1\left[\mathbf{x}(k\,\Delta t), \frac{\Delta \mathbf{x}(k\,\Delta t)}{\Delta t}, k\,\Delta t\right] \right\} \Delta t$$

and for arbitrarily small Δt, this augmented performance measure to be minimized (maximized) approaches

$$J_a = \int_0^{t_b} [f(\mathbf{x},\dot{\mathbf{x}},t) + \lambda_1(t) g_1(\mathbf{x},\dot{\mathbf{x}},t)]\, dt$$

which is of the form expressed by 3-108.

3-8c. Constraints of the Form $z_i(\mathbf{x},t) = 0$

Consider constraint equations of the form

$$z_i(\mathbf{x},t) = 0, \quad i = 1, 2, \ldots, m < n \tag{3-114}$$

Sect. 3-8   CONSTRAINTS   95

These constraints are actually a special case of those considered in subsection 3-8b. Here, the constraint equations are independent of $\dot{\mathbf{x}} = \{\dot{x}_1, \dot{x}_2, \ldots, \dot{x}_n\}$, and it may be possible to solve 3-114 for m of the x_i's in terms of t and the remaining $n - m$ of the x_i's. The x_i's thus obtained could be substituted appropriately into 3-98, and optimization could then be effected without further regard to the original constraint equations.

On the other hand, it may be difficult at the outset to solve 3-114 for some of the x_i's in terms of t and other x_i's. If so, exactly the same procedure as given in Section 3-8b can be employed to obtain the following set of Euler-Lagrange equations:

$$f_{x_i} + \sum_{k=1}^{m} \lambda_k \frac{\partial z_k}{\partial x_i} = f_{\dot{x}_i t} + \sum_{k=1}^{n} (f_{\dot{x}_i \dot{x}_k} \ddot{x}_k + f_{\dot{x}_i x_k} \dot{x}_k) \qquad (3\text{-}115)$$

where $i = 1, 2, \ldots, n$. Although the λ_k's in 3-115 are still functions of t in general, no $\dot{\lambda}_k$'s appear; and therefore the set of Equations 3-114 and 3-115 are usually somewhat easier to solve than the corresponding equations in Section 3-8b.

3-8d. Inequality Constraints

When the equality symbols in Equations 3-104, 3-107, 3-111a, and/or 3-114 are replaced by inequality symbols, the resulting relations are inequality constraints. For example, a particular inequality constraint might assume the form

$$0 \leq g(\mathbf{x}, \dot{\mathbf{x}}, t) \qquad (3\text{-}116)$$

which is an inequality constraint corresponding to the equality constraints of 3-107. Quite often, such inequality constraints define a connected region of allowable x_i's and $\dot{x}_i$'s in the space of possible x_i's and $\dot{x}_i$'s, and this region is generally a function of t. In some problems, the optimal value of the functional is obtained when the x_i's and $\dot{x}_i$'s assume values on the boundary of the allowable region for all t in the closed interval $[t_a, t_b]$, and in this case the optimum is obtained when the equality, rather than the inequality, is used in 3-116. If the preceding statement holds, the appropriate method of treating the equality constraint can be used. On the other hand, if it is questionable that the optimal x_i's and $\dot{x}_i$'s correspond to boundary values of a connected region, more devious methods must be employed. Three of these approaches are given in the following paragraphs of this section.

For differential inequality constraints of the form

$$c_1 \leq g \equiv g(\mathbf{x}, \dot{\mathbf{x}}, t) \leq c_2 \qquad (3\text{-}117)$$

or for inequality constraints of the form

$$c_1 \leq z \equiv z(\mathbf{x},t) \leq c_2 \tag{3-118}$$

the first two methods that follow are applicable, at least in theory. For illustrative purposes, only one constraint of the form given in 3-117 is considered.

The *first method* is the *slack-variable method* suggested by Valentine [3.29]. In applying this method, we introduce a new variable $x_{n+1} \equiv x_{n+1}(t)$ which is defined by

$$x_{n+1}^2 \triangleq (g - c_1)(c_2 - g) \tag{3-119}$$

where g, c_1, and c_2 are the same entities that appear in 3-117. Assuming that $x_{n+1}(t)$ is a real-valued function of t, the right-hand member of 3-119 must be non-negative over the range of t under consideration. But of course this implies that the constraint is satisfied; i.e., $g \geq c_1$ and $g \leq c_2$. Thus, Equation 3-119 is rearranged, and a new equality constraint (3-120) is defined to replace the old inequality constraint (3-117).

$$g_e \triangleq (g - c_1)(c_2 - g) - x_{n+1}^2 = 0 \tag{3-120}$$

Because this new constraint equation is of the form considered in Section 3-8b, an augmented functional J_a is formed in accord with the theory developed in Section 3-8b,

$$J_a = \int_{t_a}^{t_b} \{f + \lambda(t)[(g - c_1)(c_2 - g) - x_{n+1}^2]\} \, dt \tag{3-121}$$

from which the following Euler-Lagrange equations are derived in the ordinary way:

$$f_{x_i} + \lambda g_{x_i}(c_1 + c_2 - 2g) - \frac{d}{dt}[f_{\dot{x}_i} + \lambda g_{\dot{x}_i}(c_1 + c_2 - 2g)] = 0 \tag{3-122}$$

where $i = 1, 2, \ldots, n$; and an additional Euler-Lagrange equation is

$$-2x_{n+1}(t)\lambda(t) = 0 \tag{3-123}$$

Note from the above equation that the nonzero domains of $x_{n+1}(t)$ and $\lambda(t)$ are mutually exclusive. The problem, therefore, is to determine over which intervals of t the Lagrange multiplier $\lambda(t)$ is zero, in which case Equations 3-122 reduce to the unconstrained Euler-Lagrange equations; and over which intervals of t the variable $x_{n+1}(t)$ is zero, in which case the function g equals either c_1 or c_2. Unfortunately, the computations that are required to obtain numerical solutions to these equations are quite difficult in general. But as a means of determining properties of solutions and of solving certain special cases, the above development and similar developments are quite useful (see Problem 3.19 and references [3.3, 3.4, and 3.29]).

The *second method* to be considered here is a *penalty-function method* known as the *method of elastic stops*. This method is based on the premise that if the performance measure is severely penalized when the constraint 3-117 *is not* satisfied but the performance measure is not penalized when the same constraint *is* satisfied, then the optimal solution will be forced to satisfy the constraint. More precisely, suppose that the functional J of 3-98 is to be minimized with respect to the selection of the x_i's, and that the constraint 3-117 has to be satisfied by the optimal solution. If a function $f_1 = f_1(g,c_1,c_2)$ can be found with the property that

$$f_1 \cong 0 \quad \text{for} \quad c_1 \leq g \leq c_2 \tag{3-124}$$

and

$$f_1 \gg 0 \quad \text{for} \quad c_1 > g \text{ or } g > c_2 \tag{3-125}$$

then this function can be added to the integrand of 3-98 to obtain an augmented functional J_a,

$$J_a = \int_{t_a}^{t_b} (f + f_1)\, dt \tag{3-126}$$

and the minimum of J_a, with no further regard to the constraint, will be almost the same as the *constrained minimum* of J in 3-98.

Consider a particular f_1 function,

$$f_1 = \left(\frac{2g - c_1 - c_2}{c_2 - c_1}\right)^{2k} \tag{3-127}$$

where k is a positive integer. Typical curves of this f_1 versus g are given in Figure 3-10 for various values of k. It is observed that if k is taken sufficiently large, f_1 of 3-127 satisfies the requirements of 3-124 and 3-125.

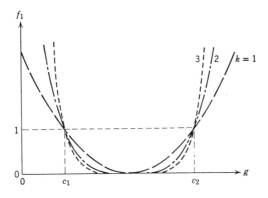

Figure 3-10. Elastic stops at $g = c_1$ and $g = c_2$.

The Euler-Lagrange equations corresponding to 3-126 are obtained in the ordinary way with f_1 as given by 3-127; thus,

$$f_{x_i} + \frac{4kg_{x_i}(2g - c_1 - c_2)^{2k-1}}{(c_2 - c_1)^{2k}} = \frac{d}{dt}\left[f_{\dot{x}_i} + \frac{4kg_{\dot{x}_i}(2g - c_1 - c_2)^{2k-1}}{(c_2 - c_1)^{2k}}\right] \quad (3\text{-}128)$$

for $i = 1, 2, \ldots, n$. Even though Equations 3-128 appear to be unwieldy, they have the desirable feature that they possess no unknown Lagrange multiplier. Once k is selected in 3-128, the problem is the same as that treated in Section 3-7. Typically, we assume a small value of k at the start to obtain an approximate solution; and then, for an accurate solution, we assume as large a value of k as can be handled with the computational facilities available.

The *third approach* of this section concerns isoperimetric inequality constraints only. To apply the method in the case of a single isoperimetric constraint, assume some particular equality which is consistent with the original inequality constraint and use this in place of the original inequality constraint. Next, solve the problem as though it were of the ordinary equality constraint type and evaluate the performance measure J. Continue by repeating the preceding step using different equalities for the equality constraint; and finally, with the aid of systematic search techniques (Chapter 6), converge to an optimal solution.

It is all too apparent that the above methods of treating inequality constraints leave much to be desired. Another approach which is useful at times is based on the parametric representation of the functions involved [3.12]. Still other approaches have been developed in the literature for certain types of inequality constraints which often occur in practice—see Chapters 7 and 8 in this regard.

Example 3-9. Reconsider Examples 3-4 and 3-6, but assume here that the input energy E_i,

$$E_i = \int_0^T i v_i \, dt = \int_0^T C \dot{v}_o v_i \, dt \quad (3\text{-}129)$$

is fixed, rather than the energy K dissipated in the resistor R of Figure 3-4. The voltages v_o and v_i in 3-129 are related by the differential equation

$$\dot{v}_o - \frac{v_i - v_o}{RC} = 0, \quad v_o(0) = 0 \quad (3\text{-}130)$$

As in Example 3-4, the problem is to maximize the average value 3-41 of the output $v_o(t)$. Here, however, constraint equations 3-129 and 3-130 are in effect. One way to proceed would be to solve 3-130 for v_i in terms of v_o and $\dot{v}_o$, in which case the result could be used to eliminate v_i from 3-129, but this approach would not illustrate application of the theory of Section

3-8b. In order to illustrate use of the theory of both Sections 3-8a and 3-8b, the approach employed here is as follows. First, an augmented functional J_a and Lagrange multipliers h_1 and $\lambda_1 \equiv \lambda_1(t)$ are introduced:

$$J_a = \int_0^T \left[\frac{v_o}{T} + h_1 C \dot{v}_o v_i + \lambda_1\left(\dot{v}_o - \frac{v_i - v_o}{RC}\right)\right] dt$$

$$= \int_0^T f_a \, dt \tag{3-131}$$

Next, v_o is associated with x_1, v_i with x_2, $[(v_o/T) + h_1 C \dot{v}_o v_i]$ with f, and $[(v_i - v_o)/RC]$ with q_1. The Euler-Lagrange equations 3-111b are applicable here and assume the particular form

$$\frac{1}{T} + \frac{\lambda_1}{RC} = h_1 C \dot{v}_i + \dot{\lambda}_1 \tag{3-132}$$

and

$$h_1 C \dot{v}_o - \frac{\lambda_1}{RC} = 0 \tag{3-133}$$

Simultaneous solution of 3-130, 3-132, and 3-133 gives

$$v_o(t) = c_1 t^2 + c_2 t \tag{3-134}$$

$$v_i(t) = c_1 t^2 + (2RCc_1 + c_2)t + RCc_2 \tag{3-135}$$

and

$$\lambda_1(t) = \frac{t}{2T} + \frac{c_2}{4c_1 T} \tag{3-136}$$

where the Lagrange multiplier h_1 has been absorbed by the constants c_1 and c_2, $c_1 = 1/(4h_1 TRC^2)$. The transversality condition associated with $v_o(T)$ and the isoperimetric constraint 3-129 are the two conditions which enable explicit determination of c_1 and c_2. With details left to the reader (Problem 3.17), the results are

$$h_1 C v_i(T) + \lambda_1(T) = 0 \tag{3-137}$$

$$c_1(T^2 + 4RCT) + c_2(T + 2RC) = 0 \tag{3-138}$$

$$c_1 = (3E_i/RC^2 T^3)^{1/2}[(T + 2RC)/(T + 8RC)]^{1/2} \tag{3-139}$$

and

$$c_2 = [-T(T + 4RC)/(T + 2RC)]c_1 \tag{3-140}$$

3-9. A GENERAL CONTROL PROBLEM

Suppose that a given system with known dynamic properties is to accomplish some task(s) in an optimal fashion. Moreover, suppose that the optimal performance corresponds to the minimum of a specified integral and that the dynamics of the system are given by a set of differential equations of the form

$$\dot{x}_i = q_i(\mathbf{x},\mathbf{m},t), \qquad i = 1, 2, \ldots, n \tag{3-141a}$$

or in matrix form,

$$\dot{\mathbf{x}} = \mathbf{q}(\mathbf{x},\mathbf{m},t) \tag{3-141b}$$

where $\mathbf{x} = [x_1(t) \quad x_2(t) \quad \cdots \quad x_n(t)]'$ is called the *state vector* of the system, and $\mathbf{m} = [m_1(t) \quad m_2(t) \quad \cdots \quad m_r(t)]'$ is called the *control vector* of the system. The components of $\mathbf{q}(\mathbf{x},\mathbf{m},t) = [q_1(\mathbf{x},\mathbf{m},t) \quad q_2(\mathbf{x},\mathbf{m},t) \quad \cdots \quad q_n(\mathbf{x},\mathbf{m},t)]'$ are expressed single-valued functions of their arguments and are of class C^2 with respect to their arguments.

It is a simple matter to reduce higher-order differential equations to the form given in 3-141. For example, consider the second-order linear differential equation

$$\ddot{x}_1 + b\dot{x}_1 + cx_1 = m_1 \tag{3-142}$$

which is equivalent to the pair

$$\dot{x}_1 = x_2, \qquad \dot{x}_2 = m_1 - bx_2 - cx_1 \tag{3-143}$$

It is desired to minimize a functional

$$J(\mathbf{m}) = \int_{t_a}^{t_b} f(\mathbf{x},\mathbf{m},t)\, dt \tag{3-144}$$

by selecting the proper control vector $\mathbf{m}^*$ for $\mathbf{m}$. As before, f is assumed to be of class C^2 with respect to its arguments. For some problems, end-point conditions on $\mathbf{x}$ are specified at both t_a and t_b; in other cases, end-point conditions are only partially specified, in which case transversality conditions (Sections 3-5 and 3-7) apply.

In the above form, this problem is formally equivalent to that considered in Section 3-8b, the only outward difference being that a control vector $\mathbf{m}$, but not $\dot{\mathbf{m}}$, has been introduced. Because $\dot{\mathbf{m}}$ does not appear, the rate of change of the components of $\mathbf{m}$ is unrestricted, and therefore, these components may be discontinuous in time. Also, if no additional restrictions on $\mathbf{m}$ are given, the "optimal" m_i's may be unbounded. For this reason, many practical control problems have constraints of the form

$$c_{1i} \leq m_i \leq c_{2i}, \qquad i = 1, 2, \ldots, r \tag{3-145}$$

Sect. 3-9　　A GENERAL CONTROL PROBLEM　　101

In theory, such constraints can be taken into account with the aid of material in Section 3-8d (see Problem 3.19), but they are incorporated more directly with the aid of dynamic programming (Chapter 7) and Pontryagin's maximum principle (Chapter 8). The following development applies only to those problems for which the m_i's that minimize J of 3-144 are bounded implicitly, without resort to constraints of the form 3-145.

Within this framework, the Euler-Lagrange equations, as formed on the basis of the material in Section 3-8b, are

$$\dot{\lambda}_i = f_{x_i} - \sum_{k=1}^{n} \frac{\partial q_k}{\partial x_i} \lambda_k \qquad (3\text{-}146)$$

for $i = 1, 2, \ldots, n$; and

$$0 = f_{m_j} - \sum_{k=1}^{n} \frac{\partial q_k}{\partial m_j} \lambda_k \qquad (3\text{-}147)$$

for $j = 1, 2, \ldots, r$. With these equations, and those of 3-141, first-variational curves may be determined. As before, the λ_k's are Lagrange multipliers; but because of the nature of the problem, they are also referred to as *costate variables*.

As with other problems of the calculus of variations, transversality conditions must be satisfied by the solution of this optimal control problem. For this problem, the integrand f_a of the augmented performance measure is

$$f_a = f(\mathbf{x},\mathbf{m},t) + \sum_{k=1}^{n} \lambda_k[\dot{x}_k - q_k(\mathbf{x},\mathbf{m},t)] \qquad (3\text{-}148)$$

The general transversality condition for this problem is obtained through substitution of f_a and $\partial f_a/\partial \dot{x}_k$ for f and $\partial f/\partial \dot{x}_k$, respectively, in 3-100. The result of this substitution is

$$\left\{ \left[f(\mathbf{x},\mathbf{m},t) - \sum_{k=1}^{n} \lambda_k \dot{x}_k \right] dt + \sum_{k=1}^{n} \lambda_k\, dx_k \right\}_{t=t_a}^{t=t_b} = 0 \qquad (3\text{-}149)$$

If $\mathbf{x}(t_a)$, $\mathbf{x}(t_b)$, t_a and t_b are all specified, the dx_k's and dt in 3-149 must be identically zero at $t = t_a$ and $t = t_b$, and 3-149 yields no useful information. On the other hand, if t_a, t_b, and $\mathbf{x}(t_a)$ are specified, but $\mathbf{x}(t_b)$ is free to be selected in an optimal manner, then 3-149 reduces to

$$\sum_{k=1}^{n} \lambda_k(t_b)\, dx_k(t_b) = 0$$

And because the $dx_k(t_b)$'s are arbitrary under the assumed conditions, it necessarily follows that

$$\lambda_k(t_b) = 0 \qquad \text{for} \qquad k = 1, 2, \ldots, n \qquad (3\text{-}150)$$

The problem of solving the Euler-Lagrange equations in this case is therefore one of solving a two-point boundary-value problem; the x_i state variables are specified at $t = t_a$, whereas the λ_i costate variables are specified at $t = t_b$. Other end conditions can be obtained on the basis of 3-149—see Problems 3.28 and 3.29, for example.

Now, in a given case, suppose that the preceding two-point boundary-value problem can be solved efficiently—this is a big assumption. Even with such a solution, it is not obvious how we should implement an optimal, stable, feedback system. Ideally, we would like a control system in which, at each instant of time, the control variables are adjusted on the basis of current state-variable information so as to minimize some suitable performance measure over the remaining time interval of interest. In special cases, this ideal is feasible (examples are given in Chapters 7 and 8). A somewhat less pure, but often acceptable approach is that of designing the system to be stable and to respond optimally under some set of representative conditions; in so doing, the hope is that the response of the system under a variety of other conditions will at least be "good" if not optimal. The following example illustrates this latter approach.

Example 3-10. Reconsider Example 3-5, but assume here that the isoperimetric constraint 3-54 does not apply; and in its place, the control signal $v \equiv m_1$ is limited by the following isoperimetric constraint

$$K = \int_0^\infty (m_1)^2 \, dt \qquad (3\text{-}151)$$

The performance measure 3-53 and the following relation carry over from Example 3-5.

$$J_m \ddot{\theta} + B_m \dot{\theta} - m_1 = 0 \qquad (3\text{-}152)$$

where $\theta(0) = 0$, $\dot{\theta}(\infty) = 0$, $\theta(\infty) = b$, and (in addition) $\dot{\theta}(0) = 0$.

To render the problem in the required form, θ_1 and θ_2 are introduced:

$$\theta_1 \triangleq \theta \quad \text{and} \quad \dot{\theta}_1 \triangleq \theta_2$$

which are used to reduce 3-152 to the pair

$$\dot{\theta}_1 = \theta_2, \qquad \theta_1(0) = 0 \quad \text{and} \quad \theta_1(\infty) = b \qquad (3\text{-}153\text{a})$$

and

$$\dot{\theta}_2 = -\frac{B_m}{J_m} \theta_2 + \frac{1}{J_m} m_1, \qquad \theta_2(0) = \theta_2(\infty) = 0 \qquad (3\text{-}153\text{b})$$

Equations 3-53, 3-151, 3-153a, and 3-153b are combined to obtain an augmented functional J_a, corresponding to that of 3-108, as follows:

Sect. 3-9   A GENERAL CONTROL PROBLEM   103

$$J_a = \int_0^\infty \left\{ (\theta_1 - b)^2 + h_1(m_1)^2 + \lambda_1(\dot{\theta}_1 - \theta_2) \right.$$

$$\left. + \lambda_2 \left[\dot{\theta}_2 + \frac{B_m}{J_m} \theta_2 - \frac{1}{J_m} m_1 \right] \right\} dt \quad (3\text{-}154)$$

At this point,[6] Equations 3-146 and 3-147 can be applied: associate x_1 with θ_1, x_2 with θ_2, q_1 with θ_2 of 3-153a, q_2 with $-(B_m/J_m)\theta_2 + (1/J_m)m_1$ of 3-153b, and f with $[(\theta_1 - b)^2 + h_1(m_1)^2]$. The resulting Euler-Lagrange equations are

$$\dot{\lambda}_1 = 2(\theta_1 - b) \quad (3\text{-}155a)$$

$$\dot{\lambda}_2 = -\lambda_1 + \frac{B_m}{J_m} \lambda_2 \quad (3\text{-}155b)$$

and

$$0 = 2h_1 m_1 - \frac{1}{J_m} \lambda_2 \quad (3\text{-}156)$$

Equations 3-153a, 3-153b, 3-155a, and 3-155b constitute four coupled first-order differential equations with four unknowns, $\theta_1(t)$, $\theta_2(t)$, $\lambda_1(t)$, and $m_1(t)$; $\lambda_2(t) = 2h_1 J_m m_1$ from 3-156, and h_1 is evaluated at the very last on the basis that 3-151 must be satisfied. Four boundary conditions are given— $\theta_1(0) = \theta_2(0) = \theta_2(\infty) = 0$, and $\theta_1(\infty) = b$—and these are used in the solution of the four first-order differential equations.

After a number of interesting intermediate steps (see Problem 3.20), the optimal response $\theta_1^* = \theta^*$ of the system is found to be

$$\theta^* = \frac{b}{\alpha_1 - \alpha_2} [\alpha_1(1 - e^{-\alpha_2 t}) - \alpha_2(1 - e^{-\alpha_1 t})] \quad (3\text{-}157)$$

where

$$\alpha_1 = \{(B_m^2/2J_m^2) + [(B_m^2/2J_m^2)^2 - (1/h_1 J_m^2)]^{1/2}\}^{1/2} \quad (3\text{-}158)$$

and

$$\alpha_2 = \{(B_m^2/2J_m^2) - [(B_m^2/2J_m^2)^2 - (1/h_1 J_m^2)]^{1/2}\}^{1/2} \quad (3\text{-}159)$$

After several additional steps (see Problem 3.21), the new compensator $G_c(s)$ of Figure 3-6 is found to be

$$G_c(s) = \alpha_1 \alpha_2 \frac{J_m s + B_m}{s + \alpha_1 + \alpha_2} \quad (3\text{-}160)$$

[6] Actually, this problem could be solved more directly by using Wiener-Hopf spectrum factorization (Chapter 4).

and the Laplace transform $V^*(s)$ of the optimal control signal $v^*(t) = m_1^*(t)$ is obtained,

$$V^*(s) = G_c(s) \frac{b - s\Theta^*(s)}{s} \qquad (3\text{-}161)$$

Because of the involved way in which h_1 enters into $V^*(s)$, h_1 should be treated as a parameter; and, for given values of J_m and B_m, a plot of K versus h_1 could be made on the basis of Equation 3-151. As in Example 3-5, the solution obtained in this example is unrealistic under certain conditions (see Problem 3.22).

•

3-10. SUFFICIENT CONDITIONS AND ADDITIONAL NECESSARY CONDITIONS

3-10a. Discussion

For the classical problem of the calculus of variations, the necessary conditions for optimality—the conditions an optimal solution must satisfy—developed in prior sections of this chapter are a subset of the necessary conditions that apply. The previously developed necessary conditions constitute a set of equations, the solutions of which are candidates for the optimum; it is true, of course, that many computational difficulties can be encountered when we attempt to obtain these solutions numerically. If only one solution satisfies these necessary conditions, either the solution is optimum or an optimal solution does not exist in the class of functions under consideration (see Problem 3.35). On the other hand, if many solutions satisfy the previously developed necessary conditions, appropriate sufficiency tests can be of value in the determination of relative maxima and minima.

In this section, the primary goal is the establishment of sufficient conditions for relative maxima and minima. These conditions are based on previously developed necessary conditions and on additional necessary conditions: the Legendre condition of Section 3-10b, the Jacobi condition of Section 3-10d, and the Weierstrass condition of Section 3-10g.

The contrast between weak and strong relative extrema is considered in Section 3-2. On the basis of their differences, it should be clear that any strong relative minimum (maximum) automatically satisfies a test which is both necessary and sufficient for weak relative minima (maxima), but that the counter statement is false. For weak relative extrema, the main sufficient condition consists of a set of subconditions, each subcondition being either a necessary condition or a strengthened version of a necessary condition.

Sect. 3-10   SUFFICIENT AND NECESSARY CONDITIONS   105

For strong relative extrema, an additional condition, the Weierstrass condition of Section 3-10g, must be satisfied.

Mathematical field concepts (Section 3-10c) are requisite for full appreciation of both the Jacobi condition and the Weierstrass condition. Similarly, Green's theorem (Section 3-10f) serves as a tool in the development of the Weierstrass condition; and, as an added bonus, Green's theorem can be utilized to advantage in the extremization of linear integrals.

3-10b. The Second Variation and the Legendre Condition

Consider first the functional J of 3-11 with fixed end conditions. As has been shown, a necessary condition for a relative minimum of $J(x)$ at $x = x_\alpha$ is that the first variation $\delta J = [\partial J(x + \epsilon\, \delta x)/\partial \epsilon]|_{\epsilon=0}\, \epsilon$ be zero at $x = x_\alpha$ for any allowable variation $\epsilon\, \delta x$; this implies that x_α is a first-variational curve,[7] a solution of the Euler-Lagrange equation (3-25).

When $x(t)$ is a first-variational curve, the increment ΔJ,

$$\Delta J = J(x + \epsilon\, \delta x) - J(x) \tag{3-162}$$

assumes the form (see Equation 3-3)

$$\Delta J = \tfrac{1}{2}\delta^2 J + Rm$$

$$= \left[\frac{\partial^2}{\partial \epsilon^2} \int_{t_a}^{t_b} f(x + \epsilon\, \delta x, \dot{x} + \epsilon\, \delta \dot{x}, t)\, dt\right]_{\epsilon=0} \frac{\epsilon^2}{2} + Rm$$

$$= \tfrac{1}{2}\int_{t_a}^{t_b} [(\epsilon\, \delta \dot{x})^2 f_{\dot{x}\dot{x}} + 2\epsilon^2\, \delta x\, \delta \dot{x}\, f_{x\dot{x}} + (\epsilon\, \delta x)^2 f_{xx}]\, dt + Rm \tag{3-163}$$

where the remainder term Rm in the above is of the order ϵ^3 if f is of class C^2 with respect to its arguments. The increment ΔJ of 3-163 is placed in a more compact form by integrating the center term of the integrand in 3-163 by parts,

$$\int_{t_a}^{t_b} 2\epsilon^2\, \delta x\, \delta \dot{x}\, f_{x\dot{x}}\, dt = 2(\epsilon\, \delta x)^2 f_{x\dot{x}}\bigg|_{t_a}^{t_b} - \int_{t_a}^{t_b} \left[2(\epsilon\, \delta x)^2 \left(\frac{d}{dt} f_{x\dot{x}}\right) + 2\epsilon^2\, \delta x\, \delta \dot{x}\, f_{x\dot{x}}\right] dt$$

in which $\delta x(t_a) = \delta x(t_b) = 0$. Thus, by identifying the right-hand member of the integrand on the right with that to the left of the equality, it follows that

$$\int_{t_a}^{t_b} 2\epsilon^2\, \delta x\, \delta \dot{x}\, f_{x\dot{x}}\, dt = -\int_{t_a}^{t_b} (\epsilon\, \delta x)^2 \left(\frac{d}{dt} f_{x\dot{x}}\right) dt$$

---

[7] As noted earlier in this chapter, first-variational curves are commonly called extremal curves in the literature.

When the above identity is substituted appropriately in 3-163, the result is

$$\Delta J \cong \tfrac{1}{2}\delta^2 J = \int_{t_a}^{t_b} [(\epsilon\,\delta\dot{x})^2 \xi + (\epsilon\,\delta x)^2 \nu]\,dt \qquad (3\text{-}164)$$

where $\delta^2 J$ is the second variation of J, and

$$\xi \triangleq \tfrac{1}{2} f_{\dot{x}\dot{x}}, \qquad \nu \triangleq \frac{1}{2}\left(f_{xx} - \frac{d}{dt} f_{x\dot{x}}\right) \qquad (3\text{-}165)$$

Suppose that the second variation $\delta^2 J$ of 3-164 were negative for some admissible $\delta x = \delta x_\beta$ and any arbitrarily small $|\epsilon|$. In that case, the value of $J(x)$ would be greater than $J(x + \epsilon\,\delta x_\beta)$ for any $|\epsilon| \neq 0$ below a certain value, and $J(x)$ would not be a relative minimum of J. It follows that *another necessary condition* for a relative minimum of $J(x)$ at $x = x_\alpha$ is that the second variation $\delta^2 J$ evaluated at $x = x_\alpha$ must be non-negative for any admissible δx.

The above requirement on the second variation is utilized in finding yet another necessary condition in terms of $\xi = \xi(t)$ in 3-165. If $\xi(t)$ is negative over a certain interval, then $(\epsilon\,\delta\dot{x})^2$ can be made large over portions of that interval—while keeping $(\epsilon\,\delta x)^2$ arbitrarily small over the same interval—and a smaller value of J could be obtained thereby. Thus, an additional *necessary condition*, known as the *Legendre condition*, for a relative minimum of $J(x)$ at $x = x_\alpha$, a first-variational curve, is that

$$f_{\dot{x}\dot{x}}|_{x=x_\alpha} \geq 0 \qquad (3\text{-}166a)$$

must be satisfied at every point t in the interval $[t_a, t_b]$. The strengthened version of this condition is

$$f_{\dot{x}\dot{x}}|_{x=x_\alpha} > 0 \qquad (3\text{-}166b)$$

Let $\xi_\alpha(t)$ and $\nu_\alpha(t)$ correspond to a first-variational curve $x_\alpha(t)$; that is,

$$\xi_\alpha \equiv \xi_\alpha(t) \triangleq \tfrac{1}{2} f_{\dot{x}\dot{x}}|_{x=x_\alpha(t)} \qquad (3\text{-}167a)$$

and

$$\nu_\alpha \equiv \nu_\alpha(t) \triangleq \frac{1}{2}\left(f_{xx} - \frac{d}{dt} f_{x\dot{x}}\right)\bigg|_{x=x_\alpha(t)} \qquad (3\text{-}167b)$$

It is easily verified that *if* ξ_α and ν_α of 3-167a and 3-167b, respectively, are positive over the interval $[t_a, t_b]$, *then* there exist constants k_1 and k_2 such that $\delta^2 J$ at $x = x_\alpha$ is greater than zero whenever $0 \neq |\epsilon\,\delta x| < k_1$ and $0 \neq |\epsilon\,\delta\dot{x}| < k_2$ over the interval defined by $t_a \leq t \leq t_b$. Hence, a *sufficient condition* for weak relative minimum of $J(x)$ at $x = x_\alpha$ is that ξ_α and ν_α be greater than zero at every point in $[t_a, t_b]$. This condition is stronger than required, however; a first-variational curve for which $\nu_\alpha(t)$, $t_a \leq t \leq t_b$, is sometimes negative may still be a weak relative minimum. Less stringent sufficient conditions are examined in subsequent subsections.

3-10c. Fields of Solutions

Concepts of mathematical fields that are of importance here are conveniently introduced in examples. Consider the differential equation

$$\ddot{x} = 1, \quad x(0) = 0 \quad \text{and} \quad \dot{x}(0) = z \quad (3\text{-}168\text{a})$$

where z is a real parameter and is independent of x and t. The solution of 3-168a is

$$x = \frac{t^2}{2} + zt, \quad t \geq 0 \quad (3\text{-}168\text{b})$$

and for a range of values of z, say, values of z between z_a and z_b, typical solutions are displayed in Figure 3-11. Also shown in Figure 3-11 is a simply connected domain D_1 in the x,t plane. *To each point of D_1, one and only one member of the family of solutions corresponds*; it is this property that is associated with a *proper field* (or, simply, a field), and solutions 3-168b are said to constitute a proper field over D_1.

To emphasize the dependence of solution 3-168b on both t and z, let x be denoted by $x(t, z)$. Also, let $s \triangleq \partial x(t,z)/\partial t$ denote the *slope function of the field* over D_1. Corresponding to 3-168b,

$$s = \frac{\partial}{\partial t}\left(\frac{t^2}{2} + zt\right)$$

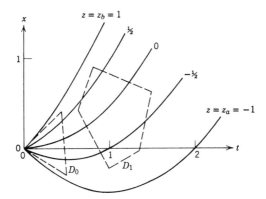

Figure 3-11. A proper field over D_1 and a central field over D_0.

and because z and t are independent variables,
$$s = t + z$$
With this result and that of 3-168b, s can be expressed explicitly in terms of x and t, as follows:
$$s = \frac{t}{2} + \frac{x}{t} \tag{3-169}$$
Thus, as anticipated, a unique slope of the field corresponds to each point of D_1.

Also shown in Figure 3-11 is the connected domain D_0. The family of solutions does not constitute a proper field over D_0 because the slope function s is not unique at the boundary point $(0,0)$ of D_0. However, because there is only one such point in D_0, the family of solutions is called a *central field* over D_0. The point $(0,0)$ is, in this example, the *pencil point* of the central field.

As an example of a family of solutions which do not form a field over a given domain, consider the equation
$$\ddot{x} + x = 0, \quad \dot{x}(0) = 0 \quad \text{and} \quad x(0) = z \tag{3-170a}$$
which has the solution
$$x = z \cos t \tag{3-170b}$$
and, although this result forms a proper field over D_1 in Figure 3-12, it does not result in a field over the D_0 shown because the slope function is not unique at two points in D_0.

The generalization of the preceding field concepts to an $(n + 1)$-dimensional space (n x_i's and t) is straightforward. Suppose a set of n second-order

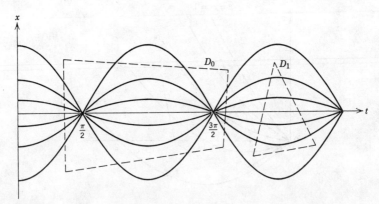

Figure 3-12. A field over D_1 but not over D_0.

differential equations in x_i's and t are solved with specific values of $x_i(t_a)$'s given at $t = t_a$, but with the $\dot{x}_i(t_a)$'s equated to real independent parameters z_i's. If each point of a connected domain D_0 of this $(n + 1)$-dimensional space is pierced by a single solution, one corresponding to a unique setting of the z_i's, the solution forms a proper field over D_0. If the initial point is included in D_0, and if the property assumed in the preceding sentence applies at all points of D_0 except the initial point, the solution forms a central field over D_0.

The graphical portrayal of a family of solutions is difficult in multivariable cases, and therefore an analytical test for a field is of value. The test to be considered next and to be applied in the following subsection is for the planar x,t case; extensions to the multivariable case are straightforward. Consider the derivative of a family $x(t,z)$ of solutions with respect to the independent parameter z. If it should occur that $\partial x/\partial z = 0$ at some point in D_0, infinitesimal changes in z at this point would result in no change in x, and therefore a unique value of z would not be associated with the point in question, in which case D_0 would not be a proper field. But, if $\partial x/\partial z$ is nonzero at every point in D_0, every infinitesimal change in z results in a corresponding change in x, so that each point in D_0 has a unique value of z associated with it, and D_0 is then a proper field.

To illustrate use of this test, consider the $\partial x/\partial z$ corresponding to 3-168b:

$$\frac{\partial x}{\partial z} = t$$

which is zero only for $t = 0$, and therefore the previously observed field aspects of 3-168b are confirmed. Similarly, corresponding to 3-170b,

$$\frac{\partial x}{\partial z} = \cos t$$

which is zero at $t = k\pi/2$, for $k = 1, 3, \ldots$; and here also, the previously noted field aspects are evident.

3-10d. The Jacobi Condition

Consider the particular functional 3-11 with t_a, t_b, and $x_a = x(t_a)$ fixed, but with $\dot{x}(t_a)$ equal to a real independent parameter z. The associated family $x(t,z)$ of first-variational curves must satisfy the Euler-Lagrange equation

$$f_x[x(t,z),\dot{x}(t,z),t] - \frac{d}{dt} f_{\dot{x}}[x(t,z),\dot{x}(t,z),t] = 0 \qquad (3\text{-}171)$$

where now $\dot{x}(t,z)$ denotes $\partial x(t,z)/\partial t$; and although the total derivative with respect to t is indicated in the second term of 3-171, it is to be understood that z is held fixed in any differentiation with respect to t.[8] Corresponding to a particular value z_0 of z, the first-variational curve $x(t,z_0)$ is to be examined. The conclusions of Section 3-10c give that $x(t,z_0)$ can be imbedded in a central field if $\gamma(t)$,

$$\left.\frac{\partial x(t,z)}{\partial z}\right|_{z=z_0} \triangleq \gamma(t) \qquad (3\text{-}172)$$

is nonzero at each t of interest. If $\gamma(t)$ is zero for some particular t_c which is greater than t_a, then t_c is said to be a *conjugate point* with respect to t_a. The Jacobi condition is conveniently stated in terms of conjugate points. That is, *the Jacobi condition is satisfied by $x(t,z_0)$ if $\gamma(t)$ contains no points conjugate to t_a in the open interval (t_a,t_b)*. The strengthened Jacobi condition is similar, but with (t_a,t_b) replaced by the interval $(t_a,t_b]$. The Jacobi condition can be shown to be a necessary one, whereas the strengthened Jacobi condition is one part of the sufficient condition of this section.

In support of the necessity of the Jacobi condition, consider the two arbitrarily close curves (ab_1cd_1 and ab_2cd_2) in Figure 3-13. Both curves are

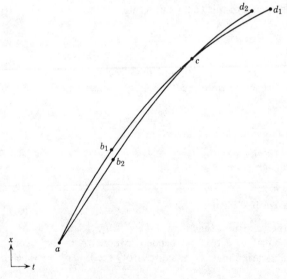

Figure 3-13. Arbitrarily close, first-variational curves which cannot be imbedded in a field of first-variational curves.

[8] To indicate partial differentiation with respect to t in 3-171 would be more misleading than is the total differentiation shown.

assumed to be first-variational curves which cannot be imbedded in a field over the interval of interest. Let J_{ab_1c} denote the value of the functional J corresponding to the path ab_1c, and let other values of J be similarly denoted. Suppose it is asserted, quite falsely, that $J_{ab_1cd_1}$ is minimum relative to all sufficiently close admissible curves that start at a and terminate at d_1, and that $J_{ab_2cd_2}$ is minimum relative to all sufficiently close admissible curves that start at a and terminate at d_2. From the first assertion,

$$J_{ab_1c} + J_{cd_1} < J_{ab_2c} + J_{cd_1}$$

which gives

$$J_{ab_1c} < J_{ab_2c} \tag{3-173}$$

But from the second assertion,

$$J_{ab_2c} + J_{cd_2} < J_{ab_1c} + J_{cd_2}$$

with the obvious simplification

$$J_{ab_2c} < J_{ab_1c} \tag{3-174}$$

which result is in contradiction with that of 3-173. Thus, at least one of the two initial assertions must be false; indeed, extensive arguments could be given to show that both are false.

The strengthened Jacobi condition is clearly satisfied by a given first-variational curve (extremal curve) $x(t,z_0)$ if the curve can be imbedded in a central field of first-variational curves. But because graphical portrayal of families of curves is generally difficult, an analytically based Jacobi test is of value. Such a test is based on $\gamma(t)$ of 3-172. To generate $\gamma(t)$, Equation 3-171 is differentiated with respect to z to obtain

$$f_{xx}\frac{\partial x(t,z)}{\partial z} + f_{x\dot{x}}\frac{\partial^2 x(t,z)}{\partial t\,\partial z} - \frac{d}{dt}\left[f_{x\dot{x}}\frac{\partial x(t,z)}{\partial z} + f_{\dot{x}\dot{x}}\frac{\partial^2 x(t,z)}{\partial t\,\partial z}\right] = 0$$

or equivalently, upon simplification,

$$\frac{d}{dt}\left[f_{\dot{x}\dot{x}}\frac{\partial}{\partial t}\frac{\partial x(t,z)}{\partial z}\right] - \left(f_{xx} - \frac{d}{dt}f_{x\dot{x}}\right)\frac{\partial x(t,z)}{\partial z} = 0$$

which, in view of identities 3-165, assumes the compact form

$$\frac{d}{dt}\left[\xi\frac{\partial}{\partial t}\frac{\partial x(t,z)}{\partial z}\right] - \nu\frac{\partial x(t,z)}{\partial z} = 0 \tag{3-175a}$$

Now, along a particular first-variational curve, say, $x_a \triangleq x(t,z_0)$, the curve $\gamma(t) = \partial x(t,z_0)/\partial z$ is to be examined, where $\gamma(t)$ is generated by use of 3-175a in the special form

$$\frac{d}{dt}(\xi_a\dot{\gamma}) - \nu_a\gamma = 0, \quad \gamma(t_a) = 0 \quad \text{and} \quad \dot{\gamma}(t_a) = 1 \tag{3-175b}$$

The functions ξ_α and ν_α are expressed in 3-167a and 3-167b. The initial value $\gamma(t_a)$ is assigned the value of zero because t_a corresponds to the pencil point of the family of curves under test; the initial slope $\dot\gamma(t_a)$ is assigned the value of 1, but other nonzero values would do as well because $\dot\gamma(t_a)$ plays the role of a scale factor for the z parameter. Generally, computer solutions of 3-175b are readily obtained, and the strengthened Jacobi condition is satisfied by a given $x_\alpha(t)$ if the corresponding $\gamma(t)$ is nonzero for all $t \in (t_a, t_b]$.

3-10e. Sufficient Condition for Weak Extrema

If both the strengthened Legendre and the strengthened Jacobi conditions are satisfied by a given first-variational curve $x_\alpha(t)$, it can be shown that the second variation of the functional J is positive for sufficiently small variations away from x_α and $\dot x_\alpha$. Thus, the functional J of 3-11 has a weak relative minimum at x equal to x_α if: (1) x_α is a first-variational curve which satisfies end conditions; (2) x_α satisfies the strengthened Legendre condition, Equation 3-166b; and (3) x_α satisfies the strengthened Jacobi condition. For a weak relative maximum, the preceding sufficiency test applies if the inequality associated with the strengthened Legendre test is reversed.

The preceding sufficiency test can be extended directly to the case of n dependent variables, in which case it is given as follows. The functional $J(\mathbf{x})$ of 3-98 has a weak relative minimum at $\mathbf{x}$ equal to

$$\mathbf{x}_\alpha = \{x_{\alpha 1}(t), x_{\alpha 2}(t), \ldots, x_{\alpha n}(t)\}$$

if $\mathbf{x}_\alpha$ is a set of admissible curves with the following properties:

1. The curves $x_{\alpha i}(t)$, $i = 1, 2, \ldots, n$, are a consistent set of first-variational curves; i.e., a solution of the Euler-Lagrange equations 3-99, and $\mathbf{x}_\alpha$ satisfies end conditions.
2. The strengthened Legendre condition for n dependent variables,

$$\begin{vmatrix} f_{\dot x_1 \dot x_1} & f_{\dot x_1 \dot x_2} & \cdots & f_{\dot x_1 \dot x_k} \\ f_{\dot x_2 \dot x_1} & & \ddots & \vdots \\ \vdots & & & \\ f_{\dot x_k \dot x_1} & \cdots & & f_{\dot x_k \dot x_k} \end{vmatrix} > 0, \qquad k = 1, 2, \ldots, n \qquad (3\text{-}176)$$

is satisfied when $\mathbf{x} = \mathbf{x}_\alpha$ for all t in $[t_a, t_b]$.
3. The half-open interval $(t_a, t_b]$ does not contain points conjugate to $t = t_a$.

SUFFICIENT AND NECESSARY CONDITIONS

For the n-dependent-variable case, a conjugate point can be defined in the following manner. Let

$$\begin{aligned} \Gamma_1 &= \{\gamma_{11}, \gamma_{12}, \ldots, \gamma_{1n}\} \\ \Gamma_2 &= \{\gamma_{21}, \gamma_{22}, \ldots, \gamma_{2n}\} \\ &\vdots \\ \Gamma_n &= \{\gamma_{n1}, \gamma_{n2}, \ldots, \gamma_{nn}\} \end{aligned} \qquad (3\text{-}177)$$

be n sets of solutions of the equations

$$\sum_{i=1}^{n} \nu_{ij}\gamma_{ki} - \frac{d}{dt}\sum_{i=1}^{n}\xi_{ij}\dot{\gamma}_{ki} = 0, \qquad \begin{Bmatrix} j = 1,2,\ldots,n \\ k = 1,2,\ldots,n \end{Bmatrix} \qquad (3\text{-}178)$$

with initial conditions

$$\gamma_{ki}(t_a) = 0, \qquad i = 1, 2, \ldots, n, \quad k = 1, 2, \ldots, n$$

and

$$\dot{\gamma}_{ki}(t_a) = \begin{cases} 1 & \text{for } k = i \\ 0 & \text{for } k \neq i \end{cases}$$

where

$$\xi_{ij} \triangleq \tfrac{1}{2} f_{\dot{x}_i \dot{x}_j}\big|_{\mathbf{x}=\mathbf{x}_a} \quad \text{and} \quad \nu_{ij} \triangleq \frac{1}{2}\left(f_{x_i x_j} - \frac{d}{dt} f_{x_i \dot{x}_j}\right)\bigg|_{\mathbf{x}=\mathbf{x}_a} \qquad (3\text{-}179)$$

By definition, if the determinant

$$\begin{vmatrix} \gamma_{11}(t) & \gamma_{12}(t) & \cdots & \gamma_{1n}(t) \\ \gamma_{21}(t) & & & \\ \vdots & & \ddots & \vdots \\ \gamma_{n1}(t) & & \cdots & \gamma_{nn}(t) \end{vmatrix}$$

vanishes at the point $t = t_g \neq t_a$, the point t_g is said to be conjugate to the point $t = t_a$.

Example 3-11. Assume a_1 and a_3 of Equation 3-27 in Example 3-1 are positive. Determine if the first-variational curve(s), which is a solution of 3-28, results in a weak relative minimum of the quadratic functional $J(x)$ of Example 3-1.

In the first place, subcondition 2 of the sufficient condition theorem is satisfied; that is, $f_{\dot{x}\dot{x}}|_{x=x_a} = 2a_1 > 0$.

In the second place, the differential equation of part 3 of the sufficient condition theorem is

$$v_\alpha \gamma - \frac{d}{dt}(\xi_\alpha \dot{\gamma}) = 0, \qquad \gamma(t_a) = 0 \quad \text{and} \quad \dot{\gamma}(t_a) = 1 \qquad (3\text{-}180)$$

where, from Equations 3-167a and 3-167b, $\xi_\alpha(t) = a_1$ and $v_\alpha(t) = a_3$. Solution of 3-180 for $\gamma(t)$ gives

$$\gamma(t) = \frac{1}{2c}[e^{c(t-t_a)} - e^{-c(t-t_a)}], \qquad c^2 = \frac{a_3}{a_1} \qquad (3\text{-}181)$$

which is clearly nonzero for all $t > t_a$. Hence, condition 3 is satisfied; therefore, any first-variational curve of the functional in Example 3-1 with fixed end points gives rise to a weak relative minimum of the functional.

•

3-10f. Green's Theorem

A form of Green's theorem, which is a well-known result in mathematical physics, is simply given in this subsection for use in the next. Consider two real-valued functions $v_1 \equiv v_1(x,s,t)$ and $v_2 \equiv v_2(x,s,t)$ where x and t are independent variables, but s is an explicit function of x and t. The value of $\Delta = J_a + J_b$ is to be found:

$$\Delta \triangleq \int_a \left(v_1 \frac{dx}{dt} + v_2\right) dt + \int_b \left(v_1 \frac{dx}{dt} + v_2\right) dt \qquad (3\text{-}182)$$

where a and b denote specific paths of integration in the x,t plane; the initial point of a is identical to the terminal point of b, and vice versa—as illustrated in Figure 3-14.

Under the condition that v_1, v_2, and s are analytic both on a and b and over the region R_{ab} between a and b, Green's theorem gives that Δ of 3-182 can be expressed as an integration over R_{ab}, namely,

$$\Delta = \iint_{R_{ab}} \left(\frac{\partial v_2}{\partial x} + \frac{\partial v_2}{\partial s}\frac{\partial s}{\partial x} - \frac{\partial v_1}{\partial t} - \frac{\partial v_1}{\partial s}\frac{\partial s}{\partial t}\right) dx\, dt \qquad (3\text{-}183)$$

Several points are worth mentioning. First, the direction of the ab path dictates the sign of the right-hand member of 3-183—a clockwise direction would have resulted in the opposite sign. Second, in the classical form of Green's theorem, s does not appear as an argument of v_1 and v_2, in which case the $\partial s/\partial x$ and $\partial s/\partial t$ terms vanish from 3-183. And third, consider the case in which the integrand of 3-183 is positive above path a, negative below path a, and zero on path a; under these conditions, J_a is the maximum

Sect. 3-10 SUFFICIENT AND NECESSARY CONDITIONS 115

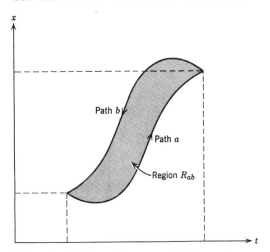

Figure 3-14. Paths of integration in the x,t plane.

value of J—an important fact which the reader can verify on the basis of the first comment below 3-183.

The last comment above forms a basis for the extremization of linear integrals (linear in dx/dt). An advantage of the approach is that constraints, ones of the form which limit the optimal trajectory to a region of the x,t plane, are easily accounted for. Now, in general the integrand of 3-183 is not zero at specified fixed end points, and the optimal trajectory may consist of connected arcs, some through the region and some on the constraint boundary. An account of the method is given by Miele [3.18]. The purpose of this section is best served by a return to questions of sufficiency at this point.

3-10g. The Weierstrass Condition and Strong Extrema

Again consider the basic functional of Lagrange, Equation 3-11, with fixed end points. Suppose that a first-variational curve $x_\alpha = x_\alpha(t)$ has been obtained and has satisfied the sufficient condition test associated with a weak relative minimum. It can therefore be imbedded in a field of first-variational curves; the slope function of this field is denoted by $s = s(x,t)$. *Question:* Does x_α also correspond to a strong relative minimum? To answer this question, an examination of ΔJ is required:

$$\Delta J = \int_\beta f(x,\dot{x},t)\,dt - \int_\alpha f(x,\dot{x},t)\,dt \qquad (3\text{-}184)$$

where the first integration is along a path labeled β in the x,t plane and the second integration is along the path α, i.e., $x_\alpha(t)$, which is under test. If $J(x_\alpha)$ is a strong relative minimum, it must be that ΔJ of 3-184 is greater than zero for any curve x_β which satisfies the fixed end points and which satisfies

$$\max_{t \in [t_a, t_b]} |x_\beta - x_\alpha| < \omega \tag{3-185}$$

for some number $\omega > 0$.

As a first step toward the answer to the basic question, a preliminary identity will be shown to hold, namely,

$$\int_\alpha f(x,\dot{x},t)\, dt \equiv \int_\beta \left[f(x,s,t) + \left(\frac{dx}{dt} - s\right) \frac{\partial f(x,s,t)}{\partial s} \right] dt$$

$$= \int_\beta \left[v_1 \frac{dx}{dt} + v_2 \right] dt \tag{3-186}$$

where $v_1 = \partial f(x,s,t)/\partial s = f_s$, and $v_2 = f - sf_s$. Equation 3-186 is obviously true when β is identical to α, but not so obvious for any other β which does not violate the boundaries of the field. To justify 3-186, Green's theorem of the preceding subsection is applied. Let Δ be defined by

$$\Delta \triangleq \int_\alpha f(x,\dot{x},t)\, dt - \int_\beta \left(v_1 \frac{dx}{dt} + v_2 \right) dt$$

$$= \int_\alpha \left(v_1 \frac{dx}{dt} + v_2 \right) dt - \int_\beta \left(v_1 \frac{dx}{dt} + v_2 \right) dt \tag{3-187a}$$

Via Green's theorem,

$$\Delta = \iint_{R_{\alpha\beta}} \left(\frac{\partial v_2}{\partial x} + \frac{\partial v_2}{\partial s} \frac{\partial s}{\partial x} - \frac{\partial v_1}{\partial t} - \frac{\partial v_1}{\partial s} \frac{\partial s}{\partial t} \right) dx\, dt \tag{3-187b}$$

and in view of the previously given identities for v_1 and v_2, $\partial v_1/\partial t = f_{st}$, $\partial v_1/\partial s = f_{ss}$, $\partial v_2/\partial x = f_x - sf_{sx}$, and $\partial v_2/\partial s = -sf_{ss}$. By appropriate substitution of these identities into 3-187b and judicious grouping of terms, the result is

$$\Delta = \iint_{R_{\alpha\beta}} \left[\frac{\partial f(x,s,t)}{\partial x} - \frac{d}{dt} \frac{\partial f(x,s,t)}{\partial s} \right]_{dx/dt = s} dt\, dx \tag{3-187c}$$

The integrand of 3-187c equals zero over $R_{\alpha\beta}$ because each member of the field is a first-variational curve which satisfies the Euler-Lagrange equation. Thus, $\Delta = 0$, and 3-186 is valid.

Identity 3-186 can be used to place ΔJ of 3-184 in the form

$$\Delta J = \int_\beta \left[f(x,\dot{x},t) - f(x,s,t) - \left(\frac{dx}{dt} - s\right) \frac{\partial f(x,s,t)}{\partial s} \right] dt \tag{3-188}$$

For convenience, a Weierstrass function (originally called an *excess* function) $E(x,\dot{x},s,t)$ is identified with the integrand of 3-188,

$$E(x,\dot{x},s,t) \triangleq f(x,\dot{x},t) - f(x,s,t) - (\dot{x} - s)\frac{\partial f(x,s,t)}{\partial s} \quad (3\text{-}189)$$

It follows that

$$\Delta J = \int_\beta E(x,\dot{x},s,t)\, dt$$

$$= \int_{t_a}^{t_b} E[x_\beta, \dot{x}_\beta, s(x_\beta,t), t]\, dt \quad (3\text{-}190)$$

Now, as previously assumed, $J(x_\alpha)$ has been validated as a weak minimum prior to testing it for properties of a strong minimum. For this reason, attention is focused here on differences between $\dot{x}_\beta$ and the slope function $s(x_\beta,t)$. Consider the set of x_β's for which

$$\max_{t \in [t_a, t_b]} |x_\beta(t) - x_\alpha(t)| < \epsilon \text{ (a small positive value)}$$

but for which $|\dot{x}_\beta - \dot{x}_\alpha|$ may vary between zero and arbitrarily large values. Corresponding to this set of x_β's, the slope function $s(x_\beta,t) = \dot{x}_\alpha(t) + o_1(\epsilon,t)$, and $x_\beta(t) = x_\alpha(t) + o_2(\epsilon,t)$, where $o_1(\epsilon,t)$ and $o_2(\epsilon,t)$ are of order ϵ at each $t \in [t_a, t_b]$. With these identities, ΔJ of 3-190 assumes the form

$$\Delta J = \int_{t_a}^{t_b} E(x_\alpha + o_2, \dot{x}_\beta, \dot{x}_\alpha + o_1, t)\, dt$$

$$= \int_{t_a}^{t_b} E(x_\alpha, \dot{x}_\beta, \dot{x}_\alpha, t)\, dt + o(\epsilon) \quad (3\text{-}191)$$

where the latter identity is obtained through appropriate Taylor's series expansion. For infinitesimal ϵ, ΔJ is effectively

$$\Delta J = \int_{t_a}^{t_b} E(x_\alpha, \dot{x}_\beta, \dot{x}_\alpha, t)\, dt \quad (3\text{-}192)$$

A strong relative minimum is assured if ΔJ is non-negative which is certainly true if $E(x_\alpha, \dot{x}_\beta, \dot{x}_\alpha, t)$ is non-negative at every $t \in [t_a, t_b]$ for arbitrary $\dot{x}_\beta$—this statement constitutes the Weierstrass condition, the key to the sufficiency test for strong relative extrema.

In summary, $x_\alpha(t)$ corresponds to a strong relative minimum of $J(x)$ in 3-11 if:

1. $x_\alpha(t)$ satisfies the requirements of a weak relative minimum.[9]

---

[9] Actually, it can be shown that the Legendre condition is satisfied whenever the Weierstrass and the Jacobi conditions are satisfied.

2. And the Weierstrass condition is satisfied; namely, at each t in $[t_a,t_b]$, $E(x_\alpha,\dot{x}_\beta,\dot{x}_\alpha,t)$ is non-negative for any value of $\dot{x}_\beta$, where

$$E(x_\alpha,\dot{x}_\beta,\dot{x}_\alpha,t) = f(x_\alpha,\dot{x}_\beta,t) - f(x_\alpha,\dot{x}_\alpha,t) - (\dot{x}_\beta - \dot{x}_\alpha)\frac{\partial f(x_\alpha,\dot{x}_\alpha,t)}{\partial \dot{x}} \quad (3\text{-}193)$$

The test for strong relative maxima is similar, but with the requirement on the Weierstrass function being that it be nonpositive.

For the case of n dependent variables, it can be shown that the Weierstrass function is

$$E(\mathbf{x}_\alpha,\dot{\mathbf{x}}_\beta,\dot{\mathbf{x}}_\alpha,t) = f(\mathbf{x}_\alpha,\dot{\mathbf{x}}_\beta,t) - f(\mathbf{x}_\alpha,\dot{\mathbf{x}}_\alpha,t) - \sum_{i=1}^{n}(\dot{x}_{\beta i} - \dot{x}_{\alpha i})\frac{\partial f(\mathbf{x}_\alpha,\dot{\mathbf{x}}_\alpha,t)}{\partial \dot{x}_i} \quad (3\text{-}194)$$

The sufficient condition associated with 3-194 is the obvious extension of that given in the preceding paragraph.

Example 3-12. Assume the conditions of Example 3-1 hold. The integrand f in that case is

$$f = a_1\dot{x}^2 + a_2\dot{x}x + a_3x^2 + g_1(t)\dot{x} + g_2(t)x \quad (3\text{-}27)$$

from which

$$\frac{\partial f}{\partial \dot{x}} = 2a_1\dot{x} + a_2x + g_1(t) \quad (3\text{-}195)$$

The E function is formed by using 3-193, 3-27, and 3-195—several terms cancel—with the final result that

$$E(x_\alpha,\dot{x}_\beta,\dot{x}_\alpha,t) = a_1(\dot{x}_\alpha - \dot{x}_\beta)^2 \quad (3\text{-}196)$$

And clearly, $E \geq 0$ if $a_1 > 0$. It is interesting to note that the results in this example and that of Example 3-11 are obtained without the explicit determination of any first-variational curve $x_\alpha(t)$. In general, however, the particular $x_\alpha(t)$ is required in a given test.

●

3-11. DIRECT METHODS

3-11a. Discussion

In many optimal design problems within the realm of the calculus of variations, the Euler-Lagrange equations are nonlinear differential equations and are subject to two-point boundary conditions. It is a difficult task in general to solve such equations for first-variational curves—often, solutions are obtained only after many iterative computations. It is desirable, therefore, to have available other techniques for solving calculus-of-variations problems, techniques known as direct methods which do not entail the solution of nonlinear differential equations with two-point boundary conditions.

Direct methods of the calculus of variations reduce the problem of extremizing a functional with respect to a function (or functions) to the problem of extremizing a function with respect to a set of parameters—albeit, the solution obtained is in general an approximate solution, unless the set of parameters is an infinite set. Two distinct forms of direct methods are considered in this section: those which are based on series representation of functions and those which are based on finite difference methods.

In the following, only the functional $J(x)$ of 3-11 is treated by direct methods. Obvious extensions are apparent for treating, among others, the functional $J(\mathbf{x})$ of 3-98. Also, although not considered here, it is interesting to note that many problems of mathematical physics (with more than one independent variable) reduce to the problem of extremizing a functional, and direct methods have been used extensively in the treatment of such problems (see references [3.12] and [3.24]).

3-11b. Series Approximations

Series approximation methods (also known as methods of Ritz, Galerkin, et al.) are based on the underlying assumption that a function $x = x(t)$ which extremizes $J(x)$ of 3-11 can be expanded in a series of the form

$$x = \sum_{k=0}^{\infty} a_k \psi_k(t) \qquad (3\text{-}197)$$

where the a_k's are independent of t, and the $\psi_k(t)$'s are known functions.

In applying series approximation methods, truncated forms of 3-197 are used. The truncated forms constitute a sequence of approximating functions x_{sn},

$$x_{sn} = \sum_{k=0}^{n} a_k \psi_k(t), \qquad n = 0, 1, 2, \ldots \qquad (3\text{-}198)$$

It is assumed that this sequence of approximating functions is *complete* for functions x of the class that minimize $J(x)$. By definition, a sequence of approximating functions is *complete* (in the strictest sense of the word) under the following condition: for any real number $\epsilon > 0$, there exists an integer j such that

$$\min_{a_k\text{'s}} [\max_{t \in [t_a, t_b]} |x_{sj}(t) - x(t)|] < \epsilon \qquad (3\text{-}199)$$

for each x in the class of x's under consideration. The importance of this definition lies in the fact that

$$\min_{x} J(x) = \min_{a_k\text{'s}} [\lim_{j \to \infty} J(x_{sj})] \qquad (3\text{-}200)$$

if $J(x)$ is strongly continuous (defined in Section 3-2) and if the sequence of functions is complete. A formal proof of the above statement is given in reference [3.12].

When an approximating sequence x_{sn} is used in place of x, the functional 3-11 assumes the approximate form $J_s(x_{sn})$:

$$J_s(x_{sn}) = \int_{t_a}^{t_b} f\left(\sum_{k=0}^{n} a_k \psi_k, \sum_{k=0}^{n} a_k \dot{\psi}_k, t\right) dt \qquad (3\text{-}201)$$

The parameters a_k are the only unknowns in the right-hand member of the above expression; and in order for $J_s(x_{sn})$ to be a relative extremum, $J_s(x_{sn})$ must be stationary with respect to each a_k, i.e.,

$$\frac{\partial J_s(x_{sn})}{\partial a_k} = \int_{t_a}^{t_b} \frac{\partial f(x_{sn}, \dot{x}_{sn}, t)}{\partial a_k} dt = 0, \qquad k = 0, 1, \ldots, n \qquad (3\text{-}202)$$

Once the indicated integrations in 3-202 are effected, the resulting set of equations can be solved for stationary points, and those stationary points which are relative maxima (minima) can be determined on the basis of the theory developed in Chapter 2; or the search techniques of Chapter 6 can be applied to 3-201 directly.

An unlimited number of approximating sequences are available for use. Typical examples are:

$$x_{sn} = \sum_{k=0}^{n} a_k t^k \qquad (3\text{-}203)$$

or

$$x_{sn} = \sum_{k=0}^{n} a_k e^{-kct} \qquad (c \text{ equals a constant}) \qquad (3\text{-}204)$$

or

$$x_{sn} = \sum_{k=0}^{n} a_k \sin(kct) \qquad (c \text{ equals a constant}) \qquad (3\text{-}205)$$

and in many cases, the problem solution is facilitated if the ψ_k's are orthonormal (see Problem 2.29). Certain rules-of-thumb should be applied in the selection of the form of ψ_k in a given instance. First, if there is a physical basis for suspecting a particular form of the extremizing curve $x^*(t)$, the first term ψ_0 should be picked as close as possible to the suspected optimal

DIRECT METHODS

curve. Second, the ψ_k's should be selected on the basis that boundary values on $x(t)$ are automatically satisfied. This is accomplished by picking $\psi_0(t)$ and a_0 such that $a_0 \psi_0(t_a) = x(t_a) = c_a$, $a_0\psi_0(t_b) = x(t_b) = c_b$, and by selecting $\psi_k(t)$ with the property $\psi_k(t_a) = \psi_k(t_b) = 0$ for $k = 1, 2, \ldots$. Finally, the ψ_k's should be such that the integration indicated in 3-202 is easily effected. This requirement is readily satisfied, for example, if the integrand of 3-11 is of the form

$$f = \sum_{j=0}^{m} \sum_{i=0}^{m} b_{ij} x^j \dot{x}^i \qquad (3\text{-}206)$$

Even in this case, however, if the b_{ij}'s are involved functions of t, it may be necessary to approximate the b_{ij}'s by a series of the form

$$b_{ij} \cong \sum_{k=0}^{n} c_{ijk} \psi_k(t) \qquad (3\text{-}207)$$

where the c_{ijk}'s are appropriate constants. Moreover, when f of 3-11 assumes a more involved form than that given by 3-206, it may be very difficult to effect the indicated integration in 3-202. For such cases, the method of finite differences (Section 3-11c) may be easier to apply.

After the ψ_k's are specified, it is usually convenient to use a low-order approximating sequence, say, with $n = 1$ or 2, to obtain a crude approximation for the optimal x. With this first approximation, a better approximation can be found by selecting a new ψ_0, which corresponds closely to the first approximation, and by proceeding as before.

It is quite difficult in general to obtain an analytical expression for the error involved in the above procedure. Usually, the best that we can do from an applied standpoint is to solve the problem by using both x_{sn} and $x_{s(n+1)}$ where n is sufficiently large so that

$$\max_{t_a \leq t \leq t_b} |x_{s(n+1)} - x_{sn}| \qquad (3\text{-}208)$$

is less than a preselected error bound.

3-11c. Finite Differences

Euler introduced the use of the method of finite differences in the solution of calculus-of-variations problems during the eighteenth century. One basic form of the method is given in this section. Again, the functional 3-11 is to be extremized. The underlying assumption here is that the optimal $x(t)$ can be approximated as close as desired by linear line-segments which are equally spaced on the closed interval $[t_a, t_b]$ (Figure 3-15). The interval

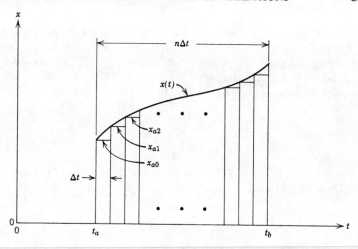

Figure 3-15. Finite difference approximation.

$[t_a, t_b]$ is divided into n adjoining increments of fixed length Δt, $n\,\Delta t = t_b - t_a$. For t in the interval defined by $t_a + k\,\Delta t \leq t < t_a + (k+1)\,\Delta t$, $x(t)$ is approximated by $x_{ak} \equiv x(t_a + k\,\Delta t)$, $\dot{x}(t)$ is approximated by $(\Delta x_{ak}/\Delta t) \equiv (x_{a,k+1} - x_{ak})/\Delta t$, and t itself is approximated by $t_a + k\,\Delta t$. The net result of these discrete approximations is that J of 3-11 assumes the approximate form J_d,

$$J_d \equiv \sum_{k=0}^{n-1} \int_{t_a + k\,\Delta t}^{t_a + (k+1)\,\Delta t} f(x_{ak}, \Delta x_{ak}/\Delta t, t_a + k\,\Delta t)\,dt$$

$$= \sum_{k=0}^{n-1} f(x_{ak}, \Delta x_{ak}/\Delta t, t_a + k\,\Delta t)\,\Delta t \qquad (3\text{-}209)$$

In the above form, J_d is a function of $n+1$ parameters, x_{a0} through x_{an}. If the boundary values x_{a0} and x_{an} are specified, J_d is a function of $n-1$ parameters, x_{a1} through $x_{a,n-1}$. Note that the parameter x_{am} appears in $f_m \equiv f(x_{am}, \Delta x_{am}/\Delta t, t_a + m\,\Delta t)$ and in $f_{m-1} \equiv f[x_{a,m-1}, \Delta x_{a,m-1}/\Delta t, t_a + (m-1)\,\Delta t]$ but in no other terms of the sum 3-209. To extremize J_d with respect to the selection of the x_{am}'s, therefore, the following equations must be satisfied,

$$\frac{\partial J_d}{\partial x_{am}} = \frac{\partial f_m}{\partial x_{am}} + \frac{\partial f_{m-1}}{\partial x_{am}} = 0, \qquad m = 1, 2, \ldots, n-1 \qquad (3\text{-}210)$$

Sect. 3-12   SENSITIVITY CONSIDERATIONS   123

and

$$\frac{\partial J_d}{\partial x_{a0}} = \frac{\partial f_0}{\partial x_{a0}} = 0, \qquad \frac{\partial J_d}{\partial x_{an}} = \frac{\partial f_{n-1}}{\partial x_{an}} = 0 \qquad (3\text{-}211)$$

If $x_{a0} = x(t_a)$ and $x_{an} = x(t_b)$ are specified, Equations 3-211 do not apply, in which case Equations 3-210 alone are solved for the x_{am}'s corresponding to stationary points. Again, those stationary points which are points of relative maximum (minimum) can be determined by use of the methods outlined in Chapters 2 and 6.

Here, as in the preceding subsection, the accuracy of the solution can be increased by increasing n, provided n is not so large that computational round-off errors are significant. Neglecting round-off errors in the computations, the optimal x_{ak} would equal the optimal value of $x(t)|_{t=t_a+k\Delta t}$ in the limit as n approaches infinity.

3-12. SENSITIVITY CONSIDERATIONS

If a system is designed under one set of assumed conditions, whereas different sets of conditions may exist for the implemented system, a question to be answered is: "How much is the performance of the system affected, that is, how sensitive is the performance of the system to changes?" In many cases the sensitivity problem is of sufficient concern to warrant inclusion of a measure of sensitivity in performance criteria. In fact, a system may be designed with minimum sensitivity to prescribed changes as a prime objective. If system performance is measured in terms of a function of system parameters, the sensitivity measures of Chapter 2 may be appropriately applied. When functionals are involved, however, the sensitivity question assumes a new dimension.

For the basic problem of Lagrange (Section 3-3), it is natural to measure sensitivity in terms of the second variation of J, Equation 3-11, about an optimal function $x^*(t)$. This is so because the first variation $\delta J(x^*) = 0$, and the increment $\Delta J(x^*)$ is approximately equal to $\delta^2 J(x^*)/2$ for small (in magnitude) variations $\epsilon\, \delta x$ (see Equation 3-3). Thus, the fractional change $\Delta J(x^*)/J(x^*)$ may be expressed as

$$\frac{\Delta J(x^*)}{J(x^*)} \simeq \frac{1}{2} \frac{\delta^2 J(x^*)}{J(x^*)} \qquad (3\text{-}212)$$

and because of the identity 3-164,

$$\frac{\Delta J(x^*)}{J(x^*)} \simeq \frac{\int_{t_a}^{t_b} [(\epsilon\, \delta \dot{x})^2 \xi^*(t) + (\epsilon\, \delta x)^2 v^*(t)]\, dt}{\int_{t_a}^{t_b} f(x^*, \dot{x}^*, t)\, dt} \qquad (3\text{-}213)$$

where $\xi^*(t)$ and $v^*(t)$ are known functions, expressed by 3-167a and 3-167b with α replaced by *. Equation 3-213 has no value if only one explicit variation $\epsilon\, \delta x(t)$ is of interest because direct and exact evaluation of $\Delta J(x^*)/J(x^*)$ could be made in such a case. The utility of 3-213 is implied in the following observation: Subintervals of $[t_a, t_b]$ over which $|\xi^*(t)|$ is relatively large (small) are subintervals where relatively large values of $|\epsilon\, \delta \dot{x}|$ would tend to yield poor (nonpoor) performance. Similarly, subintervals of $[t_a, t_b]$ over which $|v^*(t)|$ is relatively large (small) are subintervals over which relatively large values of $|\epsilon\, \delta x|$ would tend to yield poor (nonpoor) performance. Of course, δx and $\delta \dot{x}$ are related by

$$\delta x(t) = \delta x(t_a) + \int_{t_a}^{t} \delta \dot{x}(t)\, dt \qquad (3\text{-}214)$$

and this dependence cannot be ignored.

Fortunately, many combinations of variations may be disallowed by the way in which a given system is implemented. In fact, the way in which a system is implemented has a strong bearing on the resulting sensitivity of the optimum performance—this point is given consideration in Section 4-8 of the next chapter. If the first-variational curves for a given case are explicit functions of time and system parameters, then performance variations may be measured in terms of variations of system parameters. For example, the optimal controller of Example 3-10 is characterized by a Laplace-transform transfer function of the form $k_1(s + k_2)/(s + k_3)$ where k_1, k_2, and k_3 are functions of B_m and J_m. For a particular design, assume that B_m and J_m are known to within $\pm 20\%$. If k_1, k_2, and k_3 are set at design-center values, the sensitivity measures of Chapter 2 can be applied to determine the worst-case system performance corresponding to critical nondesign-center values of B_m and J_m.

Sensitivity concepts for system functionals involve many interesting facets which are outside the scope of this book. Recent research ([3.8, 3.17, 3.22, 3.28], and others) has contributed much to this area.

3-13. CONCLUSION

In this chapter, attention is centered on various forms of the problem of Lagrange in the calculus of variations. Many of the techniques developed here for the problem of Lagrange can be modified to treat other general problem forms. Examples of other general problem forms are: (1) the problems of Bolza and Mayer (Section 3-2b) in which a function of the end-point values is incorporated in the performance measure; (2) problems with several independent variables and therefore with multiple integrals in the performance

CONCLUSION

measure; and (3) problems with mixed conditions, e.g., a problem wherein the limits of integration on an isoperimetric constraint are different from those on the performance measure.

The examples presented in this chapter are elementary enough to allow closed-form solutions to be obtained. The reason for this, of course, is to illustrate use of the theory as clearly as possible, without getting involved in a maze of computations. For additional examples of use of the theory, see references [3.9, 3.16, 3.18, 3.21, 3.23, 3.25, and 3.27] which contain typical engineering applications. For problems characterized by linear time-invariant systems and by quadratic integrands of performance measures, the use of transform theory often yields more direct results (Chapter 4). For nonlinear systems and nonquadratic integrands of performance measures, we must in most cases resort to computer solutions of the two-point boundary-value problems that result from application of the methods of this chapter. The approaches to numerical solution of two-point boundary-value problems are varied (e.g., [3.5, 3.11, 3.14, 3.15, 3.26]). Search techniques (Chapter 6) are used in these iterative numerical schemes, as is evidenced in Section 8-8. Both Chapters 7 and 8 contain techniques with which problems of the calculus of variations can be treated. The techniques of these chapters are especially appropriate when inequality constraints are associated with the variables of a problem.

If both parameters and functions are to be selected to obtain the optimum of a continuously differentiable functional, necessary conditions from both Chapter 2 and Chapter 3 must be satisfied, in general, by the optimal solution. One way to show this is to embed parameter optimization in the framework of the calculus of variations. For example, consider the case in which $J(\mathbf{x})$ is to be extremized, where

$$J(\mathbf{x}) = \int_{t_a}^{t_b} f(\mathbf{x},\dot{\mathbf{x}},t) \, dt \qquad (3\text{-}98)$$

and where a particular component x_k of $\mathbf{x}$ is required to be independent of t and the remaining x_i's. In this case, $\dot{x}_k$ must equal zero, so the approach of Section 3-8b can be applied: an augmented functional $J_a(\mathbf{x})$ is formed, where

$$J_a(\mathbf{x}) = \int_{t_a}^{t_b} [f(\mathbf{x},\dot{\mathbf{x}},t) + \lambda_k(t)\dot{x}_k] \, dt \qquad (3\text{-}215)$$

and where $\lambda_k(t)$ is a Lagrange multiplier. The Euler-Lagrange equation (see Equation 3-109) corresponding to x_k is

$$\dot{\lambda}_k = \frac{\partial f(\mathbf{x},\dot{\mathbf{x}},t)}{\partial x_k} \qquad (3\text{-}216)$$

and therefore,

$$\lambda_k(t) = \lambda_k(t_a) + \int_{t_a}^{t} \left[\frac{\partial f(\mathbf{x},\dot{\mathbf{x}},t)}{\partial x_k} \right] dt \qquad (3\text{-}217)$$

Furthermore, in order for the transversality condition (3-100) to be satisfied, the *variations* $dx_k(t_a)$ and $dx_k(t_b)$ of x_k at the end points must be independent of the remaining x_i's, $i \neq k$, and of t at the end points; but since x_k is restricted to be independent of t, the variation $dx_k(t_a)$ must equal the variation $dx_k(t_b)$. Thus, the part of 3-100 associated with x_k reduces to

$$0 = \frac{\partial}{\partial \dot{x}_k} [f(\mathbf{x},\dot{\mathbf{x}},t) + \lambda_k(t)\dot{x}_k] \Big|_{t_a}^{t_b} = \lambda_k(t_b) - \lambda_k(t_a) \qquad (3\text{-}218)$$

But note that at $t = t_b$, Equation 3-217 reduces to

$$\lambda_k(t_b) = \lambda_k(t_a) + \int_{t_a}^{t_b} \frac{\partial f(\mathbf{x},\dot{\mathbf{x}},t)}{\partial x_k} dt \qquad (3\text{-}219)$$

and in view of 3-218 this is equivalent to

$$0 = \int_{t_a}^{t_b} \frac{\partial f(\mathbf{x},\dot{\mathbf{x}},t)}{\partial x_k} dt \qquad (3\text{-}220)$$

which is what would have been obtained if the theory of Section 2-8 had been applied directly.

REFERENCES

[3.1] Akhiezer, N. I. *The Calculus of Variations.* Translated from Russian by A. H. Frink, Blaisdell Publishing, New York, 1962.

[3.2] Bedrosian, E. "The Optimum Detection of Analog-Type Digital Data." *Proceedings of the IRE,* **48**, 1655–1656, Sept. 1960.

[3.3] Berkovitz, L. D. "Variational Methods in Problems of Control and Programming." *Journal of Mathematical Analysis and Applications,* **3**, 145–169, Aug. 1961.

[3.4] Berkovitz, L. D. "On Control Problems with Bounded State Variables." *Journal of Mathematical Analysis and Applications,* **5**, 488–498, Dec. 1962.

[3.5] Bingulac, S. P., and S. J. Kahne. "On the Solution of Two-Point Boundary Value Problems." *IEEE Transactions on Automatic Control,* **AC-10**, 208, April 1965.

[3.6] Bliss, G. A. *Lectures on the Calculus of Variations.* Univ. of Chicago Press, Chicago, 1946.

[3.7] Bolza, O. *Lectures on the Calculus of Variations.* Univ. of Chicago Press, Chicago, 1904.

REFERENCES

[3.8] Cruz, J. B., Jr., and W. R. Perkins. "On Invariance and Sensitivity." *IEEE International Convention Record*, **14**, part 7, 159–162, 1966.

[3.9] Drake, J. H., R. B. Mayall, L. K. Krichmayer, and H. Wood. "Optimum Operation of a Hydro-Thermal System." *Transactions of the AIEE*, **81**, part III, 242–249, 1962.

[3.10] Elsgolc, L. E. *Calculus of Variations*. Addison-Wesley, Reading, Mass. 1962.

[3.11] Fox, L. *The Numerical Solution of Two-Point Boundary Problems in Ordinary Differential Equations*. Oxford at the Clarendon Press, London, 1957.

[3.12] Gelfand, I. M., and S. V. Fomin. *Calculus of Variations*. Prentice-Hall, Englewood Cliffs, N.J., 1963.

[3.13] Hedvig, T. I. *The Determination of Optimum Paths in the Phase Plane for Dynamic Systems Using Calculus of Variations Techniques*. Ph.D. Thesis, Univ. of Illinois, Urbana, Ill., 1961, 115 pp.

[3.14] Kahne, S. J. "Note on Two-Point Boundary Value Problems." *IEEE Transactions on Automatic Control*, **AC-8**, 257–258, July 1963.

[3.15] Kenneth, P., and G. E. Taylor. "Solution of Variational Problems with Bounded Control Variables by Means of the Generalized Newton-Raphson Method." Pages 471–487 of *Recent Advances in Optimization Techniques*. A. Lavi and T. P. Vogl (Editors), Wiley, New York, 1966.

[3.16] Kliger, I. "On Closed-Loop Optimal Control." *IEEE Transactions on Automatic Control*, **AC-10**, 207, April 1965.

[3.17] Kokotovic, P. V., and R. S. Rutman. "Sensitivity of Automatic Control Systems (Survey)." *Automation and Remote Control*, **26**, 727–749, 1965.

[3.18] Miele, A. "Extremization of Linear Integrals by Green's Theorem" (Chapter 3) and "The Calculus of Variations in Applied Aerodynamics and Flight Mechanics" (Chapter 4) of *Optimization Techniques with Applications to Aerospace Systems*. G. Leitmann (Editor), Academic Press, New York, 1962.

[3.19] Mikhlin, S. G. *The Problem of the Minimum of a Quadratic Functional*. Translated from Russian by A. Feinstein, Holden-Day, San Francisco, 1965.

[3.20] Mostov, P. M., J. L. Neuringer, and D. S. Rigney. "Optimum Capacitor Charging Efficiency for Space Systems." *Proceedings of the IRE*, **49**, 941–948, May 1961.

[3.21] Paris, D. T., and F. K. Hurd. "Relaxation Properties of Fields and Circuits." *Proceedings of the IEEE*, **53**, 150–156, Feb. 1965.

[3.22] Rohrer, R. A., and M. Sobral, Jr. "Sensitivity Considerations in Optimal System Design." *IEEE Transactions on Automatic Control*, **AC-10**, 43–48, Jan. 1965.

[3.23] Sidar, M. "On Closed-Loop Optimal Control." *IEEE Transactions on Automatic Control*, **AC-9**, 292–293, July 1964.

[3.24] Sigalov, A. G. *Two-Dimensional Problems of the Calculus of Variations*. Translation No. 83, Amer. Math. Soc., 1953, 121 pp.

[3.25] Solymar, L. "A Note on the Optimum Design of Non-Uniform Transmission Lines." *Proceedings*, Institution of Electrical Engineers, London, **107**, part C, 100–104, 1960.

[3.26] Stakhovskii, R. I. "One Algorithm for Solution of Boundary-Value Problems." *Automation and Remote Control*, **24**, 879–888, 1963.
[3.27] Sugai, I. "Variationally Minimum Reflection Coefficient for Nonuniform Transmission Lines." *Proceedings of the IEEE*, **51**, 1789–1790, Dec. 1963.
[3.28] Tomovic, R. *Sensitivity Analysis of Dynamic Systems*. McGraw-Hill, New York, 1963.
[3.29] Valentine, F. A. "The Problem of Lagrange with Differential Inequalities as Added Side Conditions." Pages 403–447 of *Contributions to the Calculus of Variations*. Univ. of Chicago Press, Chicago, 1937.

PROBLEMS

3.1 A system is to be designed to generate $x(t)$ which is to be used in two different processes: process A and process B. For process A, the ideal $x(t)$ would be e^{-t} for the time interval of interest $0 \le t \le T$; for process B, the ideal value of dx/dt would be -1 for $0 \le t \le T$. Because the ideal situations cannot be realized simultaneously by the function $x(t)$, a compromise is to be made, as follows: the function $x(t)$ is to be selected on the basis that it results in a minimum of J,

$$J = \int_0^T [(x - e^{-t})^2 + (\dot{x} + 1)^2]\,dt$$

a. Find a differential equation that the optimal $x(t)$, $x^*(t)$, must satisfy and solve for this $x^*(t)$ under the conditions that $T = 1$, $x(0) = 1$, and $x(1) = 0.5$.
b. Can $x^*(t)$ contain corner points?
c. Suppose $x(T)$ must equal T which is not specified. Find the equations which enable solution of the problem.

3.2 Find $y(x)$ such that the Euclidean distance between $(x_a, y(x_a))$ and $(x_b, y(x_b))$ is a minimum, i.e., find the minimum of

$$J = \int_{x_a}^{x_b} \left[1 + \left(\frac{dy}{dx}\right)^2\right]^{1/2} dx$$

with respect to $y(x)$. (*Hint:* Find the "first integral.")

3.3 Consider the problem of finding a first-variational curve for $J(v)$:

$$J(v) = \int_0^L \left[\frac{\Gamma(x)v + \dot{v}}{\Gamma(x)v - \dot{v}}\right] dx$$

where $\Gamma(x)$ is an explicit function of x. The Euler-Lagrange equation for this problem can be reduced to a first-order differential equation in terms of $y \triangleq \dot{v}/v$. Given $v(0) = c_0$ and $v(L) = c_L$, find an expression for the unique first-variational curve $v^*(x)$. (A problem of this type is treated by Sugai [3.27].)

PROBLEMS

3.4 Find the first-variational curve which corresponds to the functional

$$\int_{-1}^{1} t^2 \dot{x}^2 \, dt$$

when $x(-1) = -1$ and $x(1) = 1$. How does this curve compare to that given in Section 3-2a for the minimum of this functional?

3.5 Consider the problem of minimizing $J(y)$:

$$J(y) = \int_0^1 \left(\frac{dy}{dx}\right)^2 dx$$

subject to satisfaction of N isoperimetric constraints:

$$0 = \int_0^1 y(x)(\cos \mu_k x) \, dx, \qquad k = 1, 2, \ldots, N$$

where the μ_k's are constants, and where the optimal y, $y^*(x)$, is to equal the constant c_a at $x = 0$ and c_b at $x = 1$. Find $y^*(x)$. (A problem of this type is treated by Solymar [3.25].)

3.6 Consider the area J between the curve $y(x)$ and the x axis:

$$J = \int_{x_a}^{x_b} y \, dx$$

where $x_a, x_b, y(x_a) = c_a$, and $y(x_b) = c_b$ are specified constants. Find an expression for the curve $y^*(x)$ which when substituted for y in the above equation yields a maximum of J, subject to the constraint that the length of the curve $y^*(x)$ is equal to a constant K:

$$K = \int_{x_a}^{x_b} [1 + \dot{y}^2]^{1/2} \, dx \bigg|_{y = y^*}$$

3.7 Consider the area J_A of a surface of revolution generated by rotating the curve $y(x)$ about the x axis:

$$J_A = 2\pi \int_{x_a}^{x_b} y[1 + \dot{y}^2]^{1/2} \, dx$$

where $x_a, x_b, y(x_a) = c_a$, and $y(x_b) = c_b$ are specified constants. Find an expression for the curve $y^*(x)$ which when substituted for y in the above equation yields a minimum of J_A.

3.8 Consider the volume J_V contained within a surface of revolution generated by rotating the curve $y(x)$ about the x axis:

$$J_V = \pi \int_{x_a}^{x_b} y^2 [1 + \dot{y}^2]^{1/2} \, dx$$

where $x_a, x_b, y(x_a) = c_a$, and $y(x_b) = c_b$ are specified constants. Find an expression for the curve $y^*(x)$ which when substituted for y in the above equation yields a maximum of J_V, under the constraint that the surface area J_A equals a specified constant K.

3.9 Repeat Example 3-4, except replace J of Equation 3-41 with

$$J = \int_0^T \dot{v}_o(t)\, dt$$

and assume $v_o(T)$ is not specified in advance.

3.10 Find an expression for the maximum of J:

$$J = \int_{t_a}^{t_b} (e^{-t} - e^{-2t})\, dt$$

with respect to the selection of t_a and t_b, but subject to the constraint that $t_b - t_a = c$, a constant. (*Hint:* Apply the transversality condition developed for Case D of Section 3-5.)

3.11 Can corner points exist for problems of the type posed in Example 3-1?

3.12 Show that the first-variational curves of Example 3-8 are of the form $c_1 t + c_0$, and find a curve which yields the minimum when $x(0) = 1$ and $x(3) = 0$.

3.13 Given to be minimized:

$$J = \int_0^1 (x_1^2 + \dot{x}_1^2 + x_2^2 + \dot{x}_2^2)\, dt$$

where $\dot{x}_1 = -x_1 + x_2$, $x_1(0) = 1$, and $x_1(1) = x_2(0) = x_2(1) = 0$. Find a set of equations which the optimal $x_1(t)$ and $x_2(t)$ must satisfy, and solve for $x_1^*(t)$ and $x_2^*(t)$.

3.14 Consider the functional $J(\mathbf{x})$ of 3-98 with t_a and $x_k(t_a)$'s fixed and with $x_k(t_b)$'s required to equal $\phi_k(t_b)$'s. The $\phi_k(t)$'s are given, differentiable functions of t. To what does the transversality condition 3-100 reduce in this case?

3.15 Consider the functional $J(\mathbf{x})$ of 3-98 with $\mathbf{x} = \{x_1, x_2\}$, t_a and $\mathbf{x}(t_a)$ fixed, and $x_1(t_b)$ required to equal a given differentiable function $\phi = \phi(x_2, t)$ at $t = t_b$. To what does the transversality condition 3-100 reduce in this case? (*Hint:* Because $x_1(t_b) = \phi(x_2, t)$ at $t = t_b$, it follows that $[(\partial\phi/\partial x_2)\, dx_2 + (\partial\phi/\partial t)\, dt]_{t=t_b}$ must equal $dx_1(t_b)$. Also, the equivalent of 3-100 for this problem contains the variations $dx_1(t_b)$, $dx_2(t_b)$, and dt_b. Between the two equations—one for each of the two preceding sentences—eliminate one of the three variations dt_b, $dx_1(t_b)$, and $dx_2(t_b)$ to obtain an equation in terms of the remaining two. In the resulting equation, the coefficients of the two remaining independent variations must be zero.)

3.16 Given J:

$$J = \int_{t_a}^{t_b} f(x, \dot{x}, \ddot{x}, t)\, dt$$

Prove that a valid Euler-Lagrange equation for this functional is

$$f_x - \frac{d}{dt} f_{\dot{x}} + \frac{d^2}{dt^2} f_{\ddot{x}} = 0$$

(*Hint:* Evaluate $[\partial J(x + \epsilon\, \delta x)/\partial \epsilon]|_{\epsilon=0} = 0$ and perform integration by parts on all terms which contain $\delta \dot{x}$ and $\delta \ddot{x}$ in the integrand.)

3.17 Supply the steps which are required to obtain the last set of equations in Example 3-9. The terminal value $v_o(T)$ of $v_o(t)$ is free.

3.18 Repeat Example 3-9, using a different performance measure, as follows:

$$J = \int_0^T v_o^2 \, dt$$

3.19 A given system is characterized by $\ddot{x}_1 + \dot{x}_1 + x_1 = m_1$ where $x_1(0) = c_1$ and $\dot{x}_1(0) \triangleq x_2(0) = c_2$ are constants, $x_1(T)$ and $\dot{x}_1(T) = x_2(T)$ are required to be zero, and T is the terminal time which is to be minimized subject to the constraint $|m_1(t)|$ being less than or equal to unity for all t. Prove that the optimal $m_1(t)$, $m_1^*(t)$, which yields the minimum of T assumes either the value $+1$ or the value -1 at all but at most a finite number of instants of time in the interval $[0,T]$. (*Hint:* Use the first method of Section 3-8d; show that the costate variables (Lagrange multipliers) corresponding to x_1 and x_2, $x_2 = \dot{x}_1$, have solutions which are nonzero, in general, at all but at most a finite number of points in $[0, T]$; use this fact in conjunction with the set of Euler-Lagrange equations to obtain the proof.)

3.20 Supply the steps which are required to obtain $\theta^*(t)$ in Example 3-10. (*Suggestion:* Laplace transform Equations 3-153a, 3-153b, 3-155a, 3-155b, and 3-156 and insert unknown initial conditions on λ_1 and λ_2; then solve for $\Theta(s)$ which is of the form of a ratio of polynomials in s, the denominator polynomial of which is $s[s^4 - (B_m/J_m)^2 s^2 + (1/h_1 J_m^2)]$. The term within the brackets evidences two roots in the left half of the s plane and two in the right half. In order for the terminal conditions on θ and $\dot{\theta}$ to be satisfied, the initial conditions $\lambda_1(0)$ and $\lambda_2(0)$, which appear in the numerator of $\Theta(s)$, must be such that zeros of the numerator polynomial exactly cancel the right-half-plane denominator factors.)

3.21 Supply the steps required to obtain $G_c(s)$ in Example 3-10, given $\theta^*(t)$.

3.22 Discuss the utility of the "optimal" controller in Example 3-10 if K is allowed to approach infinity. (*Hint:* Examine the character of the closed-loop poles as h_1 approaches zero.)

3.23 Given: $\dot{x} = m$, $J = \int_0^T \frac{1}{2}(x^2 + m^2) \, dt$, $x(0) = x_0$ (a constant), and $x(T) = 0$.
  a. Find optimal functions $x^*(t)$ and $m^*(t)$ which yield a minimum of J. These functions will be in terms of t, T, and x_0 only.
  b. Find a way of expressing $m^*(t)$ in terms of $x^*(t)$ multiplied by a time-varying function $\psi(t - T)$ which is independent of x_0.

(This problem is considered by Sidar [3.23] and Kliger [3.16].)

3.24 Repeat Problem 3.23, but with $x(T)$ unspecified.

3.25 Given are space ships A and B in free space. Space ship B is c_0 (meters) away from space ship A at time $t = 0$; and ship B is moving away from A at a constant rate c_1 (meters/second). The captain of ship A desires to catch and dock with ship B. For $t \geq 0$, the position $x(t)$ of ship A is given to be governed by

$$\ddot{x}(t) = m(t), \quad x(0) = 0, \quad \text{and} \quad \dot{x}(0) = 0$$

The position $s(t)$ of ship B, relative to the initial position of ship A, is

$$s(t) = c_0 + c_1 t$$

At the *unspecified* docking time T,

$$x(T) = s(T) = c_0 + c_1 T \quad \text{and} \quad \dot{x}(T) = c_1$$

The energy consumed by ship A in the docking maneuver is given to be proportional to J:[10]

$$J = \int_0^T m(t)^2 \, dt$$

Find the thrusting action $m^*(t)$ of $m(t)$ which yields a minimum of J while accomplishing docking. Although T is not specified, the maneuver must be accomplished by $t = T_1$, a specified time; i.e., T is constrained to be less than or equal to T_1.

3.26 Given J to be minimized:

$$J = \int_{t_a}^{t_b} (x^2 + m^2) \, dt$$

where $\dot{x} = -x + x^3 + m$. Find Euler-Lagrange equations which apply and simplify as much as possible. Use one of the following sets of boundary conditions.
a. t_a, $x(t_a)$, and t_b specified, but $x(t_b)$ free.
b. t_a, $x(t_a)$, and $x(t_b)$ specified, but t_b free.

3.27 A process characterized by $x(t)$ is to be driven from $x(0) = 1$ to $x(T) = 0$ in such a way that the terminal time T is minimized. Constraints are

$$\dot{x} = ax + m, \quad \int_0^T m^2 \, dt = K \quad \text{(a constant)},$$

$x(0) = 1$, and $x(T) = 0$; and $m(t)$ is the control function to be selected for the optimum. Give a complete set of equations from which the optimal $m(t)$, $m^*(t)$, can be determined and solve for $m^*(t)$.

3.28 For the control problem of Section 3-9, suppose that fixed end conditions are t_a and $\mathbf{x}(t_a)$, but that $\phi_1(\mathbf{x}, t)$, a given differentiable function of $\mathbf{x}$ and t, must equal zero at $t = t_b$. Find n additional conditions which the optimal solution must satisfy at $t = t_b$. (*Hint:* Make use of the transversality condition and the fact that $[(\partial \phi_1/\partial t) \, \delta t + \sum_{i=1}^n (\partial \phi_1/\partial x_i) \, \delta x_i]_{t=t_b}$ must equal zero.)

3.29 For the control problem of Section 3-9, suppose that fixed end conditions are t_a t_b, and $\mathbf{x}(t_a)$, but that $\phi_1(\mathbf{x})$ and $\phi_2(\mathbf{x})$, given differentiable functions of $\mathbf{x}$, must equal zero at $t = t_b$. Find $n - 2$ additional conditions which the optimal solution must satisfy at $t = t_b$.

3.30 Widgets are produced by company Z; rate of production dx_1/dt and rate of sales are assumed essentially equal with stored widgets negligible. The rate of sales is dependent on the rate r of advertising expenditures and time t; that is,

$$\frac{dx_1}{dt} = q_1(r, t)$$

---

[10] A more realistic criterion for Problem 3.25 is $J = \int_0^T |m(t)| \, dt$, $|m(t)| \le m_{\max}$, which is proportional to the fuel consumed by ship A in docking (see Chapter 8).

where q_1 is a known function of r and t. In addition to advertising expense, the production expenditures per unit time, dx_2/dt, equal a known function $q_2(dx_1/dt, d^2x_1/dt^2)$ of rate of production and change in rate of production; to simplify the development, observe that

$$\frac{dx_2}{dt} = q_2(dx_1/dt, d^2x_1/dt^2)$$
$$= q_2[q_1(r,t), (\partial q_1/\partial t) + (\partial q_1/\partial r)(dr/dt)]$$
$$\triangleq g_2(r,\dot{r},t)$$

Widgets are sold at c_1 dollars per unit, and net profit P is to be maximized over the time interval $[t_a, t_b]$ of operation:

$$P = \int_{t_a}^{t_b} [c_1(dx_1/dt) - (dx_2/dt) - r] dt$$
$$= \int_{t_a}^{t_b} [c_1 q_1(r,t) - g_2(r,\dot{r},t) - r] dt$$

where r is constrained to be non-negative, i.e., $r \geq 0$. Assuming t_a, t_b, and $r(t_b) = 0$ are specified, supply Euler-Lagrange equations and transversality conditions which must be satisfied by the $r(t)$ which maximizes P.

3.31 Parallel the approach of Section 3-11c with the following modifications:

$$x(t) \cong (x_{ak} + x_{a,k+1})/2 \quad \text{when } t_a + k \Delta t \leq t \leq t_a + (k+1) \Delta t$$

and

$$t \cong t_a + (k + \tfrac{1}{2}) \Delta t \quad \text{when } t_a + k \Delta t \leq t \leq t_a + (k+1) \Delta t$$

3.32 Conduct a sensitivity study on a problem from this chapter.

3.33 Formulate an approximate form of one of the problems from this chapter by use of a "direct method."

3.34 Consider the RL circuit shown. The initial current $i(0) = 0$. A maximum of $v_o(T)$ is desired, where T is specified and where

$$v_o(T) = \int_0^T R \frac{di}{dt} dt$$

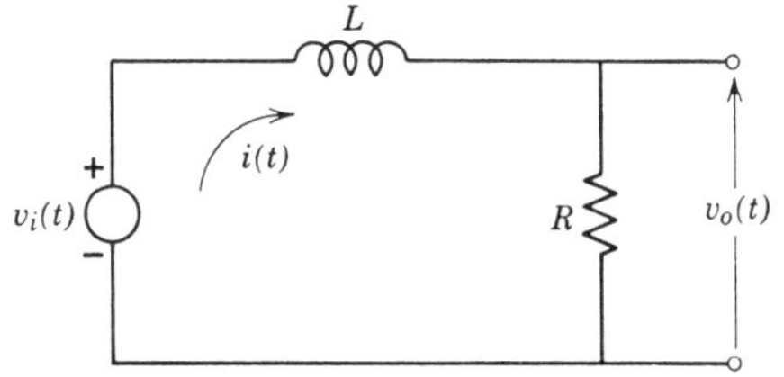

Figure 3-P34.

Constraints on the selection of the optimal $v_i(t)$ and $i(t)$ are

$$\frac{di}{dt} = \frac{1}{L} v_i - \frac{R}{L} i \quad \text{(circuit characterization)}$$

and

$$K = \int_0^T v_i i \, dt \quad \text{(an energy constraint)}$$

Within the framework of this problem statement, is it possible for jump-type discontinuities to exist in the optimal $v_i(t)$? Is it possible for the optimal $v_i(t)$ to exhibit Dirac delta functions? Obtain Euler-Lagrange equations for the problem and comment on any conditions which are implied by these equations. (The theory of Chapter 4 can be applied to obtain the solution of problems of this type.)

3.35 Given to be minimized:

$$J = \int_0^T (\dot{x}^2 - x^2) \, dt$$

with $x(0) = 0$ and $x(T) = 1$.
  a. For what values of T can first-variational curves be imbedded in a field?
  b. If $T = 3\pi/2$, does the minimum of J correspond to the first-variational curve which fits the end points? Is the minimum of J bounded for $T = 3\pi/2$?
(*Comment:* The answer to this problem indicates limited practicality for this functional.)

3.36 Given to be minimized:

$$J = \int_{t_a}^{t_b} \dot{x}^3 \, dt$$

where $x(t_a) = 0$ and $x(t_b) = b > 0$.
  a. Find the Euler-Lagrange equation and solve for a first-variational curve which satisfies the end conditions.
  b. Does the first-variational curve of Part a satisfy sufficient conditions for a weak relative minimum?
  c. Does the first-variational curve of Part a satisfy sufficient conditions for a strong relative minimum? If not, find a curve which can be placed arbitrarily close to that of Part a, and corresponding to which the value of J is smaller.

3.37 a. Corresponding to a variation $\delta \mathbf{x}$, find the first variation δJ_a of the functional

$$J_a = \int_{t_a}^{t_b} f_a(\mathbf{x}, \dot{\mathbf{x}}, t) \, dt$$

(*Hint:* Expand $\Delta J_a = J_a(\mathbf{x} + \epsilon \, \delta \mathbf{x}) - J_a(\mathbf{x})$ in a Maclaurin's series with respect to the scalar variable ϵ and evaluate the first term in the series at $\epsilon = 1$.)
  b. Find the second variation $\delta^2 J_a$ associated with Part a.

WIENER-HOPF SPECTRUM FACTORIZATION AND FREQUENCY-DOMAIN OPTIMIZATION

4

4-1. INTRODUCTION

Much of the underlying work in engineering is concerned with linear time-invariant systems, e.g., systems which are characterized by ordinary linear differential equations with constant coefficients. Even if a system is nonlinear in the large, it is often characterized by linear differential equations when operation is confined to the vicinity of a stable operating point. The analysis and design of linear time-invariant systems is expedited by the use of various transform theories, perhaps the foremost of these being the two-sided Laplace transform[1] from which the one-sided Laplace transform, the Fourier transform, and the one-sided and two-sided z-transforms can be developed as special cases. It is assumed here that the reader is familiar with basic properties of two-sided Laplace transforms (Appendix B).

To obtain optimal performance from a linear system, or any system for that matter, a performance measure should be specified. Any number of performance measures may be desirable from the practical standpoint in a given case, but only a few of these lead to closed-form solutions which are obtainable by using basic analytic techniques. Fortunately, the integral-square-error performance measure (Appendix B) and the mean-square-error performance measure (Appendix C) are reasonable measures of performance for a large class of practical systems and are easily embodied in the design of linear systems by the use of transform theory. Yet another tractable

---

[1] Also called the bilateral Laplace transform.

performance measure to be utilized in this chapter is the integral of the time-weighted output signal of a linear system when the input energy to the system is limited.

The mean-square-error performance measure is a natural one for use in conjunction with linear systems which are subjected to stationary random inputs. It is assumed in the following sections that the reader is familiar with correlation-function and power-density-spectrum concepts which are associated with stationary random signals (see Appendix C).

Sections 4-2 through 4-6 are devoted to several classes of optimization problems which can be solved by using the Wiener-Hopf spectrum-factorization technique [4.43, 4.44]. In Section 4-2, a basic optimal design problem is posed which includes filter problems, control system design problems, and predictor problems as special cases. In Section 4-3, a basic optimal pulse-shape problem is posed, and practical special cases are noted. Additional applications are to be found in the references cited at the end of this chapter and in the problems of this chapter. In all these problems, a Laplace transformable function is to be specified over a given interval of time; the function is to yield an extremum of a given performance measure, subject to specified constraints. The common features of the problems of Sections 4-2 and 4-3 are investigated in Section 4-4 in which the Wiener-Hopf spectrum-factorization technique is developed. On the basis of the general solution obtained in Section 4-4, specific solutions to the problems of Sections 4-2 and 4-3 are given in Sections 4-5 and 4-6, respectively.

Despite its many applications, the Wiener-Hopf technique cannot be applied to solve every optimal design problem which is associated with a linear system and for which the performance measure is the integral of a quadratic form. Certain general techniques are applicable, however, and these are illustrated by examples in Section 4-7 wherein the partial energy output of a general linear system is maximized, subject to the constraints that the input signal is time limited and that the input energy is fixed.

The attainment of optimal input-output relationships with a system is by itself insufficient for good system design. It may be that a system is optimum under one set of conditions, but is far from optimum under a closely related set of conditions; such a system is said to be highly sensitive and is usually most unsatisfactory. The design of systems for optimal input-output relations with sensitivity in mind is the theme of Section 4-8.

4-2. FILTER, CONTROL, AND PREDICTOR PROBLEMS

4-2a. Description of the System

Consider the block-diagram shown in Figure 4-1. In regard to the system shown, the following conditions are assumed. The functions $G_0(s)$, $G_2(s)$,

Sect. 4-2 FILTER, CONTROL, AND PREDICTOR PROBLEMS

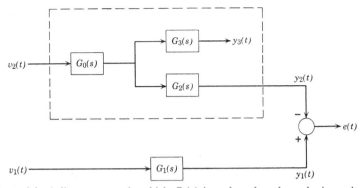

Figure 4-1. A linear system in which $G_0(s)$ is to be selected to obtain optimal performance.

and $G_3(s)$ are Laplace-transform transfer functions of linear, stable, single-input, single-output, and physically realizable systems. The functions $G_2(s)$ and $G_3(s)$ are given functions of s and characterize the *plant* of the overall system, whereas $G_0(s)$ is to be selected by the designer. The function $G_1(s)$ is a given Laplace-transform transfer function of a linear and stable, but not necessarily physically realizable system; $G_1(s)$ represents the ideal system for the process at hand with an ideal real input signal $v_1(t)$. In contrast, the actual system to be designed (that which is characterized by the blocks contained within the dashed lines of Figure 4-1) is subject to a possible nonideal real input signal $v_2(t)$ and must be physically realizable. Furthermore, the actual system may contain a signal $y_3(t)$—perhaps more than one such signal—which is known to be limited in either power or energy.

In the remainder of this section, numerous problems are described, all of which are concerned with systems which are special cases of that of Figure 4-1. The problem in each case is to select $G_0(s)$ so as to minimize either the integral-square error or the mean-square error, whichever is appropriate, of the difference $e(t)$ between the ideal real output $y_1(t)$ and the actual real output $y_2(t)$.

4-2b. Integral-Square-Error Problems

In this subsection, the input signals $v_1(t)$ and $v_2(t)$ in Figure 4-1 are assumed to be deterministic and to be Laplace transformable. It is also assumed that the integral-square values of $y_3(t)$ and $e(t)$ exist, i.e., that

$$\int_{-\infty}^{\infty} [y_3(t)]^2 \, dt < \infty \tag{4-1}$$

and
$$\int_{-\infty}^{\infty} [e(t)]^2 \, dt < \infty \tag{4-2}$$

The general problem here is to select $G_0(s)$ so as to minimize the integral-square error J_1,

$$J_1 = \int_{-\infty}^{\infty} [e(t)]^2 \, dt$$

$$= \frac{1}{2\pi j} \int_{-j\infty}^{j\infty} E(s) \, E(-s) \, ds \tag{4-3}$$

where $E(s) = \mathscr{L}\{e(t)\}$ is the (two-sided) Laplace transform of $e(t)$, and the second equality in 4-3 is a statement of Parseval's theorem (Appendix B). A given constraint on the minimization process is that the integral-square value of $y_3(t)$ must equal a constant K_1:

$$K_1 = \int_{-\infty}^{\infty} [y_3(t)]^2 \, dt$$

$$= \frac{1}{2\pi j} \int_{-j\infty}^{j\infty} Y_3(s) \, Y_3(-s) \, ds \tag{4-4}$$

Note that Equation 4-4 constitutes an isoperimetric constraint on the minimization problem (the theory associated with isoperimetric constraints is considered in Chapter 3). An equivalent problem to that stated in the preceding paragraph is therefore to find the $G_0(s)$ which yields an absolute minimum of the augmented functional J_{a1},

$$J_{a1} = \frac{1}{2\pi j} \int_{-j\infty}^{j\infty} [E(s)E(-s) + h Y_3(s) Y_3(-s)] \, ds \tag{4-5}$$

where h is a Lagrange multiplier which is independent of the variable s of integration. This Lagrange multiplier is evaluated ultimately on the basis that the "energy" constraint 4-4 must be satisfied. If additional constraints of the form 4-4 exist for a given problem, they are taken into account in the same way with additional Lagrange multipliers.

From Figure 4-1,

$$E(s) = V_1(s)G_1(s) - V_2(s)G_0(s)G_2(s) \tag{4-6}$$

and

$$Y_3(s) = V_2(s)G_0(s)G_3(s) \tag{4-7}$$

The above identities are substituted appropriately into Equation 4-5 which is then rearranged in the form

Sect. 4-2   FILTER, CONTROL, AND PREDICTOR PROBLEMS   139

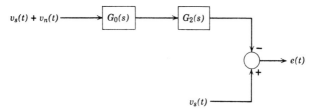

Figure 4-2a. A special case of the system of Figure 4-1; the associated problem is a filter problem.

$$J_{a1} = \frac{1}{2\pi j} \int_{-j\infty}^{j\infty} \{G_0(s)G_0(-s)V_2(s)V_2(-s)[hG_3(s)G_3(-s) + G_2(s)G_2(-s)]$$
$$- G_0(s)[G_1(-s)G_2(s)V_1(-s)V_2(s)]$$
$$- G_0(-s)[G_1(s)G_2(-s)V_1(s)V_2(-s)]$$
$$+ G_1(s)G_1(-s)V_1(s)V_1(-s)\} \, ds \quad (4\text{-}8)$$

This equation is of the general form considered in Section 4-4. The general solution obtained in Section 4-4 is applied in Section 4-5 to determine the $G_0(s)$ which yields a minimum of J_{a1} of Equation 4-8.

A few of the many special cases of this integral-square-error problem are listed next.

Case A. A filter problem in which $v_2(t) = v_s(t) + v_n(t)$, $v_n(t)$ is an undesired signal, $v_1(t) = y_1(t) = v_s(t)$ is the desired output signal, and $G_3(s) = 0$ (see Figure 4-2a).

Case B. A predictor problem in which $G_1(s) = e^{sT}$ (where T equals a positive constant), $G_2(s) = 1$, $G_3(s) = 0$, and $v_1(t) = v_2(t)$ (see Figure 4-2b).

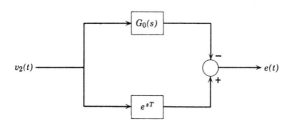

Figure 4-2b. A special case of the system of Figure 4-1; the associated problem is a predictor problem.

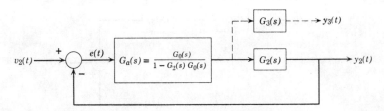

Figure 4-2c. A unity feedback control system.

Case C. A tracking control problem in which $v_2(t)$, $v_1(t)$, and $y_1(t)$ are equal, and a unity feedback control system (Figure 4-2c) is required which is equivalent, as far as input-output relations are concerned, to the system shown within the dashed lines in Figure 4-1. If feedback is employed to realize a system, the resulting system is generally less sensitive to changes in plant characteristics than is the "equivalent" open-loop system—this statement is expanded in Section 4-8.

4-2c. Mean-Square-Error Problems

In this subsection, the input signals $v_1(t)$ and $v_2(t)$ in Figure 4-1 are assumed to be stationary random signals with known correlation functions (Appendix C) which are assumed to be Laplace transformable (Appendix B).

The general problem here is to select $G_0(s)$ so as to minimize the mean-square error J_2,

$$J_2 = \lim_{T \to \infty} \frac{1}{2T} \int_{-T}^{T} [e(t)]^2 \, dt \qquad (4\text{-}9)$$

subject to the constraint that the mean-square value of $y_3(t)$ must equal a constant K_2.

$$K_2 = \lim_{T \to \infty} \frac{1}{2T} \int_{-T}^{T} [y_3(t)]^2 \, dt \qquad (4\text{-}10)$$

Note that Equation 4-10 constitutes an isoperimetric constraint on the minimization problem. An equivalent problem to that posed in the preceding paragraph is therefore to find the $G_0(s)$ which yields a minimum of the augmented functional J_{a2}:

$$J_{a2} = \lim_{T \to \infty} \frac{1}{2T} \int_{-T}^{T} [e(t)^2 + hy_3(t)^2] \, dt \qquad (4\text{-}11)$$

where h is a Lagrange multiplier which is independent of t and which is evaluated ultimately on the basis that the "power" constraint (4-10) must be satisfied.

The use of correlation-function theory, power-density-spectrum theory, and Parseval's theorem (Appendices B and C) yields the equivalent of 4-11:

$$J_{a2} = \frac{1}{2\pi j} \int_{-j\infty}^{j\infty} [\Phi_{ee}(s) + h\Phi_{y_3y_3}(s)] \, ds \tag{4-12}$$

where $\Phi_{ee}(s)$ and $\Phi_{y_3y_3}(s)$ are the power-density spectra of $e(t)$ and $y_3(t)$, respectively.

From Figure 4-1, in conjunction with the theory of Appendix C,

$$\Phi_{y_3y_3}(s) = G_0(s)G_0(-s)G_3(s)G_3(-s)\,\Phi_{v_2v_2}(s) \tag{4-13}$$

and

$$\Phi_{ee}(s) = G_0(s)G_0(-s)G_2(s)G_2(-s)\Phi_{v_2v_2}(s) + G_1(s)G_1(-s)\Phi_{v_1v_1}(s)$$
$$- \Phi_{v_2v_1}(s)G_0(-s)G_2(-s)G_1(s) - \Phi_{v_2v_1}(-s)G_0(s)G_2(s)G_1(-s) \tag{4-14}$$

When the identites given by 4-13 and 4-14 are substituted appropriately into the right-hand member of Equation 4-12, the net result is

$$J_{a2} = \frac{1}{2\pi j} \int_{-j\infty}^{j\infty} \{G_0(s)G_0(-s)\Phi_{v_2v_2}(s)[hG_3(s)G_3(-s) + G_2(s)G_2(-s)]$$
$$- G_0(s)G_2(s)G_1(-s)\Phi_{v_2v_1}(-s)$$
$$- G_0(-s)G_2(-s)G_1(s)\Phi_{v_2v_1}(s)$$
$$+ G_1(s)G_1(-s)\Phi_{v_1v_1}(s)\} \, ds \tag{4-15}$$

Observe that the above form for J_{a2} is similar to that for J_{a1} in 4-8; Equation 4-15 is generated from Equation 4-8 if the following replacements are made: $\Phi_{v_2v_2}(s)$ for $V_2(s)V_2(-s)$, $\Phi_{v_2v_1}(-s)$ for $V_2(s)V_1(-s)$, $\Phi_{v_2v_1}(s)$ for $V_2(-s)V_1(s)$, and $\Phi_{v_1v_1}(s)$ for $V_1(s)V_1(-s)$. As in the preceding subsection, therefore, the general solution of Section 4-4 is applicable. The $G_0(s)$ which yields a minimum of J_{a2} in 4-15 is given in Section 4-5.

Just as at the end of subsection 4-2b, special cases of this problem may be listed. The same examples as given there are appropriate here with obvious modifications in interpretation of the nature of the input and output signals.

4-3. A GENERAL OPTIMAL PULSE-SHAPE PROBLEM

4-3a. Description of the System

Consider a linear time-invariant low-pass system with real-valued input $v_i(t)$ and real-valued output $v_o(t)$, both of which are assumed to be Laplace transformable. The known transfer function $G(s)$ of the system is given by the ratio

$$G(s) = \frac{V_o(s)}{V_i(s)} \qquad (4\text{-}16)$$

where $V_o(s)$ and $V_i(s)$ are the two-sided Laplace transforms of $v_o(t)$ and $v_i(t)$, respectively.

It is assumed that the input $v_i(t)$ is zero outside the closed time interval $[0,T]$, T equals a constant, and that the input energy K_3 to the system is limited,

$$K_3 = \int_0^T v_i(t) i_i(t) \, dt \qquad (4\text{-}17)$$

where $i_i(t)$ is the input "current" to the system, i.e., $\mathscr{L}[i_i(t)] = I_i(s) = V_i(s) Y_i(s)$ where $Y_i(s)$ is the known input admittance to the system.

A general pulse-shape problem is this: Given that the input $v_i(t) = 0$ for all t outside the interval $[0,T]$ and that K_3 is fixed, find that $v_i(t)$ which yields a maximum of the following weighted value of the output $v_o(t)$,

$$J_3 = \int_0^\infty w_o(t) v_o(t) \, dt \qquad (4\text{-}18)$$

where the weighting factor $w_o(t)$ is assumed to have a Laplace transform $W_o(s)$.

Although initial conditions are assumed at zero in this section, they can be taken into account in a straightforward manner (see Problem 4.2). Also, linear active devices are readily incorporated in the problem formulation (see Problem 4.3).

As in the preceding section, the isoperimetric constraint 4-17 is adjoined to the performance measure 4-18 with a Lagrange multiplier h to obtain the augmented functional J_{a3},

$$J_{a3} = \int_{-\infty}^\infty [h v_i(t) i_i(t) + w_o(t) v_o(t)] \, dt \qquad (4\text{-}19)$$

Using Parseval's theorem (Appendix B) and the previously given Laplace-transform relationships results in the following frequency-domain equivalent of 4-19:

$$J_{a3} = \frac{1}{2\pi j} \int_{-j\infty}^{j\infty} [V_i(s) V_i(-s) h Y_i(s) + V_i(s) G(s) W_o(-s)] \, ds \qquad (4\text{-}20)$$

in which the Lagrange multiplier h is evaluated ultimately on the basis that the energy constraint 4-17 must be satisfied. The $V_i(s)$ is desired which results in a maximum of J_{a3} in 4-20.

Sect. 4-3   A GENERAL OPTIMAL PULSE-SHAPE PROBLEM   143

Two of the many special cases of this problem are given in the next two subsections. Solutions for these cases are obtained from the general solution of Section 4-4 and are listed in Section 4-6.

4-3b. Maximum Peak Output [4.31]

In Equation 4-18, J_3 will equal $v_o(T)$, the value of $v_o(t)$ at the end of the input-pulse period, if the weighting factor $w_o(t)$ is chosen as a Dirac delta function $\delta(t - T)$. The corresponding $W_o(s)$ is e^{-sT}.

This problem is the counterpart of the problem posed by Beattie [4.1]. Beattie considers the problem of minimizing the energy required of an input pulse so as to obtain a specified peak signal at the output of a circuit at the end of the pulse period. Numerous applications are cited by Beattie, and it is apparent that the solution obtained here is applicable to Beattie's problem with only minor modification. In Beattie's work a complicated integral equation which must be satisfied by the optimal input pulse is obtained, and particular examples are given in which the integral equation is solved. In contrast, the work in this chapter leads to a general solution which by-passes the integral equation solution.

In addition to being a problem of obvious interest to pulse-communication engineers, this problem is a generalization of a specific (linear) problem treated by Mostov, *et al.* [4.27]. Mostov, *et al.*, consider the problem of selecting the shape of a time-limited pulse which is applied at the input terminals of a network, the output element of which is effectively a capacitor which acts as an ion accelerator. The prime concern in selecting the pulse shape is that the voltage across the capacitor be as large as possible at the end of the pulse period, subject to the constraint that the energy associated with the input pulse equals a constant value K_3. Solutions obtained by Mostov, *et al.*, are for relatively simple networks and are obtained in the time domain. Here, general solutions for the linear network case are derived in the frequency domain.

The solutions of this problem to be given in Section 4-6 include the following special cases, in all of which $G(s)$ is assumed to be physically realizable.

Case A.   A fairly general case in which $Y_i(s)$ is a ratio of polynomials in s.

Case B.   The case where $Y_i(s)$ is a constant.

Case C.   The case where the time interval $[0,T]$ is replaced by the time interval $[-\infty,0]$.

Case D.   A classic case in which the conditions of both Case B and Case C are assumed.

4-3c. Maximum of the Average Output

If $w_o(t) = [u(t) - u(t - T)]/T$, where $u(t)$ is a *unit step function*,[2] the corresponding value of $W_o(s)$ is

$$W_o(s) = \lim_{\epsilon \to 0+} \frac{1 - e^{-sT}}{(s + \epsilon)T} \tag{4-21}$$

in which the positive parameter ϵ is introduced to avoid interpretational difficulties which stem from the occurrence of a zero of the denominator of $W_o(s)$ on the $j\omega$ axis of the s plane when this axis is used as the inversion contour. The Laplace transforms of time-limited functions exhibit the interesting property that they have no finite s-plane poles, e.g., $[W_o(s)_{\epsilon \equiv 0}]_{s \to 0} = 1$ in 4-21.

The effect of the weighting factor (4-21) is to make the performance measure J_3 of Equation 4-18 equal the average value of the output signal $v_o(t)$ over the period $[0,T]$. The time-domain solution of a specific problem of this kind is given in Example 3-9 of Chapter 3.

4-4. A GENERAL PROBLEM AND SOLUTION

4-4a. The Problem

The functionals of the problems of the two preceding sections have the common form J_a:

$$J_a = \frac{1}{2\pi j} \int_{-j\infty}^{j\infty} [A_1(s)F(s)F(-s) + A_2(s)F(s) + A_3(s)F(-s) + A_4(s)]\, ds \tag{4-22}$$

where each $A_i(s)$ ($i = 1, 2, 3,$ and 4) is a known, two-sided Laplace-transform function with strip of convergence which includes the $j\omega$ axis of the s plane, and the two-sided Laplace-transform $F(s)$ is to be selected to obtain an extremum of J_a. It is assumed that the inverse Laplace transform $\mathscr{L}^{-1}[F(s)] = f(t)$ is required to equal zero for all t outside a specified time interval defined by $0 \leq t \leq T_f$. Other time intervals could be assumed with only slight differences in the development that follows.

4-4b. Initial Steps to Solution

The above problem falls within the realm of the calculus of variations, but it is not exactly the type of problem which is considered in Chapter 3. As for the first few steps in the attainment of a necessary condition for an optimum, the procedure here is much the same as that of Chapter 3. Later steps deviate considerably however.

---

[2] The unit step function $u(t)$ is 0 for $t \leq 0$ and 1 for $t > 0$.

Sect. 4-4   A GENERAL PROBLEM AND SOLUTION   145

First, assume that $F^*(s)$, the optimal form of $F(s)$, exists and that any allowable $F(s)$ can be expressed as[3]

$$F(s) = F^*(s) + \epsilon\, \delta F(s) \tag{4-23}$$

where ϵ is a real variable, and $\epsilon\, \delta F(s)$ is any allowable variation of $F(s)$ from $F^*(s)$. An allowable variation is one that satisfies the condition that $\mathscr{L}^{-1}[\delta F(s)] = \delta f(t)$ equals zero outside the time interval $[0, T_f]$ because both $f(t)$ and $f^*(t)$ are so restricted.

Suppose $F^*(s)$ yields a minimum of J_a in 4-22. A typical plot of $J_a[F^*(s) + \epsilon\, \delta F(s)]$ versus ϵ, for a specific $\delta F(s)$, is depicted in Figure 4-3. Note that at the minimum of $J_a[F^*(s) + \epsilon\, \delta F(s)]$ (that is, at $\epsilon = 0$) the value of $\partial J_a/\partial \epsilon$ also equals zero; thus, a necessary condition that $F^*(s)$ must satisfy is obtained by differentiating both members of Equation 4-22 with respect to ϵ and by setting the result equal to zero at $\epsilon = 0$, as follows:

$$\left.\frac{\partial J_a}{\partial \epsilon}\right|_{\epsilon=0} = 0$$

$$= \frac{1}{2\pi j} \int_{-j\infty}^{j\infty} [A_1(s) F^*(s)\, \delta F(-s) + A_1(s) F^*(-s)\, \delta F(s)$$

$$\qquad + A_2(s)\, \delta F(s) + A_3(s)\, \delta F(-s)]\, ds$$

$$= \frac{1}{2\pi j} \int_{-j\infty}^{j\infty} \delta F(-s)[A_1(s) F^*(s) + A_3(s)]\, ds$$

$$\qquad + \frac{1}{2\pi j} \int_{-j\infty}^{j\infty} \delta F(s)[A_1(s) F^*(-s) + A_2(s)]\, ds \quad (4\text{-}24)$$

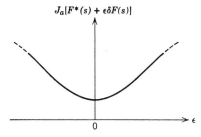

Figure 4-3. Plot of typical $J_a[F^*(s) + \epsilon\, \delta F(s)]$ versus ϵ with specific $\delta F(s)$.

---

[3] This use of "*" is not to be confused with either that in which the * signifies the complex conjugate of a function or that in which the * signifies the Laplace transform of an impulse-modulated function of time.

By Parseval's theorem, the second integral of the right-hand member of Equation 4-24 is equivalent to

$$\frac{1}{2\pi j}\int_{-j\infty}^{j\infty} \delta F(-s)[A_1(-s)F^*(s) + A_2(-s)]\,ds$$

which is substituted appropriately into 4-24 to obtain

$$\frac{1}{2\pi j}\int_{-j\infty}^{j\infty} \delta F(-s)\{[A_1(s) + A_1(-s)]F^*(s) + A_2(-s) + A_3(s)\}\,ds = 0 \quad (4\text{-}25)$$

The object here, of course, is to find $F^*(s)$ which satisfies 4-25. To this end, define $A_{11}(s)$ and $A_f(s)$ as follows:

$$A_{11} \triangleq A_1(s) + A_1(-s) \quad (4\text{-}26)$$

and

$$A_f(s) \triangleq A_{11}(s)F^*(s) + A_2(-s) + A_3(s) \quad (4\text{-}27)$$

Again, application of Parseval's theorem in conjunction with identity 4-27 and Equation 4-25 yields

$$0 = \int_{-\infty}^{\infty} \delta f(t)a_f(t)\,dt$$

$$= \int_{-\infty}^{\infty} \left\{\delta f(t)\left[a_2(-t) + a_3(t) + \int_{-\infty}^{\infty} f^*(t_1)a_{11}(t-t_1)\,dt_1\right]\right\} dt \quad (4\text{-}28a)$$

in which $\delta f(t) = \mathscr{L}^{-1}[\delta F(s)]$, and $a_f(t) = \mathscr{L}^{-1}[A_f(s)]$. Equation 4-28a is one form of the *Wiener-Hopf equation*.

To proceed at this point, recall that $\delta f(t)$ is an arbitrary Laplace transformable function in the closed interval $[0,T_f]$ and zero outside that interval. For Equation 4-28a to be satisfied, therefore, $a_f(t)$ *must equal zero for any t in the interval* $[0,T_f]$, i.e.,

$$a_2(-t) + a_3(t) + \int_{-\infty}^{\infty} f^*(t_1)a_{11}(t-t_1)\,dt_1 = 0, \quad t \in [0,T_f] \quad (4\text{-}28b)$$

which is a form of Abel's integral equation. But note that $a_f(t)$ *is not necessarily zero outside this interval*. This latter statement constitutes the heart of the problem, for because of it we cannot simply equate $A_f(s)$ equal to zero to determine $F^*(s)$.

The Wiener-Hopf spectrum-factorization technique effectively resolves the just-noted difficulty. Pertinent notation is introduced next, prior to presentation of the technique.

4-4c. Notation for Spectrum Factorization

Let $A_{11}(s)$ of 4-26 be factored as follows:

$$A_{11}(s) = A_{11}(s)^+ A_{11}(s)^- \tag{4-29}$$

where both $A_{11}(s)^+$ and $1/A_{11}(s)^+$ are analytic in finite regions which lie to the right of the $j\omega$ axis in the s plane, and where $A_{11}(s)^-$ and $1/A_{11}(s)^-$ are analytic in finite regions which lie to the left of the $j\omega$ axis in the s plane. In the case that $A_{11}(s)$ is a rational fraction function, i.e., a ratio of polynomials in s, the above statements are equivalent to requiring that $A_{11}(s)^+$ contain all the poles and zeros of $A_{11}(s)$ which lie in the left half of the s plane (to the left of the $j\omega$ axis in the s plane) and that $A_{11}(s)^-$ contain all the poles and zeros of $A_{11}(s)$ which lie in the right half of the s plane. Only the rational fraction case of $A_{11}(s)$ is considered in detail in the following. For an account of a more general case, see reference [4.32].

Next, define $A_2(-s)_{T_f+}$ and $A_2(-s)_{T_f-}$ by the pair of equations

$$A_2(-s)_{T_f+} \triangleq \mathscr{L}\{u(t - T_f)\mathscr{L}^{-1}[A_2(-s)]\} \tag{4-30}$$

$$A_2(-s)_{T_f-} \triangleq A_2(-s) - A_2(-s)_{T_f+} \tag{4-31}$$

which give that $A_2(-s)_{T_f+}$ is the Laplace transform of that part of the inverse transform of $A_2(-s)$ which exists for $t > T_f$. In the same way,

$$A_3(s)_{T_f+} = \mathscr{L}\{u(t - T_f)\mathscr{L}^{-1}[A_3(s)]\} \tag{4-32}$$

and

$$A_3(s)_{T_f-} = A_3(s) - A_3(s)_{T_f+} \tag{4-33}$$

A hypothetical $a_3(t) = \mathscr{L}^{-1}[A_3(s)]$ is given in Figure 4-4.

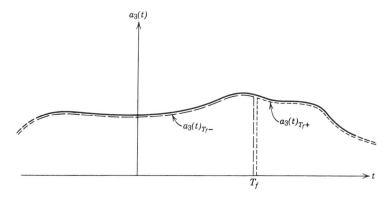

Figure 4-4. Hypothetical $a_3(t)$ with parts $a_3(t)_{T_f-}$ and $a_3(t)_{T_f+}$.

Notation similar to that of the preceding equations is

$$B(s) = B(s)_+ + B(s)_- \qquad (4\text{-}34)$$

where $B(s)$ is any Laplace transform and

$$B(s)_+ = \mathscr{L}\{u(t)\mathscr{L}^{-1}[B(s)]\} \qquad (4\text{-}35)$$

An additional relationship concerning $F^*(s)$ and $A_{11}(s)$, Equation 4-26, is established here for use in the next subsection. Consider $[A_{11}(s)F^*(s)]_{T_f+}$:

$$[A_{11}(s)F^*(s)]_{T_f+} = [A_{11}(s)_+ F^*(s) + A_{11}(s)_- F^*(s)]_{T_f+}$$
$$= [A_{11}(s)_+ F^*(s)]_{T_f+} + [A_{11}(s)_- F^*(s)]_{T_f+} \qquad (4\text{-}36)$$

The right-most term of 4-36 is identically zero; this fact can be reasoned by viewing $a_{11}(t)_- = \mathscr{L}^{-1}[A_{11}(s)_-]$ as the input to a linear system with transfer function $F^*(s)$; because $a_{11}(t)_- = 0$ for t greater than zero and because $f^*(t)$ is zero for t greater than T_f, the response $\mathscr{L}^{-1}[A_{11}(s)_- F^*(s)]$ must be identically zero for t greater than T_f; thus,

$$[A_{11}(s)F^*(s)]_{T_f+} = [A_{11}(s)_+ F^*(s)]_{T_f+} \qquad (4\text{-}37)$$

Furthermore, if $f^*(t)$ is viewed as the input to a linear system with transfer function $A_{11}(s)_+$, the response of this system for t greater than T_f is an unforced (homogeneous) response because $f^*(t)$ is identically zero for t greater than T_f. When $A_{11}(s)$ is a rational function, the preceding verbal statement, in conjunction with 4-37, is equivalent to the following analytical one:

$$[A_{11}(s)F^*(s)]_{T_f+} = e^{-sT_f} \sum_{k=1}^{n} \frac{c_k}{s - p_k} \qquad (4\text{-}38)$$

where the c_k's are constants, as yet unspecified, and the p_k's are the poles of $A_{11}(s)_+$. For notational convenience, the p_k's are assumed to be distinct; it is evident that Equation 4-38 can and should be modified when $A_{11}(s)_+$ contains poles of higher order.

In the above notation, we should carefully note the difference in meaning between the superscript "+'s and −'s" and the subscript "+'s and −'s".

For any allowable transforms $B_1(s)$ and $B_2(s)$, certain associative laws hold, as follows:

$$[B_1(s)B_2(s)]^+ = B_1(s)^+ B_2(s)^+ \qquad (4\text{-}39a)$$

$$[B_1(s)B_2(s)]^- = B_1(s)^- B_2(s)^- \qquad (4\text{-}39b)$$

$$[B_1(s) + B_2(s)]_+ = B_1(s)_+ + B_2(s)_+ \qquad (4\text{-}39c)$$

A GENERAL PROBLEM AND SOLUTION

and

$$[B_1(s) + B_2(s)]_- = B_1(s)_- + B_2(s)_- \tag{4-39d}$$

But note that, in general, $[B_1(s)B_2(s)]_+$ does *not* equal $B_1(s)_+ B_2(s)_+$; nor does $[B_1(s) + B_2(s)]^+$ equal $B_1(s)^+ + B_2(s)^+$ in general.

4-4d. Solution Using Wiener-Hopf Spectrum Factorization

Consider the function $A_f(s)_-$:

$$A_f(s)_- = A_f(s) - A_f(s)_+ \tag{4-40a}$$

and since $a_f(t) = 0$ for t in the interval $[0, T_f]$, as found in Section 4-4b, it follows that

$$A_f(s)_- = A_f(s) - A_f(s)_{T_f+} = A_f(s)_{T_f-} \tag{4-40b}$$

and by using 4-27

$$\begin{aligned}
A_f(s)_- &= [A_{11}(s)F^*(s)]_{T_f-} + A_2(-s)_{T_f-} + A_3(s)_{T_f-} \\
&= A_{11}(s)F^*(s) - [A_{11}(s)F^*(s)]_{T_f+} + A_2(-s)_{T_f-} + A_3(s)_{T_f-}
\end{aligned} \tag{4-41}$$

Furthermore, the result of Equation 4-38 is used in 4-41 to obtain

$$A_f(s)_- = A_{11}(s)F^*(s) - \left(e^{-sT_f} \sum_{k=1}^{n} \frac{c_k}{s - p_k} \right) + A_2(-s)_{T_f-} + A_3(s)_{T_f-} \tag{4-42}$$

corresponding to which $\mathscr{L}^{-1}[A_f(s)_-] = 0$ for t greater than 0.

The object of the above operation and of the steps that follow is to obtain an equation which contains $F^*(s)$ and which has the property that the inverse transform of the left-hand member of the equation is zero for $t > 0$ while that of the right-hand member is zero for $t < 0$. For such an equation, it is necessary that each member of the equation equals zero, independent of the other member. At each step of the process, care is taken to preserve $F^*(s)$ unaltered.

The left-hand member of Equation 4-42 exhibits the required property, but the right-hand member does not. To proceed, we use the fact that $A_{11}(s) = A_{11}(s)^+ A_{11}(s)^-$ to obtain

$$\frac{A_f(s)_-}{A_{11}(s)^-} = A_{11}(s)^+ F^*(s) - \left[\frac{\left(e^{-sT_f} \sum_{k=1}^{n} \frac{c_k}{s - p_k} \right) - A_2(-s)_{T_f-} - A_3(s)_{T_f-}}{A_{11}(s)^-} \right] \tag{4-43}$$

Going one step further, the bracketed term of the right-hand member of 4-43 is denoted by $B(s) = B(s)_+ + B(s)_-$ to obtain

$$\frac{A_f(s)_-}{A_{11}(s)^-} + B(s)_- = A_{11}(s)^+ F^*(s) - B(s)_+ \tag{4-44}$$

which is of the required form; the inverse transform of the left-hand member of 4-44 is zero for $t > 0$, and the inverse transform of the right-hand member is zero for $t < 0$; thus,

$$F^*(s) = \frac{B(s)_+}{A_{11}(s)^+} = \frac{\left\{ \dfrac{-A_2(-s)_{T_f-} - A_3(s)_{T_f-} + e^{-sT_f} \sum_{k=1}^{n} \dfrac{c_k}{s - p_k}}{[A_1(s) + A_1(-s)]^-} \right\}_+}{[A_1(s) + A_1(-s)]^+} \tag{4-45}$$

Equation 4-45 can be inverted to obtain $f^*(t)$ in terms of c_k ($k = 1, 2, \ldots, n$). To determine specific values for c_k's, it is only necessary to apply the condition that $f^*(t) = 0$ for $t > T_f$.

The extremizing nature of $F^*(s)$ is determined by examining the second partial derivative of J_a with respect to ϵ. In Equation 4-22, J_a is in terms of $F(s) = F^*(s) + \epsilon \, \delta F(s)$, and the second partial $\partial^2 J_a / \partial \epsilon^2$ is

$$\frac{\partial^2 J_a}{\partial \epsilon^2} = \frac{1}{2\pi j} \int_{-j\infty}^{j\infty} 2 A_1(s) \, \delta F(s) \, \delta F(-s) \, ds \tag{4-46}$$

If $\partial^2 J_a / \partial \epsilon^2$ is greater (less) than zero for all allowable values of $\delta F(s)$, a minimum (maximum) of J_a is obtained when $F(s) = F^*(s)$ given by Equation 4-45. Since only one function $F^*(s)$ satisfies the necessary condition $\partial J_a / \partial \epsilon |_{\epsilon=0} = 0$, this $F^*(s)$ yields the absolute minimum (maximum) of J_a.

An important special case of 4-45 is that in which T_f is allowed to approach infinity; in that case, $A_2(-s)_{T_f-}$ and $A_3(s)_{T_f-}$ approach $A_2(-s)$ and $A_3(s)$, respectively. Also, the inverse transforms of the terms associated with e^{-sT_f} in 4-45 are shifted to infinity as T_f approaches infinity and are therefore of no consequence in finite intervals of time; thus,

$$F^*(s) = \frac{-\left\{ \dfrac{A_2(-s) + A_3(s)}{[A_1(s) + A_1(-s)]^-} \right\}_+}{[A_1(s) + A_1(-s)]^+} \qquad \text{(for } T_f \to \infty\text{)} \tag{4-47}$$

4-5. SOLUTIONS TO FILTER, CONTROL, AND PREDICTOR PROBLEMS

Solutions to the problems described in Section 4-2 are obtained here simply by identifying the corresponding terms of the integrands of 4-8 and 4-15

Sect. 4-5 FILTER, CONTROL, AND PREDICTOR SOLUTIONS

with those of the integrand 4-22. These identities are then substituted appropriately into the solution 4-47 to obtain particular solutions to the problem at hand.

Thus, for the most general integral-square-error problem of Section 4-2b, corresponding terms of J_{a1} (Equation 4-8) and J_a (Equation 4-22) are as follows:

$$F(s) \rightarrow G_0(s)$$
$$A_1(s) \rightarrow [hG_3(s)G_3(-s) + G_2(s)G_2(-s)] V_2(s) V_2(-s)$$
$$A_2(s) \rightarrow -G_2(s)G_1(-s)V_1(-s)V_2(s)$$
$$A_3(s) \rightarrow -G_2(-s)G_1(s)V_1(s)V_2(-s)$$
$$A_4(s) \rightarrow G_1(s)G_1(-s)V_1(s)V_1(-s)$$

In this case, $A_{11}(s) = A_1(s) + A_1(-s) = 2A_1(s)$, $A_3(s) = A_2(-s)$, and $T_f \rightarrow \infty$ are properties used in conjunction with the above identities to obtain the optimal solution $G_0^*(s)$ for $G_0(s)$ from Equation 4-47:

$$G_0^*(s) = \frac{\left\{ \frac{G_1(s)G_2(-s)V_1(s)V_2(-s)}{[hG_3(s)G_3(-s) + G_2(s)G_2(-s)]^- [V_2(s)V_2(-s)]^-} \right\}_+}{[hG_3(s)G_3(-s) + G_2(s)G_2(-s)]^+ [V_2(s)V_2(-s)]^+} \quad (4\text{-}48)$$

In the right-hand member of 4-48, only the Lagrange multiplier h is unknown at this point. Corresponding to different values of h, different values of K_1, Equation 4-4, are obtained. If $G_2(s)$ and $G_3(s)$ differ by a factor which consists of a low-order transfer function, we may be able to effect the required spectrum factorization in Equation 4-48 to obtain $G_0^*(s)$ in terms of h explicitly. More often, however, $G_2(s)$ and $G_3(s)$ are such that the spectrum factorization can be performed effectively only if numerical values of h are used. If so, several values of h are tried, and corresponding values of K_1 are determined from

$$K_1 = \frac{1}{2\pi j} \int_{-j\infty}^{j\infty} V_2(s)V_2(-s)G_0^*(s)G_0^*(-s)G_3(s)G_3(-s) \, ds \quad (4\text{-}49)$$

A plot of K_1 versus h is obtained as a result; so that for a given K_1, the corresponding value of h to be used in evaluating $G_0^*(s)$ can be obtained from the plot.

The heart of the just-mentioned problem is that in many instances the numerator polynomial of

$$hG_3(s)G_3(-s) + G_2(s)G_2(-s) = hG_3(s)G_3(-s)\left[1 + \frac{1}{h}\frac{G_2(s)G_2(-s)}{G_3(s)G_3(-s)}\right] \quad (4\text{-}50)$$

cannot be factored unless h is specified numerically. If $G_2(s)$ and $G_3(s)$ are rational fraction functions, however, it is always possible to define ξ by

$\xi = -s^2$ to reduce the bracketed term in 4-50 to the form

$$1 + \frac{1}{h}\frac{N(\xi)}{D(\xi)} \tag{4-51}$$

where $N(\xi)$ and $D(\xi)$ are polynomials in ξ. The zeros of 4-51 as a function of h can be obtained by making a standard root-locus plot in the ξ plane.[4] Once the ξ-plane zeros of 4-51 are known for given values of h, the corresponding s-plane zeros of the bracketed term in Equation 4-50 are found by using the identity $s = \pm(-\xi)^{1/2}$. Because values of h are specified in the above process, corresponding values of K_1 are eventually determined. As previously noted, therefore, an iterative technique must be used to find the particular value of h which yields the allowable K_1.

The detailed examples that follow this section illustrate the many steps which are required to obtain $G_0^*(s)$ by using Equation 4-48.

Of the many special cases of the integral-square-error problem of Section 4-2b, those designated as cases A, B, and C at the end of that section have the following solutions (the reader is invited to verify these).

Case A.

$$G_0^*(s) = \frac{\left\{\frac{G_2(-s)V_s(s)[V_s(-s) + V_n(-s)]^+}{[G_2(s)G_2(-s)]^-[V_s(s) + V_n(s)]^-}\right\}_+}{[G_2(s)G_2(-s)]^+[V_s(s) + V_n(s)]^+[V_s(-s) + V_n(-s)]^+} \tag{4-52}$$

and in the special subcase where $[G_2(-s)]^- = G_2(-s)$ and

$$[V_s(-s) + V_n(-s)]^- = V_s(-s) + V_n(-s),$$

$$G_0^*(s) = \frac{[V_s(s)]_+}{G_2(s)[V_s(s) + V_n(s)]} \tag{4-53}$$

Case B.

$$G_0^*(s) = \frac{\{e^{sT}[V_1(-s)V_1(s)]^+\}_+}{[V_1(-s)V_1(s)]^+} \tag{4-54}$$

Case C.

$$G_0^*(s) = \frac{\left\{\frac{G_2(-s)[V_2(s)V_2(-s)]^+}{[hG_3(s)G_3(-s) + G_2(s)G_2(-s)]^-}\right\}_+}{[hG_3(s)G_3(-s) + G_2(s)G_2(-s)]^+[V_2(s)V_2(-s)]^+} \tag{4-55}$$

---

[4] Chang [4.5] assigns the name *root-square locus* to this particular type of root locus. Root-locus theory is presented in most basic textbooks on feedback control theory. As an alternative to the root-square locus approach, digital computer routines are generally available for finding zeros of polynomials, in particular, zeros of $hD(\xi) + N(\xi)$.

Sect. 4-5   FILTER, CONTROL, AND PREDICTOR SOLUTIONS   153

and in the special subcase (1), where $G_3(s) = 0$ and $[V_2(s)V_2(-s)]^+ = V_2(s)$,

$$G_0^*(s) = \frac{\left\{ \frac{G_2(-s)V_2(s)}{[G_2(s)G_2(-s)]^-} \right\}_+}{V_2(s)[G_2(s)G_2(-s)]^+} \quad (4\text{-}56)$$

and in the especially simple subcase (2), where $G_2(s) = [G_2(s)G_2(-s)]^+$ and the conditions of subcase (1) apply,

$$G_0^*(s) = \frac{1}{G_2(s)} \quad (4\text{-}57)$$

For the mean-square-error problem of Section 4-2c, $G_0^*(s)$ is obtained by identifying like terms in the integrands of Equations 4-15 and 4-22 and then substituting these identities into the general solution 4-47, as was done on page 151, to obtain:

$$G_0^*(s) = \frac{\left\{ \frac{G_1(s)G_2(-s)\Phi_{v_1 v_2}(-s)}{[hG_3(s)G_3(-s) + G_2(s)G_2(-s)]^- \Phi_{v_2 v_2}(s)^-} \right\}_+}{[hG_3(s)G_3(-s) + G_2(s)G_2(-s)]^+ \Phi_{v_2 v_2}(s)^+} \quad (4\text{-}58)$$

Also, as under Cases A, B, and C of this section, various special solutions can be obtained from this solution.

Example 4-1. Consider the following filter problem. In Figure 4-1, let $v_2(t) = v_s(t) + v_n(t)$ and $v_1(t) = v_s(t)$, where $v_s(t)$ is the desired signal and $v_n(t)$ is an unwanted noise signal. The autocorrelation functions of the signal and noise are, respectively,

$$\phi_{v_s v_s}(\tau) = K_s e^{-|\tau/\tau_s|} \quad (4\text{-}59)$$

and

$$\phi_{v_n v_n}(\tau) = K_n e^{-|\tau/\tau_n|} \quad (4\text{-}60)$$

where K_s, K_n, τ_s, and τ_n are positive constants.

The cross-correlation function $\phi_{v_n v_s}(\tau)$ is assumed to be zero, i.e., the signal and the noise are uncorrelated.

In this example, the transfer functions of Figure 4-1 assume the values $G_3(s) = 0$, $G_2(s) = 1$, and $G_1(s) = e^{-sT}$. The role of the constant T in the ideal transfer function $G_1(s)$ is examined in detail. If we can afford to wait for the filtering action, T can be set relatively large, and, as is evidenced, better filtering can be achieved. In many applications, especially in the areas of communication and radar, a relatively large T may be of the order of milliseconds, and no detrimental effects result from the added delay. On the other hand, in control applications, an added delay may be intolerable, and T may actually be negative to achieve some predictive effect.

As a first step in determining the filter $G_0^*(s)$ which minimizes the mean-square error $\phi_{ee}(0)$, the power-density spectra of $v_s(t)$ and $v_n(t)$ are obtained by applying the definitions of Appendix C in conjunction with Equations 4-59 and 4-60; thus,

$$\Phi_{v_s v_s}(s) = \frac{2\tau_s K_s}{(\tau_s s + 1)(-\tau_s s + 1)} \quad \text{for } \frac{-1}{\tau_s} < \text{Re}(s) < \frac{1}{\tau_s} \quad (4\text{-}61)$$

and

$$\Phi_{v_n v_n}(s) = \frac{2\tau_n K_n}{(\tau_n s + 1)(-\tau_n s + 1)} \quad \text{for } \frac{-1}{\tau_n} < \text{Re}(s) < \frac{1}{\tau_n} \quad (4\text{-}62)$$

Because the signal and the noise are assumed to be uncorrelated,

$$\Phi_{v_s v_n}(s) = 0 \quad (4\text{-}63)$$

$$\Phi_{v_1 v_2}(s) = \Phi_{v_s v_s}(s) \quad (4\text{-}64)$$

and

$$\Phi_{v_2 v_2}(s) = \Phi_{v_s v_s}(s) + \Phi_{v_n v_n}(s) \quad (4\text{-}65)$$

These identities and the previous assignment of the transfer functions are used to reduce 4-58 to

$$G_0^*(s) = \frac{\left\{ \dfrac{e^{-sT} \Phi_{v_s v_s}(s)}{[\Phi_{v_s v_s}(s) + \Phi_{v_n v_n}(s)]^-} \right\}_+}{[\Phi_{v_s v_s}(s) + \Phi_{v_n v_n}(s)]^+} \quad (4\text{-}66)$$

where, on the basis of Equations 4-61 and 4-62,

$$\Phi_{v_s v_s}(s) + \Phi_{v_n v_n}(s) = 2\frac{\tau_n K_n(\tau_s s + 1)(-\tau_s s + 1) + \tau_s K_s(\tau_n s + 1)(-\tau_n s + 1)}{(\tau_n s + 1)(\tau_s s + 1)(-\tau_n s + 1)(-\tau_s s + 1)}$$

$$= \frac{2[(\tau_n K_n + \tau_s K_s)(\tau_u s + 1)(-\tau_u s + 1)]}{(\tau_n s + 1)(\tau_s s + 1)(-\tau_n s + 1)(-\tau_s s + 1)} \quad (4\text{-}67)$$

in which

$$\tau_u \triangleq \left[\frac{\tau_n \tau_s (\tau_s K_n + \tau_n K_s)}{\tau_s K_s + \tau_n K_n} \right]^{1/2} \quad (4\text{-}68)$$

The factored parts of Equation 4-67 are

$$[\Phi_{v_s v_s}(s) + \Phi_{v_n v_n}(s)]^+ = \frac{(\tau_n K_n + \tau_s K_s)(\tau_u s + 1)}{(\tau_n s + 1)(\tau_s s + 1)} \quad (4\text{-}69)$$

and

$$[\Phi_{v_s v_s}(s) + \Phi_{v_n v_n}(s)]^- = \frac{2(-\tau_u s + 1)}{(-\tau_n s + 1)(-\tau_s s + 1)} \quad (4\text{-}70)$$

Sect. 4-5 FILTER, CONTROL, AND PREDICTOR SOLUTIONS

These identities and that of 4-61 are substituted appropriately into Equation 4-66 to obtain

$$G_0^*(s) = \frac{\tau_s K_s}{(\tau_s K_s + \tau_n K_n)} \frac{(\tau_s s + 1)(\tau_n s + 1)}{(\tau_u s + 1)} \left[\frac{(-\tau_n s + 1)e^{-sT}}{(\tau_s s + 1)(-\tau_u s + 1)} \right]_+$$

$$= \frac{\tau_s K_s (\tau_s s + 1)(\tau_n s + 1)}{(\tau_s K_s + \tau_n K_n)(\tau_s + \tau_u)(\tau_u s + 1)} \left[\frac{(\tau_n + \tau_s)e^{-sT}}{\tau_s s + 1} + \frac{(\tau_n - \tau_u)e^{-sT}}{\tau_u s - 1} \right]_+$$

(4-71)

Two cases are to be distinguished: one for $T \leq 0$ and one for $T > 0$. For $T \leq 0$, the second term within the brackets of 4-71 has an inverse transform which is zero for $t > 0$, and therefore it does not play a role in the solution. The first term within the brackets of 4-71 has the inverse

$$\mathscr{L}^{-1}\left[\frac{(\tau_n + \tau_s)e^{-sT}}{\tau_s s + 1}\right] = \frac{\tau_n + \tau_s}{\tau_s}[e^{-(t-T)/\tau_s}]u(t-T) \quad (4\text{-}72)$$

which is zero for $t < T$, where in this case T is less than or equal to zero (see Figure 4-5). The (right-sided) Laplace transform of 4-72 yields

$$\left[\frac{(\tau_n + \tau_s)e^{-sT}}{\tau_s s + 1}\right]_+ = \frac{(\tau_n + \tau_s)e^{T/\tau_s}}{\tau_s s + 1} \quad \text{for } T \leq 0 \quad (4\text{-}73)$$

which is used in 4-71 to obtain

$$G_0^*(s) = \frac{\tau_s + \tau_n}{\tau_s + \tau_u} e^{T/\tau_s} \frac{\tau_s K_s(\tau_n s + 1)}{(\tau_n K_n + \tau_s K_s)(\tau_u s + 1)} \quad \text{for } T \leq 0 \quad (4\text{-}74)$$

For $T > 0$, the first term within the brackets of 4-71 has an inverse which is zero for all t less than zero (actually, for all $t < T$ where $T > 0$), and it is

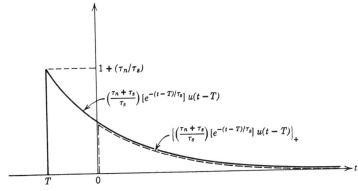

Figure 4-5. Part of the factoring process for $T \leq 0$.

therefore unaltered by the indicated factoring operation in 4-71. On the other hand, the second term $(\tau_n - \tau_u)e^{-sT}/(\tau_u s - 1)$ is inverted to obtain

$$\mathcal{L}^{-1}[(\tau_n - \tau_u)e^{-sT}/(\tau_u s - 1)] = \frac{-\tau_n + \tau_u}{\tau_u} e^{(t-T)/\tau_u} u(-t + T) \quad (4\text{-}75)$$

which is depicted in Figure 4-6.

On the basis of 4-75, the subscript $+$ of $(\tau_n - \tau_u)e^{-sT}/(\tau_u s - 1)$ is obtained in the standard way:

$$\left[\frac{(\tau_n - \tau_u)e^{-sT}}{(\tau_u s - 1)}\right]_+ = \left(\frac{\tau_u - \tau_n}{\tau_u}\right) \int_0^\infty e^{(t-T)/\tau_u} u(-t + T) e^{-st} \, dt$$

$$= \left(\frac{\tau_u - \tau_n}{\tau_u}\right) \int_0^T e^{-T/\tau_u} e^{t(-s\tau_u + 1)/\tau_u} \, dt$$

$$= (\tau_u - \tau_n) \frac{e^{-sT} - e^{-T/\tau_u}}{-\tau_u s + 1} \quad (4\text{-}76)$$

By using this result, in conjunction with the statement following 4-74, Equation 4-71 can be reduced to

$$G_0^*(s) = \frac{\tau_s K_s(\tau_s s + 1)(\tau_n s + 1)}{(\tau_s K_s + \tau_n K_n)(\tau_s + \tau_u)(\tau_u s + 1)}$$

$$\left[\frac{(\tau_n + \tau_s)e^{-sT}}{(\tau_s s + 1)} + \frac{(\tau_u - \tau_n)(e^{-sT} - e^{-T/\tau_u})}{(-\tau_u s + 1)}\right] \quad (4\text{-}77)$$

And after several intermediate steps, the above can be reduced to

$$G_0^*(s) = \frac{\tau_s K_s(\tau_n s + 1)}{(\tau_n K_n + \tau_s K_s)(\tau_u s + 1)} \left[\frac{(\tau_n s - 1)e^{-sT}}{(\tau_u s - 1)} + \frac{(\tau_u - \tau_n)(\tau_s s + 1)}{(\tau_u + \tau_s)(\tau_u s - 1)} e^{-T/\tau_u}\right] \quad (4\text{-}78)$$

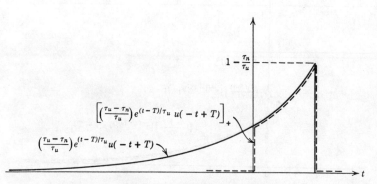

Figure 4-6. Part of the factoring process for $T > 0$.

Sect. 4-5 FILTER, CONTROL, AND PREDICTOR SOLUTIONS 157

As a special case, note that when $\tau_u \ll T$, the second term within the brackets of 4-78 constitutes an insignificant part of $G_0{}^*(s)$; thus,

$$G_0{}^*(s) = \frac{\tau_s K_s e^{-sT}(\tau_n s + 1)(\tau_n s - 1)}{(\tau_n K_n + \tau_s K_s)(\tau_u s + 1)(\tau_u s - 1)} \quad \text{for } T \gg \tau_u \quad (4\text{-}79)$$

To illustrate the effect of the added delay T on the mean-square error $\phi_{ee}(0)$, consider the special case in which τ_s and $K_s = 1$, $K_n = \frac{10}{3}$, and $\tau_n = \frac{2}{5}$. This is a very noisy case. The value of $\phi_{ee}(0)$ for $T = 0$ is to be compared with that for $T \gg \tau_u$.

From Equation 4-68, τ_u is found to equal 0.8. Equation 4-78 is used to find $G_0{}^*(s)$ with $T = 0$:

$$G_0{}^*(s) = \frac{3}{7}\frac{(0.4s + 1)}{(0.8s + 1)}\left[\frac{(0.4s - 1)}{(0.8s - 1)} + \frac{(0.4)(s + 1)}{(1.8)(0.8s - 1)}\right]$$

$$= \frac{1}{3}\frac{0.4s + 1}{0.8s + 1} \quad \text{for } T = 0 \quad (4\text{-}80)$$

With the same values of K's and τ's, Equation 4-79 yields $G_0{}^*(s)$ for $T \gg \tau_u$:

$$G_0{}^*(s) = \frac{3}{7}\frac{(0.4s + 1)(0.4s - 1)e^{-sT}}{(0.8s + 1)(0.8s - 1)} \quad \text{for } T \gg \tau_u \quad (4\text{-}81)$$

Using Equation 4-14, as modified by the particular features of this problem, gives

$$\Phi_{ee}(s) = G_0(s)G_0(-s)[\Phi_{v_s v_s}(s) + \Phi_{v_n v_n}(s)]$$
$$+ \Phi_{v_s v_s}(s)[1 - G_0(s)e^{sT} - G_0(-s)e^{-sT}] \quad (4\text{-}82)$$

where in this special case, Equations 4-61 and 4-67 are reduced to

$$\Phi_{v_s v_s}(s) = \frac{2}{(s + 1)(-s + 1)} \quad (4\text{-}83)$$

and

$$\Phi_{v_s v_s} + \Phi_{v_n v_n}(s) = \frac{(14/3)(0.8s + 1)(-0.8s + 1)}{(0.4s + 1)(-0.4s + 1)(s + 1)(-s + 1)} \quad (4\text{-}84)$$

Of course, the ideal (but unobtainable) value of $\phi_{ee}(0)$ is zero. On the other hand, if $G_0(s)$ is set equal to zero, $\Phi_{ee}(s)$ of 4-82 is reduced to $\Phi_{v_s v_s}(s)$; and therefore, by using 4-83, $\phi_{ee}(0)$ equals

$$\phi_{v_s v_s}(0) = \frac{1}{2\pi j}\int_{-j\infty}^{j\infty} \frac{2}{(s + 1)(-s + 1)} ds = 1 \quad (4\text{-}85)$$

Obviously, the optimal filter $G_0{}^*(s)$ must yield a value of $\phi_{ee}(0)$ which is less than one but greater than zero.

For $T = 0$, the identities of Equations 4-80, 4-83, and 4-84 are used in 4-82 to obtain

$$\Phi_{ee}(s) = \frac{1}{9} \frac{14/3}{(s+1)(-s+1)}$$

$$+ \frac{2}{(s+1)(-s+1)} \left[1 - \frac{1}{3} \frac{(0.4s+1)}{(0.8s+1)} - \frac{1}{3} \frac{(-0.4s+1)}{(-0.8s+1)} \right]$$

$$= \frac{(68/27)}{(s+1)(-s+1)} - \frac{(4/3)[(0.32)^{1/2}s + 1][-(0.32)^{1/2}s + 1]}{(0.8s^2 + 1.8s + 1)(0.8s^2 - 1.8s + 1)} \quad (4\text{-}86)$$

And $\phi_{ee}(0) = (1/2\pi j) \int_{-j\infty}^{j\infty} \Phi_{ee}(s) \, ds$ is evaluated by using Table B-1 of Appendix B,

$$\phi_{ee}(0) = \left(\frac{68}{27}\right)\frac{1}{2} - \left(\frac{4}{3}\right)\frac{0.32 + 0.8}{(2)(1.8)(0.8)}$$

$$= \frac{20}{27} \cong 0.741 \qquad \text{for } T = 0 \quad (4\text{-}87)$$

For $T \gg \tau_u$, the identities of Equations 4-81, 4-83, and 4-84 are used in 4-82 to obtain

$$\Phi_{ee}(s) = \frac{9}{49} \frac{(0.4s+1)^2(0.4s-1)^2}{(0.8s+1)^2(0.8s-1)^2} \left[\frac{(14/3)(0.8s+1)(-0.8s+1)}{(0.4s+1)(s+1)(-0.4s+1)(-s+1)} \right]$$

$$+ \frac{2}{(s+1)(-s+1)} \left[1 - \frac{6}{7} \frac{(0.4s+1)(0.4s-1)}{(0.8s+1)(0.8s-1)} \right]$$

$$= \frac{2}{(s+1)(-s+1)} - \frac{(6/7)(0.4s+1)(-0.4s+1)}{(0.8s^2 + 1.8s + 1)(0.8s^2 - 1.8s + 1)} \quad (4\text{-}88)$$

And $\phi_{ee}(0)$ is evaluated by using Table B-1 of Appendix B,

$$\phi_{ee}(0) = (2)\frac{1}{2} - \left(\frac{6}{7}\right)\frac{0.16 + 0.8}{(2)(1.8)(0.8)}$$

$$= \frac{5}{7} \cong 0.715 \qquad \text{for } T \gg \tau_u \quad (4\text{-}89)$$

which is an improvement over the value of $\phi_{ee}(0)$ for $T = 0$; a slight improvement if the delay of $T \gg \tau_u$ can be tolerated.

●

Example 4-2. A control system characterized by Figures 4-1 and 4-2c is to be designed on a minimum-integral-square-error basis. The signals $v_1(t)$, $v_2(t)$, and $y_1(t)$ are given as unit step functions; thus, $G_1(s) = 1$. The transfer

Sect. 4-5 FILTER, CONTROL, AND PREDICTOR SOLUTIONS 159

function $G_2(s)$ of the plant is given by

$$G_2(s) = G_{mp}(s)(\tau_1 s + 1) \tag{4-90}$$

where τ_1 may be positive or negative (both cases are considered below) and $G_{mp}(s)$ is a *minimum-phase function*; i.e., the finite poles and zeros of $G_{mp}(s)$ lie in the LHP (left half-plane), to the left of the $j\omega$ axis in the s plane. The block $G_3(s)$, from which the energy-constrained signal $y_3(t)$ emanates, is assumed to be related to $G_2(s)$ by the equation

$$G_3(s) = sG_2(s) \tag{4-91}$$

The problem is to find the physically realizable $G_0^*(s)$ which yields the minimum of the integral-square error J_1 (Equation 4-3), subject to the constraint that the integral-square value of the derivative $\dot{y}_2(t)$ of the output $y_2(t)$ is at most K_1.

Two possibilities are examined in the following: (1) τ_1 may be positive or zero in which case $G_2(s)$ is a minimum-phase function; and (2) τ_1 may be negative in which case $G_2(s)$ is a nonminimum-phase function.

$\tau_1 \geq 0$

The general solution is given by Equation 4-48. In this problem,

$$G_1(s)G_2(-s)V_1(s)V_2(-s) = \lim_{\epsilon \to 0+} \frac{G_2(-s)}{(s + \epsilon)(-s + \epsilon)} \tag{4-92}$$

$$[V_2(s)V_2(-s)]^- = \lim_{\epsilon \to 0+} \frac{1}{-s + \epsilon} \tag{4-93}$$

$$[V_2(s)V_2(-s)]^+ = \lim_{\epsilon \to 0+} \frac{1}{s + \epsilon} \tag{4-94}$$

and

$$hG_3(s)G_3(-s) + G_2(s)G_2(-s)$$
$$= \lim_{\epsilon \to 0+} h(s + \epsilon)(-s + \epsilon)G_2(s)G_2(-s)\left[1 + \frac{(1/h)}{(s + \epsilon)(-s + \epsilon)}\right]$$
$$= G_2(s)G_2(-s)(h_1 s + 1)(-h_1 s + 1) \tag{4-95}$$

in which h_1 is defined to be $h^{1/2}$ for notational convenience.

Note that because of the particularly simple relationship between $G_2(s)$ and $G_3(s)$, spectrum factorization can be effected in this example without resorting to the root-square-locus procedure described in Section 4-5. Thus, for $\tau_1 > 0$, the above equation is factored to obtain

$$[hG_3(s)G_3(-s) + G_2(s)G_2(-s)]^+ = G_2(s)(h_1 s + 1) \tag{4-96}$$

and

$$[hG_3(s)G_3(-s) + G_2(s)G_2(-s)]^- = G_2(-s)(-h_1s + 1) \qquad (4\text{-}97)$$

On the basis of the identities 4-92 through 4-97, Equation 4-48 assumes the particularly simple form

$$G_0^*(s) = \lim_{\epsilon \to 0+} \frac{[1/(-h_1s + 1)(s + \epsilon)]_+}{G_2(s)(h_1s + 1)/(s + \epsilon)}$$

$$= \frac{1}{G_2(s)(h_1s + 1)} \qquad \text{for } \tau_1 \geq 0 \qquad (4\text{-}98)$$

The Laplace transform $Y_3(s)$ of the constrained signal $y_3(t)$ is

$$Y_3 = V_2(s)G_0^*(s)G_2(s)s = \frac{1}{h_1s + 1} \qquad (4\text{-}99)$$

and the corresponding value of K_1 is

$$K_1 = \frac{1}{2\pi j}\int_{-j\infty}^{j\infty} \left(\frac{1}{h_1s+1}\right)\left(\frac{1}{-h_1s+1}\right) ds \qquad (4\text{-}100a)$$

On the basis of Table B-1 (Appendix B),

$$K_1 = \frac{1}{2h_1} = \frac{1}{2h^{1/2}} \qquad (4\text{-}100b)$$

from which

$$h^{1/2} = \frac{1}{2K_1} \qquad (4\text{-}101)$$

The Laplace transform $E(s)$ of the error signal $e(t)$ is

$$E(s) = V_2(s)[1 - G_0^*(s)G_2(s)]$$

$$= \frac{h_1}{sh_1 + 1} \qquad (4\text{-}102)$$

and the corresponding value of J_1 is

$$J_1 = \frac{h_1}{2} = \frac{1}{4K_1} \qquad \text{for } \tau_1 \geq 0 \qquad (4\text{-}103)$$

Finally, the transfer function $G_a(s)$ of the cascade controller in Figure 4-2c is

$$G_a(s) = \frac{G_0^*(s)}{1 - G_2(s)G_0^*(s)} = \frac{1}{G_2(s)}\frac{1}{h_1s}, \qquad \tau_1 \geq 0 \qquad (4\text{-}104)$$

$\tau_1 < 0$

As in the preceding case, Equations 4-92, 4-93, 4-94, and 4-95 apply. Here, however,

$$[hG_3(s)G_3(-s) + G_2(s)G_2(-s)]^+ = G_{mp}(s)(h_1s + 1)(-\tau_1 s + 1) \quad (4\text{-}105)$$

since $\tau_1 < 0$. In like manner,

$$[hG_3(s)G_3(-s) + G_2(s)G_2(-s)]^- = G_{mp}(-s)(-h_1s + 1)(\tau_1 s + 1) \quad (4\text{-}106)$$

Using the above identities in 4-48 gives

$$G_0^*(s) = \lim_{\epsilon \to 0+} \frac{[(-\tau_1 s + 1)/(-h_1 s + 1)(\tau_1 s + 1)(s + \epsilon)]_+}{G_{mp}(s)(h_1 s + 1)(-\tau_1 s + 1)/(s + \epsilon)}$$

$$= \frac{1}{G_{mp}(s)(h_1 s + 1)(-\tau_1 s + 1)} \quad \text{for } \tau_1 < 0 \quad (4\text{-}107)$$

Observe that the nonminimum-phase zero of the plant is not canceled by a corresponding pole of the optimum controller; an arbitrarily small error in any such cancellation attempt would lead to a pole of the overall system in the right half-plane, and the system would be unstable.

The Laplace transform $Y_3(s)$ of the constrained signal $y_3(t)$ is

$$Y_3(s) = V_2(s)G_0^*(s)sG_2(s)$$

$$= \frac{\tau_1 s + 1}{(h_1 s + 1)(-\tau_1 s + 1)} = \frac{\tau_1 s + 1}{-\tau_1 h_1 s^2 + (h_1 - \tau_1)s + 1} \quad (4\text{-}108)$$

and the corresponding value of K_1, Equation 4-4, as obtained by using Table B-1 (Appendix B), is

$$K_1 = \frac{\tau_1^2 - \tau_1 h_1}{2(-\tau_1 h_1)(h_1 - \tau_1)} = \frac{1}{2h_1} = \frac{1}{2h^{1/2}} \quad (4\text{-}109a)$$

from which

$$h^{1/2} = \frac{1}{2K_1} \quad (4\text{-}109b)$$

The Laplace transform $E(s)$ of the error $e(t)$ is

$$E(s) = V_2(s)[1 - G_0^*(s)G_2(s)]$$

$$= \frac{-\tau_1 h_1 s + (h_1 - 2\tau_1)}{-\tau_1 h_1 s^2 + (h_1 - \tau_1)s + 1} \quad (4\text{-}110)$$

and the corresponding value of the performance measure is

$$J_1 = \frac{(h_1 - 2\tau_1)^2 - \tau_1 h_1}{2(h_1 - \tau_1)}$$

$$= \frac{h_1 - 4\tau_1}{2} \quad (4\text{-}111)$$

Using Equation 4-109b gives

$$J_1 = \frac{1}{4K_1} - 2\tau_1 \quad \text{for } \tau_1 < 0 \qquad (4\text{-}112)$$

Again, the transfer function $G_a(s)$ of the cascade controller in the feedback configuration of Figure 4-2c is

$$G_a(s) = \frac{G_0{}^*(s)}{1 - G_2(s)G_0{}^*(s)}$$

$$= \frac{1}{G_{mp}(s)} \frac{1}{s[-\tau_1 h_1 s + (h_1 - 2\tau_1)]}, \quad \tau_1 < 0 \qquad (4\text{-}113)$$

Feedback configurations are generally preferred over the open-loop configuration of Figure 4-1 because of sensitivity considerations (Section 4-8). In Example 4-4 which follows Section 4-8, the results of the present example are examined from the standpoint of system sensitivity with respect to changes in the plant.

•

The conclusion to be drawn from the preceding example is that the presence of a right-half-plane (RHP) zero in the fixed portion (plant) of the system is usually detrimental to system performance. With no RHP zero, J_1 of Equation 4-103 approaches zero as K_1 approaches infinity. With one RHP zero associated with the plant, however, the best that can be attained for the integral-square error is given by Equation 4-112; as K_1 approaches infinity, J_1 approaches $-2\tau_1$ where $-1/\tau_1$ is the location of the RHP zero in the s plane.

4-6. SOLUTIONS TO OPTIMAL PULSE-SHAPE PROBLEMS

Solutions to the problems described in Section 4-3 are obtained here simply by identifying terms in the integrand of 4-20 with corresponding terms in the integrand of 4-22. These identities are then substituted appropriately into the general solution 4-45 to obtain particular solutions.

Corresponding terms are as follows:

$$F(s) \rightarrow V_i(s)$$
$$A_1(s) \rightarrow h Y_i(s)$$
$$A_2(s) \rightarrow G(s) W_o(-s)$$
$$A_3(s) \text{ \& } A_4(s) \rightarrow 0$$
$$T_f \rightarrow T$$

and therefore, on the basis of 4-45,

$$V_i^*(s) = \frac{\left\{\dfrac{-[G(-s)W_o(s)]_{T-} + e^{-sT}\sum_{k=1}^{n}\dfrac{c_k}{s-p_k}}{[Y_i(s)+Y_i(-s)]^-}\right\}_+}{h[Y_i(s)+Y_i(-s)]^+} \qquad (4\text{-}114)$$

where $V_i^*(s)$ denotes the optimal $V_i(s)$; the p_k's are the poles of the input admittance $Y_i(s)$; the c_k's are determined numerically from the requirement that $v_i^*(t) = 0$ for $t > T$; and the Lagrange multiplier h is selected to satisfy the constraint equation

$$K_3 = \frac{1}{2\pi j}\int_{-j\infty}^{j\infty} V_i^*(-s)V_i^*(s)Y_i(s)\,ds \qquad (4\text{-}115)$$

In the special cases of Section 4-3b, the value of the output signal $v_o(t)$ at time $t = T$ is to be maximized, and the appropriate weighting factor $W_o(s)$ is e^{-sT}.

Case A. Conditions: $G(s)$ is physically realizable, $Y_i(s)$ equals a ratio of polynomials in s, and $W_o(s) = e^{-sT}$.

$$V_i^*(s) = \frac{\left\{\dfrac{-G(-s)e^{-sT} + e^{-sT}\sum_{k=1}^{n}\dfrac{c_k}{s-p_k}}{[Y_i(s)+Y_i(-s)]^-}\right\}_+}{h[Y_i(s)+Y_i(-s)]^+} \qquad (4\text{-}116)$$

Case B. Conditions: Same as those listed under Case A, except that $Y_i(s) = K_y$, a constant.

$$V_i^*(s) = \frac{[-G(-s)e^{-sT}]_+}{2hK_y} \qquad (4\text{-}117)$$

Case C. Conditions: Same as those listed under Case A, except that the time interval $0 \le t \le T$ of the input pulse is changed to the time interval $-\infty \le t \le 0$, i.e., $v_i(t) = 0$ for $t > 0$.

$$V_i^*(s) = \frac{\left\{\dfrac{-G(-s)}{[Y_i(s)+Y_i(-s)]^+}\right\}_-}{h[Y_i(s)+Y_i(-s)]^-} \qquad (4\text{-}118)$$

The proof of this relationship is left as an exercise for the reader (Problem 4.12).

Case D. Conditions: Same as those listed under Case C, except that $Y_i(s) = K_y$, a constant.

$$V_i^*(s) = \frac{-G(-s)}{2hK_y} \qquad (4\text{-}119)$$

which is obtained from 4-118 by direct substitution of K_y for $Y_i(s)$. Equation 4-119 is a well-known result of matched-filter theory [4.42]. It relates, in essence, that under the stated conditions the optimal pulse $v_i^*(t)$ is a multiple of the mirror image of the impulse response $g(t)$ of the system.

In the special cases of Section 4-3c, the time average of the output signal $v_o(t)$ over the period $0 \leq t \leq T$ is to be maximized. The appropriate weighting factor $W_o(s)$ is given by Equation 4-21. For the case that $Y_i(s)$ is a rational fraction function and that $G(s)$ is physically realizable, the resulting $V_i^*(s)$ as obtained from 4-114 is

$$V_i^*(s) = \lim_{\epsilon \to 0+} \frac{\left\{ \dfrac{-G(-s)(1 - e^{-sT})}{T(s + \epsilon)} + e^{-sT} \sum_{k=1}^{n} \dfrac{c_k}{s - p_k} \right\}}{h[Y_i(s) + Y_i(-s)]^+} \qquad (4\text{-}120)$$

It remains to be shown that the $V_i^*(s)$'s obtained in this section yield a (constrained) maximum of the weighted output. In all cases considered, $A_1(s) = hY_i(s)$ which is used in 4-46 to obtain

$$\frac{\partial^2 J_a}{\partial \epsilon^2} = \frac{1}{2\pi j} \int_{-j\infty}^{j\infty} 2h Y_i(s)\, \delta F(s)\, \delta F(-s)\, ds \qquad (4\text{-}121)$$

Note that $\delta F(s) Y_i(s)$ represents the input "current" which stems from an allowable variation $\delta F(s)$ of the input signal. Thus, it follows that $\partial^2 J_a/\partial\epsilon^2$ of Equation 4-121 equals $2h$ times the energy supplied to the system as a result of an applied signal $\delta F(s)$; and assuming this energy is positive, $\partial^2 J_a/\partial\epsilon^2$ will be negative if h is negative. A negative value of $\partial^2 J_a/\partial\epsilon^2$ signifies that the unique $V_i^*(s)$ for a given problem yields the (constrained) absolute maximum of the weighted output for the given problem.

Example 4-3. For the circuit shown in Figure 4-7, the output voltage $v_o(t)$ at time $t = T$ is to be maximized. The initial conditions on the capacitors are assumed to be zero, and the input voltage $v_i(t)$ is assumed to be zero outside the time interval $[0,T]$. The problem is to find the $v_i(t)$, $v_i^*(t)$, which maximizes $v_o(T)$ subject to the constraint that the input energy is a constant value K_3.

Sect. 4-6 SOLUTIONS TO OPTIMAL PULSE-SHAPE PROBLEMS

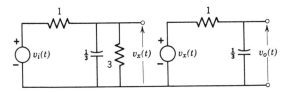

Figure 4-7. Circuit used in Example 4-3.

Equation 4-116 is applicable here, once the values of $G(-s)$ and $Y_i(s)$ have been determined. For the circuit shown,

$$G(-s) = \frac{9}{(-s + 3)(-s + 4)} \tag{4-122}$$

and

$$Y_i(s) = \frac{s + 1}{s + 4} \tag{4-123}$$

It follows that

$$Y_i(s) + Y_i(-s) = \frac{(s + 1)(-s + 4) + (-s + 1)(s + 4)}{(s + 4)(-s + 4)}$$

$$= \frac{2(s + 2)(-s + 2)}{(s + 4)(-s + 4)} \tag{4-124}$$

which is factored to obtain

$$[Y_i(s) + Y_i(-s)]^+ = \frac{2(s + 2)}{(s + 4)} \tag{4-125}$$

and

$$[Y_i(s) + Y_i(-s)]^- = \frac{(-s + 2)}{(-s + 4)} \tag{4-126}$$

The identities given by 4-122, 4-123, 4-125, and 4-126 are substituted appropriately into 4-116 to obtain

$$V_i^*(s) = -\frac{1}{2h} \frac{s + 4}{s + 2} \left[\frac{\frac{9}{(-s + 3)(-s + 4)} - \frac{c_1}{s + 4}}{\frac{e^{sT}(-s + 2)}{(-s + 4)}} \right]_+$$

$$= -\frac{1}{2h} \frac{s + 4}{s + 2} \left[\frac{-4c_1 e^{-sT}}{3(s + 4)} - \frac{9e^{-sT}}{-s + 3} + \frac{\left(9 - \frac{c_1}{3}\right)e^{-sT}}{-s + 2} \right]_+ \tag{4-127}$$

In Equation 4-127, the bracketed term with the subscript + is replaced by the transform of that portion of its inverse which is zero for $t < 0$, as follows:

$$V_i^*(s) = -\frac{1}{2h}\frac{s+4}{s+2}\left[\frac{-4c_1 e^{-sT}}{3(s+4)} + \frac{9(e^{-3T} - e^{-sT})}{-s+3} + \frac{\left(9 - \frac{c_1}{3}\right)(e^{-sT} - e^{-2T})}{-s+2}\right]$$
(4-128)

and the inverse transform of this equation is found—after several intermediate steps—to be

$$v_i^*(t) = \frac{1}{2h}[u(t) - u(t-T)]\left\{\left[\left(4.5 - \frac{c_1}{6}\right)e^{-2T} - 3.6e^{-3T}\right]e^{-2t}\right.$$
$$\left. + 12.6e^{3(t-T)} - (13.5 - 0.5c_1)e^{2(t-T)}\right\} \quad (4-129)$$

provided that

$$c_1 = \frac{(5.4 - 27e^{-4T} + 21.6e^{-5T})}{(9 - e^{-4T})}$$
(4-130)

The transform of the input current corresponding to the $V_i^*(s)$ in Equation 4-128 is

$$I_i(s) = V_i^*(s) Y_i(s)$$

$$= V_i^*(s) \frac{s+1}{s+4}$$

$$= -\frac{1}{2h}\frac{s+1}{s+2}\left[\frac{-4c_1 e^{-sT}}{3(s+4)} + \frac{9(e^{-3T} - e^{-sT})}{-s+3} + \left(9 - \frac{c_1}{3}\right)\frac{e^{-sT} - e^{-2T}}{-s+2}\right]$$
(4-131)

and this is inverted to obtain

$$i_i(t) = \frac{1}{2h}\left\{\left[1.8e^{-3T} - \left(2.25 - \frac{c_1}{12}\right)e^{-2T}\right]e^{-2t}\right.$$
$$\left. + 7.2e^{3(t-T)} + (6.75 - 0.25c_1)e^{2(t-T)}\right\} \quad (4-132)$$

which applies for the time interval $0 \leq t \leq T$.

The energy K_3 supplied by the input pulse is

$$K_3 = \int_0^T v_i(t) i_i(t)\, dt$$
(4-133)

Sect. 4-6   SOLUTIONS TO OPTIMAL PULSE-SHAPE PROBLEMS

In this example, the optimal value of $v_i(t)$ given by 4-129 and the corresponding optimal value of $i_i(t)$ given by 4-132 are substituted appropriately into 4-133, and the indicated integration is effected to obtain

$$K_3 = \frac{1}{4h^2}[0.125(a_1{}^2 - a_0{}^2)(1 - e^{-4T}) + 15.12(1 - e^{-6T})$$
$$+ 0.9a_0 e^{-2T}(1 - e^{-T}) + 2.7a_1(1 - e^{-5T})] \quad (4\text{-}134)$$

in which

$$a_0 \triangleq \left(4.5 - \frac{c_1}{6}\right)e^{-2T} - 3.6e^{-3T} \quad (4\text{-}135)$$

and

$$a_1 \triangleq -13.5 + 0.5c_1 \quad (4\text{-}136)$$

where c_1 is given by 4-130.

Equation 4-134 is rearranged to obtain the Lagrange multiplier h in terms of the input energy K_3 and the pulse length T:

$$h = -\frac{1}{2K_3^{1/2}}[0.125(a_1{}^2 - a_0{}^2)(1 - e^{-4T}) + 15.12(1 - e^{-6T})$$
$$+ 0.9a_0 e^{-2T}(1 - e^{-T}) + 2.7a_1(1 - e^{-5T})]^{1/2} \quad (4\text{-}137)$$

and a plot of $-2K_3^{1/2}h$ versus T is given in Figure 4-8.

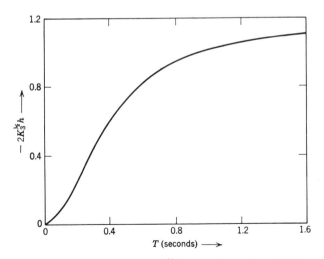

Figure 4-8. A plot of $-2K_3^{1/2}h$ versus the pulse width T.

The data concerning h is used in 4-129, and plots of $v_i^*(t)/K_3^{1/2}$ versus $t - T$ are given in Figure 4-9 for $T = 0.2$, $T = 0.4$, $T = 0.8$, and $T = \infty$.

It is expected that for a given value of K_3, the value of the output voltage at the terminal time $t = T$ should be greatest for the case that $T = \infty$. This is verified for this example by obtaining the inverse transform of $V_o(s)$ evaluated at $t = T$:

$$v_o(T) = \mathscr{L}^{-1}[V_i^*(s)G(s)]_{t=T}$$
$$= -\frac{1}{2h}[-13.5e^{-6T} + (32.4 - 0.6c_1)e^{-5T}$$
$$+ (0.75c_1 - 20.25)e^{-4T} + 1.35 - 0.15c_1] \quad (4\text{-}138)$$

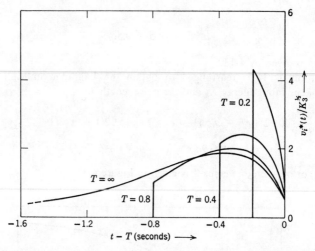

Figure 4-9. Plots of $v_i^*(t)/K_3^{1/2}$ versus $t - T$ for various values of T.

where the details of obtaining 4-138 are straightforward. A plot of $v_o(T)/K_3^{1/2}$ versus T is given in Figure 4-10.

For comparative purposes, a rectangular pulse of width T is supplied to the input terminals of the circuit shown in Figure 4-7. A summary of results is given here:

$$V_i(s) = \frac{c_r(1 - e^{-sT})}{s} \quad (4\text{-}139)$$

$$I_i(s) = c_r \frac{(s+1)(1-e^{-sT})}{(s+4)s} = \frac{c_r}{4}\left[\frac{1}{s} + \frac{3}{s+4}\right](1 - e^{-sT}) \quad (4\text{-}140)$$

$$K_3 = \int_0^T v_i(t)i_i(t)\, dt = \frac{c_r^2}{16}(4T + 3 - 3e^{-4T}) \quad (4\text{-}141)$$

Sect. 4-6 SOLUTIONS TO OPTIMAL PULSE-SHAPE PROBLEMS

$$c_r = \frac{4K_3^{1/2}}{(4T + 3 - 3e^{-4T})^{1/2}} \tag{4-142}$$

$$V_o(s) = c_r \frac{9(1 - e^{-sT})}{s(s + 3)(s + 4)} \tag{4-143}$$

and

$$v_o(T) = \frac{c_r}{4}(3 - 12e^{-3T} + 9e^{-4T}) \tag{4-144}$$

Plots of $v_i(t)/K_3^{1/2}$ versus $t - T$ are given in Figure 4-11 for $T = 0.2$, $T = 0.4$, $T = 0.8$, and $T = 1.6$; these curves are to be compared with the optimal

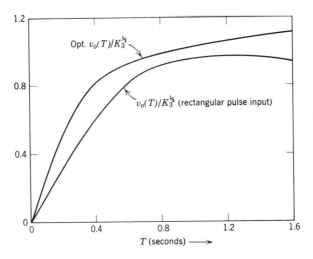

Figure 4-10. Plots of $v_o(T)/K_3^{1/2}$ versus T.

curves in Figure 4-9. Values of $v_o(T)/K_3^{1/2}$ are given in Figure 4-10 for both the case of the rectangular pulse input and the case of the optimal pulse input, and the superiority of the optimal pulse input is thereby displayed.

In many designs, it is desirable to use rectangular or other easily obtained pulse shapes, provided that the resulting $v_o(T)$ is not too much less than the optimum $v_o(T)$ which could be obtained. This is illustrated in this example in that the $v_o(T)$ obtained as a result of a rectangular pulse input is greater than 88% of the $v_o(T)$ obtained as a result of the optimal pulse input for any value of T between 0.6 second and 1.4 seconds.

●

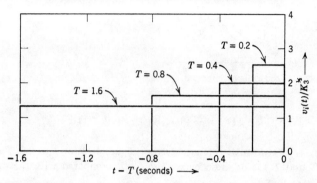

Figure 4-11. Plots of $v_i(t)/K_3^{1/2}$ versus $t - T$ where $v_i(t)$ is a nonoptimal rectangular shaped pulse.

4-7. NON-WIENER-HOPF-TYPE FREQUENCY-DOMAIN PROBLEMS

4-7a. General Comments

Although the Wiener-Hopf technique encompasses a broad class of problems concerned with optimal design of linear systems, it does not form a basis for the solution of numerous other optimization problems in which frequency-domain analysis plays a leading role. Examples of these include optimization problems relating to both antenna design [4.2, 4.24] and communication theory [4.8, 4.40, 4.41]. In most cases, however, the basic variational operations of the calculus of variations can be used to find necessary conditions which must be satisfied by the optimal solution. The problem treated in the following subsection exemplifies the basic approach and evidences the varied form of solution which may result in a non-Wiener-Hopf-type frequency-domain problem.

4-7b. Pulse Shape for Maximum Output Energy [4.30]

Consider a linear system with transfer function $G(s)$. The input $v_i(t)$ to the system is zero outside the time interval $[0,T_i]$, whereas the output energy of interest has associated with it the time interval $[0,T_o]$; in general, $T_i \neq T_o$.

The problem to be solved in this section is the following: Given that the input "energy" is fixed, find the optimal form of the input pulse $v_i(t)$ which results in the maximization of the partial "energy" output, where

$$K_4 = \int_0^{T_i} [v_i(t)]^2 \, dt \tag{4-145}$$

is the normalized input energy, and

$$J_4 = \int_0^{T_o} [v_o(t)]^2 \, dt \tag{4-146}$$

is the normalized output energy over the period defined by $0 \leq t \leq T_o$.

The input energy given by 4-145 equals that of 4-17 if $Y_i(s)$ in Section 4-3 is set equal to unity. A more general formulation of the problem treated here would enfold $Y_i(s)$, rather than a constant input admittance.

It is obvious that the problem, as stated, also characterizes the case in which $v_i(t)$ represents the envelope of an amplitude-modulated pulse. In such a case, $G(s)$ is the transfer function which relates the output envelope to that of the input.

The problem of determining the most desirable envelope of a modulated pulse has been considered by Chalk [4.4]. In Chalk's work, the prime concern is that of minimizing the energy which is transmitted to channels which are adjacent to the excited channel, and therefore Chalk considers only the case where $T_o = \infty$ in 4-146.

In many applications, however, the output energy resulting from a pulse-shaped input is essentially useless unless most of the energy is confined to a given interval of time. This is especially true when a sequence of pulses is sent at a rapid rate. Hence, it is assumed here that the usable output is confined to the time interval $[0, T_o]$ when the input signal is in the form of a single pulse which is initiated at time $t = 0$ and is terminated at time $t = T_i$. The resulting solution is an effective compromise between that obtained on the basis of the complete elimination of intersymbol interferences [4.9, 4.12] and that obtained on the basis of the absolute minimization of interchannel interference [4.4].

As a preliminary step in the solution of this problem, the usable output energy J_4 is expressed as

$$J_4 = \int_0^{T_i} \int_0^{T_i} v_i(t_1) v_i(t_2) \gamma(t_1, t_2) \, dt_1 \, dt_2 \tag{4-147}$$

in which

$$\gamma(t_1, t_2) = \frac{1}{(2\pi j)^2} \int_{-j\infty}^{j\infty} \int_{-j\infty}^{j\infty} \frac{G(s_1) G(s_2) e^{-s_1 t_1} e^{-s_2 t_2} [e^{(s_1 + s_2) T_o} - 1]}{s_1 + s_2} \, ds_1 \, ds_2 \tag{4-148a}$$

and if $T_o \to \infty$,

$$\gamma(t_1, t_2)|_{T_o = \infty} = \frac{-1}{(2\pi j)^2} \int_{-j\infty}^{j\infty} \int_{-j\infty}^{j\infty} \frac{G(s_1) G(s_2) e^{-s_1 t_1} e^{-s_2 t_2}}{s_1 + s_2} \, ds_1 \, ds_2 \tag{4-148b}$$

Equations 4-147 and 4-148 are obtained by using the Laplace-transform relationship between $V_i(s)$ and $V_o(s)$: $v_o(t)^2$ is replaced by

$$\mathscr{L}^{-1}[V_i(s_1)G(s_1)]\mathscr{L}^{-1}[V_i(s_2)G(s_2)]$$

in 4-146; this is followed by a change in the order of integration and an integration to eliminate t; followed by replacement of $V_i(s_1)V_i(s_2)$ by the equivalent of $\mathscr{L}[v_i(t_1)]\mathscr{L}[v_i(t_2)]$ in terms of s_1, s_2, t_1, and t_2; followed by a final interchange in the order of integration (see Appendix II of reference [4.30]).

Consider the quantity J_{a4},

$$J_{a4} = J_4 + hK_4 \tag{4-149}$$

where h is a negative multiplier which is to be evaluated so as to satisfy the constraint equation. Substituting the right-hand members of 4-145 and 4-147 into 4-149 yields:

$$J_{a4} = \int_0^{T_i}\int_0^{T_i} v_i(t_1)v_i(t_2)\gamma(t_1,t_2)\,dt_1\,dt_2 + h\int_0^{T_i} v_i^2(t_1)\,dt_1 \tag{4-150}$$

An optimal value of $v_i(t)$ is assumed to exist and is denoted by $v_i^*(t)$. In general, $v_i(t)$ equals $v_i^*(t) + \epsilon\,\delta v_i(t)$; ϵ is a real variable, $\epsilon\,\delta v_i(t)$ is an arbitrary variation from $v_i^*(t)$ in the interval $[0,T_i]$, and $\delta v_i(t) = 0$ outside said interval.

As in Section 4-4, the basic variational technique of the calculus of variations is applied, as follows:

$$\left.\frac{\partial J_{a4}}{\partial \epsilon}\right|_{\epsilon=0} = 0$$

$$= \int_0^{T_i}\int_0^{T_i} \delta v_i(t_1)v_i^*(t_2)\gamma(t_1,t_2)\,dt_1\,dt_2$$

$$+ \int_0^{T_i}\int_0^{T_i} \delta v_i(t_2)v_i^*(t_1)\gamma(t_1,t_2)\,dt_1\,dt_2$$

$$+ h\int_0^{T_i} 2\delta v_i(t_1)v_i^*(t_1)\,dt_1 \tag{4-151}$$

It is apparent from Equation 4-148 that $\gamma(t_1,t_2) = \gamma(t_2,t_1)$; therefore the first and second terms of the right-hand member of 4-151 are equal, and an equation which is equivalent to 4-151 is

$$0 = 2\int_0^{T_i} \delta v_i(t_1)\left[hv_i^*(t_1) + \int_0^{T_i} v_i^*(t_2)\gamma(t_1,t_2)\,dt_2\right]dt_1 \tag{4-152}$$

But since $\delta v_i(t_1)$ is arbitrary in the interval defined by $0 \le t_1 \le T_i$, it follows from 4-152 that

$$v_i^*(t_1) = \mu \int_0^{T_i} v_i^*(t_2)\gamma(t_1,t_2)\,dt_2 \qquad (4\text{-}153)$$

for $0 \le t_1 \le T_i$ where $\mu = -1/h$. Values of μ for which solutions to 4-153 exist are called *characteristic values*.

Expression 4-153 is a homogeneous linear integral equation of the second kind [4.39]. The *kernel* $\gamma(t_1,t_2)$ of 4-153 is symmetric; that is, $\gamma(t_1,t_2) = \gamma(t_2,t_1)$, in which case at least one characteristic value μ is known to exist for which a nonzero $v_i^*(t)$ satisfies 4-153. Thus, the existence of a solution is assured although it might not be a unique solution in a given instance.

If more than one characteristic value μ exists, the particular characteristic value μ_o which results in the absolute maximum of the output energy J_4 is determined by noting that if $-h > 0$ is made as large as possible in 4-149, J_4 will be as great as possible for a given value of K_4. It follows that the maximum value of J_4 occurs when $v_i(t)$ is proportional to any one of the characteristic functions $v_i^*(t)$ corresponding to the smallest positive characteristic value $\mu_o = 1/(-h)$ of μ for which 4-153 is satisfied.

The solution of 4-153 with $\gamma(t_1,t_2)$ specified is not as formidable as we might at first imagine. A very practical approach to the solution of 4-153 is based on the following discrete approximation:

$$v_i^*(k\rho) = \mu\rho \sum_{j=0}^{N-1} v_i^*(j\rho)\gamma(k\rho,j\rho) \qquad (4\text{-}154)$$

where $N\rho = T_i$, and $k = 0, 1, \ldots, N-1$. If $[v_i^*(0) \; v_i^*(\rho) \; \cdots \; v_i^*(N\rho - \rho)]'$ is viewed as an unknown column vector, the solution of 4-154 is equivalent to the solution of an eigenvalue problem of matrix theory. Corresponding to the largest value of the eigenvalue $1/\mu\rho = -h/\rho$ which satisfies 4-154, the solution can be found by using a straightforward iterative technique [4.11, pp. 367–368] which is easily programmed for digital computation.[5] The amplitude coefficient of $v_i^*(t)$ is then determined by the requirement that Equation 4-145 be satisfied.

For an account of the more classical methods of solution of integral equations of the above type, the interested reader is referred to the work of Tricomi [4.39].

4-8. SENSITIVITY CONSIDERATIONS IN DESIGN

In the single-loop system of Figure 4-12, the transfer function $G_2(s)$ is that of the plant of the system, $G_a(s)$ is the transfer function of the *cascade*

---

[5] The convenience of this approach was suggested to the author by Dr. C. K. Rushforth.

controller, and $G_b(s)$ is the transfer function of the *feedback* controller. Because two essentially independent controllers are contained in the system, the system is said to exhibit two *degrees of control freedom*. The problem considered in this section is that of specifying these two controllers to accomplish two essentially independent control objectives: (1) the attainment of optimal input-output relations, and (2) the attainment of a system which is insensitive to relatively large (but slowly varying) changes in plant parameters. The input-output transfer function $W_2(s) \triangleq Y_2(s)/V_2(s)$ of the system is

$$W_2(s) = \frac{G_a(s)G_2(s)}{1 + G_a(s)G_b(s)G_2(s)}$$

$$= \frac{1}{G_b(s)} \frac{L(s)}{1 + L(s)} \qquad (4\text{-}155)$$

where $L(s) \triangleq G_a(s)G_b(s)G_2(s)$ is called the *loop transmission*.

Suppose, as is often the case, that the controllers of the system can be built so that $G_a(s)$ and $G_b(s)$ exhibit only nominal change over operational limits of the system, whereas the properties of the plant transfer function $G_2(s)$ are known to a lesser degree or are subject to relatively large, but slowly varying, changes over a period of time. Measures of sensitivity of $W_2 = W_2(s)$ with respect to changes (or uncertainties) in $G_2 = G_2(s)$ are of interest.

Consider the following Taylor's series expansion of W_2 about a design-center form W_2° of W_2:

$$W_2 = W_2^\circ + \frac{\partial W_2}{\partial G_2}\bigg|_\circ (G_2 - G_2^\circ) + \tfrac{1}{2} \frac{\partial^2 W_2}{\partial G_2^2}\bigg|_\circ (G_2 - G_2^\circ)^2 + \cdots \qquad (4\text{-}156)$$

where $(\partial W_2/\partial G_2)|_\circ$ denotes the evaluation of $(\partial W_2/\partial G_2)$ at the design-center form G_2° of G_2. With the particular W_2 given by 4-155, we can effect the

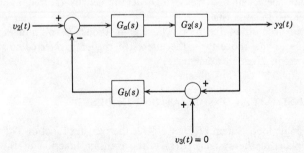

Figure 4-12. A control system with two degrees of control freedom.

Sect. 4-8   SENSITIVITY CONSIDERATIONS IN DESIGN   175

indicated differentiations in 4-156 to obtain

$$\frac{W_2 - W_2{}^\circ}{W_2{}^\circ} = \left(\frac{1}{1 + L^\circ}\right)\left(\frac{G_2 - G_2{}^\circ}{G_2{}^\circ}\right) + \frac{-L^\circ}{(1 + L^\circ)^2}\left(\frac{G_2 - G_2{}^\circ}{G_2{}^\circ}\right)^2 + \cdots \quad (4\text{-}157)$$

In keeping with the sensitivity development in Chapter 2, Equation 4-157 is expressed as

$$\frac{W_2 - W_2{}^\circ}{W_2{}^\circ} \cong \left(\frac{G_2 - G_2{}^\circ}{G_2{}^\circ}\right) S_{G_2}^{W_2} + \left(\frac{G_2 - G_2{}^\circ}{G_2{}^\circ}\right)^2 Sq_{G_2}^{W_2} \quad (4\text{-}158)$$

where the *linear sensitivity* (or simply, sensitivity) $S_{G_2}^{W_2}$ of W_2 with respect to G_2 is given by

$$S_{G_2}^{W_2} \triangleq \left.\frac{\partial W_2}{\partial G_2}\right|_0 \frac{G_2{}^\circ}{W_2{}^\circ} = \frac{1}{(1 + L^\circ)} \quad (4\text{-}159)$$

and the *square sensitivity* $Sq_{G_2}^{W_2}$ is given by

$$Sq_{G_2}^{W_2} \triangleq \frac{-L^\circ}{(1 + L^\circ)^2} \quad (4\text{-}160)$$

Of course, both $S_{G_2}^{W_2}$ and $Sq_{G_2}^{W_2}$ are functions of the Laplace variable s; this fact is used in design to determine the sensitivity of W_2 with respect to G_2 over critical values of s.

An alternative way of viewing the sensitivity of W_2 with respect to G_2 is of interest. On the basis of 4-157, the ratio $[(W_2 - W_2{}^\circ)/W_2{}^\circ]/[G_2 - G_2{}^\circ)/G_2{}^\circ]$ should be close to $S_{G_2}^{W_2}$ when G_2 is close to $G_2{}^\circ$, that is,

$$\left(\frac{W_2 - W_2{}^\circ}{W_2{}^\circ}\right)\left(\frac{G_2{}^\circ}{G_2 - G_2{}^\circ}\right) = \frac{[L/(1 + L)] - [L^\circ/(1 + L^\circ)]}{[L^\circ/(1 + L^\circ)]} \frac{L^\circ}{(L - L^\circ)}$$

$$= \frac{(1 + L^\circ)L - (1 + L)L^\circ}{(1 + L)(L - L^\circ)}$$

$$= \frac{1}{1 + L} \quad (4\text{-}161)$$

This equation gives an exact relationship between the fractional change of W_2 and that of G_2; but observe that 4-161 is in terms of L and not L°. Typically, a design-center form of L° can be determined, and therefore no problem is encountered in the evaluation of $1/(1 + L^\circ)$. On the other hand, if a worst-case form of L is known, this form can be used in 4-161 to obtain a *worst-case sensitivity function*.

The sensitivity $S_{G_2}^{W_2}$ is a function of the complex variable s. A logical question to ask is: "Over what values of s are the values of $S_{G_2}^{W_2}$ most critical in regard to system performance?" A good way to answer this question is to examine the nature of the transfer function $W_2{}^o(s)$ along the particular Laplace-transform inversion contour $s = j\omega$ where ω is "real" frequency in radians per second. Assuming $W_2{}^o(j\omega)$ characterizes a low-pass system, $|W_2{}^o(j\omega)|$ approaches zero as ω approaches infinity; and for some particular frequency ω_1,

$$|W_2{}^o(j\omega)| \ll 1 \quad \text{for } \omega \geq \omega_1 \qquad (4\text{-}162)$$

Consider the case in which $|S_{G_2}^{W_2}|$ is designated to be much less than unity for $\omega < \omega_1$, but $|S_{G_2}^{W_2}|$ equals approximately unity for $\omega > \omega_2 > \omega_1$. In this case, relatively large differences between G_2 and $G_2{}^o$ for $\omega < \omega_1$ do not result in corresponding differences between W_2 and $W_2{}^o$. On the other hand, for $\omega > \omega_2$ relatively large differences between G_2 and $G_2{}^o$ result in corresponding differences between W_2 and $W_2{}^o$, but with $|W_2(j\omega)| \ll 1$—see 4-162—so that only nominal change is exhibited in the inversion integral, i.e., for a large class of $V_2(j\omega)$'s,

$$\frac{1}{2\pi}\int_{-\infty}^{\infty} V_2(j\omega)W_2(j\omega)e^{j\omega t}\,d\omega \cong \frac{1}{2\pi}\int_{-\infty}^{\infty} V_2(j\omega)W_2{}^o(j\omega)e^{j\omega t}\,d\omega \qquad (4\text{-}163)$$

In design, therefore, it is desirable to make $|S_{G_2}^{W_2}|$ small over the bandwidth associated with $W_2{}^o(j\omega)$.

Suppose that the optimal transfer function $W_2{}^o$ between the input $v_2(t)$ and the output $y_2(t)$ is given by

$$W_2{}^o = G_0{}^*(s)G_2{}^o(s) \qquad (4\text{-}164)$$

If the open-loop configuration of Figure 4-1 is used to realize $W_2{}^o$, the sensitivity of the resulting system with respect to G_2 is

$$S_{G_2}^{W_2} = \left[\frac{\partial(G_0{}^*G_2)}{\partial G_2}\bigg|_{G_2=G_2{}^o}\right]\left(\frac{G_2{}^o}{G_0{}^*G_2{}^o}\right) = 1 \quad \text{(open loop)} \qquad (4\text{-}165)$$

If the feedback configuration of Figure 4-12 is used to realize $W_2{}^o(s)$, but with $G_b(s)$ set equal to unity, the sensitivity $S_{G_2}^{W_2}$ is a special case of 4-159; namely,

$$S_{G_2}^{W_2} = \frac{1}{1 + G_a G_2{}^o} \quad \text{(closed loop with } G_b = 1\text{)} \qquad (4\text{-}166)$$

And finally, if the feedback configuration of Figure 4-12 is used, the sensitivity $S_{G_2}^{W_2}$ of 4-159 applies.

In the preceding paragraph, all three realizations of $W_2{}^o(s)$ give the same design-center input-output relationship $G_0{}^*G_2{}^o$, but all three have different sensitivities. The controller in the open-loop realization of $W_2{}^o$ is completely specified once $G_0{}^*(s)$ is specified; no other adjustment can be made to in-

Sect. 4-8   SENSITIVITY CONSIDERATIONS IN DESIGN   177

fluence other factors of interest, e.g., the sensitivity of the system with respect to changes in plant parameters and the effect of noise which might enter the plant via a different path than that shown in Figure 4-1.

Even with the feedback configuration of Figure 4-2c, we still have only one controller (one degree of control freedom) which, if specified solely on the basis of a specific performance measure of response, could result in an overall system that is very sensitive to certain parameter changes or that does not adequately suppress noise from sources in the plant and feedback path of the system. Nevertheless, it is often true that the feedback configuration of Figure 4-2c is less sensitive to changes in plant parameters than is the configuration of Figure 4-1. This point is illustrated in the example which concludes this section.

Of particular interest here is the realization of $W_2{}^o(s)$ with two degrees of control freedom. If the configuration of Figure 4-12 is employed, the right-hand members of 4-155 and 4-164 can be equated to obtain

$$G_0{}^* = \frac{G_a}{1 + L^o} \qquad (4\text{-}167)$$

and because $1/(1 + L^o) = S_{G_2}^{W_2}$, Equation 4-167 is equivalent to

$$G_a = \frac{G_0{}^*}{S_{G_2}^{W_2}} \qquad (4\text{-}168)$$

which gives G_a in terms of $G_0{}^*$ and $S_{G_2}^{W_2}$. Similarly, for G_b, Equation 4-167 can be rearranged to obtain

$$G_b = \frac{1 - S_{G_2}^{W_2}}{G_0{}^* G_2{}^o} \qquad (4\text{-}169)$$

which relates G_b to $S_{G_2}^{W_2}$, $G_0{}^*$, and $G_2{}^o$. It would appear on the basis of 4-168 and 4-169 that $G_0{}^*$ and $S_{G_2}^{W_2}$ can be assigned independently to obtain the optimum $W_2{}^o$ and a very insensitive system—$|S_{G_2}^{W_2}|$ much less than 1 over the bandwidth associated with $W_2{}^o(j\omega)$—but the following other factors must be considered.

In order for $|S_{G_2}^{W_2}| \ll 1$, the magnitude $|L^o|$ of the loop transmission must be much greater than 1 (see Equation 4-159). In some cases, it is economically disadvantageous to require the use of very large open-loop gain.

Furthermore, suppose a nonzero noise signal $v_3(t)$ enters the feedback path as shown in Figure 4-12. The transfer function between $v_3(t)$ and $y_2(t)$ is

$$\frac{Y_2(s)}{V_3(s)} = \frac{-L^o}{1 + L^o} \qquad \text{for } V_2(s) = 0 \qquad (4\text{-}170)$$

To obtain effective rejection of the noise associated with $V_3(j\omega)$, therefore, $|L^o(j\omega)|$ should be much less than 1 over the frequency range associated with $V_3(j\omega)$. But for sensitivity purposes, as previously observed, $|L^o(j\omega)|$ should be much greater than 1 over the bandwidth associated with $W_2{}^o(j\omega)$. Thus, a compromise must be reached. Typically, $|L^o(j\omega)|$ is designed so as to approach zero at a rapid rate as ω exceeds the bandwidth corresponding to $W^o(j\omega)$. This "rapid rate" is also limited, however, by stability requirements of the system; Example 4-4 which concludes this section illustrates this point.

Yet another limitation is illustrated in Example 4-4; it is shown that $|L^o(j\omega)|$ is limited in magnitude by requirements of stability in the case that $G_2{}^o(s)$ contains a right-half-plane zero.

Conceivably, the difficulties noted in the preceding paragraphs could be by-passed if more than two degrees of control freedom are available to the designer. A fundamental limitation in this regard is noted by Horowitz [4.13]: if the plant of the system has but one input and one measurable output (as in Figure 4-12), no more than two degrees of control freedom can be achieved, regardless of the complexity of the configuration designed around the plant.

Example 4-4. The optimal input-output relations of Example 4-2 can be realized by any one of the systems represented in Figures 4-1, 4-2c, and 4-12. The sensitivity of these realizations is to be investigated. In the first part of Example 4-2, the plant of the system is characterized by a minimum-phase function $G_2(s)$; in the second part, the plant is characterized by the non-minimum-phase function $G_2(s) = G_{mp}(s)(\tau_1 s + 1)$ where τ_1 is less than zero and $G_{mp}(s)$ is a minimum-phase function. The two parts of this example are extensions of the corresponding parts in Example 4-2. In both parts that follow, the sensitivity of the open-loop realization (Figure 4-1) of $W_2{}^o(s)$ is known to be unity on the basis of 4-165.

$\tau_1 \geq 0$

On the basis of Equation 4-98, the optimal overall transfer function $W_2{}^o(s)$ is

$$W_2{}^o(s) = G_0{}^*(s) G_2{}^o(s) = \frac{1}{h_1 s + 1} \qquad (4\text{-}171)$$

A plot of 20 log $|W_2{}^o(j\omega)|$ versus ω on a logarithmic scale is one of the curves given in Figure 4-13.

If the unity-feedback configuration of Figure 4-2c is used to realize $W_2{}^o$, G_a is given by 4-104, and the expression for sensitivity $S_{G_2}^{W_2}$ in Equation 4-166 applies, as follows:

$$S_{G_2}^{W_2} = \frac{1}{1 + G_a G_2{}^o} = \frac{h_1 s}{h_1 s + 1} \qquad (4\text{-}172)$$

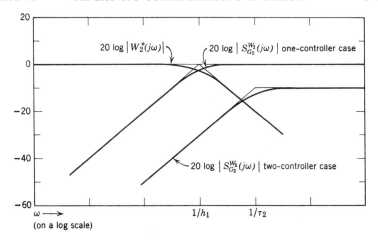

Figure 4-13. Typical Bode diagrams for the one- and two-controller, minimum-phase cases.

By use of this result with $s = j\omega$, $20 \log |S_{G_2}^{W_2}(j\omega)|$ is plotted versus $\log \omega$ as one of the curves in Figure 4-13. It is to be observed that the magnitude of the sensitivity $S_{G_2}^{W_2}(j\omega)$ is less than 1, the value which holds for the open-loop case, for all finite ω: it approaches zero as ω approaches zero, it equals $(1/2)^{1/2}$ at the critical corner frequency $\omega = 1/h_1$ of $W_2^o(j\omega)$, and it approaches unity as ω approaches infinity.

For the two-controller configuration, Equations 4-168 and 4-169 apply:

$$G_a(s) = \frac{1}{G_2^o(s)(h_1 s + 1)S_{G_2}^{W_2}} \qquad (4\text{-}173)$$

and

$$G_b(s) = (1 - S_{G_2}^{W_2})(h_1 s + 1) \qquad (4\text{-}174)$$

The sensitivity $S_{G_2}^{W_2}$ remains to be specified. Observe that poles of $S_{G_2}^{W_2}$ cannot be located in the right half-plane; poles of $S_{G_2}^{W_2}$ are both zeros of $G_a(s)$ and poles of $1 - S_{G_2}^{W_2}$ and, therefore, poles of $G_b(s)$; in practice, the inherent inexact cancellation of right-half-plane poles and zeros in the loop transmission $L(s) = G_a(s)G_b(s)G_2(s)$ would result in closed-loop poles in the right half-plane, with instability the result (we can reason this on the basis of root-locus theory, there being a closed-loop pole on the locus between the inexactly canceled pole-zero pair).

From the one-controller case, it is known that $S_{G_2}^{W_2}(j\omega)$ can be made to equal zero at $\omega = 0$. Thus, for this two-controller case, let $S_{G_2}^{W_2}$ assume the form

$$S_{G_2}^{W_2} = \frac{\alpha s}{\tau_2 s + 1} \quad (4\text{-}175)$$

where α and τ_2 are positive parameters to be specified by the designer to obtain a required degree of insensitivity. More complicated forms of $S_{G_2}^{W_2}$ could be assumed, but this particular form is sufficient for the present sensitivity illustration, as is to be observed in Figure 4-13.

With the particular form of $S_{G_2}^{W_2}$ in 4-175, $G_a(s)$ and $G_b(s)$ of 4-173 and 4-174 are, respectively,

$$G_a(s) = \frac{(\tau_2 s + 1)}{G_2^o(s)(h_1 s + 1)s\alpha} \quad (4\text{-}176)$$

and

$$G_b(s) = \frac{[(\tau_2 - \alpha)s + 1](h_1 s + 1)}{(\tau_2 s + 1)} \quad (4\text{-}177)$$

Typically, the loop transmission $L(j\omega)$ should approach zero as ω approaches infinity; this requires that $S_{G_2}^{W_2}$ approach unity as ω approaches infinity (see Equation 4-159). With this restriction, α is required to equal τ_2, and a zero in the right-hand member of 4-177 is eliminated.

$\tau_1 < 0$

On the basis of Equation 4-107, the optimal transfer function $W_2^o(s)$ is

$$W_2^o(s) = G_0^*(s)G_2^o(s) = \frac{(\tau_1 s + 1)}{(h_1 s + 1)(-\tau_1 s + 1)} \quad (4\text{-}178)$$

A plot of $20 \log |W_2^o(j\omega)|$ versus $\log \omega$ is one of the curves given in Figure 4-14.

If the unity-feedback configuration of Figure 4-2c is used to realize W_2^o, G_a is given by 4-113, and the expression for sensitivity $S_{G_2}^{W_2}$ in Equation 4-166 applies, as follows:

$$S_{G_2}^{W_2} = \frac{s(-h_1\tau_1 s + h_1 - 2\tau_1)}{-h_1\tau_1 s^2 + (h_1 - \tau_1)s + 1}$$

$$= \frac{s\left(s - \frac{1}{\tau_1} + \frac{2}{h_1}\right)}{\left(s - \frac{1}{\tau_1}\right)\left(s + \frac{1}{h_1}\right)} \quad (4\text{-}179)$$

Two representative cases of this sensitivity are given in Figure 4-14; one for which $-1/\tau_1 = -1/\tau_1'$ is greater than $1/h_1$ and one for which $-1/\tau_1''$ is less than $1/h_1$. In both cases, the sensitivity is poorer than that which is obtained in the corresponding minimum-phase case of Figure 4-13.

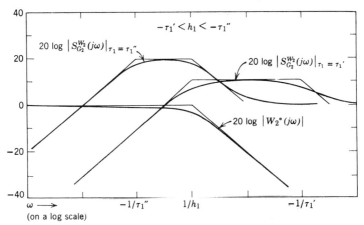

Figure 4-14. Typical Bode diagrams for the one-controller, nonminimum-phase case.

For the two-controller configuration, Equations 4-168 and 4-169 apply:

$$G_a(s) = \frac{1}{G_{mp}^\circ(s)(h_1 s + 1)(-\tau_1 s + 1) S_{G_2}^{W_2}} \tag{4-180}$$

and

$$G_b(s) = \frac{(1 - S_{G_2}^{W_2})(h_1 s + 1)(-\tau_1 s + 1)}{(\tau_1 s + 1)} \tag{4-181}$$

To avoid a right-half-plane cancellation attempt between $G_b(s)$ and $G_2^\circ(s)$, $S_{G_2}^{W_2}$ must be specified such that $1 - S_{G_2}^{W_2}$ has a zero at $s = -1/\tau_1$, with the result that $G_b(s)$ in 4-181 has no pole at $s = -1/\tau_1$. Also, the limit as $|s|$ approaches infinity of $S_{G_2}^{W_2}$ is taken as unity in order that the loop transmission $L^\circ(j\omega)$ approach zero as ω approaches infinity.

A form of $S_{G_2}^{W_2}$ which complies with the above-noted requirements is

$$S_{G_2}^{W_2} = \frac{s(s + z_1)}{(s + p_1)(s + p_2)} \tag{4-182}$$

in which the parameters z_1, p_1, and p_2 are related by

$$\frac{(-1/\tau_1)[(-1/\tau_1) + z_1]}{[(-1/\tau_1) + p_1][(-1/\tau_1) + p_2]} = 1, \qquad \tau_1 < 0 \tag{4-183}$$

because $G_b(s)$ is to have no pole at $s = -1/\tau_1$. Both p_1 and p_2 must be

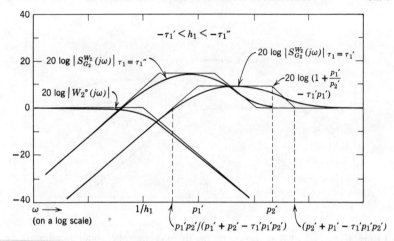

Figure 4-15. Typical Bode diagrams for the two-controller, nonminimum-phase case.

greater than zero in order that right-half-plane pole-zero cancellation is not attempted between G_a and G_b in $L°$. It follows from 4-183 that

$$z_1 = p_1 + p_2 - \tau_1 p_1 p_2, \qquad \tau_1 < 0 \qquad (4\text{-}184)$$

which is satisfied only if z_1 is greater than both p_1 and p_2. Without loss of generality, it is assumed that

$$0 < p_1 < p_2 < z_1 \qquad (4\text{-}185)$$

Values of p_1 and p_2 can be specified by the designer, and the value of z_1 determined by using 4-184. Typical plots of $20 \log |S_{G_2}^{W_2}(j\omega)|$ versus $\log \omega$ are given in Figure 4-15. It is to be observed that the corner frequency $1/h_1$ of $W_2°(j\omega)$ can be less than the 0-db cross-over frequency

$$p_1 p_2/(p_1 + p_2 - \tau_1 p_1 p_2)$$

of the asymptote of $S_{G_2}^{W_2}$ only if h_1 is greater than $-\tau_1$. Thus, if the corner frequency associated with the RHP zero of the plant is within the bandwidth of this closed-loop system, the resulting system is relatively poor from the standpoint of sensitivity.

●

4-9. CONCLUSION

The subject of this chapter is the optimal design of linear time-invariant systems by using transform concepts and Wiener-Hopf spectrum factorization. Filter, control, predictor, and pulse-shape problems which can be treated by this approach are given in Section 4-2 and 4-3. In Section 4-4, a general problem is expressed and solved by use of spectrum factorization. Basic techniques of variational calculus and spectrum factorization are applied to find the Laplace transform $F(s)$ which results in an extremum of the inversion integral

$$J_a = \frac{1}{2\pi j} \int_{-j\infty}^{j\infty} [A_1(s)F(s)F(-s) + A_2(s)F(s) + A_3(s)F(-s) + A_4(s)] \, ds \tag{4-186}$$

The general solution so obtained is used in Sections 4-5 and 4-6 to obtain solutions for the previously mentioned problems.

As noted in Section 4-2, the selection of $F(s)$ in 4-186 may involve one or more constant Lagrange multipliers which are specified so that corresponding isoperimetric constraints are satisfied. Only in a limited class of problems can factorization be effected without the Lagrange multipliers being specified numerically in advance. Thus, in general, a search must be conducted to determine values of Lagrange multipliers which result in satisfaction of isoperimetric constraints. The content of Chapter 6 is of use in this regard, as also is the work of Sancho and Roberts [4.35] in which computer oriented methods are presented for treating the case of two Lagrange multipliers.

Numerous extensions and generalizations of the basic Wiener-Hopf approach are available in the literature. For multivariable systems, performance measures may contain more than one function to be selected for the optimum, in which case the factorization of a "spectral" matrix is required [4.7, 4.14, 4.23, 4.33, 4.45]. For systems in which some signals are sampled, z-transform theory is often of use in the factorization process [4.5, 4.16, 4.18, 4.34, 4.36, 4.38]. For most distributed-parameter systems and systems with time delay, the factorization process can be readily effected only after suitable approximations have been made [4.22, 4.29, 4.32]. For linear systems with time-varying parameters, the transform approach is replaced by a parallel time-domain approach [4.3, 4.19]. Nonlinear systems also are treated by use of spectrum factorization, but in a very limited sense [4.20, 4.28, 4.29]. And notably, for a general class of filter and estimation problems, the solutions of Kalman and Bucy are known to reduce to the Wiener solution in the steady state [4.15]; in fact, Kalman filters are alternatively called Kalman-Bucy filters or Kalman-Bucy-Wiener filters.

The sensitivity of performance of linear time-invariant systems is conveniently examined in terms of frequency response concepts. These concepts are used in Section 4-8 to determine the practicality of different ways of implementing an overall optimal solution. For additional background on these aspects of sensitivity, the reader is referred to studies by Horowitz [4.13] and Merchav [4.26].

Finally, the pulse-shape problem of Section 4-7 illustrates the fact that spectrum-factorization techniques are not applicable to all optimization problems associated with linear time-invariant systems. Indeed, many such problems remain to be solved.

REFERENCES

[4.1] Beattie, L. A. "Minimum Energy Triggering Signals." *Proceedings of the IRE*, **46**, 751–757, April 1958.

[4.2] Brown, W. M. *Analysis of Linear Time-Invariant Systems*. McGraw-Hill, New York, 1963.

[4.3] Booton, R. C., Jr. "An Optimization Theory for Time-Varying Linear Systems with Nonstationary Statistical Inputs." *Proceedings of the IRE*, **40**, 977–981, Aug. 1952.

[4.4] Chalk, J. H. H. "The Optimum Pulse-Shape for Pulse Communication." *Proceedings of the Institution of Electrical Engineers*, London, **97**, part 3, 88–92, 1950.

[4.5] Chang, S. S. L. *Synthesis of Optimum Control Systems*. McGraw-Hill, New York, 1961.

[4.6] Cochran, W. T., J. J. Downing, D. L. Favin, H. D. Helms, R. A. Kaenel, W. W. Lang, and D. E. Nelson. "Burst Measurements in the Frequency Domain." *Proceedings of the IEEE*, **54**, 830–841, June 1966.

[4.7] Davis, M. C. "Factoring the Spectral Matrix." *IEEE Transactions on Automatic Control*, **AC-8**, 296–305, Oct. 1963.

[4.8] Franks, L. E. "A Method for Optimizing Performance of Pulse Transmission Systems." *IEEE International Convention Record*, **12**, part 5, 279–290, March 1964.

[4.9] Gerst, I., and J. Diamond. "The Elimination of Intersymbol Interference by Input Signal Shaping." *Proceedings of the IRE*, **49**, 1195–1203, July 1961.

[4.10] Gupta, S. C., and R. J. Solem. "Optimum Filters for Second- and Third-Order Phase-Locked Loops by an Error Function Criterion." *IEEE Transactions on Space Electronics and Telemetry*, **SET-11**, 54–66, June 1965.

[4.11] Hamming, R. W. *Numerical Methods for Scientists and Engineers*. McGraw-Hill, New York, 1962.

[4.12] Hancock, J. C., H. Schwarzlander, and R. E. Totty. "Optimization of Pulse Transmission." *Proceedings of the IRE*, **50**, 2136, Oct. 1962.

[4.13] Horowitz, I. M. *Synthesis of Feedback Systems*. Academic Press, New York, 1963.

[4.14] Hsieh, H. C., and C. T. Leondes. "On the Optimum Synthesis of Multipole Control Systems in the Wiener Sense." *IRE Transactions on Automatic Control*, **AC-4**, 16–29, Nov. 1959.

[4.15] Hutchinson, C. E. "An Example of the Equivalence of the Kalman and Wiener Filters." *IEEE Transactions on Automatic Control*, **AC-11**, 324, April 1966.

REFERENCES

[4.16] Jackson, R. "Optimum Sampled-Data Control." *Proceedings of the Institution of Electrical Engineers*, London, **108**, part C, 309–316, 1961.

[4.17] Jaffe, R., and E. Rechtin. "Design and Performance of Phase-Lock Circuits Capable of Near-Optimum Performance over a Wide Range of Input Signal and Noise Levels." *IRE Transactions on Information Theory*, **IT-1**, 66–76, March 1955.

[4.18] Jury, E. I. *Theory and Application of the z-Transform Method.* Wiley, New York, 1964.

[4.19] Johnson, E. P., Jr. "Optimization of Linear Systems." Chapter 12 of *Control Systems Engineering*, W. W. Seifert and C. W. Steeg, Jr. (Editors). McGraw-Hill, New York, 1960.

[4.20] Katzenelson, J., and L. A. Gould. "A Spectrum Factorization Method for the Calculation of Nonlinear Filters of the Volterra Type." *Information and Control*, **8**, 239–250, June 1965.

[4.21] Knapp, C. H. "Optimum Linear Systems with Unstable Plants." *IEEE Transactions on Automatic Control*, **AC-8**, 258–259, July 1963.

[4.22] Kolb, R. C., and D. A. Pierre. "Averaged Integral-Square Error and an Approximation to Optimal Control of Distributed-Parameter Systems." *Preprint Volume*, Joint Automatic Control Conference, 811–822, August 1966.

[4.23] MacCracken, L. C. "An Extension of Wiener Theory to Multivariable Controls." *IRE International Convention Record*, **9**, part 4, 56–60, 1961.

[4.24] MacPhie, R. H. *On Maximizing the Signal-to-Noise Ratio of a Linear Receiving Antenna Array.* Antenna Laboratory Technical Report, No. 65, Electrical Engineering Research Laboratory, Univ. of Illinois, Urbana, Ill., Jan. 1963, 44 pp.

[4.25] Meer, S. A. "A Class of Wiener Filters Useful in PLL Applications." *Proceedings of the IEEE*, **53**, 2121, Dec. 1965.

[4.26] Merchav, S. J. "Compatibility of a Two-Degree-of-Freedom System with a Set of Independent Specifications." *IEEE Transactions on Automatic Control*, **AC-7**, 67–72, Jan. 1962.

[4.27] Mostov, P. M., J. L. Neuringer, and D. S. Rigney. "Optimum Capacitor Charging Efficiency for Space Systems." *Proceedings of the IRE*, **49**, 941–948, May 1961.

[4.28] Newton, G. C., Jr. "Compensation of Feedback Control Systems Subject to Saturation." *Journal of the Franklin Institute*, **254**, 281–296 and 391–413, Oct. and Nov. 1952.

[4.29] Newton, G. C., Jr., L. A. Gould, and J. F. Kaiser. *Analytical Design of Linear Feedback Controls.* Wiley, New York, 1957.

[4.30] Pierre, D. A. *Optimal Pulse-Shapes for Energy-Constrained Inputs to Linear Systems.* IEEE Region 6 Annual Conference, April 1964, 23 pp.

[4.31] Pierre, D. A. "Optimal Time-Limited Energy-Constrained Inputs to Linear Systems." *Proceedings*, Columbia University and IEEE Symposium on Signal Transmission and Processing, 32–41, May 1965.

[4.32] Pierre, D. A. "Minimum Mean-Square-Error Design of Distributed-Parameter Control Systems." *ISA Transactions*, **5**, 263–271, July 1966.

[4.33] Riddle, A. C., and B. D. Anderson. "Spectral Factorization Computational

Aspects." *IEEE Transactions on Automatic Control*, **AC-11**, 764–765, Oct. 1966.

[4.34] Robins, H. M. "An Extension of Wiener Filter Theory to Partly Sampled Systems." *IRE Transactions on Circuit Theory*, **CT-6**, 362–370, Dec. 1959.

[4.35] Sancho, N. G. F., and A. P. Roberts. "Use of a Digital Computer for the Optimization of Automatic Control Systems with Random Inputs." *International Journal of Control*, **1**, 501–518, June 1965.

[4.36] Tou, J. T. *Digital and Sampled-Data Control Systems.* McGraw-Hill, New York, 1959.

[4.37] Tozer, R. F., and D. W. Tufts. "Experimental Simulation of Optimal State Variable Estimation." *Proceedings of the IEEE*, **53**, 498–499, May 1965.

[4.38] Tretter, S. A., and K. Steiglitz. "Some Properties of Minimum Mean-Square-Error Filters for Sampled-Data Reconstruction." *IEEE International Convention Record*, **13**, part 7, 134–146, 1965.

[4.39] Tricomi, F. G. *Integral Equations.* Interscience, New York, 1957.

[4.40] Tufts, D. W. *Nyquist's Problem in Pulse Transmission Theory.* Technical Report, No. 425, Cruft Laboratory, Harvard Univ., Cambridge, Mass., Sept. 1963, 22 pp.

[4.41] Tufts, D. W. "Nyquist's Problem—The Joint Optimization of Transmitter and Receiver in Pulse Amplitude Modulation." *Proceedings of the IEEE*, **53**, 248–259, March 1965.

[4.42] Turin, G. L. "An Introduction to Matched Filters." *IEEE Transactions on Information Theory*, **IT-6** (special issue on matched filters), 311–329, June 1960.

[4.43] Wiener, N. "Generalized Harmonic Analysis." *Acta Mathematica*, **55**, 117–258, 1930.

[4.44] Wiener, N. *Extrapolation, Interpolation, and Smoothing of Stationary Time Series.* The M.I.T. Press, Cambridge, Mass.; first printing, August 1949; first paperback edition, Feb. 1964.

[4.45] Youla, D. C. "On the Factorization of Rational Matrices." *IRE Transactions on Information Theory*, **IT-7**, 172–189, July 1961.

[4.46] Zotov, M. G. "A Special Case of the Wiener Equation." *Engineering Cybernetics*, 80–85, Sept.–Oct. 1964.

PROBLEMS

4.1 Suppose that the blocks within the dashed lines of Figure 4-1 are replaced by the blocks within the dashed lines of Figure 4-P1. The signal $v_n(t) = \mathscr{L}^{-1}[V_n(s)]$ is an undesired disturbance. Find $E(s)$ of Figure 4-P1 as a linear function of $G_0(s)$, $G_0(s) \triangleq G_a(s)/[1 + G_a(s)G_{p1}(s)G_{p2}(s)]$, and use this $E(s)$ in Equation 4-5. Identify $F(s)$ of 4-22 with $G_0(s)$ and express $A_1(s)$ through $A_4(s)$ of 4-22 in terms of V_s, V_n, V_1, G_{p1}, G_{p2}, G_1, G_3, and h. (*Hint:* $Y_2(s) = V_s(s)G_0(s)G_{p1}(s)G_{p2}(s) + V_n(s)G_{p2}(s)[1 - G_0(s)G_{p1}(s)G_{p2}(s)]$.)

A special case of this problem is considered by Zotov [4.46].

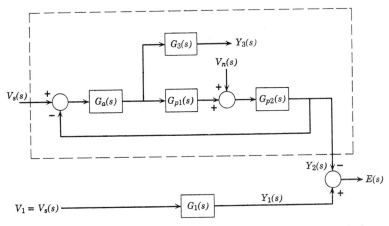

Figure 4-P1. A unity feedback control system with a disturbance $v_n(t)$ in the loop.

4.2 Consider Figure 4-P2. The operational impedances of the circuit components are expressed in terms of the Laplace variable s, and initial conditions associated with the capacitor and the inductor are represented by equivalent

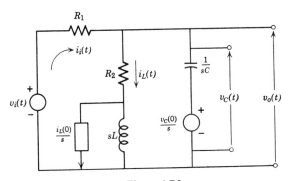

Figure 4-P2.

voltage and current sources, respectively. The transform $V_o(s)$ of the output $v_o(t)$ is of the form

$$V_o(s) = V_i(s)G_1(s) + v_C(0)G_2(s) + i_L(0)G_3(s)$$

and the transform $I_i(s)$ of the input current $i_i(t)$ is of the form

$$I_i(s) = V_i(s)G_4(s) + v_C(0)G_5(s) + i_L(0)G_6(s)$$

where the G_k's are functions of s and the circuit parameters.
    a. If the augmented performance measure to be maximized is that of Equation 4-19, find the frequency-domain equivalent of 4-19 in terms of $V_i(s)$ which

applies here. Identify $V_i(s)$ with $F(s)$ of 4-22 and find expressions for $A_1(s)$ through $A_4(s)$ which apply here.

b. Evaluate $G_1(s)$ through $G_6(s)$ in terms of s and the circuit parameters.

4.3 Given the circuit diagram of Figure 4-P3 and the augmented performance measure of 4-19, place this performance measure in the standard form of 4-22 with $V_i(s)$ identified with $F(s)$.

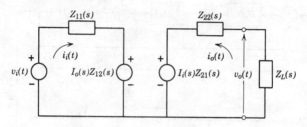

Figure 4-P3. An active network: $V_i(s) = Z_{11}(s)I_i(s) + Z_{12}(s)I_o(s)$, $V_o(s) = Z_{21}(s)I_i(s) + Z_{22}(s)I_o(s)$, and $V_o(s) = -I_o(s)Z_L(s)$.

4.4 Consider the block diagram of Figure 4-P4. The transfer function $G_p(s)$ is assumed known; $v_s(t)$ is a stationary random signal with known power-density spectrum $\Phi_{v_s v_s}(s)$; $v_n(t)$ is a stationary random noise with known power-density spectrum $\Phi_{v_n v_n}(s)$; the output $x_1(t)$ of the block associated with the known $G_{11}(s)$ is a state variable of the system characterized by $G_p(s)$; the output $x_2(t)$ of the block associated with the known $G_{12}(s)$ is a second state variable associated with $G_p(s)$; and so forth. The blocks on the right-hand side of Figure 4-P4 represent an ideal system for determination of the state variables, but this ideal system

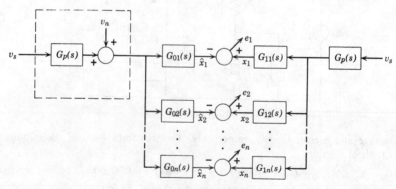

Figure 4-P4.

cannot be realized because of measurement noise in the actual system depicted by the left-hand blocks in the figure. For each integer i, $i = 1, 2, \ldots, n$, the problem is to determine the optimal form $G_{0i}^*(s)$ of $G_{0i}(s)$ which yields a minimum of the

PROBLEMS 189

mean-square value of the error $e_i(t)$ between the actual ith state variable $x_i(t)$ and the estimate $\hat{x}_i(t)$ of the ith state variable.

    a. Find an expression for the power density spectrum $\Phi_{e_i e_i}(s)$.

    b. Find a general expression for $G_{0i}^*(s)$ which yields a minimum of J_i, where

$$J_i = \frac{1}{2\pi j}\int_{-j\infty}^{j\infty} \Phi_{e_i e_i}(s)\,ds$$

4.5 Supply the steps required to obtain 4-52 and 4-53 from 4-48.

4.6 Supply the steps required to obtain 4-54 from 4-48.

4.7 Supply the steps required to obtain 4-55, 4-56, and 4-57 from 4-48.

4.8 In Figure 4-1, assume that $G_1(s) = G_2(s) = 1$, $G_3(s) = 0$, that $v_1(t)$ has an associated autocorrelation function $\phi_{v_1 v_1}(\tau) = e^{-\alpha|\tau|}$ where α is a constant, and that $v_2(t)$ has an associated autocorrelation function $e^{-\alpha|\tau|} + \delta(\tau)$ where $\delta(\tau)$ is the Dirac delta function. Determine the $G_0(s)$ which yields the minimum of J_{a2} in 4-12, where $g_0(t) = \mathcal{L}^{-1}[(G_0(s)]$ equals zero for $t < 0$. Assume $\Phi_{v_1 v_2} = \Phi_{v_1 v_1}$.

4.9 In Figure 4-1, assume that $G_1(s) = G_2(s) = 1$, $G_3(s) = s$, and that $V_1(s) = \mathcal{L}[v_1(t)]$ and $V_2(s) = \mathcal{L}[v_2(t)]$ equal $\lim_{\epsilon \to 0+} 1/(s + \epsilon)$. Find the $G_0(s)$ which results in the minimum of J_{a1} in 4-8, subject to the constraint that $g_0(t) = 0$ for $t < 0$ and subject to the additional constraint that $g_0(t) = 0$ for $t > 1$ second.

4.10 Repeat Example 4-1, but with $K_n = \tau_n = 1/3$ and $K_s = \tau_s = 1$.

4.11 Derive Equations 4-116 and 4-117 from Equation 4-114.

4.12 Derive Equation 4-118.

4.13 Consider the network of Figure 4-P13:

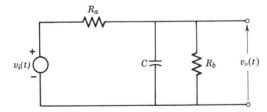

Figure 4-P13.

    a. Find the particular $G(s)$ and $Y_i(s)$ of Section 4-3 which apply here. (*Hint:* To simplify notation, let $a \triangleq (R_a + R_b)/R_a R_b C$ and $b^2 \triangleq R_a/(R_a + R_b)$.)

    b. Find the $v_i^*(t)$ corresponding to 4-116 in terms of R_a, R_b, C, h, T, and t.

4.14 Solve Problem 4.13 under the simplifying condition that R_b approaches infinity.

4.15 In Figure 4-1, assume the following: the plant of the system is characterized by $G_2(s) = e^{-sD}G_{mp}(s)$ where D is a positive time-delay constant and $G_{mp}(s)$ is a minimum-phase function; the ideal transfer function is e^{-sD}; $G_3(s) = sG_2(s)$; the ideal input is a unit step function initiated at $t = 0$ for which $V_1(s) = 1/s$; the actual input rises exponentially from zero for $t \leq 0$ to the unity level as t approaches infinity, for which $V_2(s) = (1/s) - [1/(s + a)] = a/(s^2 + sa)$ where a is a constant. Evaluate G_0^* of 4-48 under these conditions.

4.16 In Figure 4-1, assume that $G_1(s) = 1$, $G_3(s) = 0$, $V_1(s) = V_2(s) = \lim_{\epsilon \to 0+} 1/(s + \epsilon)$, and $G_2(s) = G_{mp}(s)(\tau_1 s + 1)$ where $G_{mp}(s)$ is a minimum-phase function and τ_1 is a real constant. For rejection of noise, it is generally desirable to have a small bandwidth associated with the transfer function $W_2(s) = G_0(s)G_2(s)$. For zero steady-state error, to a step input, it is required that $W_2(0) = 1$. For a low-pass system with the property that $W_2(0) = 1$, a performance measure which is indicative of bandwidth is J_b:

$$J_b = \frac{1}{2\pi j} \int_{-j\infty}^{j\infty} W_2(s)W_2(-s)\,ds$$

A constraint on the minimization of J_b is that the integral-square error J_1 equals a specified constant value.

a. Find the constrained minimum of J_b and the corresponding G_0^* in the case that $\tau_1 \geq 0$.
b. Find the constrained minimum of J_b and the corresponding G_0^* in the case that $\tau_1 < 0$.

(*Hint:* Compare the form of this problem with that of Example 4-2.)

4.17 As a special case of Problem 4.4, consider the following: the number n of state variables equals 2; $G_p(s) = 1/(s^2 + 0.2s + 1)$; $G_{11}(s) = 1$; $G_{12}(s) = s$; $\Phi_{v_s v_s}(s) = 10$; and $\Phi_{v_n v_n}(s) = 0.164$. Find $G_{01}^*(s)$ and $G_{02}^*(s)$. (A version of this problem is treated by Tozer and Tufts [4.37] by use of Kalman filter theory. Their problem is more general in that initial conditions, $x_1(0)$ and $x_2(0)$, and initial probabilistic data concerning $x_1(0)$ and $x_2(0)$ are taken into account. The filter which results in that case is a time-varying one and one which is obtained on the basis of numerical solution of a set of first-order time-variant differential equations.)

4.18 Consider the block diagram of Figure 4-P18. Under the condition that $v_n(t)$ equals identically zero and $v_s(t)$ equals $u(t)$, a unit step function, the integral-

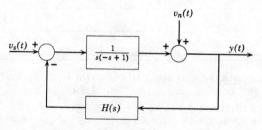

Figure 4-P18.

square value of $v_s(t) - y(t)$ is to be less than or equal to a given constant K. But under the condition that $v_s(t)$ equals identically zero and the disturbance $v_n(t)$ equals $u(t)$, a minimum of the integral-square value of $y(t)$ is desired. Find a physically realizable $H(s)$ which results in a stable closed-loop system and which satisfies the noted specifications. (*Hint:* No attempt should be made to exactly cancel the right-half-plane pole at $s = 1$ with a right-half-plane zero of $H(s)$; express pertinent equations in terms of $F_1(s) \triangleq 1/[s(-s+1) + H(s)]$ and determine the optimum $F_1^*(s)$ which is required to be analytic in the right half-plane. Problems similar to this are considered by Knapp [4.21].)

4.19 With regard to the circuit diagram of Figure 4-P19, the average value

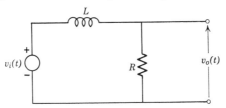

Figure 4-P19.

of $v_o(t)$ is to be maximized over the time interval $[0,T]$. At $t = 0-$, the value of $v_o(0-)$ is zero. The input *energy* is constrained to be less than K. Find the optimal form $V_i^*(s)$ of $V_i(s)$ and the corresponding $v_i^*(t)$. (*Hint:* Dirac delta functions will be included in your solution, but c_1 of 4-120 will be forced to be zero in order that only finite energy be supplied.)

4.20 It is given that $v_i(t) = 0$ for all t outside the closed interval $[0,T]$ and that $v_i(t)$ is constrained by

$$\int_0^T [v_i(t)]^2 \, dt = K \quad \text{(constant)}$$

The maximum of P is to be determined, where

$$P = \int_0^\infty [v_o(t)]^2 \, dt$$

and where $V_o(s) = \mathscr{L}[v_o(t)]$ and $V_i(s) = \mathscr{L}[v_i(t)]$ are related by $V_o(s) = V_i(s)[u(\omega_o + s/j) - u(-\omega_o + s/j)]$. The constant ω_o is the bandwidth of an ideal filter and the u's are standard unit step functions. Find the expression for $\gamma(t_1, t_2)$ of Equation 4-148 which applies to this problem. (An equivalent problem is considered in reference [4.6].)

4.21 Derive Equations 4-147 and 4-148.

4.22 For the system depicted by the block diagram of Figure 4-P22, find $S_{G_2}^W$ and express G_a and G_b in terms of W, G_2, and $S_{G_2}^W$.

4.23 Repeat Example 4-4 using the configuration of Figure 4-P22 in place of that in Figure 4-12.

4.24 With $V_n(s) = 0$ in Figure 4-P1, find the linear sensitivity function corresponding to the sensitivity of $E(s)/V_s(s)$ with respect to $G_{p2}(s)$.

4.25 Find the linear sensitivity function corresponding to the sensitivity of $V_o(s)/V_i(s)$ in Figure 4-P3 with respect to:
a. $Z_{11}(s)$
b. $Z_{12}(s)$
c. $Z_{21}(s)$
d. $Z_{22}(s)$

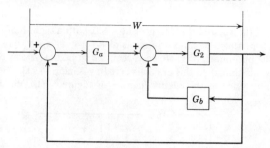

Figure 4-P22.

4.26 Assume that the parameters of the plant in Figure 4-P26 vary slowly in comparison with the time constants of the system. Describe how the adaptive

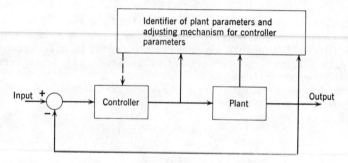

Figure 4-P26.

loop might be implemented to obtain optimal input-output relationships over an extended period of time. What effect does the adaptive loop have on the sensitivity of the system?

4.27 Compare the results of the first part of Example 4-2 with the results of Example 3-5. What problems might be associated with the exact implementation of the results of Example 4-2.

4.28 Prove that the Lagrange multiplier approach used in Section 4-7 is valid.

4.29 Consider minimization of J:

$$J = \int_0^\infty (m^2 - x^2)\, dt$$

where

$$\dot{x} = m, \quad x(0) = x_0$$

What peculiar result is obtained through the formal application of spectrum factorization on this problem? How do you explain this result? (*Hint:* Reconsider Problem 3.35.)

THE SIMPLEX TECHNIQUE AND LINEAR PROGRAMMING

5

5-1. INTRODUCTION

The *linear* in "linear programming" indicates that only linear equations are involved. The *programming* in "linear programming" indicates that various variables are to be programmed—programmed in the sense of being scheduled or selected—to optimize a linear performance measure. In the usual linear programming problem (Section 5-2), the performance measure is a specified linear algebraic equation, and other linear algebraic equations act as constraints on the optimization process. Exceptions to the usual linear programming problem are manifest (Sections 5-12 and 5-13).

The simplex technique (Sections 5-4, 5-5, and 5-6) is an efficient method for solving linear programming problems which are of a general form. Certain modifications of the simplex technique are particularly well-suited for digital computer solution, and routines for these are commonly available at digital computer centers. Because of this, the theory associated with the technique and the procedure best suited for manual computations are stressed here, and so are applications (Sections 5-3 and 5-12) from many areas.

Whether manual computations (for small problems) or computer computations (for large problems) are used, scaling and initializing of problems (Section 5-7) must be effected to minimize harmful effects of round-off errors.

For every linear programming problem, a dual problem exists which may be easier to solve. The solution to the original (or primal) problem can be obtained directly from the solution of the dual when the simplex technique is used (Section 5-9). Once an optimal solution has been obtained, it is of practical interest to study closely related problems and solutions, closely related in the sense that various coefficients are changed. Sensitivity analysis (Section 5-10) is a tool of use in making such studies. When solutions are obtained on analog computers (Section 5-11), sensitivity analysis is very easy to perform; but both the size of problem treated and the accuracy of solution obtained are limited on an analog computer, more so than on the average general-purpose digital computer.

The first seven sections of this chapter should be read in order. Sufficient background is gained from these first sections to enable the assimilation of Sections 5-8 through 5-13 in any order.

5-2. THE GENERAL PROBLEM AND ITS STANDARD FORM

Particular linear programming problems may appear to be quite different in nature at first glance (as will be observed in the next section and in Section 5-12). Despite this fact, essentially all linear programming problems can be placed in a common (standard) form. Methods of solving a standard form of linear programming problem may be used to solve any linear programming problem provided it is restated in the standard form.

In this book, the standard form adopted for linear programming problems is as follows.

The column matrix $\mathbf{x} = [x_1 \quad x_2 \quad \cdots \quad x_n]'$ of real parameters x_i is to be specified to obtain the maximum of the performance measure P:

$$P = P(\mathbf{x}) = \mathbf{c}'\mathbf{x} \tag{5-1}$$

where $\mathbf{c}' = [c_1 \quad c_2 \quad \cdots \quad c_n]$ is a row matrix, the entries c_i of which are real and are called *value coefficients*; and where the entries of $\mathbf{x}$ are constrained by the following two sets of equations.

$$\mathbf{x} \geq [0 \quad 0 \quad \cdots \quad 0]' \triangleq \mathbf{0} \tag{5-2}$$

and

$$\mathbf{Ax} = \mathbf{b} \tag{5-3}$$

in which $\mathbf{A}$ is an $m \times n$ (m rows by n columns) matrix, $m < n$, and $\mathbf{b}$ is an $m \times 1$ matrix.[1] The matrices $\mathbf{A}$ and $\mathbf{b}$ are expressed in terms of known a_{ij} and b_i real values:

---

[1] Column matrices are sometimes referred to as column vectors and may also be called "points" in multidimensional Euclidean space, depending on the context in which they are used.

$$A = \begin{bmatrix} a_{11} & a_{12} & \cdots & a_{1n} \\ a_{21} & & \ddots & \vdots \\ \vdots & & & \\ a_{m1} & & \cdots & a_{mn} \end{bmatrix} \quad (5\text{-}4)$$

$$b = \begin{bmatrix} b_1 \\ b_2 \\ \vdots \\ b_m \end{bmatrix} \quad (5\text{-}5)$$

5-3. CONVERSION TO THE STANDARD FORM

That many seemingly different types of problems can be placed in the preceding standard form is shown by the following cases.

Case A. (Minimal $P(x)$ desired.)

If the minimum of $P(x)$ is desired instead of the maximum, we need only define $P_1(x)$ by

$$P_1(x) \triangleq -P(x) = -c'x \quad (5\text{-}6)$$

and proceed to obtain the x that maximizes $P_1(x)$. When the maximum of $P_1(x)$ is obtained, the desired minimum of $P(x)$ is found from the elementary relationship $\min P(x) = -\max P_1(x)$.

Case B. (Constraints of the form $d_i \leq x_i$.)

In the standard form, constraints of the type $0 \leq x_i$ are considered. To account for constraints of the type $d_i \leq x_i$, which is equivalent to $0 \leq x_i - d_i$, we define $x_{\alpha i} \triangleq x_i - d_i$ and replace x_i with $x_{\alpha i} + d_i$ in all equations. In the altered equations, $x_{\alpha i} \geq 0$ is a constraint of the form which is taken into account by 5-2.

Case C. (Constraints of the form $d_i \geq x_i$, and x_i not necessarily greater than zero.)

In a manner similar to that of Case B above, define $x_{\beta i} \triangleq d_i - x_i \geq 0$ and replace x_i with $d_i - x_{\beta i}$ in all equations to obtain an altered set of equations in which $x_{\beta i}$ is constrained to be greater than or equal to zero.

Case D. (x_i unrestricted in sign.)

Define $x_{i+} - x_{i-} \triangleq x_i$ where x_{i+} and x_{i-} are both greater than or equal to zero, as required by the standard form, and replace x_i with $x_{i+} - x_{i-}$ in all equations. In the final solution, at least one of x_{i+} and x_{i-} will equal zero, depending on whether the optimal x_i is positive, negative, or zero.

Case E. (Constraints of the form $d_{i1} \leq x_i \leq d_{i2}$.)

This constraint is equivalent to the constraint $0 \leq x_i - d_{i1} \leq d_{i2} - d_{i1}$. Let $x_{\alpha i} = x_i - d_{i1}$, and it is clear that $x_{\alpha i}$ conforms to the constraints of 5-2. Thus, replace x_i with $x_{\alpha i} + d_{i1}$ in all equations in which it appears, but note that $x_{\alpha i}$ must be less than or equal to $d_{i2} - d_{i1}$. The problem this poses can be handled in two ways: The most efficient way is called *upper-bounding* and is presented in Section 5-8; if only a few such constraints exist, we may simply add a *slack variable* (a positive quantity which, literally, takes up the slack) x_{si} to $x_{\alpha i}$ to obtain the following equality:

$$x_{\alpha i} + x_{si} = d_{i2} - d_{i1} \tag{5-7}$$

where both $x_{\alpha i}$ and x_{si} are greater than or equal to zero. Because Equation 5-7 is of the form of those in 5-3, it is adjoined to this set to render the problem in standard form.

Slack variables play a leading role in the cases that follow.

Case F. (Constraints of the form $a_1 x_1 + a_2 x_2 + \cdots + a_k x_k \leq b$.)

This constraint is placed in the form of those in 5-3 by the addition of a unique non-negative slack variable $x_s \geq 0$ to the left-hand member, as follows:

$$a_1 x_1 + a_2 x_2 + \cdots + a_k x_k + x_s = b \tag{5-8}$$

Case G. (Constraints of the form $a_1 x_1 + a_2 x_2 + \cdots + a_k x_k \geq b$.)

To place this constraint in standard form, subtract a non-negative slack variable x_s from the left-hand member of the inequality. This action yields

$$a_1 x_1 + a_2 x_2 + \cdots + a_k x_k - x_s = b \tag{5-9}$$

Case H. (Constraints of the form $b_1 \leq a_1 x_1 + a_2 x_2 + \cdots + a_k x_k \leq b_2$.)

Replace this constraint with two constraint equations which form an equivalent constraint:

$$a_1 x_1 + a_2 x_2 + \cdots + a_k x_k - x_{s1} = b_1 \tag{5-10}$$

$$a_1 x_1 + a_2 x_2 + \cdots + a_k x_k + x_{s2} = b_2 \tag{5-11}$$

where the slack variables x_{s1} and x_{s2} are greater than or equal to zero. Equations 5-10 and 5-11 are of the standard form 5-3.

Sect. 5-3   CONVERSION TO THE STANDARD FORM   197

Case I. (Constraints of the form $|a_1 x_1 + a_2 x_2 + \cdots + a_k x_k| \leq b$.)
This constraint is actually a special case of that given under Case H with $b = b_2 = -b_1$.

Case J. (Minimum of $|P(\mathbf{x})| = |c_1 x_1 + c_2 x_2 + \cdots + c_k x_k|$ desired.)
We solve this problem by solving two ordinary linear programming problems, at least one of which yields the desired solution. The two problems are given in the following subcases.

Subcase J1. Assume that $P(\mathbf{x}) = \mathbf{c}'\mathbf{x}$ is non-negative at the minimum value of $|P(\mathbf{x})|$. The equivalent problem under this condition is to maximize $P_1(\mathbf{x}) = -\mathbf{c}'\mathbf{x}$—as in Case A—subject to the additional constraint $c_1 x_1 + c_2 x_2 + \cdots + c_k x_k - x_s = 0$ where x_s is a unique non-negative slack variable.

Subcase J2. Assume that $P(\mathbf{x})$ is nonpositive at the minimum value of $|P(\mathbf{x})|$. In this case, the equivalent problem is to maximize $P(\mathbf{x}) = \mathbf{c}'\mathbf{x}$ subject to the additional constraint that $\mathbf{c}'\mathbf{x} \leq 0$ or $c_1 x_1 + c_2 x_2 + \cdots + c_k x_k + x_s = 0$ where x_s is a unique non-negative slack variable.

Since either subcase J1 or subcase J2 (or both if optimum $\mathbf{c}'\mathbf{x} = 0$) must apply to the problem posed in Case J, the procedure is to solve both subcase problems and to use the solution to which corresponds the minimum of $|P(\mathbf{x})|$.

Other cases of this type can be reduced to the standard form which is characterized by Equations 5-1, 5-2, and 5-3—see Problems 5.1 through 5.6. Note, however, that many of the cases cited are treated in the literature with less effort when the associated problem exhibits special features.

Example 5-1. Consider the problem of a manager who wishes to schedule most profitably the production of two types of integrated circuit. Three machines are used in the process: machine A, an integrated circuit former with 50 hours of machine-time available; machine B, an incaser with 35 hours of machine-time available; and machine C, a tester and packager with 80 hours of machine-time available. The manufacturing time per lot of each integrated circuit is listed, as follows:

Integrated Circuit	Time on A	Time on B	Time on C
type 1	10 hrs/lot	5 hrs/lot	5 hrs/lot
type 2	5 hrs/lot	5 hrs/lot	15 hrs/lot

Profit on the integrated circuits is $100 per lot for type 1 and $80 per lot for type 2. The manager wishes to maximize the total profit.

These verbal statements must be restated algebraically. Let x_1 equal the lots of type 1 to be produced and x_2 equal the lots of type 2 to be produced. Total profit P is then

$$P = 100x_1 + 80x_2 \quad \text{(dollars)} \tag{5-12}$$

Of course, x_1 and x_2 are restricted to be greater than or equal to zero; and from the available machine-time limitations,

$$10x_1 + 5x_2 \leq 50 \quad \text{(machine A)} \tag{5-13}$$

$$5x_1 + 5x_2 \leq 35 \quad \text{(machine B)} \tag{5-14}$$

$$5x_1 + 15x_2 \leq 80 \quad \text{(machine C)} \tag{5-15}$$

The maximum value of P for this problem is found by using the simplex technique in Section 5-6.

•

Example 5-2. Consider the "curve-tracking" control problem characterized by

$$\frac{dy}{dt} = -ay + m(t) \tag{5-16}$$

where $y(0) \equiv y_0$ is specified and where $m(t)$ is constrained by

$$0 \leq m(t) \leq m_{\max} \tag{5-17}$$

It is desired that a minimum of the performance measure J be attained:

$$J = \int_0^T \gamma(t)[y_d(t) - y(t)] \, dt \tag{5-18}$$

where the response $y(t)$ is constrained to be less than or equal to the desired response $y_d(t)$ at each instant of time. (This is known as overcontrol elimination; whether it is useful or not from the practical standpoint depends on the particular application; constraining $y(t)$ to be less than or equal to $y_d(t)$ does simplify the computations in the linear programming solution.) The terminal time T is fixed, and the function $\gamma(t)$ in Equation 5-18 is selected by the designer to weight the error most heavily at specific critical values of t.

As in Example 5-1, the purpose of this example is to illustrate the steps in the process of reducing the problem to one of linear programming form.

First, Equation 5-16 is replaced by the following discrete approximation (more accurate discrete approximations could be employed with no essential change in the development).

$$\frac{y_{k+1} - y_k}{(T/K)} = -ay_k + m_k \tag{5-19}$$

Sect. 5-3 CONVERSION TO THE STANDARD FORM 199

for $k = 0, 1, 2, \ldots, K - 1$ and where $y_k \triangleq y(kT/K)$ and $m_k \triangleq m(kT/K)$. The larger the value of K in Equation 5-19, the more accurate the approximation—assuming that K is not so large that accumulated round-off errors in the computations are significant. Equation 5-19 is rewritten here in more suitable form:

$$y_{k+1} = \frac{K - aT}{K} y_k + \frac{T}{K} m_k, \quad k = 0, 1, \ldots, K - 1 \quad (5\text{-}20)$$

From this and the given initial condition y_0, the values y_1 through y_K are found in terms of the control sequence. Thus, let $(K - aT)/K \triangleq \beta$ and $(T/K) \triangleq \rho$ to obtain

$$\begin{aligned}
y_1 &= \beta y_0 + \rho m_0 \leq y_{d1} \\
y_2 &= \beta^2 y_0 + \beta \rho m_0 + \rho m_1 \leq y_{d2} \\
y_3 &= \beta^3 y_0 + \beta^2 \rho m_0 + \beta \rho m_1 + \rho m_2 \leq y_{d3} \\
&\vdots \qquad\qquad\qquad\qquad \vdots \\
y_K &= \beta^K y_0 + \rho \sum_{j=0}^{K-1} m_j \beta^{K-1-j} \leq y_{dK}
\end{aligned} \quad (5\text{-}21)$$

The performance measure 5-18 is replaced by the discrete approximation P:

$$P \triangleq \sum_{k=1}^{K} \gamma_k (y_{dk} - y_k) \rho \quad (5\text{-}22)$$

Because ρ, γ_k, and y_{dk} are known values, minimization of the performance measure P of Equation 5-22 can be replaced by the maximization of P_0, where

$$P_0 \triangleq \sum_{k=1}^{K} \gamma_k y_k \quad (5\text{-}23)$$

subject to the constraints previously given. At this point, the problem is in a linear programming form, but the following manipulations result in the elimination of variables and, thereby, reduced computational requirements.

By using the equations listed in 5-21, Equation 5-23 is rewritten as

$$\begin{aligned}
P_0 &= y_0 \sum_{k=1}^{K} \gamma_k \beta^k + \rho m_0 \sum_{k=0}^{K-1} \gamma_{k+1} \beta^k \\
&\quad + \rho m_1 \sum_{k=0}^{K-2} \gamma_{k+2} \beta^k + \cdots + \rho m_{K-1} \sum_{k=0}^{K-K} \gamma_{k+K} \beta^k \\
&= y_0 \sum_{k=1}^{K} \gamma_k \beta^k + \rho \sum_{i=0}^{K-1} \left[m_i \sum_{k=0}^{K-1-i} \gamma_{k+i+1} \beta^k \right]
\end{aligned} \quad (5\text{-}24)$$

Note that the first summation in the right-hand member of 5-24 is a constant. Because of this fact, we need only maximize P_1:

$$P_1 \triangleq \sum_{i=0}^{K-1} c_i m_i \qquad (5\text{-}25)$$

where

$$c_i \triangleq \sum_{k=0}^{K-1-i} \gamma_{k+i+1} \beta^k \qquad (5\text{-}26)$$

In addition to the constraints 5-21, the upper-bound constraints corresponding to 5-17 must be satisfied, i.e., $0 \le m_k \le m_{\max}$ for all k. The number of constraints is therefore $2n$; but n of these, the upper-bound constraints, are rather easily handled by the upper-bounding procedure of Section 5-8. Thus, the problem is reduced to a well-defined linear programming problem.

•

5-4. ANALYTICAL BASIS

5-4a. Prelude

The constraint Equations 5-3 correspond geometrically to hyperplanes (generalized planes in n dimensions) in the n-dimensional space of the x_i's, $i = 1, 2, \ldots, n$. The intersection of these hyperplanes is another hyperplane, one of lower dimension, and the values of **x** which lie on this hyperplane and which satisfy 5-2 form a *convex set* R_c of points, as is shown in the following Section 5-4b. This convex set R_c is generally a polyhedron, and the vertices of R_c have at most m nonzero components, where m is the number of equations in 5-3; this is shown in Section 5-4c. *The vertices of R_c are of interest because the optimal value of P in Equation 5-1 is obtained at a vertex of R_c*, as is shown in Section 5-4d.

In view of the above geometrical interpretation, interest is focused on those solutions of Equations 5-3 which are obtained by setting $n - m$ of the x_i's equal to zero. Such solutions are called *basic solutions*. If the remaining m of the x_i's of a basic solution are forced to be nonzero, the solution is called a *nondegenerate basic solution*; but if one or more of the remaining x_i's can assume the zero value, the solution is called a *degenerate basic solution*. Furthermore, if a basic solution contains no negative x_i's, it is called a *basic feasible solution*—feasible in the sense that it satisfies the constraint set 5-2. A nondegenerate basic feasible solution is necessarily a vertex of R_c.

To obtain a particular basic solution, we first set $n - m$ of the x_i's equal to zero in 5-3. In the **A** matrix of 5-3, those columns associated with the zero x_i's are deleted, and the resulting m equations (which must be independent if the solution is to be nondegenerate) are solved for the remaining m of the x_i's.

In the following subsections, basic solutions of 5-3 are assumed to be nondegenerate unless stated otherwise. This is not too restrictive an assumption. Llewellyn [5.44] and others report that degenerate solutions are rarely, if ever, found in practice. In case one should appear, moreover, we can always perturb the coefficients of the **A** matrix slightly to eliminate it. This process is illustrated in Example 5-3 and is further discussed in Section 5-7a. Wolfe [5.69] gives systematic methods with which degeneracy can be avoided in computer solutions of linear programming problems.

5-4b. Convexity

Consider initially the definition of a *nonconvex set*: A set R_X of points in n-dimensional Euclidean space is a nonconvex set if two points of R_X, say, points $\mathbf{x}_a$ and $\mathbf{x}_b$, exist with the property that at least one of the points on the straight line segment joining $\mathbf{x}_a$ and $\mathbf{x}_b$ is not in R_X. Thus, let $\mathbf{x}_c = \mathbf{x}_c(\alpha)$ denote the straight line segment between the two points $\mathbf{x}_a$ and $\mathbf{x}_b$ in R_X:

$$\mathbf{x}_c = \alpha \mathbf{x}_a + (1 - \alpha)\mathbf{x}_b, \qquad 0 \leq \alpha \leq 1 \tag{5-27}$$

If there exists some α, $0 \leq \alpha \leq 1$, and some $\mathbf{x}_a$ and $\mathbf{x}_b$ in R_X such that the corresponding $\mathbf{x}_c$ is not in R_X, then R_X is nonconvex. But, if no such α, $\mathbf{x}_a$, and $\mathbf{x}_b$ exist, R_X is a convex set, i.e., R_X is a *convex set* if it is not nonconvex.

Examples of convex sets and nonconvex sets are displayed in Figures 5-1 and 5-2 for the case that $n = 2$. Convex sets of the form given in Figure

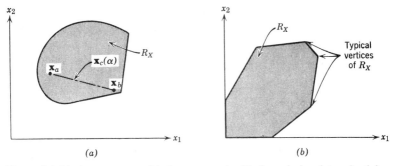

Figure 5-1 (a) A convex set. (b) A convex set with boundaries determined by "hyperplanes."

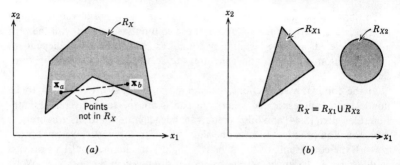

Figure 5-2 (a) A nonconvex set with boundaries determined by hyperplanes. (b) Two convex sets, the union of which is nonconvex.

5-1b are of particular interest in this chapter because the boundaries of such sets are defined by linear algebraic equations.

Let the set of points that satisfy constraints 5-2 and 5-3 be denoted by R_c. To show that R_c is convex, consider first the constraints of 5-2. Given two column matrices $\mathbf{x}_a$ and $\mathbf{x}_b$, each of which has no negative components, it necessarily follows that

$$\mathbf{x}_c = \alpha \mathbf{x}_a + (1 - \alpha)\mathbf{x}_b \quad \text{for } 0 \leq \alpha \leq 1 \tag{5-28}$$

has no negative components.

Consider next the constraints 5-3. Again, given any two column matrices $\mathbf{x}_a$ and $\mathbf{x}_b$, each of which has the property

$$\mathbf{A}\mathbf{x}_a = \mathbf{b} \quad \text{and} \quad \mathbf{A}\mathbf{x}_b = \mathbf{b} \tag{5-29}$$

it follows that

$$\begin{aligned} \mathbf{A}\mathbf{x}_c &= \mathbf{A}[\alpha\mathbf{x}_a + (1 - \alpha)\mathbf{x}_b] \\ &= \alpha\mathbf{A}\mathbf{x}_a + (1 - \alpha)\mathbf{A}\mathbf{x}_b \\ &= \alpha\mathbf{b} + (1 - \alpha)\mathbf{b} = \mathbf{b} \end{aligned} \tag{5-30}$$

Thus R_c is convex.

Note that if the performance measure P in Equation 5-1 is treated as an additional variable, say, $x_{n+1} = P$, Equations 5-1 and 5-3 correspond to hyperplanes in $(n + 1)$-dimensional space. From this point of view, the set R_p of points which satisfy 5-1, 5-2, and 5-3 is also a convex set—as the proof of this fact parallels the proof of the preceding paragraph, it is omitted here. Because R_p is convex, the point in R_p with the greatest (least) value of $x_{n+1} = P$ corresponds to an absolute maximum (minimum) of P, i.e., there exists only one relative maximum (minimum) of P, and this maximum (minimum) is the absolute maximum (minimum).

5-4c. Extreme Point and Verticy Properties

By definition, a point $\mathbf{x}_v$ of a convex set R_X is an *extreme point* if no straight line segment completely in R_X of finite length contains the point $\mathbf{x}_v$ internally; analytically, $\mathbf{x}_v$ is an extreme point of R_X if $\mathbf{x}_c$,

$$\mathbf{x}_c = \beta \mathbf{x}_v + (1 - \beta)\mathbf{x}_a \qquad \text{for } \mathbf{x}_a \neq \mathbf{x}_v \tag{5-31}$$

is not an element of R_X for any $\beta > 1$ and for any $\mathbf{x}_a \in R_X$. In the particular case that the convex set R_X is the previously defined polyhedron R_c, the extreme points of R_c are appropriately called *vertices*.

Theorem 5-1. *The vertices of R_c have at least $n - m$ zero components, where R_c is the set of points which satisfies constraints 5-2 and 5-3.*

Proof: Assume the contrary, i.e., that $\mathbf{x}_q$ is a vertex of R_c which has less than $n - m$ zero components. Pick any $\mathbf{x}_a \in R_c$ which does not equal $\mathbf{x}_q$ but which contains the same zero components. Many such $\mathbf{x}_a$'s exist in general because Equations 5-3 are underspecified when less than $n - m$ x_i's are set equal to zero.

With the above assumptions in mind, consider the column vector $\mathbf{x}_c$:

$$\mathbf{x}_c = \beta \mathbf{x}_q + (1 - \beta)\mathbf{x}_a \qquad \text{for } \beta > 1 \tag{5-32}$$

and note that if β is sufficiently small, but yet greater than one, $\mathbf{x}_c$ is nonnegative because $\mathbf{x}_q$ and $\mathbf{x}_a$ have the same positive components, i.e., $\mathbf{x}_c$ satisfies 5-2. It is equally true that $\mathbf{x}_c$ satisfies 5-3 because

$$\begin{aligned}
\mathbf{A}\mathbf{x}_c &= \mathbf{A}[\beta\mathbf{x}_q + (1 - \beta)\mathbf{x}_a] \\
&= \beta \mathbf{A}\mathbf{x}_q + (1 - \beta)\mathbf{A}\mathbf{x}_a \\
&= \beta \mathbf{b} + (1 - \beta)\mathbf{b} = \mathbf{b}
\end{aligned} \tag{5-33}$$

Thus, $\mathbf{x}_c$ is an element of R_c for some sufficiently small $\beta > 1$, but from the definition of a vertex, this latter statement contradicts the original assumption that $\mathbf{x}_q$ in Equation 5-32 is a vertex of R_c.

On the other hand, if $n - m$ x_i's are zero, the remaining nonzero x_i's, of which there are m, can be determined uniquely in general from the set of m Equations 5-3. If the x_i's of this unique solution are positive, they constitute a *vertex* $\mathbf{x}_v$ because any other $\mathbf{x}_a \in R_c$ cannot contain all of the same zero components. Thus, at least one component of $\mathbf{x}_d$,

$$\mathbf{x}_d \triangleq \beta \mathbf{x}_v + (1 - \beta)\mathbf{x}_a \qquad \text{for } \beta > 1 \tag{5-34}$$

must be negative, i.e., $\mathbf{x}_d \notin R_c$ for any $\mathbf{x}_a \in R_c$ which concludes the proof of Theorem 5-1.

●

5-4d. Optimal P at a Vertex

Theorem 5-2. *Assuming a finite maximum (minimum) of the P of Equation 5-1 exists, it is attained at one (at least) of the vertices of the convex set R_c which is defined by 5-2 and 5-3.*

Proof: Suppose R^* is the set of all points of R_c at which P assumes its maximum (minimum) value P^*. This set R^* is convex because, for $\mathbf{x}_c$ equal to $\alpha \mathbf{x}_a + (1 - \alpha)\mathbf{x}_b$ where $\mathbf{x}_a$ and $\mathbf{x}_b$ are elements of R^*,

$$P(\mathbf{x}_c) = \mathbf{c}'[\alpha \mathbf{x}_a + (1 - \alpha)\mathbf{x}_b]$$
$$= \alpha P^* + (1 - \alpha)P^* = P^* \tag{5-35}$$

Let $\mathbf{x}_v$ be any point of R^* which satisfies the properties of a vertex for R^* and assume that this vertex of R^* is not a vertex of R_c. Then $\mathbf{x}_c = \beta \mathbf{x}_v + (1 - \beta)\mathbf{x}_a$ must be an element of R_c for some $\beta > 1$ and for some $\mathbf{x}_a \in R_c$. Because this $\mathbf{x}_c$ is assumed not to be an element of R^*, but is an element of R_c, it apparently follows that

$$\mathbf{c}'\mathbf{x}_c < P^* \quad \text{(assuming } P^* \text{ is \textit{the} maximum of } P) \tag{5-36}$$

But in truth,

$$\mathbf{c}'\mathbf{x}_c = \beta \mathbf{c}'\mathbf{x}_v + (1 - \beta)\mathbf{c}'\mathbf{x}_a$$
$$= P^* + (\beta - 1)(P^* - \mathbf{c}'\mathbf{x}_a) \geq P^* \tag{5-37}$$

Hence, a contradiction is revealed which can be resolved only if the assumption which follows Equation 5-35 is changed: A vertex of R^* must also be a vertex of R_c.

The practical value of this theorem is that it allows us to restrict the search for the maximum (minimum) of P to a search over the vertices of which there are a finite number.

●

5-5. SIMPLEX ALGORITHM THEORY

Having shown that the search for a maximum of P can be restricted to a search over the vertices of R_c, we find that the effort required to solve the problem is greatly reduced. Still, it may be expensive and time-consuming to test each vertex and to pick the one which yields the optimal P. For example, because $\mathbf{A}$ of Equation 5-3 is an $m \times n$ matrix with $n > m$, the number N of distinct basic solutions, of which the vertices of R_c are some subset, could be as great as $n!/m!(n - m)!$, i.e., as great as the number of n things taken m at a time. If $\mathbf{A}$ is 50×100—which corresponds to a very reasonable linear programming problem for computer solution—N could be as large as $100!/(50!)^2 > 10^{29}$.

What is desired, therefore, is an effective technique or algorithm with which to progress from one basic feasible solution to another basic feasible solution so that the second is guaranteed in advance to have a greater value of P. The simplex technique (also called the simplex algorithm) is such a procedure.

The original form of the simplex algorithm was developed by George B. Dantzig in 1947 and was formally published in 1951 [5.19]. Many variations of the original technique have been developed since, but it is significant that the original simplex algorithm is still the best procedure for the solution of the general linear programming problem when manual computations are used [5.4]. Certain other revised simplex algorithms are computationally advantageous when the solution is effected with a digital computer.

In this section, theoretical justification is presented for the simplex algorithm of linear programming theory.

A few remarks are in order concerning the suitability of the word "simplex" in the name "simplex algorithm." In general mathematical terms, a *k-dimensional* simplex (sometimes called a *closed simplex*) is a set which consists of the following: (1) $k + 1$ linearly independent points $x_1, x_2, \ldots, x_{k+1}$ of a Euclidean space of dimension greater than k; and (2) all points of the type

$$x = \alpha_1 x_1 + \alpha_2 x_2 + \cdots + \alpha_{k+1} x_{k+1} \qquad (5\text{-}38)$$

where

$$\sum_{i=1}^{k+1} \alpha_i = 1 \qquad (5\text{-}39)$$

and where all α_i's are constrained by $\alpha_i \geq 0$.

Several examples of simplexes are depicted in Figure 5-3. Note that the linearly independent points mentioned in part 1 of the preceding definition are vertices of the corresponding simplex. Also, it can easily be shown that any simplex is convex, as defined in Section 5-4. Because these general properties of simplexes are likewise satisfied by the set R_c of points defined by 5-2 and 5-3, the phrases "simplex algorithm" and "simplex technique" are suggestive of the nature of the problem; consequently these names are used interchangeably in the literature in reference to the solution procedure that follows.

To use the simplex technique, an initial vertex (an initial basic feasible solution) must be found. In the cases where slack variables are added to the left-hand member of each pertinent constraint of a given problem to obtain Equations 5-3, the slack variables by themselves (with all other variables equated to zero) form an obvious but usually far-from-optimal basic feasible

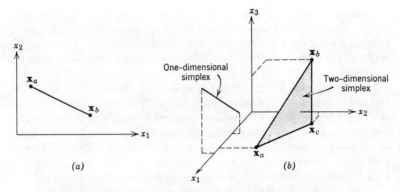

Figure 5-3 (a) A one-dimensional simplex in two-dimensional space. (b) Examples of simplexes in three-dimensional space.

solution. In other cases, more devious techniques may be required to obtain an initial basic feasible solution; these are considered in Section 5-7.

Once an initial vertex is available, one iteration of the simplex technique is used to find a neighboring vertex with which is associated a greater value of P. Relatively little effort is required to effect one iteration. Assuming that the hyperspace R_c is bounded, repetition of the process leads sequentially to the maximum of P because the number of vertices is finite. The number of iterations required is generally between m and $2m$ [5.44].

Notation is introduced as required in the following development. Consider, first, the columns of the **A** matrix in 5-3. The first column is denoted by $\mathbf{a}_1$, the second by $\mathbf{a}_2$, and so on to $\mathbf{a}_n$. Equations 5-3 can now be rewritten in the form

$$x_1\mathbf{a}_1 + x_2\mathbf{a}_2 + \cdots + x_m\mathbf{a}_m + \cdots + x_n\mathbf{a}_n = \mathbf{b} \qquad (5\text{-}40)$$

Assume that an initial vertex is found on the basis of m of the x_i's. These m positive x_i's are labeled $x_1^1, x_2^1, \ldots, x_m^1$ where the superscript 1's denote that these x_i's are associated with the first vertex. Note that x_i^1 *need not* be identified with x_i. After $(N-1)$ iterations of the simplex algorithm, the Nth vertex $\mathbf{x}^N$ is obtained. For this Nth vertex $\mathbf{x}^N$, the value of $P(\mathbf{x}^N)$ of Equation 5-1 is expressed by

$$\begin{aligned} P(\mathbf{x}^N) &= (\mathbf{c}^N)'\mathbf{x}^N \\ &= c_1^N x_1^N + c_2^N x_2^N + \cdots + c_m^N x_m^N \end{aligned} \qquad (5\text{-}41)$$

where c_i^N is one of the original value coefficients, the one associated with x_i^N. The parameters x_{m+1}^N through x_n^N are zero, and therefore are not entered in Equation 5-41.

For the Nth vertex, 5-3 assumes the form

$$x_1^N \mathbf{a}_1^N + x_2^N \mathbf{a}_2^N + \cdots + x_m^N \mathbf{a}_m^N = \mathbf{b}^N \tag{5-42}$$

where $\mathbf{a}_i^N$ is the column of $\mathbf{A}$ associated with x_i^N. The column matrices $\mathbf{a}_1^N$ through $\mathbf{a}_m^N$, which are also referred to as vectors, are assumed to be linearly independent; that is, these vectors form a basis for an m-dimensional space of reals, and therefore any m-dimensional real vector can be expressed as a linear combination of the $\mathbf{a}_i^N$ vectors, $i = 1, 2, \ldots, m$. In particular, the column matrix $\mathbf{a}_j^N$ is expressible in the form

$$\mathbf{a}_j^N = y_{1j}^N \mathbf{a}_1^N + y_{2j}^N \mathbf{a}_2^N + \cdots + y_{mj}^N \mathbf{a}_m^N \tag{5-43}$$

for $j = m + 1, m + 2, \ldots, n$ where the y_{ij}^N's are appropriate constants. By using these same constants y_{ij}^N, the scalars s_j^N are defined for $j = m + 1, m + 2, \ldots, n$ as follows:

$$s_j^N \triangleq c_1^N y_{1j}^N + c_2^N y_{2j}^N + \cdots + c_m^N y_{mj}^N \tag{5-44}$$

where the c_i^N's are the same as those found in 5-41. The importance of this definition will soon be apparent.

Suppose θ^N is a positive parameter, and suppose we multiply Equations 5-43 and 5-44 by θ^N and subtract the resulting equations from 5-42 and 5-41, respectively, to obtain

$$(x_1^N - \theta^N y_{1j}^N)\mathbf{a}_1^N + (x_2^N - \theta^N y_{2j}^N)\mathbf{a}_2^N + \cdots + (x_m^N - \theta^N y_{mj}^N)\mathbf{a}_m^N + \theta^N \mathbf{a}_j^N = \mathbf{b}^N \tag{5-45}$$

and

$$(x_1^N - \theta^N y_{1j}^N)c_1^N + \cdots + (x_m^N - \theta^N y_{mj}^N)c_m^N + \theta^N c_j^N = P(\mathbf{x}^N) + \theta^N(c_j^N - s_j^N) \tag{5-46}$$

for $j = m + 1, m + 2, \ldots, n$.

Next, suppose there exists a j for which $c_j^N - s_j^N > 0$. For this value of j, θ^N is assigned according to the following equation:

$$\theta^N = \underset{i, y_{ij}^N > 0}{\text{minimum}} \frac{x_i^N}{y_{ij}^N} \triangleq \frac{x_u^N}{y_{uj}^N} \tag{5-47}$$

This process of selecting θ^N is known as *the θ rule*. The value of θ^N obtained thereby reduces the coefficient[2] of $\mathbf{a}_u^N$ to zero in Equation 5-45 and results in the introduction of a new positive $x_u^{N+1} = \theta^N$. The net result is a new vertex of R_c, a vertex $\mathbf{x}^{N+1}$ which by Equation 5-46 has a new, improved value of P:

$$P(\mathbf{x}^{N+1}) = P(\mathbf{x}^N) + \theta^N(c_j^N - s_j^N), \quad c_j^N - s_j^N > 0 \tag{5-48}$$

---

[2] If two or more values of i yield the same minimum value of θ^N, a degenerate basic feasible solution is obtained for $\mathbf{x}^{N+1}$. The procedure used in such a case is given under Case C in Section 5-6.

In general, there may be several values of j which satisfy the condition assumed after Equation 5-46. It would be plausible to pick the particular j which yields the greatest increase in P, i.e., to pick j so as to maximize the increase ΔP in P,

$$\max \Delta P = \max_j \min_i \frac{x_i^N}{y_{ij}^N}(c_j^N - s_j^N), \quad c_j^N - s_j^N \text{ and } y_{ij}^N > 0 \quad (5\text{-}49)$$

This procedure is not followed in the simplex technique. Instead, we simply *select j on the basis that the difference $c_j^N - s_j^N$ is maximized.* This procedure is sometimes referred to as the *rule of steepest ascent*, because the rate of change in P is maximized by its use.

After a particular value of j is selected, the parameter θ^N is computed from the equation

$$\theta^N = \underset{i, y_{ij}^N > 0}{\operatorname{minimum}} \frac{x_i^N}{y_{ij}^N} = \frac{x_u^N}{y_{uj}^N} \quad (5\text{-}50)$$

and the components of the new, improved vertex $\mathbf{x}^{N+1}$ are obtained from Equation 5-45 as follows:

$$\left.\begin{aligned}
x_1^{N+1} &= x_1^N - \theta^N y_{1j}^N & (\mathbf{a}_1^{N+1} &= \mathbf{a}_1^N) \\
x_2^{N+1} &= x_2^N - \theta^N y_{2j}^N & (\mathbf{a}_2^{N+1} &= \mathbf{a}_2^N) \\
&\vdots & &\vdots \\
x_{u-1}^{N+1} &= x_{u-1}^N - \theta^N y_{(u-1)j}^N & (\mathbf{a}_{u-1}^{N+1} &= \mathbf{a}_{u-1}^N) \\
x_u^{N+1} &= \theta^N & (\mathbf{a}_u^{N+1} &= \mathbf{a}_j^N) \\
x_{u+1}^{N+1} &= x_{u+1}^N - \theta^N y_{(u+1)j}^N & (\mathbf{a}_{u+1}^{N+1} &= \mathbf{a}_{u+1}^N) \\
&\vdots & &\vdots \\
x_m^{N+1} &= x_m^N - \theta^N y_{mj}^N & (\mathbf{a}_m^{N+1} &= \mathbf{a}_m^N)
\end{aligned}\right\} \quad (5\text{-}51)$$

and

$$x_{m+1}^{N+1} = x_{m+2}^{N+1} = \cdots = x_n^{N+1} = 0 \quad (5\text{-}52)$$

At this point, the process could be repeated, provided the optimum is not obtained at the $(N + 1)$st vertex—the optimum is obtained when $c_j^{N+1} - s_j^{N+1} \leq 0$ for all j. However, rather than use the simplex technique exactly as outlined above, which is somewhat awkward, the process is greatly simplified if the **b** column and the columns of the **A** matrix are appropriately modified from one vertex to the next. The modification is easily effected as shown next.

To initiate the process, suppose that the initial matrix equation $\mathbf{A}^1 \mathbf{x}^1 = \mathbf{b}^1$ is of the form

$$\begin{bmatrix} 1 & 0 & 0 & 0 & \cdots & 0 & a^1_{1,m+1} & \cdots & a^1_{1,n} \\ 0 & 1 & 0 & & & & a^1_{2,m+1} & & \\ 0 & & \cdot & & & & \vdots & & \\ \vdots & & & \cdot & & & \vdots & & \vdots \\ & & & & \cdot & 0 & & & \\ 0 & \cdots & & 0 & 0 & 1 & a^1_{m,m+1} & \cdots & a^1_{m,n} \end{bmatrix} \begin{bmatrix} x_1^1 \\ x_2^1 \\ \vdots \\ x_m^1 \\ \vdots \\ x_n^1 \end{bmatrix} = \begin{bmatrix} b_1^1 \\ b_2^1 \\ \vdots \\ b_m^1 \end{bmatrix} \quad (5\text{-}53)$$

in which $x^1_{m+1} = x^1_{m+2} = \cdots = x_n^1 = 0$, and in which all b_i^1's are greater than zero. If $\mathbf{Ax} = \mathbf{b}$ is not initially in this form, it can be reduced to this form by one of several techniques (see Section 5-7). The convenience of this form lies in the fact that $y_{ij}^1 = a_{ij}^1$, as is observed by comparing Equations 5-43 and 5-53 with $N = 1$. These values of y_{ij}^1 are used in conjunction with 5-44 to obtain s_j^1's in a straightforward manner. In general,

$$s_j^N = c_1^N a_{1j}^N + c_2^N a_{2j}^N + \cdots + c_m^N a_{mj}^N \quad (5\text{-}54)$$

Thus, all the factors which are required to determine the next improved vertex are at hand. Once the improved vertex is found, simple linear operations on the matrix equation are used to reduce it to the form of Equation 5-53, and the process is repeated as necessary. At the Nth vertex, the equations assume the form

$$\begin{bmatrix} 1 & 0 & 0 & \cdots & 0 & a^N_{1,m+1} & \cdots & a^N_{1,n} \\ 0 & 1 & & & & & & \\ \vdots & & \ddots & & \vdots & \vdots & & \vdots \\ & & & & 0 & & & \\ 0 & \cdots & & 0 & 1 & a^N_{m,m+1} & \cdots & a^N_{m,n} \end{bmatrix} \begin{bmatrix} x_1^N \\ x_2^N \\ \vdots \\ x_n^N \end{bmatrix} = \begin{bmatrix} b_1^N \\ b_2^N \\ \vdots \\ b_m^N \end{bmatrix} \quad (5\text{-}55)$$

For manual computation, the individual steps of the simplex technique are most easily effected when the matrix coefficients are placed in a tabular array which is called a simplex tableau. As will be observed in the next section, the rules for finding the entries in the simplex tableau stem directly from the preceding theory.

5-6. SIMPLEX ALGORITHM MECHANICS: THE SIMPLEX TABLEAU

The simplex tableau (the plural of which is tableaux) is simply a tabular way of effecting the details of the simplex technique which is derived in the preceding section. The essential features of the simplex tableau are illustrated by the following example.

It is desired to maximize P:

$$P = c_1 x_1 + c_2 x_2 \qquad (5\text{-}56)$$

subject to the constraints

$$\left.\begin{aligned} a_{11}x_1 + a_{12}x_2 &\leq b_1 \\ a_{21}x_1 + a_{22}x_2 &\leq b_2 \\ a_{31}x_1 + a_{32}x_2 &\leq b_3 \end{aligned}\right\} \qquad (5\text{-}57)$$

and

$$x_1 \text{ and } x_2 \geq 0 \qquad (5\text{-}58)$$

where b_1, b_2, and b_3 are assumed to be positive in this example.

Slack variables x_3, x_4, and x_5 are added to the left-hand members of the constraints in 5-57 to obtain

$$\begin{bmatrix} a_{11} & a_{12} & 1 & 0 & 0 \\ a_{21} & a_{22} & 0 & 1 & 0 \\ a_{31} & a_{32} & 0 & 0 & 1 \end{bmatrix} \begin{bmatrix} x_1 \\ x_2 \\ x_3 \\ x_4 \\ x_5 \end{bmatrix} = \begin{bmatrix} b_1 \\ b_2 \\ b_3 \end{bmatrix} \qquad (5\text{-}59)$$

and these data, as well as those contained in Equation 5-56, are arranged in a simplex tableau (Table 5-1). Note that the value coefficients c_3, c_4, and c_5 are assigned zero values in Table 5-1 because the slack variables x_3, x_4, and x_5 do not contribute to the performance measure P.

TABLE 5-1. A SIMPLEX TABLEAU

	$c_j \rightarrow$		c_1	c_2	0	0	0		
(Current vertex) x_i	c_i	**b**	**a**$_1$	**a**$_2$	**a**$_3$	**a**$_4$	**a**$_5$	Check	Possible θ's
x_3	0	b_1	a_{11}	a_{12}	1	0	0		
x_4	0	b_2	a_{21}	a_{22}	0	1	0		
x_5	0	b_3	a_{31}	a_{32}	0	0	1		
Current $P = 0$	s_j	0	0	0	0	0			
	$c_j - s_j$		c_1	c_2	0	0	0		

Sect. 5-6   SIMPLEX ALGORITHM MECHANICS   211

The initial vertex (the initial basic feasible solution) is selected in Table 5-1 to correspond to the *unit vectors* $\mathbf{a}_3$, $\mathbf{a}_4$, and $\mathbf{a}_5$; that is, the initial vertex is at $x_3 = b_1$, $x_4 = b_2$, $x_5 = b_3$, $x_1 = 0$, and $x_2 = 0$. In general, the vertex of any tableau corresponds to positive unit vectors from the columns of the modified **A** matrix. Also, all **b**-column entries *must be positive* in tableaux if the simplex algorithm is to be applied in a straightforward manner. If a given **b**-column entry is nonpositive, appropriate steps (which are described toward the end of this section and in the next) must be taken to satisfy the above-noted requirement.

The current value of P in Table 5-1 is found by summing 0 times b_1 plus 0 times b_2 plus 0 times b_3 in the tableau (the sum of the products of the paired numbers in the c_i and **b** columns). The values of s_j^1 are computed by using Equation 5-54, i.e.,

$$s_j^1 = c_1^1 a_{1j}^1 + c_2^1 a_{2j}^1 + c_3^1 a_{3j}^1 \tag{5-60}$$

where $c_1^1 = c_3 = 0$, $c_2^1 = c_4 = 0$, and $c_3^1 = c_5 = 0$. In this case, therefore, all s_j^1's are zero, and the last two rows of the first tableau are correct as shown in Table 5-1.[3]

Suppose c_1 is greater than $c_2 > 0$. In this case, the $\mathbf{a}_1$ column is the *key column in the first tableau* because it possesses the maximum $c_j - s_j > 0$, and x_1 is to be assigned a positive value at the next vertex. The entries in the key column are used to compute possible θ's as indicated by Equation 5-50. If all a_{i1}'s are greater than zero, the possible θ's are as follows: b_1/a_{11}, the entry in the first row of the current **b** column divided by the entry in the first row of the key column; b_2/a_{21}, the entry in the second row of the current **b** column divided by the entry in the second row of the key column; and b_3/a_{31}, the entry in the third row of the current **b** column divided by the entry in the third row of the key column. On the other hand, if a_{21} were negative or zero, only b_1/a_{11} and b_3/a_{31} would be possible θ's.

Based on 5-50, θ^1 is selected such that

$$\theta^1 = \underset{a_{i1} > 0}{\text{minimum}} \; (b_1/a_{11}, b_2/a_{21}, b_3/a_{31}) \tag{5-61}$$

and the row which corresponds to θ^1 is designated as the *key row* of the first tableau. The particular x_i^1 in the left-hand column of the key row is reduced to zero at the next vertex, to be displaced by a positive value of $x_1 = x_i^2$ in this case.

---

[3] Superscripts are not used in tableaux because the arrangement of the variables in the tableaux and the order of the tableaux are clearly indicative of the order of the vertices obtained. The left-hand x_i and c_i columns are in fact x_i^N and c_i^N columns. Note that the order of the $\mathbf{a}_j$ columns, and of the associated x_j's, remains unaltered throughout the sequence of tableaux.

To proceed at this point, it is convenient to assign numbers to the a_{ij}'s, b_i's, and c_i's. Let $a_{11} = 10$, $a_{12} = 5$, $a_{21} = 5$, $a_{22} = 5$, $a_{31} = 5$, $a_{32} = 15$, $b_1 = 50$, $b_2 = 35$, $b_3 = 80$, $c_1 = 10$, and $c_2 = 8$ (see Example 5-1). With these specifications, the first tableau (Table 5-1) assumes the form shown in Table 5-2.

TABLE 5-2. A PARTICULAR SIMPLEX TABLEAU

	$c_j \rightarrow$		10	8	0	0	0			
(Current vertex)				(Current)					Possible θ's	
x_i	c_i	b	a_1	a_2	a_3	a_4	a_5	Check		
x_3	0	50	⑩	5	1	0	0	66	5	Key row
x_4	0	35	5	5	0	1	0	46	7	
x_5	0	80	5	15	0	0	1	101	16	
Current $P = 0$ / s_j		0	0	0	0	0				
$c_j - s_j$		10	8	0	0	0				

Key column

The numbers in the *check column* of this tableau require an explanation. As its name implies, the check column affords a check on certain of the computations that are to follow. A given entry in the check column is obtained by summing all numbers in the row to the left of the check number within the double lines—more on this later.

Note that the x_i associated with the key column is the x_i which is to be increased from a zero value at the old vertex to a positive value at the new vertex; in this case x_1 is to be greater than zero at the second vertex. Also note that the key row dictates which of the original nonzero vertex components is to be reduced to zero at the new vertex; the x_3 in the current-vertex column of the key row indicates that the variable x_3 is to be reduced to zero at the second vertex.

The entry which appears at the intersection of the key row and the key column is the *pivotal entry*. This pivotal entry is used to find the coordinates of the next vertex in two easy steps: *Rule 1*, divide all entries in the key row within the double lines by the pivotal entry to obtain an altered key row; and *Rule 2*, subtract an appropriate multiple of the altered key row from

Sect. 5-6   SIMPLEX ALGORITHM MECHANICS   213

each of the other rows within the double lines so as to reduce all entries (except the pivotal entry) to zero in the key column within the double lines. The results of these two operations are placed in the next tableau (Table 5-3).

TABLE 5-3. SECOND TABLEAU: AN UPDATED VERSION OF THE TABLEAU IN TABLE 5-2

	$c_j \rightarrow$	10	8	0	0	0			
(Current vertex) x_i	c_i	b	$\mathbf{a}_1$	(Current) $\mathbf{a}_2$	$\mathbf{a}_3$	$\mathbf{a}_4$	$\mathbf{a}_5$	Check	Possible θ's
x_1	10	5	1	0.5	0.1	0	0	6.6	10
x_4	0	10	0	(2.5)	-0.5	1	0	13	4	Key row
x_5	0	55	0	12.5	-0.5	0	1	68	4.4
Current $P = 50$ / s_j		10	5	1	0	0			
/ $c_j - s_j$		0	3	-1	0	0			

Key column

Note that the entry in a given row of the check column in the second tableau is obtained by exactly the same linear combinational operations as used to obtain the other entries in the same row within the double lines. Thus, to check the accuracy of these linear combinational operations, the numbers in a given row within the double lines to the left of the check entry are summed, and the summed result is compared with that of the check-column entry. If the two numbers are not equal, an arithmetic error has been made.

To find a third vertex with a greater value of P, the s_j's and $(c_j - s_j)$'s are computed as before: $s_1 = 10 \times 1 + 0 \times 0 + 0 \times 0$; $s_2 = 10 \times 0.5 + 0 \times 2.5 + 0 \times 12.5$; $s_3 = 10 \times (0.1) + 0 \times (-0.5) + 0 \times (-0.5)$; etc. At this point, $j = 2$ yields the only $c_j - s_j$ which is greater than zero in Table 5-3. The corresponding column $\mathbf{a}_2$ is therefore the key column of the second tableau. Possible θ's are generated by dividing greater-than-zero entries in the key column into corresponding entries in the current **b** column, with the results as shown in the second tableau of Table 5-3. The minimal θ is $\theta = 4$, and therefore the corresponding center row is the key row of the second tableau. The fact that the key column is $\mathbf{a}_2$ dictates that the variable

x_2 is positive at the third vertex; the fact that x_4 is in the current-vertex column of the key row dictates that the slack variable x_4 is reduced to zero in obtaining the third vertex. Rules 1 and 2, given just prior to the second tableau (Table 5-3), are used again, this time on the second tableau, to obtain the third tableau (Table 5-4).

TABLE 5-4. THIRD TABLEAU: AN UPDATED VERSION OF THE TABLEAU IN TABLE 5-3

	$c_j \rightarrow$		10	8	0	0	0		
x_i	c_i	b	a_1	a_2	a_3	a_4	a_5	Check	θ's
x_1	10	3	1	0	0.2	-0.2	0	4	
x_2	8	4	0	1	-0.2	0.4	0	5.2	
x_5	0	5	0	0	2	-5	1	3	
Current $P = 62$ / s_j		10	8	0.4	1.2	0		
$c_j - s_j$		0	0	-0.4	-1.2	0		

Because all $c_j - s_j$ are less than or equal to zero in the third tableau, the optimum has been attained, and the results are read directly from the tableau: $P^* = 62$ which results from setting $x_1 = 3$, $x_2 = 4$, $x_3 = 0$, $x_4 = 0$, and $x_5 = 5$.

It is clear that the same approach as outlined above can be used for larger tableaux. There are, however, several critical cases which require investigation.

Case A. (Unbounded solution.)

Suppose that at some point in the computations a basic feasible solution is found with the following property: for some $x_j^N = 0$, $c_j^N - s_j^N$ is positive, and all elements of the corresponding $\mathbf{a}_j^N$ are nonpositive; *then there is no upper bound on the value of the performance measure P*. To prove this statement, consider Equations 5-45 and 5-46. Observe that all x_i^{N+1}'s, which are the coefficients of the $\mathbf{a}_i^N$'s in Equation 5-45, remain non-negative for arbitrarily large values of θ^N because the a_{ij}^N's = y_{ij}^N's are assumed to be nonpositive in this case. Thus, θ^N can be made as large as desired, which means that the performance measure $P(\mathbf{x}^{N+1}) = P(\mathbf{x}^N) + \theta^N(c_j^N - s_j^N)$ could be made as large as desired in Equation 5-48.

Case B. (Maximal $c_j - s_j$ not unique.)

If it should occur that two (or more) values of $c_j - s_j$ are equal and larger than any other $(c_j - s_j) > 0$, we must choose between the two in selecting the key column. Actually, either may be selected with no deleterious effects, but an overall solution may be obtained quicker if the choice is made by computing possible θ's for each of the columns in question and by using that value of j with which is associated the largest actual value of θ. In this way, a greater increase in P is obtained, as is indicated by Equation 5-48.

Case C. (Avoiding degeneracy resulting from equal θ's.)

If two (or more) values of i result in θ^N,

$$\theta^N = \minimum_{i, a_{ij}^N > 0} (x_i^N / a_{ij}^N) \qquad (5\text{-}62)$$

then two (or more) of the nonzero x_i's at the current vertex are reduced to zero at the next vertex (observe Equation 5-45), whereas only one of the x_i's which is zero at the current vertex is made positive at the next vertex. Hence, more than $n - m$ x_i's will be zero at the next vertex, and this corresponds to a degenerate basic feasible solution. To avoid the difficulties stemming from this occurrence, we need only perturb the **b**-column entry (picked at random) which corresponds to one of these θ's by an amount $-\epsilon$ where ϵ is a small positive number. The particular θ^N that results will be a unique minimum, and the simplex technique is applied as before. The details of this procedure are illustrated in the following example.

Example 5-3. Given:

$$x_1 + x_2 + 2x_3 \leq 5 \qquad (5\text{-}63)$$

$$3x_1 + x_2 + 4x_3 \leq 10 \qquad (5\text{-}64)$$

$$x_1, x_2, \text{ and } x_3 \geq 0 \qquad (6\text{-}65)$$

Maximize:

$$P = x_1 + 1.5x_2 + 2x_3 \qquad (5\text{-}66)$$

The steps in the solution are contained in the following tableaux (Table 5-5).

•

5-7. INITIALIZING AND SCALING

5-7a. Avoiding Initial Degeneracy

In Case C of Section 5-6, a method is given with which we can avoid degeneracy encountered in progressing from one basic feasible solution to another. If the initial basic feasible solution is degenerate, however, a slightly different approach must be used.

TABLE 5-5. SOLUTION TABLEAUX FOR EXAMPLE 5-3

(Current vertex) x_i	c_i	c_j → b	1 a_1	1.5 a_2	2 a_3	Slack variables 0 a_4	0 a_5	Check	Possible θ's
x_4	0	$5 - \epsilon$	1	1	(2)	1	0	$10 - \epsilon$	$2.5 - 0.5\epsilon$
x_5	0	10	3	1	4	0	1	19	2.5
		$P = 0$ / s_j	0	0	0	0	0		
		$c_j - s_j$	1	1.5	2	0	0		
x_3	2	$\dfrac{5-\epsilon}{2}$	0.5	(0.5)	1	0.5	0	$\dfrac{10-\epsilon}{2}$	5
x_5	0	2ϵ	1	-1	0	-2	1	$-1 + 2\epsilon$	—
		$P = 5 - \epsilon$ / s_j	1	1	2	1	0		
		$c_j - s_j$	0	0.5	0	-1	0		
x_2	1.5	$5 - \epsilon$	1	1	2	1	0	$10 - \epsilon$	
x_5	0	$5 + \epsilon$	2	0	2	-1	1	$9 + \epsilon$	
		max $P = 7.5$ / s_j	1.5	1.5	3	1.5	0		
		$c_j - s_j$	-0.5	0	-1	-1.5	0	All nonpositive	

The initial basic feasible solution is degenerate if and only if one or more entries in the initial **b** column are zero. A negative entry in the **b** column should be converted to a positive entry by reversing the signs of all coefficients in the corresponding equation, and therefore such entries cause no particular difficulties.

One procedure for avoiding initial degeneracy is as follows. Assume that k of the entries in the initial **b** column are zero. Select a small positive number ϵ on the basis that $k\epsilon \ll 1$. Using this value of ϵ, replace the first zero entry with ϵ, replace the second zero entry with 2ϵ, replace the third zero entry with 3ϵ, and so on. The resulting initial basic feasible solution is then nondegenerate. The details of this procedure are illustrated in Example 5-4 which follows Section 5-7c.

5-7b. Generating an Initial Basic Feasible Solution

When some of the constraint equations are originally in equality form or when some of the equality constraint equations have been obtained by the process of subtracting slack variables from left-hand members, an initial basic feasible solution will not be clearly evident. There is a straightforward way, however, to generate one in every case.

Consider Equations 5-3 and assume that appropriate steps have been taken (Section 5-7a) to make all entries of the **b** column positive. If a particular equation in 5-3 does not contain a unique slack variable with positive coefficient, the procedure is to add a unique *artificial variable* to the left-hand member of that equation. Of course, if any of the artificial variables happened to be nonzero in the final solution, the desired equality constraints (those existing before the addition of artificial variables) would not be satisfied. To avoid the possibility of a nonzero artificial variable in the final solution, the quantity M times the sum of the artificial variables is subtracted from the original performance measure P. This procedure, first suggested by Charnes [5.15], results in an augmented performance measure P_a:

$$P_a = P - M \sum_k x_{ak} \tag{5-67}$$

where the x_{ak}'s are artificial variables and where M is much greater than the magnitudes of other c_i's. The value coefficient of each artificial variable x_{ak} is $-M$ and is called a *penalty coefficient* for the obvious reason that any feasible solution which contains a positive x_{ak} is so severely penalized that it cannot possibly be the optimal solution.

The initial basic feasible solution which results from this process consists of the artificial variables and the slack variables with positive coefficients equated to the corresponding **b**-column entries, with all other x_i's equated to zero. The augmented performance measure P_a is used in place of P to assure eventual reduction of the artificial variables to zero. If for any reason these cannot be driven to zero in a given problem, the original constraints imposed on the problem are inconsistent.

If a given problem contains many variables and constraints, the use of a large M as described above may lead to round-off errors in key computations, and incorrect answers could thereby result. Another method of treating artificial variables, one which avoids the complications developed by large

M's, is credited to Dantzig, Orden, and others [5.23]. The method is called the "two-phase" method. One version of the method is as follows.

In Phase I, value coefficients of -1, rather than $-M$, are assigned to all artificial variables. To all other variables, value coefficients of zero are assigned, regardless of the value coefficients in the original problem. The modified performance measure P_{av} equals the negative of the sum of the artificial variables, and its minimal value is zero, which value is attained when all artificial variables are driven to zero by use of the standard simplex technique.

At the end of Phase I, three possibilities exist:

1. max $P_{av} < 0$, in which case no feasible solution exists for the original problem.
2. max $P_{av} = 0$, and the final **b**-column entries are associated with non-artificial variables. This is a basic feasible solution to the original problem.
3. max $P_{av} = 0$, but one or more entries in the final **b** column are zero and are equal to artificial variables. This is also a basic feasible solution to the original problem, but there may be redundancy in the original constraint equations. (See Hadley [5.31] for an extensive treatment of this possibility.)

If possibility 1 applies, no feasible solution exists for the original problem. Otherwise, if possibility 2 applies, Phase II is brought to bear.

The final tableau of Phase I is used to start Phase II with the following modifications (only possibility 2 is considered here): the value coefficients of the original problem are inserted appropriately into the tableau; the columns corresponding to the artificial variables may be deleted from the tableau; and new values of $c_j - s_j$ are computed and are inserted in place of the old values. The remainder of Phase II consists of determining the optimal value of the original performance measure by use of the simplex technique.

5-7c. Scaling

A linear programming problem is scaled correctly when essentially all of the nonzero entries (exceptions to be noted later) in the **A**, **b**, and **c** matrices are of the same order of magnitude. Because the coefficients of the nonzero variables in the basic feasible solution are unity, the order of 10^0 is used for all coefficients.

To attempt to solve a linear programming problem without scaling is to run the risk of introducing round-off errors which alter the original problem and which may result in a false optimum being designated. Round-off errors are most critical in two particular simplex computations: (1) if certain non-zero numbers in a given row of a simplex tableau are significantly greater in magnitude than others, the number in the check column may be dominated by the large numbers—round-off errors in the large numbers may be greater than the small numbers in which case the check column is useless; and (2) if certain nonzero numbers in a column of the tableau are significantly greater in magnitude than others in the same column, the corresponding values of s_j (by the use of which $c_j - s_j$ and the key column are determined) may be dominated by the large numbers—again, seemingly small round-off errors in the large numbers may eventually mask the effect of the small numbers and thereby lead to inaccurate if not completely misleading results.

To scale a linear programming problem, two different operations are used. One operation is to multiply both sides of a particular equation from the set 5-1 and 5-3 by an appropriate positive number, thereby modifying the coefficients in the equation—this is called a *row-scaling operation*. The second operation is to replace the variable x_i, as it appears in both 5-1 and 5-3, with the product of α_i and a scaled x_i where α_i is an appropriate positive number—this is called a *column-scaling operation*.

A systematic way to apply the above two operations is by using the following *scaling table* (Table 5-6a) of coefficients from Equations 5-1 and 5-3.

TABLE 5-6a. INITIAL SCALING TABLE

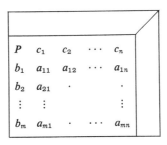

The objective in scaling is to reduce all nonzero coefficients to the same order of magnitude, if possible, by allowable row-scaling and column-scaling operations. Toward this end, row-scaling operations are performed initially with the sole purpose of making the nonzero members in each column of the same order of magnitude (not necessarily of magnitude 10^0). The result is the following updated version of Table 5-6a. The constants $\beta_p, \beta_1, \beta_2, \ldots, \beta_m$ are selected in Table 5-6b to obtain the same order of magnitude within the individual columns.

TABLE 5-6b. SCALING COMPLETED BY ROWS

$\beta_p P$	$\beta_p c_1$	$\cdots$	$\beta_p c_n$	β_p
$\beta_1 b_1$	$\beta_1 a_{11}$	$\cdots$	$\cdot$	β_1
$\vdots$	$\vdots$		$\vdots$	$\vdots$
$\beta_m b_m$	$\cdot$	$\cdots$	$\beta_m a_{mn}$	β_m

The second step is to apply column-scaling operations on Table 5-6b to reduce the nonzero members in each row to the same order of magnitude, order of magnitude 10^0. The result is updated Table 5-6c.

TABLE 5-6c. SCALING COMPLETED BY ROWS AND COLUMNS

α_p	α_1	α_2	$\cdots$	α_n	
$\alpha_p \beta_p P$	$\alpha_1 \beta_p c_1$	$\alpha_2 \beta_p c_2$	$\cdots$	$\alpha_n \beta_p c_n$	β_p
$\alpha_p \beta_1 b_1$	$\alpha_1 \beta_1 a_{11}$	$\cdot$	$\cdots$	$\cdot$	β_1
$\vdots$	$\vdots$			$\vdots$	$\vdots$
$\alpha_p \beta_m b_m$	$\cdot$	$\cdot$	$\cdots$	$\alpha_n \beta_m a_{mn}$	β_m

The coefficients in Table 5-6c are used in the simplex tableau in place of the original coefficients. After finding the optimal solution in terms of the scaled coefficients, the actual solution is found by use of straightforward conversion equations:

$$P = (\text{scaled } P)/\alpha_p \beta_p \qquad (5\text{-}68)$$

and

$$x_i = (\text{scaled } x_i)(\alpha_i/\alpha_p) \quad \text{for } i = 1, 2, \ldots, n \qquad (5\text{-}69)$$

Three types of coefficients should be ignored (left unchanged) in the scaling operation. One type is obviously a coefficient of zero which is unaltered by the scaling operations. A second type is an ϵ coefficient entered in the **b** column to avoid initial degeneracy; the ϵ coefficient should be treated as zero as far as scaling is concerned. The third type is a $-M$ coefficient which acts as the penalty coefficient for an artificial variable.

Example 5-4. Consider the set of constraints:

$$\left. \begin{array}{r} -20x_1 + 100x_2 + 400x_3 \geq 0 \\ x_1 - 10x_2 + 100x_3 \leq 0 \\ 100x_1 + 1{,}000x_2 + 8{,}000x_3 \leq 50{,}000 \end{array} \right\} \qquad (5\text{-}70)$$

Sect. 5-7   INITIALIZING AND SCALING   221

And consider the performance measure P:

$$P = 0.2x_1 + x_2 + 4x_3 \quad (x_1, x_2, \text{ and } x_3 \geq 0) \quad (5\text{-}71)$$

The performance measure P is to be maximized subject to satisfaction of the constraints.

With the appropriate addition of slack and artificial variables, the constraints and the augmented performance measure assume the form

$$\left.\begin{array}{r}-20x_1 + 100x_2 + 400x_3 - x_4 + x_5 = \epsilon \\ x_1 - 10x_2 + 100x_3 + x_6 = 2\epsilon \\ 100x_1 + 1{,}000x_2 + 8{,}000x_3 + x_7 = 50{,}000\end{array}\right\} \quad (5\text{-}72)$$

and

$$P_a = 0.2x_1 + x_2 + 4x_3 - Mx_5 \quad (5\text{-}73)$$

The variables x_4, x_6, and x_7 are slack variables; the variable x_5 is an artificial variable; the quantity ϵ is a small positive number introduced to avoid initial degeneracy; and the penalty coefficient M is a large positive number introduced to insure the eventual reduction to zero of the artificial variable x_5.

The scaling tables (Tables 5-7a, 5-7b, and 5-7c) are obtained in accordance with the theory outlined in Section 5-7c.

TABLE 5-7a. UNSCALED COEFFICIENTS

—	0.2	1	4	0	$-M$	0	0	
ϵ	-20	100	400	-1	1	0	0	
2ϵ	1	-10	100	0	0	1	0	
50,000	100	1,000	8,000	0	0	0	1	

TABLE 5-7b. ROW-SCALING OPERATIONS COMPLETED

—	2	10	40	0	$-M$	0	0	10
ϵ	-2	10	40	-0.1	0.1	0	0	10^{-1}
2ϵ	1	-10	100	0	0	1	0	1
500	1	10	80	0	0	0	0.01	10^{-2}

TABLE 5-7c. ROW AND COLUMN SCALING OPERATIONS COMPLETED

10^{-2}	1	2×10^{-1}	0.5×10^{-1}	10	10	1	100	
—	2	2	2	0	$-M$	0	0	10
ϵ	-2	2	2	-1	1	0	0	10^{-1}
2ϵ	1	-2	5	0	0	1	0	1
5	1	2	4	0	0	0	1	10^{-2}

The coefficients in Table 5-7c are used in the first simplex tableau, and the remaining simplex tableaux are obtained in the standard way (see Table 5-8).

Finally, the actual optimal values P^*, x_1^*, and x_2^* are obtained from the scaled values of the last tableau by using Equations 5-68 and 5-69. The pertinent scale factors are found in the final scaling table: $\alpha_p = 10^{-2}$, $\beta_p = 10$, $\alpha_1 = 1$, and $\alpha_2 = 2 \times 10^{-1}$. Thus,

$$\left.\begin{aligned} x_1^* &= (1 \times 1.667)/10^{-2} = 166.7 \\ x_2^* &= (2 \times 10^{-1} \times 1.667)/10^{-2} = 33.3 \\ x_3^* &= 0 \end{aligned}\right\} \qquad (5\text{-}74)$$

and

$$P^* = 6.67/(10^{-2} \times 10) = 66.7 \qquad (5\text{-}75)$$

Before leaving the example, two important features should be noticed. In the first place, observe that once the artificial variable x_5 is reduced to zero in a basic feasible solution, it and its associated column are no longer considered; the sole purpose of the artificial variable is to establish an initial basic feasible solution. When the x_5 column is ignored in the second tableau, however, the check column has to be reinitialized.

A second, less obvious feature illustrated by this example is the fact that only when the entries in a given key column are nonpositive in those positions which are associated with entries of order ϵ in the **b** column does the performance measure increase by more than an order of ϵ (observe the key column in the second tableau). The reason for this, of course, is that the increase in the performance measure is proportional to θ which in turn is of order ϵ as long as any one of the possible θ's is of order ϵ. This suggests that we should modify the rule for picking the key column when ϵ coefficients are present in the **b** column: select the jth column as the key column if $(c_j - s_j) > 0$ and if none of the corresponding possible θ's are of order ϵ; but if no such columns exist, proceed according to the previously established rule.

TABLE 5-8. SIMPLEX TABLEAUX WITH SCALED ENTRIES

		c_j	2	2	2	2	0	$-M$	0	0		
	c_i	b	a_1	a_2	a_3	a_4	a_5	a_6	a_7	Check	θ's
x_5	$-M$	ϵ	-2	②	2	-1	1	0	0	$2+\epsilon$	$\epsilon/2$
x_6	0	2ϵ	1	-2	5	0	0	1	0	$5+2\epsilon$	—
x_7	0	5	1	2	4	0	0	0	1	13	5/2
$P_a = -M\epsilon$		s_j	$2M$	$-2M$	$-2M$	M	$-M$	0	0		
		$c_j - s_j$	$2-2M$	$2+2M$	$2+2M$	$-M$	0	0	0		
x_2	2	$\epsilon/2$	-1	1	1	$-1/2$	—	0	0	$(1+\epsilon)/2$	—
x_6	0	3ϵ	-1	0	7	-1	—	1	0	$6+3\epsilon$	—
x_7	0	$5-\epsilon$	③	0	2	1	—	0	1	$12-\epsilon$	$(\tfrac{5}{3})-o(\epsilon)$
$P_a = \epsilon$		s_j	-2	2	2	-1	—	0	0		
		$c_j - s_j$	4	0	0	1	—	0	0		
x_2	2	5/3	0	1	5/3	$-1/6$	—	0	1/3	4.5	
x_6	0	5/3	0	0	23/3	$-2/3$	—	1	1/3	10	
x_1	2	5/3	1	0	2/3	1/3	—	0	1/3	4	
$P_a = 20/3$		s_j	2	2	14/3	1/3	—	0	4/3		
		$c_j - s_j$	0	0	$-8/3$	$-1/3$	—	0	$-4/3$		

5-8. UPPER-BOUNDING ALGORITHM

Consider a linear programming problem in which each x_i, $x_i \geq 0$, is bounded from above by a constant d_i. As noted under Case E of Section 5-3, each such upper-bound constraint could be placed in standard form by adding a row and a column to the simplex tableau. A larger, more cumbersome tableau would result, however, and if a digital computer is used, increased storage space is required.

As an alternative to enlarging the tableau, upper-bounding additions to the simplex technique have been developed by Charnes, Lemke, Dantzig, and others. The upper-bounding procedure given here is a modification of those previously developed.

The essential steps in the upper-bounding modification are displayed in the flow diagram of Figure 5-4. Notation used is as follows:

x_i^N = the particular x_j, where $j \neq i$ in general, which equals the ith entry of the **b** column in the Nth tableau for $i \leq m$ and which equals zero for $i > m$.

d_i^N = the upper-bound constraint on x_i^N (note that certain d_i^N may be infinite).

c_i^N = the value coefficient of x_i^N.

a_{ik}^N = the a entry associated with the ith row and with the same column as x_k^N (not necessarily the kth column of the **A** matrix).

The first tableau is arranged just as though no upper bounds exist. In block A of Figure 5-4, the ordinary simplex technique is effected to determine that the column associated with x_k^N, $N = 1$, is the key column. In block B, the first θ, θ^1, is determined in the ordinary way, but the effect of the d_i's is also considered. For each nonzero x_i^N (i.e., for each x_i^N, $i \leq m$), it is possible that x_i^{N+1} will be driven beyond its upper bound d_i^N if the next tableau is generated in the ordinary way—the reader may verify that x_i^N would be driven beyond its upper bound d_i^N if

$$\frac{d_i^N - b_i^N}{-a_{ik}^N} < \theta^N \quad \text{for} \quad a_{ik}^N < 0 \tag{5-76}$$

Thus, in block B, we test for this possibility.

It is also apparent that the value assumed by x_k^N in the next tableau would be θ^N if the ordinary simplex technique is followed. If θ^N is greater than d_k^N, however, the upper-bound constraint on x_k^N would be violated by the standard approach. In blocks C_a and C_b, therefore, θ^N is compared with d_k^N and with

$$\frac{d_r^N - b_r^N}{-a_{rk}^N} \triangleq \min_{1 \leq i \leq m} \frac{d_i^N - b_i^N}{-a_{ik}^N} \quad \text{for } a_{ik}^N < 0 \tag{5-77}$$

UPPER-BOUNDING ALGORITHM

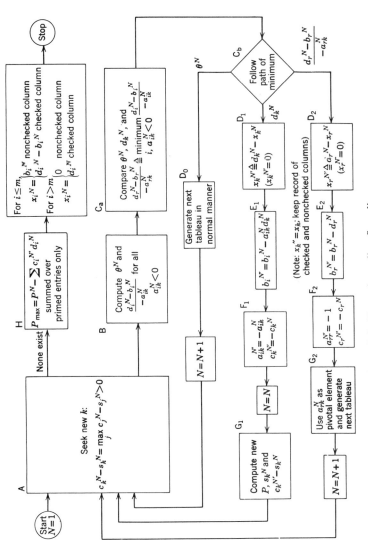

Figure 5-4. Upper-bounding flow diagram.

and a minimum of the three is found. Corresponding to this minimum, one of three paths is taken in the flow diagram.

In the first place, if θ^N is the minimum, the next tableau is generated in the ordinary way because no upper-bound constraint is violated (block D_0).

If d_k^N is the minimum, however, the procedure is to set x_k^N at its upper bound d_k^N—but note that this action results in more than m nonzero x_i's in the solution. To offset this fact and to render the problem in a form to which the ordinary simplex method is again applicable, the variable $x_k^{N'}$ is defined by $d_k^N - x_k^N$ (block D_1) and $d_k^N - x_k^{N'}$ is substituted for x_k^N in the tableau. The result is that $b_i^{N'} = b_i^N - d_k^N a_{ik}^N$ replaces b_i^N (block E_1) followed by a_{ik}^N being replaced by $a_{ik}^{N'} = -a_{ik}^N$, and c_k^N being replaced by $c_k^{N'} = -c_k^N$ (block F_1). The $\mathbf{a}_k^N$ column is now associated with the variable $x_k^{N'}$, and this fact is recorded, say, by a check at the head of the $\mathbf{a}_k^N$ column (if this same column is checked once again in a later tableau, it is converted back to the unchecked state). Because $x_k^N = d_k^N$, $x_k^{N'} = 0$, and $x_k^{N'}$ is bounded from above by d_k^N. New values of P, s_k^N, and $c_k^{N'} - s_k^N$ are computed at this point (block G_1); and the procedure is repeated starting at block A, with a tableau which is only slightly altered.

The third alternative is that $(d_r^N - b_r^N)/(-a_{rk}^N)$ is the minimum found in block C_b. In this case, x_r^N is set to its upper bound d_r^N, and for the same reason as given in the preceding paragraph, x_r^N is replaced in the tableau by $d_r^N - x_r^{N'}$. Thus, blocks D_2, E_2, and F_2 are the counterparts of blocks D_1, E_1, and F_1 previously considered. Block E_2 requires only the one computation $b_r^{N'} = b_r^N - d_r^N$ because $a_{ir}^N = 1$ for $i = r$, and equals zero otherwise. Similarly, only $a_{rr}^N = 1$ is replaced by -1 in block F_2. The variable $x_r^{N'}$ is zero in the new solution, whereas the variable x_k^N is increased from its zero value. The next tableau is generated therefore by using the entry a_{rk}^N as the pivotal entry (block G_2).

At the maximum, all $c_j^N - s_j^N$ are less than or equal to zero in block A. In general the value of P^N given in the final tableau is not the maximal value, however, because of the $x_i^{N'}$ substitutions. For one such primed variable, say, $x_i^{N'}$, P^N contains the term

$$c_i^{N'} x_i^{N'} = -c_i^N(d_i^N - x_i^N) = c_i^N x_i^N - c_i^N d_i^N$$

The term $c_i^N d_i^N = -c_i^{N'} d_i^N$ must therefore be added to P^N, as shown in block H.

Finally, the optimal x_i's are computed: (1) for $i \leq m$,

$$x_i^N = \begin{cases} b_i^N \text{ if the } \mathbf{a}_i^N \text{ column is not checked} \\ d_i^N - b_i^N \text{ if the } \mathbf{a}_i^N \text{ column is checked} \end{cases} \quad (5\text{-}78)$$

and (2) for $i > m$,

$$x_i^N = \begin{cases} 0 \text{ if the } \mathbf{a}_i^N \text{ column is not checked} \\ d_i^N \text{ if the } \mathbf{a}_i^N \text{ column is checked} \end{cases} \quad (5\text{-}79)$$

5-9. DUAL PROBLEMS

5-9a. Duals in General

If a problem has a dual, the dual problem is posed in terms of the same types of variables as the original (*primal*) problem, but with the roles of certain variables being interchanged. Thus, for electrical duals, current is the dual of voltage and vice versa; for mechanical duals, force is the dual of velocity and vice versa. The solution of a dual problem is related in some well-defined way to the solution of the primal problem; and in certain cases, the solution of the dual problem is easier to obtain. It is for this reason, primarily, that dual problems are considered.

The existence of duals for linear programming problems was first proposed by von Neumann in 1947 [5.63]. Much of the initial work on duality properties can be traced to Gale, Kuhn, and Tucker [5.27]. Of the known dual problems of linear programming, one of fundamental importance is the symmetric dual problem (Section 5-9b) from which other useful dual problems can be developed (Section 5-9c). The goal here is the utilization of duality relations to solve problems, and therefore the reader who desires a detailed derivation of the symmetric dual relationships is referred to other sources (e.g., [5.21, 5.31, 5.44, 5.59]).

5-9b. Symmetric Duals

Consider these two problems:

Problem S.1.

$$\text{Constraints} \begin{cases} \mathbf{Ax} \leq \mathbf{b} \text{ (A is assumed to be } m \times k \text{ in this section)} \\ \mathbf{x} \geq \mathbf{0} \end{cases} \quad (5\text{-}80)$$

Performance measure P_{S1} to be maximized:

$$P_{S1} = \mathbf{c'x} \quad (5\text{-}81)$$

Problem S.2.

$$\text{Constraints} \begin{cases} \mathbf{A'w} \geq \mathbf{c} \\ \mathbf{w} \geq \mathbf{0} \end{cases} \quad (5\text{-}82)$$

Performance measure P_{S2} to be minimized:

$$P_{S2} = \mathbf{b'w} \quad (5\text{-}83)$$

Problems S.1 and S.2 are (symmetric) duals of one another. The maximum of P_{S1} equals the minimum of P_{S2}. As to whether one is called the primal problem and the other is called the dual problem depends on the particular application. Note that if the number k of x_i variables in Problem S.1 is much less than the number m of constraints, the number m of w_i variables in Problem S.2 is much greater than the number k of associated constraints. Thus, when $m \gg k$, Problem S.2 is easier to solve since the number of iterations of the simplex technique is related (loosely) to the number of constraint equations; but when $m \ll k$, Problem S.1 is solved with less effort.

Once a dual problem has been solved, it is desirable to find the solution of the corresponding primal problem. As previously stated, the performance measures P_{S1} and P_{S2} have identical optimal values. Furthermore, the values of the primal variables at the optimum can be obtained from the entries in the final $c_j - s_j$ row in the simplex solution of the dual.

Assume that either Problem S.1 or S.2 is the dual problem and has been solved by using the simplex technique, as described in Section 5-6. *If Problem S.2 has been solved, it is assumed that the tableaux were arranged to find the maximum of* $-P_{S2}$. Rules for determining the optimal primal variables *from the final tableau* of the dual are:

1. *Negate the entries* in the $c_j - s_j$ row *under the slack columns* to obtain the optimal values of the *structural variables* of the primal problem—structural variables are all nonslack and nonartificial variables. The slack column corresponding to the first constraint of the dual yields the first structural variable of the primal; the slack column corresponding to the second constraint of the dual yields the second structural variable of the primal; and so forth.

2. *Negate the entries* in the $c_j - s_j$ row *under the structural columns* to obtain the optimal slack values of the primal problem. The column corresponding to the first structural variable yields the slack variable associated with the first constraint of the primal problem; the column corresponding to the second structural variable yields the slack variable associated with the second constraint of the primal problem; and so forth.

3. If S.1 is solved, the optimal value of P_{S2} equals the value of P_{S1} found in the last tableau; if S.2 is solved, the value max P_{S1} = min P_{S2} equals the negative of max $(-P_{S2})$, which is found in the last tableau.

Rules 1 and 2 are independent of whether S.1 or S.2 is the primal problem, except that $(-b_j - s_j)$ replaces $(c_j - s_j)$ when S.2 is solved. All three rules remain unchanged even if certain entries in the **A**, **b**, and **c** matrices are negative. This fact is of importance when we desire the dual of a problem which contains both types of inequality constraints: constraints of the form 5-84 and constraints of the form 5-86. Thus, for example, consider a given problem which contains one constraint of the form

$$a_{i1}x_1 + a_{i2}x_2 + \cdots + a_{ik}x_k \geq b_i \qquad (5\text{-}84)$$

Sect. 5-9   DUAL PROBLEMS   229

where b_i is a positive constant. All other constraints are assumed to be of the form 5-80. To place the problem in form S.1, we multiply both sides of 5-84 by -1:

$$-a_{i1}x_1 - a_{i2}x_2 - \cdots - a_{ik}x_k \leq -b_i \qquad (5\text{-}85)$$

A dual problem S.2 can now be formed in the straightforward manner.

5-9c. Other Duals

Dantzig and Orden [5.22] present an unsymmetric dual problem pair, one of the problems of which is in the standard form given by Equations 5-1, 5-2, and 5-3. The unsymmetric dual of a problem in standard form is generally more difficult to solve than is the problem in standard form. *It is advisable, therefore, to form the dual of a given problem prior to the addition (subtraction) of slack and artificial variables.*

The unsymmetric dual problem is a special case of the *mixed dual problem* which is developed here. Mixed dual problems contain both equality and inequality constraints and are shown to be special cases of symmetric dual problems.

Suppose for the moment that a problem is in the form S.1 except that the mth (and only the mth) constraint of the set 5-80 is an equality constraint rather than an inequality constraint. This problem is called a *modified S.1 problem*, as are others which are of the form S.1 but contain equality constraints.

Consider the inequality constraints

$$a_{m1}x_1 + \cdots + a_{mk}x_k \leq b_m \qquad (5\text{-}86)$$

and

$$-a_{m1}x_1 - \cdots - a_{mk}x_k \leq -b_m \qquad (5\text{-}87)$$

which can both be true, obviously, only if equality holds. By using 5-86 and 5-87 in place of the equality constraint, the problem is placed in the form of S.1 from which a dual of form S.2 can be obtained. But in the dual so obtained, several novel features arise. Note that each constraint equation of the dual is of the form

$$a_{1j}w_1 + \cdots + a_{mj}w_{m+} - a_{mj}w_{m-} \geq c_j \qquad (5\text{-}88)$$

where w_{m+} and w_{m-} stem from constraints 5-86 and 5-87, respectively, and the performance measure is

$$P_{S2} = b_1w_1 + b_2w_2 + \cdots + b_mw_{m+} - b_mw_{m-} \qquad (5\text{-}89)$$

with all w's greater than or equal to zero. In all these relationships, the variables w_{m+} and w_{m-} appear with coefficients equal in magnitude but opposite in sign. It is possible therefore—see Case D of Section 5-3—to replace variables w_{m+} and w_{m-} with one variable w_m which is *unrestricted in sign*. In general, it follows that if r primal constraints are equality constraints, the corresponding r variables of the dual are unrestricted in sign.

By working the above argument in reverse, it is easily established that if a given problem contains r variables which are unrestricted in sign, the corresponding r constraints of the dual problem are equality constraint equations. These statements apply equally well if the primal problem is of the modified S.2 form.

When equality constraints and/or unrestricted variables exist in the primal problem, additional rules are required to determine the solution of the primal from the final tableau of the dual. Only the equality constraints and the unrestricted variables are involved in these additional rules.

Consider first the case that the ith constraint of the primal is an equality constraint. The corresponding variable in the dual problem can assume both positive and negative values, which fact leads to the use of two non-negative variables in the dual as discussed under Case D of Section 5-3. Since the equality constraint in the primal allows no slack, the $c_j - s_j$ value under the columns of these two variables must equal zero in the final tableau of the dual.

Finally, if the ith variable of the primal is unrestricted in sign, the ith constraint equation of the dual is an equality constraint in which no slack variable appears. An artificial variable can be added to the left-hand member of this equation, however, to enable solution by using the standard simplex technique. The column associated with this artificial variable should be carried in all tableaux of the dual solution in order to obtain the optimal value of the ith primal variable: this equals the value found in the column of the artificial variable in the s_j row of the final tableau of the dual.

Example 5-5. Consider the following primal problem.
Maximize P:

$$P = 2x_1 + x_2 - 3x_3 \tag{5-90}$$

Subject to:

$$5x_1 + 2x_2 + 4x_3 \leq 20 \tag{5-91}$$

$$x_1 + x_2 + 2x_3 \leq 8 \tag{5-92}$$

$$4x_1 + 3x_2 + 6x_3 \geq 5 \tag{5-93}$$

$$x_1 + x_2 + x_3 = 6 \tag{5-94}$$

Sect. 5-9   DUAL PROBLEMS   231

TABLE 5-9. SIMPLEX TABLEAUX FOR EXAMPLE 5-5

	$-b_j$		-20	-8	5	-6	6	0	0	$-M$	$-M$	$-M$		
w_i	$-b_i$	c	w_1	w_2	w_3	w_{4+}	w_{4-}	w_5	w_6	w_7	w_8	w_9	Check	θ's
w_7	$-M$	2	⑤	1	-4	1	-1	-1	0	1	0	0	4	0.4
w_8	$-M$	1	2	1	-3	1	-1	0	-1	0	1	0	1	0.5
w_9	$-M$	3	-4	-2	6	-1	1	0	0	0	0	1	4	—
$-P_{S2} = -6M$		s_j	$-3M$	0	M	$-M$	M	M	M	$-M$	$-M$	$-M$		
		$-b_j - s_j$	$3M$	-8	$-M$	M	$-M$	$-M$	$-M$	0	0	0		
w_1	-20	0.4	1	0.2	-0.8	0.2	-0.2	-0.2	0		0	0	0.6	—
w_8	$-M$	0.2	0	0.6	-1.4	0.6	-0.6	0.4	-1		1	0	-0.2	—
w_9	$-M$	4.6	0	-1.2	②.⑧	-0.2	0.2	-0.8	0		0	1	6.4	1.642
$-P_{S2} = -4.8M$		s_j	-20	$0.6M$	$-1.4M$	$-0.4M$	$0.4M$	$0.4M$	M		$-M$	$-M$		
		$-b_j - s_j$	0	$-0.6M$	$1.4M$	$0.4M$	$-0.4M$	$-0.4M$	$-M$		0	0		
w_1	-20	1.715	1	-0.143	0	0.1428	-0.1428	-0.4285	0		0	0.286	2.43	12
w_8	$-M$	2.5	0	0	0	⑤	-0.5	0	-1		1	0.5	3.00	5
w_3	5	1.642	0	-0.429	1	-0.0714	0.0714	-0.286	0		0	0.357	2.284	—
$-P_{S2} = -2.5M$		s_j	—	—	—	$-0.5M$	$0.5M$	—	M		$-M$	$-0.5M$		
		$-b_j - s_j$	—	—	—	$0.5M$	$-0.5M$	—	$-M$		0	$-0.5M$		
w_1	-20	1	1	-0.143	0	0	0	-0.429	0.286		0	0.1432		
w_{4+}	-6	5	0	0	0	1	-1	0	-2		1	1		
w_3	5	2	0	-0.429	1	0	0	-0.286	-0.143		0	0.4284		
$-P_{S2} = -40$		s_j	-20	0.715	5	-6	6	7.14	5.574		$-M$	-6.722		
		$-b_j - s_j$	0	-7.285	0	0	0	-7.14	-5.574		0	$-M + 6.722$		

and

$$x_1 \geq 0, \quad x_2 \geq 0 \tag{5-95}$$

Note that x_3 is unrestricted in sign. To obtain a modified S.1 form, constraint 5-93 is replaced by

$$-4x_1 - 3x_2 - 6x_3 \leq -5 \tag{5-96}$$

The dual constraint corresponding to the x_1 variable is

$$5w_1 + w_2 - 4w_3 + w_4 \geq 2 \tag{5-97}$$

That corresponding to x_2 is

$$2w_1 + w_2 - 3w_3 + w_4 \geq 1 \tag{5-98}$$

And that corresponding to x_3 is

$$4w_1 + 2w_2 - 6w_3 + w_4 = -3 \tag{5-99}$$

Since the fourth constraint 5-94 of the primal is an equality constraint, the fourth variable w_4 of the dual is unrestricted in sign. The performance measure P_{S2} of the dual is $P_{S2} = 20w_1 + 8w_2 - 5w_3 + 6w_4$, and a maximum of $-P_{S2}$ is to be found.

After appropriate addition of artificial and slack variables, the constraints of the dual are placed in the first simplex tableau (Table 5-9). From the final simplex tableau of the dual, the solution for the primal problem is found to be $\max P = P^* = 40$, $x_1^* = 7.14$, $x_2^* = 5.57$, $x_3^* = -6.72$, (slack) $x_4^* = 0$, (slack) $x_5^* = 7.28$, and (slack) $x_6^* = 0$.

●

5-10. SENSITIVITY ANALYSIS

Sensitivity analysis is the ever-present after-product of optimization. Given an optimal solution which corresponds to fixed **A**, **b**, and **c** matrices (Equations 5-1 and 5-3), the immediate practical questions to be answered are: "How is the value of P affected if a suboptimal feasible solution is used rather than the optimal solution? And how does the nature of the optimal solution change if various coefficients in the **A**, **b**, and **c** matrices are changed?"

The first question is easily answered by direct evaluation of the difference $P_{\text{opt}} - P_{\text{subopt}}$. As to the second question, no single answer applies. It may be that individual probability distributions are given for certain a, b, or c coefficients, and that the expected values of the coefficients are used in the equations. This approach is unsatisfactory, however, if the distributions are relatively broad and if the constraint equations are of the "hard" variety, that is, they are not to be violated (much). Alternatively, it may be that the various coefficients are placed at worst-case limits in the equations, in which case optimization yields the optimal x_i's for the worst case (see Section 5-12d). Again, it may be that the coefficients change in some known way, perhaps as a function of a given parameter (the term parametric programming is then used), and it is desired to determine over what range the coefficients can change before a change occurs in the nature of the optimal solution, i.e., before one (or more) of the nonzero (*basis*) variables in the current optimal solution is replaced by a variable which is zero in the current optimal solution.

The range of values that any single coefficient can assume without causing a change in optimal basis variables is known as the *sensitivity range* of the coefficient. Note that within the sensitivity range of a given coefficient, the optimal value of the performance measure may change, as well as the optimal values of the basis variables, but the x_i's which are zero at the initial point in the range remain at zero throughout the sensitivity range.

To illustrate the sensitivity range concept, consider the case in which the sensitivity range of each value coefficient c_i is to be determined independently [5.32]. In the first place, assume that the coefficient c_i is associated with a variable x_i which is zero in the optimal solution and assume that $\bar{c}_i$ is the value of c_i which was used in determining the optimal solution. It follows from the simplex technique of Section 5-6 that such a c_i can be decreased from $\bar{c}_i$ by any amount without affecting the optimal solution; however, it can be increased only by an amount which does not exceed the value $|\bar{c}_i - s_i|$ which is found in the last tableau, because the solution is no longer optimal if c_i is increased to the point that $c_i - s_i$ is positive. Thus, the sensitivity range on this type of c_i is given by

$$-\infty \le c_i - \bar{c}_i \le |\bar{c}_i - s_i| \tag{5-100}$$

Next, assume that a given coefficient c_i is associated with a variable which is in the optimal basis. In this case, any variation of c_i from $\bar{c}_i$ will affect the value of P and the values of certain s_j's. The sensitivity range of c_i is determined by computing $(c_j - s_j)$'s as a function of $c_i - \bar{c}_i$ and by determining the range of $c_i - \bar{c}_i$ for which all $(c_j - s_j)$'s remain nonpositive. Note that all $(c_j - s_j)$'s which correspond to positive x_j's in the optimal solution are zero, independent of the change in c_i, because only those s_j's for which the corresponding a_{ij} entries are nonzero are affected by a change in c_i (see Equation 5-54). Thus, a zero value of $\bar{a}_{ij}$—the coefficient which appears under the column of the jth variable and in *the row of* the ith variable in the final tableau—renders the value $c_j - s_j$ impervious to changes in c_i. For all other $(c_j - s_j)$'s, the following equation applies:

$$c_j - s_j = (c_j - \bar{s}_j) - (c_i - \bar{c}_i)\bar{a}_{ij} \tag{5-101}$$

The value of $c_i - \bar{c}_i$ which yields a zero value of $c_j - \bar{s}_j$ is of interest. Thus, for nonzero $\bar{a}_{ij}$'s, $(c_j - \bar{s}_j) = 0$ when

$$c_i - \bar{c}_i = \frac{c_j - \bar{s}_j}{\bar{a}_{ij}} \qquad (5\text{-}102)$$

and provided there exists at least one $\bar{a}_{ij} < 0$ for some j and one $\bar{a}_{ij} > 0$ for another j, $j \neq i$, the sensitivity range on c_i is determined by

$$\operatorname*{maximum}_{\substack{j, j \neq i \\ \bar{a}_{ij} > 0}} \left\{ \frac{c_j - \bar{s}_j}{\bar{a}_{ij}} \right\} \leq c_i - \bar{c}_i \leq \operatorname*{minimum}_{\substack{j \\ \bar{a}_{ij} < 0}} \left\{ \frac{c_j - \bar{s}_j}{\bar{a}_{ij}} \right\} \qquad (5\text{-}103)$$

However, if all $\bar{a}_{ij}$'s ($j = 1, 2, \ldots$, but $j \neq i$) are non-negative (nonpositive), the upper (lower) limit on $c_i - \bar{c}_i$ is infinity (minus infinity), a fact which can be verified on the basis of 5-101.

Example 5-6. In the final tableau of Example 5-3, the sensitivity range of c_1 is $-\infty \leq c_1 - 1 \leq 0.5$, or $-\infty \leq c_1 \leq 1.5$; the sensitivity range of c_3 is $-\infty \leq c_3 - 2 \leq 1$, or $-\infty \leq c_3 \leq 3$; and the sensitivity range of c_4 is $-\infty \leq c_4 \leq 1.5$. For the variable x_2 which is in the basis, the sensitivity range is determined by using 5-103 and the statement that follows 5-103; in this case, $\max\{-0.5, -0.5, -1.5\} \leq c_2 - 1.5 \leq \infty$, or $1 \leq c_2 \leq \infty$. Similarly, for x_5, $\max\{-0.25, -0.5\} \leq c_5 \leq \min\{1.5\}$, or $-0.25 \leq c_5 \leq 1.5$.

•

5-11. ANALOG SOLUTIONS

5-11a. Analogies

Any physical system which is characterized (in the steady state) by a set of linear algebraic constraint equations and by a linear algebraic performance measure can, in theory, be used to obtain solutions to certain linear programming problems. Many such analog approaches to solutions are impractical because of difficulties encountered both in constructing the analogs and in obtaining accurate results from the analogs. In certain analog schemes, however, the above-noted difficulties are sufficiently minimized, and the use of the analog is warranted because it enables sensitivity analysis to be effected in the easiest of ways, e.g., changing component values to obtain closely related optimal solutions.

Perhaps the most promising analog schemes are those which are restricted to particular types of linear programming problems. For example, consider the "minimum-chain problem" which is a generalized version of the "minimum-time problem" of Section 5-12c. A chain network consists of nodes (tie-points) and chain links between certain nodes. If two particular nodes are connected by a chain link, the "length" of this direct chain link could be proportional to different things in different problems: e.g., it could be proportional to time of transit in going from the one node to the other without passing through any node; it could be proportional to the direct distance between the nodes; or it could be proportional to the cost of constructing telephone wires between nodes. The problem is to find the path which minimizes the connected-chain length between two widely separated nodes. Two of the analog schemes [5.49, 5.55] for solving this problem are as follows.

The simplest analog for the chain network consists of connected pieces of unstretchable string, the length of the piece of string between two given nodes being proportional to the length of the corresponding direct chain link. To use this analog of strings as an analog computer, pick up the two nodes between which the minimum connected chain is desired, one in each hand, and stretch as far as possible without breaking. The taut path is the minimum connected chain. To find the next-best minimum connected chain—if one of the chain links in the optimal path should fail—cut the corresponding string and stretch as before.

Electrical analogs are also available for the above problem. In one such analog [5.49], each chain link is replaced by a gas tube (or a symmetrical Zener diode) which exhibits a break-down voltage proportional to the length of the chain link. A current source is then applied between the two nodes in question, and the minimum connected chain is found to correspond to the path of current flow.

For the general linear programming problem, Dennis [5.24] shows that any linear programming problem can be modeled by an electrical network consisting of voltage sources, current sources, ideal diodes, and ideal transformers. The latter element is not required in many well-known linear programming problems, e.g., the transportation problem. Dennis stresses the fact that the electrical analogy permits novel interpretations of algorithms for obtaining solutions of programming problems and that these analogies often point the way to the formulation of new analytical techniques.

5-11b. Linear Programming on the General-Purpose Analog Computer

Consider a linear programming problem which is characterized by the following normalized constraints and performance measure P:[4]

---

[4] The approach of this section is applicable if **Ac** is greater than **0**. This statement is clarified in the development. Also, inequality constraints of the form $a_1 x_1 + a_2 x_2 + \cdots + a_k x_k \leq -1$ can be treated by making minor changes in the computer circuits (see Problem 5.22).

$$\mathbf{Ax} \leq [1 \quad 1 \quad 1 \quad \cdots \quad 1]' \quad (\mathbf{A} \text{ is assumed to be } m \times k \text{ in this section}) \quad (5\text{-}104)$$

$$\mathbf{x} \geq [0 \quad 0 \quad 0 \quad \cdots \quad 0]' = \mathbf{0} \tag{5-105}$$

and

$$P = \mathbf{c}'\mathbf{x} \tag{5-106}$$

where P is to be maximized.

Pyne [5.54] gives a method for solving this problem on a general-purpose analog computer. Diodes, resistors, capacitors, and operational amplifiers are required. The number of required operational amplifiers is bounded between $m + 2k$ and $2m + 3k$, where k equals the number of x_j's and m equals the number of constraint equations. If all a_{ij}'s of 5-104 are positive, the lower limit holds; but if each row and each column of the $\mathbf{A}$ matrix contains one or more negative a_{ij}'s, the upper limit holds.

It is illuminating to examine the geometrical motivation for the method. The constraints 5-104 and 5-105 define an allowable region in the k-dimensional space of the x_j's. It is known (Section 5-4) that this region is convex and that the unique maximum of P is attained at some point, or on some connected set of points, on the boundary. Consider the *gradient* ∇P of the performance measure P:

$$\text{grad } P = \nabla P = [c_1 \quad c_2 \quad \cdots \quad c_k]' \tag{5-107}$$

Grad P is a column vector which points in the direction of greatest increase for P (a proof of this statement is given in the next chapter). If the direction ∇P is followed from some starting point in the allowed region, a boundary is ultimately reached. At the boundary, a new direction must be taken in order to stay inside the allowed region. The question is how to determine the new direction.

Suppose the boundary corresponding to $a_{i1}x_1 + a_{i2}x_2 + \cdots + a_{ik}x_k = 1$ is reached for some specific integer i. A column vector which is orthogonal to this boundary and which points toward the allowed region is given by

$$\mathbf{n}_i = -[a_{i1} \quad a_{i2} \quad \cdots \quad a_{ik}]' \tag{5-108}$$

If this vector is added to an appropriate multiple of ∇P, the resulting vector points back into the allowed region. One could force $\mathbf{x}$ to move in this composite-vector direction for a short distance, until safely in the allowed region, and then return to movement in the gradient direction. This approach is illustrated in Figure 5-5.

The preceding statements are formalized as follows: Let the column vector $\mathbf{f}$ be defined by

$$\mathbf{f} \triangleq K \nabla P + \sum_{i=1}^{m} \delta_i \mathbf{n}_i \tag{5-109}$$

Sect. 5-11   ANALOG SOLUTIONS   237

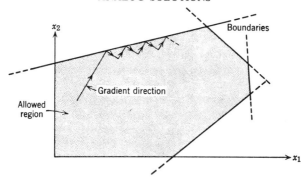

Figure 5-5. Search in the gradient direction.

where K is a constant (to be specified shortly), and where δ_i is given by

$$\delta_i = \begin{cases} 0 & \text{when } \sum_{j=1}^{k} a_{ij}x_j \leq 1 \\ 1 & \text{when } \sum_{j=1}^{k} a_{ij}x_j > 1 \end{cases} \quad (5\text{-}110)$$

The vectors ∇P and $\mathbf{n}_i$ are as previously defined.

Let $\mathbf{x} = [x_1 \ x_2 \ \cdots \ x_k]'$ denote the location of the x_j's at any given instant of time. The vector $d\mathbf{x}/dt = [dx_1/dt \ dx_2/dt \ \cdots \ dx_k/dt]'$ then equals the rate of change of $\mathbf{x}$. Suppose $d\mathbf{x}/dt$ is equated to $\gamma \mathbf{f}$ where γ is a positive (time-scaling) constant. Components of the resulting vector are

$$\frac{dx_j}{dt} = \gamma \left(Kc_j - \sum_{i=1}^{m} \delta_i a_{ij} \right), \quad j = 1, 2, \ldots, k \quad (5\text{-}111)$$

When none of the constraints are violated (when all δ_i's equal zero), the x_j's which satisfy 5-111 increase in the direction of ∇P; when a particular constraint is violated, the corresponding δ_i assumes unity value, and the solution is driven back into the allowed region.[5] Typically, the path along a constraint boundary is a series of small zig-zags, back and forth across the boundary, but always in the general direction of increasing P. In the limit as $t \to \infty$ the optimal solution is attained, provided of course it is bounded.

---

[5] We might rightly conjecture that appropriate δ's for the constraints 5-105 should be included in Equations 5-109 and 5-111. In actual implementation, however, it is easier to take this set of constraints into account by using limiting diodes on voltages which correspond to x_j's.

238   THE SIMPLEX TECHNIQUE AND LINEAR PROGRAMMING   Ch. 5

To insure that the solution of 5-111 cannot break through any boundary, we must set the constant K such that the magnitude of the orthogonal component of $-K\nabla P$ on each boundary is less than the magnitude of $\mathbf{n}_i$ for all i. This is fulfilled if the following dot-product inequalities (in matrix form) hold for all i. (We assume that $\nabla P'\mathbf{n}_i < 0$ for all i. See Problem 5.23.)

$$-K\nabla P'\mathbf{n}_i \leq \mathbf{n}_i'\mathbf{n}_i \tag{5-112}$$

or equivalently,

$$K \leq \min_i \left\{ \frac{\sum_{j=1}^{k} a_{ij}^2}{\sum_{j=1}^{k} c_j a_{ij}} \right\} \tag{5-113}$$

To attain a solution in the least possible time, K should be set as close to the minimum indicated in 5-113 as tolerances will allow.

To program the set of Equations 5-111, the following are required: (1) k integrators, the outputs of which are the x_j's (Figure 5-6); (2) m summing amplifiers and m switching amplifiers to form the δ_i's (Figure 5-7); and (3) at most $m + k$ sign-changing amplifiers to obtain $-x_j$ and $-\delta_i$, as necessary. In Figures 5-6 and 5-7, all proportionality constants β are the same and should be selected on the basis of the particular computer used. The voltage V_0 should be set to obtain the solution in the linear range of amplifier operation. The integrating unit and the summing unit are standard analog computer configurations (see Jackson [5.34], for example), except that the diode on the integrating unit holds the variable x_j to non-negative values. The switching unit operates as follows: as long as v_i is greater than zero—that is, as long as $V_0 \sum_j a_{ij}x_j$ is less than V_0—diode d_1 is forward biased and the output $V_0\delta_i$ equals zero; but when v_i is less than zero, $V_0 \sum_j a_{ij}x_j$ is greater than V_0, and diode d_2 (rather than d_1) is forward biased, yielding an output $\delta_i V_0 = V_0$.

To obtain accurate solutions on an analog computer, the original problem

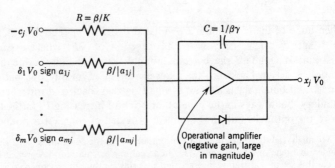

Figure 5-6. Integrating unit.

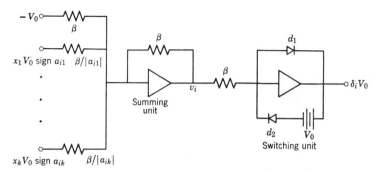

Figure 5-7. Summing and switching units.

should be scaled (see Section 5-7). Also, the number of inputs that can be applied to an operational amplifier is limited by practical considerations, so that additional summing amplifiers are required to solve a large problem.

As for the time required to obtain the desired steady-state solution, the factor γ is of importance. In the unconstrained case, the variable x_r which changes the fastest is determined by the rate-of-change $|\gamma K c_r| = \max_j |\gamma K c_j|$ (see Equation 5-111). A given computer has a prescribed range for the rate of integration, and since c_j's are known and K is determined from 5-113, γ can be specified so that $|\gamma K c_r|$ is at the upper end of the allowed range.

5-12. APPLICATIONS

Numerous applications of linear programming are considered in this section, any part of which can be omitted by the person who has a definite application in mind. In addition to the applications considered here, we can observe in the literature ([5.3, 5.13, 5.28, 5.33], for example) applications where nonlinear problems have been placed in linear programming form by using piecewise linear approximations to nonlinear equations; linear programming is applied over a restricted range of the variables, and the linear model is updated periodically.

5-12a. Problems of Economics

The first significant problems of linear programming were posed during and immediately following World War II. Most of these problems were centered around scheduling (programming) of resources to satisfy demands in such a way that the overall efficiency of operation was optimized. Current uses include public utility systems planning [5.10] and the establishment of economical power-generating schedules [5.33]. Masse and Gibrat [5.48] use linear programming to determine optimal investment policies in the electrical industry. More recently, Asher [5.5] and Dov [5.25] proposed linear programming models for the optimal allocation of research and development efforts.

Specific economic uses of linear programming can be broadly categorized according to type [5.16]:

1. Optimal selection of alternatively available materials, resources, supplies, and/or machines.
2. Optimal use and scheduling of existing tools, machines, and resources (production scheduling).
3. Optimal establishment of inventories.
4. Minimization of transportation and distribution costs.

In a specific problem, aspects of each of these types may be involved. Standard linear programming books [e.g., 5.15, 5.21, 5.31, 5.44] may be consulted for detailed examples of such problems.

5-12b. Control Problems

Linear programming techniques can be applied in various ways to solve optimal control problems. Zadeh [5.71] traces the early history of proposed control applications, the first of which was by Krasovskii [5.38] in 1957. In addition to being applicable to certain problems posed for discrete (sampled-data) control systems [5.37, 5.52, 5.53, 5.61, 5.64], linear programming theory has also been proposed for use on suitable discrete approximations of continuous linear systems, including distributed-parameter control systems [5.45, 5.57].

Many of the proposed applications deal with a system which is characterized (in part) by the discrete state equation

$$\mathbf{y}_k = \mathbf{G}\mathbf{y}_{k-1} + \mathbf{h}m_{k-1}, \qquad k = 1, 2, \ldots \qquad (5\text{-}114)$$

where $\mathbf{y}_k = [y_1(k) \quad y_2(k) \quad \cdots \quad y_n(k)]'$ is a column vector, the entries of which equal state variables at the kth equally-spaced sampling instant; $\mathbf{G}$ is an $n \times n$ matrix of known constant entries; $\mathbf{h}$ is a column vector of constants; and $m_k \triangleq m(k)$ is the scalar control function at the kth sampling instant. Equation 5-114 is readily solved for $\mathbf{y}_k$ in terms of the initial condition $\mathbf{y}_0$ and in terms of the controlling actions, as follows:

$$\begin{aligned}\mathbf{y}_k &= (\mathbf{G})^k \mathbf{y}_0 + (\mathbf{G})^{k-1}\mathbf{h}m_0 + (\mathbf{G})^{k-2}\mathbf{h}m_1 + \cdots + \mathbf{G}\mathbf{h}m_{k-2} + \mathbf{h}m_{k-1} \\ &= (\mathbf{G})^k \mathbf{y}_0 + \sum_{i=0}^{k-1} (\mathbf{G})^i \mathbf{h} m_{k-1-i} \end{aligned} \qquad (5\text{-}115)$$

Sect. 5-12   APPLICATIONS   241

where $(\mathbf{G})^i\mathbf{h}$, $i = 0, 1, \ldots$, is the discrete weighting function of the system.

Given an initial condition $\mathbf{y}_0$ and given the fact that m_k is constrained by the relationship

$$m_{\min} \leq m_k \leq m_{\max} \tag{5-116}$$

numerous linear programming problems can be cited [5.64, 5.71], the most interesting of which are:

Terminal Error Problem 1.

For the system characterized by 5-114, 5-115, and 5-116, find a control sequence $m_0, m_1, \ldots, m_{K-1}$ which yields the minimum of P:

$$P \triangleq \sum_{i=1}^{j} |e_i| \tag{5-117}$$

where the e_i is the ith entry of the column vector $\mathbf{e}$:

$$\mathbf{e} \triangleq \mathbf{Q}\mathbf{y}_d - \mathbf{Q}\mathbf{y}_K \tag{5-118}$$

in which $\mathbf{y}_d$ is the desired terminal state at the specified terminal time which is denoted by the integer K, and $\mathbf{Q}$ is a specified $j \times n$ matrix of constants. When $j = n$, and $\mathbf{Q}$ equals an $n \times n$ identity matrix, P reduces to the sum of the absolute values of the terminal errors in the states.

This problem may include constraints of the form

$$L(\mathbf{y}_1, \mathbf{y}_2, \ldots, \mathbf{y}_K) \leq \mathbf{d} \tag{5-119}$$

where $L = L(\mathbf{y}_1, \mathbf{y}_2, \ldots, \mathbf{y}_K)$ is a linear vector function of $\mathbf{y}_1, \mathbf{y}_2, \ldots, \mathbf{y}_K$, and $\mathbf{d}$ is a constant column vector. Also, a constraint of the form

$$\sum_{i=0}^{K-1} |m_i| \leq c \quad \text{(a constant)} \tag{5-120}$$

can be treated (see Problem 5.4).

Terminal Error Problem 2.

With conditions as described under 1 above, replace 5-117 with

$$P = (\max_i |e_i|) \tag{5-121}$$

where a minimum of P is desired (see Problem 5.5).

Terminal Error Problem 3.

Again, with conditions as described under Terminal Error Problem 1, replace 5-117 with

$$P = \sum_{i=1}^{j} e_i \tag{5-122}$$

with the additional constraint that $e_i \geq 0$ for $i = 1, 2, \ldots, j$. The latter restriction is referred to as *overcontrol elimination*.

Minimal Action Problem 1. (A Minimal Fuel Problem)

Given the system characterized by 5-114, 5-115, and 5-116, find a control sequence (i.e., find $m_0, m_1, \ldots, m_{K-1}$) which drives the system from an initial state $\mathbf{y}_0$ to a desired state $\mathbf{y}_d$ at time K (actually, at the time corresponding to the integer K) and which results in a minimum of P:

$$P = \sum_{i=0}^{K-1} |m_i| \qquad (5\text{-}123)$$

Constraints of the form 5-119 may also be included in this problem.

Minimal Action Problem 2.

Here the same conditions as described under Minimal Action Problem 1 apply, but 5-123 is replaced by

$$P = (\max_i |m_i|) \qquad (5\text{-}124)$$

where a minimum of P is desired.

Minimal Time Problem.

For the system characterized by 5-114, 5-115, and 5-116, find a control sequence $m_0, m_1, \ldots, m_{K-1}$ which drives the system from an initial state $\mathbf{y}_0$ to a desired state $\mathbf{y}_d$ in minimal time K. If more than one such sequence exists for the minimal K, select that sequence which results in the minimum of P given by Equation 5-123. This problem can be solved in several ways [5.61], one of which is to solve the Minimal Action Problem 1 repeatedly for increasingly large values of K. The first value (smallest value) of K for which a solution exists is also the solution to the Minimal Time Problem.

In other control applications of linear programming, Porcelli [5.52] gave a method to determine the digital compensator of a closed-loop system to minimize the time-weighted sum of the error between output $c(k)$ and input $r(k)$, i.e., a minimum is attained for P:

$$P = \sum_k k|r(k) - c(k)| \qquad (5\text{-}125)$$

where $r(k)$ is a deterministic input as a function of integer values of k. Saturation constraints on the error signal are easily incorporated in the design.

Sakawa [5.57] determined the control action which minimizes the performance measure

$$P = \int_0^L |y_d(x) - y(x,T)| \, dx \tag{5-126}$$

where $y(x)$ is the desired distribution of a distributed state variable $y(x,t)$ at time T; the particular distributed-parameter system treated by Sakawa is characterized by

$$\frac{\partial^2 y(x,t)}{\partial x^2} = \frac{\partial y(x,t)}{\partial t} \tag{5-127}$$

subject to the initial conditions $y(x,0) = 0$, and subject to the boundary conditions

$$\left.\frac{\partial y(x,t)}{\partial x}\right|_{x=1} = 0, \quad \left.\frac{\partial y(x,t)}{\partial x}\right|_{x=0} = \alpha[y(0,t) - v(t)] \tag{5-128}$$

where α is a constant, and $v = v(t)$ is the temperature associated with a fuel flow $m(t)$:

$$\frac{dv}{dt} + v = m(t), \quad 0 \le m(t) \le 1 \tag{5-129}$$

Lorchirachoonkul [5.45] generalized Sakawa's approach [5.57] by introducing a performance measure with absolute values of errors weighted both in space and time; Lorchirachoonkul's work is generally applicable to those linear distributed-parameter systems in which control actions are constant between sampling instants, and may also be applied to suitably approximated continuous control problems.

For process control applications, Kuehn and Porter [5.40] introduced the concept of *move penalties*; these account for the cost of changing the settings on a steady-state industrial process subject to control. Certain equality constraints which are "hard constraints" (must be closely satisfied)[6] whereas others which are "soft constraints" (need not be closely satisfied) can be treated by appropriate use of slack variables in the associated equation in conjunction with penalty coefficients in the performance measure—large penalty coefficients for hard constraints and small penalty coefficients for soft constraints. Because most variables in industrial control processes are

---

[6] The terms "hard constraint" and "soft constraint" have origins in mechanics. A hard constraint or limit can be likened to a stiff spring; if a stiff spring (hard constraint) is to be stretched (violated) a significant amount, a relatively great force (expense) is required. Similarly, a soft constraint can be likened to a weak spring; the stretching (violation) of such a spring (constraint) requires relatively little proportional force (expense).

bounded from above, Kuehn and Porter advocate the use of the upper-bounding addition (Section 5-8) to the simplex technique.

5-12c. Communications Problems

A communications network consists of nodes (stations or switching centers) and branches (links between the nodes). Numerous problems of optimization have been posed for such networks: Solutions based on linear programming theory are common, although dynamic programming theory has also been applied in cases (as is evidenced in Chapter 7).

Kalaba and Juncosa [5.36] and Kalaba [5.35] have given an excellent account of the early history of such problems. Two major categories of the problems are to be distinguished: problems of optimal utilization and problems of optimal design. The fairly general problems described by Kalaba and Juncosa are presented next, to be followed by various other problem descriptions and special cases.

The general features of the system are:

1. A set of N stations at which messages are received, sent, and relayed.
2. A set of links connected from station to station. Each station may be connected to every other by distinct links.
3. A set of messages.

Terminology used is:

$T =$ An interval of time in which messages are sent, relayed, and received. It may be of the order of one second or one hour depending on the problem.

$a_{ij} =$ Demand coefficients representing the number of messages originating at station i during a certain time interval T and destined for another station j. By definition, a_{ii} equals identically zero for any i. A message which originates at station i is one which is brought there by means other than the links in question and one which is to be delivered, if possible, over the links to a station other than i.

$c_{ij} =$ Capacity of the direct link from station i to another station j during the time interval T.

$d_{ij} =$ All inclusive total cost per unit of link capacity c_{ij}. This implies a linear relation between link capacity and total link cost—an assumption not altogether justified in many applications but one which is useful if linearization of the actual cost equation can be effected in the vicinity of the optimal solution.

$s_i =$ Switching capacity at station i during the time interval T. Both the number of messages entering and the number of those leaving the ith node during T cannot exceed s_i.

Sect. 5-12   APPLICATIONS   245

d_i = All inclusive total cost per unit of switching capacity s_i. (Same comment as given with d_{ij} applies here.)

x_{ijk} = Number of messages sent from station i (not necessarily originating at station i) over one link to another station j during T and destined for $k \neq i$. By definition, x_{iik} equals identically zero for any i,k combination, as also does x_{iji} for any i,j combination.

P_1 = Ratio of the number of delivered messages to the number of originating messages during T.

Consider, first, the particular systems-utilization problem known as the *optimal routing problem*. The problem is to deliver as many messages as possible in the time interval T; thus the performance measure to be maximized is P_1, as defined above, which is

$$P_1 = \frac{\sum_{i,j} x_{ijj}}{\sum_{i,j} a_{ij}} \tag{5-130}$$

where the constraints to be satisfied are

$$\sum_m x_{imj} - \sum_m x_{mij} \leq a_{ij} \quad \text{for all } i,j; i \neq j \tag{5-131}$$

$$\sum_k x_{ijk} \leq c_{ij} \quad \text{for all } i,j; i \neq j \tag{5-132}$$

$$\sum_{j,k} x_{ijk} \leq s_i \quad \text{for all } i \tag{5-133}$$

and

$$\sum_{j,k} x_{kij} \leq s_i \quad \text{for all } i \tag{5-134}$$

The constraints 5-131, of which there are $N^2 - N$, stem from the demand limitations; constraints 5-132, of which there are $N^2 - N$, are the link-capacity constraints; constraints 5-133, of which there are N, are the switching-center-capacity constraints for messages leaving nodes; and constraints 5-134, of which there are N, are the switching-center-capacity constraints for messages entering nodes. Figure 5-8 illustrates a communications network for the special case of $N = 3$ (see Problem 5.7).

It is clear that expressions 5-130 through 5-134 are of a form to which linear programming theory can be applied. Kalaba and Juncosa have noted, however, that in a system such as the Western Union System, which has some 15 regional switching centers all connected to each other, an optimal routing problem of this type has $2 \times 15^2 = 450$ constraints, in addition to the non-negative constraints on the variables of which there are around 3,000. If

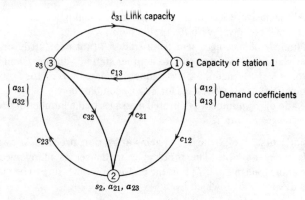

Figure 5-8. A communications network.

solved by using the standard simplex technique, this would be at the threshold of the capacity of most digital computers. Fortunately, most communications problems of the above type have simplifying features which enable special solution algorithms to be developed. In fact, Kalaba [5.35] developed such an algorithm for this problem.

The optimal routing problem just given is easily modified to include cost coefficients which may be associated with messages sent. The performance measure P_1 in that case would include weighting coefficients multiplying each x_{ijj}.

To implement the above system in practice, we would survey the demands a_{ij} during typical time intervals (rush hours, slack periods, etc.). The optimization problem would then be solved by using these several different sets of a_{ij}'s, and the routing procedure changed from time interval to time interval in accordance with the solution obtained for the corresponding time interval.

Essentially the same terminology as used in the preceding routing problem is used by Kalaba and Juncosa [5.36] to describe a communications system *design problem*. Additional notation is required to denote existing link and switching capacities: c_{ij}^e is the existing link capacity and s_i^e is the existing switching capacity in a given system. The quantities c_{ij}^e and s_i^e are taken as zero in the special case that the system is not yet in existence. The problem to be solved is as follows.

Minimize the cost P_2 of adding capacity to the system:

$$P_2 = \sum_{\substack{i,j \\ i \neq j}} d_{ij}(c_{ij} - c_{ij}^e) + \sum_{i} d_i(s_i - s_i^e) \qquad (5\text{-}135)$$

where the c_{ij}'s and the s_i's are the parameters to be selected. As in the preceding problem, constraints 5-131 through 5-134 apply; and in addition, the following constraints must be satisfied:

$$c_{ij} \geq c_{ij}^e, \qquad s_i \geq s_i^e \qquad (5\text{-}136)$$

and

$$\frac{\sum_{i,j} x_{ijj}}{\sum_{i,j} a_{ij}} \geq P_1 \qquad (5\text{-}137)$$

where, in this case, P_1 is specified in the closed interval [0, 1]. Note that this last constraint insures that at least a specified fraction P_1 of the messages will be delivered during T. In practice, the problem would be solved for several values of P_1, and a policy decision based on the results would be made to decide the trade-off point between system performance (characterized by P_1) and the cost P_2 of increased capacity. As before, the problem is one to which linear programming theory is applicable. Wing and Chien [5.68] and Gomory and Hu [5.29] have treated problems which are very similar to the preceding problems.

Kalaba and Juncosa conclude their paper [5.36] by describing an approach which can be used to preschedule increases in capacity for future years. In essence, discrete probabilities are assigned, on the basis of comprehensive economic studies, to probable future demands a_{ij} and to probable future costs d_{ij} and d_i. Because of the probabilistic nature of the problem, we then minimize the expected value of the cost of adding capacity to the system at some future date(s).

Many other communications problems can be cited. Kruskal's work [5.39] applies to the problem of a TV broadcasting company which wishes to lease video links so that its stations in various cities may be formed into a connected network. Each city is to have at most one incoming link, and the cost of leasing links between cities is known. Because of its relatively simple nature, this problem can be solved by several straightforward methods [5.35]. Kruskal's method of solution is the following: among the links not yet included in the connecting net, pick the lowest priced link which does not form any loops with the links already chosen. The proof that this method leads to minimal cost is readily established by assuming the contrary.

A *minimum-time problem* is reviewed by Kalaba [5.35]. Here the problem is to route a message through a network along the path which yields the least time-delay within the network (the delay includes switching time, travel time, etc.). Novel analogs [5.35, 5.49, 5.55] are available for the solution of this problem (see Section 5-11).

Pollack [5.51] considers the problem of finding the *maximum capacity route* through a network. For any given route, the capacity of the route is defined by Pollack as the minimum of the individual link capacities. The capacity of a particular link in one direction may differ from that in the opposite direction.

Kalaba [5.35] considers the problem of determining the path between two given nodes which has the greatest probability of being available for service. Along the same lines, Amara [5.2] considers probabilistic communications networks with nodes and links which are subject to random failure or destruction. Frisch [5.26] considers a communications system in which there is associated with each link a link capacity and a link reliability. Frisch uses an algorithm developed by Wing [5.67] to find routes between a given pair of stations with maximum capacity and with reliability not less than some prescribed minimal value.

5-12d. Circuit Design Applications

In designing an electronic circuit, the question is more than just "how to design the circuit," it is "how to design the circuit so that it operates satisfactorily when it is both mass-produced and used under field conditions." A given circuit is characterized in general by many circuit parameters and circuit constants, each of which has associated tolerances. Tolerances stem from initial manufacture, temperature operating range, aging, and various other reasons in specific cases. For example, a resistor may have an initial $\pm 10\%$ tolerance, plus a $\pm 5\%$ tolerance because of the specified temperature range, plus a $\pm 5\%$ tolerance over the useful life of the circuit; the resulting tolerance is $\pm 20\%$, assuming that the individual tolerances are independent.

Knowledge of the tolerances associated with the components of a circuit enables us to determine worst-case limits under which the circuit should remain operative. The process of designing a circuit to meet performance specifications under extremes of tolerance variations is known as *worst-case design*.

Typically, certain major components of a given circuit are selected solely on the basis of required performance, e.g., power output, gain-bandwidth product, etc. These components are called the "fixed" parts of the circuit. Moreover, the topology of the circuit is usually selected on the basis of its ability to perform the desired circuit operation. In this task, the experience and skill of a circuit designer are of the utmost importance—a poor choice here will yield a poor design in spite of subsequent optimization.

With the fixed parts and the topology specified (for the most part), the *terminal design* task remains. This is the task of selecting the mean values and the tolerances of other assorted circuit components. The terminal design philosophy followed here is that of worst-case design, with the additional feature of optimizing a performance measure under worst-case conditions.

Constraints on the optimization process stem from the equations of performance of the circuit. If the equations of performance or the performance measure are nonlinear, linearization must be performed in the vicinity of the operating point if linear programming theory is to be applied; otherwise, nonlinear or dynamic programming might be used. The linearization approach has been applied to computer circuit design by Goldstick and Mackie [5.28]. Brown and Yang [5.13] have also incorporated linear programming in a closely related approach.

An alternative terminal design philosophy to that given here is embodied in the work of Clunies-Ross and Husson [5.17]. In their work, linear programming theory is used to select parameter values which yield the maximum of the expected number of operable circuits, from the mass production of a circuit, given the continuous probability distributions of various parameter values in the circuit; the distribution about a mean value is assumed known for each parameter, but the mean value itself is selected in the process of optimization.

The following elementary example illustrates the terminal design steps; more involved examples are to be found in the literature [5.13, 5.28].

Example 5-7. Consider the voltage-divider circuit shown in Figure 5-9a. It is assumed that $E_1 \geq V_0 \geq E_2$ over the entire range of V_0; thus, current flow in all branches is positive as referenced to the arrows. It is required that V_0 be bounded inside the specified closed interval $[V_1, V_2]$. The design-center values of E_1 and E_2 are constants which are given—they have been fixed by other considerations—tolerances on E_1 and E_2 are also assumed known. The current to be drawn from the divider is at most $I_{\max}$ and could be as small as $I_{\min}$ at times.

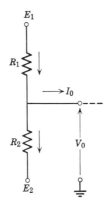

Figure 5-9a. A voltage-divider circuit.

250   THE SIMPLEX TECHNIQUE AND LINEAR PROGRAMMING   Ch. 5

The standard worst case design problem would be to determine design-center values of R_1 and R_2 such that the voltage constraint on V_0 is satisfied at all times. For optimal worst-case design, an additional feature is incorporated in the problem; here the worst-case value R_{max} of the output impedance is to be minimized with respect to the selection of R_1 and R_2, but subject to satisfaction of the previously noted constraints, where

$$R_{max} = \frac{R_{1max}R_{2max}}{R_{1max} + R_{2max}}$$

$$= \frac{1}{G_{1min} + G_{2min}} \qquad (5\text{-}138)$$

and where $G_{1min} = 1/R_{1max}$ and $G_{2min} = 1/R_{2max}$. It is clear that R_{max} is minimized if the following performance measure P is maximized:

$$P = G_{1min} + G_{2min} = (1 - \epsilon_1)\bar{G}_1 + (1 - \epsilon_2)\bar{G}_2 \qquad (5\text{-}139)$$

where ϵ_1 is the assumed-known fractional tolerance of G_1, ϵ_2 is the assumed-known fractional tolerance of G_2, and $\bar{G}_1$ and $\bar{G}_2$ are design-center values.

From the equivalent circuit shown in Figure 5-9b, it is seen that

$$V_0 = \frac{E_1 G_1 + E_2 G_2 - I_0}{G_1 + G_2} \qquad (5\text{-}140)$$

Note that

$$\frac{\partial V_0}{\partial G_1} = \frac{G_2(E_1 - E_2) + I_0}{(G_1 + G_2)^2} > 0 \qquad (5\text{-}141)$$

and

$$\frac{\partial V_0}{\partial G_2} = \frac{-G_1(E_1 - E_2) + I_0}{(G_1 + G_2)^2} < 0 \qquad (5\text{-}142)$$

These inequalities indicate that V_0 assumes its least value when G_1 assumes

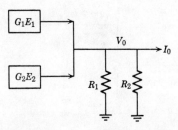

Figure 5-9b. An equivalent circuit.

its minimum value and G_2 assumes its maximum value. Thus, the constraint that V_0 be greater than V_1 is

$$\frac{E_{1\min}G_{1\min} + E_{2\min}G_{2\max} - I_{0\max}}{G_{1\min} + G_{2\max}} \geq V_1 \tag{5-143}$$

and similarly, in order for V_0 to be less than V_2,

$$\frac{E_{1\max}G_{1\max} + E_{2\max}G_{2\min} - I_{0\min}}{G_{1\max} + G_{2\min}} \leq V_2 \tag{5-144}$$

Inequalities 5-143 and 5-144 are rearranged to obtain

$$(1 - \epsilon_1)(E_{1\min} - V_1)\bar{G}_1 + (1 + \epsilon_2)(E_{2\min} - V_1)\bar{G}_2 \geq I_{0\max} \tag{5-145}$$

and

$$(1 + \epsilon_1)(E_{1\max} - V_2)\bar{G}_1 + (1 - \epsilon_2)(E_{2\max} - V_2)\bar{G}_2 \leq I_{0\min} \tag{5-146}$$

where $\bar{G}_1$ and $\bar{G}_2$ are greater than or equal to zero. The attainment of maximal P in 5-139 subject to these constraints constitutes a well-defined linear programming problem.

●

5-12e. Field Problems

In this subsection, a "field" problem which reduces to a linear programming problem is posed. The problem has direct analogs in the areas of fluid flow and heat conduction.

Consider a two-dimensional conducting-sheet field problem with specified boundary conditions (see Figure 5-10, for example). The black areas on the

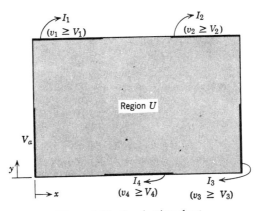

Figure 5-10. Conducting sheet.

boundary of the conducting sheet in Figure 5-10 are areas of essentially perfect conduction and therefore are areas of uniform potential. Voltage V_a is an applied potential and is fixed. Currents $I_1, I_2, \ldots$ ($I_i \geq 0$) are parameters to be specified such that a maximum of P is obtained:

$$P = \sum_i I_i \tag{5-147}$$

subject to the restriction that

$$v_i \geq V_i \quad \text{for } i = 1, 2, \ldots \tag{5-148}$$

where v_i is the voltage at terminal i and V_i is a given lower bound imposed on the potential v_i.

Potentials within the region U satisfy Laplace's equation

$$\frac{\partial^2 v(x,y)}{\partial x^2} + \frac{\partial^2 v(x,y)}{\partial y^2} = 0 \tag{5-149}$$

and a discrete approximation to this equation is

$$\frac{v(x_{i-1},y_j) - 2v(x_i,y_j) + v(x_{i+1},y_j)}{(\Delta x)^2} + \frac{v(x_i,y_{j-1}) - 2v(x_i,y_j) + v(x_i,y_{j+1})}{(\Delta y)^2} = 0 \tag{5-150}$$

Let $\Delta x = \Delta y$ and let $v(x_i,y_j) \triangleq v(i,j)$. An equivalent to Equation 5-150 is then

$$v(i,j-1) + v(i,j+1) + v(i-1,j) + v(i+1,j) - 4v(i,j) = 0 \tag{5-151}$$

which equation applies for all i,j combinations which correspond to interior nodes.

In Figure 5-11 a coarse approximation is depicted for the conducting sheet of Figure 5-10. The resistance R in ohms of each unmarked resistor is the resistance across opposite edges of a square section of the sheet; this resistance is independent of the size of the square—an interesting property. Note that the resistors along the edge of the nonconducting boundary are assigned values of $2R$ because they represent only half of a square area. For example, at node (6,3) the characterizing equation is

$$2v(5,3) + v(6,4) + v(6,2) - 4v(6,3) = 0 \tag{5-152}$$

and since $v(6,4) = v_2$ and $v(6,2) = v_3$, this is equivalent to

$$2v(5,3) + v_2 + v_3 - 4v(6,3) = 0 \tag{5-153}$$

At the conducting boundaries the equations for the voltages are governed by Kirchhoff's current law; at boundary number 4, for example,

$$v(1,0) + 2v(2,1) + 2v(3,1) + 2v(4,1) + v(5,0) - 8v_4 = 2I_4 R \tag{5-154}$$

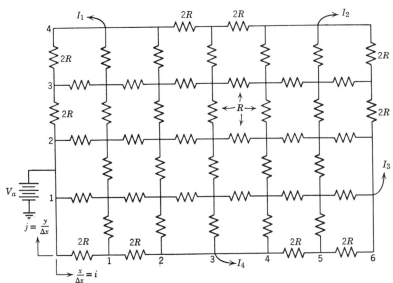

Figure 5-11. An approximation to a conducting sheet.

Similar equations to those obtained above can be obtained for each of the unknown node voltages. The problem is to maximize P given by 5-147 without violating the lower-bound constraints on the voltages v_1, v_2, v_3, and v_4.

Although the problem is in a linear programming form as developed above, the optimization process is simplified considerably if we solve for v_1, v_2, v_3, and v_4 in terms of the current sources I_1, I_2, I_3, and I_4 and in terms of the applied voltage V_a—this is most easily accomplished by using the principle of superposition, perhaps in conjunction with relaxation methods [5.58]. The resulting equations are still linear, obviously, and the number of unknowns is greatly reduced by this approach.

5-12f. Other Applications

A brief listing of other applications of linear programming is given in this subsection. Applications that are significantly different from those described in all of Section 5-12 are to be found in bibliographical sources [e.g., 5.1, 5.4, 5.56]. Without doubt, numerous additional applications will be considered in future literature. In fact, the reader may have already conceived such applications.

Methods of linear programming have been considered for use in optimal logic design and related areas [5.11, 5.12]. In these applications, variables are usually restricted to be Boolean, so that the simplex technique of this chapter is not directly applicable. Similarly, linear programming problems in which some or all of the variables are restricted to integral values may be solved by use of *integer linear programming* techniques [5.8, 5.60].

Other proposed applications include the static design of multicoordinate systems [5.30] and the development of schemes for pattern recognition [5.47, 5.50]. Pursuit of these topics is left to readers with associated interests.

5-13. CONCLUSION

Linear programming is a relatively well-developed area of knowledge. For additional reading, standard works are available at various levels of depth [5.15, 5.21, 5.31, 5.44], and bibliographical sources are available for those seeking additional applications and/or methods [5.1, 5.4, 5.56].

In Section 5-2, a standard form of linear programming problem is posed, and the degree of generality of the problem is exhibited in Section 5-3. Although quite general, this standard form excludes integer linear programming problems [5.7, 5.8, 5.18, 5.60, 5.70], problems with linear integral performance measures [5.41, 5.43, 5.62], stochastic linear programming problems [5.9, 5.20, 5.46, 5.66], and others.

The simplex technique for solving linear programming problems is developed in Sections 5-4 through 5-6. This technique and its variations are considered to be the best general approach to the solution of standard linear programming problems. Other techniques, too numerous to list here, are available, however; and for problems which are less general in form or which exhibit special features, other techniques may afford computational savings [5.6, 5.21, 5.37].

Scaling, initializing, and upper-bounding are important considerations in the solution of any linear programming problem and are given detailed attention in Sections 5-7 and 5-8. As for Sections 5-9 and 5-10, much more space than is given could be fruitfully devoted to the subjects of dual problems [5.42, 5.44] and sensitivity analysis [5.14, 5.65].

Analog methods are considered in Section 5-11. These methods have been less developed in the literature than numerical methods, but the analog methods proposed to date are promising and appear to be especially useful for solving special types of linear programming problems. Relatively little detail is reported in the literature on actual linear programming problems solved by use of analog methods.

REFERENCES

[5.1] *Linear Programming*. OTS Selective Bibliography SB-407, Revised, 1950 to June 1963, Office of Technical Services, U.S. Dept. of Commerce, Washington, D.C.

[5.2] Amara, R. C. "Computer Design and Control of Probabilistic Communication Networks." *IEEE Transactions on Communication Systems*, **CS-11**, 30–36, March 1963.

[5.3] Andreyev, N. I. "Method of Solving Certain Problems of Nonlinear Programming." *Engineering Cybernetics*, 13–20, Jan.–Feb. 1963.

[5.4] Arnoff, E. L., and S. S. Sengupta. "Mathematical Programming." Chapter 4 of *Progress in Operations Research*, vol. 1, R. L. Ackoff (Editor). Wiley, New York, 1961.

[5.5] Asher, D. T. "A Linear Programming Model for the Allocation of R and D Efforts." *IRE Transactions on Engineering Management*, **EM-9**, 154–157, Dec. 1962.

[5.6] Bakes, M. D. "Solution of Special Linear Programming Problems with Additional Constraints." *Operations Research Quarterly*, **17**, 425–445, Dec. 1966.

[5.7] Beale, E. M. L. *A Method of Solving Linear Programming Problems when Some but not All of the Variables Must Take Integral Values*. Technical Report No. 19, Statistical Techniques Research Group, Princeton University, Princeton, N.J., 1958.

[5.8] Beale, E. M. L. "Survey of Integer Programming." *Operations Research Quarterly*, **16**, 219–228, June 1965.

[5.9] Bereanu, B. "On Stochastic Linear Programming, the Laplace Transform of the Distribution of the Optimum and Applications." *Journal of Mathematical Analysis and Applications*, **15**, 280–294, Aug. 1966.

[5.10] Bergsman, J. "Electric Power Systems Planning Using Linear Programming." *IEEE Transactions on Military Electronics*, **MIL-8**, 59–62, April 1964.

[5.11] Breuer, M. A. *The Use of Mathematical Programming in the Implementation of Boolean Switching Functions*. Ph.D. Thesis, Univ. of California, Berkeley, Calif., 1965, 120 pp.

[5.12] Breuer, M. A. "Implementation of Threshold Nets by Integer Linear Programming." *IEEE Transactions on Electronic Computers*, **EC-14**, 950–952, Dec. 1965.

[5.13] Brown, A., and H. C. Yang. "Optimal Worst-Case Circuit Design." *Proceedings of the National Electronics Conference*, **21**, 737–742, 1965.

[5.14] Charnes, A., W. W. Cooper, and B. Mellon. "A Model for Programming and Sensitivity Analysis in Integrated Oil Company." *Econometrica*, **22**, 193–217, April 1954.

[5.15] Charnes, A., W. W. Cooper, and A. Henderson. *An Introduction to Linear Programming*. Wiley, New York, 1953.

[5.16] Cheney, L. K. "The Significance of Mathematical Programming in the Business World." *Symposium on Mathematical Programming*, RAND Corporation, Santa Monica, Calif., March 16, 1959.

[5.17] Clasing, R. ???, C., and S. S. Haddam. "Statistical Techniques in Circuit Optimization." *Proceedings of the National Electronics Conference*, **18**, 325–334, 1962.

[5.18] Cook, R. A. *An Algorithm for Integer Linear Programming.* Ph.D. Thesis, Washington University, St. Louis, 1966, 163 pp.

[5.19] Dantzig, G. B. "Maximization of a Linear Function of Variables Subject to Linear Inequalities." Chapter 21 of *Activity Analysis of Production and Allocation*, T. C. Koopmans (Editor). Cowles Commission Monograph No. 13, Wiley, New York, 1951.

[5.20] Dantzig, G. B. "Linear Programming Under Uncertainty." *Management Science*, **1**, 197–206, April–July 1955.

[5.21] Dantzig, G. B. *Linear Programming and Extensions.* Princeton Univ. Press, Princeton, N.J., 1963.

[5.22] Dantzig, G. B., and A. Orden. "A Duality Theorem Based on the Simplex Method." Pages 51–55 of *Symposium on Linear Inequalities*, A. Orden and L. Goldstein (Editors). Project SCOOP Publication No. 10, Directorate of Management Analysis Service, Washington, D.C., 1952.

[5.23] Dantzig, G. B., A. Orden, and P. Wolfe. "The Generalized Simplex Method for Minimizing a Linear Form under Linear Inequality Restraints." *Pacific Journal of Mathematics*, **5**, 183–195, June 1955.

[5.24] Dennis, J. B. *Mathematical Programming and Electrical Networks.* The M.I.T. Press, Cambridge, Mass., 1959.

[5.25] Dov, A. G. B. "Optimal Assignment of Research and Development Projects in a Large Company Using an Integer Programming Model." *IEEE Transactions on Engineering Management*, **EM-12**, 138–142, Dec. 1965.

[5.26] Frisch, I. T. "Optimum Routes in Communication Systems with Channel Capacities and Channel Reliabilities." *IEEE Transactions on Communication Systems*, **CS-11**, 241–245, June 1963.

[5.27] Gale, D., H. W. Kuhn, and A. W. Tucker. "Linear Programming and the Theory of Games." Chapter 19 of *Activity Analysis of Production and Allocation*, T. C. Koopmans (Editor). Cowles Commission Monograph No. 13, Wiley, New York, 1951.

[5.28] Goldstick, G. H., and D. G. Mackie. "Design of Computer Circuits Using Linear Programming Techniques." *IRE Transactions on Electronic Computers*, **EC-11**, 518–530, Aug. 1962.

[5.29] Gomory, R. E., and T. C. Hu. "An Application of Generalized Linear Programming to Network Flows." *Journal of the Society for Industrial and Applied Mathematics*, **10**, 260–283, June 1962.

[5.30] Gurevich, A. M. "Application of Mathematical Programming to the Static Design of Multicoordinate Linear Systems." *Automation and Remote Control*, **26**, 291–296, 1965.

[5.31] Hadley, G. *Linear Programming.* Addison-Wesley, Reading, Mass., 1962.

[5.32] Hanson, G. E. *Linear Programming Sensitivity Analysis.* Report R63CD12, General Electric Computer Department, Phoenix, Ariz., April 1964, 32 pp.

[5.33] Hovanessian, S. A., and T. M. Stout. "Optimum Fuel Allocation in Power Plants." *IEEE Transactions on Power Apparatus and Systems*, **82**, 329–335, June 1963.

REFERENCES

[5.34] Jackson, A. S. *Analog Computation*. McGraw-Hill, New York, 1960.

[5.35] Kalaba, R. "On Some Communication Network Problems." *Proceedings, Symposium in Applied Mathematics* (April 1958), American Math. Society, **10**, 261–280, 1960.

[5.36] Kalaba, R., and M. Juncosa. "Optimal Design and Utilization of Communication Networks." *Management Science*, **3**, 33–44, Oct. 1956.

[5.37] Kim, M. "On the Minimum Time Control of Linear Sampled-Data Systems." *Proceedings of the IEEE*, **53**, 1263–1264, Sept. 1965.

[5.38] Krasovskii, N. N. "On the Theory of Optimal Regulation." *Automatika i Telemekhanika*, **18**, pp. 960–970, Nov. 1957.

[5.39] Kruskal, J. B., Jr. "On the Shortest Spanning Subtree of a Graph and the Traveling Salesman Problem." *Proceedings, American Math. Society*, **7**, 48–50, 1956.

[5.40] Kuehn, D. R., and J. Porter. "The Application of Linear Programming Techniques in Process Control." *IEEE Transactions on Industrial Electronics and Control Instrumentation*, **IECI-11**, 64–70, Feb. 1964.

[5.41] Lehman, R. S. *On the Continuous Simplex Method*. Research Memo. RM-1386, RAND Corporation, Santa Monica, Calif., Dec. 1954, 55 pp.

[5.42] Lemke, C. E. "The Dual Method of Solving the Linear Programming Problem." *Naval Research Logistics Quarterly*, **1**, 36–47, 1954.

[5.43] Levinson, N. "A Class of Continuous Linear Programming Problems." *Journal of Mathematical Analysis and Applications*, **16**, 73–83, Oct. 1966.

[5.44] Llewellyn, R. W. *Linear Programming*, Holt, Rinehart, and Winston, New York, 1964.

[5.45] Lorchirachoonkul, V., and D. A. Pierre. "Optimal Control of Multi-Variable Distributed-Parameter Systems Through Linear Programming." *Preprint Volume*, Joint Automatic Control Conference, 702–710, June 1967.

[5.46] Madansky, A. *Methods of Solution of Linear Programs Under Uncertainty*. Research Memo. RM-2752, RAND Corporation, Santa Monica, Calif., April 1961, 20 pp.

[5.47] Mangasarian, O. L. "Linear and Nonlinear Separation of Patterns by Linear Programming." *Operations Research*, **13**, 444–452, May–June 1965.

[5.48] Masse, P., and R. Gibrat. "Application of Linear Programming to Investments in the Electric and Power Industry." *Management Science*, **3**, 149–166, Jan. 1957.

[5.49] Miehle, W. "Link-Length Minimization in Networks." *Operations Research*, **6**, 232–243, March–April 1958.

[5.50] Pervozvanskiy, A. A. "Recognition of Abstract Patterns as a Problem of Linear Programming." *Engineering Cybernetics*, 38–41, July–Aug. 1965.

[5.51] Pollack, M. "The Maximum Capacity Route Through a Network." *Operations Research*, **8**, 733–736, Sept.–Oct. 1960.

[5.52] Porcelli, G. "Linear Programming Design of Digitally Compensated Systems." *Preprint Volume*, Joint Automatic Control Conference, 412–421, June 1964.

[5.53] Propoi, A. I. "Use of Linear Programming Methods for Synthesizing Sampled-Data Automatic Systems." *Automation and Remote Control*, **24**, 837–844, 1963.

[5.54] Pyne, I. B. "Linear Programming on an Electronic Analogue Computer." *AIEE Transactions*, Part I, Communications and Electronics, **75**, 139–143, May 1956.

[5.55] Rapaport, H., and P. Abramson. "An Analog Computer for Finding an Optimum Route Through a Communication Network." *IRE Transactions on Communication Systems*, **CS-7**, 37–41, May 1959.

[5.56] Riley, V., and S. I. Gass. *Linear Programming and Associated Techniques: A Comprehensive Bibliography on Linear, Nonlinear, and Dynamic Programming.* Johns Hopkins Press, Baltimore, Md., 1958.

[5.57] Sakawa, Y. "Solution of an Optimal Control Problem in a Distributed-Parameter System." *IEEE Transactions on Automatic Control*, **AC-9**, 420–426, Oct. 1964.

[5.58] Shaw, F. S. *An Introduction to Relaxation Methods.* Dover Publications, New York, 1953.

[5.59] Sreedharan, V. P. "A Short Proof of the Duality Theorem of Linear Programming." *Journal of the Society for Industrial and Applied Mathematics*, **13**, 423–424, June 1965.

[5.60] Srinivasan, A. V. "An Investigation of Some Computational Aspects of Integer Programming." *Journal of the Association for Computing Machinery*, **12**, 525–535, Oct. 1965.

[5.61] Torng, H. C. "Optimization of Discrete Control Systems Through Linear Programming." *Journal of the Franklin Institute*, **277**, 28–44, July 1964.

[5.62] Tyndall, W. F. "A Duality Theorem for a Class of Continuous Linear Programming Problems." *Journal of the Society for Industrial and Applied Mathematics*, **13**, 644–666, Sept. 1965.

[5.63] von Neumann, J. "Discussion of a Maximum Problem." Institute for Advanced Study, Princeton, N.J., 1947.

[5.64] Whalen, B. H. *Linear Programming for Optimal Control.* Ph.D. Thesis, Univ. of California, Berkeley, Calif., 1963, 69 pp.

[5.65] Williams, A. C. "Marginal Values in Linear Programming." *Journal of the Society for Industrial and Applied Mathematics*, **11**, 82–94, March 1963.

[5.66] Williams, A. C. "On Stochastic Linear Programming." *Journal of the Society for Industrial and Applied Mathematics*, **13**, 927–940, Dec. 1965, and **15**, 228, Jan. 1967.

[5.67] Wing, O. "Algorithms to Find the Most Reliable Path in a Network." *IRE Transactions on Circuit Theory*, **CT-8**, 78–79, March 1961.

[5.68] Wing, O., and R. T. Chien. "Optimum Synthesis of Communication Networks." *IRE Transactions on Circuit Theory*, **CT-8**, 44–48, March 1961.

[5.69] Wolfe, P. "A Technique for Resolving Degeneracy in Linear Programming." *Journal of the Society for Industrial and Applied Mathematics*, **11**, 205–211, June 1963.

[5.70] Young, R. D. "A Primal (All-Integer) Integer Programming Algorithm." *Journal of Research of the National Bureau of Standards*, **69B**, section B, 213–250, July–Sept. 1965.

[5.71] Zadeh, L. A. "On Optimal Control and Linear Programming." *IRE Transactions on Automatic Control*, **AC-7**, 45–46, July 1962.

PROBLEMS

5.1 Consider the following linear programming problem.
Maximize P:
$$P = |x_1| + x_2 + x_3$$
Subject to:
$$x_2 \geq 0 \text{ and } x_3 \geq 0$$
$$x_1 + 2x_2 + x_3 = 8$$
$$2x_1 + x_2 + x_3 \geq 10$$
$$x_1 - x_2 + x_3 \leq 5$$

Place this problem in standard linear programming form.

5.2 Given P to be minimized:
$$P = 1/(c_1 x_1 + c_2 x_2 + \cdots + c_n x_n)$$
where all c_i's and x_i's are assumed non-negative. Constraints of the form 5-2 and 5-3 are assumed to apply. Give an equivalent linear programming problem in standard form.

5.3 Three men are to be scheduled to perform six different tasks. The ith man can perform a_{ij} units of the jth task in one hour and is paid c_{ij} dollars for one hour's work on the jth task. The total number of hours which the ith man can work is b_{1i}, and the number of units required to complete the jth task is b_{2j}. A minimum of the cost P is desired:
$$P = \sum_{i=1}^{3} \sum_{j=1}^{6} c_{ij} x_{ij}$$
where x_{ij} is the number of hours that the ith man spends on the jth task. Place this problem in a standard linear programming form.

5.4 Place the performance measure of Equation 5-123 in the standard linear programming form of Equation 5-1. (*Hint:* Follow an approach similar to that given under Case D of Section 5-3.)

5.5 Place the performance measure of Equation 5-124, or that of 5-121, in the standard linear programming form of Equation 5-1. (*Hint:* Introduce a new variable z with the property that $z \geq |m_i|$ for all i; convert the latter set of constraints to standard form, and minimize z.)

5.6 Parallel the development given in Example 5-2, using the following performance measure in place of 5-18:
$$J = \int_0^T y(t) |y_d(t) - y(t)| \, dt$$
in which $y(t)$ need not be less than the desired response $y_d(t)$.

5.7 As a specific example of the Kalaba and Juncosa routing problem (Section 5-12c), consider the three-station case depicted in Figure 5-8. Give the performance measure and the constraint equations which correspond to the following coefficients:

$$s_1 = 5, s_2 = 4, s_3 = 6, a_{12} = 2, a_{13} = 1, a_{23} = 2, a_{31} = 2, a_{21} = 1$$
$$a_{32} = 2, c_{12} = 3, c_{13} = 3, c_{21} = 1, c_{23} = 2, c_{31} = 2, \text{ and } c_{32} = 2$$

5.8 Consider Figure 5-P8: A given set of N data points (y_k, i_k)'s, where y_k is a given function of i_k, are to be fitted approximately by a continuous function

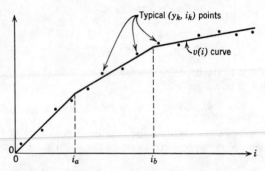

Figure 5-P8. A best fit of line segments to data points.

$v \equiv v(i)$ which is linearly related to i between the given values $i = 0$, i_a, i_b, and i_c. Notationally,

$$v_k \stackrel{\Delta}{=} v(i_k) = \begin{cases} \beta_1 i_k & \text{for } 0 \leq i_k \leq i_a \\ \beta_2 i_k + \beta_0 & \text{for } i_a \leq i_k \leq i_b \\ \beta_4 i_k + \beta_5 & \text{for } i_b \leq i_k \end{cases}$$

For continuity,
$$\beta_1 i_a = \beta_2 i_a + \beta_3$$
and
$$\beta_2 i_b + \beta_3 = \beta_4 i_b + \beta_5$$

Values of the β_j's are to be selected.

a. Frame the problem in terms of a standard linear programming form, given that
$$P = \sum_{k=1}^{N} |v_k - y_k|$$
is to be minimized.

b. Frame the problem in terms of a standard linear programming form, given that
$$P = \max |v_k - y_k|$$
is to be minimized. (See the hint given under Problem 5.5.)

c. Suppose the v versus i characteristic of Figure 5-P8 is to be implemented with a network of diodes, resistors, and voltage sources: i is the input current and v is the voltage across the input terminals. Describe such a network.

5.9 Draw an x_1, x_2 coordinate system and show the convex region which satisfies the constraints that follow:

$$x_2 \geq 0 \quad \text{and} \quad 0 \leq x_1 \leq 3$$
$$-x_1 + x_2 \leq 1$$
$$x_1 + x_2 \leq 4$$

Of the points in this region, find the point (or set of points) which yields the maximum of:

a. $P = 2x_1 + x_2$
b. $P = x_1 + x_2$
c. $P = x_1 + 2x_2$

5.10 Which of the following characterize convex sets?

a. $x_1^2 + x_2 \leq 1$
b. $x_1^2 + x_2 \geq 1$
c. $x_1 + x_2 + x_3 = 2$
   $x_1 + 2x_2 = 1$
d. $x_1 + x_2 + x_3 = 2$
   $x_1 + 2x_2 \leq 1$
e. $x_1^2 + x_2^2 \leq 1$
   $x_1 - x_2 \geq 1$
f. The intersection of two convex sets.

5.11 Determine the extreme points of the convex sets in Problem 5.10.

5.12 Use the simplex technique to find the x_i's which yield the maximum of P,

$$P = 2x_1 + 3x_2 + x_3$$

where x_1, x_2, and x_3 are required to be non-negative and to satisfy the constraints that follow:

$$x_1 + x_2 + 2x_3 \leq 200$$
$$3x_1 + 2x_2 + x_3 \leq 500$$
$$x_1 + 2x_2 + 2x_3 \leq 300$$

5.13 Assuming optimal values P^*, x_1^*, x_2^*, and x_3^* corresponding to Problem 5.12 are known, express the solution of the following problem in terms of these known values.

Maximize P_1,

$$P_1 = 2x_1 + 1.5x_2 + x_3$$

where the non-negative parameters x_1, x_2, and x_3 must satisfy

$$x_1 + 0.5x_2 + 2x_3 \leq 2$$
$$3x_1 + x_2 + x_3 \leq 5$$
$$x_1 + x_2 + 2x_3 \leq 3$$

(*Hint:* Note that Problem 5.13 is a column-scaled version of Problem 5.12.)

5.14 Consider the problem of maximizing P,

$$P = 2x_1 + 0.1x_2 + 10x_3$$

where x_1, x_2, and x_3 are constrained by $x_1, x_2, x_3 \geq 0$ and

$$-x_1 + 0.2x_2 + 10x_3 \geq 1$$
$$200x_1 - 10x_2 + 2{,}000x_3 \leq 0$$
$$10x_1 + x_2 + 100x_3 \leq 40$$

a. Scale and initialize this problem.
b. Solve the scaled problem by using the simplex algorithm and determine the optimal values of the unscaled variables.

5.15 A maximum of P is desired, where

$$P = 3x_1 + 2x_2 - 2x_3$$

Constraints which must be satisfied are:

$$4x_1 + 2x_2 + 2x_3 \leq 20$$
$$2x_1 + 2x_2 + 4x_3 \geq 6$$

and $x_1 \geq 0$, $x_2 \geq 0$, but x_3 is unrestricted in sign.
a. Solve the problem.
b. Give the dual problem and obtain the solution of the dual problem from the last tableau of the primal solution.

5.16 Repeat the running Example of Section 5-6, but with an upper bound of 4 on x_1.

5.17 Consider the following abbreviated simplex tableau:

i	Current **b**	Key column k	Possible θ's	d_i
1	10	2	5	$d_1 = 20$
2	8	-3	—	$d_2 = $ (as specified in problem statement)
3	6	3	2	$d_3 = 12$

The d_i's in the right-hand column are upper bounds on the corresponding x_i's. Designate the path to be followed in the flow diagram of Figure 5-4 under each of the following sets of conditions:
a. The upper bound d_k on the variable associated with the key column equals 4, and d_2 equals 17.
b. The upper bound d_k on the variable associated with the key column equals 4, and d_2 equals 11.
c. The upper bound d_k on the variable associated with the key column equals 1, and d_2 equals 17.

5.18 Give the dual problem of Problem 5.1 and solve the dual to obtain a solution of the primal.

5.19 Consider the following primal problem: A maximum of P is desired, where

$$P = 1.1x_1 + 1.2x_2 + x_3$$

Constraints which must be satisfied are:

$$2x_1 + 2x_2 + 2x_3 \leq 10$$
$$x_1 + 3x_2 + x_3 \leq 10$$
$$4x_1 + x_2 + x_3 \leq 10$$
$$3x_1 + x_2 + 3x_3 \leq 10$$
$$x_1 + 2x_2 + 3x_3 \leq 10$$
$$3x_1 + 2x_2 + x_3 \leq 10$$

where all x_i's are constrained to be non-negative.
  a. Give the dual problem.
  b. Solve the dual problem.
  c. Obtain the solution of the primal problem from the last tableau of the dual solution.

5.20 Find the sensitivity ranges associated with c_1, c_2, c_3, and c_4 in Table 5-8.

5.21 Construct an analog computer diagram for the solution of the following problem:

$$P = 10x_1 + 8x_2 \text{ (to be maximized)}$$

subject to satisfaction of $x_1, x_2, x_3 \geq 0$ and

$$10x_1 + 5x_2 \leq 50$$
$$5x_1 + 5x_2 \leq 35$$
$$5x_1 + 15x_2 \leq 80$$

5.22 Derive an analog computer circuit that can be used to account for a constraint of the form

$$a_1 x_1 + a_2 x_2 + \cdots + a_k x_k \leq -1$$

5.23 Consider maximization of P,

$$P = x_1 + 0.3x_2$$

subject to

$$x_1 - \tfrac{1}{3}x_2 \leq 1 \quad \text{and} \quad -x_1 + x_2 \leq 1$$

  a. Show that this problem does not satisfy the assumption made above inequality 5-112 (see also the footnote on page 235). Give a graphical portrayal of the solution, as in Figure 5-5.
  b. Modify the approach of Section 5-11b to apply to this problem. Make sure that a constraint breakthrough cannot occur.

5.24 Use a linear programming computer program on a general-purpose digital computer to solve a problem of your own choosing.

SEARCH TECHNIQUES AND NONLINEAR PROGRAMMING

6

6-1. INTRODUCTION

The process of searching for a solution to a problem is one which is basic to all mankind. When many solutions to the same problem exist, the search for the best solution, best in some sense, is just as basic. Now, just as problems appear in ever-varied form, so also do search techniques. A given search technique may be superior to all others in a limited domain, but we cannot hope that it will cope with all problems equally well. It is for this reason that a variety of search techniques, along with underlying principles, is presented in this chapter.

The primary problem considered in this chapter is that of finding the extrema of a performance measure which is a nonlinear real-valued function of n parameters. The function may be given analytically or it may be determined experimentally; noise and experimental error may or may not be associated with the function; the function may or may not exhibit discontinuities; and constraint equations may exist which limit the arguments of the performance measure. In the latter case, the problem is called the *nonlinear programming* problem, in analogy with the naming of linear programming (Chapter 5).

Search techniques are called *direct methods* of solving problems of the type posed in the preceding paragraph. This is in contrast to the methods of Chapter 2 which are indirect methods, ones based solely on differential properties that certain classes of functions exhibit at points of extrema. The set of search techniques may be subdivided in many ways: discrete search versus continuous search; nonsequential search versus sequential search; local search versus global search; search with quadratic convergence versus search without quadratic convergence; and so forth. Specific methods which fall within any one of these categories have merit for certain problems. Given a particular optimization problem, a searcher's prime concern is to utilize a search technique which not only solves the problem, but also solves it efficiently; we might say that a searcher seeks the search technique which is "optimum" for his optimization problem.

The efficiency of a given search technique is affected by certain global and local properties of functions (Section 6-2). For functions of one variable (Section 6-3), five one-dimensional search techniques are presented, the first two of which are more appropriately used on analytically given functions while the remaining three may be more useful when data are obtained experimentally. The importance of efficient one-dimensional search is heightened by the fact that many n-dimensional search techniques incorporate a sequence of one-dimensional[1] searches in n space (Section 6-3f).

Nonsequential search methods (Section 6-4) are generally inefficient, but are useful under important special conditions. Univariate (one variable at a time) search and relaxation search techniques (Section 6-5) are often convenient when data are obtained experimentally, but these techniques do not provide rapid convergence to the optimum of most analytically given functions. Basic gradient methods (Section 6-6) are perhaps the best-known search techniques; continuous gradient search (Section 6-6b) can be programmed on general-purpose analog computers; best-step steepest ascent (Section 6-6c) is a stepping-stone to the more efficient acceleration-step search (Section 6-7); and Newton search (Section 6-6d) is a generalization of the one-dimensional Newton-Raphson search technique (Section 6-3a). The method of parallel tangents (Section 6-7b) and conjugate-direction search techniques (Section 6-8) are based in part on the gradient concept, and exhibit the desirable property of quadratic convergence (defined in Section 6-3a). That the methods of Sections 6-7 and 6-8 have proved to be highly efficient is shown by an overall comparison (Section 6-12).

The first application of gradient search was given by Cauchy [6.16]; he outlined a procedure for solving a set of simultaneous algebraic equations by using search techniques. This procedure (Section 6-10) is incorporated in the penalty-function method of solving the general nonlinear programming problem (Section 6-11). Once a performance measure is augmented by a penalty function, any of the known search methods may be used in obtaining the optimum. Also examined in Section 6-11 is Everett's theorem which brings the Lagrange multiplier once more to the fore, this time for the solution of very general nonlinear programming problems.

[1] A one-dimensional search in an n-dimensional space is not to be confused with a one-variable (univariate) search in an n-dimensional space. The former is more general than the latter.

The search methods of Section 6-9 have novel facets which render them worthy of consideration. Pattern search is a special case of a general class of search techniques called direct search techniques (Section 6-9b); bunny-hop search is a special case of search by directed array (Section 6-9c); creeping random methods (Section 6-9d) are sequential random methods; and centroid methods (Section 6-9e) are based on the density concept of a function.

An illustrative problem is treated by several different methods (Examples 6-2, 6-3, 6-4, 6-5, and 6-6), and the overall philosophy of search is summarized in Section 6-13. As for applications of search techniques, the following list is suggestive: simultaneous solution of nonlinear algebraic equations [6.6]; combined use of direct methods of the calculus of variations (Section 3-11) and search techniques for systems design [6.18, 6.47]; filter and circuit design [6.1, 6.75, 6.86, 6.87]; control systems [6.42, 6.50, 6.62]; optimal scheduling of hydrothermal systems [6.92]; determination of the parameters of a dynamic process [6.108]; root-locus plotting by continuous search [6.59, 6.67]; adaptive control [6.17, 6.95]; and adaptive pattern recognition [6.5]. It should be evident too that the applications noted in Chapter 5 are also relevant in this chapter; problems that can be solved by using linear programming methods are a subclass of those that can be solved by using nonlinear programming methods.

6-2. GEOMETRICAL INTERPRETATION AND SCALING

All search techniques involve some form of comparison; e.g., a function $f(\mathbf{x})$ is evaluated at various allowed values of $\mathbf{x} = \{x_1, x_2, \ldots, x_n\}$, and certain values of $f(\mathbf{x})$ are compared in order to find an estimate of the optimum $f(\mathbf{x})$. It is helpful, therefore, to examine typical properties of functions in the large and to introduce pertinent concepts at this point for later use.

6-2a. Local Properties

Certain local properties of functions are examined in Chapter 2. In Section 2-2, the focal point in regard to local properties centers on the concept of a stationary point. A given stationary point is either a point of local maximum, point of local minimum, or a saddle point. In this chapter, the scope of local properties is broadened to include local properties of real-valued functions in the vicinity of arbitrary points, in addition to stationary points.

A useful local property, if it exists, is the gradient $\nabla f(\mathbf{x})$ of a real-valued function $f = f(\mathbf{x})$:

$$\nabla f(\mathbf{x}) \triangleq [\partial f/\partial x_1 \quad \partial f/\partial x_2 \quad \cdots \quad \partial f/\partial x_n]'$$
$$\equiv \mathbf{g}(\mathbf{x}) \tag{6-1}$$

Sect. 6-2   GEOMETRICAL INTERPRETATION AND SCALING

which is expressed here as a vector in column matrix form. Its usefulness is shown by considering the Taylor's series expansion of $f(\mathbf{x})$ about an arbitrary point $\mathbf{x} = \mathbf{x}_a$:

$$f(\mathbf{x}) = f(\mathbf{x}_a) + (\mathbf{x} - \mathbf{x}_a)'\mathbf{g}(\mathbf{x}_a) + \cdots \qquad (6\text{-}2)$$

where $(\mathbf{x} - \mathbf{x}_a)'$ is the row vector $[x_1 - x_{1a} \quad x_2 - x_{2a} \quad \cdots \quad x_n - x_{na}]$ and where $\mathbf{g}(\mathbf{x}_a)$ equals $\mathbf{g}(\mathbf{x})$ evaluated at $\mathbf{x} = \mathbf{x}_a$. If the right-hand member of 6-2 is truncated after the first term, the resulting equation $f(\mathbf{x}) = f(\mathbf{x}_a)$ is generally satisfied by a set of values of $\mathbf{x}$. The set of such values of $\mathbf{x}$ generally defines an $(n - 1)$-dimensional hypersurface in an n-dimensional Euclidean space; a value $\mathbf{x}_b$ of $\mathbf{x}$ lies on this $f(\mathbf{x}_a)$ hypersurface if and only if $f(\mathbf{x}_b) = f(\mathbf{x}_a)$. At point $\mathbf{x}_a$, the gradient vector $\mathbf{g}(\mathbf{x}_a)$ is *normal* to the $f(\mathbf{x}_a)$ hypersurface. Note that if the second term in the series of 6-2 is equated to zero, the resulting linear equation

$$(\mathbf{x} - \mathbf{x}_a)'\mathbf{g}(\mathbf{x}_a) = 0 \qquad (6\text{-}3)$$

defines a hyperplane in the n-dimensional Euclidean space of the x_i's. This hyperplane passes through the point $\mathbf{x}_a$ and has a normal vector $\mathbf{g}(\mathbf{x}_a)$. Thus, the hyperplane defined by 6-3 is tangent at $\mathbf{x}_a$ to the $f(\mathbf{x}_a)$ hypersurface.

In three-dimensional space, a two-dimensional hypersurface conforms to the every-day concept of surface. In two-dimensional space, however, the equation $f(x_1,x_2) = f(x_{1a},x_{2a})$ defines a line which is appropriately called an *isoline* of $f(x_1,x_2)$; a one-dimensional hypersurface is simply an isoline.

6-2b. Regional Properties

In addition to local properties, *global properties* and *regional properties* of functions $f(\mathbf{x})$ are of interest. Global properties of $f(\mathbf{x})$ are those properties which are independent of particular values of $\mathbf{x}$. For example, if the components of ∇f exist and are continuous for all finite $\mathbf{x}$, $f(\mathbf{x})$ has the global property of being of class C^1. Regional properties of $f(\mathbf{x})$ are those which are valid for any $\mathbf{x}$ in a specified region of the $\mathbf{x}$ domain. Hence, a global property is a regional property for which the defined region is the entire $\mathbf{x}$ domain.

Many regional properties are easily visualized in the case that $f(\mathbf{x})$ is a function of two variables x_1 and x_2. Consider x_1 and x_2 as Cartesian coordinates, and assign $f(x_1,x_2)$ to a third coordinate, one orthogonal to the first two. The coordinate $f(x_1,x_2)$ corresponds, say, to elevation on a mountain range (Figure 6-1a).

In addition to peaks (relative maxima), valleys (relative minima), and passes (saddle points), the mountain range may include *ridges* and *ravines*. These are conveniently visualized on a topographic map (Figure 6-1b). The contours of equimagnitude in Figure 6-1b are isolines of $f(x_1,x_2)$.

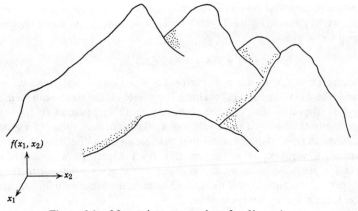

Figure 6-1a. Mountain-range analogy for $f(x_1, x_2)$.

With the indirect methods, so called, of Chapter 2, ravines and ridges play no role whatsoever. When using direct methods, however, the presence of ravines and ridges causes considerable inconvenience; search points must be carefully selected if we are to avoid a false assignment of the minimum (maximum) to a nonoptimal point on a ravine (ridge). Certain techniques of Sections 6-7, 6-8, and 6-9 are especially useful in this regard.

It is convenient at times to view $f(\mathbf{x})$ as a density function in the n-dimensional space of the x_i's. This is especially true for $n = 3$, in which case $f(x_1, x_2, x_3)$ can be visualized as the density of a piece of material. The object of the search is to find the densest part of the material, when a maximum is

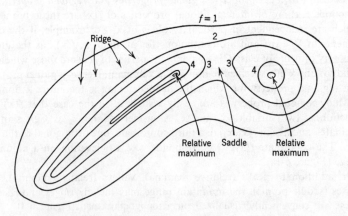

Figure 6-1b. Topographic map of $f(x_1, x_2)$.

Sect. 6-2 GEOMETRICAL INTERPRETATION AND SCALING

desired. This viewpoint is stressed in the centroid methods of Section 6-9e.

In some instances, reasoning based on physical principles gives *a priori* that a certain function $f(\mathbf{x})$ has the regional property of being *unimodal*, meaning that $f(\mathbf{x})$ has but one relative maximum (or relative minimum), which is therefore *the* maximum (minimum). Such advance information can be used to simplify the search technique employed, and the use of some search techniques is inappropriate unless this property holds (Fibonacci search, Section 6-3d, for example).

An even stronger piece of *a priori* information is that $f(\mathbf{x})$ is either *convex* or *concave*; a function $f(\mathbf{x})$ is *convex* over n-dimensional Euclidean space if the following condition is satisfied:

$$(1 - \alpha)f(\mathbf{x}_a) + \alpha f(\mathbf{x}_b) \geq f[(1 - \alpha)\mathbf{x}_a + \alpha \mathbf{x}_b] \qquad (6\text{-}4)$$

for any $\mathbf{x}_a, \mathbf{x}_b$ pair of points and any α in the interval $[0,1]$. If the inequality in 6-4 can be replaced by a strict inequality sign over the open interval $(0,1)$, $f(\mathbf{x})$ is said to be *strictly convex*. Furthermore, it may be that $f(\mathbf{x})$ is convex only *over some convex set of points* R_c (as defined in Section 5-4b) in the n-dimensional space of the x_i's. In such a case, the point-pairs $\mathbf{x}_a, \mathbf{x}_b$ of 6-4 are restricted to R_c. Finally, for a concave function, the inequality in 6-4 is reversed.

As an example of convex functions in one dimension, consider functions $f_1(x)$ and $f_2(x)$ in Figure 6-2. The function $f_1(x)$ is strictly convex over the closed convex set $[x_c, x_d]$ whereas $f_2(x)$ is, simply, convex over the same set. The line denoted by $(1 - \alpha)f_1(x_a) + \alpha f_1(x_b)$ in the figure is representative. An interesting property, which the reader is invited to verify (Problem 6.1), is that the sum $f_1(x) + f_2(x)$ is also convex over $[x_c, x_d]$. In general, the sum

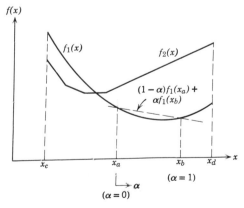

Figure 6-2. Convex functions.

270 SEARCH TECHNIQUES AND NONLINEAR PROGRAMMING Ch. 6

of convex functions is also convex. Additional convexity concepts and relations have been examined by Ponstein [6.78].

6-2c. Scaling and Change of Variables

The rate of convergence of any search technique is highly dependent on the given function $f(\mathbf{x})$—it is for this reason that there is no optimal optimum-seeking method for arbitrary $f(\mathbf{x})$. For most search techniques, however, the rate of convergence is enhanced if *interaction* between the variables can be eliminated (or at least reduced) by making simple changes of variable. A function $f(\mathbf{x})$ is said to be *noninteracting* or *separable* in its variables if it can be placed in the form

$$f(\mathbf{x}) = \sum_{i=1}^{n} q_i(x_i) \qquad (6\text{-}5)$$

where each $q_i(x_i)$ is a function of x_i only. In the case of Equation 6-5, it is apparent that $f(\mathbf{x})$ is maximized if all $q_i(x_i)$'s are maximized independently. Hence, the one n-dimensional search problem is reduced to n one-dimensional search problems—a saving of considerable value when univariate search (Section 6-5) is utilized.

In addition to being dependent on interaction of variables, the efficiency of search techniques is usually dependent on the choice of scale used. An example serves to illustrate these points.

Example 6-1. Consider $f(x_1, x_2)$:

$$f(x_1, x_2) = x_1^2 + 10 x_1 x_2 + 100 x_2^2 = [x_1 \ x_2] \begin{bmatrix} 1 & 5 \\ 5 & 100 \end{bmatrix} \begin{bmatrix} x_1 \\ x_2 \end{bmatrix} \qquad (6\text{-}6a)$$

which is to be minimized with respect to x_1 and x_2. Note that this $f(x_1, x_2)$ is characterized by a narrow ravine and that the minimum of $f(x_1, x_2)$ is easily seen by analytical methods to be zero. Ascertainment of this latter fact by using search techniques is expedited by the following operations.

Let $z_1 = x_1 + a x_2$, and let $z_2 = x_2/b$ where a and b are constants. Alternatively, $x_2 = b z_2$, and $x_1 = z_1 - a b z_2$. These identities are substituted into the original $f(x_1, x_2)$ function to obtain

$$f = z_1^2 + (-2ab + 10b) z_1 z_2 + (-10 a b^2 + a^2 b^2 + 100 b^2) z_2^2 \qquad (6\text{-}6b)$$

This equation is made noninteracting if $a = 5$, for then $-2ab + 10b = 0$; and the constant b is specified to obtain a unity coefficient for z_2^2:

$$-(10)(5) b^2 + 100 b^2 + 5^2 b^2 = 1$$

from which

$$b = \pm(1/75)^{1/2}$$

with the result that

$$f = z_1^2 + z_2^2 \qquad (6\text{-}6c)$$

This f is noninteracting in z_1 and z_2, and is characterized by circular contours in the z_1,z_2 plane. All of the better-known search techniques are highly efficient in finding the minimum of the f in Equation 6-6c.

•

From this discussion, it is evident that two preliminary procedures should be effected for efficient search: (1) reduce interaction between variables as much as possible by defining new variables in terms of old variables where necessary; and (2) change the scale of variables to obtain equimagnitude contours which are circular in shape.

When $f(\mathbf{x})$ is a very complicated function of its variables, or when $f(\mathbf{x})$ is not known explicitly (i.e., when $f(\mathbf{x})$ is determined experimentally), the above general rules may be difficult, if not impossible, to effect with certainty. If any form of sequential search is used, however, the data obtained during the initial phases of the search may point to appropriate scale and variable changes to be made in the latter phases of the search.

6-2d. Noise Considerations

When the performance measure $f(\mathbf{x})$ is evaluated experimentally, there exist experimental errors from measurements and from noise in the physical system under study. Under such conditions, exact expressions are not obtainable for the function $f(\mathbf{x})$ or for the gradient of $f(\mathbf{x})$. It is possible, however, to obtain *expected values* of these by using the following equations:

$$E[f(\mathbf{x}_a)] \simeq \text{Average} f(\mathbf{x})|_{\mathbf{x}=\mathbf{x}_a} \qquad (6\text{-}7)$$

and

$$E\left[\frac{f(x_1,x_2,\ldots,x_i+\epsilon,\ldots,x_n) - f(\mathbf{x})}{\epsilon}\right]_{\mathbf{x}=\mathbf{x}_a} \simeq \text{Avg.} \left.\frac{\partial f}{\partial x_i}\right|_{\mathbf{x}=\mathbf{x}_a} \qquad (6\text{-}8)$$

where $E[f]$ denotes the expected value of f, and ϵ is a relatively small positive number. In practice, we usually measure the value of $f(\mathbf{x}_a)$ a number of times and take the average value of the readings obtained as the expected value, assuming, of course, that the function $f(\mathbf{x})$ does not change significantly during the time required for measurements and that the experimental and noise errors are not biased. Geometrically, the response hypersurface can be visualized as varying randomly about some mean value. The work of Box and Wilson [6.7] is classic in the use of this type of approach. Estimates based on the minimum of squared error often prove to be more accurate than those obtained by simple averaging. Yet other techniques of statistical estimation theory can be employed if probability density functions for the noise are available.

6-2e. Constraint Geometry

The real arguments $\{x_1, x_2, \ldots, x_n\}$ of a real-valued function $f(\mathbf{x})$ constitute an n-dimensional space. The effect of an equality constraint of the form $f_i(\mathbf{x}) = c_i$ in this n-dimensional space is generally to restrict the optimizing $\mathbf{x}$ to lie on an $(n-1)$-dimensional *hypersurface* in n-dimensional space, a surface defined by $f_i(\mathbf{x}) = c_i$ where c_i is a real constant. If k such constraints exist, i.e., $f_i(\mathbf{x}) = c_i$ for $i = 1, 2, \ldots, k$, the region of n-dimensional space which satisfies all k constraints is referred to as a *manifold*,[2] a manifold of dimension $n - k$ (typically), or as simply a hypersurface of dimension $n - k$ (typically).

The effect of an inequality constraint of the form $f_i(\mathbf{x}) \leq c_i$ in the n-dimensional space of x_k's is to restrict the allowed optimum to lie within a given n-dimensional region, the boundary of which is given by $f_i(\mathbf{x}) = c_i$. The effect of constraints in search operations is considered in detail in Section 6-11.

6-3. ONE-DIMENSIONAL SEARCH

Five one-dimensional search techniques are presented in this section. The first two techniques (Sections 6-3a and 6-3b) are basically analytical while the latter three are appropriate when the gradient is not readily available, e.g., when the performance measure is evaluated experimentally. The importance of these search techniques goes beyond one-dimensional problems in that many n-dimensional search techniques utilize a sequence of one-dimensional searches. The basics of one-dimensional search in n-dimensional space are given in Section 6-3f. Also, certain of the techniques given in this section have extensions to higher dimensions, and these are duly noted.

6-3a. Newton-Raphson Search

Consider a given function $f(x)$ of one real variable x. The function $f = f(x)$ is assumed to be of class C^2. The stationary points of f are to be found, that

---

[2] In general, the term *manifold* can be used to denote any collection or set of objects; but more restricted meanings of the word exist [6.52].

is, the values of x which satisfy

$$\frac{df}{dx} \triangleq g(x) = 0 \tag{6-9}$$

are desired.

Newton-Raphson search is an iterative technique which can be used to find zeros of $g(x)$. The search consists of a series of trials, and each trial consists of the following steps:

1. Take the value x^k of x obtained from the kth trial (x^0 is picked by a judicious guess) and evaluate $g(x^k)$ and $\dot{g}(x^k)$.[3]
2. Fit a straight line to the data obtained in Step 1.
3. Determine the point x^{k+1} at which the straight-line approximation intersects the x axis. This point x^{k+1} is an approximation to a stationary point and is used as the initial point for a new trial.

These steps are now expressed analytically for convenience of computation. In Figure 6-3, it is to be observed that

$$\dot{g}(x^k)(x^k - x^{k+1}) = g(x^k) \tag{6-10}$$

from which

$$x^{k+1} = x^k - \frac{g(x^k)}{\dot{g}(x^k)}$$

$$= x^k - \frac{f(x^k)}{\ddot{f}(x^k)} \tag{6-11}$$

An important property of this technique is that it yields *quadratic*

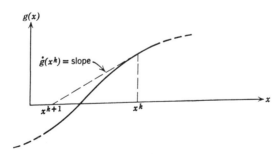

Figure 6-3. Newton-Raphson search.

---

[3] Note that x^k does not denote x raised to the kth power in this instance. This superscript notation is used to denote a sequence of events. We might be tempted to use x_k to denote the kth trial point, but it is possible to confuse this with the kth entry of the matrix $\mathbf{x} = [x_1 \cdots x_n]'$. In instances where confusion might result, x raised to the kth power is denoted by $(x)^k$ rather than x^k in this chapter.

convergence. Quadratic convergence is that property of a search technique which enables a minimum (maximum) of a quadratic function to be located exactly in a finite number (usually a number equal to the number of variables) of steps, subject to round-off errors.

To prove quadratic convergence in this one-dimensional case, assume that f is of the quadratic form

$$f = a(x)^2 + bx + c \tag{6-12}$$

By employing Equation 6-11, it follows that

$$x^1 = x^0 - \frac{2ax^0 + b}{2a}$$

$$= \frac{-b}{2a} \tag{6-13}$$

which the reader may readily verify to be the exact minimum point (if $a > 0$).

In general, the property of quadratic convergence is an important attribute for a search technique. To substantiate this statement, recall that any differentiable function $f(\mathbf{x})$ can be expanded in a Taylor's series about a given relative extremum point. The first three terms of such a series, the terms which form a quadratic, are dominant for any $\mathbf{x}$ in the near vicinity of the extremum point. Thus, using a quadratic-convergent search technique insures rapid convergence once the near vicinity of the extremum point is reached.

The number of trials required in the Newton-Raphson process is typically small if we are dealing with a well-behaved function; the degree and range of linearity of $g(x)$ in the vicinity of the extremum point are the determining factors in this regard. If more than one extremum point exists for $f(x)$, the initial choice x^0 of x determines which of the points is obtained by $\lim_{k \to \infty} x^k$, if any—it is possible for the search to be trapped in a cycle and to oscillate back and forth.

One additional fact should be considered before progressing to other methods. Note that in the vicinity of an extremum point, the second derivative $\dot{g}(x)$ of $f(x)$ is constant in sign, whereas the first derivative $g(x)$ of $f(x)$ changes in sign from one side of the extremum point to the other. The first derivative $f(x)$ is a one-dimensional gradient of $f(x)$ and, depending on the sign of the gradient $f(x^k)$, the $(k + 1)$st value x^{k+1} of x is either smaller or larger than the kth value. Thus, the gradient of f determines the *direction* of change in x. This method is therefore related to n-dimensional gradient methods of search (Sections 6-6, 6-7, and 6-8).

6-3b. Cubic-Convergent Search without Second Derivatives

The search technique in this subsection is a variation of one which has been used with success by Davidon [6.24] and others. It consists of three distinct phases, the first and third of which are quadratic convergent and cubic convergent, respectively. In analogy to the definition of a quadratic-convergent search, a *cubic-convergent search* is one with which the relative minimum (maximum) of a cubic polynomial—one with a well-defined relative minimum (see Problem 2.4)—is located exactly after a finite number of search iterations, aside from round-off error.

As is to be observed, each evaluation of f and $\dot{f}$ is used to the utmost in the procedure; a relatively large amount of computational effort is devoted to interpolation phases of the search. It is to be expected, therefore, that this procedure is of most value when f and $\dot{f}$ are complicated functions which require a proportionally larger amount of time for evaluation than that used in the interpolation phases.

Phase I.

Given: x^0, $f(x^0)$, $\dot{f}(x^0) = g(x^0)$, f_{est} (the estimated minimum of $f(x)$), and r^0 (a parameter which assumes a significant role in n-dimensional search—see Sections 6-3f and 6-8).

Find: d_s, where d_s is the true minimum point of $y(\alpha) = f(x^0 + \alpha r^0)$ *if* $y(\alpha)$ is a quadratic with a well-defined minimum and *if* f_{est} is the true minimum of $y(\alpha)$.

Thus, if
$$y(\alpha) = a(\alpha)^2 + b\alpha + c, \qquad a > 0 \qquad (6\text{-}14)$$
and if
$$f_{est} = c - \frac{b^2}{4a} \quad \text{at } \alpha = d_s = \frac{-b}{2a} \qquad (6\text{-}15)$$
the given data $y(0) = f(x^0) = c$ and $\dot{y}(0) = r^0 g(x^0) = b$ are used in 6-15 to obtain
$$d_s = \frac{-b}{2a} = 2\frac{f_{est} - f(x^0)}{r^0 g(x^0)} \qquad (6\text{-}16)$$

In general, of course, $y(\alpha)$ is not quadratic, but Equation 6-16 is used regardless (see Figure 6-4).

Phase II.

The value of f_{est} is always adjusted to be less than or equal to $f(x^0)$. If f_{est} is also less than the actual f_{min}, $|d_s|$ becomes excessively large when x^0 is near the minimum point—see Equation 6-16—because $g(x^0)$ approaches zero at that point. To offset this possibility, a step of value d_0, rather than d_s, is taken in α; where
$$d_0 = \begin{cases} d_s, & |d_s| < |r^0| \\ |r^0| \text{ sign } d_s, & \text{otherwise} \end{cases} \qquad (6\text{-}17)$$

276   SEARCH TECHNIQUES AND NONLINEAR PROGRAMMING   Ch. 6

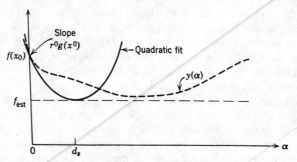

Figure 6-4a. A representative case of Phase I of the search of Section 6-3b.

With d_0 so obtained, the value of $f(x^0 + \alpha r^0)$ is evaluated at $\alpha = d_0, 2d_0, 2^2 d_0, 2^3 d_0, \ldots, d_1, d_2, d_3$, where the terminal value d_3 is determined on the basis that

$$f(x^0 + d_3 r^0) \geq f(x^0 + d_2 r^0) \tag{6-18}$$

The next step concerns $\dot{y}(d_2) = r^0 g(x^0 + d_2 r^0)$. With $d_2 > 0$, if $\dot{y}(d_2) < 0$, interpolation between the points $x^0 + d_2 r^0$ and $x^0 + d_3 r^0$ is effected; but if $\dot{y}(d_2) > 0$, interpolation between the points $x^0 + d_1 r^0$ and $x^0 + d_2 r^0$ is effected. (If $\dot{y}(d_2) = 0$, the one-dimensional search is ended.) In any case, search in Phase III that follows is conducted between two points:

$$x^0 + d_a r^0 \leq x \leq x^0 + d_b r^0 \tag{6-19}$$

where

$$d_a = \begin{cases} d_1, & \dot{y}(d_2) > 0 \\ d_2, & \dot{y}(d_2) < 0 \end{cases} \quad \text{for } d_2 > 0 \tag{6-20}$$

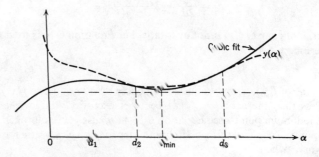

Figure 6-4b. A representative case of Phases II and III of the search of Section 6-3b.

and

$$d_b = \begin{cases} d_2, & \dot{y}(d_2) > 0 \\ d_3, & \dot{y}(d_2) < 0 \end{cases} \text{ for } d_2 > 0 \qquad (6\text{-}21)$$

Phase III

Phase III consists of a cubic interpolation between two points, d_a and d_b, to locate an approximate minimum of $y(\alpha)$. Values of $y(d_a)$, $\dot{y}(d_a)$, $y(d_b)$, and $\dot{y}(d_b)$ are computed, and an *estimate* $d_{\min}$ of the minimum point is determined by the use of Davidon's interpolation equations:

$$d_{\min} = d_b - (d_b - d_a)\left[\frac{\dot{y}(d_b) + u_2 - u_1}{\dot{y}(d_b) - \dot{y}(d_a) + 2u_2}\right] \qquad (6\text{-}22)$$

where

$$u_1 \equiv \dot{y}(d_a) + \dot{y}(d_b) + 3\frac{y(d_a) - y(d_b)}{d_b - d_a} \qquad (6\text{-}23)$$

and

$$u_2 \equiv [u_1{}^2 - \dot{y}(d_a)\dot{y}(d_b)]^{\frac{1}{2}} \qquad (6\text{-}24)$$

The above equations are a very compact representation of the relative minimum point $d_{\min}$ of a cubic of the form

$$y(\alpha) = a(\alpha - d_a)^3 + b(\alpha - d_a)^2 + c(\alpha - d_a) + d \qquad (6\text{-}25)$$

where the known data—$y(d_a) = d$, $\dot{y}(d_a) = c$, $y(d_b) = a(d_b - d_a)^3 + b(d_b - d_a)^2 + c(d_b - d_a) + d$, and $\dot{y}(d_b) = 3a(d_b - d_a)^2 + 2b(d_b - d_a) + c$—are sufficient to determine the constants a, b, c, and d; and by using ordinary min-max theory, the minimum point $d_{\min}$ can be shown to be

$$d_a - \left(\frac{b}{3a}\right) + \left[\left(\frac{b}{3a}\right)^2 - \frac{c}{3a}\right]^{\frac{1}{2}} \text{ sign } (a)$$

If neither $y(d_a)$ nor $y(d_b)$ is less than $y(d_{\min}) = f(x^0 + d_{\min}r^0)$, $d_{\min}$ may be assumed to be a sufficiently accurate estimate of a relative minimum point. On the other hand, Phase III may be repeated over one of the following two intervals: $(d_a, d_{\min})$ or $(d_{\min}, d_b)$. The former interval is used if $\dot{y}(d_{\min}) > 0$; the latter is used if $\dot{y}(d_{\min}) < 0$. These statements hold when $d_{\min} > 0$.

6-3c. Quadratic-Convergent Search without Derivatives

Powell's search procedure [6.79], as amended by Zangwill [6.109], does not involve the use of derivatives, but is quadratic convergent all the same, even in the n-dimensional case. One part of Powell's procedure involves minimization along a straight line; a version of this part is considered in this subsection.

As in the preceding subsection, assume a function $y(\alpha) = f(x^0 + \alpha r^0)$ is to be minimized with respect to α. Given are the point x^0 and the "direction" r^0—this use of the word "direction" has significance in the n-dimensional case where a set $\mathbf{x}$ of n real variables, rather than x, is the argument of f.

Also assumed given are the following: an upper bound d_{ub} on steps in α; a starting step magnitude d_0 for α, $d_0 < d_{ub}$; and a value d_{ac} which is used in determining the accuracy of the assumed minimum which is obtained.

The procedure is begun with a comparison of $f(x^0)$ and $f(x^0 + d_0 r^0)$. If $f(x^0)$ is less than $f(x^0 + d_0 r^0)$, $f(x^0 - d_0 r^0)$ is computed; otherwise $f(x^0 + 2d_0 r^0)$ is computed. In either case, three evaluations of the function are obtained; these are labeled $y(d_1) = f(x^0 + d_1 r^0)$, $y(d_2) = f(x^0 + d_2 r^0)$, and $y(d_3) = f(x^0 + d_3 r^0)$; and for convenience they are ordered such that $d_3 > d_2 > d_1$. The reader may verify (Problem 6.4) that a quadratic which is fitted to these data has a stationary point d_s:

$$d_s = \frac{1}{2} \frac{d_{23}^s y(d_1) + d_{31}^s y(d_2) + d_{12}^s y(d_3)}{d_{23} y(d_1) + d_{31} y(d_2) + d_{12} y(d_3)} \tag{6-26}$$

where

$$d_{23} \equiv d_2 - d_3, \quad d_{31} \equiv d_3 - d_1, \quad d_{12} \equiv d_1 - d_2$$

and

$$d_{23}^s \equiv d_2^2 - d_3^2, \quad d_{31}^s \equiv d_3^2 - d_1^2, \quad d_{12}^s \equiv d_1^2 - d_2^2$$

The stationary point d_s is a minimum of the quadratic, provided that

$$-2 \frac{d_{23} y(d_1) + d_{31} y(d_2) + d_{12} y(d_3)}{d_{23} d_{31} d_{12}} \triangleq D > 0 \tag{6-27}$$

If the inequality in 6-27 is not satisfied, one of two actions is taken: when $y(d_3) > y(d_1)$, (1) d_3 is redefined to be the old value of d_2, d_2 is redefined to be the old value of d_1, and d_1 is redefined to be (old) $d_1 - d_{ub}$; or when $y(d_3) \leq y(d_1)$, (2) d_1 is redefined to be the old value of d_2, d_2 is redefined to be the old value of d_3, and d_3 is redefined to be (old) $d_3 + d_{ub}$. Three-point interpolation is repeated with the new data points (see Figure 6-5).

If 6-27 is satisfied, however, one of two conditions prevails: case A, $\min\{|d_1 - d_s|, |d_3 - d_s|\} > d_{ub}$; or case B, $d_{ub} \geq \min\{|d_1 - d_s|, |d_3 - d_s|\}$. In case A, action (1) of the preceding paragraph is taken if $d_s < d_1 - d_{ub}$, or action (2) is taken if $d_s > d_3 + d_{ub}$. In case B, the values $|d_s - d_1|$, $|d_s - d_2|$, and $|d_s - d_3|$ are compared with the accuracy constant d_{ac}: if d_{ac} is greater than any one of these three values, the point d_s is accepted as a winning point in the search for a relative minimum point of $y(\alpha)$; but if d_{ac} is less than all three values, the point d_s replaces one of the three data points,

Sect. 6-3   ONE-DIMENSIONAL SEARCH

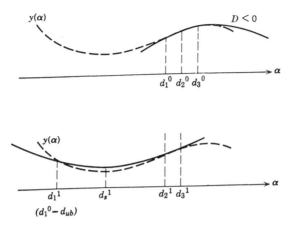

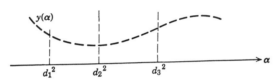

Figure 6-5. A representative case of the search of Section 6-3c.

normally[4] the one with the greatest value of f, the points are reordered so that $d_1 < d_2 < d_3$, and the interpolation 6-26 and 6-27 is repeated as before.

Terminal cycles of the search can be simplified by using the following approximation D to the second derivative of $f(x^0 + \alpha r^0)$ with respect to α:

$$D \cong \frac{\partial^2 f(x^0 + \alpha r^0)}{\partial \alpha^2}$$

$$= -2 \frac{d_{23} y(d_1) + d_{31} y(d_2) + d_{12} y(d_3)}{d_{23} d_{31} d_{12}} \tag{6-28}$$

In the vicinity of a relative minimum point, the second derivative remains relatively constant. Thus, the value D can be used, in conjunction with just two data points (d_1's and d_2's), to determine corresponding d_s's.

---

[4] If the minimum point does not lie between data points which result from this normal procedure, whereas it would be bracketed by replacing a different data point by d_s, the latter procedure is applied.

The procedure to be followed is: first, make repeated use of the interpolation Equations 6-26 and 6-27, at least until a positive value of D (Equation 6-27) is obtained; and second, compute d_s by use of the equation[5]

$$d_s = \frac{d_1 + d_2}{2} - \frac{y(d_2) - y(d_1)}{(d_2 - d_1)D} \qquad (6\text{-}29)$$

where d_1 and d_2 are two data points, one or perhaps both of which are previously obtained values of d_s. As before, the search is terminated when d_{ac} is greater than either $|d_s - d_1|$ or $|d_s - d_2|$.

Powell suggests the following modification of the above. Note that Equation 6-29 is equivalent to

$$D^{1/2} d_s = \frac{d_1 D^{1/2} + d_2 D^{1/2}}{2} - \frac{y(d_2) - y(d_1)}{d_2 D^{1/2} - d_1 D^{1/2}} \qquad (6\text{-}30)$$

Consider the introduction of (new d_i's) $\triangleq$ (old d_i's) $D^{1/2}$ and $r^1 =$ (new r^0) $\triangleq$ (old r^0)/$D^{1/2}$ where the latter identity is made to obtain $y(\text{new } d_i) = y(\text{old } d_i)$. In terms of the new d_i's, the scaled equivalent of 6-30 is

$$d_s = \frac{d_1 + d_2}{2} - \frac{y(d_2) - y(d_1)}{d_2 - d_1} \qquad (6\text{-}31)$$

6-3d. Fibonacci Search

Consider an industrial process. The performance of the process is measured in terms of a function $f(x)$, the form of which is not known explicitly but which can be evaluated experimentally. The argument x of $f(x)$ could be a valve or potentiometer setting, for example. From the physical nature of the process, it is known (or at least strongly suspected) that $f(x)$ is unimodal. It is also suspected that $f(x)$ is discontinuous so that the quadratic-convergent search of the preceding subsection is deemed unsuitable. A maximum of $f(x)$ is to be located with x in the interval $c_1 \leq x \leq c_2$, but each test point that is checked expends a certain amount of time and performance—a change in the setting of x may require some settling time before $f(x)$ can be measured. Thus, only a limited number N of tests are to be made before the final set-point is picked. It is assumed that $f(x)$ itself does not change appreciably with time. (Note that this list of assumptions greatly restricts the set of problems for which Fibonacci search is "best.")

A search technique is required with which to determine the *smallest possible interval* of uncertainty, in which the maximum lies, after N tests are

---

[5] Equation 6-29 is obtained by assuming the quadratic form $y(\alpha) = a(\alpha - d_1)^2 + b(\alpha - d_1) + c$ where $y(d_1) = c$, $y(d_2) = a(d_2 - d_1)^2 + b(d_2 - d_1) + c$, and $\ddot{y}(\alpha) = 2a = D$. The minimum occurs at $d_s = d_1 - b/2a$.

completed. The technique must guarantee the size of the uncertainty interval independent of the particular function $f(x)$ of the class described. The search technique used to perform this feat is called Fibonacci search and was developed by Kiefer [6.57].

Preliminary notation for Fibonacci search is as follows:

N = number of tests to be made.
d_1 = width of the initial interval of uncertainty $(c_2 - c_1)$.
d_k = width of interval of uncertainty after k tests have been made.

The test points are to be selected sequentially so that the value of d_N is guaranteed to be as small as possible, independent of the outcome of any given test.[6]

The simplest case is treated first. If only two tests, $N = 2$, are allowed, it is evident that the best that can be done is to eliminate one-half of the initial interval by placing the points as close as experimentally possible[7] in the center of the interval, as shown in Figure 6-6a. In this figure, it is assumed that test point 1 results in the greater value of f and, therefore, the maximum must lie to the right of center in the interval.

Next, consider the case of three allowed tests. In this case it is required that wherever the first two test points are placed, the final test will be made in an optimal manner; that is, the final test will be made as close as possible to the best of the first two test points so as to eliminate one-half of the next-to-last interval of uncertainty. In Figure 6-6b, therefore, test point 2 is placed half way between test point 1 and the left-hand boundary, and test point 1 is placed half way between test point 2 and the right-hand boundary. The result is that $d_2 = \frac{2}{3} d_1$, and either 1 or 2 can accept the role of "center" in the optimal two-test search that remains. If it happens that test point 1 yields a greater value of $f(x)$ than does test point 2—this is assumed in Figure 6-6b—then the maximum of f must lie in the interval from 2 to the right-hand boundary, and the next test point 3 is placed as close as possible to 1 as was done in the optimal two-test case. The final uncertainty interval has width $d_3 = \frac{1}{3} d_1$ for $N = 3$.

Going one step farther, for $N = 4$ the goal is to select the first two of the four test points so that d_4 is minimized by using an optimal three-test search after the first two tests, independent of the "winner" of the first two tests. This is depicted in Figure 6-6c. The final interval has width $d_4 = \frac{1}{5} d_1$.

Similarly, the arrangement when $N = 5$ is depicted in Figure 6-6d. In general, it is determined by this chain of reasoning that for an N test-point

---

[6] It is interesting to note that we could formulate the minimization of d_N as a linear programming problem. The dynamic programming approach used here is patterned after that of Johnson [6.54]; it yields the results more directly.

[7] Wilde and Oliver [6.105] specify a width ϵ for this separation; but the results obtained differ only by the order of ϵ, and therefore for convenience in this derivation all d_k are assumed to be much greater than ϵ.

282   SEARCH TECHNIQUES AND NONLINEAR PROGRAMMING   Ch. 6

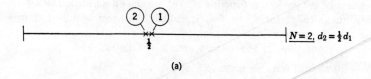

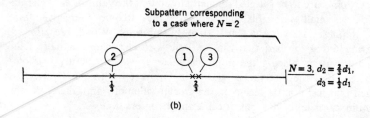

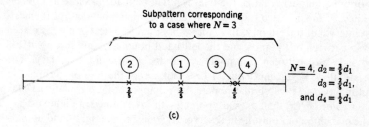

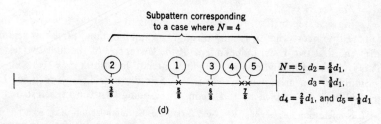

Figure 6-6. Four cases of Fibonacci search corresponding to four different values of N.

case, the width d_N of the interval of uncertainty in which the maximum lies at the end of the N tests is given by

$$d_N = \frac{d_1}{F_N} \tag{6-32}$$

where the integer F_N is governed by the difference equation

$$F_N = F_{N-1} + F_{N-2}, \quad F_0 = F_1 = 1 \tag{6-33}$$

The numbers generated by the above difference equation have an interesting historical background. They were first noticed by Leonardo of Pisa (nicknamed Fibonacci, hence the naming of the method) in 1202 A.D. By using these numbers, now called Fibonaccian numbers, Fibonacci was able to characterize the reproduction of rabbits. The interested reader is referred to Wilde's humorous discussion [6.105] of this.

The difference equation 6-33 can be solved by classical techniques, or by z-transform methods, with the result that

$$F_N = \frac{(0.5)^{N+1}}{\sqrt{5}} [(-1)^N(\sqrt{5} - 1)^{N+1} + (\sqrt{5} + 1)^{N+1}] \tag{6-34}$$

which is a rather unwieldy, but exact relationship.

Return now to the problem of placing the test points. For $N = 2$, the two test points are placed at a distance $\frac{1}{2}d_1 = (F_1/F_2)d_1$ from the ends of the interval. For $N = 3$, the first two test points are placed at a distance $\frac{2}{3}d_1 = (F_2/F_3)d_1$ from the ends of the initial interval, and the final uncertainty interval is $d_3 = \frac{1}{3}d_1 = (F_1/F_3)d_1$. For $N = 4$, the uncertainty interval d_2 after the first two tests equals $\frac{3}{5}d_1 = (F_3/F_4)d_1$; the uncertainty interval d_3 remaining after the third test equals $\frac{2}{5}d_1 = (F_2/F_4)d_1$; and the final uncertainty interval d_4 equals $\frac{1}{5}d_1 = (F_1/F_4)d_1$.

In general, if N test points are to be used, the first two test points are located at distances of $(F_{N-1}/F_N)d_1$ from the ends of the initial interval. Corresponding to the winner of the first two tests, an uncertainty interval $d_2 = (F_{N-1}/F_N)d_1$ is determined. The third test point is placed so as to obtain an assured uncertainty interval of width $d_3 = (F_{N-2}/F_N)d_1$; and in general, the kth test point is placed so as to obtain an assured uncertainty interval of width $d_k = (F_{N-k+1}/F_N)d_1$.

The first twenty Fibonaccian numbers and the reduction ratio d_N/d_1 are given in Table 6-1.

An n-dimensional extension of Fibonaccian search has been proposed by Sugie [6.94]. The extension requires the application of many one-dimensional Fibonaccian searches in sequence, each one-dimensional Fibonaccian search carried out to such a degree that the maximum for the one-dimensional search is effectively obtained. This technique has not been compared with other

TABLE 6-1. FIBONACCIAN NUMBERS AND REDUCTION RATIOS

N	FIBONACCIAN NUMBER F_N	REDUCTION RATIO d_N/d_1
1	1	1.
2	2	0.5
3	3	0.333
4	5	0.2
5	8	0.125
6	13	0.077
7	21	0.0476
8	34	0.0294
9	55	0.0182
10	89	0.0112
11	144	0.00694
12	233	0.00429
13	377	0.00265
14	610	0.00164
15	987	0.00101
16	1597	0.000625
17	2584	0.000387
18	4181	0.000240
19	6765	0.000148
20	10946	0.0000914

n-dimensional search methods; but even without computational experience with typical problems, it can be asserted that the proposed extension, which requires a relatively large number of test points for accuracy, is inefficient in comparison to the efficiency of many other methods when applied to problems of practical interest.

6-3e. Search by Golden Section

The same conditions apply here as are given at the start of Section 6-3d, with the exception that the number N of tests is not specified in advance. We may choose to specify N only after a certain number of tests have been made to determine the relative smoothness of $f(x)$. In this case, it is logical to assume initially that a great number of tests will be made and to choose d_2 on the following basis:

$$d_2 = \lim_{N \to \infty} \frac{F_{N-1}}{F_N} d_1 \qquad (6\text{-}35)$$

ONE-DIMENSIONAL SEARCH

In turn,

$$d_3 = \lim_{N \to \infty} \frac{F_{N-2}}{F_N} d_1$$

$$= \lim_{N \to \infty} \frac{F_{N-2}}{F_{N-1}} \frac{F_{N-1}}{F_N} d_1$$

$$= \lim_{N \to \infty} \left(\frac{F_{N-1}}{F_N}\right)^2 d_1 \qquad (6\text{-}36)$$

And in general, by following this line of reasoning,

$$d_k = \lim_{N \to \infty} \left(\frac{F_{N-1}}{F_N}\right)^{k-1} d_1 \qquad (6\text{-}37)$$

To use the above expression for d_k, the limit as N approaches infinity of F_{N-1}/F_N must be determined. Let Υ be defined by

$$\Upsilon \triangleq \lim_{N \to \infty} \frac{F_N}{F_{N-1}} \qquad (6\text{-}38)$$

By using Equation 6-33, F_N/F_{N-1} is replaced by $1 + (F_{N-2}/F_{N-1})$ in 6-38 to obtain

$$\Upsilon = \lim_{N \to \infty} \left(1 + \frac{F_{N-2}}{F_{N-1}}\right)$$

$$= 1 + \lim_{N \to \infty} \frac{F_{N-2}}{F_{N-1}}$$

$$= 1 + \frac{1}{\Upsilon} \qquad (6\text{-}39)$$

Hence,

$$\Upsilon^2 - \Upsilon - 1 = 0 \qquad (6\text{-}40)$$

which yields

$$\Upsilon = \frac{1 + 5^{1/2}}{2} \cong 1.618 \qquad (6\text{-}41)$$

Equation 6-37 is updated by including this result, as follows:

$$d_k = \left(\frac{1}{\Upsilon}\right)^{k-1} d_1$$

$$= (0.618)^{k-1} d_1 \qquad (6\text{-}42)$$

These values of d_k are used to find the test points, as explained in the preceding section.

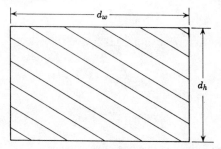

Figure 6-7. Golden-section design.

An interesting historic note is that the number Υ was known to the ancients as the *golden-section* ratio. Some of the early Greek architecture was designed with the ratio of the width-plus-height to the width of a building equal to the ratio of the width to the height of the building. In terms of Figure 6-7,

$$\frac{d_w + d_h}{d_w} = \frac{d_w}{d_h} \tag{6-43}$$

where $d_w + d_h$ equals a constant. The resulting ratio $d_w/d_h = \Upsilon$; a fact which the reader is invited to verify.

6-3f. One-Dimensional Search in n-Dimensional Space

To define a straight line in n-dimensional Euclidean space E^n, either two points or a point and a direction must be specified. If $\mathbf{x}^a$ and $\mathbf{x}^b$ are specified points, then $\mathbf{x}(\alpha) = (1 - \alpha)\mathbf{x}^a + \alpha\mathbf{x}^b$, $-\infty < \alpha < \infty$, defines a straight line which passes through points $\mathbf{x}^a = \mathbf{x}(0)$ and $\mathbf{x}^b = \mathbf{x}(1)$. On the other hand, if $\mathbf{x}^a$ is a specified point and $\mathbf{r} = [r_1 \ r_2 \ \cdots \ r_n]'$ is a specified direction vector, then $\mathbf{x}(\alpha) = \mathbf{x}^a + \alpha\mathbf{r}$ is a line which passes through the point $\mathbf{x}^a$ in the $\mathbf{r}$ direction. That these two representations of a line are equivalent is shown by identifying $\mathbf{r}$ with $\mathbf{x}^b - \mathbf{x}^a$ to obtain $\mathbf{x}(\alpha) = \mathbf{x}^a + \alpha(\mathbf{x}^b - \mathbf{x}^a) = \mathbf{x}^a + \alpha\mathbf{r}$.

Consider $y(\alpha) \equiv f(\mathbf{x}^a + \alpha\mathbf{r})$. The first derivative $\dot{y}(\alpha)$ of $y(\alpha)$ is expressed as

$$\frac{dy(\alpha)}{d\alpha} = \left[\frac{\partial f(\mathbf{x})}{\partial x_1}\frac{\partial x_1}{\partial \alpha} + \frac{\partial f(\mathbf{x})}{\partial x_2}\frac{\partial x_2}{\partial \alpha} + \cdots + \frac{\partial f(\mathbf{x})}{\partial x_n}\frac{\partial x_n}{\partial \alpha}\right]_{\mathbf{x} = \mathbf{x}^a + \alpha\mathbf{r}}$$

$$= \left[r_1\frac{\partial f}{\partial x_1} + r_2\frac{\partial f}{\partial x_2} + \cdots + r_n\frac{\partial f}{\partial x_n}\right]_{\mathbf{x} = \mathbf{x}^a + \alpha\mathbf{r}}$$

$$= \mathbf{r}'\mathbf{g}(\mathbf{x}^a + \alpha\mathbf{r}) \tag{6-44}$$

where $\mathbf{g}(\mathbf{x})$ is the gradient ∇f of f. In like manner, the second derivative $\ddot{y}(\alpha)$

of $y(\alpha)$ is obtained:

$$\frac{d^2 y(\alpha)}{d\alpha^2} = \frac{d}{d\alpha}[\mathbf{r}'\mathbf{g}(\mathbf{x}^a + \alpha\mathbf{r})]$$

$$= \sum_{i=1}^{n} r_i \frac{d}{d\alpha}\left[\left(\frac{\partial f}{\partial x_i}\right)_{\mathbf{x}=\mathbf{x}^a+\alpha\mathbf{r}}\right]$$

$$= \sum_{i=1}^{n} r_i \sum_{k=1}^{n} r_k \left.\frac{\partial^2 f}{\partial x_k \partial x_i}\right|_{\mathbf{x}=\mathbf{x}^a+\alpha\mathbf{r}}$$

$$= \mathbf{r}'\mathbf{A}\mathbf{r} \tag{6-45}$$

where

$$\mathbf{A} = \begin{bmatrix} \frac{\partial^2 f}{\partial x_1^2} & \frac{\partial^2 f}{\partial x_1 \partial x_2} & \cdots & \frac{\partial^2 f}{\partial x_1 \partial x_n} \\ \vdots & & \ddots & \vdots \\ \frac{\partial^2 f}{\partial x_n \partial x_1} & & \cdots & \frac{\partial^2 f}{\partial x_n^2} \end{bmatrix}_{\mathbf{x}=\mathbf{x}^a+\alpha\mathbf{r}} \tag{6-46}$$

Because of the $n(n + 1)/2$ functions that must be evaluated to obtain $\mathbf{A}$ of 6-46, it is not surprising that the more efficient n-dimensional search techniques make no direct use of the second derivative.

As an example of one-dimensional search in n-dimensional space, consider the use of Davidon's one-dimensional search (Section 6-3b) to obtain the minimum of $f(\mathbf{x}^k + \alpha\mathbf{r}^k)$ with respect to α. Equation 6-16 assumes the form

$$d_s = 2\left(\frac{f_{\text{est}} - f(\mathbf{x}^k)}{\mathbf{r}^{k'}\mathbf{g}(\mathbf{x}^k)}\right) \tag{6-47}$$

where the superscript k denotes that k one-dimensional searches in sequence, each with a different value of $\mathbf{r}$ in general, have been used to locate $\mathbf{x}^k$. Note the use of Equation 6-44 in obtaining the denominator $\mathbf{r}^{k'}\mathbf{g}(\mathbf{x}^k)$ of Equation 6-47. The initial step size d_0 is determined by a generalization of Equation 6-17, as follows:

$$d_0 = \begin{cases} d_s, & |d_s| \leq (\mathbf{r}^{k'}\mathbf{r}^k)^{1/2} \\ (\mathbf{r}^{k'}\mathbf{r}^k)^{1/2} \operatorname{sign} d_s, & \text{otherwise} \end{cases} \tag{6-48}$$

The remaining steps of the search are identical to those given in Section 6-3b and are therefore omitted here.

6-4. NONSEQUENTIAL METHODS

Nonsequential search techniques are one-shot affairs; all data are obtained simultaneously (at least from the viewpoint of the person conducting the search), and therefore no information from past data is used to determine new data points. We might surmise that such techniques are less efficient on the average than sequential search techniques, and indeed this has been shown experimentally by Brooks [6.11] for a set of typical problems. Why then should we consider the use of nonsequential search techniques? The answer to this is simply that certain types of problems are amenable only to nonsequential search. One such type is that in which all available data are given; much data may have been accumulated concerning an industrial process which has operated under various conditions. A search of this data may logically reveal certain controller settings which resulted in superior performance.

Another type of problem which necessitates nonsequential search is one for which each data point requires considerable time to obtain. Example: the productivity of a certain plant under various fertilizer and soil-moisture conditions.

Additional reasons for using nonsequential search techniques are: They may be used in the early stages of a search, perhaps on data recorded for other purposes, to find regions which are likely to contain the absolute maximum (minimum); more efficient sequential search techniques can then be started in these restricted regions. Also, nonsequential search techniques do not require assumptions concerning the abruptness of changes in $f(\mathbf{x})$ whereas most sequential search techniques are inefficient if $f(\mathbf{x})$ exhibits radical vacillations.

6-4a. Nonsequential Random Search

A *nonsequential random search* for an optimum of $f(\mathbf{x})$ is one in which a given number of values of $\mathbf{x}$ are generated at random—within some domain of interest—and the particular $\mathbf{x}$ for which $f(\mathbf{x})$ is best is denoted the winner. This type of nonsequential search, sometimes referred to as Monte Carlo search, has been shown by Brooks [6.11] to be somewhat less efficient in general than that of nonsequential factorial search which is considered in the next subsection. Random search has, however, the desirable feature of being completely unbiased in the selection of search points; as such, while often inefficient, it could conceivably lead to the detection of an absolute optimum which would be missed otherwise.

6-4b. Nonsequential Factorial Search

Nonsequential factorial search is a search which is conducted over evenly spaced points in a simply connected region of a Euclidean space. Each of the x_i coordinates is assigned a set of evenly spaced points, called *grid points*, and only the values of x_i at these grid points are used. The function $f(\mathbf{x})$ is evaluated

Sect. 6-4 NONSEQUENTIAL METHODS 289

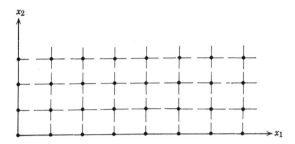

Figure 6-8. Factorial search points.

for all possible combinations of the grid points (Figure 6-8), and the grid-value of **x** which yields the best $f(\mathbf{x})$ is deemed the winner.

The selection of the spacing of the grid points for each coordinate requires judgment on the part of the searcher, hopefully judgment based on the particular performance measure $f(\mathbf{x})$. As for the number of data points required, if each x_i coordinate is assigned k evenly spaced points, the total number of data points is k^n where, as before, n is the number of entries in **x**.

Example 6-2. Here a problem is posed, and preliminary results are obtained. Later examples in this chapter feature search solutions to the problem.

Consider the performance measure $f = f(x_1, x_2)$:

$$f = (1 + a - bx_1 - bx_2)^2 + (b + x_1 + ax_2 - bx_1x_2)^2 \qquad (6\text{-}49)$$

where a and b are specified positive constants, and x_1 and x_2 are non-negative parameters to be selected to attain the absolute minimum of f. The two-dimensional nature of the problem permits visual display of interesting features of search techniques; for solutions to more complicated problems, the use of computers is in order.

Actually, as a challenging exercise (Problem 2.11), the reader may prove by methods of Chapter 2 that the true optimal values x_1^* and x_2^* of x_1 and x_2 are

$$x_1^* = \begin{cases} \dfrac{a}{b} + \left(\dfrac{a}{b^2} - 1\right)^{1/2} & \text{for } a > b^2 > 0 \\ \dfrac{a}{b} & \text{for } 0 < a \leq b^2 \end{cases} \qquad (6\text{-}50)$$

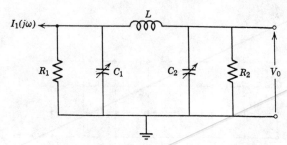

Figure 6-9. Pi circuit.

and

$$x_2^* = \begin{cases} \frac{1}{b} + \left(\frac{a}{b^2} - 1\right)^{1/2} & \text{for } a > b^2 > 0 \\ \frac{1}{b} & \text{for } 0 < a \le b^2 \end{cases} \quad (6\text{-}51)$$

Motivation for the performance measure 6-49 stems from the following circuit problem (see Figure 6-9).

This circuit could represent the interstage coupling of a tuned amplifier; similar circuits are used as output networks of automatically tuned transmitters in which C_1, the tuning capacitor, and C_2, the loading capacitor, are adjusted with semi-independent control loops. With a given carrier frequency ω, the purpose of such a tuning system is usually to obtain 180° phase shift between $I_1(j\omega)$ and the voltage across C_1, and to obtain a certain level of power output to an antenna system. Rather than pursue the foregoing objective, in which tuned values of C_1 and C_2 are constrained to satisfy a 180° phase-shift requirement, the objective of this example is simply to find values of C_1 and C_2 such that maximum power is delivered to R_2.

In Figure 6-9, $I_1(j\omega)$ is the complex-number representation of the sinusoidal steady-state driving current with frequency ω radians/second; the resistors R_1 and R_2 are fixed, as is the inductor L; and the capacitors C_1 and C_2 are to be adjusted to maximize the power output P_0:

$$P_0 = \frac{|V_0(j\omega)|^2}{R_2} \quad (6\text{-}52)$$

Let $\tau_1 \triangleq R_1 C_1$ and $\tau_2 \triangleq R_2 C_2$; circuit analysis gives

$$V_0(s) = \frac{-R_1 R_2 I_1(s)}{s^3 (L\tau_1\tau_2) + s^2 L(\tau_1 + \tau_2) + s(L + R_1\tau_2 + R_2\tau_1) + R_1 + R_2} \quad (6\text{-}53)$$

Sect. 6-4   NONSEQUENTIAL METHODS   291

from which

$$|V_0(j\omega)| = \frac{I_1 R_1}{(\text{Re}^2 + \text{Im}^2)^{1/2}}, \qquad I_1 \equiv |I_1(j\omega)| \qquad (6\text{-}54)$$

where

$$\text{Re} = 1 + \frac{R_1}{R_2} - \frac{\omega^2 L \tau_1}{R_2} - \frac{\omega^2 L \tau_2}{R_2} \qquad (6\text{-}55)$$

and

$$\text{Im} = \frac{\omega L}{R_2} + \frac{R_1 \omega \tau_2}{R_2} + \omega \tau_1 - \frac{\omega^3 L \tau_1 \tau_2}{R_2} \qquad (6\text{-}56)$$

To simplify notation, the following dimensionless parameters are introduced:

$$a = R_1/R_2$$
$$b = \omega L/R_2$$
$$x_1 = \omega \tau_1$$

and

$$x_2 = \omega \tau_2$$

By substituting the above identities appropriately into Equations 6-54, 6-55, and 6-56, and by noting that P_0 is maximized when $1/|V_0(j\omega)|^2$ is minimized, it follows that Equation 6-49 is an appropriate measure of performance for this circuit problem. The values of a and b must be specified prior to applying search techniques.

Let $a = 10$ and $b = 1$ in Equation 6-49; thus,

$$f = (11 - x_1 - x_2)^2 + (1 + 10x_2 + x_1 - x_1 x_2)^2 \qquad (6\text{-}57)$$

Table 6-2 contains values of f obtained by a factorial search, the results of which are intended to suggest fruitful starting points for more comprehensive search techniques.

TABLE 6-2. RESULTS OF FACTORIAL SEARCH

x_2 \ x_1	0	5	10	15
10	10,202	3,152	202	1,352
5	2,637	962	137	162
0	122	72	122	272

●

6-5. UNIVARIATE AND RELAXATION SEARCH

With the methods considered in this section, the x_i variables are adjusted one at a time; these methods date back to Gauss. Given certain initial settings of the variables, the methods tend to lead to relative extrema, but no assurance is given that absolute extrema will be obtained. When $f(\mathbf{x})$ is given analytically, the procedures are generally less efficient than those which are considered in succeeding sections. When $f(\mathbf{x})$ is obtained experimentally, however, these methods may have a distinct advantage.

6-5a. Univariate Search

Univariate search has been employed by everyone who has had occasion to tune an electrical circuit by adjusting several parameters. First, one parameter is adjusted, using a one-dimensional search, until no further improvement is gained; then another parameter is adjusted until no additional improvement results, and so on. After each parameter has been adjusted once, the process is repeated by returning to the first parameter and proceeding as before. If the parameters are noninteracting, which is not very likely, once through the above cycle is enough. If strong interaction and ridges or ravines exist, many cycles may be required; and we may actually be deceived into stopping the process before the optimum is reached.

In using univariate search, it is wise to alter the sequence of adjusting the x_i's now and then so as to add some *dither* to the process. Dither tends to offset slowly converging cycles which might otherwise be established.

Depending on the nature of a particular problem—whether it be analytical or experimental, whether $f(\mathbf{x})$ be continuous or discontinuous, whether noise is involved or not—a particular one-dimensional search procedure (Section 6-3) may be more appropriate than others for the individual steps of a univariate cycle.

6-5b. Southwell's Relaxation Search

Southwell's work on relaxation methods in theoretical physics [6.93] is well known, and the adaptation of relaxation procedures to methods of search is straightforward. Basically, relaxation search is a univariate search in which a particular x_i is picked for relaxation (adjustment to find a relaxed, i.e., optimal condition) if it yields the maximum rate of change of $f(\mathbf{x})$ at a given point $\mathbf{x}^k$. That is, coordinate x_i is picked for search on the basis that

$$\left|\frac{\partial f}{\partial x_i}\right|_{\mathbf{x}=\mathbf{x}^k} \geq \left|\frac{\partial f}{\partial x_j}\right|_{\mathbf{x}=\mathbf{x}^k} \qquad (6\text{-}58)$$

for $j = 1, 2, \ldots, n$. A one-dimensional search is then effected in the x_i direction. At the end of this one-dimensional search, a point $\mathbf{x} = \mathbf{x}^{k+1}$ is located and the partial derivatives are again computed at $\mathbf{x} = \mathbf{x}^{k+1}$. A new x_i is picked, as before, for another one-dimensional search; and the process is continued until all $\partial f/\partial x_j$ are essentially zero at the final test point.

As noted in Section 6-5a, many different one-dimensional search techniques could be used. One procedure used by Bromberg [6.8] is a sequence of Newton-Raphson searches. In this case, a step equal to $-(\partial f/\partial x_i)/(\partial^2 f/\partial x_i^2)|_{\mathbf{x}=\mathbf{x}^k}$ is taken along the x_i coordinate, i.e.,

$$x_i^{k+1} = x_i^k - \left.\frac{\partial f/\partial x_i}{\partial^2 f/\partial x_i^2}\right|_{\mathbf{x}=\mathbf{x}^k} \tag{6-59}[8]$$

and

$$x_j^{k+1} = x_j^k \qquad \text{for } j \neq i \tag{6-60}$$

New values of $\partial f/\partial x_j$, $j = 1, 2, \ldots, n$, are computed at $\mathbf{x}^{k+1}$, and the process is repeated as before. This procedure works relatively well when the response surface $f(\mathbf{x})$ is approximately quadratic; however, relaxation search is not quadratic convergent (quadratic convergence is defined in Section 6-3a).

6-5c. Southwell-Synge Search

A variation of Southwell's relaxation search was proposed by Synge [6.97]. In this variation, a step is taken away from a given point $\mathbf{x}^k$ along the coordinate x_i for which

$$\left|\frac{\partial f/\partial x_i}{\partial^2 f/\partial x_i^2}\right|_{\mathbf{x}=\mathbf{x}^k} \geq \left|\frac{\partial f/\partial x_j}{\partial^2 f/\partial x_j^2}\right|_{\mathbf{x}=\mathbf{x}^k} \tag{6-61}$$

for all j. As before, the step size is given by 6-59, and therefore the largest possible step, consistent with the Newton-Raphson method, is taken at each iteration.

A serious disadvantage of this method is that all $\partial^2 f/\partial x_j^2$'s must be computed at each search point. For performance measures which are evaluated experimentally, finite-difference approximations of the form

$$\frac{\partial^2 f}{\partial x_j^2} \simeq \frac{f(x_1, \ldots, x_j + \epsilon, \ldots, x_n) - 2f(\mathbf{x}) + f(x_1, \ldots, x_j - \epsilon, \ldots, x_n)}{\epsilon^2}$$

would have to be used. Thus, $2n + 1$ function evaluations would be required prior to each major step of the search.

---

[8] Note that the sign of $\partial^2 f/\partial x_i^2$ is indicative of the curvature of the quadratic fit; if a minimum of f is desired, but the sign of $\partial^2 f/\partial x_i^2$ is negative at point $\mathbf{x}^k$, the step suggested by 6-59 should not be taken.

Example 6-3. The problem of Example 6-2 is to be solved by using univariate search. The performance measure is

$$f = (11 - x_1 - x_2)^2 + (1 + x_1 + 10x_2 - x_1x_2)^2 \quad (6\text{-}62)$$
$$= \text{Re}^2 + \text{Im}^2$$

where x_1 and x_2 are restricted to be non-negative. It follows that

$$(\partial f/\partial x_1) = -2\,\text{Re} + 2(1 - x_2)\,\text{Im} \quad (6\text{-}63)$$

$$(\partial f/\partial x_2) = -2\,\text{Re} + 2(10 - x_1)\,\text{Im} \quad (6\text{-}64)$$

$$(\partial^2 f/\partial x_1^2) = 2 + 2(1 - x_2)^2 \quad (6\text{-}65)$$

$$(\partial^2 f/\partial x_2^2) = 2 + 2(10 - x_1)^2 \quad (6\text{-}66)$$

and

$$(\partial^2 f/\partial x_1 \partial x_2) = 2(1 - \text{Im}) + 2(1 - x_2)(10 - x_1) \quad (6\text{-}67)$$

Suppose that the first starting point is selected at $\mathbf{x}^0 = [0\ \ 0]'$ and that the initial one-dimensional search is conducted along the x_1 coordinate—this selection is a reasonable consequence of the factorial search of Example 6-2. It follows that $\text{Re}^0 = 11$ and $\text{Im}^0 = 1$, by the use of which $(\partial f/\partial x_1)^0 = -20$, and $(\partial^2 f/\partial x_1^2)^0 = 4$. A Newton-Raphson search (Equation 6-59) yields

$$x_1^1 = x_1^0 - \left.\frac{\partial f/\partial x_1}{\partial^2 f/\partial x_1^2}\right|_{\mathbf{x}=\mathbf{x}^0} - 0\ \ (-20/4) = 5$$

Note that $f(x_1, 0)$ is a quadratic in x_1; this single iteration of the Newton-Raphson procedure is therefore sufficient to locate the minimum point $x_1^1 = 5$ along the line defined by $x_2 = 0$.

Next, the search should be conducted in the x_2 direction, while holding x_1 equal to 5. But with $\mathbf{x}^1 = [5\ \ 0]'$, $(\partial f/\partial x_2)$ is found to equal 48—increasing x_2 away from zero would result in an increase in f! Because x_1 and x_2 are restricted to be non-negative, it would appear that a constrained minimum of f occurs at $\mathbf{x} = [5\ \ 0]'$. Indeed, by closely checking f it is found that $f(5,0) = 72$ is the best that can be done in the vicinity of $\mathbf{x} = [5\ \ 0]'$.

Well, should the search be continued? The coarse factorial search of Example 6-2 gives no hint of better things to be obtained, but perhaps more scrutiny is required. Let the new initial search point be given by $\mathbf{x}^0 = [5\ \ 1]'$. As before,

$$x_1^1 = 5 - (-10/2) = 10$$

and at the point $\mathbf{x}^1 = [10\ \ 1]'$, both $\partial f/\partial x_1$ and $\partial f/\partial x_2$ equal zero! Also at $\mathbf{x}^1 = [10\ \ 1]'$, $\partial^2 f/\partial x_1^2$ and $\partial^2 f/\partial x_2^2$ equal 2, so it would appear that $f(10,1) =$

Sect. 6-5 UNIVARIATE AND RELAXATION SEARCH 295

121 is another relative minimum—*but is it*? Perhaps the test associated with inequality 2-21 should be applied, as follows:

$$\frac{\partial^2 f}{\partial x_1^2} = 2 > 0$$

and

$$\begin{vmatrix} \dfrac{\partial^2 f}{\partial x_1^2} & \dfrac{\partial^2 f}{\partial x_1 \partial x_2} \\ \dfrac{\partial^2 f}{\partial x_1 \partial x_2} & \dfrac{\partial^2 f}{\partial x_2^2} \end{vmatrix} = \begin{vmatrix} 2 & -20 \\ -20 & 2 \end{vmatrix} = -396 < 0$$

and therefore $[10 \ \ 1]'$ is *a saddle point, and not a local minimum point*.

It is the nature of certain saddle points that they lie between two distinct points of relative minima. To determine if $[10 \ \ 1]'$ is such a saddle point for f, let a new starting point $\mathbf{x}^0 = [18 \ \ 3]'$. Repeated application of Newton-Raphson univariate search gives

$$x_1^1 = 18 - (40/10) = 14, \qquad \mathbf{x}^1 = [14 \ \ 3]'$$
$$x_2^2 = 3 - (-12/34) = 3.353, \qquad \mathbf{x}^2 = [14 \ \ 3.353]'$$

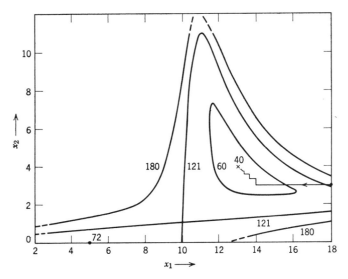

Figure 6-10. Contours of $(11 - x_1 - x_2)^2 + (1 + x_1 + 10x_2 - x_1 x_2)^2$ and a univariate search for the minimum.

and similarly,

$$\mathbf{x}^3 = [13.62 \quad 3.353]'$$
$$\mathbf{x}^4 = [13.62 \quad 3.584]'$$
$$\mathbf{x}^5 = [13.33 \quad 3.584]'$$
$$\mathbf{x}^6 = [13.33 \quad 3.733]'$$
$$\vdots \quad \vdots \quad \vdots$$
$$\mathbf{x}^\infty = [13 \quad 4]'$$

These results are sketched in Figure 6-10. The absolute minimum is $f(13,4) = 40$ which is a significant improvement over the relative minimum of $f(5,0) = 72$ previously obtained.

●

6-6. BASIC GRADIENT METHODS

6-6a. Common Features

Common to all gradient search techniques is the use of the gradient $\nabla f = \mathbf{g} = [\partial f/\partial x_1 \quad \partial f/\partial x_2 \quad \cdots \quad \partial f/\partial x_n]'$. In the case that $f(\mathbf{x})$ is determined experimentally, a discrete approximation to the gradient is used; and in the work of Box and Wilson [6.7], statistical procedures are used to estimate the gradient $\mathbf{g}$ from experimentally obtained noisy data.

All gradient methods are governed, at least in part, by the following equation:

$$\mathbf{x}^{k+1} = \mathbf{x}^k - \alpha \mathbf{H}\mathbf{g}|_{\mathbf{x}=\mathbf{x}^k} \qquad (6\text{-}68)$$

in which $\mathbf{x}^k$ is the "old" value of $\mathbf{x}$, $\mathbf{x}^{k+1}$ is the "new" value of $\mathbf{x}$, $\mathbf{g} = \nabla f$ is the gradient of f in column-vector form, $\mathbf{H}$ is an $n \times n$ matrix, and α is a real number. Gradient methods differ in the way in which $\mathbf{H}$ and α are selected at $\mathbf{x} = \mathbf{x}^k$.

For example, if $\mathbf{H}$ is taken to be the matrix which contains all zero entries, except for a single 1 in the ith row on the main diagonal, and if α is taken to be $1/(\partial^2 f/\partial x_i^2)$, then Equation 6-68 reduces to 6-59 and 6-60, and the result is Southwell's relaxation search. In a limited sense, therefore, Southwell's relaxation search is a form of gradient search.

In the following subsections, more appropriate forms for $\mathbf{H}$ are considered; and in addition, the case in which α is infinitesimal is treated. This latter leads to a continuous search scheme, one which can be programmed on a general-purpose analog computer.

Because the gradient of $f(\mathbf{x})$ is generally affected by a change in scale of any given x_i, it is not surprising that the rate of convergence of basic gradient methods depends on the scale used [6.22]. Another drawback of basic gradient methods is that they are relatively inefficient when ridges or ravines are salient. Extensions of the basic gradient methods alleviate these faults to a great extent, as is evidenced in Sections 6-7 and 6-8.

6-6b. Continuous Steepest Ascent [Descent]

Cauchy first introduced the concept of steepest descent in 1847 [6.16] for use in the simultaneous solution of equations, and the method has enjoyed popularity ever since. The key idea supporting continuous steepest ascent—to keep the presentation unencumbered, only the ascent case is given—is that a maximum is sought by always proceeding in the direction which yields the greatest rate of increase of $f(\mathbf{x})$. It is as a blindfolded man who strives to reach the top of a hill by always climbing the steepest slope.

To use the method, the following question must be resolved. Given an initial starting point $\mathbf{x}^0 = [x_1^0 \ x_2^0 \ \cdots \ x_n^0]'$, in what direction in the n-dimensional Euclidean space of the x_i's from $\mathbf{x} = \mathbf{x}^0$ does $f(\mathbf{x})$ tend to increase the most? It is shown in the next few paragraphs that the gradient direction $\mathbf{g}(\mathbf{x}^0)$ yields the greatest incremental increase of $f(\mathbf{x})$ for a fixed incremental distance moved from $\mathbf{x} = \mathbf{x}^0$. The derivation follows that of Kelley [6.55].

Let the fixed incremental distance moved from $\mathbf{x} = \mathbf{x}^0$ be denoted by ϵ. By Pythagoras's theorem,

$$\epsilon^2 = (\Delta x_1^0)^2 + (\Delta x_2^0)^2 + \cdots + (\Delta x_n^0)^2 \qquad (6\text{-}69)$$

Let $\mathbf{x}^1$ denote the value of $\mathbf{x}$ obtained by an incremental move:

$$\begin{aligned}\mathbf{x}^1 &= [x_1^0 + \Delta x_1^0 \ \ x_2^0 + \Delta x_2^0 \ \ \cdots \ \ x_n^0 + \Delta x_n^0]' \\ &= \mathbf{x}^0 + \Delta \mathbf{x}^0 \end{aligned} \qquad (6\text{-}70)$$

In seeking a maximum of $f(\mathbf{x})$ by steepest ascent search, the object is to maximize $f(\mathbf{x}^0 + \Delta \mathbf{x}^0)$ by appropriate selection of the Δx_i^0's. Recall, however, that the Δx_i^0's are constrained by Equation 6-69. Thus, the Lagrange multiplier technique of Chapter 2 is applicable: the augmented function f_a is defined by

$$f_a = f(\mathbf{x}^0 + \Delta \mathbf{x}^0) + h \sum_{j=1}^{n} (\Delta x_j^0)^2 \qquad (6\text{-}71)$$

and the necessary condition for a maximum of f_a is

$$\frac{\partial f_a}{\partial \Delta x_j^0} = 0 \qquad (6\text{-}72)$$

or

$$\left.\frac{\partial f}{\partial x_j}\right|_{\mathbf{x}=\mathbf{x}^0+\Delta\mathbf{x}^0} + 2h\Delta x_j^0 = 0 \qquad (6\text{-}73)$$

for all j. Hence,

$$\Delta x_j^0 = \frac{-1}{2h} \left.\frac{\partial f(\mathbf{x})}{\partial x_j}\right|_{\mathbf{x}=\mathbf{x}^0+\Delta\mathbf{x}^0} \qquad (6\text{-}74a)$$

which in matrix form is

$$\Delta \mathbf{x}^0 = \frac{-1}{2h} \nabla f(\mathbf{x}^0 + \Delta \mathbf{x}^0) = \frac{-1}{2h} \mathbf{g}(\mathbf{x}^0 + \Delta \mathbf{x}^0) \qquad (6\text{-}74\text{b})$$

The Lagrange multiplier h is evaluated by using the result given by 6-74a in Equation 6-69:

$$\epsilon^2 = \frac{1}{4h^2} \sum_{j=1}^{n} \left(\frac{\partial f}{\partial x_j}\right)^2 \bigg|_{\mathbf{x}=\mathbf{x}^0+\Delta\mathbf{x}^0} = \frac{1}{4h^2} \mathbf{g}(\mathbf{x}^0 + \Delta\mathbf{x}^0)' \mathbf{g}(\mathbf{x}^0 + \Delta\mathbf{x}^0)$$

from which it follows that

$$\frac{-1}{2h} = [\mathbf{g}(\mathbf{x}^0 + \Delta\mathbf{x}^0)' \mathbf{g}(\mathbf{x}^0 + \Delta\mathbf{x}^0)]^{-\frac{1}{2}} \epsilon$$

This result is used to eliminate $-1/(2h)$ from Equation 6-74b:

$$\Delta \mathbf{x}^0 = \epsilon \mathbf{g}(\mathbf{x}^0 + \Delta\mathbf{x}^0)/[\mathbf{g}(\mathbf{x}^0 + \Delta\mathbf{x}^0)' \mathbf{g}(\mathbf{x}^0 + \Delta\mathbf{x}^0)]^{\frac{1}{2}} \qquad (6\text{-}75)$$

If the incremental distance ϵ is sufficiently small, then so also is each Δx_i; and $\mathbf{g}(\mathbf{x}^0 + \Delta\mathbf{x}^0)$ can be safely replaced simply by $\mathbf{g}(\mathbf{x}^0)$. Also note that the denominator of the right-hand member of 6-75 is a positive number, except at stationary points in which case it equals zero. It is convenient to lump this term with ϵ, i.e., to form a new increment

$$\Delta \tau \triangleq \epsilon/[\mathbf{g}(\mathbf{x}^0)' \mathbf{g}(\mathbf{x}^0)]^{\frac{1}{2}}$$

and rather than view $\Delta \tau$ as a function of $\mathbf{x}^0$, $\Delta \tau$ is held fixed by assigning an appropriate value of the increment ϵ for each value of $\mathbf{x}$. In other words $\epsilon \triangleq \Delta\tau[\mathbf{g}(\mathbf{x})' \mathbf{g}(\mathbf{x})]^{\frac{1}{2}}$ where $\Delta \tau$ is constant. The net result is that

$$\Delta x_i|_{\mathbf{x}=\mathbf{x}^0} = \Delta\tau[\partial f/\partial x_i]_{\mathbf{x}=\mathbf{x}^0} \qquad i = 1, 2, \ldots, n \qquad (6\text{-}76\text{a})$$

or in column-vector form

$$\Delta \mathbf{x}^0 = \Delta\tau \mathbf{g}(\mathbf{x}^0) \qquad (6\text{-}76\text{b})$$

for steepest ascent search.

Up to this point, $\Delta \tau$ and ϵ have been assumed small increments, small enough so that $\nabla f(\mathbf{x}^0 + \Delta\mathbf{x}^0) \simeq \nabla f(\mathbf{x}^0)$. If $\Delta \tau$, and therefore also Δx_i, is now allowed to approach $0+$, Equations 6-76 reduce to a set of first-order differential equations:

$$\frac{dx_i}{d\tau} = \frac{\partial f}{\partial x_i} \qquad i = 1, 2, \ldots, n \qquad (6\text{-}77)$$

By assigning τ the role of time, this set of equations can be programmed on a general-purpose analog computer [6.2, 6.25, 6.41, 6.66, and 6.84]. The final values of the x_i's in the analog computer solution correspond to stationary values of $f(\mathbf{x})$ because at stationary values, $\partial f/\partial x_i$ and therefore dx_i/dt equal zero. As with any analog computer problem, amplitude and time scaling are required. The interested reader is referred to standard analog computer texts [6.51 or 6.66, for example] for computer programming details.

It is conceivable that one might unknowingly start the solution of 6-77 with initial conditions at a relative minimum or at a saddle point of $f(\mathbf{x})$. In such a case, all $\partial f/\partial x_i$ are initially zero, and theoretically the search would not begin. In practice, however, noise in the system is sufficient to deviate the solution from either a minimum or a saddle point; and once away, the solution diverges from these unstable equilibrium points. In fact, Zellnick et al. [6.110] found it difficult to determine the character of functions in the vicinity of saddle points because their search techniques lead them abruptly away.

6-6c. Discrete Steepest Ascent [Descent]

A discrete version of steepest ascent search is obtained from Equation 6-76b, namely,

$$\mathbf{x}^{k+1} - \mathbf{x}^k = \Delta\tau\mathbf{g}|_{\mathbf{x}=\mathbf{x}^k} \qquad (6\text{-}78)$$

which is rearranged to correspond in form to gradient search in general, Equation 6-68, as follows:

$$\mathbf{x}^{k+1} = \mathbf{x}^k + \Delta\tau\mathbf{I}\mathbf{g}|_{\mathbf{x}=\mathbf{x}^k} \qquad (6\text{-}79)$$

In this special case, the matrix $\mathbf{H}$ of Equation 6-68 equals the identity matrix $\mathbf{I}$, the nonzero entries of which are "ones" on the major diagonal, and α of Equation 6-68 equals $-\Delta\tau$.

The step size $\Delta\tau$ remains to be determined. Note that by replacing $\mathbf{x}$ in $f(\mathbf{x})$ with $\mathbf{x}^k + \Delta\tau\mathbf{g}(\mathbf{x}^k)$, the function $f[\mathbf{x}^k + \Delta\tau\mathbf{g}(\mathbf{x}^k)]$ is a function of the parameter $\Delta\tau$ only, i.e., the function is in a parametric form. There are an unlimited number of ways in which $\Delta\tau$ can be selected, all of which correspond to some form of one-dimensional search (see Section 6-3). Lapidus et al. [6.63] compare six of these, each but slightly different, by applying them to a common problem.

Because of the computations involved in evaluating the gradient of $f(\mathbf{x})$ at a given point, it is usually advantageous to make the most of each gradient computation before making another; that is, to search in the direction of the gradient until $\partial f[\mathbf{x}^k + \Delta\tau\mathbf{g}(\mathbf{x}^k)]/\partial\Delta\tau = 0$ for some $\Delta\tau$. This approach is referred to in the literature [6.12] as the method of *optimum steepest ascent*, an unfortunate designation in that quite often improvements in efficiency can be made by incorporating additional features, as is done in Section 6-7. In this work, therefore, the phrase "*best-step steepest ascent*" is used in place of "optimum steepest ascent."

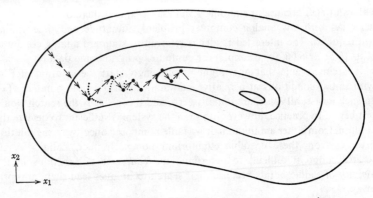

Figure 6-11a. Example of best-step steepest ascent search.

Figure 6-11a depicts a search conducted by best-step steepest ascent when $\mathbf{x} = \{x_1, x_2\}$. Note that if a different scale is employed for the x_1 coordinate, as in Figure 6-11b, the number of iterations is changed considerably. Hence, the desirability of proper scaling is clearly evident.

Example 6-4. Again, the problem of Example 6-2 is considered. The fumbling perpetrated in the initial phases of Example 6-3 is by-passed here, however, and the initial search point $\mathbf{x}^0 = [18 \quad 3]'$ is selected to correspond to that $\mathbf{x}^0$ finally used in Example 6-3. The solution is obtained by using discrete gradient search for the minimum. For the one-dimensional part of the search, phase I of the procedure in Section 6-3b is used with slight modifications (see also Equations 6-47 and 6-48), and this is followed by a simple quadratic interpolation.

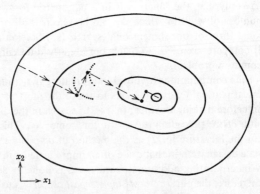

Figure 6-11b. Example of best-step steepest ascent search with proper scaling.

A computer flow diagram for the procedure is given in Figure 6-12. The flow diagram is generally applicable to $f(\mathbf{x})$'s of class C^1. It is also to be observed that the direction vector $\mathbf{r}^k$ is identified with $-\mathbf{g}^k$ only in block number 5 which is contained within the dashed lines—the remainder of the flow diagram is generally applicable for any $\mathbf{r}^k$. Thus, other means of specifying $\mathbf{r}^k$ could be incorporated in block number 5. As presented in Sections 6-7 and 6-8, highly efficient search techniques utilize certain previously computed gradients, direction vectors, and "best-step" points to obtain the current direction vector $\mathbf{r}^k$; means of storing the required past history must be added to the flow diagram of Figure 6-12 to render it applicable to more general cases.

The sequence of computational events is as follows:

1. Explicit forms of $f(\mathbf{x})$ and $\nabla f(\mathbf{x}) = \mathbf{g}(\mathbf{x})$ are programmed, and values of f_{est}^0 and $\mathbf{x}^0$ are specified.
2. k is equated to zero initially.
3. Values are computed for $y(0) = f(\mathbf{x}^k) \equiv f^k$ and $\mathbf{g}(\mathbf{x}^k) \equiv \mathbf{g}^k$.
4. A halt condition is indicated if the value of $(\mathbf{g}^k)'\mathbf{g}^k$ is less than 10^{-14} (nominally), and control is transferred to point B (block 18) in the flow diagram.
5. The direction vector $\mathbf{r}^k$ is generated.
6. $\dot{y}(0) = (\mathbf{r}^k)'\mathbf{g}^k$ is evaluated.
7. If $|\dot{y}(0)|$ is less than 10^{-14} (nominally), $\mathbf{r}^k$ is redefined as $-\mathbf{g}^k$, and $\dot{y}(0)$ is redefined as $-(\mathbf{g}^k)'\mathbf{g}^k$.
8. $d_s = -2(f^k - f_{\text{est}}^k)/\dot{y}(0)$ is computed, as in Equation 6-47.
9. d_0 is computed, as in Equation 6-48.
10. An index j is equated to unity for later use.
11. To compute the next estimate d_1 of a minimum point of $y(\alpha)$, the three values $y(0)$, $\dot{y}(0)$, and $y(d_0)$ are fitted with a quadratic which has a well-defined minimum, provided that u is greater than zero, where

$$u \triangleq y(d_0) - y(0) - d_0 \dot{y}(0) \qquad (6\text{-}80)$$

12. If u is greater than zero, d_1 is equated to the minimum point of the quadratic, which point the reader may show to be located at $\alpha = -\dot{y}(0)(d_0)^2/2u$.
13. If u is less than zero, however, d_1 is equated to twice d_0.[9]
14. In either of the preceding two cases, $y(d_1)$ is evaluated and compared to $y(d_0)$ and $y(0)$.
15. In the case that $y(0)$ is less than or equal to $y(d_0)$ and $y(d_1)$, the index j

---

[9] From the way in which $\dot{y}(0)$ and d_0 are formed, they must be opposite in sign. Because of this fact, $y(d_0)$ must be less than $y(0)$ if u in 6-80 is less than zero. This property is utilized in step 13.

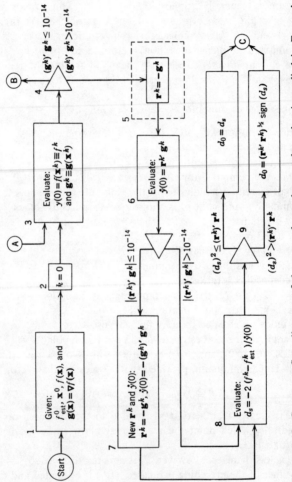

Figure 6-12. Computer flow diagram of the general search procedure that is applied in Example 6-4. (The figure is continued on the following page.)

Sect. 6-6 BASIC GRADIENT METHODS

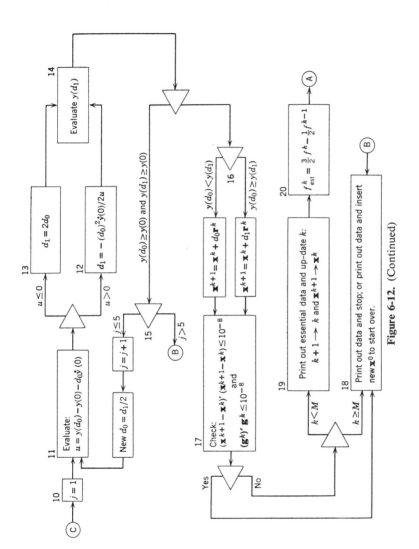

Figure 6-12. (Continued)

started in step 10 is compared to 20 (nominally)—if j is less than 20, j is incremented up by one, d_0 is reset to $d_1/2$, and steps 11 through 13 are repeated to obtain new values of $y(d_0)$, d_1, and $y(d_1)$; but if j is greater than 20, a halt condition is signified because the current one-dimensional search has been unsuccessful in twenty attempts to obtain some value of α for which $y(\alpha) < y(0)$.

16. In the case that at least one of $y(d_0)$ and $y(d_1)$ is less than $y(0)$, the lesser of the two values is identified with $f(\mathbf{x}^{k+1})$, and $\mathbf{x}^{k+1}$ is thereby defined.
17. If both $(\mathbf{x}^{k+1} - \mathbf{x}^k)'(\mathbf{x}^{k+1} - \mathbf{x}^k)$ and $(\mathbf{r}^k)'\mathbf{r}^k$ are less than 10^{-8} (nominally), a halt condition is indicated, and control is transferred to point B in the flow diagram.
18. If the halt condition in step 17 is not satisfied, however, k is compared to M, a specified integer; and if $k \geq M$, a halt condition is indicated.
19. But if $k < M$, intermediate search results can be recorded, as desired, and the value of k is then updated: $k + 1$ replaces k, and $\mathbf{x}^{k+1}$ replaces $\mathbf{x}^k$.
20. A new value of f_{est}^k is computed, $f_{est}^k = (3/2)f^k - (1/2)f^{k-1}$, which is necessarily less than f^k.
21. And finally, control is transferred to step 3 (point A in the flow diagram), and a new one-dimensional search is begun. After at most M such one-dimensional searches (the first of which begins at $\mathbf{x}^0$), a "winning" point, if not an optimum point, is reached.

For the present example, relationships are as follows:

$$f_{est}^0 = 72 \tag{6-81}$$

which is the smallest value of f obtained in the factorial search of Example 6-2;

$$\mathbf{x}^0 = \begin{bmatrix} 18 \\ 3 \end{bmatrix} \tag{6-82}$$

which corresponds to the value of $\mathbf{x}^0$ finally used in Example 6-3;

$$y(\alpha) = f(\mathbf{x}^k + \alpha \mathbf{r}^k) \\ = \text{Re}^2 + \text{Im}^2 \tag{6-83}$$

where

$$\text{Re} = 11 - x_1^k - \alpha r_1^k - x_2^k - \alpha r_2^k \tag{6-84a}$$

and

$$\text{Im} = 1 + x_1^k + \alpha r_1^k + 10(x_2^k + \alpha r_2^k) - (x_1^k + \alpha r_1^k)(x_2^k + \alpha r_2^k) \tag{6-84b}$$

Sect. 6-6 BASIC GRADIENT METHODS

which stem from Example 6-3; and finally,

$$\mathbf{r}^k = -\mathbf{g}^k$$

$$= \begin{bmatrix} 2\,\text{Re} - 2(1 - x_2^k)\,\text{Im} \\ 2\,\text{Re} - 2(10 - x_1^k)\,\text{Im} \end{bmatrix}_{\alpha=0} \tag{6-85}$$

The first one-dimensional search proceeds as follows: with $\mathbf{x}^0 = [18\ \ 3]'$, $f^0 = 125$ (by using Equation 6-83), $\mathbf{g}^0 = [40\ \ 100]'$ (by using Equation 6-85), and $\mathbf{g}^{0\prime}\mathbf{g}^0 = 11{,}600$; step 8 yields $d_s = 0.91 \times 10^{-2}$; step 9 yields $d_0 = 0.91 \times 10^{-2}$; Equation 6-83 yields $y(d_0) = 83.4$; Equation 6-80 yields $u = 64.3$; step 12 yields $d_1 = 0.75 \times 10^{-2}$; Equation 6-83 yields $y(d_1) = 82.0$; and finally, step 16 yields $\mathbf{x}^1 = \mathbf{x}^0 - d_1 \mathbf{g}^0 = [17.7\ \ 2.25]'$. At this point, $\mathbf{g}^1$ is determined, f_{est}^1 is set at $(3/2)f^1 - (1/2)f^0 = 60.48$, and the cycle is repeated. Table 6-3 contains pertinent results, and Figure 6-13 displays these graphically. Note that the initial one-dimensional searches are not of the best-step variety, but that successive one-dimensional searches near the minimum are characterized by orthogonal directions and are therefore "best-step" searches.

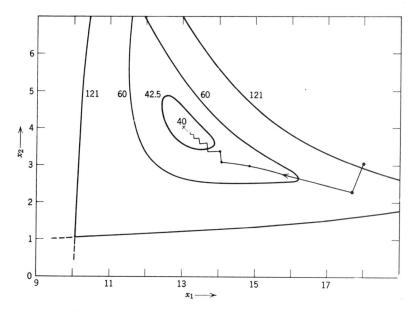

Figure 6-13. Discrete gradient search of Example 6-4.

TABLE 6-3. RESULTS OF DISCRETE GRADIENT SEARCH

k	x_1	x_2	$g'g$	d_0	d_1	$y(d_0)$	$y(d_1)$	f_{est}
0	18	3	11,600.000	0.009138	0.007523	83.380	81.988	72
1	17.699	2.248	220.530	0.195040	0.421985	48.916	62.025	60.482
2	14.887	2.943	62.186	0.531819	0.111595	94.648	44.365	32.379
3	14.015	3.059	96.923	0.046950	0.032197	43.132	42.815	42.090
4	13.986	3.374	29.773	0.052073	0.061167	41.924	41.902	42.040
5	13.654	3.345	33.775	0.027038	0.039293	41.303	41.235	41.445
6	13.631	3.572	14.974	0.044482	0.055107	40.838	40.822	40.902
7	13.418	3.553	12.985	0.031845	0.044809	40.555	40.530	40.615
8	13.403	3.714	7.126	0.040931	0.051060	40.356	40.348	40.384
9	13.267	3.702	5.083	0.035844	0.049010	40.233	40.223	40.257
10	13.257	3.812	3.320	0.038621	0.048849	40.148	40.145	40.161
.	.	.	.					.
19	13.028	3.967	0.051	0.044749	0.057048	40.00259	40.00252	40.0028

6-6d. Newton Search

The n-dimensional counterpart of one-dimensional Newton-Raphson search (Section 6-3a) is a form of gradient search. It has the desirable attribute of quadratic convergence, but it requires evaluation of second partial derivatives and inverse matrices, and it is therefore well-suited for end-game search only.

Suppose that in the vicinity of the current search point $\mathbf{x} = \mathbf{x}^k$, $f(\mathbf{x})$ can be represented adequately by the truncated Taylor's series expansion $f_s(\mathbf{x})$:

$$f_s(\mathbf{x}) = f(\mathbf{x}^k) + \sum_{i=1}^{n} \frac{\partial f(\mathbf{x}^k)}{\partial x_i}(x_i - x_i^k) + \tfrac{1}{2} \sum_{i=1}^{n} \sum_{j=1}^{n} a_{ij}(x_i - x_i^k)(x_j - x_j^k)$$

$$\cong f(\mathbf{x}) \tag{6-86}$$

where

$$a_{ij} = a_{ji} = \left.\frac{\partial^2 f}{\partial x_i \partial x_j}\right|_{\mathbf{x}=\mathbf{x}^k} \tag{6-87}$$

The stationary points of the quadratic function $f_s(\mathbf{x})$ are obtained by the classical approach:

$$\frac{\partial f_s(\mathbf{x})}{\partial x_m} = \frac{\partial f(\mathbf{x}^k)}{\partial x_m} + \tfrac{1}{2} \sum_{i=1}^{n} a_{mi}(x_i - x_i^k) + \tfrac{1}{2} \sum_{j=1}^{n} a_{mj}(x_j - x_j^k)$$

$$= \frac{\partial f(\mathbf{x}^k)}{\partial x_m} + \sum_{j=1}^{n} a_{mj}(x_j - x_j^k) \tag{6-88}$$

$$= 0 \quad \text{for } m = 1, 2, \ldots, n$$

and in matrix form, the set of equations represented by 6-88 is expressed by

$$\mathbf{A}^k(\mathbf{x} - \mathbf{x}^k) = -\nabla f(\mathbf{x}^k) = -\mathbf{g}^k \tag{6-89}$$

where

$$\mathbf{A}^k = \begin{bmatrix} a_{11} & a_{12} & \cdots & a_{1n} \\ a_{21} & & & \\ \vdots & & \ddots & \vdots \\ a_{n1} & & \cdots & a_{nn} \end{bmatrix} \tag{6-90}$$

Assuming that the inverse matrix $(\mathbf{A}^k)^{-1}$ of $\mathbf{A}^k$ exists,[10] the $\mathbf{x}$, $\mathbf{x} = \mathbf{x}^{k+1}$,

---

[10] The inverse $\mathbf{A}^{-1}$ of $\mathbf{A}$ exists if the determinant $|\mathbf{A}|$ of $\mathbf{A}$ is nonzero; and $\mathbf{A}^{-1}$ equals *adj* $\mathbf{A}/|\mathbf{A}|$ where the (i, j)th element c_{ij} of *adj* $\mathbf{A}$ equals $(-1)^{i+j}|\mathbf{M}_{ji}|$, where $\mathbf{M}_{ji}$ is the $(n-1) \times (n-1)$ matrix obtained by deleting the jth row and the ith column of $\mathbf{A}$. This inversion procedure is relatively inefficient in comparison to certain others given in the references for Appendix A. Also, $|\mathbf{A}|$ may be nonzero but yet so close to zero as to lead to meaningless numerical results; in such cases, $\mathbf{A}$ is said to be *ill-conditioned*.

which results in a stationary value of $f_s(\mathbf{x})$ is obtained by premultiplying both sides of Equation 6-89 by $(\mathbf{A}^k)^{-1}$; thus,

$$\mathbf{x}^{k+1} = \mathbf{x}^k - (\mathbf{A}^k)^{-1}\mathbf{g}^k \tag{6-91}$$

and this equation corresponds in form to Equation 6-68 which expresses gradient search in general. Actually, prior to the inversion of $\mathbf{A}^k$, $\mathbf{A}^k$ should be tested for positive (negative) definiteness, as described in Section 2-4, to determine if point $\mathbf{x}^k$ is "near" a local minimum (maximum); if not, Newton search may diverge.

Consider the effort involved in one iteration of Newton search: First, the n components of the gradient must be computed at $\mathbf{x} = \mathbf{x}^k$; second, the number of distinct a_{ij}'s is $n(n + 1)/2$ and these correspond to second partial derivatives of $f(\mathbf{x})$ evaluated at $\mathbf{x} = \mathbf{x}^k$; and third, the inverse of the $\mathbf{A}^k$ matrix must be obtained. Many authors have suggested that the effort expended above could be put to better use on other approaches; indeed, this is self-evident when $\mathbf{x}^k$ is far from an extremum point of a nonquadratic function. However, when $\mathbf{x} = \mathbf{x}^k$ is near an extremum point—that is, when the search is near completion—Newton's method affords a rate of convergence which leaves little room for improvement. Also, as noted by Crockett *et al.* [6.21] and others, the matrix $\mathbf{A}^k$ of second derivatives remains fairly constant near the optimum and can be held fixed for several iterations.

Example 6-5. For the problem of Example 6-2,

$$\mathbf{A} = \begin{bmatrix} 2 + 2(1 - x_2)^2 & 2 - 2\,\text{Im} + 2(10 - x_1)(1 - x_2) \\ 2 - 2\,\text{Im} + 2(10 - x_1)(1 - x_2) & 2 + 2(10 - x_1)^2 \end{bmatrix} \tag{6-92}$$

and

$$\mathbf{g} = [-2\,\text{Re} + 2(1 - x_2)\,\text{Im} \quad -2\,\text{Re} + 2(10 - x_1)\,\text{Im}]' \tag{6-93}$$

as obtained from Equations 6-62 through 6-67. With $\mathbf{x}^0 = [18 \quad 3]'$, which corresponds to the $\mathbf{x}^0$ used in the preceding example, $\mathbf{A}^0$ and $\mathbf{g}^0$ are found:

$$\mathbf{A}^0 = \begin{bmatrix} 10 & 44 \\ 44 & 130 \end{bmatrix}, \quad \mathbf{g}^0 = \begin{bmatrix} 40 \\ 100 \end{bmatrix}$$

The inverse of $\mathbf{A}^0$ is obtained:

$$(\mathbf{A}^0)^{-1} = \frac{1}{|\mathbf{A}^0|} \begin{bmatrix} 130 & -44 \\ -44 & 10 \end{bmatrix} = \frac{-1}{636} \begin{bmatrix} 130 & -44 \\ -44 & 10 \end{bmatrix}$$

Note that the determinant of $\mathbf{A}^0$ is -636; the fact that $|\mathbf{A}^0|$ is negative suggests that the point $\mathbf{x}^0$ is not in the immediate vicinity of a minimum point (see inequality 2-21). If this observation is ignored, however, Equation 6-91 is used to obtain $\mathbf{x}^1$:

$$\mathbf{x}^1 = \begin{bmatrix} 18 \\ 3 \end{bmatrix} + \frac{1}{636}\begin{bmatrix} 800 \\ -760 \end{bmatrix} = \begin{bmatrix} 19.26 \\ 1.8 \end{bmatrix}$$

which is a change in the opposite direction to that which is desired (see Figure 6-13)!

To illustrate the better qualities of Newton search, a point $\mathbf{x}^1$ which is closer to the minimum point is used for the next iteration. Let $\mathbf{x}^1 = [14 \ 3]'$; this is the same point $\mathbf{x}^1$ resulting from the univariate search of Example 6-3. Again,

$$\mathbf{A}^1 = \begin{bmatrix} 10 & 12 \\ 12 & 34 \end{bmatrix}, \quad \mathbf{g}^1 = [0 \ -12]'$$

and

$$(\mathbf{A}^1)^{-1} = \frac{1}{196}\begin{bmatrix} 34 & -12 \\ -12 & 10 \end{bmatrix}$$

from which,

$$\mathbf{x}^2 = \begin{bmatrix} 14 \\ 3 \end{bmatrix} - \frac{1}{196}\begin{bmatrix} 144 \\ -120 \end{bmatrix} = \begin{bmatrix} 13.265 \\ 3.613 \end{bmatrix}$$

The superiority of this result is clearly evident when compared with the $\mathbf{x}^2$ obtained by univariate search (Figure 6-10) and with that obtained by discrete gradient search (Figure 6-13).

•

6-7. ACCELERATION-STEP SEARCH

6-7a. Two-Dimensional Case

In 1951, Forsythe and Motzkin [6.39] advanced a procedure, called acceleration-step search, for expediting the rate of convergence of the best-step version of steepest ascent when $\mathbf{x} = \{x_1, x_2\}$. Surprisingly, not only the number of required one-dimensional searches is reduced in general by using acceleration-step search, but also the gradient need not be computed at the start of each one-dimensional search. If $f(x_1, x_2)$ is a quadratic function with a well-defined minimum, the procedure requires the use of three one-dimensional searches in sequence to locate the optimum exactly; thus, the procedure is quadratic convergent (the proof of this fact is contained in the more general proofs of Theorems 6-1 and 6-2 which are given in Section 6-8).

One version of the method is as follows. Starting at an initial point $\mathbf{x} = \mathbf{x}^0$ the gradient $\mathbf{g}(\mathbf{x}^0)$ is evaluated, and a one-dimensional search for a maximum is conducted in the gradient direction. (If a minimum of $f(\mathbf{x})$ is desired, the preceding one-dimensional search is conducted in the negative gradient direction.) At the winning point $\mathbf{x}^1$, the gradient $\mathbf{g}(\mathbf{x}^1)$ is evaluated and a second one-dimensional search is conducted, as before. The winner of this second search is designated $\mathbf{x}^2$. At this stage of the process, $\nabla f(\mathbf{x}^2)$ *is not* computed, rather a one-dimensional search is conducted along the straight line which connects the initial point $\mathbf{x}^0$ and the point $\mathbf{x}^2$. In terms of $\mathbf{x}$, this latter *acceleration-step* search is conducted along the line $\mathbf{x} = \beta \mathbf{x}^2 + (1 - \beta)\mathbf{x}^0$ where β is the search parameter. If $f(x_1,x_2)$ is quadratic, this third, best-step search results in the optimal value of $f(x_1,x_2)$; this is illustrated in Figure 6-14. Notice that in the quadratic case, the scale employed *is not* critical when acceleration-step search is used; this is not the case for best-step gradient search and univariate search.

In the practical case where $f(x_1,x_2)$ is not quadratic, acceleration-step search requires use of additional iterations. The process is continued by using at least one best-step steepest ascent search before each acceleration-step. The acceleration-step search is conducted along the line which connects the most recent best-step winner $\mathbf{x}^k$ to the point $\mathbf{x}^{k-2}$ which was obtained as a result of the $(k - 2)$nd one-dimensional search. An example of this is depicted in Figure 6-15.

Example 6-6. Consider one acceleration-step search using the points $\mathbf{x}^2$ and $\mathbf{x}^4$ of Example 6-4. The direction $\mathbf{r}^4$ defined by $\mathbf{x}^4 - \mathbf{x}^2$ is obtained from tabulated values (Table 6-3):

$$\mathbf{r}^4 = \mathbf{x}^4 - \mathbf{x}^2 = \begin{bmatrix} 13.986 \\ 3.374 \end{bmatrix} - \begin{bmatrix} 14.887 \\ 2.943 \end{bmatrix} = \begin{bmatrix} -0.901 \\ 0.431 \end{bmatrix}$$

and the minimum of $f(\mathbf{x}^4 + \alpha \mathbf{r}^4)$ is desired. For this one dimensional search, the method of Section 6-3c is employed with the initial step size d_0 equal to 0.1. Thus for $\alpha = 0$, Table 6-3 gives $f(\mathbf{x}^4) = 42.815$, and this is compared with $f(\mathbf{x})$ evaluated at $\mathbf{x}^4 + 0.1\mathbf{r}^4$:

$$\mathbf{x}^4 + 0.1\mathbf{r}^4 = \begin{bmatrix} 13.986 \\ 3.374 \end{bmatrix} + \begin{bmatrix} -0.0901 \\ 0.0431 \end{bmatrix} = \begin{bmatrix} 13.896 \\ 3.417 \end{bmatrix}$$

With this, Equation 6-57 yields $f(\mathbf{x}^4 + 0.1\mathbf{r}^4) = 42.360$ which is less than $f(\mathbf{x}^4) = 42.815$; therefore, a larger step is taken in the same direction with the result that $f(\mathbf{x}^4 + 0.2\mathbf{r}^4) = f(13.806, 3.460) = 41.943$.

Now, let $d_1 = 0$, $d_2 = 0.1$, and $d_3 = 0.2$; the corresponding values of y are $y(d_1) = 42.815$, $y(d_2) = 42.360$, and $y(d_3) = 41.943$. The identities which follow 6-26 are evaluated: $d_{23} = -0.1$, $d_{31} = 0.2$, $d_{12} = -0.1$, $d_{23}^2 = -0.03$,

Sect. 6-7 ACCELERATION-STEP SEARCH 311

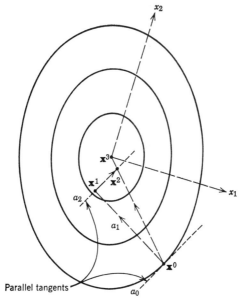

Figure 6-14. Two-dimensional acceleration-step search for the maximum of a quadratic.

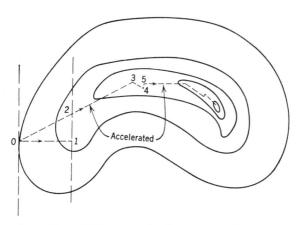

Figure 6-15. An acceleration-step search.

$d_{31}^s = 0.04$, and $d_{12}^s = -0.01$. These results are substituted into Equation 6-27:

$$\frac{d_{23}y(d_1) + d_{31}y(d_2) + d_{12}y(d_3)}{d_{23}\,d_{31}\,d_{12}} = \frac{-0.0038}{0.002}$$

Because this result is less than zero, Equation 6-26 is evaluated to obtain $d_s = 1.25$ which is used to find the estimate $\mathbf{x}^5$ of the minimum point:

$$\mathbf{x}^5 = \mathbf{x}^4 + 1.25\mathbf{r}^4 = [12.860 \quad 3.913]'$$

(Compare this result with that obtained by discrete gradient search in Figure 6-13.)

•

6-7b. n-Dimensional Case: PARTAN

In their original memorandum, Forsythe and Motzkin [6.39] suggested the possible fruitful use of acceleration steps in the n-dimensional case. That is, they suggested alternating between steepest ascent (descent) search and search along the line connecting the most recent winner $\mathbf{x}^k$ of best-step steepest ascent (descent) search to a corresponding winner $\mathbf{x}^{k-2}$. In this search, the optimum of $f[\beta \mathbf{x}^k + (1 - \beta)\mathbf{x}^{k-2}]$ is located with respect to β, $0 \le \beta \le \infty$. Finkel [6.33] found this approach to be relatively successful for the n-dimensional case. Later, Shah et al. [6.14, 6.89] conducted a rigorous study of this approach, defining several new variations known as methods of *parallel tangents* (in short, partan). It is also of interest to note that the acceleration-step concept can be used to advantage in conjunction with univariate search procedures; Pinsker and Tseitlin [6.77] have made a formal study of this.

The name partan has no significance as far as the mechanics of the search procedure are concerned; however, the name has an interesting geometrical origin which is shown in the two-dimensional case of Figure 6-14. The line labeled a_0 in Figure 6-14 is tangent to an equimagnitude contour of $f(\mathbf{x})$ at the original search point $\mathbf{x}^0$; the line labeled a_2 is tangent to an equimagnitude contour at the search point $\mathbf{x}^2$; and the line labeled a_1 is perpendicular to both a_0 and a_2; thus, lines a_0 and a_2 are *parallel tangents*. Note that the acceleration step from $\mathbf{x}^0$ through $\mathbf{x}^2$ to $\mathbf{x}^3$ is taken through the two points $\mathbf{x}^0$ and $\mathbf{x}^2$ at which the two parallel lines a_0 and a_2 are tangent to equimagnitude contours. This feature is common to all partan methods; and as is shown in Section 6-8, the directions of search generated by the procedure are "conjugate directions" with which the minimum of a quadratic can be located in a finite number of iterations.

Sect. 6-7   ACCELERATION-STEP SEARCH

The strong point of this common feature is that it enables us to traverse swiftly along straight and narrow ridges (ravines). To see this, consider any two lines in the x_1,x_2 plane which are parallel and which intersect a straight ridge of $f(x_1,x_2)$ (Figure 6-16). Observe that the points of tangency define a line which parallels the ridge. Hence, by searching along the parallel ridge line, we effectively follow the ridge. For curved ridges, partan search is not quite so efficient, but it is invariably much better than gradient search alone.

Shah *et al.* found that the particular version of partan called *continued partan* worked best for most of the problems they considered, and they proved that the optimum of $f(\mathbf{x})$ is attained after at most $2n - 1$ best-step one-dimensional searches if $f(\mathbf{x})$ is a quadratic with a well-defined optimum. The general procedure for continued partan search is as follows: (1) starting at the initial point $\mathbf{x}^0$, search in the direction defined by $\nabla f(\mathbf{x}^0)$ until extremal point $\mathbf{x}^1$ is found; (2) search for the extremal point $\mathbf{x}^2$ which lies along the line defined by $\mathbf{x}^1$ and $\nabla f(\mathbf{x}^1)$; (3) search for the extremal point $\mathbf{x}^3$ which corresponds to the optimum of $f[\mathbf{x}^2 + \beta(\mathbf{x}^2 - \mathbf{x}^0)]$ with respect to β; and (4) alternate between best-step gradient search and acceleration steps. Best-step steepest ascent (descent) searches are used to find $\mathbf{x}^1, \mathbf{x}^2, \mathbf{x}^4, \mathbf{x}^6, \ldots$, and acceleration steps are used to locate $\mathbf{x}^3, \mathbf{x}^5, \mathbf{x}^7, \mathbf{x}^9, \ldots$. With continued partan, the acceleration steps are conducted through the following pairs of points: $(\mathbf{x}^0,\mathbf{x}^2)$, $(\mathbf{x}^1,\mathbf{x}^4)$, $(\mathbf{x}^3,\mathbf{x}^6), \ldots, (\mathbf{x}^{2k-3},\mathbf{x}^{2k}), \ldots$. Note that, except for the first point pair,

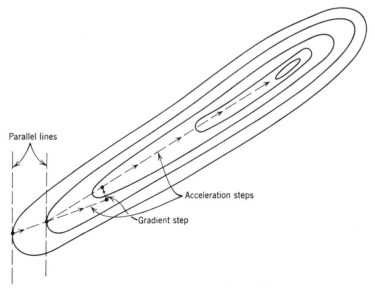

Figure 6-16. A partan search on a ridge.

the accleration steps are conducted through points which differ in index by 3. If $f(\mathbf{x})$ is a quadratic with a well-defined optimum, the exact optimal point is located after $2n - 1$ one-dimensional searches [6.89], except for round-off error.

6-8. CONJUGATE-DIRECTION METHODS

6-8a. Conjugate Directions

The conjugate-direction methods described in this section have proved to be highly efficient search techniques. The techniques presented here fall under the general heading of gradient methods, but this is not true of all conjugate-direction methods; e.g., the extension [6.109] of Powell's method [6.79] is a conjugate-direction method which does not even require evaluation of derivatives, but in general it is not as efficient as the methods given in this section, provided derivatives are available. All conjugate-direction methods possess the desirable property of quadratic convergence.

Conjugate-direction methods are conceptually related to the conjugate-gradient method of Hestenes and Stiefel [6.46] which was developed in 1952 for the solution of a set of simultaneous linear equations having a symmetric positive-definite matrix with real entries. (Beckman [6.4] gives an excellent account, including computer flow charts, of the conjugate-gradient method.) The relationship is evidenced by examining the quadratic form $f_s(\mathbf{x})$:

$$f_s(\mathbf{x}) = f(\mathbf{x}_a) + (\mathbf{x} - \mathbf{x}_a)'\mathbf{b} + \tfrac{1}{2}(\mathbf{x} - \mathbf{x}_a)'\mathbf{A}(\mathbf{x} - \mathbf{x}_a) \qquad (6\text{-}94)$$

where $\mathbf{A}$ is an $n \times n$ symmetric matrix of known real values, and $\mathbf{b}$ and $\mathbf{x}_a$ are specified $n \times 1$ matrices of real values. The stationary value $\mathbf{x}_s$ of $\mathbf{x}$ is obtained from $\nabla f_s(\mathbf{x}) = \mathbf{0}$, i.e.,

$$\mathbf{b} = -\mathbf{A}(\mathbf{x}_s - \mathbf{x}_a)$$

which is a set of linear equations, the solution of which is

$$\mathbf{x}_s = \mathbf{x}_a - \mathbf{A}^{-1}\mathbf{b} \qquad (6\text{-}95)$$

and $\mathbf{x}_s$ is a minimum point provided that $\mathbf{A}$ is a positive-definite matrix. ($\mathbf{A}$ is a positive-definite matrix if, and only if, the test associated with inequality 2-21 is satisfied.)

Given a positive-definite $n \times n$ matrix $\mathbf{A}$, a column vector $\mathbf{r}_1$ is said to be *conjugate* (with respect to $\mathbf{A}$) to a column vector $\mathbf{r}_2$ if and only if

$$\mathbf{r}_1'\mathbf{A}\mathbf{r}_2 = 0 \qquad (6\text{-}96)$$

When $\mathbf{r}_1$ and $\mathbf{r}_2$ are interpreted as directions in an n-dimensional space, they are said to be $\mathbf{A}$-*conjugate directions* if they satisfy 6-96.

Sect. 6-8   CONJUGATE-DIRECTION METHODS   315

As a prelude to the presentation of conjugate-direction search techniques, important properties of conjugate directions are examined in this subsection.

Theorem 6-1. *If n nontrivial $\mathbf{r}_i$'s, $i = 0, 1, \ldots, n - 1$, are known to be A-conjugate to one another, it necessarily follows that these $\mathbf{r}_i$'s are linearly independent. The matrix $\mathbf{A}$ is assumed to be positive-definite.*

A proof by contradiction is obtained by assuming that a particular $\mathbf{r}_i$, say $\mathbf{r}_k$, is linearly dependent on the remaining $\mathbf{r}_i$'s, i.e.,

$$\mathbf{r}_k = \sum_{\substack{i \\ i \neq k}} \beta_i \mathbf{r}_i$$

where at least one of the real β_i's, say, β_j, is nonzero. By the assumed conjugate property,

$$\mathbf{r}_j' \mathbf{A} \mathbf{r}_k = 0 \tag{6-97a}$$

But by the supposed linear dependence property,

$$\mathbf{r}_j' \mathbf{A} \mathbf{r}_k = \mathbf{r}_j' \mathbf{A} \left(\sum_{\substack{i \\ i \neq k}} \beta_i \mathbf{r}_i \right) = \beta_j \mathbf{r}_j' \mathbf{A} \mathbf{r}_j \tag{6-97b}$$

where $\mathbf{r}_j' \mathbf{A} \mathbf{r}_j$ is nonzero because of the positive-definite nature of $\mathbf{A}$. It is apparent that 6-97a and 6-97b are in contradiction, and therefore the supposed dependence of $\mathbf{r}_k$ on the remaining $\mathbf{r}_i$'s must be false.

●

Theorem 6-2. *If nontrivial directions $\mathbf{r}_i$ ($i = 0, 1, \ldots, n - 1$) are mutually conjugate with respect to a positive-definite $\mathbf{A}$ matrix in Equation 6-94, the exact minimum of f_s in Equation 6-94 can be obtained by the following sequence of n one-dimensional searches: starting at the point $\mathbf{x}^0$, $\mathbf{x}^1$ is determined from $f_s(\mathbf{x}^1) = \min_{\alpha_0} f_s(\mathbf{x} + \alpha_0 \mathbf{r}_0)$; $\mathbf{x}^2$ is determined from $f_s(\mathbf{x}^2) = \min_{\alpha_1} f_s(\mathbf{x}^1 + \alpha_1 \mathbf{r}_1)$; and so forth, the final result being $\mathbf{x}_{\min} = \mathbf{x}^n$ which is extracted from $f_s(\mathbf{x}^n) = \min_{\alpha_{n-1}} f_s(\mathbf{x}^{n-1} + \alpha_{n-1} \mathbf{r}_{n-1})$.*

Proof: Consider the linear transformation

$$\mathbf{x} = \sum_{i=0}^{n-1} \alpha_i \mathbf{r}_i = [\mathbf{r}_0 \quad \mathbf{r}_1 \quad \cdots \quad \mathbf{r}_{n-1}] \begin{bmatrix} \alpha_0 \\ \alpha_1 \\ \vdots \\ \alpha_{n-1} \end{bmatrix} \tag{6-98}$$

Equation 6-94 assumes the form

$$f_s\left(\sum_{i=0}^{n-1} \alpha_i \mathbf{r}_i\right) = f(\mathbf{x}_a) - \mathbf{x}_a'\mathbf{b} + \tfrac{1}{2}\mathbf{x}_a'\mathbf{A}\mathbf{x}_a$$
$$+ \sum_{i=0}^{n-1} [\alpha_i \mathbf{r}_i'(\mathbf{b} - \mathbf{A}\mathbf{x}_a) + \tfrac{1}{2}\alpha_i^2 \mathbf{r}_i'\mathbf{A}\mathbf{r}_i] \quad (6\text{-}99)$$

in which all $\mathbf{r}_i'\mathbf{A}\mathbf{r}_j$ terms, for $i \neq j$, have been deleted because of the assumed conjugate nature of the $\mathbf{r}_i$'s. We need only observe that Equation 6-99 is noninteracting (see Section 6-2) in the α_i variables to substantiate the theorem.

•

In terms of matrix theory (Appendix A), conjugate directions are used to diagonalize the positive-definite $\mathbf{A}$ matrix. The linear transformation of 6-98 converts the quadratic term $\mathbf{x}'\mathbf{A}\mathbf{x}$ to

$$\mathbf{x}'\mathbf{A}\mathbf{x} = \left(\mathbf{R}\begin{bmatrix}\alpha_0 \\ \alpha_1 \\ \vdots \\ \alpha_{n-1}\end{bmatrix}\right)'\mathbf{A}\mathbf{R}\begin{bmatrix}\alpha_0 \\ \alpha_1 \\ \vdots \\ \alpha_{n-1}\end{bmatrix}$$

$$= [\alpha_0 \ \alpha_1 \ \cdots \ \alpha_{n-1}](\mathbf{R}'\mathbf{A}\mathbf{R})\begin{bmatrix}\alpha_0 \\ \alpha_1 \\ \vdots \\ \alpha_{n-1}\end{bmatrix}$$

where $\mathbf{R}$ is identically equal to the $n \times n$ matrix $[\mathbf{r}_0 \ \mathbf{r}_1 \ \cdots \ \mathbf{r}_{n-1}]$, the columns of which are mutually conjugate directions corresponding to $\mathbf{A}$. The matrix $\mathbf{R}$ is used in the formation of $\mathbf{D}$:

$$\mathbf{D} \triangleq \mathbf{R}'\mathbf{A}\mathbf{R}$$

$$= \begin{bmatrix}\mathbf{r}_0'\mathbf{A}\mathbf{r}_0 & \mathbf{r}_0'\mathbf{A}\mathbf{r}_1 & \cdots & \mathbf{r}_0'\mathbf{A}\mathbf{r}_{n-1} \\ \mathbf{r}_1'\mathbf{A}\mathbf{r}_0 & & & \vdots \\ \vdots & & \ddots & \\ \mathbf{r}_{n-1}'\mathbf{A}\mathbf{r}_0 & & \cdots & \mathbf{r}_{n-1}'\mathbf{A}\mathbf{r}_{n-1}\end{bmatrix}$$

With $\mathbf{R}$ nonsingular, a fact which follows from the linear independence of the $\mathbf{r}_i$'s, the matrices $\mathbf{A}$ and $\mathbf{D}$ are said to be *congruent*.

Note that the diagonal matrix $\mathbf{D}$ and the $\mathbf{r}_i$ conjugate directions are not uniquely determined by the given matrix $\mathbf{A}$. Because the intent is to make

Sect. 6-8 CONJUGATE-DIRECTION METHODS 317

zero all terms of $\mathbf{D} = \mathbf{D}'$ above the major diagonal, there exists a set of $n(n-1)/2$ equations to be satisfied; namely,

$$\mathbf{r}_i'\mathbf{A}\mathbf{r}_j = \mathbf{r}_j'\mathbf{A}\mathbf{r}_i = 0, \qquad i \neq j$$

But n^2 is the number of elements in the $\mathbf{r}_i$'s, so they can be specified in many ways. For example, one procedure for generation of a set of $\mathbf{A}$-conjugate $\mathbf{r}_i$'s is the following:

Any desired nontrivial $\mathbf{r}_0$ is allowed, and the row vector

$$\mathbf{r}_0'\mathbf{A} = [c_{01} \quad c_{02} \quad \cdots \quad c_{0n}]$$

can be generated. $\mathbf{r}_1$ is then generated on the basis that

$$\mathbf{r}_0'\mathbf{A}\mathbf{r}_1 = \sum_{i=1}^{n} c_{0i}r_{i1} = 0 \tag{6-100}$$

and $n-1$ degrees of freedom generally exist in the satisfaction of 6-100 by specific values of r_{i1}'s. With a particular nontrivial $\mathbf{r}_1$ which satisfies 6-100, $\mathbf{r}_1'\mathbf{A} = [c_{11} \quad c_{12} \quad \cdots \quad c_{1n}]$ is formed, and both

$$\mathbf{r}_0'\mathbf{A}\mathbf{r}_2 = \sum_{i=1}^{n} c_{0i}r_{i2} = 0 \tag{6-101a}$$

and

$$\mathbf{r}_1'\mathbf{A}\mathbf{r}_2 = \sum_{i=1}^{n} c_{1i}r_{i2} = 0 \tag{6-101b}$$

are to be satisfied by the nontrivial $\mathbf{r}_2$ chosen. Next,

$$\mathbf{r}_2'\mathbf{A} = [c_{21} \quad c_{22} \quad \cdots \quad c_{2n}]$$

is formed, and $\mathbf{r}_3$ is chosen on the basis that it is $\mathbf{A}$-conjugate to $\mathbf{r}_0$, $\mathbf{r}_1$, and $\mathbf{r}_2$. The process is continued in like manner; ultimately $\mathbf{r}_{n-1}$ is formed on the basis that it is $\mathbf{A}$-conjugate to the previously formed vectors $\mathbf{r}_0, \mathbf{r}_1, \ldots, \mathbf{r}_{n-2}$.

Because the n $\mathbf{r}_i$'s which are generated are known to be linearly independent (Theorem 6-1), they span the n-dimensional space in question; any other real n-dimensional vector can be expressed as a linear combination of these $\mathbf{r}_i$ vectors. Thus, if an additional nontrivial vector $\mathbf{r}_n$ is proposed, it can be expressed as a linear combination of the n $\mathbf{r}_i$'s, and by Theorem 6-1 this $\mathbf{r}_n$ vector cannot be $\mathbf{A}$-conjugate to each vector of the set of n mutually $\mathbf{A}$-conjugate vectors.

The definitive criterion for two directions $\mathbf{r}_1$ and $\mathbf{r}_2$ to be conjugate is, of course, $\mathbf{r}_1'\mathbf{A}\mathbf{r}_2 = 0$. However, if $\mathbf{A}$ is not known explicitly or if the $f(\mathbf{x})$ to be minimized equals $f_s(\mathbf{x})$ plus higher-order terms, other relations must be used to determine conjugate directions—one such relation stems from the following theorem.

Theorem 6-3. *Consider Equation 6-94 and suppose that $\partial f_s(\alpha \mathbf{r}_1)/\partial \alpha$ equals zero at $\alpha = 0$. Also, suppose that $\partial f_s(\mathbf{x}_a - \mathbf{x}_b + \alpha \mathbf{r}_1)/\partial \alpha$ equals zero at $\alpha = 0$, where $\mathbf{r}_1$ and $\mathbf{x}_a - \mathbf{x}_b \neq \mathbf{0}$ are linearly independent. Under these conditions, it is necessarily true that the direction $\mathbf{r}_1$ is $\mathbf{A}$-conjugate to the direction $\mathbf{r}_2 \equiv \mathbf{x}_a - \mathbf{x}_b$.*

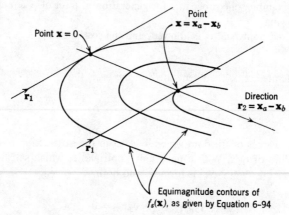

Figure 6-17. A-conjugate directions $\mathbf{r}_1$ and $\mathbf{r}_2$.

Proof: The assumed conditions are depicted in Figure 6-17. By assumption,

$$\frac{\partial f_s(\alpha \mathbf{r}_1)}{\partial \alpha} = 0 \quad \text{at } \alpha = 0$$

In terms of Equation 6-94, this is equivalent to

$$\mathbf{r}_1'(\mathbf{b} - \mathbf{A}\mathbf{x}_a) = 0 \tag{6-102a}$$

Similarly,

$$\frac{\partial}{\partial \alpha} f_s(\mathbf{x}_a - \mathbf{x}_b + \alpha \mathbf{r}_1) = 0 \quad \text{at } \alpha = 0$$

which is equivalent to

$$\mathbf{r}_1'(\mathbf{b} - \mathbf{A}\mathbf{x}_b) = 0 \tag{6-102b}$$

And by subtracting corresponding members of Equations 6-102a and 6-102b, the desired result is obtained:

$$\mathbf{r}_1'\mathbf{A}(\mathbf{x}_a - \mathbf{x}_b) = \mathbf{r}_1'\mathbf{A}\mathbf{r}_2 = 0 \tag{6-103}$$

●

This theorem clearly suggests that the partan methods (Section 6-7) are a class of conjugate-direction methods; for example, observe that the points of tangency of the parallel lines in Figure 6-14 define a search direction which, according to Theorem 6-3, is conjugate to the direction of the parallel lines. The theorem also suggests a way of generating conjugate directions without using the gradient of $f(\mathbf{x})$ [6.79, 6.109].

6-8b. Method of Fletcher and Reeves

In the method of Fletcher and Reeves [6.36], the conjugate directions $\mathbf{r}^i$ are obtained in sequence by use of the following computations:

$$\mathbf{r}^0 = -\mathbf{g}(\mathbf{x}^0) = -\mathbf{g}^0 \qquad (6\text{-}104a)$$

and

$$\mathbf{r}^i = -\mathbf{g}^i + \frac{\mathbf{g}^{i\prime}\mathbf{g}^i}{\mathbf{g}^{i-1\prime}\mathbf{g}^{i-1}} \mathbf{r}^{i-1}, \qquad i = 1, 2, \ldots, n \qquad (6\text{-}104b)$$

where $\mathbf{g}^i$ equals the gradient of $f(\mathbf{x})$ at the point $\mathbf{x}^i$, and $\mathbf{x}^i$ is determined from the relationship

$$f(\mathbf{x}^i) = \min_{\alpha} f(\mathbf{x}^{i-1} + \alpha \mathbf{r}^{i-1}) \qquad (6\text{-}105)$$

Fletcher and Reeves show that the directions which are generated in this way are **A**-conjugate.

Fletcher and Reeves also give computer programming details for their method. Based on practical experience with various nonquadratic functions, they suggest that the direction $\mathbf{r}^i$ be reinitialized to the gradient direction periodically at multiples of $n + 1$ one-dimensional searches, i.e., let

$$\mathbf{r}^i = -\mathbf{g}^i \qquad \text{for } i = j(n+1) \text{ where } j = 0, 1, 2, \ldots \qquad (6\text{-}106a)$$

and

$$\mathbf{r}^i = -\mathbf{g}^i + \frac{\mathbf{g}^{i\prime}\mathbf{g}^i}{\mathbf{g}^{i-1\prime}\mathbf{g}^{i-1}} \mathbf{r}^{i-1} \qquad (6\text{-}106b)$$

otherwise.

The type of one-dimensional search employed is not central to the method; however, it should be of high accuracy in the vicinity of the optimum to insure that conjugate directions are generated—the technique of Sections 6-3b and 6-3f is recommended.

As for criteria for stopping the search, it is recommended that at least n one-dimensional searches be effected, regardless of other criteria, and after that, the search be terminated either when a specified number N of one-dimensional searches have been completed or when both the value of $\mathbf{g}^{k\prime}\mathbf{g}^k$

and the value of $(\mathbf{x}^{k+1} - \mathbf{x}^k)'(\mathbf{x}^{k+1} - \mathbf{x}^k)$ are diminished below preassigned levels.

The reader should compare the method of this subsection with the DFP method of the next. Representative of the examples given by Fletcher and Reeves is one for which their method requires 27 one-dimensional searches whereas the DFP method requires but 18 for the same accuracy of search. This is not as detrimental to the method of Fletcher and Reeves as might be supposed, however, because the DFP method requires more computer memory and more computations per one-dimensional search.

6-8c. Davidon's Method via Fletcher and Powell [The DFP Method]

Davidon's method [6.24], as described and extended by Fletcher and Powell [6.35], is one of the most sophisticated search techniques available. It is summarily agreed by leading workers in the field that the DFP method is the best single method available to date for the following type of problem: (1) $f(\mathbf{x})$ is given analytically and is of class C^2 with respect to its arguments; and (2) evaluations of $f(\mathbf{x})$ and $\nabla f(\mathbf{x})$ require a relatively large amount of computation time as compared to that required for the matrix manipulations which are associated with the method. These assertions are justified on the basis of considerable experimental evidence (Section 6-12).

In Section 6-6, it is shown that the minimum of a quadratic function could be obtained in one iteration, provided that the inverse $\mathbf{A}^{-1}$ of the $\mathbf{A}$ matrix of second partial derivatives was available. In the DFP method, the $\mathbf{A}^{-1}$ matrix is generated after n one-dimensional searches, *without the use of second partial derivatives.*

The essential steps in the method are depicted in the flow diagram of Figure 6-18. In block A of this figure, the search directions defined by the $-\mathbf{H}^k \mathbf{g}^k$'s assume the role of the $\mathbf{r}^k$'s of the preceding subsection. The following statements hold when the DFP method is accurately applied to the quadratic form of Equation 6-94: (1) $(\mathbf{H}^i \mathbf{g}^i)' \mathbf{A} (\mathbf{H}^j \mathbf{g}^j) = 0$ for $i \neq j$; (2) $\mathbf{H}^n = \mathbf{A}^{-1}$; and (3) $\mathbf{g}^n = 0$. Even if the function in question is nonquadratic, the matrix $\mathbf{H}^j$ remains positive-definite, and therefore the one-dimensional search in block A of Figure 6-18 need be conducted over positive α only. The interested reader is referred to the work of Fletcher and Powell [6.35] for justification of these assertions.

In block B of Figure 6-18, $\Delta \mathbf{x}_k = \mathbf{x}^{k+1} - \mathbf{x}^k$ and $\Delta \mathbf{g}_k = \mathbf{g}^{k+1} - \mathbf{g}^k$ are computed, and these are used in blocks C_1 and C_2 to compute $\mathbf{M}_1$ and $\mathbf{M}_2$—note that the numerators of $\mathbf{M}_1$ and $\mathbf{M}_2$ are $n \times n$ matrices, whereas the denominators are scalars. A new $n \times n$ matrix $\mathbf{H}^{k+1} = \mathbf{H}^k + \mathbf{M}_1{}^k + \mathbf{M}_2{}^k$ is evaluated in block D of the figure, and the cycle is repeated at least n times (block E) to guarantee the formation of $\mathbf{A}^{-1}$ in the quadratic case. During

Sect. 6-8 CONJUGATE-DIRECTION METHODS 321

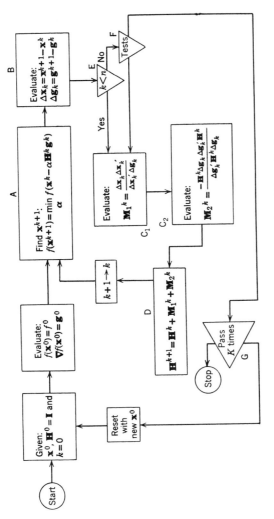

Figure 6-18. Flow diagram for the DFP method.

some cycle k, for $k > n$, a relative minimum is assumed to be obtained when either of the following two tests (block F) are satisfied: (1) k exceeds a specified integer N; or (2) $\Delta \mathbf{x}_k{'} \Delta \mathbf{x}_k$ and $\mathbf{g}^{k+1}{'}\mathbf{g}^{k+1}$ decrease below preassigned levels. At this point (block G), either the search is stopped or it is reinitiated with a new value of $\mathbf{x}^0$.

Fletcher and Powell applied this method to typical test problems and found it to be completely satisfactory, even in a case where 100 variables were involved. However, Volz [6.99] found that the method could be hampered if the one-dimensional searches are of limited accuracy, especially if the initial search point $\mathbf{x}^0$ is far from a relative minimum. Since Volz's refinement [6.99] of the method is unnecessary in most cases, it is left for the interested reader to pursue.

6-9. OTHER SEARCH METHODS

6-9a. Discussion

It requires but little imagination to realize at this point that the number of and variations of search techniques are unlimited. The search techniques presented in the following subsections are a representative sample of additional ones to be found in the literature. Each of these has its own ingenious features, enabling good results to be obtained under certain conditions; but for the most part, they are not as efficient as the methods of Sections 6-7 and 6-8, provided these are applicable.

The search techniques to be considered are more "global" than "local" in character. As such, most of them are categorized under the heading of *direct search*. Hooke and Jeeves [6.49] coined the term "direct search" to describe sequential examination of trial solutions involving comparison of each trial solution with the best obtained up to that time, together with a strategy for determining what the next trial will be as a function of earlier results. Among the techniques which are included in the broad scope of this definition, illustrative ones are given in Sections 6-9b and 6-9c. The techniques considered in Sections 6-9d and 6-9e have some of the features of direct search but also have features which are common to gradient search methods.

6-9b. Pattern Search

As devised by Hooke and Jeeves [6.49], pattern search is a direct search method which results in relatively efficient search along straight ridges or ravines. We attempt with pattern search to establish the "pattern" of successful search points in the immediate past from which plausible future

Sect. 6-9   OTHER SEARCH METHODS   323

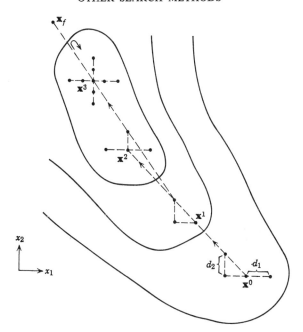

Figure 6-19. Example of a pattern search.

search points are predicted. The following example illustrates the method for the two-dimensional case.

Example 6-7. In Figure 6-19, the search progresses from an initial base point $\mathbf{x}^0$, and a minimum of $f(\mathbf{x})$ is to be found without the use of derivatives of $f(\mathbf{x})$. A small displacement d_1 from $\mathbf{x}^0$ in the x_1 direction is effected, and $f(x_1^0 + d_1, x_2^0)$ is evaluated and compared with $f(x_1^0, x_2^0)$. If the latter is smaller than the former, a small displacement in the $-x_1$ direction is effected, and $f(x_1^0 - d_1, x_2^0)$ is compared with $f(x_1^0, x_2^0)$. If the latter is greater than the former, a small displacement d_2 is made from $(x_1^0 - d_1, x_2^0)$ in the x_2 direction, and $f(x_1^0 - d_1, x_2^0)$ is compared with $f(x_1^0 - d_1, x_2^0 + d_2)$. Assuming the latter is the smaller of the two, a *pattern move* is made in the direction established by the preceding successful moves to the point $\mathbf{x}^1$ in Figure 6-19. At $\mathbf{x}^1$, the small moves which proved to be successful around $\mathbf{x}^0$ are repeated and if successful lead to a second pattern move to $\mathbf{x}^2$. From $\mathbf{x}^2$, small displacements of magnitude d_1 along the x_1 coordinate prove to be unsuccessful, whereas the displacement d_2 yields $f(x_1^2, x_2^2 + d_2) < f(x_1^2, x_2^2)$. A pattern

324   SEARCH TECHNIQUES AND NONLINEAR PROGRAMMING   Ch. 6

move is made therefore along the line which passes through the points $(x_1^1 - d_1, x_2^1 + d_2)$ and $(x_1^2, x_2^2 + d_2)$.

Note that the step size of the pattern search is made progressively larger as the general directions dictated by the pattern prove successful. When failure occurs, as at point $\mathbf{x}_f$ in Figure 6-19, the step size of the pattern move is reduced as depicted, to point $\mathbf{x}^3$.

At $\mathbf{x}^3$, the displacements d_1 from $\mathbf{x}^3$ along the x_1 coordinate fail to improve $f(\mathbf{x})$, as do displacements d_2 from $\mathbf{x}^3$ along the x_2 coordinate. Reductions in d_1 and d_2 are therefore required at this point,[11] and the process starts anew. Ultimately both the displacement and the pattern step sizes are reduced below some preassigned limit, and the search is terminated.

The principles underlying the above procedure have been mechanized for process control by the Westinghouse Electric Corporation in the form of a controller called "Opcon" [6.111]. Opcon has been used to optimize automatically a pilot plant of the Dow Chemical Company [6.111] and to optimize operation of a distillation column [6.104]. Weisman *et al.* [6.103] give a Fortran computer program for pattern search.

●

6-9c. Search by Directed Array

Search by directed array is this author's conception of a direct search method in which the sequential examination of trial solutions is made in a definite

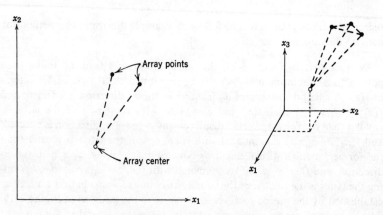

Figure 6-20. Typical directed arrays.

---

[11] These indicated reductions in d_1 and d_2 could conceivably fail to yield an improvement if $\mathbf{x}^3$ is a saddle point; a case in point is encountered in Example 6-3.

directed array from the current search point. As used here, a directed array consists of a set of points, finite in number, which are equidistant from a given point called the array center. An unlimited number of arrays is possible, and many different strategies may be employed to select both the array center of the next array and the step size of the array.

Typically, the number of points in a given directed array equals the dimension n of $\mathbf{x}$. Figure 6-20 depicts typical arrays for $n = 2$ and $n = 3$. The search procedure presented in the following example illustrates the general concepts involved in a directed array search.

Example 6-8. (Bunny-Hop Search[12])

Suppose a certain function $f(x_1, x_2)$ describes the altitude of some unimodal portion of land; x_1 represents the distance east of a fixed origin, and x_2 represents the distance north of this origin. The problem is to determine the location of the lowest point in the region in which the representation $f(x_1, x_2)$ is valid; this point is *the bunny's home*.

In bunny-hop search, the directed array consists of a pair of radial line segments (*the bunny's feet!*) separated by a fixed acute angle θ. The line segments of the kth array are of length d_k. The *clockwisemost* of the pair is terminated by point $\mathbf{a}_k$ and the other by the point $\mathbf{b}_k$. The vertex of the array is located at a point called a center; the kth array center is at the point $\mathbf{c}_k$. To orient the kth array, two factors are taken into account: the orientation of the $(k-1)$st array and the point in the $(k-1)$st array which was chosen for the kth array center $\mathbf{c}_k$. The relationships are illustrated in Figure 6-21.

Analytically,

$$\mathbf{c}_k = \begin{cases} \mathbf{a}_{k-1} & \text{if } f(\mathbf{c}_{k-1}) > f(\mathbf{a}_{k-1}) \leq f(\mathbf{b}_{k-1}) \\ \mathbf{b}_{k-1} & \text{if } f(\mathbf{c}_{k-1}) > f(\mathbf{b}_{k-1}) < f(\mathbf{a}_{k-1}) \\ \mathbf{c}_{k-1} & \text{if } f(\mathbf{a}_{k-1}) \geq f(\mathbf{c}_{k-1}) \leq f(\mathbf{b}_{k-1}) \end{cases} \quad (6\text{-}107)$$

and

$$\phi_k = \begin{cases} \phi_{k-1} - \theta & \text{if } \mathbf{c}_k = \mathbf{a}_{k-1} \\ \phi_{k-1} + \theta & \text{if } \mathbf{c}_k = \mathbf{b}_{k-1} \\ \phi_k & \text{if } \mathbf{c}_k = \mathbf{c}_{k-1} \end{cases} \quad (6\text{-}108)$$

where ϕ_k is the angle that the $\mathbf{c},\mathbf{a}$ line segment of the kth array makes with the x_1 axis. If both $f(\mathbf{b}_{k-1})$ and $f(\mathbf{a}_{k-1})$ are greater than or equal to $f(\mathbf{c}_{k-1})$,

---

[12] The details of this delightful search technique were developed by James W. Leggate, Donald A. Rudberg, and the author in the spring of 1965.

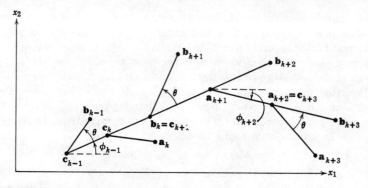

Figure 6-21. A bunny-hop search.

Equation 6-108 does not give a way of finding ϕ_k; a procedure to be used in that case is given three paragraphs hence.

Observe from the preceding that the bunny tends to bias the orientation of her feet in the direction of search which has proved successful in past hops. She is therefore an adaptive bunny and can follow a curved ravine with considerable agility. To enhance the bunny's adaptive character, a reward is given her each time she completes a successful hop (the $k - 1$st hop is successful if either $\mathbf{c}_k = \mathbf{a}_{k-1}$ or $\mathbf{c}_k = \mathbf{b}_{k-1}$). If the $k - 1$st hop is successful, d_k is set equal to γd_{k-1}, where γ is a constant greater than unity; and the bunny hops with greater vigor on her kth hop. Similarly, the bunny is punished if she makes a bad hop; if $\mathbf{c}_k = \mathbf{c}_{k-1}$, as provided by Equation 6-108, the step size is attenuated by a fraction λ; in which case $d_k = \lambda d_{k-1}$.

Once the array center $\mathbf{c}_k = (x_{1c}^k, x_{2c}^k)$ is located by using Equation 6-107, the coordinates (x_{1a}^k, x_{2a}^k) and (x_{1b}^k, x_{2b}^k) of the array points $\mathbf{a}_k$ and $\mathbf{b}_k$ are determined analytically by

$$\begin{aligned} x_{1a}^k &= x_{1c}^k + d_k \cos \phi_k \\ x_{2a}^k &= x_{2c}^k + d_k \sin \phi_k \end{aligned} \tag{6-109}$$

and

$$\begin{aligned} x_{1b}^k &= x_{1c}^k + d_k \cos (\phi_k + \theta) \\ x_{2b}^k &= x_{2c}^k + d_k \sin (\phi_k + \theta) \end{aligned} \tag{6-110}$$

where d_k is the step size in the kth array. The sine and cosine terms in the above equations need not be evaluated directly; if these require a relatively large time for direct digital computation, they can be evaluated by using

Sect. 6-9   OTHER SEARCH METHODS   327

previously computed values. The appropriate identities are obtained from the relationship of Equation 6-108, as follows:

$$\cos \phi_k = \begin{cases} \cos \theta \cos \phi_{k-1} + \sin \theta \sin \phi_{k-1} & \text{if } \mathbf{c}_k = \mathbf{a}_{k-1} \\ \cos \theta \cos \phi_{k-1} - \sin \theta \sin \phi_{k-1} & \text{if } \mathbf{c}_k = \mathbf{b}_{k-1} \end{cases} \quad (6\text{-}111)$$

$$\sin \phi_k = \begin{cases} \cos \theta \sin \phi_{k-1} - \sin \theta \cos \phi_{k-1} & \text{if } \mathbf{c}_k = \mathbf{a}_{k-1} \\ \cos \theta \sin \phi_{k-1} + \sin \theta \cos \phi_{k-1} & \text{if } \mathbf{c}_k = \mathbf{b}_{k-1} \end{cases} \quad (6\text{-}112)$$

$$\cos (\phi_k + \theta) = \cos \phi_k \cos \theta - \sin \phi_k \sin \theta \quad (6\text{-}113)$$

and

$$\sin (\phi_k + \theta) = \sin \phi_k \cos \theta + \sin \theta \cos \phi_k \quad (6\text{-}114)$$

If the bunny's hop from $\mathbf{c}_{k-1}$ is unsuccessful, the angular direction ϕ_k of the kth hop must be determined. There are many ways to select such a ϕ_k; the one used here is based on the negative of the gradient of f at $\mathbf{c}_k$. Thus

$$\begin{bmatrix} \cos \phi_k \\ \sin \phi_k \end{bmatrix} \triangleq [-\nabla f/(\nabla f' \, \nabla f)^{1/2}]_{\mathbf{x} = \mathbf{c}_k} \quad \text{if } \mathbf{c}_k = \mathbf{c}_{k-1} \quad (6\text{-}115)$$

Equation 6-115 is also used to compute the initial trigonometric identities $\cos \phi_0$ and $\sin \phi_0$, which correspond to the initial array center $\mathbf{c}_0$.

A computer flow diagram for realizing bunny-hop search is given in Figure 6-22. In addition to the search program, the computer is supplied with the following initial data: the functions f, $\partial f/\partial x_1$, and $\partial f/\partial x_2$; the values of $\cos \theta$ and $\sin \theta$; the initial step size d_0; the initial array center $\mathbf{c}_0$ (or perhaps several such initial points for several independent searches); the reward factor γ; the punishment factor λ; a limit N on the number of hops allowed; a lower limit d_L on the allowed step size; and a lower limit g_L on allowed values of $(\nabla f' \, \nabla f)^{1/2}$. The blocks within the dashed lines in Figure 6-22 constitute the gradient computation: if $(\nabla f' \, \nabla f)^{1/2} > g_L$, the bunny hops, if not, the bunny stops. If the kth hop is a good hop, the bunny tests to see if $k < N$; if yes, she is not tired and goes on to be rewarded in step size by factor γ; if no, she is tired and must stop. After being rewarded, the bunny reorients her feet and proceeds with the next hop. If the kth hop is unsuccessful, the bunny is punished in step size by the factor λ, $d_{k+1} = \lambda d_k$, and she tests to see if she is home; yes, she is home if $d_{k+1} < d_L$; no, she is not home if $d_{k+1} \geq d_L$, and so onward she hops.

The often-used test function $f(x_1, x_2) = 100(x_1^2 - x_2)^2 + (x_1 - 1)^2$ was used to evaluate the performance of bunny-hop search. The starting point $\mathbf{x}^0 = (-1.2, 1.0)$ was used in all tests. Based on several trials, the following values were found to be most efficient in locating the minimum: $\theta = 30°$,

328 SEARCH TECHNIQUES AND NONLINEAR PROGRAMMING Ch. 6

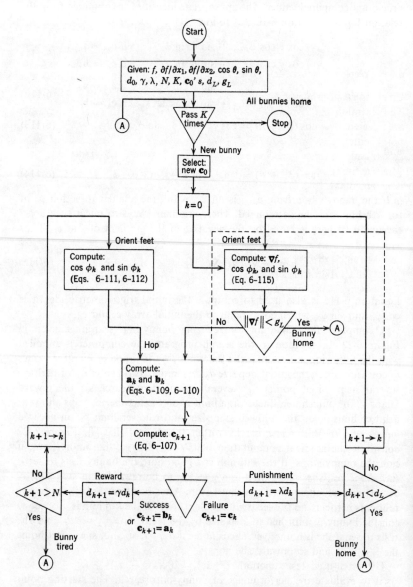

Figure 6-22. Flow diagram for bunny-hop search.

$\gamma = 2$, and $\lambda = 0.1$. The efficiency of this method is compared with that of other methods in Section 6-12.

•

6-9d. Creeping Random Methods

Creeping random methods have been considered at length by Brooks [6.10]. The central theme to all such methods is that $f(\mathbf{x})$ is evaluated at points which are selected at random from a set of points in the vicinity of a current search-point center. Two main arteries of search strategies stem from this base: one global and the other local in character.

In the first place, if the point which is the temporary winner of the search is assigned the role of new search-point center from which the process is repeated, the search is a global search in the sense defined in Section 6-2. On the other hand, if a number of search points are selected at random from the immediate vicinity of the associated search-point center, the temporary winner of this search defines an approximate gradient direction. A step may be taken in the direction so found to locate the next search-point center.

The efficiency of either of the above processes is admittedly poor [6.48]. The principal attribute of random search is that it has no innate bias which is found in nonrandom methods. Because of this fact, it has been suggested by Bromberg [6.8] and others that a random search be used, every so often, in conjunction with nonrandom methods to provide dither to an overall search process. Also, use of random search around a supposed optimal point is an effective way to expose it if it is not optimum.

6-9e. Centroid Methods

Fend and Chandler [6.28] use the density concept—treating $f(\mathbf{x})$ as a density function in the n-dimensional space of the x_i's—as the basis for various optimum-seeking methods. These methods are designed to seek out the centroid, the effective "mass" center, of n-dimensional solids which have surfaces defined in terms of the x_i's and which are characterized by the density function $f(\mathbf{x})$.[13] The boundaries of any one of these n-dimensional solids may enclose either a small part or a large part of the $\mathbf{x}$ space under consideration.

In one form or another, all centroid methods utilize an approximation of the following equation to determine the coordinates $\bar{\mathbf{x}}^k = (\bar{x}_1^k, \bar{x}_2^k, \ldots, \bar{x}_n^k)$

---

[13] The use of the term "centroid" in this regard is not completely consistent with standard terminology [6.52]; the phrase "center of mass" is preferred when $f(\mathbf{x})$ is nonuniform in 6-116, whereas the term "centroid" strictly applies when $f(\mathbf{x})$ equals a constant.

of the centroid of the kth n-dimensional object, the boundaries of which are defined by the limits of integration L_k:

$$\bar{x}_i^k = \frac{L_k^{\int x_i f(\mathbf{x})\,d\mathbf{x}}}{L_k^{\int f(\mathbf{x})\,d\mathbf{x}}}, \qquad i = 1, 2, \ldots, n \tag{6-116}$$

The given integrals denote multiple integrals. An important property of this kth centroid is that its location with respect to the kth n-dimensional object is invariant under linear translations of the origin of the orthogonal x_i coordinate axes.

Typically, centroid methods are sequential in nature; on the basis of the location of the centroid associated with the kth n-dimensional object, a $(k + 1)$st n-dimensional object is defined. In locating the $(k + 1)$st n-dimensional object, the goal is to bring the centroid of the $(k + 1)$st n-dimensional object closer to the optimal value of $\mathbf{x}$ than is the centroid of the kth n-dimensional object.

One form of centroid search is the *moment-rosetta search* of Fend and Chandler [6.28]. By definition, a rosetta is an ornament, object, or arrangement shaped like a rose. In the remainder of this subsection, a version of moment-rosetta search is described in which the rosetta consists of the set of $2n$ points which are centered on the $2n$ faces of a hyper-rectangle. These faces are aligned with the orthogonal x_i coordinate axes; and using the fact that the relative location of a centroid is independent of the location of the coordinate origin, the centroid is computed in terms of Δx_i variations away from the current center $\mathbf{x}^k$ of the kth hyper-rectangle.

A first-order approximation of Equation 6-116 is

$$\overline{\Delta x_i}^k = \frac{d_i^k [f(x_i^k + d_i^k, \mathbf{x}^k) - f(x_i^k - d_i^k, \mathbf{x}^k)]}{\sum_{j=1}^{n} [f(x_j^k + d_j^k, \mathbf{x}^k) + f(x_j^k - d_j^k, \mathbf{x}^k)]} \tag{6-117}$$

in which d_i^k is one-half the width of the kth hyper-rectangle in the x_i direction and where $f(x_i^k + d_i^k, \mathbf{x}^k)$ is shorthand notation for

$$f(x_1^k, x_2^k, \ldots, x_{i-1}^k, x_i^k + d_i^k, x_{i+1}^k, \ldots, x_n^k).$$

An interesting observation can be made at this point. If all d_i^k's are selected to be both equal and small, the incremental changes $\overline{\Delta x_i}^k$ of Equation 6-117 are in the gradient direction associated with the point $\mathbf{x}^k$. Hence, moment-rosetta search has much in common with discrete gradient search.

A typical moment-rosetta search might proceed on the following basis: (1) A region of $\mathbf{x}$ space in which $f(\mathbf{x})$ is unimodal is isolated by one means or another, possibly by using other search techniques; (2) scaling of the x_i's is

Sect. 6-10 COMBINED USE OF INDIRECT AND DIRECT METHODS 331

then effected so that the "width" of the above isolated region is the same for all x_i coordinates; (3) because of the above scaling, it is deemed appropriate to let $d_k \triangleq d_1{}^k = d_2{}^k \cdots = d_n{}^k$; (4) for the first rosetta, pick d_0 equal to one-half the width of the isolated region and locate the center $\mathbf{x}^0$ of the first rosetta at the center of the isolated region, and compute the centroid $\bar{\mathbf{x}}^0 = \mathbf{x}^0 + \overline{\Delta \mathbf{x}^0}$ by using Equation 6-117; (5) for the second rosetta, select $d_1 = \lambda d_0$ where λ is a positive fraction, locate the center $\mathbf{x}^1$ of the second rosetta at the centroid $\bar{\mathbf{x}}^0$ of the first rosetta (let $\mathbf{x}^1 = \bar{\mathbf{x}}^0$), and compute the centroid $\bar{\mathbf{x}}^1$ by using Equation 6-117; (6) in general, select $d_k = \lambda d_{k-1}$, let $\mathbf{x}^k = \bar{\mathbf{x}}^{k-1}$, and solve for $\bar{\mathbf{x}}^k$ by using 6-117.

The unimodal property of $f(\mathbf{x})$ is necessary to insure convergence of the above process. From computational experience, Fend and Chandler [6.28] found that the rate of convergence of moment-rosetta search is relatively swift if the optimal $f(\mathbf{x})$ remains located within the "branches" of the rosetta but is relatively slow if the optimal $f(\mathbf{x})$ is located outside the branches of the rosetta. For this reason, it is suggested that moment-rosetta search be used only as an end-of-overall-search method to pinpoint the optimum to a higher degree of accuracy.

6-10. COMBINED USE OF INDIRECT AND DIRECT METHODS

6-10a. Equation Solution by Search

Cauchy's original application [6.16] of steepest descent search is devoted to the simultaneous solution of sets of algebraic equations. To solve such equations by using steepest descent search, or by using any other search technique for that matter, we must first formulate a minimization problem, the solutions of which are forced to be identical to the solutions of the set of equations. This can be accomplished with relative ease, as follows.

Suppose real values of $\mathbf{x} = [x_1 \ x_2 \ \cdots \ x_n]'$ which satisfy the set of equations

$$f_i(\mathbf{x}) = 0, \quad i = 1, 2, \ldots, n \quad (6\text{-}118)$$

are to be found. Each function $f_i(\mathbf{x})$ is assumed to be real-valued and to be of class C^1.

The key step in forming an equivalent minimization problem is to define $f(\mathbf{x})$:

$$f(\mathbf{x}) \triangleq \sum_{i=1}^{n} [f_i(\mathbf{x})]^2 \quad (6\text{-}119)$$

Note that $f(\mathbf{x})$ is always non-negative, and therefore, if values of $\mathbf{x}$ exist for which $f(\mathbf{x}) = 0$, these must be minimum points for $f(\mathbf{x})$. When $f(\mathbf{x}) = 0$, moreover, each $f_i(\mathbf{x})$ must also be zero. Thus, the solutions of the set of Equations 6-118 are obtained by searching for the minima of Equation 6-119.

6-10b. Reduction of Dimensionality

In Chapter 2, the problem of finding extrema of an analytic function $f(\mathbf{x})$ is reduced to the solution of the set of equations

$$\frac{\partial f(\mathbf{x})}{\partial x_i} = 0, \qquad i = 1, 2, \ldots, n \tag{6-120}$$

If $f(\mathbf{x})$ is a complicated function of the x_i's, it is seldom possible to solve analytically for the x_i's which satisfy 6-120. It is often the case, however, that one or more of the x_i's which satisfy 6-120 can be obtained in terms of the other x_i's. When this is true, the solution of the original problem may be obtained more easily by applying search techniques to solve equivalent formulations of the original problem. Two such formulations are given in the next few paragraphs.

To illustrate the approach, suppose that Equations 6-120 can be rearranged (or solved) to yield the set of equations

$$f_i(\mathbf{x}_r) = 0, \qquad i = 1, 2, \ldots, n-1 \tag{6-121}$$

and

$$x_n = f_n(\mathbf{x}_r) \tag{6-122}$$

where the set $\mathbf{x}_r = \{x_1, x_2, \ldots, x_{n-1}\}$ is the *reduced* set of variables.

One approach is to replace x_n in the original function $f(x_1, x_2, \ldots, x_n)$ with $f_n(\mathbf{x}_r)$:

$$f_0(\mathbf{x}_r) \triangleq f[x_1, x_2, \ldots, x_{n-1}, f_n(\mathbf{x}_r)] \tag{6-123}$$

It is clear that $f_0(\mathbf{x}_r) = f(\mathbf{x})$ at the stationary points of $f(\mathbf{x})$, but that $f_0(\mathbf{x}_r) \neq f(\mathbf{x})$ in general. A second important property of $f_0(\mathbf{x}_r)$ is that its stationary points include those of $f(\mathbf{x})$; this fact follows from the identity

$$\frac{\partial f_0(\mathbf{x}_r)}{\partial x_i} = \left[\frac{\partial f(\mathbf{x})}{\partial x_i} + \frac{\partial f(\mathbf{x})}{\partial x_n}\frac{\partial f_n(\mathbf{x}_r)}{\partial x_i}\right]_{x_n = f_n(\mathbf{x}_r)}, \qquad i = 1, 2, \ldots, n-1 \tag{6-124}$$

Note that all $\partial f_0(\mathbf{x}_r)/\partial x_i$ equal zero when all $\partial f(\mathbf{x})/\partial x_i$ equal zero. Thus, the relative maxima and minima of $f_0(\mathbf{x}_r)$ include those of $f(\mathbf{x})$. If a comprehensive search is conducted in the reduced $(n-1)$-dimensional space to obtain the extrema of $f_0(\mathbf{x}_r)$, the values of $\mathbf{x}_r$ so obtained, in conjunction with Equation 6-122, can be examined to obtain the values of $\mathbf{x}$ which extremize $f(\mathbf{x})$.

Computational savings are generally obtained by conducting the search in the least-possible number of dimensions.

An alternative approach to that given in the preceding paragraph is to solve the reduced set of equations 6-121 by using the procedure described in Section 6-10a, i.e., minimize the function $f_{n+1}(\mathbf{x}_r)$ where

$$f_{n+1}(\mathbf{x}_r) \triangleq \sum_{i=1}^{n-1} [f_i(\mathbf{x}_r)]^2 \qquad (6\text{-}125)$$

An advantage of this approach is that the minima of $f_{n+1}(\mathbf{x}_r)$ are known to be zero, and a given search procedure may incorporate this fact to advantage. On the other hand, the fact that $f_{n+1}(\mathbf{x}_r)$ is very close to zero for a particular $\mathbf{x}_r$ does not guarantee the immediate proximity of an extremum: a case in point is a "ravine" along which $f(\mathbf{x})$ is a slowly decreasing function of $\mathbf{x}$; on such a ravine, the derivatives 6-120 from which $f_{n+1}(\mathbf{x}_r)$ is obtained are "close" to zero, and yet $f(\mathbf{x})$ may be far from optimum.

Example 6-9. Consider the case in which $f(\mathbf{x})$ is of the form

$$f(\mathbf{x}) = x_n{}^2 q_1(\mathbf{x}_r) + x_n q_2(\mathbf{x}_r) + q_3(\mathbf{x}_r) \qquad (6\text{-}126)$$

where $f(\mathbf{x})$ is assumed to be of class C^1. The stationary values of $\mathbf{x}$ for $f(\mathbf{x})$ are solutions of the set of Equations 6-127 and 6-128:

$$\frac{\partial f(\mathbf{x})}{\partial x_i} = 0, \quad i = 1, 2, \ldots, n-1 \qquad (6\text{-}127)$$

and

$$\frac{\partial f(\mathbf{x})}{\partial x_n} = 2 x_n q_1(\mathbf{x}_r) + q_2(\mathbf{x}_r) = 0 \qquad (6\text{-}128)$$

From Equation 6-128,

$$x_n = \frac{-q_2(\mathbf{x}_r)}{2 q_1(\mathbf{x}_r)} \qquad (6\text{-}129)$$

Equations 6-127 and 6-129 are easily arranged in the form of Equations 6-121 and 6-122, and either of the two approaches given in Section 6-10b is applicable at this point. In a given case, the functions $q_1(\mathbf{x}_r)$, $q_2(\mathbf{x}_r)$, and $q_3(\mathbf{x}_r)$ are specified, and one of the two approaches may appear to be more appropriate.

●

6-11. CONSTRAINTS

6-11a. The Nonlinear Programming Problem

Consider the problem of extremizing the nonlinear performance measure P:

$$P = f_0(\mathbf{x}) \qquad (6\text{-}130)$$

where the real x_i's of $\mathbf{x} = [x_1 \; x_2 \; \cdots \; x_n]'$ are constrained by the following sets of equations:

$$f_i(\mathbf{x}) = c_i, \qquad i = 1, 2, \ldots, k < n \qquad (6\text{-}131)$$

and

$$f_i(\mathbf{x}) \leq c_i, \qquad i = k + 1, k + 2, \ldots, m \qquad (6\text{-}132)$$

All c_i are constants, and all $f_i(\mathbf{x})$ are assumed to be of class C^1 and real-valued over the set of $\mathbf{x}$'s defined by 6-131 and 6-132; if not, only those search techniques which are nongradient methods may be applicable. This general problem is called the *nonlinear programming problem* because of its similarity of appearance to that of linear programming (Chapter 5).

The gradient projection method [6.80, 6.81] can be used to solve nonlinear programming problems. This method is convenient when the constraint equations are linear algebraic equations, or nearly so; but it is considerably less attractive in the more general case. Other methods have been developed which apply if Equations 6-130, 6-131, and 6-132 assume special forms; e.g., the *quadratic programming problem* [6.45, Chapter 7] in which the performance measure is quadratic but the constraints are linear. Wolfe [6.106] reviews several of these search techniques. Also of value for the solution of nonlinear programming problems are necessary-and-sufficient-condition tests such as those presented in Chapter 2 and those developed by Fritz John [6.53, 6.71], Kuhn and Tucker [6.60], King [6.58], and Fiacco [6.32].

It is evident that the constraints in 6-132 can be placed in the form of those in 6-131 by the use of slack variables, as is done in Chapters 2 and 5; however, this incorporation of additional search variables is unnecessary when the penalty-function approach is used to solve nonlinear programming problems. General forms of this approach are considered in Sections 6-11b through 6-11e; and in Sections 6-11f and 6-11g, the versatility of the Lagrange multiplier is again demonstrated.

6-11b. Outside Penalty Functions for Inequality Constraints

The concept of a penalty function is one of those which is so basic that it would be facetious to attribute its initial use to any single person. For purposes of nonlinear programming, *an outside penalty function is one which is added to the performance measure and which detracts from good performance, as measured by the performance measure, when an associated constraint is violated.* The following is a particularly useful penalty function for the inequality constraints 6-132:

$$p_i(\mathbf{x}) \triangleq [f_i(\mathbf{x}) - c_i]^2 \delta_i, \qquad i = k + 1, k + 2, \ldots, m \qquad (6\text{-}133)$$

where

$$\delta_i \triangleq \begin{cases} 0 \text{ for } f_i(\mathbf{x}) \leq c_i \\ 1 \text{ for } f_i(\mathbf{x}) > c_i \end{cases}$$

The advantage of this form of penalty function over certain others that might be used is that it is of class C^1, provided $f_i(\mathbf{x})$ is of class C^1. Thus, it can be used in conjunction with gradient techniques. Figure 6-23 depicts the effect of such a penalty function in a particular case. Observe in Figure 6-23 that the constrained minimum of $f_0(\mathbf{x})$ is approximated closely by the "unconstrained" minimum of $f_0(\mathbf{x})$-plus-penalty-function.

6-11c. Penalty Functions for Equality Constraints

Typically, Equations 6-131 define an $(n - k)$-dimensional manifold (sometimes termed an $(n - k)$-fold) in the n-dimensional space of the x_i's. A penalized performance measure is to be formed by augmenting the given performance measure with penalty functions; those points and only those points off the $(n - k)$-fold are to be penalized.

The same penalty functions as used in the preceding subsection are appropriate here, but without the δ_i's. Thus, corresponding to equality constraint equations 6-131, penalty functions are

$$p_i(\mathbf{x}) \triangleq [f_i(\mathbf{x}) - c_i]^2, \quad i = 1, 2, \ldots, k \tag{6-134}$$

The effect of this type of penalty function is illustrated in Figure 6-24. For the case shown, the penalty function forces the two-dimensional search to be conducted primarily along the line defined by $p_1(\mathbf{x}) = 0$. In general, such a penalty function in an n-dimensional search forces the search to be conducted primarily on an $(n - 1)$-dimensional surface in n-dimensional space.

6-11d. Minimization of the Penalized Performance Measure

The penalized performance measure P_p is defined as the unconstrained performance measure augmented by the sum of all associated penalty functions, each of which is weighted by a coefficient w_i, as follows:

$$P_p \triangleq f_0(\mathbf{x}) + \sum_{i=1}^{m} w_i p_i(\mathbf{x}) \tag{6-135a}$$

where the $p_i(\mathbf{x})$'s stem from both equality and inequality constraints, in general, as considered in the preceding subsections. All w_i's are positive if a minimum of P_p is desired, but they are negative if a maximum is desired. Only the minimization case is considered here because the significant features of

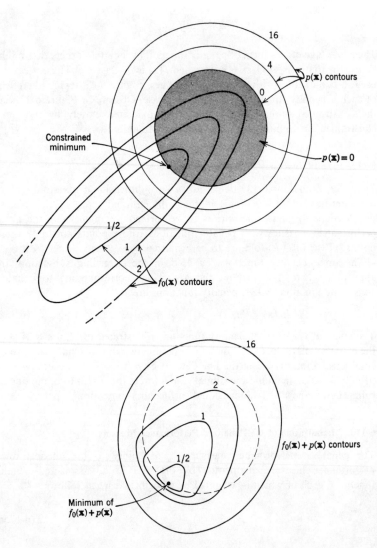

Figure 6-23. An outside penalty function $p(\mathbf{x})$ for an inequality constraint, and its effect on the performance measure $f_0(\mathbf{x})$.

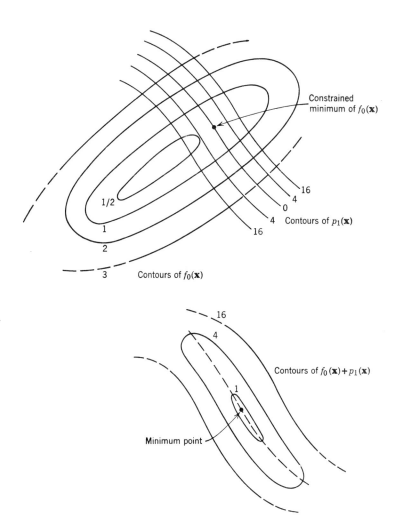

Figure 6-24. A penalty function $p_1(\mathbf{x})$ for an equality constraint, and its effect on the performance measure $f_0(\mathbf{x})$.

both cases are the same, and only outside penalty functions are considered—the approach to be used with inside penalty functions is indicated in Section 6-11e.

Rather than weight each penalty function with a distinct w_i, as in 6-135a, it is sufficient to use but one w to weight the sum of the penalty functions, for if the sum of the penalty functions is forced to zero, so also must each member of the sum be forced. Thus, P_p, as given by

$$P_p = f_0(\mathbf{x}) + w \sum_{i=1}^{m} p_i(\mathbf{x}) \qquad (6\text{-}135\text{b})$$

is a somewhat less general penalized performance measure, but the slight loss in generality is more than offset by the simplicity gained. Note the similarity between the summation in 6-135b and that in 6-119: in each case, a zero solution of the sum is desired. A word of caution, however, is in order: Certain $p_i(\mathbf{x})$'s may be so large as to obscure the effects of others in Equation 6-135b if scaling is not performed. It is assumed, therefore, that each $p_i(\mathbf{x})$ is scaled initially by an appropriate multiplicative factor to insure that the effect of violating a given constraint by a specified amount (metric distance) is of the same order of magnitude as the effect of violating any other constraint by the same amount.

The penalized performance measure P_p 6-135b has one encumbering feature which distinguishes it from the original performance measure P: P_p contains the w weighting factor. It is not enough, therefore, to say simply that a certain search technique will be used to minimize P_p—the following must also be specified: the initial w, the conditions under which w is to be changed (increased), the factor by which w is increased, and criteria for terminating increases in w.

The examples previously considered in Figures 6-23 and 6-24 correctly suggest that *constrained minima of P, Equation* 6-130, *are identical to unconstrained minima* of $\lim_{w \to \infty} P_p$. It should also be clear, however, that immensely large values of w should not be used at the start of a search; for then steep ravines, such as depicted in Figure 6-24, would exist with the initial search point $\mathbf{x}^0$ far from the minimal point, in general. On the other hand, a w which is too small could conceivably lead to an unbounded minimum of P_p. Hence, a compromise is required in choosing the initial w. The following is suggested when the gradient of P_p, or an approximation thereto, is available.

Define

$$p_{m+1} \triangleq \sum_{i=1}^{m} p_i(\mathbf{x})$$

Sect. 6-11   CONSTRAINTS   339

and at the initial search point $\mathbf{x}^0$, compute $p_{m+1}(\mathbf{x}^0)$. If $p_{m+1}(\mathbf{x}^0)$ equals essentially zero, conduct a search for a minimum of $f_0(\mathbf{x})$, but stop this search at the first search point for which p_{m+1} exceeds zero by a preassigned amount; the designation $\mathbf{x}^0$ is reassigned to this point. Starting from an $\mathbf{x}^0$ for which p_{m+1} is not zero, compute $\nabla f_0(\mathbf{x}^0)$ and $\nabla p_{m+1}(\mathbf{x}^0)$, and assign the initial weighting w^0 by using the equation

$$w^0 = [\nabla f_0(\mathbf{x}^0)' \, \nabla f_0(\mathbf{x}^0) / \nabla p_{m+1}(\mathbf{x}^0)' \, \nabla p_{m+1}(\mathbf{x}^0)]^{1/2} \qquad (6\text{-}136)$$

This assignment assures that the gradient of the constraining penalty $w^0 p_{m+1}(\mathbf{x}^0)$ equals $\nabla f_0(\mathbf{x}^0)$ in magnitude, thereby minimizing (but not completely eliminating) the possibility of a constraint breakthrough. A search is then initiated for the minimum of $P_p = f_0(\mathbf{x}) + w^0 p_{m+1}(\mathbf{x})$.

A mandatory situation in which to increase w is when a constraint breakthrough is apparent: when p_{m+1} continually increases, and $f_0(\mathbf{x})$ decreases at a faster rate. In addition, there are several alternative points in a search at which we may update w: (1) at the end of one iteration of a basic search procedure; (2) at the end of n iterations of a basic search procedure (n equals the dimension of $\mathbf{x}$); or (3) at the end of an overall search with w fixed throughout. The second of the above would appear to be a reasonable compromise and is the one advocated here. The factor by which w is incremented depends somewhat on the particular search technique employed. It could be made to depend on previously computed values of f_0, p_{m+1}, ∇f_0, and ∇p_{m+1}, but this practice is not common. Typically, w^{k+1} is related to w^k by a factor γ: $w^{k+1} = \gamma w^k$ where typical values of γ are between 4 and 10, depending on the particular search technique employed. In fact, we might wish to change γ at some point in an overall search on the basis of information gained in the search.

Once all the penalty functions $p_i(\mathbf{x})$ decrease below small preassigned values, the above criteria for increasing w are by-passed, i.e., w is held fixed at the last used value, and the particular minimization procedure employed runs its ultimate course, unless any one of the $p_i(\mathbf{x})$'s exceeds its preassigned limit at some point, in which case w is increased as before.

An important attribute of the preceding penalty-function approach is that the initial search point $\mathbf{x}^0$ need not satisfy the constraints. A simultaneous solution to the constraint equations is attained concurrently to the attainment of constrained relative minima.

6-11e. Inside Penalty Functions

Penalty functions other than those suggested in the preceding subsections are of value. Carroll [6.15] introduced a particularly interesting penalty-function method which has been developed extensively by Fiacco and

McCormick [6.29, 6.30, 6.31]. Versions of the method have been applied to a variety of problems, including problems of network synthesis [6.88, 6.100] and optimal control [6.101, 6.102]—these control problems have given rise to generalizations of the method because the associated performance measures are functionals rather than ordinary functions. The version of the method developed by Fiacco and McCormick has been named SUMT, standing for "sequential unconstrained minimization technique."

To illustrate the basic idea of Carroll, consider one inequality constraint of the form 6-132, namely,

$$0 \le c_m - f_m(\mathbf{x}) \tag{6-137}$$

It is assumed that 6-137 defines a connected region in n-dimensional space. A minimum of $f_0(\mathbf{x})$ in 6-130 is desired, subject to satisfaction of 6-137. If the performance measure $f_0(\mathbf{x})$ is augmented by $w/[c_m - f_m(\mathbf{x})]$, the result is a new penalized performance measure P_p:

$$P_p = f_0(\mathbf{x}) + \frac{w}{c_m - f_m(\mathbf{x})} \tag{6-138}$$

Suppose a search for the minimum of P_p in 6-138 is started with some positive value of w and with an initial search point $\mathbf{x}^0$ which satisfies 6-137. Any value of $\mathbf{x}$ that is sufficiently near, but inside, the constraint boundary results in a relatively large value of $w/[c_m - f_m(\mathbf{x})]$, and $-\nabla P_p(\mathbf{x})$ points into the allowed region for any such $\mathbf{x}$. The inside penalty function sets up a barrier inside the allowed region; the barrier cannot be crossed by any continuous search for the constrained minimum. With a discrete search, on the other hand, the barrier can be inadvertently crossed; but this fact can be detected immediately from the fact that $w/[c_m - f_m(\mathbf{x})]$ is negative for any $\mathbf{x}$ that does not satisfy 6-137—a negative value of $w/[c_m - f_m(\mathbf{x})]$ indicates that the search should revert to the preceding value of $\mathbf{x}$, which is necessarily inside the allowed region, and a new search tack should be taken from that point. After a minimum has been obtained for one value of w, a new and *smaller* (but positive) value of w is used in the next search. It is easily seen that each constrained relative minimum of $f_0(\mathbf{x})$ is approached asymptotically by a relative minimum of

$$\lim_{w \to 0+} \left[f_0(\mathbf{x}) + \frac{w}{c_m - f_m(\mathbf{x})} \right], \qquad w \not\equiv 0$$

For detailed accounts of interior penalty methods of the preceding type, we can refer to the previously noted works. A logarithmic interior penalty function is considered by Lootsma [6.68].

6-11f. Equality Constraints and Classical Lagrange Multipliers

Forsythe [6.38] has examined the role of classical Lagrange multipliers (Sections 2-5 and 2-6) when they are used in conjunction with gradient-type search techniques. Consider, for example, the problem of maximizing P:

$$P = f_0(\mathbf{x}) \tag{6-139}$$

where allowed values of the column matrix $\mathbf{x}$ are constrained to satisfy

$$f_1(\mathbf{x}) = c_1 \quad \text{(a constant)} \tag{6-140}$$

The real-valued functions f_0 and f_1 are assumed to be of class C^1 with respect to the x_i entries of $\mathbf{x}$. The classical approach is that of maximizing an augmented but unconstrained performance measure P_a:

$$P_a = f_0(\mathbf{x}) + hf_1(\mathbf{x}) \tag{6-141}$$

where h is a Lagrange multiplier which, if the original problem is to be solved, must be specified to obtain equality in 6-140. To obtain the desired solution, one approach that might be tried is as follows: First, a value of h is specified; second, a search is conducted to obtain the maximum of P_a with respect to the x_i's; third, the "optimal" $\mathbf{x}$ obtained in the second step is used to determine how close Equation 6-140 is to being satisfied; and fourth, the procedure is repeated with more appropriate values of h. Caution is in order here, however, for it can easily occur that the values of h selected will lead to senseless answers—Forsythe gives a specific example (see Problem 6.28) in which, depending only on the value of h, either a saddle point or a local minimum or a local maximum corresponds to the stationary value of a particular P_a.

An approach which is better than that of the preceding paragraph is: First, a composite function P_c is formed,

$$P_c = (\nabla P_a)' \nabla P_a + [f_1(\mathbf{x}) - c_1]^2 \tag{6-142}$$

where ∇P_a is the gradient with respect to $\mathbf{x}$ of P_a in 6-141; second, an $(n + 1)$-dimensional search (with respect to $\mathbf{x}$ *and* h) is conducted for all relative *minima* of P_c; third, those relative minima which do not correspond to $P_c = 0$ are discarded; and fourth, the remaining solutions, ones corresponding to absolute minima of P_c, are examined to determine the one (or more) which yields the largest value of $f_0(\mathbf{x})$. Note that this approach is an application of Cauchy's procedure (Section 6-10a) for the simultaneous solution of sets of equations. The approach is easily extended to account for more than one equality constraint, but the disadvantage is that the number of search parameters is increased by the number of Lagrange multipliers.

6-11g. General Constraints and Lagrange Multipliers

Everett [6.27] gave a basic theorem which has promise of utilization in a variety of optimization techniques. Slightly extended versions of the theorem are given in this subsection; the extensions involve consideration of equality constraints in addition to general inequality constraints. The results obtained relate two equivalent problems, which are given next.

For the first problem, consider the augmented performance measure $P_a \equiv P_a(\mathbf{x},\mathbf{h})$:

$$P_a = \mathbf{h}'\mathbf{f}(\mathbf{x}) \tag{6-143}$$

where $\mathbf{h}' \equiv [h_0 \ h_1 \ \cdots \ h_m]$ is a row matrix of Lagrange multipliers, and $\mathbf{f}(\mathbf{x}) \equiv [f_0 \ f_1 \ \cdots \ f_m]'$ is a column matrix of given real-valued functions of the n entries of $\mathbf{x}$. Note that *no assumptions in regard to the continuity of the f_i's are made*. Without loss of generality, $h_0 \equiv 1$ is assumed in the following development.

Suppose a subset R_I of the n-dimensional space of x_i's is of interest. There are no particular restrictions on R_I except that it is assumed closed; it could, for example, be a set of discrete points. For some given values of the h_i's, it is assumed that a maximum P_a^* of P_a exists over R_I; that is,

$$P_a^* = \max_{\mathbf{x} \in R_I} P_a(\mathbf{x},\mathbf{h})$$
$$= P_a[\mathbf{x}^*(\mathbf{h}),\mathbf{h}] \tag{6-144}$$

where the dependence of the optimal $\mathbf{x}$ on $\mathbf{h}$ is shown explicitly as $\mathbf{x}^*(\mathbf{h})$.

For the second problem, suppose a maximum of the unaugmented performance measure P is desired:

$$P = f_0(\mathbf{x}) \tag{6-145}$$

where, in addition to being constrained to be an element of R_I, the optimal $\mathbf{x}$, in this case $\mathbf{x}^o(\mathbf{h})$, is required to satisfy

$$f_i(\mathbf{x}) \leq f_i[\mathbf{x}^*(\mathbf{h})] \quad \text{if} \quad h_i < 0 \tag{6-146a}$$

and

$$f_i(\mathbf{x}) \geq f_i[\mathbf{x}^*(\mathbf{h})] \quad \text{if} \quad h_i > 0 \tag{6-146b}$$

for $i = 1, 2, \ldots, m$, where $\mathbf{x}^*(\mathbf{h})$ is the optimal $\mathbf{x}$ of the preceding problem, and the $f_i[\mathbf{x}^*(\mathbf{h})]$'s assume specific values for given $\mathbf{h}$. Constraints 6-146 define a subset of the n-dimensional space of the x_i's; the intersection of this subset with R_I is denoted by $R_{\mathbf{x}(\mathbf{h})}$; that is $R_{\mathbf{x}(\mathbf{h})}$ denotes the set of those points, and only those points, which satisfy 6-146 and which are contained in R_I.

The constrained maximum P^o of P for this problem is assumed to exist and is expressed as

$$P^o = \max_{\mathbf{x} \in R_{\mathbf{x}(\mathbf{h})}} f_0(\mathbf{x})$$
$$= f_0[\mathbf{x}^o(\mathbf{h})] \qquad (6\text{-}147)$$

The general Lagrange multiplier theorem is now in order.

Theorem 6-4. *The $f_0[\mathbf{x}^o(\mathbf{h})]$ of 6-147 equals $f_0[\mathbf{x}^*(\mathbf{h})]$, where $\mathbf{x}^*(\mathbf{h})$ yields P_a^* in 6-144; and if unique absolute maxima exist, it necessarily follows that $\mathbf{x}^o(\mathbf{h}) = \mathbf{x}^*(\mathbf{h})$.*

Proof: It is clear that

$$f_0[\mathbf{x}^o(\mathbf{h})] \geq f_0[\mathbf{x}^*(\mathbf{h})] \qquad (6\text{-}148)$$

because $\mathbf{x}^*(\mathbf{h})$ is one of the elements of $R_{\mathbf{x}(\mathbf{h})}$ from which $\mathbf{x}^o(\mathbf{h})$ is selected. It only remains to show that the statement "$f_0[\mathbf{x}^o(\mathbf{h})] > f_0[\mathbf{x}^*(\mathbf{h})]$" leads to a contradiction. From 6-146, it follows that $h_i f_i[\mathbf{x}^o(\mathbf{h})] \geq h_i f_i[\mathbf{x}^*(\mathbf{h})]$ for all i. With the preceding fact, and with the (false) assumption $f_0[\mathbf{x}^o(\mathbf{h})] > f_0[\mathbf{x}^*(\mathbf{h})]$, it would appear that

$$f_0[\mathbf{x}^o(\mathbf{h})] + \sum_{i=1}^{m} h_i f_i[\mathbf{x}^o(\mathbf{h})] > f_0[\mathbf{x}^*(\mathbf{h})] + \sum_{i=1}^{m} h_i f_i[\mathbf{x}^*(\mathbf{h})] \qquad (6\text{-}149)$$

But inequality 6-149 contradicts the known fact that $\mathbf{x}^*(\mathbf{h})$ yields the maximum of P_a over those $\mathbf{x}$'s contained in R_I. Thus, equality must hold in 6-148, and the theorem is justified.

●

Consider the direct (but often frustrating) use that may be made of Theorem 6-4: A given problem could be that of maximizing $f_0(\mathbf{x})$ in 6-145 subject to sets of constraints of the form

$$f_i(\mathbf{x}) \leq c_i, \quad i = 1, 2, \ldots, k \qquad (6\text{-}150a)$$

and

$$f_i(\mathbf{x}) \geq c_i, \quad i = k+1, \ldots, m \qquad (6\text{-}150b)$$

In place of the preceding problem, we might be able to solve the problem of maximizing

$$P_a = f_0(\mathbf{x}) + \sum_{i=1}^{m} h_i f_i(\mathbf{x}) \qquad (6\text{-}151)$$

where the first k h_i's are assigned negative values, and the remaining h_i's are assigned positive values. The corresponding value $\mathbf{x}^*$ of $\mathbf{x}$ could be used to obtain values $f_i(\mathbf{x}^*)$ which would determine bounds, as in 6-146. What we would like to happen is that these $f_i(\mathbf{x}^*)$'s would equal the given c_i's, so that 6-150a and 6-150b are satisfied; and if so, Theorem 6-4 gives the desired solution. The roadblock at this point, therefore, is that we do not know what values to assign the h_i's to obtain a match between the c_i's and corresponding $f_i[\mathbf{x}^*(\mathbf{h})]$'s. Indeed, such h_i's may not exist, even in relatively simple cases, as is demonstrated in the example that follows this section. Under special conditions, linear programming can be applied to obtain the match [6.9].

The problem of existence of appropriate h_i's is linked, in part, with the fact that the optimal $\mathbf{x}$ designated by Theorem 6-4 lies on the boundary of the region defined by constraints 6-146. In fact, a weaker version of Theorem 6-4 is:

Theorem 6-5. *Let the inequalities in 6-146 be replaced by equalities, but let the remaining conditions associated with Theorem 6-4 be unchanged. It is still true that $f_0[\mathbf{x}^o(\mathbf{h})] = f_0[\mathbf{x}^*(\mathbf{h})]$.*

The proof of Theorem 6-5 parallels that of Theorem 6-4. Theorem 6-5 justifies the statement that optimal x_i^o's of Theorem 6-4 yield equalities in 6-146. It is clear, however, that optimal solutions to inequality constrained optimization problems seldom result in each inequality constraint being pushed to the limit. Thus, although the preceding suggested use of Theorem 6-4 may yield considerable insight about the form of the solution, it generally does not suffice for the complete solution. For special problem forms, ingenuity coupled with Theorems 6-4 and 6-5 may lead to many valuable results.

Example 6-10. A problem with an obvious answer is given here to illustrate the previously noted limitations of Theorem 6-4. Given to be maximized is

$$P = -(x_1 - 1)^2 - (x_2 + 2)^2 \tag{6-152}$$

subject to the constraints that both x_1 and x_2 are to be non-negative. (Here R_I is taken to be all of two-dimensional space.) Optimal values are, of course, $x_1^o = 1$, $x_2^o = 0$, and $P^o = -4$.

To apply Theorem 6-4, P_a is formed:

$$P_a = -(x_1 - 1)^2 - (x_2 + 2)^2 + h_1 x_1 + h_2 x_2 \tag{6-153}$$

And the optimal value P_a^* of this P_a can be found in the classical way:

$$\left.\frac{\partial P_a}{\partial x_1}\right|_{\mathbf{x}=\mathbf{x}^*} = -2(x_1^* - 1) + h_1 = 0$$

and

$$\left.\frac{\partial P_a}{\partial x_2}\right|_{x=x^*} = -2(x_2^* + 2) + h_2 = 0$$

from which

$$x_1^* = (h_1/2) + 1 \qquad (6\text{-}154a)$$

and

$$x_2^* = (h_2/2) - 2 \qquad (6\text{-}154b)$$

For positive h_1 and h_2, Equations 6-154 in conjunction with Theorem 6-4 give the following:

$$x_1 \geq (h_1/2) + 1 \qquad (6\text{-}155a)$$

and

$$x_2 \geq (h_2/2) - 2 \qquad (6\text{-}155b)$$

And subject to constraints 6-155, the x_1^o and x_2^o which yield a maximum of P in Equation 6-152 are

$$x_1^o = (h_1/2) + 1 \qquad (6\text{-}156a)$$

and

$$x_2^o = (h_2/2) - 2 \qquad (6\text{-}156b)$$

Note that the right-hand member of 6-155b is zero, as desired, if h_2 is selected at 4; however, the right-hand member of 6-155a cannot possibly be zero, which is the value of the specified c_1, for any positive h_1. Thus, the direct application of the theorem does not yield the known solution to the original problem. Nevertheless, it does result in the solution of related problems.

•

6-12. COMPARISON OF TECHNIQUES

Given a specific search technique, we can generally find a performance measure and an initial search point for which the given search technique performs as well as, if not better than, any other search technique. It is this characteristic which makes search techniques so difficult to compare objectively. For example, consider the use of continuous on-line search in a specific industrial process. If the performance measure of the process has a response surface which is relatively well-known in shape, it would be poor engineering, from the standpoint of computational facilities required, to demand the use of the most general search algorithm available; an economic compromise must be made between the all-encompassing cost-per-unit-time C of the search and the extra profit-per-unit-time P which the search extracts from the process—the search technique which is best in this case is the one which results in the maximum of $P - C$, assumed greater than zero, where both P and C are, in general, different for different search techniques.

In selecting one search technique from many for a given application, the following questions merit attention:

1. How much computational equipment is required?
2. Has the search technique proved to be completely successful on similar types of performance measures?
3. What accuracy is required of the search?
4. What is a fair measure of the cost of the search?
5. How will the time spent in evaluating the performance measure and its derivatives, if used, compare with the time spent on other aspects of the search?

In general, these questions can be answered independently only in special cases. In particular, of much interest are efficient, all-purpose search schemes to be used in conjunction with general-purpose digital computers[14] (the answer to question one is apparent in this case). The answer to question two is obtained by applying promising search techniques to various difficult test functions which have been advanced in the literature. For this all-purpose case, best-possible accuracy is desired (this answers question three). And the cost of a given search (question four) is proportional to the time required to effect the search on the computer. When the computer time spent in evaluating the performance measure and its derivatives is much greater than the computer time used on other aspects of the search (question five), the number of evaluations of functions is a fair indicator of the cost. Note that one evaluation of the performance measure $f(\mathbf{x})$ requires the evaluation of but one function whereas one evaluation of the gradient of the performance measure requires the evaluation of n functions, the n components of $\nabla f(\mathbf{x})$.

Representative test functions are described in Tables 6-4 and 6-5. (The reader is urged to examine these before proceeding.) As is to be expected, a given search technique may work well on a given test function, but not so well on another. Also, a given search technique may readily lead to the optimum when one initial search point is used, but not so readily when another is used. Initial search points other than those listed in Table 6-4 have been suggested in the literature, but somewhat surprising are the results of Leon's study [6.65] which strongly suggests the following: With a given

---

[14] General-purpose computer installations usually have several "all-purpose" search routines available.

TABLE 6-4. TEST FUNCTIONS: DEFINITIONS

NAME	TEST FUNCTION	OPTIMUM POINT	TYPICAL STARTING PT.
Rosie (banana valley)	$100(x_1^2 - x_2)^2 + (1 - x_1)^2$	(1, 1)	(−1.2, 1)
Shalow	$(x_1^2 - x_2)^2 + (1 - x_1)^2$	(1, 1)	(−2, −2)
Strait	$(x_1^2 - x_2)^2 + 100(1 - x_1)^2$	(1, 1)	(2, −2)
Cube	$100(x_1^3 - x_2)^2 + (1 - x_1)^2$	(1, 1)	(−1.2, 1)
Beale	$\sum_{i=1}^{3} U_i^2$, where $U_i = c_i - x_1(1 - x_2^i)$ and $c_1 = 1.5$, $c_2 = 2.25$, $c_3 = 2.625$	(3, 0.5)	(5, 0.8)
Helical	$100[(x_3 - 10\theta)^2 + (r - 1)^2] + x_3^2$ where $x_1 = r \cos 2\pi\theta$ and $x_2 = r \sin 2\pi\theta$, $-\pi/2 < 2\pi\theta < 3\pi/2$	(1, 0, 0)	(−1, 0, 0)
Powell	$(x_1 + 10x_2)^2 + 5(x_3 - x_4)^2 + (x_2 - 2x_3)^4 + (10x_1 - x_4)^4$	(0, 0, 0, 0)	(3, −1, 0, 1)
Multi-trig (n up to 100)	$\sum_{i=1}^{n} \left\{ e_i - \sum_{j=1}^{n} (a_{ij} \sin x_j + b_{ij} \cos x_j) \right\}^2$ (The a_{ij}'s and b_{ij}'s are generated at random integer values between −100 and +100; and $e_i \triangleq \sum_{j=1}^{n} (a_{ij} \sin x_j^* + b_{ij} \cos x_j^*)$ where the x_j^*'s are generated at random between $-\pi$ and π.)	x_j^*'s (plus the likelihood of many other relative minima)	$(x_j^* + 0.1\delta_j)$'s (where δ_j's are generated at random between $-\pi$ and π)

test function, a given set of search techniques, and a given set of starting points, the search technique of the set which does best for one starting point usually does well, if not best, for any other starting point of the set. This is especially true of test functions Rosie, Strait, and Cube, each of which was tested by Leon with eight different search techniques and five different starting points.

TABLE 6-5. TEST FUNCTIONS: ORIGIN AND CHARACTER

NAME	ORIGIN AND CHARACTER
Rosie	Introduced by Rosenbrock [6.83]. Has a steep (banana-shaped) valley along $x_2 = x_1^2$. Most widely used of the test functions [6.34, 6.35, 6.36, 6.65, 6.79, 6.89, 6.105, 6.107].
Shalow	Introduced by Witte *et al.* Similar to Rosie but with shallow valley. Used by Leon [6.65].
Strait	Introduced by Witte *et al.* Has a steep, straight valley along $x_1 = 1$. Used by Leon [6.65].
Cube	Introduced by Witte *et al.* Has a steep cubic valley along $x_2 = x_1^3$. Used by Leon [6.65].
Beale	Introduced by Beale. Has a narrow curving valley which approaches the line $x_2 = 1$. Used by Shah *et al.* [6.89], and by Leon [6.65].
Helical	Introduced by Fletcher and Powell [6.35]. A three-variable test function which has a helical valley in the x_3 direction with pitch 10 and radius 1. Used also by Fletcher [6.34].
Powell	Introduced by Powell [6.79]. A four-variable test function having the property that the second partial derivatives do not form a positive-definite matrix at the minimum point. Test designed especially for quadratic-convergent search [6.34, 6.35, 6.79].
Multi-trig	Introduced by Fletcher and Powell [6.35]. Test designed for multi-variable search. Function exhibits many relative minima which are not known in advance [6.35, 6.79].

It is logical to divide search techniques into two categories for general comparison purposes: category 1, those techniques which utilize derivatives of the performance measure; and category 2, those techniques which do not. In general, the best sequential search techniques are more efficient than the best nonsequential ones, and the best sequential methods which utilize the gradient are more efficient than those which do not. The comparison which follows is primarily qualitative—detailed quantitative results are to be found in references as noted. The reader should be on guard in reading the references however, in that the essential terminology of one author may differ from that of another. To one author, the evaluation of the gradient may "mean" one function evaluation, whereas to another it means n function evaluations (this author's preference), and to a third, something different still; to one author, an "iteration" may mean a single one-dimensional search in an n-dimensional space (this author's preference if the technique employs such searches), and to another it may mean n such searches; likewise, the term "cycle" may be used to describe n one-dimensional searches (this author's preference when the technique employs such searches), but the term has many other interpretations in the literature. Then too, a comparison of computing times is misleading unless the same computer is used on all techniques being compared, or unless relative computer speeds are incorporated in the results. Similarly, efficient use of a given computer is an important factor; programming skill is required to transcribe a given search procedure to a flow diagram and to a computer program which makes the best use of computer hardware and software.

For gradient-based search techniques, Leon's study [6.65] shows that best-step gradient search and three other search techniques (ones not mentioned in this book) are generally inferior to continued partan search (Section 6-7b) and Davidon's search [6.24], the second of which is sometimes called the "variable metric method" and which is the precursor of the DFP method (Section 6-8c). Of the methods not included in Leon's study, the DFP method and the method of Fletcher and Reeves (Section 6-8b) merit attention. Newton's method (Section 6-6d) might rightfully be included here also; but as evidenced in Example 6-5, its sterling advantages are apparent only when the search is employed in the vicinity of a relative optimum point.

As stated in Section 6-8c, DFP search is currently without peer when $f(\mathbf{x})$ and $\nabla f(\mathbf{x})$ are typical functions of the sort that require a relatively large amount of computer time to evaluate. Noteworthy is the fact that success of DFP search appears to be independent of the number of variables involved (it encountered no difficulty with Multi-trig, Tables 6-4 and 6-5, with $n = 100$). For cases where $f(\mathbf{x})$ and $\nabla f(\mathbf{x})$ are less involved (e.g., when $f(\mathbf{x})$ is a multivariable polynomial), the method of Fletcher and Reeves and the continued partan method appear to be of comparable value and to be more efficient than DFP search. These methods also require less computer storage space than does DFP search.

Of the search methods which do not require evaluation of derivatives, Fletcher's study [6.34] indicates that Powell's method [6.79] is generally better than Smith's method [6.91] or the DSC method [6.96]; but Powell's method encountered some difficulty with Multi-trig with $n = 20$—Zangwill's modification [6.109] of Powell's method may alleviate this difficulty. Leon's comparison [6.65] shows that "look" [6.107], a variation of pattern search, is much more efficient than univariate search, as would be expected; but of interest is the fact that, for reasons unexplained, look search stopped prematurely on several occasions, before the proximity of the optimum was attained, whereas univariate search was not so afflicted. A limited (very, very limited) cross comparison of various studies can be made on the basis of Rosie with initial search point $(-1.2, 1)$. To reduce Rosie to within 10^{-3} of the optimal value (zero), the number of function evaluations are as follows: Powell's method requires 125 [6.34]; the DSC method [6.96] (an extension of Rosenbrock's method [6.83]) requires 150 [6.34]; Smith's method requires 172 [6.34]; pattern search requires somewhere in the neighborhood of 150 [6.65] to 235 [6.107] (the range is due to the fact that numerous versions of pattern search exist); bunny-hop search of Section 6-9c requires approximately 250; and univariate search requires on the order of 500 [6.65].

When comparing highly accurate search schemes, it must be noted that the numerical accuracy used in a search can be, and usually is, an important consideration. For example, we may program a problem to be solved with a specific search technique that uses 12 significant figures in all computations, and then compare the solution so obtained with one obtained by using 24 significant figures but with other factors unchanged, and find that the paths by which the two searches finally converge to the optimum are considerably different.

6-13. CONCLUSION

The principal search techniques of this chapter are designed to locate the extrema of performance measures which are well-behaved functions of n variables. When constraint equations are associated with a given problem, (Section 6-11), the performance measure may be augmented by penalty functions, and the search techniques of this chapter may yet be applied with modifications. Also considered in Section 6-11 are Lagrange multipliers. Methods that combine Lagrange multiplier approaches with penalty-function approaches have been developed and generally are much more effective than penalty methods alone. It is significant that the Lagrange multiplier theorems of Section 6-11g are valid for integer (discrete) programming problems as well as for other programming problems. It is observed in comparing various techniques (Section 6-12) that no one search technique is optimal for all search problems, but certain guidelines aid in the selection of specific search techniques for specific problems.

Prior to conducting a search, or after an initial search phase, scaling of variables is advisable to reduce the harmful effects of ridges and ravines

(Section 6-2). A nonsingular linear transformation of variables may also be appropriate if one can be found which reduces the interaction of variables in the performance measure. In using sequential search, promising initial search points must be specified. These may be selected on the basis of initially available data and/or a preliminary nonsequential search (e.g., the random or the factorial search of Section 6-4). If the extrema of a specified differentiable function are desired, a gradient-based technique (Sections 6-6, 6-7, and 6-8) is appropriate for the main phase of the search. If the performance measure is evaluated experimentally, however, methods of direct search (Section 6-9) or methods such as that of Powell [6.79, 6.109] are advantageous. In either case, one-dimensional search algorithms (Section 6-3) play an important role in many multidimensional search schemes. One-dimensional search in n-dimensional space is not to be confused with univariate search (Section 6-5) which is one of the least efficient sequential search procedures. Terminal phases of a search—terminal in the sense of searching in the vicinity of an extremum point—should be conducted with a quadratic-convergent search algorithm (e.g., one of those in Sections 6-6d, 6-7, and 6-8). Local random methods (Section 6-9d) and centroid methods (Section 6-9e) are of value during final phases of experimentally conducted search.

In the process of conducting a search, data are generated with which we may ascertain the relative flatness of the performance measure about a given optimal point. Even so, the sensitivity measures of Chapter 2 may be of value in determining the sensitivity of the performance measure with respect to parameter changes about the optimal point. If the DFP method (Section 6-8c) is used, an approximation to the matrix of second partial derivatives is available, and this may be used to evaluate a square sensitivity matrix Sq_x^f of the type defined in Chapter 2.

Most sequential search routines are adaptive in character because both step size and direction of search are affected by previously computed data (e.g., bunny-hop search of Section 6-9c). Flood and Leon [6.37] extend the degree of adaptation further by programming several search schemes and by then programming the computer to decide, on the basis of relative success during a search, which of the search algorithms to use most often during the given search. The main deterrent to this sophisticated approach is that a relatively large amount of computer storage is required for such a general program.

Although many search techniques are directly applicable to nonlinear search problems, piecewise linear approximations that are periodically updated are still much used ([6.3, 6.43]; see also Section 5-12); the reason for this being the facility with which linear programming techniques are able to incorporate a multitude of constraints. Also, just as linear programming problems have related dual problems (Section 5-9), so do certain nonlinear programming problems—the interested reader is referred to pertinent literature [6.23, 6.70, 6.72, 6.74].

Certain important classes of search algorithms have not been given consideration in this chapter. Principal among these are ones which apply to the extremization of functionals, such as those in Chapter 3. Work on variational search techniques was done as early as 1908 by Hadamard [6.20, 6.44]. For the most part, these techniques (e.g., [6.13, 6.56, 6.73, 6.82, 6.90]) are extensions and generalizations of the gradient-based techniques of this chapter. Some of the more important search techniques for the extremization of functionals are considered in Chapter 8. Also note that functionals which are treated by the use of the direct methods of the calculus of variations (Section 3-11) revert to problems of the type considered in this chapter; numerous examples of this direct approach are to be found in the literature (e.g., [6.18, 6.47]).

Other types of search algorithms which deserve more attention than given here are the following: search algorithms designed especially to combat noise [6.61, 6.69, 6.76]; quadratic programming algorithms [6.40, 6.45, 6.98]; algorithms valid for convex (concave) performance measures and constraints [6.19, 6.45, 6.60]; algorithms for geometric programming problems [6.26];[15] and nonlinear integer programming algorithms [6.64, 6.85]—dynamic programming of Chapter 7 is useful in the solution of some of these problems.

One final point: a given search may isolate many relative extrema of a performance measure, and a search of these must be made to find the absolute extrema; this task may be quite tedious. If dynamic programming (Chapter 7) is applicable to the given problem, however, relative extrema are automatically by-passed.

REFERENCES

[6.1] Aaron, M. R. "The Use of Least Squares in System Design." *IRE Transactions on Circuit Theory*, **CT-3**, 224–231, Dec. 1956.

[6.2] Ablow, C. M., and G. Brigham. "An Analog Solution of Programming Problems." *Operations Research*, **3**, 388–394, Nov. 1955.

[6.3] Andreyev, N. I. "Method of Solving Certain Problems of Nonlinear Programming." *Engineering Cybernetics*, 13–20, Jan.–Feb. 1963.

[6.4] Beckman, F. S. "The Solution of Linear Equations by the Conjugate Gradient Method." Chapter 4 of *Mathematical Methods for Digital Computers*, A. Ralston and H. S. Wilf (Editors). Wiley, New York, 1964 (4th printing).

---

[15] Geometric programming problems are those nonlinear programming problems in which all functions can be expressed as weighted sums of products of variables which are raised to real powers.

REFERENCES

[6.5] Bishop, A. B. *Adaptive Pattern Recognition*. WESCON Paper, 1963, 10 pp.

[6.6] Booth, A. D. "An Application of the Method of Steepest Descent to the Solution of Systems of Non-Linear Simultaneous Equations." *Quarterly Journal of Mechanics and Applied Mathematics*, **2**, 460–468, 1949.

[6.7] Box, G. E. P., and K. B. Wilson. "On the Experimental Attainment of Optimum Conditions." *Journal of the Royal Statistical Society*, Series B, **13**, 1–45, 1951.

[6.8] Bromberg, N. S. "Maximization and Minimization of Complicated Multivariate Functions." *AIEE Transactions*, part I, Communications and Electronics, **80**, 725–730, Dec. 1961.

[6.9] Brooks, R., and A. Geoffrion. "Finding Everett's Lagrange Multipliers by Linear Programming." *Operations Research*, **14**, 1149–1153, Nov.–Dec. 1966.

[6.10] Brooks, S. H. "A Discussion of Random Methods for Seeking Maxima." *Operations Research*, **6**, 244–251, March 1958.

[6.11] Brooks, S. H. "A Comparison of Maximum-Seeking Methods." *Operations Research*, **7**, 430–457, July–Aug. 1959.

[6.12] Brown, R. R. "A Generalized Computer Procedure for the Design of Optimum Systems—Parts I and II." *AIEE Transactions*, part I, Communications and Electronics, **78**, 285–293, July 1959.

[6.13] Bryson, A. E., and W. F. Denham. "A Steepest-Ascent Method for Solving Optimum Programming Problems." *Transactions of the ASME*, series E (*Journal of Applied Mechanics*), **29**, 247–257, 1962.

[6.14] Buehler, R. J., B. V. Shah, and O. Kempthorne. *Some Properties of Steepest Ascent and Related Procedures for Finding Optimum Conditions*. Report, Iowa State Univ. Statistical Laboratory, Ames, Iowa, April 1961.

[6.15] Carroll, C. W. "The Created-Response-Surface Technique for Optimizing Nonlinear Restrained Systems." *Operations Research*, **9**, 169–184, March–April 1961.

[6.16] Cauchy, A. "Méthode Générale pour la Résolution des Systémes d' Équations Simultanées." *Comptes Rendus Hebdomadaires des Séances de l'Académie des Sciences*, Paris, **25**, 536–538, Oct. 1847.

[6.17] Centner, R. M., and J. M. Idelsohn. "A Milestone in Adaptive Machine Control." *Control Engineering*, **11**, 92–94, Nov. 1964.

[6.18] Chang, C. S., and P. M. DeRusso. "An Approximate Method for Solving Optimal Control Problems." *IEEE Transactions on Automatic Control*, **AC-9**, 554–555, Oct. 1964.

[6.19] Charnes, A. "The SUMT Method for Convex Programming: Some Discussion and Experience." Pages 215–216 of *Recent Advances in Optimization Techniques*, A. Lavi and T. P. Vogl (Editors). Wiley, New York, 1966.

[6.20] Courant, R. "Variational Methods for the Solution of Problems of Equilibrium and Vibrations." *Bulletin of the American Mathematical Society*, **49**, 1–23, Jan. 1943.

[6.21] Crockett, J. B., and H. Chernoff. "Gradient Methods of Maximization." *Pacific Journal of Mathematics*, **5**, 33–50, March 1955.

[6.22] Curry, H. B. "The Method of Steepest Descent for Nonlinear Minimization Problems." *Quarterly of Applied Mathematics*, **2**, 258–261, Oct. 1944.

[6.23] Dantzig, G. B., E. Eisenberg, and R. W. Cottle. "Symmetric Dual Nonlinear Programs." *Pacific Journal of Mathematics*, 15, 809–812, Sept. 1965.
[6.24] Davidon, W. C. *Variable Metric Method for Minimization.* A.E.C. Research and Development Report, ANL-5990, 1959, 21 pp.
[6.25] DeLand, E. C. *Continuous Programming Methods on an Analog Computer.* Paper P-1815, RAND Corporation, Santa Monica, Calif., 1959, 10 pp.
[6.26] Duffin, R. J., E. L. Peterson, and C. Zener. *Geometric Programming: Theory and Application.* Wiley, New York, 1967.
[6.27] Everett, H. "Generalized Lagrange Multiplier Method for Solving Problems of Optimum Allocation of Resources." *Operations Research*, 11, 399–431, May–June 1963.
[6.28] Fend, F. A., and C. B. Chandler. *Numerical Optimization for Multidimensional Problems.* GE Report No. 61GL78, March 1961, 34 pp.
[6.29] Fiacco, A. V., and G. P. McCormick. "The Sequential Unconstrained Minimization Technique for Nonlinear Programming, A Primal-Dual Method." *Management Science*, 10, 360–366, Jan. 1964.
[6.30] Fiacco, A. V., and G. P. McCormick. "Computational Algorithm for the Sequential Unconstrained Minimization Technique for Nonlinear Programming." *Management Science*, 10, 601–617, July 1964.
[6.31] Fiacco, A. V., and G. P. McCormick. "Extensions of SUMT for Nonlinear Programming: Equality Constraints and Extrapolation." *Management Science*, 12, 816–829, July 1966.
[6.32] Fiacco, A. V. "Second Order Sufficient Conditions for Weak and Strict Constrained Minima." *SIAM Journal on Applied Mathematics*, 16, 105–108, Jan. 1968.
[6.33] Finkel, R. W. "The Method of Resultant Descents for the Minimization of an Arbitrary Function." Paper 71, *Preprints of the 14th National Meeting of the Association for Computing Machinery*, 1959.
[6.34] Fletcher, R. "Function Minimization Without Evaluating Derivatives—A Review." *The Computer Journal*, 8, 33–41, April 1965.
[6.35] Fletcher, R., and M. J. D. Powell. "A Rapidly Convergent Descent Method for Minimization." *The Computer Journal*, 6, 163–168, July 1963.
[6.36] Fletcher, R., and C. M. Reeves. "Function Minimization by Conjugate Gradients." *The Computer Journal*, 7, 149–154, July 1964.
[6.37] Flood, M. M., and A. Leon. *A Universal Adaptive Code for Optimization.* Internal Working Paper No. 19, Space Sciences Laboratory, Univ. of California Berkeley, Calif., Aug. 1964, 109 pp.
[6.38] Forsythe, G. E. "Computing Constrained Minima with Lagrange Multipliers." *Journal of the Society for Industrial and Applied Mathematics*, 3, 173–178, Dec. 1955.
[6.39] Forsythe, G. E., and T. S. Motzkin. "Acceleration of the Optimum Gradient Method—Preliminary Report (Abstract)." *Bulletin of the American Mathematical Society*, 57, 304–305, July 1951.
[6.40] Frank, M., and P. Wolfe. "An Algorithm for Quadratic Programming." *Naval Research Logistics Quarterly*, 3, 95–110, March and June 1956.
[6.41] Gal'perin, M. V., G. I. Korotkevich, I. N. Minsker, and V. I. Rybasov.

"Solving a Nonlinear Mathematical Programming Problem with One and Many Extrema on an Analog Computer." *Engineering Cybernetics*, 75–83, July–Aug. 1964.

[6.42] Gnoyenskiy, L. S., and S. M. Movshovich. "Application of Mathematical Programming Methods to the Optimal Regulation Problem." *Engineering Cybernetics*, 12–26, Sept.–Oct. 1964.

[6.43] Griffith, R. E., and R. A. Stewart. "A Nonlinear Programming Technique for Optimization of Continuous Processing Systems." *Management Science*, 7, 379–392, July 1961.

[6.44] Hadamard, J. "Mémoire sur le Problème d'Analyse Relatif à l'Équilibre des Plaques Élastiques Encastrées." *Mem. Pres. Acad. Sci. France*, 33, No. 4, 1908.

[6.45] Hadley, G. *Nonlinear and Dynamic Programming*. Addison-Wesley, Reading, Mass., 1964.

[6.46] Hestenes, M., and E. Stiefel. *Method of Conjugate Gradients for Solving Linear Systems*. Report 1659, National Bureau of Standards, 1952.

[6.47] Ho, Y. C. "Solution Space Approach to Optimal Control Problems." *Transactions of the ASME*, series D (*Journal of Basic Engineering*), 83, 53–58, March 1961.

[6.48] Hooke, R., and T. A. Jeeves. "Comments on Brooks' Discussion of Random Methods." *Operations Research*, 6, 881–882, Nov. 1958.

[6.49] Hooke, R., and T. A. Jeeves. "Direct Search Solution of Numerical and Statistical Problems." *Journal of the Association for Computing Machinery*, 8, 212–229, April 1961.

[6.50] Ivakhnenko, A. G. "On Constructing an Extremum Controller without Hunting Oscillations." *IEEE Transactions on Automatic Control*, **AC-12**, 154–161, April 1967.

[6.51] Jackson, A. S. *Analog Computation*. McGraw-Hill, New York, 1960.

[6.52] James, G., and R. C. James (Editors). *Mathematics Dictionary*. Van Nostrand, Princeton, N.J., 2nd ed., 1959.

[6.53] John, F. "Extremum Problems with Inequalities as Side Conditions." Pages 187–204 of *Studies and Essays*, Courant Anniversary Volume, K. O. Friedrichs, O. E. Neugebauer, and J. J. Stoker (Editors). Wiley, New York, 1948.

[6.54] Johnson, S. M. *Best Exploration for Maximum is Fibonaccian*. Paper P-856, RAND Corporation, Santa Monica, Calif., May 1956.

[6.55] Kelley, H. J. "Methods of Gradients." Chapter 6, pp. 205–254 of *Optimization Techniques with Applications to Aerospace Systems*, G. Leitmann (Editor). Academic Press, New York, 1962.

[6.56] Kenneth, P., and G. E. Taylor. "Solution of Variational Problems with Bounded Control Variables by Means of the Generalized Newton-Raphson Method." Pages 471–487 of *Recent Advances in Optimization Techniques*, A. Lavi and T. P. Vogl (Editors). Wiley, New York, 1966.

[6.57] Kiefer, J. "Sequential Minimax Search for a Maximum." *Proceedings of the American Mathematical Society*, 4, 502–506, June 1953.

[6.58] King, R. P. "Necessary and Sufficient Conditions for Inequality

Constrained Extreme Values." *Industrial and Engineering Chemistry Fundamentals*, 5, 484–489, Nov. 1966.

[6.59] Kokotovic, P., and D. D. Siljak. "Automatic Analog Solution of Algebraic Equations and Plotting of Root Loci by Generalized Mitrovic's Method." *IEEE International Convention Record*, 12, part 1, 47–54, 1964.

[6.60] Kuhn, H. W., and A. W. Tucker. "Nonlinear Programming." *Proceedings, Second Berkeley Symposium on Mathematical Statistics and Probability*, Univ. of California Press, Berkeley, Calif., pp. 481–492, 1951.

[6.61] Kushner, H. J. "A New Method of Locating the Maximum Point of an Arbitrary Multipeak Curve in the Presence of Noise." *Preprint Volume*, Joint Automatic Control Conference, 69–79, June 1963.

[6.62] Kuznetsov, Yu. K. "The Algorithm for Finding Optimal Parameters of Control Systems by the Gradient Method." *Engineering Cybernetics*, 75–81, March–April 1964.

[6.63] Lapidus, L., E. Shapiro, S. Shapiro, and R. E. Stillman. "Optimization of Process Performance." *AIChE Journal*, 7, 288–294, June 1961.

[6.64] Lawler, E. L., and M. D. Bell. "A Method for Solving Discrete Optimization Problems." *Operations Research*, 14, 1098–1112, Nov.–Dec. 1966.

[6.65] Leon, A. "A Comparison Among Eight Known Optimizing Procedures." Pages 23–46 of *Recent Advances in Optimization Techniques*, T. P. Vogl and A. Lavi (Editors). Wiley, New York, 1966.

[6.66] Levine, L. *Methods for Solving Engineering Problems Using Analog Computers*. McGraw-Hill, New York, 1964.

[6.67] Levine, L., and H. F. Meissinger. "An Automatic Analog Computer for Solving Polynomials and Finding Root Loci." *IRE National Convention Record*, 5, part 4, 164–172, 1957.

[6.68] Lootsma, F. A. "Logarithmic Programming—A Method of Solving Nonlinear Programming Problems." *Philips Research Reports*, 22, 329–344, June 1967.

[6.69] Magee, E. J. *An Empirical Investigation of Procedures for Locating the Maximum Peak of a Multiple-Peak Regression Function*. Report 22 G-0046, Lincoln Laboratory, M.I.T., Cambridge, Mass., Oct. 1960, 31 pp.

[6.70] Mangasarian, O. L. "Duality in Nonlinear Programming." *Quarterly of Applied Mathematics*, 20, 300–302, Oct. 1962; see also same journal, 21, 252, Oct. 1963.

[6.71] Mangasarian, O. L., and S. Fromovitz. "The Fritz John Necessary Optimality Conditions in the Presence of Equality and Inequality Constraints." *Journal of Mathematical Analysis and Applications*, 17, 37–47, Jan. 1967.

[6.72] Mangasarian, O. L., and J. Ponstein. "Minmax and Duality in Nonlinear Programming." *Journal of Mathematical Analysis and Applications*, 11, 504–518, July 1965.

[6.73] McReynolds, S. R., and A. E. Bryson, Jr. "A Successive Sweep Method for Solving Optimal Programming Problems." *Preprint Volume*, Joint Automatic Control Conference, 551–555, 1965.

[6.74] Mond, B. "A Symmetric Dual Theorem for Nonlinear Programs." *Quarterly of Applied Mathematics*, 23, 265–269, Oct. 1965.

[6.75] Murata, T. "The Use of Adaptive Constrained Descent in System Design." *IEEE International Convention Record*, **12**, part 1, 296–306, 1964.
[6.76] Norris, R. C. *A Method for Locating the Maximum of a Function of a Single Variable in the Presence of Noise.* Report 22 G-0035, Lincoln Laboratory, M.I.T., Cambridge, Mass., Oct. 1960, 36 pp.
[6.77] Pinsker, I. Sh., and B. M. Tseitlin. "A Nonlinear Optimization Problem." *Automation and Remote Control*, **23**, 1510–1518, 1962.
[6.78] Ponstein, J. "Seven Kinds of Convexity." *SIAM Review*, **9**, 115–119, Jan. 1967.
[6.79] Powell, M. J. D. "An Efficient Method for Finding the Minimum of a Function of Several Variables without Calculating Derivatives." *The Computer Journal*, **7**, 155–162, July 1964.
[6.80] Rosen, J. B. "The Gradient Projection Method for Nonlinear Programming, Part 1: Linear Constraints." *Journal of the Society for Industrial and Applied Mathematics*, **8**, 181–217, March 1960.
[6.81] Rosen, J. B. "The Gradient Projection Method for Nonlinear Programming, Part 2: Nonlinear Constraints." *Journal of the Society for Industrial and Applied Mathematics*, **9**, 514–532, Dec. 1961.
[6.82] Rosenbloom, P. C. "The Method of Steepest Descent." *Numerical Analysis*, Proceedings of the Sixth Symposium in Applied Mathematics, McGraw-Hill, New York, pp. 127–176, 1956.
[6.83] Rosenbrock, H. H. "An Automatic Method for Finding the Greatest or the Least Value of a Function." *The Computer Journal*, **3**, 175–184, Oct. 1960.
[6.84] Rybashov, M. V. "On a Certain Method for Solving the Global Problem of Searching for the Roots of Finite Equations by Means of an Electronic Simulator." *Automation and Remote Control*, **23**, 1311–1313, 1962.
[6.85] Saaty, T. L. "On Nonlinear Optimization in Integers." *Naval Research Logistics Quarterly*, **15**, 1–22, March 1968.
[6.86] Sakrison, D. J. "Application of Stochastic Approximation Methods to Optimum Filter Design." *IRE International Convention Record*, **9**, part 4, 127–135, 1961.
[6.87] Sakrison, D. J. "Iterative Design of Optimum Filters for Non-Mean-Square-Error Performance Criteria." *IEEE Transactions on Information Theory*, **IT-9**, 161–166, July 1963.
[6.88] Scheibe, P. O., and E. A. Huber. "The Application of Carroll's Optimization Technique to Network Synthesis." *Proceedings*, Third Annual Allerton Conference on Circuit and System Theory, Dept. of Elect. Engr., Univ. of Illinois, Urbana, Ill., pp. 182–191, Oct. 1965.
[6.89] Shah, B. V., R. J. Buehler, and O. Kempthorne. "Some Algorithms for Minimizing a Function of Several Variables." *Journal of the Society for Industrial and Applied Mathematics*, **12**, 74–92, March 1964.
[6.90] Sinnott, J. F. Jr., and D. G. Luenberger. "Solution of Optimal Control Problems by the Method of Conjugate Gradients." *Preprint Volume*, Joint Automatic Control Conference, 566–574, June 1967.
[6.91] Smith, C. S. *The Automatic Computation of Maximum Likelihood Estimates.* N.C.B. Scientific Dept. Report S.C. 846/MR/40, 1962.

[6.92] Sokkappa, B. G. "Optimum Scheduling of Hydrothermal Systems—A Generalized Approach." *IEEE Transactions on Power Apparatus and Systems*, **82**, 97–104, 1963.

[6.93] Southwell, R. V. *Relaxation Methods in Theoretical Physics*. Oxford Univ. Press, New York, 1946.

[6.94] Sugie, N. "An Extension of Fibonaccian Searching to Multidimensional Cases." *IEEE Transactions on Automatic Control*, **AC-9**, 105, Jan. 1964.

[6.95] Suzuki, T., and B. Kondo. "Adaptive Control of Gradient-Type Optimizing Control System." *Electrical Engineering in Japan*, **86**, 67–76, Oct. 1966.

[6.96] Swann, W. H. *Report on the Development of a New Direct Search Method of Optimization*. I.C.I. Ltd., Central Instrument Laboratory Research Note 64/3, 1964.

[6.97] Synge, J. L. "A Geometrical Interpretation of the Relaxation Method." *Quarterly of Applied Mathematics*, **2**, 87–89, April 1944.

[6.98] Van De Panne, C., and A. Whinston. "The Simplex and the Dual Method for Quadratic Programming." *Operations Research Quarterly*, **15**, 355–388, Dec. 1964.

[6.99] Volz, R. A. "The Minimization of a Function by Weighted Gradients." *Proceedings of the IEEE*, **53**, 646–647, June 1965.

[6.100] Waren, A. D., and L. S. Lasdon. "Practical Filter Design Using Mathematical Optimization." *Proceedings*, Third Annual Allerton Conference on Circuit and System Theory, Dept. of Elect. Engr., Univ. of Illinois, Urbana, Ill., pp. 677–689, Oct. 1965.

[6.101] Waren, A. D., L. S. Lasdon, and R. K. Rice. "A Penalty Method for Optimal Control." *Proceedings of the IEEE*, **55**, 115–116, Jan. 1967.

[6.102] Waren, A. D., L. S. Lasdon, and R. K. Rice. "An Interior Penalty Method for Inequality Constrained Optimal Control Problems." *Preprint Volume*, Joint Automatic Control Conference, 538–548, June 1967.

[6.103] Weisman, J., C. F. Wood, and L. Rivlin. "Optimal Design of Chemical Process Systems." A. I. Ch. E. Symposium Series, *Process Control and Applied Mathematics*, **61**, 50–63, 1965.

[6.104] Weiss, E. A., D. H. Archer, and D. A. Burt. "Computer Sets Tower for Best Run." *Hydrocarbon Processing and Petroleum Refiner*, **40**, 169–174, Oct. 1961.

[6.105] Wilde, D. J. *Optimum Seeking Methods*. Prentice Hall, Englewood Cliffs, N.J., 1964.

[6.106] Wolfe, P. "Methods of Nonlinear Programming." Pages 67–86 of *Recent Advances in Mathematical Programming*, R. L. Graves and P. Wolfe (Editors). McGraw-Hill, New York, 1963.

[6.107] Wood, C. F. *Recent Developments in 'Direct Search' Techniques*. Westinghouse Research Report 62-159-522-R1, 1962, 32 pp.

[6.108] Young, P. C. "The Determination of the Parameters of a Dynamic Process." *The Radio and Electronic Engineer*, **29**, 345–362, June 1965.

[6.109] Zangwill, W. I. "Minimizing a Function Without Calculating Derivatives." *The Computer Journal*, **10**, 293–296, Nov. 1967.

[6.110] Zellnick, H. E., N. E. Sondak, and R. S. Davis. "Gradient Search Optimization." *Chemical Engineering Progress*, **58**, 35–41, Aug. 1962.
[6.111] "Progress Report on Opcon." *Control Engineering*, **6**, 124, Nov. 1959.

PROBLEMS

6.1 a. Prove that the sum $f_1(x) + f_2(x)$ is convex if both $f_1(x)$ and $f_2(x)$ are convex.

b. Prove that

$$\sum_{i=1}^{K} |a_{i1}x_1 + a_{i2}x_2 + \cdots + a_{in}x_n + b_i|$$

is a convex function of **x**, where the a_{ij}'s and b_i's are real constants.

6.2 Is the sum $f_1(x) + f_2(x)$ unimodal if both $f_1(x)$ and $f_2(x)$ are unimodal?

6.3 Given that $df_1(x)/dx$ is strictly monotonic increasing and $-df_2(x)/dx$ is strictly monotonic decreasing; show that the sum $f_1(x) + f_2(x)$ is unimodal.

6.4 a. Verify Equations 6-26 and 6-27.

b. Draw a flow diagram for the one-dimensional search procedure of Section 6-3c.

6.5 Apply appropriate search techniques from the left-hand column (see Section 6-3) to the maximization of the functions listed in the right-hand column. (Initial search points, etc., are to be supplied as necessary.)

a. Newton-Raphson
b. Cubic-Convergent
c. Search Without Derivatives
d. Fibonacci
e. Golden Section

f. $-(\sin x)/x$, $\pi \leq x \leq 2\pi$
g. $-x^4 + 10x^3 - 35x^2 + 40x$
h. $5(1 - e^{-x} - xe^{-x}) - x$, $x \geq 0$
i. Problem 6.6
j. Any one-variable maximization problem from Chapter 2

6.6 Company Z plans to borrow x dollars this year for plant expansion and to pay the money back in equal yearly installments over the next K years. Company Z finds that it will have to pay a yearly interest rate $r_1 = k_0 + k_1 x$ that depends on the amount of money borrowed; k_0 and k_1 are given constants. Money earned as a result of the expansion can be invested by company Z at a fixed interest rate of r_2 per annum. The total expected return from the expansion as reflected to the Kth year, is $c_1(1 - e^{-c_2 x})$, where c_1 and c_2 are constants. Company Z's annual payment to the lending agency is

$$\left[\frac{r_1(1 + r_1)^K}{(1 + r_1)^K - 1}\right]x$$

The total of these payments, as reflected to the end of the Kth year at interest rate r_2, is

$$\left[\frac{(1 + r_2)^K - 1}{r_2}\right]\left[\frac{r_1(1 + r_1)^K}{(1 + r_1)^K - 1}\right]x$$

To be maximized is total profit P:

$$P = c_1(1 - e^{-c_2 x}) - \left[\frac{(1 + r_2)^K - 1}{r_2}\right]\left[\frac{r_1(1 + r_1)^K}{(1 + r_1)^K - 1}\right]x$$

Note that considerations of income tax are ignored in this problem. Use a search technique to find the maximum of P with respect to x (dollars) when

$K = 10$ $k_0 = 0.05$

$c_1 = 4 \times 10^5$ dollars $k_1 = \dfrac{2}{10^7 \text{ dollars}}$

$c_2 = \dfrac{1}{10^5 \text{ dollars}}$ $r_2 = 0.05$

Suggest a scale change for x and P.

6.7 Local approximation of a function $y(\alpha) = f(\mathbf{x}^0 + \alpha \mathbf{r}^0)$ by a quadratic is to be made. Find the stationary point (under what conditions is it a maximum point?) of the quadratic in terms of one of the following sets of data.

a. $y(0) = y^0$, $\dot{y}(0) = \dot{y}^0$, and $y(d_1) = y^1$.
b. $\dot{y}(0) = \dot{y}^0$, $y(d_1) = y^1$, and $y(d_2) = y^2$.

6.8 Local approximation of a function $y(\alpha) = f(\mathbf{x}^0 + \alpha \mathbf{r}^0)$ by a cubic equation is to be made. Find the local minimum of the cubic equation in terms of the following set of data: $y(0) = y^0$, $\dot{y}(0) = \dot{y}^0$ which is less than zero, $y(d_1) = y^1$ where $d_1 > 0$, and $y(d_2) = y^2$ where both $d_2 > d_1$ and $y^2 > y^1$.

6.9 Conduct a five-point Fibonacci search on an experimentally evaluated unimodal function $f(x)$. (If you have no such function on hand, do the following: each time a comparison is to be made between the value of $f(x)$ at two points, flip a coin—if it comes up heads, assume the right-hand point is the one with the greater value of f; if tails, the left-hand point is the one with the greater value.) Give the results of your search in the same form as those in Figure 6-6.

6.10 Repeat Problem 6.9, but use search by golden section.

6.11 A one-dimensional search for a minimum of $f(\mathbf{x})$ is to be conducted along a line defined by two points $\mathbf{x}^a$ and $\mathbf{x}^b$ in n-dimensional space. The procedure of Section 6-3c is to be appropriately modified and applied. Give the required modifications.

6.12 As a result of a factorial search for the minimum of $f(x_1, x_2)$, the point x_1^b, x_2^b is the best yet obtained (see Figure 6-P12). Values of f at surrounding points have been generated in the factorial search; six of these values—$(n + 1)(n + 2)/2$ of these values with $n = 2$—are to be fitted with a quadratic function q:

$$q = b_0 + b_1 x_1 + b_2 x_2 + \tfrac{1}{2}\mathbf{x}'\mathbf{A}\mathbf{x}$$

where $\mathbf{A} = \mathbf{A}'$ is a 2×2 matrix. The minimum point associated with q is to be found.

a. Find an expression for the minimum point associated with q.
b. Find expressions for the b_i's and the a_{ij}'s as functions of $f(x_1^a, x_2^a) \equiv f^{aa}$, $f(x_1^a, x_2^b) \equiv f^{ab}$, $f(x_1^b, x_2^a) \equiv f^{ba}$, f^{bb}, f^{bc}, f^{cb}, and the corresponding points in the x_1, x_2 plane.

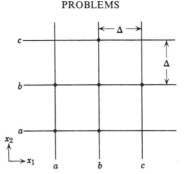

Figure 6-P12.

6.13 More than a dozen multivariable search techniques are considered in this chapter. Apply one or more of these to the problems listed in Parts a through f. Suggest scale changes as deemed appropriate.

    a. Minimize $f(x_1,x_2) = 100e^{-x_1 x_2} + x_1 + 10x_2$ subject to constraints $x_1 \geq 0$ and $x_2 \geq 0$. Use initial search point $x_1 = 1$ and $x_2 = 1$.
    b. Use a test problem from Table 6-4.
    c. Problem 6.15.
    d. Problem 6.28 by use of the approach associated with Equation 6-142.
    e. Problem 6.31.
    f. Problem 6.33.

6.14 Use a standard digital computer routine for solution of differential equations to implement continuous gradient search. Comment on the computational merits of this approach.

6.15 In planning for production of two products, company Z estimates expected profit to equal P:

$$P = \alpha_1(1 - e^{-\beta_1 x_1} - \beta_1 x_1 e^{-\beta_1 x_1}) +$$
$$\alpha_2(1 - e^{-\beta_2 x_2} - \beta_2 x_2 e^{-\beta_2 x_2}) +$$
$$\alpha_3(1 - e^{-\beta_3 x_1 x_2}) - x_1 - x_2$$

where x_1 is the amount of money spent to produce and advertise product 1, x_2 is the amount of money spent to produce and advertise product 2, and the α_i's and β_i's are specified constants. P, x_1, and x_2 are in units of 10^5 dollars. Find the maximum of P and the optimal values of x_1 and x_2 under the following conditions:
    a. $\alpha_1 = 3$, $\alpha_2 = 4$, $\alpha_3 = 1$, $\beta_1 = 1.2$, $\beta_2 = 1.5$, and $\beta_3 = 1$.
    b. $\alpha_1 = 3$, $\alpha_2 = 4$, $\alpha_3 = -1$, $\beta_1 = 1.2$, $\beta_2 = 1.5$, and $\beta_3 = 1$.

6.16 Vector $\mathbf{r}_1$ is known to be conjugate (with respect to an $n \times n$ symmetric matrix $\mathbf{A}$) to vectors $\mathbf{r}_2$ and $\mathbf{r}_3$. Is $\mathbf{r}_2$ necessarily conjugate to $\mathbf{r}_3$? Is $\mathbf{r}_1$ necessarily conjugate to $a_2 \mathbf{r}_2 + a_3 \mathbf{r}_3$ (where a_2 and a_3 are arbitrary real constants)?

6.17 A given quadratic function $f(\mathbf{x})$ is

$$f(\mathbf{x}) = f(\mathbf{x}_a) + (\mathbf{x} - \mathbf{x}_a)'\mathbf{b} + \tfrac{1}{2}(\mathbf{x} - \mathbf{x}_a)'\mathbf{A}(\mathbf{x} - \mathbf{x}_a)$$

For arbitrary $\mathbf{x}^k$ and $\mathbf{x}^{k+1}$, show that $\mathbf{g}^{k+1} - \mathbf{g}^k = \mathbf{A}(\mathbf{x}^{k+1} - \mathbf{x}^k)$.

6.18 Prove that a given direction vector $\mathbf{r}^k$ is orthogonal to the gradient $\mathbf{g}(\mathbf{x}^{k+1})$ of f if $\mathbf{x}^{k+1} = \mathbf{x}^k + \alpha \mathbf{r}^k$ where α is selected to minimize $f(\mathbf{x}^k + \alpha \mathbf{r}^k)$.

6.19 Consider an industrial process in which performance is characterized by

$$P = [(x_1 - x_1^*)\ (x_2 - x_2^*)]\begin{bmatrix} 2 & 1 \\ 1 & 2 \end{bmatrix}\begin{bmatrix} (x_1 - x_1^*) \\ (x_2 - x_2^*) \end{bmatrix}$$

where a minimum of P is desired, x_1 and x_2 represent angular positions of control shafts, and x_1^* and x_2^* are optimal values of x_1 and x_2, respectively. Over a period of time, x_1^* and x_2^* tend to drift because of various environmental factors, and x_1 and x_2 must be periodically adjusted to compensate for these changes in x_1^* and x_2^*. However, rather than adjusting x_1 and x_2 separately in a univariate search for the optimum, a gear drive is to be designed so that x_1 and x_2 are linearly related to two independent shafts characterized by y_1 and y_2; that is,

$$\begin{bmatrix} y_1 \\ y_2 \end{bmatrix} = \begin{bmatrix} w_{11} & w_{12} \\ w_{21} & w_{22} \end{bmatrix}\begin{bmatrix} x_1 \\ x_2 \end{bmatrix}$$

Find values of w_{ij}'s which reduce P to a noninteracting performance measure in terms of $y_1 - y_1^*$ and $y_2 - y_2^*$.

6.20 Find linearly independent vectors $\mathbf{r}_1$ and $\mathbf{r}_2$ which are A-conjugate to each other and to $\mathbf{r}_0$.

a. $\mathbf{r}_0' = [1\ \ 0\ \ 0]$

$$\mathbf{A} = \begin{bmatrix} 2 & 1 & 0 \\ 1 & 2 & 1 \\ 0 & 1 & 2 \end{bmatrix}$$

b. $\mathbf{r}_0' = [1\ \ 0\ \ 1]$

$$\mathbf{A} = \begin{bmatrix} 2 & 1 & 1 \\ 1 & 3 & 2 \\ 1 & 2 & 2 \end{bmatrix}$$

6.21 The search procedure of Example 6-4 is to be improved by use of the method of Fletcher and Reeves (Section 6-8b). Indicate appropriate changes in the flow diagram of Figure 6-12.

6.22 In regard to the running example of Examples 6-2 through 6-6, use starting point $\mathbf{x}^0 = [14\ \ 3]'$ and apply the following search techniques to find values of $\mathbf{x}^1$ and $\mathbf{x}^2$.

a. The method of Fletcher and Reeves (Section 6-8b).
b. The DFP method (Section 6-8c).

6.23 Use the inverse of Cauchy's procedure (Section 6-10a) to locate those real x's which result in the minimum of f.

a. $f = (x_1 + x_2 + 2)^2 + (x_1^2 - 2x_1x_2 + x_2^2)^2$
b. $f = (x_1 + 10x_1x_2 + 100x_1x_2x_3)^2 + (1 - x_1^2 + 50x_2^2 - 100x_3^2)^2$

6.24 Consider the equations

$$x_1^2 + 2x_1 + 3 + x_2 = 0$$

and

$$x_1 + x_2 + 1 = 0$$

Can Cauchy's approach of Section 6-10a be applied to simultaneous solution of these equations? Suggest a procedure with which nonreal roots of equations can be determined by use of search techniques.

6.25 Perform a dimensionality reduction (Section 6-10) as applicable to the extremization of $f(\mathbf{x})$.

a. $f(\mathbf{x}) = e^{-x_n} q_1(\mathbf{x}_r) + x_n q_2(\mathbf{x}_r) + q_3(\mathbf{x}_r), c_1 \le x_n \le c_2$
b. $f(\mathbf{x}) = \dfrac{1}{x_n} q_1(\mathbf{x}_r) + (\ln x_n) q_2(\mathbf{x}_r) + x_n q_3(\mathbf{x}_r), 0 \le c_1 \le x_n \le c_2$

6.26 Consider an equality constraint

$$f_1(\mathbf{x}) = c_1 + \alpha, \qquad -\epsilon \le \alpha \le \epsilon$$

for fixed ϵ. Suggest an inside penalty function to account for this constraint in a minimization problem.

6.27 A given function $f(\mathbf{x})$ represents a characteristic of a system where $\mathbf{x}^o$ is the design-center $\mathbf{x}$ vector. The actual values of the x_i parameters are known to be constrained by

$$|x_i - x_i^o| \le \epsilon_i \text{ (a specified constant)}, \qquad i = 1, 2, \ldots, n$$

Outline a procedure for the determination of the worst-case deviations in $f(\mathbf{x})$ from $f(\mathbf{x}^o)$; that is, the constrained maximum and minimum of $f(\mathbf{x})$ are to be found. Use functions of the form $1/(\epsilon_i - |x_i - x_i^o|)$ to generate an inside penalty function.

6.28 Consider extremization of

$$f_0(x_1, x_2) = x_2 - x_1^2$$

subject to satisfaction of the constraint

$$f_1(x_1, x_2) = x_1^2 + x_2^2 = c_1 \text{ (a positive constant)}$$

A Lagrange multiplier approach is to be applied. Use a representative set of values of the Lagrange multiplier h, $-\infty < h < \infty$, and determine the nature of the corresponding stationary points of the augmented function. What is the true constrained minimum of $f_0(x_1, x_2)$? What is the true constrained maximum of $f_0(x_1, x_2)$? (A problem of this type is considered by Forsythe [6.38].)

6.29 Given are functions $f_0(\mathbf{x})$ and $f_1(\mathbf{x})$: the domain R_I of interest is $\mathbf{x} \ge \mathbf{0}$; $f_0(\mathbf{x})$ is strictly concave over R_I and is a monotonic increasing function of each

x_i entry of $\mathbf{x}$; whereas $f_1(\mathbf{x})$ is convex over R_I and is a strictly monotonic increasing function of each x_i entry of $\mathbf{x}$. Given any real constant $c > f_1(\mathbf{0})$, prove the existence of a general Lagrange multiplier $h < 0$ such that $f_1[\mathbf{x}^*(h)] = c$, where $\mathbf{x}^*(h)$ yields a maximum of $f_0 + hf_1$.

With the existence of an appropriate Lagrange multiplier assured by your proof, the Lagrange multiplier approach of Section 6-11g can be used to find the maximum of $f_0(\mathbf{x})$, subject to the constraint that $f_1(\mathbf{x}) \leq c$.

6.30 Use the approach suggested in Problem 6.29 to find the maximum of

$$f_0(\mathbf{x}) = (1 - e^{-x_1}) + (1 - e^{-x_2})$$

subject to constraints

$$x_1 + 2x_2 \leq 3, \qquad x_1 \text{ and } x_2 \geq 0$$

6.31 Consider the circuit diagram of Figure 6-P31. The ratio $|V_2(j\omega)/V_1(j\omega)|$ is of interest at two given frequencies $\omega = \omega_1$ and $\omega = \omega_2$; $|V_2(j\omega_1)/V_1(j\omega_1)|$ is to be maximized subject to the constraint that $|V_2(j\omega_1)/V_1(j\omega_1)| = |V_2(j\omega_2)/V_1(j\omega_2)|$. Components R_1 and R_2 are fixed; C_1, C_2, L_1, and L_2 are to be selected to give the constrained optimum. Additional constraints are $L_1 \geq 0$, $L_2 \geq 0$, $C_1 \geq C_{1\,\text{min}}$, and $C_2 \geq C_{2\,\text{min}}$.

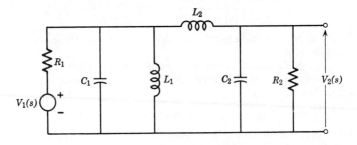

Figure 6-P31.

Express $|V_2(j\omega)/V_1(j\omega)|$ as a real-valued function, and outline a detailed procedure for computer solution of this problem. (*Suggestion:* Use a Lagrange multiplier procedure to account for the equality constraint, and inside penalty functions to account for the inequality constraints.)

6.32 Consider the circuit diagram of Figure 6-P32. Initial voltages and currents are assumed to be zero. The input voltage $v_1(t)$ is a rectangular pulse,

$$v_1(t) = u(t) - u(t - 10^{-5})$$

$R_1 = 100$ kilohms, $R_2 = 10$ kilohms, and $C = 100$ picofarads. The mutual inductance M is approximately $(L_1 L_2)^{1/2}$; positive values of L_1 and L_2 are to be specified to obtain a maximum of $v_2(10^{-5})$.

a. Obtain a set of differential equations that characterize the circuit of Figure 6-P32.
b. Suggest appropriate scale changes in the variables.

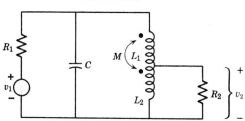

Figure 6-P32.

c. Suggest a search approach for obtaining a maximum of $v_2(10^{-5})$. Assume that you are to use a general-purpose digital computer to effect the search and the iterative solution of the differential equations.

6.33 Consider the block diagram of Figure 6-P33. With the system initially at rest, a step change in reference input $r(t)$ is applied at $t = 0$. Values of K_1, K_2, and a are to be set such that a minimum of $f(K_1, K_2, a)$ is obtained,

$$f(K_1, K_2, a) = \int_0^5 t|e(t)|\, dt$$

subject to constraints: $K_1 + (K_2/a) = 100$, $K_1 > 0$, $a > 0$, and the relationships indicated in Figure 6-P33. Outline a search procedure for solution of this problem.

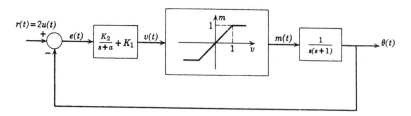

Figure 6-P33.

6.34 Figure 6-P34 depicts a maximum-seeking regulator. The signal y is an unregulated input which changes relatively slowly in comparison to the response time of the system at frequency ω_0. Explain how this system seeks out the peak of z.

6.35 Consider minimization of J:

$$J = \int_0^T \tfrac{1}{2}(x^2 + \dot{x}^2)\, dt$$

where $x(0) = 1$ and $x(T) = 1$ are specified. The optimal $x(t)$, $x^*(t)$, is known to satisfy the Euler-Lagrange equation

$$\ddot{x} - x = 0$$

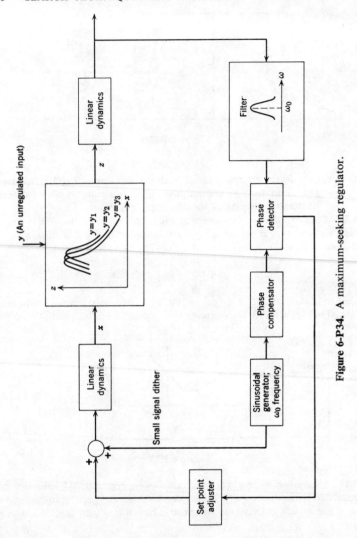

Figure 6-P34. A maximum-seeking regulator.

If $x^*(0) = 1$ and $\dot{x}^*(0) = c$, a constant, result in satisfaction of $x(T) = 1$, what value of $x(T)$ is obtained by use of $x(0) = 1$ and $\dot{x}(0) = c + \epsilon$? Under the conditions that ϵ represents an error of numerical approximation and T is relatively large, comment on the effectiveness of a search over initial values of $\dot{x}$ to obtain the desired terminal value of x.

A PRINCIPLE OF OPTIMALITY AND DYNAMIC PROGRAMMING

7

7-1. INTRODUCTION

If we were to order the many optimization theories on the basis of overall fertility, dynamic programming would, without doubt, be at the top of the list. It has evolved primarily on the basis of Richard Bellman's works on the subject (typified by references [7.4] through [7.22]) which date back to the early 1950's. The modifier "dynamic" of dynamic programming is particularly appropriate. It is suggestive of the fact that a given problem to be solved is placed in a dynamic framework; the problem is *imbedded* in a class of similar problems, the solutions of which are both logically related and scaled in degree of difficulty. The phrase *multistage decision process* can be associated with all problems which are solvable by using dynamic programming. The phrase implies that a problem which is amenable to dynamic programming can be viewed as a succession of decision problems, each one building on the last, until the problem is solved.

The variables associated with dynamic programming problems can be grouped into two categories: state variables and decision variables. State variables are difficult to describe in terms broad enough to cover the spectrum of problems to which dynamic programming can be applied, but, basically, they are a measure of the conditions that exist at the beginning of any stage of solution. Decision variables (which may also be referred to as control variables or policy variables) are directly specified to obtain an optimal solution at each stage of a dynamic programming solution.

Unlike linear programming, which is focused on the optimization of a linear algebraic performance measure subject to satisfaction of linear algebraic constraint equations, dynamic programming can be applied to a wide variety of different problems, both linear and nonlinear, but ones that can generally be characterized in terms of a Markovian process; a *Markovian process* has the property that after any number of decisions, say k, the effect of the remaining stages of the decision process upon the total return depend only upon the state of the system at the end of the kth decision and the subsequent decisions [7.12].

Unlike nonlinear programming, which is focused on the extremization of a nonlinear function subject to satisfaction of constraint equations that may contain nonlinear functions, dynamic programming is focused on an approach to solution. This approach to solution is associated with a *principle of optimality*, i.e., an optimal policy has the property that whatever the initial state and the initial decision are, the remaining decisions must constitute an optimal policy with regard to the state resulting from the first decision [7.12]. Statements of this principle, in the context of specific problem types, are given in Section 7-8.

The idea of building a solution by *imbedding* a problem in a class of similar problems is a very old one. Bellman and Kalaba [7.21] have investigated the interconnections between invariant imbedding and dynamic programming. Imbedding concepts have also been used in the study of continued fractions [7.94] and they play an important role in the development of rational fraction approximation techniques for certain classes of nonrational fraction functions [7.79]. The original aspect which appears in dynamic programming is the combined utilization of optimization theory and imbedding procedures to obtain *recurrence relations* which relate one optimal solution to the next one down the line.

The unfolding of dynamic programming in Sections 7-2 through 7-8 is pedagogically oriented; the prime reason for the particular problems and topics selected and their order of presentation is to enable the uninitiated to learn the basics of dynamic programming. Allocation problems are considered in Sections 7-2, 7-3, and 7-4. A class of minimal chain problems is considered in Section 7-5; with this type of problem, forward and backward dynamic programming algorithms are clearly distinguishable. The control problem of Section 7-6 incorporates features of both allocation problems and minimal chain problems. The common attribute of all these problems is summarized by way of Bellman's principle of optimality in Section 7-8. As with other programming procedures, numerical convergence and accuracy of dynamic programming algorithms must be considered (Section 7-7).

In the initial sections (Sections 7-2 through 7-9) of this chapter, problems with only one state variable are considered. Most problems with one state variable have multivariable counterparts, and two of these are examined in Section 7-10a. For one-state-variable problems of a general nature, clear-cut numerical procedures—hereafter referred to as *straightforward procedures* of dynamic programming—are employed in obtaining dynamic programming solutions. These straightforward methods of dynamic programming can be applied to two-state-variable problems without undue difficulty, and to three-state-variable problems in some cases; but for four or more state variables, additional computational aids (Sections 7-10, 7-11, and 7-12) are generally required. When the straightforward dynamic programming approach is employed, both the number of computations required and the amount of rapid-access storage space required increase in an exponential fashion as a function of the number n of state variables.

For some problems, the corresponding multistage decision process is infinite in extent—such problems are imbedded in an infinite number of problems. In these cases (e.g., the control problem of Section 7-11a), approximations in function space and policy space (Sections 7-11b and 7-11c) may be employed to obtain solutions. In contrast to the problem of Section 7-11, with its infinite number of discrete stages, the general problem of Section 7-12 is a continuous decision problem with a continuum of stages. This problem can be approximated by one with a finite number of discrete stages; and as shown in Section 7-12, the way in which such approximations are made can have considerable bearing on the computational facilities required for solution.

For continuous decision problems, dynamic programming theory is used in Sections 7-12c and 7-13a to obtain continuous recurrence relations. In turn, continuous recurrence relations are used in Section 7-13 to obtain fundamental results usually associated with classical calculus of variations (Chapter 3) and with Pontryagin's maximum principle (Chapter 8). The general applicability of these methods and of continuous recurrence relations is illustrated by the solutions obtained for the control problems of Section 7-14.

The problems solved by using dynamic programming in this chapter are only a representative sample of those which can be viewed as multistage decision problems. An appreciation for the diverse nature of problems which can be solved by dynamic programming may be obtained by comparing the transmission-line tower-placement problem of Section 7-9 and the stochastic decision problems of Sections 7-15 and 7-16. A listing of other problems, with references, is given in the conclusion of this chapter, Section 7-17.

7-2. ALLOCATION PROBLEMS

7-2a. Problem Statement and Applications

Consider a noninteracting performance measure P:[1]

$$P = \sum_{i=1}^{K} f_i(x_i) \qquad (7\text{-}1)$$

[1] The definition of a *noninteracting function* is given in Section 6-2c.

where K is a specified integer and where the set of x_i's is constrained by

$$\sum_{i=1}^{K} g_i(x_i) \le b \tag{7-2}$$

The functions $f_i(x_i)$ and $g_i(x_i)$ are given (either graphically or analytically) real-valued functions of real arguments, and b is a specified constant. A maximum of P is desired.

No assumptions in regard to the continuity of $f_i(x_i)$ and $g_i(x_i)$ need be made, although it will be shown (Section 7-7) that simplifications accrue if these functions are continuous. From the dynamic programming point of view, the computations are also simplified if the x_i's and $g_i(x_i)$'s can assume only non-negative integer values. This is in contrast to what we would expect from classical min-max theory which provides no efficient mechanism for treating integer programming problems.

The connection between the above problem statement and the word *allocation* is to be found in the following applications:

Application 1: A quantity b (dollars) is to be invested in K ventures; x_i is the number of dollars to be *allocated to* the ith venture which yields a return $f_i(x_i)$ in dollars. In this case, constraint 7-2 assumes the simple form $\sum_{i=1}^{K} x_i \le b$.

Application 2: A capacity b (cubic meters) is available on a cargo ship which is to transport goods to Gazue; x_i is the number of items of type i to be shipped with an anticipated profit $f_i(x_i)$. The volume of a single unit of the ith type is a_i (cubic meters) so constraint 7-2 assumes the form $\sum_{i=1}^{K} a_i x_i \le b$. Here the volume $a_i x_i$ is *allocated* to the ith item.

Application 3: A power consumption b (kilowatts) is associated with an industrial complex. There exist K power generating units; $f_i(x_i)$ is the cost per unit time of supplying x_i (kilowatts) from the ith unit to the industrial complex. In this case, the performance measure P is to be minimized by properly *allocating* the power generation among the units such that $\sum_{i=1}^{K} x_i = b$.

Application 4: Consider a system (simply an amplifier, perhaps) which is characterized by a function q (the gain of the amplifier) of K *independent* parameters $r_1, r_2, \ldots, r_K$ (resistors, active-element parameters, etc.). It is assumed that the design-center values $\mu_1, \mu_2, \ldots, \mu_K$ of the parameters have been selected in a design phase. In mass production, it is known that the parameters used in a given unit vary statistically from the design-center values—a *normal* probability density function for each parameter is assumed. The variance σ_i^2 of the density function for the ith parameter can be selected, within limitations, but the cost-per-parameter $f_i(\sigma_i^2)$ increases in some known fashion as σ_i^2 is decreased. Recall that the variance of a probability

Sect. 7-2   ALLOCATION PROBLEMS   371

density function is normally a measure of its "width." As a first-order approximation, the variance σ_s^2 of the system function q can be shown to equal the following (see Problem 7.4):

$$\sigma_s^2 = \sum_{i=1}^{K} a_i \sigma_i^2 \qquad (7\text{-}3)$$

where

$$a_i = (\partial q/\partial r_i)^2|_{r_i\text{'s}=\mu_i\text{'s}} \qquad (7\text{-}4)$$

Let b equal the maximum allowable variance of q and let $\sigma_i^2 \equiv x_i$. The objective then is to minimize the cost P,

$$P = \sum_{i=1}^{K} f_i(x_i) \qquad (7\text{-}1)$$

subject to the constraint that the variance σ_s^2 does not exceed b, i.e.,

$$\sum_{i=1}^{K} a_i x_i \leq b \qquad (7\text{-}5)$$

Additional details and extensions of the above applications and others are to be found in the literature [7.16, 7.46, 7.49, 7.72].

7-2b. Dynamic Programming Approach to Solution

Attention is now focused on the dynamic programming approach to solution of the allocation problem of 7-1 and 7-2 when P is to be maximized. Consider a slightly generalized form of constraint 7-2, namely,

$$\sum_{i=1}^{K} g_i(x_i) \leq \beta_K \leq b, \qquad x_i\text{'s} \geq 0 \qquad (7\text{-}6)$$

where β_K is the *resource variable* (also called a state variable) which is to be allocated. It is assumed that the x_i's are constrained to be non-negative and that the $g_i(x_i)$'s cannot assume negative values. Subject to satisfaction of 7-6, let $\mathscr{F}_K(\beta_K)$ denote the maximum of P in 7-1:

$$\mathscr{F}_K(\beta_K) = \underset{x_1,\ldots,x_K}{\text{maximum}} \left[f_K(x_K) + \sum_{i=1}^{K-1} f_i(x_i) \right] \qquad (7\text{-}7)$$

Note that $\mathscr{F}_K(\beta_K)$ is a function of β_K; $\mathscr{F}_K(\beta_K)$ is called a *maximum-return function*.[2] For the particular problem associated with 7-1 and 7-2, the value $\mathscr{F}_K(b)$ is to be determined.

[2] In the literature, the phrase "maximum-return function" is often abbreviated to "return function." Such functions are commonly called "minimum-cost functions" in cases where a minimum of P is desired.

The immediate goal is to relate the maximum-return function $\mathscr{F}_K(\beta_K)$ to the solution of an easier problem. To this end, let $\hat{x}_K(\beta_K)$ denote the *policy function* associated with x_K: for any given value of β_K, the corresponding value of $\hat{x}_K(\beta_K)$ is an optimal value of x_K. The particular policy $\hat{x}_K(b)$ will be denoted by $x_K{}^*$. With this notation, Equation 7-7 can be placed in the form

$$\mathscr{F}_K(\beta_K) = f_K[\hat{x}_K(\beta_K)] + \underset{x_1,\ldots,x_{K-1}}{\text{maximum}} \sum_{i=1}^{K-1} f_i(x_i) \tag{7-8}$$

in which maximization is constrained by

$$\sum_{i=1}^{K-1} g_i(x_i) \le \beta_K - g_K[\hat{x}_K(\beta_K)] \tag{7-9}$$

Note that the maximization in 7-8 is less involved than the maximization in 7-7 because one less variable is involved. Furthermore, constraint 7-9 is of the same form as the original constraint 7-6. Thus, the solution of the easier problem can be expressed in the same notational framework as used in the original problem, namely,

$$\mathscr{F}_{K-1}\{\beta_K - g_K[\hat{x}_K(\beta_K)]\} = \underset{x_1,\ldots,x_{K-1}}{\text{maximum}} \sum_{i=1}^{K-1} f_i(x_i) \tag{7-10}$$

in which maximization is over those x_i's that satisfy 7-9.

On the basis of 7-8 and 7-10, it follows that

$$\mathscr{F}_K(\beta_K) = f_K[\hat{x}_K(\beta_K)] + \mathscr{F}_{K-1}\{\beta_K - g_K[\hat{x}_K(\beta_K)]\} \tag{7-11a}$$

Equivalently,

$$\mathscr{F}_K(\beta_K) = \max_{x_K} \{f_K(x_K) + \mathscr{F}_{K-1}[\beta_K - g_K(x_K)]\} \tag{7-11b}$$

subject to $g_K(x_K) \le \beta_K$.

For the sake of completeness, let β_{K-1} be defined to equal $\beta_K - g_K(x_K)$. It follows that

$$\mathscr{F}_{K-1}(\beta_{K-1}) = \underset{x_1,\ldots,x_{K-1}}{\text{maximum}} \sum_{i=1}^{K-1} f_i(x_i) \tag{7-12a}$$

wherein the maximization process is constrained by

$$\sum_{i=1}^{K-1} g_i(x_i) \le \beta_{K-1} \tag{7-12b}$$

In considering the result of 7-11b, observe that the solution of the original problem—which is called a *K-stage problem* because K x_i's are involved—has

Sect. 7-2 ALLOCATION PROBLEMS 373

been related to the solution of a $(K-1)$-stage problem. The solution to the K-stage problem is relatively simple to obtain, provided that the $(K-1)$-stage problem has been solved. But in turn, the solution of the $(K-1)$-stage problem (as embodied in 7-12a and 7-12b) is related to that of a $(K-2)$-stage problem, and the character of the relationship is the same as that in Equation 7-11b. In general, this line of reasoning provides the following *recurrence relation*:

$$\mathscr{F}_k(\beta_k) = \max_{x_k} \{f_k(x_k) + \mathscr{F}_{k-1}[\beta_k - g_k(x_k)]\}, \qquad g_k(x_k) \leq \beta_k \quad (7\text{-}13)$$

for $k = 1, 2, \ldots, K$ with the convention that $\mathscr{F}_0(\beta_0) = 0$.

The first solution we obtain is $\mathscr{F}_1(\beta_1)$:

$$\mathscr{F}_1(\beta_1) = \max_{x_1} [f_1(x_1)]$$

subject to the constraint $g_1(x_1) \leq \beta_1$. With this and Equation 7-13, $\mathscr{F}_2(\beta_2)$ can be found, next $\mathscr{F}_3(\beta_3)$, and so on; a sequence of K one-stage *decision* problems replaces the original K-stage problem.

To illustrate the above points completely, consider the case in which

$$g_i(x_i) = a_i x_i \quad (7\text{-}14)$$

and assume that b and each a_i are specified positive integers. Assume also that each x_i is restricted to be a non-negative integer. In this case,

$$\mathscr{F}_1(\beta_1) = \max_{x_1} f_1(x_1), \qquad 0 \leq a_1 x_1 \leq \beta_1 \leq b, \quad x_1 \text{ integer}$$

or equivalently,

$$\mathscr{F}_1(\beta_1) = \max_{x_1} f_1(x_1), \qquad 0 \leq \text{integer } x_1 \leq [\beta_1/a_1] \quad (7\text{-}15)$$

where β_1 assumes values $0, 1, 2, \ldots, b$ and where $[\beta_1/a_1]$ denotes the largest integer which is less than or equal to β_1/a_1. The sequence of computations required to compute the function $\mathscr{F}_1(\beta_1)$ is as follows: (1) $\beta_1 = 0$ is specified initially; (2) with β_1 specified, the maximization in 7-15 is effected with respect to x_1; (3) the value (or values) of x_1 that yields the maximum is designated as $\hat{x}_1(\beta_1)$; (4) β_1 is incremented up by one; and (5) the procedure, starting with step 2 above, is repeated until $\beta_1 = b$. The maximum-return function $\mathscr{F}_1(\beta_1)$ for this first one-stage decision problem is tabulated along with those values of x_1, $\hat{x}_1(\beta_1)$, which correspond to $\mathscr{F}_1(\beta_1)$—see Table 7-1. Observe that $\hat{x}_1$ is a function of the resource β_1; it is the "policy" that would be adopted for x_1, given β_1, and for this reason, is called a policy function.

TABLE 7-1. TABULATED FORM OF A DYNAMIC PROGRAMMING SOLUTION WITH ONE RESOURCE (STATE) VARIABLE

$k \rightarrow$	1		2		$\cdots$	i		$\cdots$	K	
β_k	$\mathscr{F}_1(\beta_1)$	$\hat{x}_1(\beta_1)$	$\mathscr{F}_2(\beta_2)$	$\hat{x}_2(\beta_2)$	$\cdot$	$\mathscr{F}_i(\beta_i)$	$\hat{x}_i(\beta_i)$	$\cdot$	$\mathscr{F}_K(\beta_K)$	$\hat{x}_K(\beta_K)$
0	$\mathscr{F}_1(0)$	$\hat{x}_1(0)$	$\mathscr{F}_2(0)$	$\hat{x}_2(0)$	$\cdot$	$\mathscr{F}_i(0)$	$\hat{x}_i(0)$	$\cdot$	$\mathscr{F}_K(0)$	$\hat{x}_K(0)$
1	$\mathscr{F}_1(1)$	$\hat{x}_1(1)$	$\mathscr{F}_2(1)$	$\hat{x}_2(1)$	$\cdot$			$\cdot$	$\cdot$	$\cdot$
2	$\mathscr{F}_1(2)$	$\hat{x}_1(2)$	$\cdots$	$\cdots$	$\cdot$			$\cdot$	$\cdot$	$\cdot$
$\cdots$	$\cdots$	$\cdots$			$\cdot$	$\cdot$	$\cdot$	$\cdot$	$\cdots$	$\cdots$
b	$\mathscr{F}_1(b)$	$\hat{x}_1(b)$	$\cdot$	$\cdot$	$\cdot$	$\cdot$	$\cdot$	$\cdot$	$\mathscr{F}_K(b)$	$\hat{x}_K(b)$

The next solution, $\mathscr{F}_2(\beta_2)$ and $\hat{x}_2(\beta_2)$, is obtained by using the tabulated values of $\mathscr{F}_1(\beta_1)$ and the recurrence relation 7-13 which, for this case, assumes the form

$$\mathscr{F}_2(\beta_2) = \max_{x_2} [f_2(x_2) + \mathscr{F}_1(\beta_2 - a_2 x_2)], \qquad 0 \le x_2 \le [\beta_2/a_2]$$

This also is tabulated, and the process is continued in like manner,

$$\mathscr{F}_k(\beta_k) = \max_{x_k} [f_k(x_k) + \mathscr{F}_{k-1}(\beta_k - a_k x_k)], \qquad 0 \le x_k \le [\beta_k/a_k] \qquad (7\text{-}16)$$

until for $k = K$, only $\mathscr{F}_K(b)$ and $\hat{x}_K(b)$ need be generated to obtain the absolute maximum of the original problem.

To find an optimal policy, i.e., a set of x_i's that yields the absolute maximum, we need merely work backward through the tabulated values of $\hat{x}_i$'s. Assume for the present that the optimal policy (optimal set of x_i's) is unique. In this case, $\hat{x}_K(b)$ is unique and is the optimal value x_K^* of x_K. Having designated x_K^*, the optimal value β_{K-1}^* of β_{K-1} is found by using the relationship $\beta_{K-1}^* = \beta_K^* - a_K x_K^* = b - a_K x_K^*$, and the entry in the $\hat{x}_{K-1}(\beta_{K-1}^*)$ position of the table is x_{K-1}^*, the optimal value of x_{K-1}. In general, $\beta_{k-1}^* = \beta_k^* - a_k x_k^*$ with which x_{k-1}^* is equated to the entry in the $\hat{x}_{k-1}(\beta_{k-1}^*)$ position of the table.

If it should occur that two (or more) $\hat{x}_i$'s correspond to a particular β_i^*, then the optimal set of x_i's is not unique—either entry in the $\hat{x}_i(\beta_i^*)$ position may be designated as x_i^* to obtain *a* set of optimal x_i's.

Example 7-1. (This example illustrates the theory of Section 7-2. Certain particulars of the example deviate from those in the section, but these are easily resolved.)

A performance measure P is to be *minimized*:

$$P = \sum_{i=1}^{3} f_i(x_i), \qquad x_i \text{ integer} \qquad (7\text{-}17)$$

subject to the constraint that

$$\sum_{i=1}^{3} x_i = b, \qquad b \text{ integer} \qquad (7\text{-}18)$$

Solutions are to be obtained for all integer values of b in the range 1 through 15. The functions $f_i(x_i)$ are tabulated in Table 7-2.

To solve the problem via dynamic programming, the first order of business is to find the appropriate recurrence relations. The reader should verify that

TABLE 7-2. PIECEWISE LINEAR PERFORMANCE FUNCTIONS OF EXAMPLE 7-1

$f_i(x_i)$	$x_i = 0$	$1 \leq x_i \leq 5$	$5 \leq x_i \leq 15$
$f_1(x_1) =$	0	$5 + 10x_1$	$15 + 8x_1$
$f_2(x_2) =$	0	$2 + 8x_2$	$-18 + 12x_2$
$f_3(x_3) =$	0	$1 + 9x_3$	$-4 + 10x_3$

these are analogous to Equation 7-13 and that they assume the following form in this example:

$$\mathscr{F}_3(b) = \min_{x_3} [f_3(x_3) + \mathscr{F}_2(b - x_3)], \qquad 0 \leq x_3 \leq b \qquad (7\text{-}19)$$

$$\mathscr{F}_2(\beta_2) = \min_{x_2} [f_2(x_2) + \mathscr{F}_1(\beta_2 - x_2)], \qquad 0 \leq x_2 \leq \beta_2 \qquad (7\text{-}20)$$

TABLE 7-3. DYNAMIC PROGRAMMING SOLUTION TABLE FOR THE ALLOCATION EXAMPLE

$k \rightarrow$	1		2		3		
β_k	$\mathscr{F}_1(\beta_1)$	$\hat{x}_1(\beta_1)$	$\mathscr{F}_2(\beta_2)$	$\hat{x}_2(\beta_2)$	$\mathscr{F}_3(b)$	$\hat{x}_3(b)$	b
0	0	0	0	0	0	0	0
1	15	1	10	1	10	$\{0, 1\}$	1
2	25	2	18	2	18	0	2
3	35	3	26	3	26	0	3
4	45	4	34	4	34	0	4
5	55	5	42	5	42	0	5
6	63	6	54	6	52	1	6
7	71	7	66	7	61	2	7
8	79	8	77	5	70	3	8
9	87	9	87	$\{0, 5\}$	79	4	9
10	95	10	95	0	88	5	10
11	103	11	103	0	98	6	11
12	111	12	111	0	108	7	12
13	119	13	119	0	118	8	13
14	127	14	127	0	127	0	14
15	135	15	135	0	135	0	15

and

$$\mathcal{F}_1(\beta_1) = \min_{x_1} f_1(x_1), \qquad x_1 = \beta_1$$

$$= f_1(\beta_1) \tag{7-21}$$

Note that in this last relationship, x_1 equals β_1, rather than $x_1 \leq \beta_1$ as in the related case of Equation 7-15. The reason for this is that the sum of the x_i's must equal b in this example, that is, $x_1 = (b - x_3) - x_2$ which equals $\beta_2 - x_2$ which equals β_1 by definition.

Table 7-3 contains the results generated by using Equations 7-21, 7-20, and 7-19. For example, consider determination of the minimum cost $\mathcal{F}_3(7)$ and the policy $\hat{x}_3(7)$ by the use of Equation 7-19:

$$\mathcal{F}_3(7) = \min_{x_3} [f_3(x_3) + \mathcal{F}_2(7 - x_3)], \qquad 0 \leq x_3 \leq 7$$

$$= \min \{66, 64, 61, 62, 63, 64, 66, 66\} = 61$$

The minimum of 61 for $b = 7$ corresponds to $x_3 = \hat{x}_3(7) = x_3^* = 2$, $\beta_2^* = 7 - x_3^* = 7 - 2 = 5$, $\hat{x}_2(5) = x_2^* = 5$, $\beta_1^* = \beta_2^* - x_2^* = 5 - 5 = 0$, and $\hat{x}_1(0) = x_1^* = 0$. A summary of optimal values for other values of b is given in Table 7-4. The reader is urged to verify the results in Tables 7-3 and 7-4.

TABLE 7-4. RESULTS OF THE ALLOCATION EXAMPLE

b	$P_{\min}$	x_3^*	x_2^*	x_1^*
1	10	$\{0, 1\}$	$\{1, 0\}$	0
2	18	0	2	0
3	26	0	3	0
4	34	0	4	0
5	42	0	5	0
6	52	1	5	0
7	61	2	5	0
8	70	3	5	0
9	79	4	5	0
10	88	5	5	0
11	98	6	5	0
12	108	7	5	0
13	118	8	5	0
14	127	0	0	14
15	135	0	0	15

7-3. EFFICIENCY COMPARISON

A comparison of the above approach and that of factorial search (Section 6-4b) is in order. To simplify the comparison, assume that $a_i = 1$ for all i and that $\sum_{i=1}^{K} x_i$ must equal the integer b rather than be less than or equal to b. Of interest is the number N_1 of different integer combinations of the x_i which satisfy

$$\sum_{i=1}^{K} x_i = b, \qquad x_i \geq 0 \text{ for all } i \tag{7-22}$$

This number N_1 is the number of integer grid points which would necessarily be examined in a factorial search for the maximum of P. Observe from Equation 7-22 that N_1 equals the number of ways of allocating b identical objects among K different activities, that is,

$$N_1 = \frac{(K + b - 1)!}{b!(K - 1)!} \tag{7-23}$$

For $K = 10$ and $b = 20$, $N_1 = 10{,}015{,}005$.

On the other hand, with the dynamic programming approach, $\mathscr{F}_1(\beta_1)$ is found on the basis of $b + 1$ evaluations of $f_1(x_1)$. For a particular value of β_k, with $k \neq 1$, $\mathscr{F}_k(\beta_k)$ is determined on the basis of $\beta_k + 1$ evaluations, so the column $\mathscr{F}_k(\beta_k)$ is generated on the basis of the following number of evaluations:

$$\sum_{\beta_k=0}^{b} (\beta_k + 1) = \left(1 + \frac{b}{2}\right)(1 + b)$$

In the final column, only $\mathscr{F}_K(b)$ is required and it is obtained on the basis of $b + 1$ evaluations. Thus, the dynamic programming solution of this problem requires N_2 function evaluations in total, where

$$N_2 = (1 + b) + (K - 2)(1 + b)\left(1 + \frac{b}{2}\right) + (1 + b)$$

$$= (1 + b)\left[K\left(1 + \frac{b}{2}\right) - b\right] \tag{7-24}$$

Again, in the special case that $K = 10$ and $b = 20$, $N_2 = 1{,}890$ which is to be compared with the much larger value of function evaluations $N_1 = 10{,}015{,}005$ which are required in the related factorial search.

7-4. REDUNDANCY TO IMPROVE RELIABILITY [7.14]

The recurrence relation 7-13 is but one type, albeit a very common type, of those that arise in dynamic programming solutions. The problem of this section results in another.

Sect. 7-4 REDUNDANCY TO IMPROVE RELIABILITY

Consider a system that contains K critical subunits. If any one of these subunits fails, the system fails. The subunits are assumed independent with p_j being the known probability that the jth subunit will fail during the specified operational lifetime T of the system. It is assumed that spare subunits can be incorporated in the system in the following way: when the jth subunit fails, a spare of type j automatically replaces it in the system, provided all such spares have not been used. If x_j is the number of spare subunits of type j, the cost of the system is *increased* by an amount $g_j(x_j)$, where $g_j(0) = 0$.

It is further assumed that the probability of failure of a subunit of type j is p_j regardless of whether it is initially in the system or is used as a spare. The probability that *all* $1 + x_j$ of the jth subunits will fail is then $p_j^{1+x_j}$. It follows that unity minus this probability is the probability, call it $f_j(x_j)$, of having *at least one* of the jth subunits operational throughout T, i.e.,

$$f_j(x_j) = 1 - p_j^{1+x_j} \tag{7-25}$$

The system remains operational only if one or more of each subunit is functional. Thus the probability P that the system remains operational is

$$P = \prod_{j=1}^{K} f_j(x_j) \tag{7-26}$$

If an unlimited number of spares could be used, P could be made arbitrarily close to unity. However, it is given that the added cost of the spares cannot exceed b (dollars) if the system is to be economically feasible. Thus,

$$\sum_{j=1}^{K} g_j(x_j) \le b \tag{7-27}$$

As in the preceding section, let the maximum-return function $\mathscr{F}_K(\beta_K)$ equal the absolute maximum of P when

$$\sum_{j=1}^{K} g_j(x_j) \le \beta_K \le b \tag{7-28}$$

That is,

$$\begin{aligned} \mathscr{F}_K(\beta_K) &= \operatorname*{maximum}_{x_1,\ldots,x_K} \left[f_K(x_K) \cdot \prod_{j=1}^{K-1} f_j(x_j) \right] \\ &= \max_{x_K} \left[f_K(x_K) \operatorname*{maximum}_{x_1,\ldots,x_{K-1}} \prod_{j=1}^{K-1} f_j(x_j) \right] \\ &= \max_{x_K} \{ f_K(x_K) \mathscr{F}_{K-1}[\beta_K - g_K(x_K)] \} \end{aligned} \tag{7-29}$$

in which

$$\beta_K - g_K(x_K) = \beta_{K-1} \geq \sum_{j=1}^{K-1} g_j(x_j) \qquad (7\text{-}30)$$

The reader may wish to supply the verbal description, parallel to that given between Equations 7-7 and 7-12, which accompanies the steps given in Equation 7-29.

In general, reasoning analogous to that given above renders the following recurrence relation:

$$\mathscr{F}_k(\beta_k) = \max_{x_k} \{f_k(x_k)\mathscr{F}_{k-1}[\beta_k - g_k(x_k)]\} \qquad (7\text{-}31)$$

for $k = 1, 2, \ldots, K$. For this problem, $\mathscr{F}_0(\beta_0) \triangleq 1$ is used in the determination of $\mathscr{F}_1(\beta_1)$ which is used in evaluation of $\mathscr{F}_2(\beta_2)$, and so forth. As in the preceding two sections, maximization is effected with x_k in the interval defined by $0 \leq g_k(x_k) \leq \beta_k \leq b$. The mechanics of the solution are identical to those given in Section 7–2.

7-5. MINIMAL CHAIN PROBLEMS

7-5a. Chain Networks

The minimal chain problem is introduced in Section 5-11a. It falls under the heading of "graph" problems and, in particular cases, it may be referred to as a minimal time problem or a minimal path problem. An example (Example 7-2 which is interwoven throughout this section) serves to illustrate the dynamic programming approach to solution of oriented minimal chain problems. Different solutions for the example are obtained for several different *end* conditions. These solutions are differentiated by their *forward* or *backward* character—whereas the concepts of forward dynamic programming solutions and backward dynamic programming solutions have no significance for the allocation problems of the preceding sections, these concepts are of prime importance in regard to the control problems of the next.

Consider the chain network shown in Figure 7-1 which consists of *nodes* and *links* (or branches) between nodes. The *length* assigned to each link could represent, say, a normalized cost of sending a message between nodes. A minimum-length connected chain (or simply stated, a minimal chain) between $x_2 = 0$ and $x_2 = 6$ is desired. (Note that here x_1 and x_2 are not simply parameters to be selected in an optimal manner.) It is assumed that the minimal chain between $x_2 = 0$ and $x_2 = 6$ is *oriented* in a special way: It is assumed that x_1 is a single-valued function of x_2 along the minimal

Sect. 7-5   MINIMAL CHAIN PROBLEMS   381

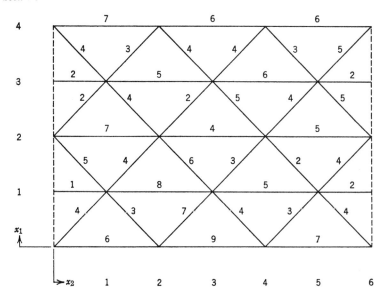

Figure 7-1. A chain network with specified link lengths.

chain; the standard notation $x_1(x_2)$ is used to express the relationship between x_1 and x_2 along such a connected chain. The reason that this single-valued assumption is made is to enable the problem to be viewed as a sequence of one-stage problems to which dynamic programming applies. If the above assumption is not satisfied, the results obtained by the specific approach given in this section correspond to a false minimal chain (but the best one of those which exhibit the single-valued-function property). A basic method for solving nonoriented minimal chain problems is given in Section 7-11d.

Several solutions are obtained in this section; these correspond to different conditions imposed on $x_1(0)$ and $x_1(6)$.

7-5b. Forward Solution I

Forward solution I applies to the case where no restrictions are imposed on the value of x_1 at $x_2 = 0$; $x_1(0)$ is to be selected in an optimal manner; this is appropriately designated as a "free" end condition. On the other hand, it is specified that a set of minimal chains must be found, one minimal chain for each of the "fixed" integer values of $x_1(6)$ in the interval $0 \leq x_1(6) \leq 4$.

Of course, by comparing the solutions obtained, an absolute minimal chain can be found between $x_2 = 0$ and $x_2 = 6$.

The stages of forward solution I are shown in Figure 7-2. Let the minimum-length function $\mathscr{F}_k(\beta_k)$ denote the minimal chain length between $x_2 = 0$ and $x_2 = k$ given only that $x_1(k) = \beta_k$. The letter $\mathscr{F}$ used here is indicative of the *forward* manner in which the solution will be generated, forward in the sense of starting with stage 1 at the left and moving from left to right as stages of the problem are progressively solved.

In the first place, consider node (1,1) of stage 1. *Question:* What is the shortest link of those that tie to node (1,1) from the left? The answer is obviously the link from (1,0) to (1,1) with length 1. These facts are recorded at node (1,1) by placing a "1" in the circle above the node, indicating the minimal chain length $\mathscr{F}_1(1)$ to node (1,1), and by placing a "$\wedge$" on the link that results in $\mathscr{F}_1(1) = 1$. Similarly, the value $\mathscr{F}_1(3) = 2$ is placed in the circle above node (3,1); and $\wedge$'s are placed on the links which tie (2,0) to (3,1) and (3,0) to (3,1), either of which result in $\mathscr{F}_1(3) = 2$. Note that $\mathscr{F}_1(0)$, $\mathscr{F}_1(2)$, and $\mathscr{F}_1(4)$ do not exist for the given problem.

For the second stage,

$$\mathscr{F}_2(0) = \min\{6, \mathscr{F}_1(1) + 3\} = \min\{6, 4\} = 4$$
$$\mathscr{F}_2(2) = \min\{\mathscr{F}_1(1) + 4, 7, \mathscr{F}_1(3) + 4\} = \min\{5, 7, 6\} = 5 \quad (7\text{-}32)$$

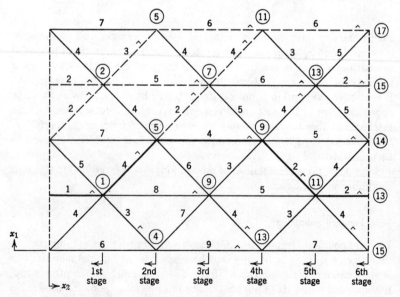

Figure 7-2. Forward dynamic programming solution with free initial condition.

and

$$\mathscr{F}_2(4) = \min\{\mathscr{F}_1(3) + 3, 7\} = \min\{5, 7\} = 5 \tag{7-33}$$

Observe that the results of stage 1 are conveniently used in the above tabulation of results for stage 2; but in addition, use is made of those link lengths that tie directly from $x_2 = 0$ to $x_2 = 2$. In general, it is to be observed that both $\mathscr{F}_{k-1}(\beta_{k-1})$ and $\mathscr{F}_k(\beta_k)$ are of use in the evaluation of the minimum-length function $\mathscr{F}_{k+1}(\beta_{k+1})$ in this problem, e.g., for $\mathscr{F}_3(3)$,

$$\begin{aligned}\mathscr{F}_3(3) &= \min\{\mathscr{F}_2(2) + 2, \mathscr{F}_1(3) + 5, \mathscr{F}_2(4) + 4\} \\ &= \min\{7, 7, 9\} = 7\end{aligned} \tag{7-34}$$

The computations are continued in this manner with the final result that $\mathscr{F}_6(0) = 15$, $\mathscr{F}_6(1) = 13$, $\mathscr{F}_6(2) = 14$, $\mathscr{F}_6(3) = 15$, and $\mathscr{F}_6(4) = 17$. The absolute minimal chain corresponds to $\mathscr{F}_6(1) = 13$.

To trace out the links of the minimal chain corresponding to $\mathscr{F}_6(1)$, we start at node (1,6) and follow the path which is marked by $\wedge$'ed links to the left: at each node along the path, there exists at least one (in this case, only one) $\wedge$'ed link that ties to the node from the left, and this link is part of a minimal chain. Note that several minimal chains may correspond to a given value of $x_1(6)$; e.g., the dashed lines in Figure 7-2 afford five different minimal chains corresponding to $\mathscr{F}_6(4) = 17$.

7-5c. Backward Solution I

Backward solution I applies to the case where no restrictions are imposed on the value of x_1 at $x_2 = 6$, that is, $x_1(6)$ is free to be selected in an optimal manner. On the other hand, it is specified that a set of minimal chains must be found, one minimal chain for each of the "fixed" integer values of $x_1(0)$ in the interval $0 \leq x_1(0) \leq 4$.

For backward solution I, the notation $\mathscr{B}_k(\beta_{K-k})$ is used to designate the minimal chain length of those chains which start at node $(\beta_{K-k}, K - k)$ and terminate at any node where $x_2 = K$. (In the running example, $K = 6$.) In the first place, minimum-length function $\mathscr{B}_1(\beta_{K-1})$ is obtained as a result of the *first* stage of a K-stage (backward) decision process. For the chain network of Figure 7-1, $\mathscr{B}_1(\beta_5)$ is used in the evaluation of $\mathscr{B}_2(\beta_4)$ which is used in turn in the evaluation of $\mathscr{B}_3(\beta_3)$ and so forth until $\mathscr{B}_6(\beta_0)$ is found. It is left to the reader to develop the iterative backward solution of this particular problem. (*Hint:* Invert the book and use the algorithm of Forward Solution I!) The solution is given in Figure 7-3.

7-5d. Backward Solution II

As the name implies, backward solution II is closely related to backward solution I; the designation of these as being "I" or "II" is artificial—the

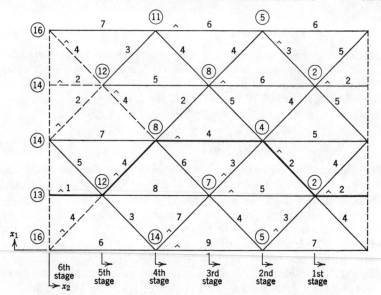

Figure 7-3. Backward dynamic programming solution with free terminal condition.

important common denominator is the phrase "backward solution." As with backward solution I, a minimal chain is to be found for each "fixed" value of $x_1(0)$. In this case, however, the right-hand node of the chain is restricted to be an element of a given proper subset of the range of values of $x_1(6)$. The particular case where this proper subset consists of a single node, here node (0,6), is treated in detail in the following paragraphs.

If the right-hand node *must be* (0,6), then only those nodes that tie to node (0,6) from $x_2 = 5$ need be considered in the first stage of this backward solution. Only node (1,5) satisfies this condition, and therefore $\mathscr{B}_1(\beta_5) = \mathscr{B}_1(1) = 4$ which is the minimal length (the only length) of chain link from (1,5) to (0,6). For the second stage, $\mathscr{B}_2(0)$ and $\mathscr{B}_2(2)$, but not $\mathscr{B}_2(4)$, need to be evaluated. $\mathscr{B}_2(4)$ does not exist because there is no allowed path (single-valued path) from node (4,4) to the node (0,6). Thus,

$$\mathscr{B}_2(0) = \min\{7, 3 + \mathscr{B}_1(1)\} = \min\{7, 7\} = 7 \qquad (7\text{-}35)$$

and

$$\mathscr{B}_2(2) = \min\{2 + \mathscr{B}_1(1)\} = \min\{6\} = 6 \qquad (7\text{-}36)$$

Sect. 7-5 MINIMAL CHAIN PROBLEMS 385

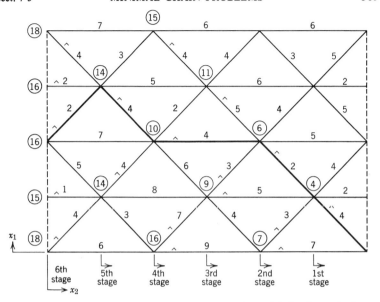

Figure 7-4. Backward dynamic programming solution with fixed terminal condition.

These values and the corresponding links are recorded, with the previously established conventions, in Figure 7-4. In like manner,

$$\mathscr{B}_3(1) = \min \{4 + \mathscr{B}_2(0), 5 + \mathscr{B}_1(1), 3 + \mathscr{B}_2(2)\} = 9 \quad (7\text{-}37)$$

and

$$\mathscr{B}_3(3) = \min \{5 + \mathscr{B}_2(2)\} = 11 \quad (7\text{-}38)$$

The process is continued iteratively to the end that $\mathscr{B}_6(0) = 18$, $\mathscr{B}_6(1) = 15$, $\mathscr{B}_6(2) = 16$, $\mathscr{B}_6(3) = 16$, and $\mathscr{B}_6(4) = 18$. The minimal chain shown in Figure 7-4 corresponds to the fixed end conditions (2,0) and (0,6). It is obtained by starting at node (2,0) and proceeding forward over the $\wedge$'ed path as shown.

7-5e. Comparison of Forward and Backward Solutions

For problems of the type posed in Section 7-5a, and for those that are presented in the next section, there are basically two types of dynamic programming approaches to solution: the backward solution and the forward solution. The same type of algorithm is used with both approaches; the choice of one over the other is based both on the information given concerning initial and terminal conditions and on the sensitivity information that one would like to extract from the solution.

For the running example of this section, the value of $x_1(0)$ for a given chain is termed the *initial* value of the chain. Similarly, the value of $x_1(6)$ for a given chain is called the *terminal* value of the chain. Suppose that both the initial value and the terminal value of the "minimal" chain are specified constants. In this case, either backward solution II or forward solution II—the counterpart of forward solution I—can be generated with equal ease. Why then should one approach be preferred over the other? The answer to this is that certain sensitivity information may be obtained from one solution, but not the other. For example, if minimal chains corresponding to different, but fixed, initial values are desired, the backward approach should be used because it supplies this information directly. If we were to use the forward approach in such a case, the problem would have to be solved anew to determine the effect of any change in the assumed initial condition. On the other hand, if minimal chains corresponding to different, but fixed, terminal values are desired, the forward approach should be used because it supplies the requisite information directly, whereas the backward solution does not.

For certain types of problems, the foregoing statements are inappropriate. Only the backward solution procedure is applicable to a class of on-line control problems (e.g., that of Section 7-11) and to a class of stochastic optimization problems (e.g., that of Section 7-15). The reasons for this are made clear in subsequent sections.

7-6. A CONTROL PROBLEM

7-6a. Statement of the Problem

Consider a sampled-data control system which is characterized (perhaps only approximately) by the first-order difference equation

$$x(k+1) = q[x(k),m(k),k], \qquad k = 0, 1, 2, \ldots, K-1 \qquad (7\text{-}39)$$

where $q[x(k),m(k),k]$ is a specified real- and single-valued function of the state variable $x(k)$, the control variable (or decision variable or policy variable) $m(k)$, and the integer k which is indicative of, but not necessarily directly proportional to, time. For notational convenience, let $x(k) = x_k$ and $m(k) = m_k$. State equation 7-39 is then of the form

$$x_{k+1} = q(x_k,m_k,k), \qquad k = 0, 1, 2, \ldots, K-1 \qquad (7\text{-}40)$$

It is specified that the control variable m_k is constrained to be an element of a given set U_k; e.g., the set U_k could consist simply of the closed interval defined by $|m_k| \leq m_{\max}$, where $m_{\max}$ is a specified constant. With these restrictions in mind, the m_k's are to be selected to minimize the performance measure P:

$$P = \sum_{i=1}^{K} f_i(x_i, m_{i-1}) \tag{7-41}$$

where $f_i(x_i, m_{i-1})$ is a specified real- and single-valued function of its arguments. An optimal set of m_k's is called an optimal policy.

7-6b. Backward Solution

As observed in Section 7-5, backward dynamic programming solutions allow one to ascertain the effect of changes in initial condition x_0 with relative ease, but any change in the assumed terminal condition x_K requires that a backward solution be completely rerun. Two backward solutions are developed in the following paragraphs: In one of the solutions, the terminal state x_K is selected in an optimal manner; and in the other, the terminal state is assumed to be fixed in advance.

For both solutions, let the minimum-cost function $\mathcal{B}_K(\beta_0)$ denote the constrained minimum of P when $x_0 = \beta_0$ is an initial state:

$$\mathcal{B}_K(\beta_0) = \min \sum_{k=1}^{K} f_k(x_k, m_{k-1}) \tag{7-42}$$

in which minimization is effected with respect to constrained m_k's,

$$m_k \in U_k, \quad k = 0, 1, \ldots, K-1 \tag{7-43}$$

and where

$$x_{k+1} = q(x_k, m_k, k), \quad k = 0, 1, \ldots, K-1 \tag{7-44}$$

with $x_0 = \beta_0$.

Observe that x_1 is a function of β_0 and m_0 only:

$$x_1 = q(\beta_0, m_0, 0) \tag{7-45}$$

This fact allows the minimization in Equation 7-42 to be ordered as follows:

$$\mathcal{B}_K(\beta_0) = \underset{m_0 \in U_0}{\text{minimum}} \left\{ f_1[q(\beta_0, m_0, 0), m_0] + \underset{\substack{m_k's \in U_k's \\ k \neq 0}}{\text{minimum}} \sum_{k=2}^{K} f_k(x_k, m_{k-1}) \right\} \tag{7-46}$$

A parallel is to be observed between the above step and the corresponding one in Section 7-2—see the material between 7-6 and 7-12. The parallel continues through the remainder of this section.

Let the minimum within the braces of 7-46 be denoted by $\mathscr{B}_{K-1}(\beta_1)$:

$$\mathscr{B}_{K-1}(\beta_1) = \underset{\substack{m_k's \in U_k's \\ k \neq 0}}{\text{minimum}} \sum_{k=2}^{K} f_k(x_k, m_{k-1}) \qquad (7\text{-}47)$$

where $\beta_1 = x_1$ assumes various values, but is held fixed at a value of interest each time minimization is effected in 7-47. Thus, $\mathscr{B}_{K-1}(\beta_1)$ is truly a function of β_1 and is the result of a $(K-1)$-stage decision process involving the selection of the $(K-1)$ control variables $m_1, m_2, \ldots, m_{K-1}$, where

$$x_{k+1} = q(x_k, m_k, k), \qquad k = 1, 2, \ldots, K - 1 \qquad (7\text{-}48)$$

with $x_1 = \beta_1$.

Suppose that the $(K-1)$-stage problem of determining $\mathscr{B}_{K-1}(\beta_1)$ has been solved for numerous values of β_1 and that these values of $\mathscr{B}_{K-1}(\beta_1)$ have been tabulated. To relate these solutions to $\mathscr{B}_K(\beta_0)$, recall that

$$\beta_1 = q(\beta_0, m_0, 0) \qquad (7\text{-}49)$$

which follows from Equation 7-44. This fact and Equations 7-46 and 7-47 are used to obtain the desired relationship which follows.

$$\mathscr{B}_K(\beta_0) = \underset{m_0 \in U_0}{\text{minimum}} \{f_1[q(\beta_0, m_0, 0), m_0] + \mathscr{B}_{K-1}[q(\beta_0, m_0, 0)]\} \qquad (7\text{-}50)$$

Now, the $(K-1)$-stage problem associated with Equations 7-47 and 7-48 is of exactly the same form as that of the original K-stage problem, so it too can be related to a simpler problem, a $(K-2)$-stage problem. In general, this reasoning gives the following recurrence relation which relates the k-stage problem to the $(k-1)$-stage problem:

$$\mathscr{B}_k(\beta_{K-k}) = \underset{m_{K-k} \in U_{K-k}}{\text{minimum}} \{f_{K-k+1}[q(\beta_{K-k}, m_{K-k}, K-k), m_{K-k}]$$
$$+ \mathscr{B}_{k-1}[q(\beta_{K-k}, m_{K-k}, K-k)]\} \qquad (7\text{-}51)$$

for $k = 2, 3, \ldots, K$. The case where $k = 1$ requires special treatment, for it is this case that must be solved first and is the one in which the terminal condition is incorporated.

For the one-stage problem,

$$\mathscr{B}_1(\beta_{K-1}) = \underset{m_{K-1} \in U_{K-1}}{\text{minimum}} f_K(x_K, m_{K-1}) \qquad (7\text{-}52)$$

in which account must be made of the fact that

$$x_K = q(\beta_{K-1}, m_{K-1}, K - 1) \qquad (7\text{-}53)$$

Two common terminal conditions are considered next.

Sect. 7-6 A CONTROL PROBLEM

Terminal Condition I. Assume that no restrictions are imposed on x_K; x_K is free to be selected in an optimal manner. In this case, the right-hand member of Equation 7-53 is substituted for x_K in Equation 7-52, and the following result is generated for a set of values of β_{K-1}:

$$\mathcal{B}_1(\beta_{K-1}) = \underset{m_{K-1} \in U_{K-1}}{\text{minimum}} f_K[q(\beta_{K-1}, m_{K-1}, K-1), m_{K-1}] \quad (7\text{-}54)$$

Terminal Condition II. In this case, assume that x_K must equal a given constant c_1; x_K is fixed and cannot be selected in the optimization process. This fact must be accounted for in Equations 7-52 and 7-53 with the result that

$$\mathcal{B}_1(\beta_{K-1}) = \underset{m_{K-1} \in U_{K-1}}{\text{minimum}} f_K(c_1, m_{K-1}) \quad (7\text{-}55)$$

in which the minimization is constrained by the relationship

$$c_1 = q(\beta_{K-1}, m_{K-1}, K-1), \qquad m_{K-1} \in U_{K-1} \quad (7\text{-}56)$$

For a given value of β_{K-1}, say, $\beta_{K-1} = c_2$, one of the following statements can be made on the basis of Equation 7-56: (1) It may be that no m_{K-1} which is an element of the constraint set U_{K-1} satisfies Equation 7-56, in which case the value c_2 of β_{K-1} cannot possibly be the value of x_{K-1} in an optimal solution which satisfies the given boundary condition, and therefore $\beta_{K-1} = c_2$ need not be considered in the dynamic programming solution. (2) It may be that a unique value of $m_{K-1} \in U_{K-1}$ satisfies Equation 7-56 for $\beta_{K-1} = c_2$ in which case that value, and only that value, is used to obtain the value of $\mathcal{B}_1(c_2)$ in Equation 7-55. (3) It may be that there are multiple solutions for m_{K-1} in Equation 7-56, in which case minimization is effected in Equation 7-55 over these allowed values of m_{K-1}. If $q(x_{K-1}, m_{K-1}, K-1)$ is linear in m_{K-1}, case 3 above cannot possibly apply.

Other terminal conditions could be imposed (see Problem 7.13) and could be incorporated in the backward solution in much the same way as is done in this section.

As with previous dynamic programming solutions of this chapter, the first step in the actual solution is to obtain and store the results of the one-stage problem, which in this case means values of the minimum-cost function $\mathcal{B}_1(\beta_{K-1})$ and corresponding values of the policy function $\hat{m}_{K-1}(\beta_{K-1})$ for the control variable m_{K-1}. Equation 7-51 is then used recursively to obtain $\{\mathcal{B}_2(\beta_{K-2}), \hat{m}_{K-2}(\beta_{K-2})\}, \{\mathcal{B}_3(\beta_{K-3}), \hat{m}_{K-3}(\beta_{K-3})\}, \ldots, \{\mathcal{B}_K(\beta_0), \hat{m}_0(\beta_0)\}$. If x_0 is a specified constant, then only the corresponding values of $\mathcal{B}_K(x_0)$ and $\hat{m}_0(x_0) = m_0^*$ need be tabulated in the final two columns of the dynamic programming solution.

In principle, the method of generating a dynamic programming table, and of then working backward through the table to find the optimal control policy, is the same for this control problem as for the problems discussed in Sections 7-2 and 7-4. In point of fact, however, values of the state variable and the policy variable in this control problem are from the set of real numbers, whereas only integer values have been considered in the preceding sections. This gives rise to some inconvenience because interpolation must be used both to generate the values of $\mathcal{B}_k(\beta_{K-k})$ and to find the optimal controlling actions after the table is generated. These points are given consideration in Example 7-3 and in Section 7-7.

7-6c. Forward Solutions

A forward solution algorithm for the control problem of Section 7-6a is given below. The reader is invited to derive the algorithm in detail (Problem 7.14).

Let $\mathcal{F}_K(\beta_K)$ be the minimum of P (Equation 7-41), given that $x_K = \beta_K$. An additional assumption to be used here is that the state equation $x_k = q(x_{k-1}, m_{k-1}, k-1)$ can be rearranged to obtain x_{k-1} as an explicit single-valued function v of x_k, m_{k-1}, and $k-1$:

$$x_{k-1} = v(x_k, m_{k-1}, k-1) \tag{7-57}$$

With this, it can be shown that

$$\mathcal{F}_K(\beta_K) = \underset{m_{K-1} \in U_{K-1}}{\text{minimum}} \{f_K(\beta_K, m_{K-1}) + \mathcal{F}_{K-1}[v(\beta_K, m_{K-1}, K-1)]\} \tag{7-58}$$

And in general, for $k = 2, 3, \ldots, K$,

$$\mathcal{F}_k(\beta_k) = \underset{m_{k-1} \in U_{k-1}}{\text{minimum}} \{f_k(\beta_k, m_{k-1}) + \mathcal{F}_{k-1}[v(\beta_k, m_{k-1}, k-1)]\} \tag{7-59}$$

To evaluate $\mathcal{F}_1(\beta_1)$, assumptions in regard to initial conditions must be made.

Initial Condition I. If it is specified that x_0 is free to be selected in an optimal manner, it can be shown that

$$\mathcal{F}_1(\beta_1) = \underset{m_0 \in U_0}{\text{minimum}} f_1(\beta_1, m_0) \tag{7-60}$$

Initial Condition II. If it is specified that x_0 is a constant value c_1, independent of the optimization process, it can be shown that

$$\mathcal{F}_1(\beta_1) = \text{minimum} f_1(\beta_1, m_0) \tag{7-61}$$

in which the minimization is constrained by the relationship

$$c_1 = v(\beta_1, m_0, 0), \qquad m_0 \in U_0 \tag{7-62}$$

For a given value of β_1, $\beta_1 = c_2$, one of the following statements can be made on the basis of Equation 7-62: (1) It may be that no m_0 which is an element of the constraint set U_0 satisfies Equation 7-62, in which case the value c_2 of β_1 cannot possibly be the value of x_1 in an optimal solution which satisfies the given boundary condition, and therefore $\beta_1 = c_2$ need not be considered in the dynamic programming solution. (2) It may be that a unique value of $m_0 \in U_0$ satisfies Equation 7-62 for $\beta_1 = c_2$ in which case that value, and only that value, is used to obtain the value $\mathscr{F}_1(c_2)$ in Equation 7-61. (3) It may be that there are multiple solutions for m_0 in Equation 7-62, in which case minimization is effected in Equation 7-61 over these allowed values of m_0. If $v(x_1,m_0,0)$ is linear in m_0, case 3 above cannot possibly apply.

Example 7-3. Consider the first-order differential equation

$$\dot{x}(t) + ax(t) = bm(t) \tag{7-63}$$

where a and b are constants. If $x(t_k) \equiv x_k$ is the value of $x(t)$ at time t_k, then $x(t)$ for $t \geq t_k$ is given by

$$x(t) = e^{-a(t-t_k)} x_k + \int_{t_k}^{t} e^{-a(t-\tau)} bm(\tau)\, d\tau \tag{7-64}$$

Suppose that the control action $m(t)$ is generated by a digital system and that $m(t)$ equals a constant m_k in the time interval defined by $t_k \leq t < t_{k+1}$. It follows from 7-64 that $x(t_{k+1}) \equiv x_{k+1}$ is given by

$$x_{k+1} = e^{-a(t_{k+1}-t_k)} x_k + \left[\int_{t_k}^{t_{k+1}} e^{-a(t_{k+1}-\tau)} b\, d\tau \right] m_k \tag{7-65a}$$

which is in the form of Equation 7-40. In a particular case, suppose 7-65a reduces to

$$x_{k+1} = 0.8 x_k + m_k, \quad k = 0, 1, 2 \tag{7-65b}$$

where $x_0 \triangleq 0$ and m_k is constrained by

$$-0.5 \leq m_k \leq 1, \quad k = 0, 1, 2$$

The constraint set U_k is therefore the closed interval $[-0.5, 1]$ for $k = 0, 1, 2$.

A control policy is desired which will force the state variable to follow closely the sampled response shown in Figure 7-5.

It is given that initial errors are less serious than are terminal errors, so a time-weighted absolute-value performance measure P is deemed appropriate:

$$P = \sum_{i=1}^{3} i |x_i - x_{id}| \tag{7-66}$$

where the desired values x_{id} of x_i are (see Figure 7-5) $x_{1d} = 1$, $x_{2d} = 1$, and $x_{3d} = 0$.

Either linear programming (Chapter 5) or dynamic programming theory can be used to solve this problem, but only the dynamic programming

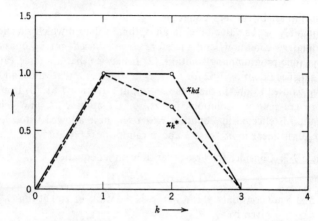

Figure 7-5. Desired sampled response x_{kd} and optimal sampled response $x_k{}^*$ for the problem of Example 7-3. Only the response at integer values of k is well-defined in this figure.

solution is considered here. For the backward dynamic programming solution, $x_K = x_3$ is not specified so that Equation 7-54 applies for the initial one-stage problem:

$$\mathscr{B}_1(\beta_2) = \underset{m_2 \in U_2}{\text{minimum}} f_3[q(\beta_2, m_2, 2), m_2] \tag{7-67a}$$

in which $q(\beta_2, m_2, 2) = 0.8\beta_2 + m_2$ (from Equation 7-65b) and $f_3(x_3, m_2) = 3|x_3|$ (from Equation 7-66); thus,

$$\mathscr{B}_1(\beta_2) \triangleq \underset{m_2 \in U_2}{\text{minimum}} \Omega_1(\beta_2, m_2)$$

$$= \underset{m_2 \in U_2}{\text{minimum}} 3|0.8\beta_2 + m_2| \tag{7-67b}$$

Similarly, Equation 7-51 is used to obtain

$$\mathscr{B}_2(\beta_1) \triangleq \underset{m_1 \in U_1}{\text{minimum}} \Omega_2(\beta_1, m_1)$$

$$= \underset{m_1 \in U_1}{\text{minimum}} [2|0.8\beta_1 + m_1 - 1| + \mathscr{B}_1(0.8\beta_1 + m_1)] \tag{7-68}$$

and

$$\mathscr{B}_3(\beta_0)|_{x_0 = \beta_0 = 0} \triangleq \underset{m_0 \in U_0}{\text{minimum}} \Omega_3(0, m_0)$$

$$= \underset{m_0 \in U_0}{\text{minimum}} [|m_0 - 1| + \mathscr{B}_2(m_0)] \tag{7-69}$$

Sect. 7-6   A CONTROL PROBLEM   393

To solve Equation 7-67b numerically,[3] discrete values of β_2 must be selected. Observe that the desired state response is in the range [0, 1]; it is reasonable to expect that x_2^* will also be in this range, or at least not too far outside this range, and probably closer to 1 than to 0. Thus, β_2 is incremented from 0 to 1 in steps of 0.1 in Table 7-5, and values of $\mathscr{B}_1(\beta_2)$ and $\hat{m}_2(\beta_2)$ are obtained by straightforward solution of Equation 7-67b.

TABLE 7-5. DYNAMIC PROGRAMMING TABLE FOR THE PROBLEM OF EXAMPLE 7-3

$j \rightarrow$	2		1		0	
β_j	$\mathscr{B}_1(\beta_2)$	$\hat{m}_2(\beta_2)$	$\mathscr{B}_2(\beta_1)$	$\hat{m}_1(\beta_1)$	$\mathscr{B}_3(\beta_0)$	$\hat{m}_0(\beta_0)$
0.0	0.00	0.00	0.75	0.625	0.75	1
0.1	0.00	−0.08	0.75	0.545		
0.2	0.00	−0.16	0.75	0.465		
0.3	0.00	−0.24	0.75	0.385		
0.4	0.00	−0.32	0.75	0.305		
0.5	0.00	−0.40	0.75	0.225		
0.6	0.00	−0.48	0.75	0.145		
0.7	0.18	−0.50	0.75	0.065		
0.8	0.42	−0.50	0.75	−0.015		
0.9	0.66	−0.50	0.75	−0.095		
1.0	0.90	−0.50	0.75	−0.175		

The next function to be evaluated is of course $\mathscr{B}_2(\beta_1)$, but here a slight inconvenience is encountered. To be specific, consider the evaluation of $\mathscr{B}_2(0.8)$:

$$\mathscr{B}_2(0.8) = \min_{m_1 \in U_1} \Omega_2(0.8, m_1)$$

$$= \min_{m_1 \in U_1} [2|m_1 - 0.36| + \mathscr{B}_1(m_1 + 0.64)] \qquad (7\text{-}70)$$

Note that the $\mathscr{B}_1(\beta_2)$ column in Table 7-5 is of direct use in evaluating $\Omega_2(0.8, m_1)$ only at $m_1 = 0.36, 0.26, 0.16, \ldots, -0.54$, and -0.64. The values of most interest are $\Omega_2(0.8, 0.26) = 0.86$, $\Omega_2(0.8, 0.16) = 0.82$, $\Omega_2(0.8, 0.06) = 0.78$, $\Omega_2(0.8, -0.04) = 0.80$, and $\Omega_2(0.8, -0.14) = 1.0$. The apparent minimum of these values is 0.78, but is this close to the value of $\mathscr{B}_2(0.8)$? To answer this question, interpolation is required. A projected linear interpolation will provide the minimum $\mathscr{B}_2(0.8)$ because $\Omega_k(\beta_{K-k}, m_{K-k})$ is

---
[3] The solution to Equation 7-67 could be deduced analytically, but to do this at this point would be to defeat the purpose of the example.

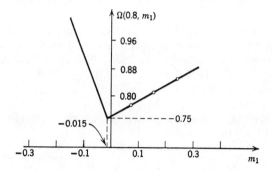

Figure 7-6. A typical projected linear interpolation.

piecewise linear and continuous for this simple problem. The indicated projected linear interpolation is sketched in Figure 7-6.

Figure 7-6 gives

$$\hat{m}_1(0.8) = -0.015 \quad \text{and} \quad \mathscr{B}_2(0.8) = 0.75$$

The remainder of Table 7-5 is generated in like fashion. From the generated table, $\hat{m}_0(0) = m_0^*$ is found to equal 1, and the minimum $\mathscr{B}_3(0)$ of the performance measure P is seen to be 0.75. Equation 7-65b gives $x_1^* = 0.8x_0^* + m_0^* = m_0^* = 1$. Thus, the use of $\beta_1 = 1$ in Table 7-5 gives $\hat{m}_1(1) = m_1^* = -0.175$. Similarly, $x_2^* = 0.8x_1^* + m_1^* = 0.8 - 0.175 = 0.625$. And finally, projected linear interpolation in the $\hat{m}_2(\beta_2)$ column gives $m_2(0.625) = m_2^* = -0.5$. The resulting optimal response is shown in Figure 7-5.

●

7-7. NUMERICAL CONSIDERATIONS

The dynamic programming tables of Section 7-2 and 7-6 are similar in form, but they differ in one essential regard: Table 7-3 was generated on the basis of integer values without the need of interpolation, whereas Table 7-5 was generated by considering real values with interpolation between discrete values being required both to generate the table and to find the optimal policy once the table was completed. When interpolation is used in solving a problem numerically, the accuracy of the numerical solution is directly related to the accuracy of the interpolation. Because of the repetitive nature of dynamic programming solutions, moreover, the extent of accumulated interpolation error is difficult to predict in practical problems; it should be noted that this type of error, in conjunction with round-off errors, may critically affect the solution and even invalidate the final solution of problems with many stages. Examples where this problem arises are considered by Guignabodet [7.43, 7.44] who gives guideline methods of estimating the accumulated error.

It is easy to show that interpolation should *not* be used in cases where the function under consideration is known or suspected to have discontinuities. For example, if real values rather than just integer values had been allowed in Example 7-1 of Section 7-2, interpolation in the policy function $\hat{x}_i(\beta_i)$ columns of Table 7-3 would be unsafe because of apparent discontinuities. In particular, if the resource (state) $b = 13.5$, and if linear interpolation is used in Table 7-3, the value of $\hat{x}_3(13.5)$ would be estimated at 4—halfway between 8 and 0—which would lead to $\hat{x}_2(9.5) = 0$ and $\hat{x}_1(9.5) = 9.5$; the corresponding value of the performance measure P, Equation 7-17, is 128. On the other hand, either $x_3 = 8.5$ or $x_3 = 0$ results in a value of P equal to 123 which is significantly less than the supposed minimum of 128 obtained by linear interpolation.

In general, it is important that we know which functions, if any, of a dynamic programming solution are continuous and are therefore more likely to lead to accurate interpolation. To this end, theorems of this section supply sufficient conditions under which the minimum-cost functions $\mathscr{F}_k(\beta_k)$'s of the control problem in Section 7-6 are continuous functions of β_k. After studying this section, the reader should be able to develop analogous theorems for other problems of Sections 7-2, 7-4, and 7-6; the theorems are analogous because of the similarity of the solutions (compare Equations 7-13, 7-31, 7-51, and 7-59).

It will be observed that the $\mathscr{F}_k(\beta_k)$'s are continuous under less restrictive conditions than are the $\hat{m}_k(\beta_k)$'s. Because of this, it is often advisable to store only the $\mathscr{F}_k(\beta_k)$'s in high-speed computer storage, thereby eliminating storage requirements for the $\hat{m}_k(\beta_k)$'s, and to compute m_k^* values by using interpolated values of $\mathscr{F}_k(\beta_k)$'s. The procedure is as follows: Assume $\beta_K = x_K^*$ is specified and that $\mathscr{F}_K(x_K^*)$ has been generated. Then m_{K-1}^* is found by determining the value of m_{K-1} which minimizes the right-hand member of 7-71:

$$\mathscr{F}_K(x_K^*) = \underset{m_{K-1} \in U_{K-1}}{\text{minimum}} \{f_K(x_K^*, m_{K-1}) + \mathscr{F}_{K-1}[v(x_K^*, m_{K-1}, K-1)]\} \quad (7\text{-}71)$$

Note that interpolation in the $\mathscr{F}_{K-1}(\beta_{K-1})$ column is required, in general, to obtain the optimal value

$$\mathscr{F}_{K-1}[v(x_K^*, m_{K-1}^*, K-1)]$$

of $\mathscr{F}_{K-1}[v(x_K{}^*,m_{K-1},K-1)]$, and that $x_{K-1}^* = v(x_K{}^*,m_{K-1}^*,K-1)$. Having determined x_{K-1}^*, we can find m_{K-2}^* in the same way as m_{K-1}^*, by finding the value of m_{K-2} which minimizes the right-hand member of 7-72:

$$\mathscr{F}_{K-1}(x_{K-1}^*) = \underset{m_{K-2} \in U_{K-2}}{\text{minimum}} \{f_{K-1}(x_{K-1}^*,m_{K-2}) + \mathscr{F}_{K-2}[v(x_{K-1}^*,m_{K-2},K-2)]\} \quad (7\text{-}72)$$

The process is continued in like fashion; m_{k-1}^* is determined on the basis that it yields the minimum of the right-hand member of 7-73:

$$\mathscr{F}_k(x_k{}^*) = \underset{m_{k-1} \in U_{k-1}}{\text{minimum}} \{f_k(x_k{}^*,m_{k-1}) + \mathscr{F}_{k-1}[v(x_k{}^*,m_{k-1},k-1)]\} \quad (7\text{-}73)$$

where interpolation is required, in general, to obtain

$$\mathscr{F}_{k-1}[v(x_k{}^*,m_{k-1}^*,k-1)]$$

from which $x_{k-1}^* = v(x_k{}^*,m_{k-1}^*,k-1)$ is also obtained.

In the theorems that follow, the set U_k of allowed values of the control variable m_k is assumed to be defined by a closed interval of the form

$$(m_k)_{\min} \leq m_k \leq (m_k)_{\max}, \quad k = 0, 1, \ldots, K-1 \quad (7\text{-}74)$$

It is left to the reader (and the researcher) to establish related theorems for other U_k forms.

Theorem 7-1. *If $f_1(\beta_1,m_0)$ is a continuous function of β_1 for each $m_0 \in U_0$ and if the function $\mathscr{F}_1(\beta_1)$ is given by*

$$\mathscr{F}_1(\beta_1) = \underset{m_0 \in U_0}{\text{minimum}} f_1(\beta_1,m_0) \quad (7\text{-}75)$$

where U_0 is a given set of real values, then $\mathscr{F}_1(\beta_1)$ is a continuous single-valued function of β_1.

Proof: This theorem is directly applicable to the first stage of the forward solution (with free initial condition) of the control problem in Section 7-6c. Pictorial justification of the theorem is obtained on the basis of Figure 7-7. Fixed values of β_1 shown in Figure 7-7 are β_{1a}, β_{1b}, and β_{1c}. Because $f_1(\beta_1,m_0)$ is assumed continuous in β_1 for any allowed value of m_0, "small" changes in β_1, as between β_{1a} and β_{1b}, result in "small" changes in the minimum of $f_1(\beta_1,m_0)$ with respect to m_0. The curve of $\mathscr{F}_1(\beta_1)$ is therefore a continuous single-valued function of β_1.

●

Note that for the particular case shown in Figure 7-7, the policy function $\hat{m}_0(\beta_1)$ is discontinuous at $\beta_1 = \beta_{1b}$, i.e.,

$$\lim_{\epsilon \to 0+} \hat{m}_0(\beta_{1b} - \epsilon) = k_2 \quad (7\text{-}76)$$

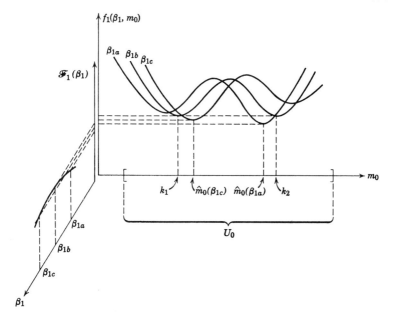

Figure 7-7. Illustration of a continuous $\mathscr{F}_1(\beta_1)$ function for Theorem 7-1.

whereas

$$\lim_{\epsilon \to 0+} \hat{m}_0(\beta_{1b} + \epsilon) = k_1 \neq k_2 \tag{7-77}$$

If we were to attempt interpolation to obtain $\hat{m}_0(\beta_{1b})$ from knowledge of $\hat{m}_0(\beta_{1b} - \epsilon) \cong k_2$ and of $\hat{m}_0(\beta_{1b} + \epsilon) \cong k_1$, the interpolated value $(k_1 + k_2)/2$ would be a poor choice indeed.

Theorem 7-2. *Consider the following recurrence relation:*

$$\mathscr{F}_k(\beta_k) = \underset{m_{k-1} \in U_{k-1}}{\text{minimum}} \{f_k(\beta_k, m_{k-1}) + \mathscr{F}_{k-1}[v(\beta_k, m_{k-1}, k-1)]\} \tag{7-78}$$

with $2 \leq k \leq K$. If the functions $f_k(\beta_k, m_{k-1})$, $v(\beta_k, m_{k-1}, k-1)$, and $\mathscr{F}_{k-1}(\beta_{k-1})$ are continuous single-valued functions of their respective arguments, $\mathscr{F}_k(\beta_k)$ is a continuous single-valued function of β_k.

Proof: This theorem is directly applicable to stages 2 through K of the forward solution (with free initial condition) of the control problem in Section 7-6c. To initiate proof of the theorem, let $\Omega_k(\beta_k, m_{k-1})$ denote the function

$$f_k(\beta_k, m_{k-1}) + \mathscr{F}_{k-1}[v(\beta_k, m_{k-1}, k-1)]$$

It is well known that a continuous single-valued function, $\mathscr{F}_{k-1}(\beta_{k-1})$, of a second continuous function, $\beta_{k-1} = v(\beta_k, m_{k-1}, k-1)$, is still continuous and single-valued in β_k and m_{k-1}. It is equally true that the sum $\Omega_k(\beta_k, m_{k-1})$ of two continuous single-valued functions,

$$f_k(\beta_k, m_{k-1}) \quad \text{and} \quad \mathscr{F}_{k-1}[v(\beta_k, m_{k-1}, k-1)],$$

is continuous and single-valued in its arguments. Thus the theorem is valid if it can be shown that $\mathscr{F}_k(\beta_k)$ is continuous and single-valued when

$$\Omega_k(\beta_k, m_{k-1})$$

is continuous and single-valued where

$$\mathscr{F}_k(\beta_k) = \underset{m_{k-1} \in U_{k-1}}{\text{minimum}} \Omega_k(\beta_k, m_{k-1}) \tag{7-79}$$

But this is similar to the result proved in Theorem 7-1. The proof of Theorem 7-2 is therefore completed by simple notational substitutions in the proof of Theorem 7-1 as follows: β_k for β_1, m_{k-1} for m_0, U_{k-1} for U_0, $\Omega_k(\beta_k, m_{k-1})$ for $f_1(\beta_1, m_0)$, and $\mathscr{F}_k(\beta_k)$ for $\mathscr{F}_1(\beta_1)$.

●

Theorem 7-3. *Consider the relation*

$$\mathscr{F}_1(\beta_1) = \underset{m_0}{\text{minimum}} f_1(\beta_1, m_0) \tag{7-80}$$

in which minimization is constrained by

$$c_1 = v(\beta_1, m_0, 0), \qquad (m_0)_{\min} \leq m_0 \leq (m_0)_{\max} \tag{7-81}$$

If both $f_1(\beta_1, m_0)$ and $v(\beta_1, m_0, 0)$ are continuous in β_1 and m_0 and if $v(\beta_1, m_0, 0)$ is a strictly monotonic function[4] of both β_1 and m_0, the function $\mathscr{F}_1(\beta_1)$ is continuous over the interval of β_1 in which solutions to 7-81 exist.

Proof: This theorem applies to stage 1 of the forward solution (with fixed initial condition) of the control problem in Section 7-6c. For convenience in establishing the proof, assume that $v(\beta_1, m_0, 0)$ is a monotonically decreasing function of m_0 and is a monotonically increasing function of β_1. Figure 7-8 is suggestive of the assumed conditions.

With small changes in β_1, as for example the change between β_{1a} and β_{1b}, the change in the m_0 values which satisfy 7-81 is correspondingly small, i.e., $\hat{m}_0(\beta_{1b}) - \hat{m}_0(\beta_{1a})$ is "small"; therefore, the change in $\mathscr{F}_1(\beta_1)$ is small,

---

[4] A function $g(t)$ is said to be a *monotonic function* of t if one, and only one, of the following conditions applies: (1) for every $t_2 > t_1$, $g(t_2) \geq g(t_1)$; or (2) for every $t_2 > t_1$, $g(t_2) \leq g(t_1)$. If the equality signs are disallowed between $g(t_1)$ and $g(t_2)$ in the above definition, then the phrase *strictly monotonic function* applies.

Sect. 7-7   NUMERICAL CONSIDERATIONS   399

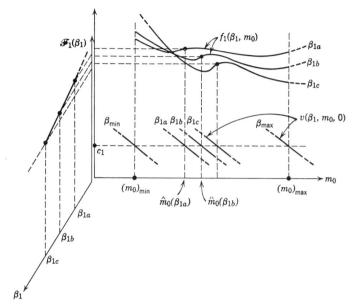

Figure 7-8. Illustration of a continuous $\mathscr{F}_1(\beta_1)$ function for Theorem 7-3.

i.e., $|f_1[\beta_{1a}, \hat{m}_0(\beta_{1a})] - f_1[\beta_{1b}, \hat{m}_0(\beta_{1b})]|$ is small because of the assumed continuity of f_1 in its arguments, and therefore $\mathscr{F}_1(\beta_1)$ is continuous in β_1.

•

It is easy to give examples in which the assumed monotonic character of $v(\beta_1, m_0, 0)$ is not satisfied and in which $\mathscr{F}_1(\beta_1)$ is discontinuous (see Problem 7.15). But it is just as easy to give examples in which the assumed monotonic character of $v(\beta_1, m_0, 0)$ is not satisfied and in which $\mathscr{F}_1(\beta_1)$ is continuous. The nonmonotonic forms of $v(\beta_1, m_0, 0)$ require individual attention.

Theorem 7-4. *Consider the following recurrence relation:*

$$\mathscr{F}_k(\beta_k) = \underset{m_{k-1}}{\text{minimum}} \{f_k(\beta_k, m_{k-1}) + \mathscr{F}_{k-1}[v(\beta_k, m_{k-1}, k-1)]\} \quad (7\text{-}82)$$

for $2 \leq k \leq K$, in which minimization is constrained both by

$$(m_{k-1})_{\min} \leq m_{k-1} \leq (m_{k-1})_{\max} \quad (7\text{-}83)$$

and by

$$(\beta_{k-1})_{\min} \leq v(\beta_k, m_{k-1}, k-1) \leq (\beta_{k-1})_{\max} \quad (7\text{-}84)$$

where it is assumed that $\mathscr{F}_{k-1}(\beta_{k-1})$ exists only on the closed interval $[(\beta_{k-1})_{\min},(\beta_{k-1})_{\max}]$. The claim is that the function $\mathscr{F}_k(\beta_k)$ exists and is continuous over a single closed interval, $(\beta_k)_{\min} \leq \beta_k \leq (\beta_k)_{\max}$, if the following conditions are satisfied:

1. Both $f_k(\beta_k,m_{k-1})$ and $v(\beta_k,m_{k-1},k-1)$ are continuous in β_k and m_{k-1}.
2. The function $v(\beta_k,m_{k-1},k-1)$ is strictly monotonic in both β_k and m_{k-1}.
3. $\mathscr{F}_{k-1}(\beta_{k-1})$ is a continuous function of β_{k-1} over the given closed interval $(\beta_{k-1})_{\min} \leq \beta_{k-1} \leq (\beta_{k-1})_{\max}$.

Proof: This theorem is directly applicable to stages 2 through K of the forward solution (with fixed initial condition) of the control problem in Section 7-6c. For convenience in establishing the proof, assume that

$$v(\beta_k,m_{k-1},k-1)$$

is a monotonic decreasing function of m_{k-1} and is a monotonic increasing function of β_k. Figure 7-9 depicts a representative constraint region defined by conditions 7-83 and 7-84. Observe that for a given value of β_k, say, $\beta_k = \beta_{ka}$, the curve $v(\beta_{ka},m_{k-1},k-1)$ crosses the constraint boundary at two points; the m_{k-1} coordinates of these two points bound a subset $U_{k-1}(\beta_{ka})$ of the set U_{k-1}, and the control variable may assume values only in this subset when $\beta_k = \beta_{ka}$. For a value of β_k which differs only slightly from β_{ka}, say, $\beta_k = \beta_{kb}$, the upper (lower) bound of the subset $U_{k-1}(\beta_{kb})$ can differ but slightly from the upper (lower) bound of the subset $U_{k-1}(\beta_{ka})$.

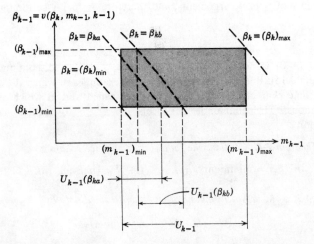

Figure 7-9. Constraint region defined by conditions 7-83 and 7-84.

Suppose that $\mathscr{F}_k(\beta_{ka})$ and $\mathscr{F}_k(\beta_{kb})$ are evaluated:

$$\mathscr{F}_k(\beta_{ka}) = \min_{m_{k-1} \in U_{k-1}(\beta_{ka})} \{f_k(\beta_{ka},m_{k-1}) + \mathscr{F}_{k-1}[v(\beta_{ka},m_{k-1},k-1)]\}$$

and

$$\mathscr{F}_k(\beta_{kb}) = \min_{m_{k-1} \in U_{k-1}(\beta_{kb})} \{f_k(\beta_{kb},m_{k-1}) + \mathscr{F}_{k-1}[v(\beta_{kb},m_{k-1},k-1)]\}$$

Because f_k and $\mathscr{F}_{k-1}$ are assumed continuous in their respective arguments and because the upper (lower) bound of $U_{k-1}(\beta_{ka})$ can differ but slightly from the upper (lower) bound of $U_{k-1}(\beta_{kb})$ if $\beta_{kb} - \beta_{ka}$ is small, $\mathscr{F}_k(\beta_{ka})$ and $\mathscr{F}_k(\beta_{kb})$ can differ by only a "small" amount, and therefore $\mathscr{F}_k(\beta_k)$ is continuous over the interval in which it is defined.

For the conditions assumed in Figure 7-9, the interval of definition for β_k is given by

$$(\beta_{k-1})_{\min} = v[(\beta_k)_{\min},(m_{k-1})_{\min},k-1] \tag{7-85}$$

and

$$(\beta_{k-1})_{\max} = v[(\beta_k)_{\max},(m_{k-1})_{\max},k-1] \tag{7-86}$$

Relations 7-85 and 7-86 are placed in more appropriate form by using Equation 7-40; thus,

$$(\beta_k)_{\max} = q[(\beta_{k-1})_{\max},(m_{k-1})_{\max},k-1] \tag{7-87}$$

and

$$(\beta_k)_{\min} = q[(\beta_{k-1})_{\min},(m_{k-1})_{\min},k-1] \tag{7-88}$$

where the set of β_k's for which $\mathscr{F}_k(\beta_k)$ exists is given by

$$(\beta_k)_{\min} \leq \beta_k \leq (\beta_k)_{\max} \tag{7-89}$$

•

In control problems such as the one considered, as well as in other problems, there is a tendency for $(\beta_k)_{\max}$ and $-(\beta_k)_{\min}$ to increase rapidly as a function of k. In such cases, the number of grid points used per stage must also be increased, and we might surmise that computational difficulty would result for large k. The factor which lessens this difficulty is that additional constraints are usually imposed on the region of state space in which trajectories are allowed;[5] when the values computed by using 7-87 and 7-88 exceed the

---

[5] When magnitude bounds are imposed on both x_k and m_k, it may be that no allowed trajectories exist between a given set S_k of allowed values of x_k at an instant k and a specified desired value of x_K at an instant K, $K > k$. The set S_k is said to be a set of *immobial states* at instant k and need not be considered in a dynamic programming solution. The theory of mobial and immobial states is considered by Kahne [7.51, 7.52].

allowed range, appropriate boundary values of the allowed range are used in place of $(\beta_k)_{max}$ and $(\beta_k)_{min}$.

The four theorems of this section supply general conditions under which the minimum-cost function of a dynamic programming solution is continuous. The continuity of such functions is desirable from the standpoint of interpolation accuracy; and, in addition, we may be able to use low-order series approximations, perhaps a low-order polynomial, to accurately represent a continuous minimum-cost function as a function of the state variable(s). Approximations of this kind are referred to in the literature as *function approximations in state space* (or sometimes simply approximations in function space). By storing the coefficients of the series rather than numerous values of the function, rapid-access computer storage space is conserved. This sort of conservation is most important when problems with several state variables are considered, as will be observed in Sections 7-10 through 7-16.

7-8. A PRINCIPLE OF OPTIMALITY

Bellman's principle of optimality [7.5] is a concise description of the phenomenon which enables problems—problems amenable to dynamic programming solution—to be viewed as a sequence of simpler problems. Depending on the context in which it is stated, this principle of optimality may assume various forms. By reviewing that which is presented in the preceding sections, the reader should now be in a position to summarize certain key results—to get the principle, as it were.

In Sections 7-2, 7-3, and 7-4, a resource is to be allocated among various activities. The return from a given activity is dependent only on the amount of resource allocated to it. If we allocate a specific amount of the resource to a given activity, e.g., activity number 1, we have a chance of obtaining an overall optimal return only if the remaining amount of available resource is allocated in an optimal fashion among the remaining activities. *An optimal policy (an optimal selection of the x_i's) is therefore one having the property that whatever the initial state (resource) and the initial decision (allocation) are, the remaining decisions (allocations) constitute an optimal policy with respect to the state (remaining resource) resulting from the initial decision*—this is Bellman's principle of optimality.

In Section 7-5, a chain network is defined with chain links between adjacent nodes. Each link of a given chain network is assigned a "length"; the problem is to find the minimum-length connected chain (the minimal chain) between an initial set of specified nodes (the set may consist of a single node) and another set, a terminal set, of specified nodes (again, this set may consist of a single node). If we arbitrarily pick a node which is directly adjacent (i.e., coupled by a direct link) to a node of the initial set and use it as a node of a connected chain between the initial and terminal sets, we have a chance of obtaining an overall minimal chain only if the net length of the remaining links in the connected chain is the least that can be obtained. In other words, *an optimal policy has the property that whatever the initial state (a node of the initial set) and the initial decision (the selection of a particular node which is adjacent to that node of the initial set) are, the remaining decisions (adjacent node selections along the chain) constitute an optimal policy (a minimal chain) with respect to the state (the node) resulting from the initial decision*—again, Bellman's principle of optimality filters through.

In Section 7-6, the state of a sampled-data system at time t_{k+1} is a given function of both a control variable and the state of the system at time t_k. The control is to be selected from an allowed set of real values at K discrete instants of time; the goal is to minimize a performance measure which is a sum of n functions, the kth such function being dependent only on the kth state and the $(k-1)$st value of the control variable. If we arbitrarily pick the value of the control variable, say, at the initial time, then we have a chance of obtaining the minimum of the performance measure only if the remaining control actions are selected in an optimal manner with respect to the state which results from the initial value of the control variable. Here too, *an optimal policy (an optimal set of control actions) has the property that whatever the initial state and the initial decision (initial value of the control variable) are, the remaining decisions (control values) constitute an optimal policy with respect to the state of the system resulting from the initial decision.* The fact that the initial decision corresponds to the control applied at the initial time in the above statement is not crucial in many applications; the initial decision could correspond to the last control action to be applied in time. Moreover, Bellman's principle of optimality may be viewed as expressing the fact that any portion of an optimal trajectory[6] is also optimal (optimal for a smaller problem, a problem with end conditions corresponding to appropriate points on the optimal trajectory).

In Sections 7-2 through 7-4, the variable β_k denotes the resource variable (state variable) associated with a k-stage decision problem. In Section 7-5, β_k denotes the state variable x_1 at $x_2 = k$. In Section 7-6, β_k denotes the state variable x_k at time t_k. Although somewhat awkward at times, the common use of β_k in the preceding sections serves to highlight the actual similarities of the problems and their dynamic programming solutions. In many cases, e.g., the problem of the following section, the reader will find that it is notationally expedient to deviate from this β_k convention.

---

[6] Within the framework of the problem of Section 7-6, an optimal trajectory is a time sequence of optimal state values.

7-9. PLACEMENT OF TRANSMISSION-LINE TOWERS [7.87]

The selection and placement of transmission-line towers along a fixed right-of-way is a novel multistage decision problem. Initially, a survey is conducted to obtain the profile $y(x)$ of the terrain along the right-of-way x (this profile may be plotted by a computer if the surveyors use mark-sense cards to record their data). The profile in Figure 7-10a represents a segment of a hypothetical transmission-line right-of-way. It is assumed that the two end-locations shown are fixed tower locations: they could represent initial or terminal points of a line; they could represent points at which the transmission-line right-of-way bends at a fixed angle; or they could represent any other point at which a tower is required for a practical reason.

The designer has available m different tower types, each of which has a designated cost, including installation costs which may depend on proposed locations of the tower. The towers differ in height and/or load-carrying ability. By reading the surveyors' notes and by viewing the profile of Figure 7-10a, the designer marks regions on the profile where towers cannot be located—e.g., on roads, in rivers, on loose foundations, etc. The designer also notes regions of the profile which are likely choices for locating a tower—e.g., near the top of a ridge.

With these conditions in mind, the designer proceeds to use a dynamic

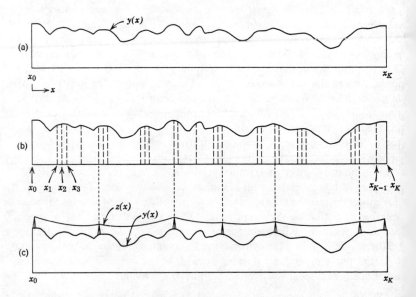

Figure 7-10. Transmission-line right-of-way profile.

Sect. 7-9 PLACEMENT OF TRANSMISSION-LINE TOWERS

programming approach by designating $K - 1$ points, labeled x_1 through x_{K-1}, between x_0 and x_K, each of which is a *potential* tower position. Observe in Figure 7-10b that many more points than are ultimately used for tower positions are selected. Relatively closely placed points are designated in regions which are likely to contain a tower in the final design; a moderate number of points are designated in level regions; only a few points are designated in regions which are unlikely to contain a tower in the final design; and no points are designated in regions which cannot support a tower.

At a given point x_k, it is *feasible* to erect at least one (perhaps any one) of m tower types; a tower of type j is denoted by T_j. If the designer considers locating a tower of type j at point x_k, the notation T_{jk} is used to describe this *tower-position pair*.

Consider two points x_k and x_r with $r < k$. If a tower of type j is located at x_k, then it may or may not be feasible to have a directly adjacent tower-position pair T_{ir}. It is feasible to span the transmission line between T_{ir} and T_{jk} if, and only if, the following conditions are simultaneously satisfied:

1. T_{ir} is a feasible tower-position pair.
2. The difference between the elevation $z(x)$ of the line and that of the ground (including surface obstruction) $y(x)$ is greater than or equal to a designated clearance height $h = h(x)$ at all points between towers.
3. The vertical and horizontal loading capacities of the towers are not exceeded.

There are subtle points involved in incorporating precise loading capabilities of towers in the design via dynamic programming, even more subtle than is indicated in the literature [7.87]. As a first-order approximation, the designer could use the known weight of the lines to specify a maximum span distance d_j for each tower T_j; the loading capabilities of tower T_j are not exceeded if the linear distance between x_j and the adjacent tower position on either side does not exceed d_j. This simplifying assumption is used in the following analysis; more accurate ones could be used.

For each feasible tower-position pair T_{jk}, there exists a set S_{jk} of other tower-position pairs T_{ir} with the following properties:

1. $0 \leq r < k \leq K$.
2. T_{ir} is a feasible tower-position pair.
3. $\{(x_k - x_r)^2 + [y(x_k) - y(x_r)]^2\}^{1/2}$ is less than both d_j and d_i.
4. $z(x) - y(x) \geq h(x)$ for x in the interval $x_r \leq x \leq x_k$ when T_{ir} and T_{jk} are adjacent tower-position pairs.

This set S_{jk} is therefore a set of tower-position pairs all of which lie to the left of x_k and any one of which could feasibly be coupled directly with T_{jk} in the final design.

The design objective, of course, is to select a subset of the set of all feasible tower-position pairs to obtain the least-possible cost, subject to satisfying the loading and clearance constraints. By using dynamic programming in fulfilling this objective, many inconsequential combinations of tower-position pairs are automatically by-passed. Either a forward or a backward dynamic programming approach can be used on this problem. The following forward solution illustrates the pertinent concepts.

Let $f(T_{jk})$ denote the cost of procuring and erecting tower T_j at point x_k, and let the minimum-cost function $\mathscr{F}(T_{jk})$ denote the *minimal* cost of a line which starts at x_0 and terminates at x_k, given that tower T_j is used at x_k. Of course, $\mathscr{F}(T_{j0}) = f(T_{j0})$ is a trivial case. As for $\mathscr{F}(T_{j1})$,

$$\mathscr{F}(T_{j1}) = \underset{T_{i0} \in S_{j1}}{\text{minimum}} [f(T_{j1}) + f(T_{i0})]$$

$$= f(T_{j1}) + \underset{T_{i0} \in S_{j1}}{\text{minimum}} \mathscr{F}(T_{i0}) \tag{7-90}$$

which is the minimal cost of a line which starts at x_0 and terminates at x_1, given that T_j is used at x_1. If the type of tower at x_0 is fixed—that is, if the designer must use a particular tower-position pair T_{v0}—then the set S_{j1} is restricted to contain, at most, the element T_{v0}. Note that if any one of the conditions associated with set S_{j1} is not satisfied by T_{v0}, S_{j1} is empty and $\mathscr{F}(T_{j1})$ must be assigned an exorbitant cost.

The object is now to build on the solutions $\mathscr{F}(T_{ir})$, for $r = 0, 1$ and for $i = 1, 2, \ldots, m$, by using these to obtain $\mathscr{F}(T_{j2})$, $j = 1, 2, \ldots, m$. The minimal cost $\mathscr{F}(T_{j2})$ can be viewed as consisting of two parts: a fixed cost $f(T_{j2})$ of the tower T_j at x_2, and a cumulative cost of the towers in the line up to, but excluding, T_{j2}. Of the tower-position pairs in S_{j2}, one—denoted by $\hat{T}(T_{j2})$—is coupled directly with T_{j2} to obtain the minimal cost $\mathscr{F}(T_{j2})$. In order for $\mathscr{F}(T_{j2})$ to be optimal, it must be that

$$\mathscr{F}(T_{j2}) = f(T_{j2}) + \mathscr{F}[\hat{T}(T_{j2})] \le f(T_{j2}) + \mathscr{F}(T_{ir}) \tag{7-91}$$

for all T_{ir} contained in S_{j2}. Equivalently,

$$\mathscr{F}(T_{j2}) = f(T_{j2}) + \underset{T_{ir} \in S_{j2}}{\text{minimum}} \mathscr{F}(T_{ir}) \tag{7-92}$$

By coupling $\hat{T}(T_{j2})$ to T_{j2}, the corresponding cost $\mathscr{F}(T_{j2})$ is the minimal cost of constructing a line from x_0 to x_2, given that T_j is used at x_2.

In general, this reasoning results in the recurrence relation

$$\mathscr{F}(T_{jk}) = f(T_{jk}) + \underset{T_{ir} \in S_{jk}}{\text{minimum}} \mathscr{F}(T_{ir}), \qquad k = 1, 2, \ldots, K \tag{7-93}$$

Sect. 7-9 PLACEMENT OF TRANSMISSION-LINE TOWERS

where the final value $\mathscr{F}(T_{jK})$ is the minimal cost of a feasible design between x_0 and x_K, assuming T_j is used at x_K.

In analogy with the dynamic programming solutions of Sections 7-2, 7-3, 7-4, and 7-6, a table of the minimum-cost functions $\mathscr{F}(T_{jk})$'s and the policy functions $\hat{T}(T_{jk})$'s is generated recursively. The distinguishing feature of this problem is that the entry $\mathscr{F}(T_{jk})$ is obtained by considering not only the $(k-1)$st $\mathscr{F}$ column, but also all other $\mathscr{F}$ columns that contain entries associated with the set S_{jk}.

To compute $\mathscr{F}(T_{jk})$, the number of comparisons required in Equation 7-93 equals the number $N(S_{jk})$ of elements in S_{jk}. The total number of such comparisons required to compute the kth column is therefore $\sum_{j=1}^{m} N(S_{jk})$. To generate the complete table, the total number N_T of these comparisons is

$$N_T = \sum_{k=1}^{K} \sum_{j=1}^{m} N(S_{jk}) \quad (7\text{-}94)$$

For example, if the average $N(S_{jk}) = 10m$, then $N_T = 10Km^2$. Furthermore, if $K = 100$ and $m = 10$, $N_T = 10^5$, which is a reasonable number for high-speed digital computation.

More time-consuming than the above comparisons is the determination of each feasibility set S_{jk}. All those feasible T_{ir}'s, $r < k$, for which either d_j or d_i is less than $\{(x_k - x_r)^2 + [y(x_k) - y(x_r)]^2\}^{1/2}$, cannot be contained in S_{jk} because of the loading restrictions. Let the set of feasible tower-position pairs which do satisfy the above loading restrictions be denoted by S'_{jk}. The set S_{jk} is a subset of S'_{jk} and is obtained by considering the ground clearance restriction.

Assume that the towers are numbered so that the height of T_{j+1} is greater than or equal to the height of T_j for all j. To determine if T_{ir}, a given member of S'_{jk}, is also a member of S_{jk}, the ground curve $y(x)$ must be subtracted from the curve which characterizes the transmission line between T_{ir} and T_{jk} (Stagg and Watson [7.87] use a parabolic fit for the transmission line). We need only compute this difference at a suitably dense set of points between towers, e.g., 30 such points; if the difference is greater than $h(x)$, T_{ir} is contained in S_{jk}. Note that because of the assumed ordering of towers in height, $T_{i+1,r}$ is an element of S_{jk} if both $T_{ir} \in S_{jk}$ and $T_{i+1,r} \in S'_{jk}$. Furthermore, any tower-position pair which is contained both in S_{jk} and in $S'_{j+1,k}$ is also contained in $S_{j+1,k}$. These relationships reduce significantly the number of clearance trials that must be effected.

Once the table of $\mathscr{F}$'s and $\hat{T}$'s is generated (Table 7-6), the optimal solution is obtained by working backwards through the table; the optimal tower T^*_{iK} at x_K is the one corresponding to $\mathscr{F}(T^*_{iK})$:

$$\mathscr{F}(T^*_{iK}) = \min \mathscr{F}(T_{iK}) \quad (7\text{-}95)$$

TABLE 7-6. TRANSMISSION-LINE TOWER SOLUTION TABLE

$k \rightarrow$		0	1		$\cdots$		K	
j	T_{jk}	$\mathscr{F}(T_{j0})$	$\mathscr{F}(T_{j1})$	$\hat{T}(T_{j1})$	$\cdots$		$\mathscr{F}(T_{jK})$	$\hat{T}(T_{jK})$
1	T_{1k}	$f(T_{10})$	$\mathscr{F}(T_{11})$	$\hat{T}(T_{11})$			$\mathscr{F}(T_{1K})$	$\hat{T}(T_{1K})$
2	T_{2k}	$f(T_{20})$	$\mathscr{F}(T_{21})$	$\hat{T}(T_{21})$			.	.
3	T_{3k}	$f(T_{30})$	(Evaluated by using Equation 7-90)	.			(Evaluated by using Equation 7-93)	
$\vdots$	$\vdots$	$\vdots$		$\vdots$				$\vdots$
m	T_{mk}	$f(T_{m0})$	$\mathscr{F}(T_{m1})$	$\hat{T}(T_{m1})$	$\cdots$		$\mathscr{F}(T_{mK})$	$\hat{T}(T_{mK})$

The entry adjacent to $\mathscr{F}(T_{iK}^*)$ in the table is $\hat{T}(T_{iK}^*) = T_{ab}^*$ which is the tower-position pair that is coupled to T_{iK}^* in the optimal design. In turn, the entry $\hat{T}(T_{ab}^*)$ equals a certain T_{cd}^* which is coupled to T_{ab}^* in the optimal design. This process is continued until the complete set of optimal tower-position pairs is obtained.

When designing transmission lines which are to extend for a considerable distance over rolling or mountainous terrain, we can expect a dynamic programming approach to yield a significantly smaller cost than is obtained by the seat-of-the-pants design approach which is based on the designer's skill in picking appropriate tower-position pairs and in using standard line-sag curves to insure adequate ground clearance. A test case [7.87] was run on an existing transmission line which traverses 18 miles over irregular terrain; the data included 97 different structure types and 18 fixed angle locations. For the optimized line, 74 towers were indicated versus the 76 towers used in the actual line, a saving of 2.6% in the number of structures. More impressive than this, however, is that the cost of the optimally selected structures was more than 5% less than the cost of the actual structures in the line.

7-10. n STATE VARIABLES: DISCRETE PROCESSES

7-10a. Problems and Difficulties

Examples of multistage decision problems with n state variables are given in this subsection. These problems are matrix extensions of the one-state-variable multistage decision problems of the preceding sections; because of this, we obtain recurrence relations which are identical in form to those of the preceding sections. It becomes quite apparent, however, that the "straightforward" solution of these recurrence relations is computationally prohibitive, in general, if the number n of state variables is large ($n \geq 4$). Thus, more devious solution schemes are examined in subsequent subsections.

Allocation Problems. For allocation problems of the type posed in Section 7-2, the number of state variables equals the number of constraints of the form shown in 7-2. In particular, consider a manufacturing process in which n machines are used to manufacture K different products. Profit from the production of x_j units of the jth product is given by a known real-valued function $f_j(x_j)$. The amount of time required to process x_j units of type j on the ith machine is given by a known real-valued function $g_{ij}(x_j)$, which generally includes a setup time. Over a specified future time period, it is given that the ith machine will be available for operation for a total of b_i hours. The order in which the machines are used to produce the jth product is nonarbitrary, in general, but considerations such as this are ignored in this introductory development of the problem.

Obviously, the maximum of the profit P is desired, where

$$P = \sum_{j=1}^{K} f_j(x_j) \tag{7-96}$$

and where the optimal solution must satisfy

$$\sum_{j=1}^{K} g_{ij}(x_j) \leq b_i, \qquad i = 1, 2, \ldots, n \tag{7-97}$$

In matrix notation, let $\mathbf{b} = [b_1 \ b_2 \ \cdots \ b_n]'$ and let

$$\mathbf{g}_j(x_j) = [g_{1j}(x_j) \ g_{2j}(x_j) \ \cdots \ g_{nj}(x_j)]'$$

Inequality 7-97 reduces to

$$\sum_{j=1}^{K} \mathbf{g}_j(x_j) \leq \mathbf{b} \tag{7-98}$$

The form of 7-96 and 7-98 is identical to that of 7-1 and 7-2; the development immediately following 7-6 leads to a recurrence relation, Equation 7-13. The reader may easily parallel that development to obtain the following n-state counterpart of Equation 7-13:

$$\mathscr{F}_k(\mathbf{b}_k) = \max_{x_k} \{f_k(x_k) + \mathscr{F}_{k-1}[\mathbf{b}_k - \mathbf{g}_k(x_k)]\} \tag{7-99}$$

for $k = 1, 2, \ldots, K$. Constraints on the maximization in 7-99 are

$$x_k \geq 0, \qquad k = 1, 2, \ldots, K \tag{7-100}$$

and

$$\mathbf{g}_k(x_k) \leq \mathbf{b}_k \triangleq [\beta_{1k} \ \beta_{2k} \ \cdots \ \beta_{nk}]' \tag{7-101}$$

The maximum profit equals $\mathscr{F}_K(\mathbf{b})$. The interpretation of the maximum-return function $\mathscr{F}_k(\mathbf{b}_k)$ is this: the "$\mathscr{F}$" of $\mathscr{F}_k(\mathbf{b}_k)$ denotes the forward nature of the solution; the subscript k of $\mathscr{F}_k$ denotes that only the first k stages (first k products) are considered; the argument $\mathbf{b}_k = [\beta_{1k} \ \beta_{2k} \ \cdots \ \beta_{nk}]'$ is the state matrix at stage k, here β_{ik} equals the number of hours available on the ith machine for processing products numbered 1 through k. Each value of the state matrix $\mathbf{b}_k$ results, via Equation 7-99, in a corresponding value of the maximum-return function $\mathscr{F}_k(\mathbf{b}_k)$.

At this point, difficulties arise: the argument $\mathbf{b}_k$ of the return function $\mathscr{F}$ in 7-99 is an $n \times 1$ column matrix. In the straightforward dynamic programming approach, $\mathscr{F}_{k-1}(\mathbf{b}_{k-1})$ is computed and stored in order to be used in the evaluation of $\mathscr{F}_k(\mathbf{b}_k)$, Equation 7-99. But if each component of $\mathbf{b}_{k-1}$ may assume ξ distinct values, the total number N of values of $\mathscr{F}_{k-1}$ which need be stored is ξ^n—with $\xi = 100$ and $n = 5$, for example, $N = 10^{10}$, a stupendous number!

Control Problems. For control problems of the type posed in Section 7-6, the number of state variables equals the number of first-order state equations, in the form of 7-40, which are necessary to characterize the system. In matrix form,

$$\mathbf{x}_{k+1} = \mathbf{q}(\mathbf{x}_k, \mathbf{m}_k, k), \qquad k = 0, 1, \ldots, K-1 \qquad (7\text{-}102)$$

in which

$$\mathbf{x}_k = [x_{1k} \ x_{2k} \ \cdots \ x_{nk}]' \qquad (7\text{-}103)$$

$$\mathbf{m}_k = [m_{1k} \ m_{2k} \ \cdots \ m_{rk}]' \qquad (7\text{-}104)$$

and

$$\mathbf{q}(\mathbf{x}_k, \mathbf{m}_k, k) = [q_1(\mathbf{x}_k, \mathbf{m}_k, k) \ q_2(\mathbf{x}_k, \mathbf{m}_k, k) \ \cdots \ q_n(\mathbf{x}_k, \mathbf{m}_k, k)]' \qquad (7\text{-}105)$$

The state matrix $\mathbf{x}_k$ denotes the state of the system at "time" k; $\mathbf{m}_k$ is an $r \times 1$ control matrix; and $q_i(\mathbf{x}_k, \mathbf{m}_k, k)$, $i = 1, 2, \ldots, n$, is a known real-valued function of its arguments.

As in Section 7-6, a performance measure P is to be minimized:

$$P = \sum_{k=1}^{K} f_k(\mathbf{x}_k, \mathbf{m}_{k-1}) \qquad (7\text{-}106)$$

where $f_k(\mathbf{x}_k, \mathbf{m}_{k-1})$ is a known real-valued function of its arguments and where $\mathbf{m}_k$ is constrained to be an element of a specified closed set U_k. Again, Equations 7-102 and 7-106 are identical in form to their one-state-variable counterparts, Equations 7-40 and 7-41; thus, the proofs and results in Section

7-6 carry over essentially unchanged. For example, the backward dynamic programming algorithm of 7-51 extends to

$$\mathscr{B}_{K-k}(\mathbf{x}_k) = \underset{\mathbf{m}_k \in U_k}{\text{minimum}} \{f_{k+1}[\mathbf{q}(\mathbf{x}_k,\mathbf{m}_k,k),\mathbf{m}_k] + \mathscr{B}_{K-k-1}[\mathbf{q}(\mathbf{x}_k,\mathbf{m}_k,k)]\} \quad (7\text{-}107)$$

for $k = K - 2, K - 3, \ldots, 2, 1, 0$. The minimum of P for any initial state $\mathbf{x}_0$ equals $\mathscr{B}_K(\mathbf{x}_0)$. The interpretation of $\mathscr{B}_{K-k}(\mathbf{x}_k)$ is as follows: the script $\mathscr{B}$ denotes the backward nature of the solution; the subscript $K - k$ of $\mathscr{B}_{K-k}(\mathbf{x}_k)$ denotes that a $(K - k)$-stage decision process is used to obtain $\mathscr{B}_{K-k}(\mathbf{x}_k)$; and the argument $\mathbf{x}_k$ of $\mathscr{B}_{K-k}(\mathbf{x}_k)$ is indicative of the state of the system at "time" k.

As with the preceding allocation problem, the $n \times 1$ state matrix of the system necessitates the use of ξ^n storage locations if all of the values of $\mathscr{B}_{K-k}(\mathbf{x}_k)$ are to be retained when each state variable (each component of $\mathbf{x}_k$) is allowed to assume ξ values. An additional complication is noticeable, moreover: minimization in 7-107 must be conducted with respect to r control variables. If r is large, considerable effort may be required to obtain the minimum in 7-107 for even one specific value of $\mathbf{x}_k$.

7-10b. Series Approximations

Rather than store all computed values of a maximum-return function, as is done in a straightforward dynamic programming solution, it may be feasible to generate a finite series approximation to a maximum-return function, a function approximation in state space, and to store numerical values for the coefficients in the series along with a program for computing the series approximation at any point in state space. Many types of series approximations may be used, the most commonly used being polynomial approximation [7.9, 7.15, 7.37, 7.63, 7.75]. A significant reduction in high-speed storage requirements is attained if the number of evaluations of the return function is significantly greater than the number of coefficients in the series plus the number of storage locations occupied by the program which must be used to evaluate the series. Two factors tend to limit use of this approach: (1) the accuracy of the series approximation; and (2) the increase in computation time due to evaluation of coefficients in the series and to evaluation of the series itself.

A typical approach is the following: A function, say, $\mathscr{B}(\mathbf{x})$, is to be evaluated at ρ points in state space. For notational convenience, let $\mathscr{B}(\mathbf{x}^i)$ denote the ith such evaluation of $\mathscr{B}(\mathbf{x})$ and let $\mathscr{B}_s(\mathbf{x}^i,\mathbf{a})$ denote the corresponding series evaluation, where $\mathscr{B}_s(\mathbf{x},\mathbf{a})$ is a linear function of the series coefficients, the components of $\mathbf{a} = [a_1 \; a_2 \; \cdots \; a_\zeta]'$. In order for $\mathscr{B}_s(\mathbf{x},\mathbf{a})$ to accurately approximate $\mathscr{B}(\mathbf{x})$, the following summation I must be small.

$$I = \sum_{i=1}^{\rho} [\mathscr{B}_s(\mathbf{x}^i,\mathbf{a}) - \mathscr{B}(\mathbf{x}^i)]^2 \tag{7-108}$$

From classical min–max theory, I may be minimized with respect to the a_i's if the following necessary conditions are satisfied.

$$\sum_{i=1}^{\rho} [\mathscr{B}_s(\mathbf{x}^i,\mathbf{a}) - \mathscr{B}(\mathbf{x}^i)][\partial \mathscr{B}_s(\mathbf{x}^i,\mathbf{a})/\partial a_j] = 0, \qquad j = 1, 2, \ldots, \zeta \tag{7-109}$$

This set of ζ equations is linear in the ζ a_i's because of the assumed linearity of $\mathscr{B}_s(\mathbf{x},\mathbf{a})$ in the a_i's.

Because rapid-access storage space is to be conserved, the values of $\mathscr{B}(\mathbf{x}^i)$ are not simultaneously available. To save storage space, each time a new $\mathscr{B}(\mathbf{x}^k)$ is generated, it is incorporated in a partial sum $\sum_{jk}$:

$$\sum_{jk} \triangleq \sum_{i=1}^{k} [\mathscr{B}_s(\mathbf{x}^i,\mathbf{a}) - \mathscr{B}(\mathbf{x}^i)][\partial \mathscr{B}_s(\mathbf{x}^i,\mathbf{a})/\partial a_j]$$

$$= \sum_{j,k-1} + [\mathscr{B}_s(\mathbf{x}^k,\mathbf{a}) - \mathscr{B}(\mathbf{x}^k)][\partial \mathscr{B}_s(\mathbf{x}^k,\mathbf{a})/\partial a_j] \tag{7-110}$$

for $j = 1, 2, \ldots, \zeta$. Only the most recent set of partial sums $\sum_{jk}, j = 1, 2, \ldots, \zeta$, need be stored. These are stored, of course, in terms of the coefficients of the a_i's and in terms of a constant for each equation of the set. Once the last set of partial sums $\sum_{j\rho}, j = 1, 2, \ldots, \zeta$, has been generated, the resulting set of linear equations

$$\sum_{j\rho} = 0, \qquad j = 1, 2, \ldots, \zeta \tag{7-111}$$

is solved for the a_i's which minimize 7-108, and these are stored, along with a program for the evaluation of $\mathscr{B}_s(\mathbf{x},\mathbf{a})$.

7-10c. Lagrange Multipliers

In some cases, Lagrange multipliers may be used to offset dimensionality difficulties [7.4, 7.46, 7.82]. For example, consider the problem of maximizing $P(\mathbf{x})$,

$$P(\mathbf{x}) = \sum_{i=1}^{K} f_i(x_i) \tag{7-112}$$

subject to the following constraints:

$$x_i \geq 0, \qquad i = 1, 2, \ldots, K \tag{7-113}$$

$$\sum_{i=1}^{K} g_{1i}(x_i) = b_1 \quad \text{(constant)} \tag{7-114}$$

and
$$\sum_{i=1}^{K} g_{2i}(x_i) = b_2 \quad \text{(constant)} \quad (7\text{-}115)$$

This problem is a two-state-variable version of the problem treated in Section 7-2; note, however, that here the x_i's are not restricted to integer values. If we have an application for which integer values of x_i's are required, the Lagrange multiplier approach given here is incomplete, and additional literature [7.50, 7.82] is pertinent.

In place of the two-state-variable problem given above, consider the following one-state-variable problem: Find the maximum of $P_a(\mathbf{x},h)$ with respect to $\mathbf{x}$, where

$$P_a(\mathbf{x},h) \triangleq \sum_{i=1}^{K} [f_i(x_i) - hg_{2i}(x_i)] \quad (7\text{-}116)$$

subject only to satisfaction of constraints 7-113 and 7-114. The parameter h is a Lagrange multiplier. For a specific value of h (and given functions $f_i(x_i)$, $g_{1i}(x_i)$, and $g_{2i}(x_i)$) this problem can be solved by using the one-state-variable procedures of this chapter.

Theorem 7-5. *Assume that $\mathbf{x}^*$ is a value of $\mathbf{x}$ which yields a maximum of $P(\mathbf{x})$ in the original two-state-variable problem, and assume that $\mathbf{x}^o(h)$ is a value of $\mathbf{x}$ which yields a maximum of $P_a(\mathbf{x},h)$ in the one-state-variable problem. If a value h^o of h exists such that*

$$\sum_{i=1}^{K} g_{2i}[x_i^o(h^o)] = b_2 \quad (7\text{-}117)$$

then it necessarily follows that $P(\mathbf{x}^*) = P[\mathbf{x}^o(h^o)]$.

Proof: According to assumptions,

$$\sum_{i=1}^{K} \{f_i[x_i^o(h^o)] - h^o g_{2i}[x_i^o(h^o)]\} \geq \sum_{i=1}^{K} [f_i(x_i^*) - h^o g_{2i}(x_i^*)] \quad (7\text{-}118)$$

or equivalently,

$$P[\mathbf{x}^o(h^o)] - h^o b_2 \geq P(\mathbf{x}^*) - h^o b_2 \quad (7\text{-}119)$$

Thus,
$$P[\mathbf{x}^o(h^o)] \geq P(\mathbf{x}^*) \quad (7\text{-}120)$$

But according to the supposition that $P(\mathbf{x}^*)$ is the constrained maximum of $P(\mathbf{x})$,

$$P(\mathbf{x}^*) \geq P[\mathbf{x}^o(h^o)] \quad (7\text{-}121)$$

Obviously, 7-120 and 7-121 are simultaneously satisfied only if

$$P(\mathbf{x}^*) = P[\mathbf{x}^o(h^o)] \tag{7-122}$$

and the desired result is obtained.

•

In practice, the one-state-variable problem is solved initially with a value h^1 of h picked by guesswork. If the corresponding value of

$$\sum_{i=1}^{K} g_{2i}[x_i{}^o(h^1)] \tag{7-123}$$

is greater (less) than b_2, a smaller (larger) value h^2 of h is selected, and therewith the one-state-variable problem is solved again. A third value h^3 of h may be picked by extrapolation or interpolation, whichever is appropriate, and this procedure is continued until the summation is sufficiently close to b_2 (see Figure 7-11).

As in the preceding subsection, a trade-off between rapid-access storage requirements and computation time is evident here. Instead of solving a two-state-variable problem once, several (on occasion, many) one-state-variable problems are solved in sequence. The Lagrange multiplier procedure of this section can be extended as far as to solve four-state-variable problems by replacing two of the four state variables with two Lagrange multipliers; but for more complicated problems than this, additional techniques are generally required to obtain numerical solutions with the aid of present computer facilities.

7-10d. Region-Limiting Strategies and Iterated Dynamic Programming

There are numerous ways in which iterative techniques can be used in conjunction with dynamic programming to solve problems with many state

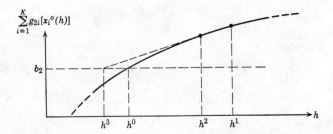

Figure 7-11. Search for the appropriate Lagrange multiplier.

Sect. 7-10   STATE VARIABLES: DISCRETE PROCESSES   415

variables. The drawback to these techniques, in contrast to dynamic programming alone, is that they do not necessarily yield the absolute maximum (minimum) of a performance measure; thus, these techniques are used primarily when computational limitations do not permit the use of the more conventional dynamic programming approach. The iterative techniques to be considered in this subsection contain two basic phases: a dynamic programming phase and a strategy phase. A dynamic programming phase consists of conventional dynamic programming applied over a restricted domain, a restricted domain of either policy space or state space or both and one which is changed from one stage of a dynamic programming solution to the next, in general. A strategy phase is one in which the limited domains for the next dynamic programming phase are selected. By alternating between strategy phases and dynamic programming phases, better and better solutions are obtained, and these, hopefully but not necessarily, converge to the optimal solution.

To illustrate the essential concepts involved, consider the control problem of Section 7-10a with the initial state $\mathbf{x}_0$ of the system fixed at some value. It is usually easy to rule out state-space regions that could not possibly contain the value of $\mathbf{x}_1$ which results in a minimum of P, Equation 7-96. In fact, a region R_k of state space may be identified for each $\mathbf{x}_k$; in region R_k, the optimal value of $\mathbf{x}_k$ is known to lie with some degree of confidence. Let $R_k{}^0$ denote the initial selection of R_k, and let the "center" point of this region $R_k{}^0$ be denoted by $\mathbf{x}_k{}^0$. The sequence of points $\mathbf{x}_0{}^0, \mathbf{x}_1{}^0, \ldots, \mathbf{x}_K{}^0$ is called the *initial*

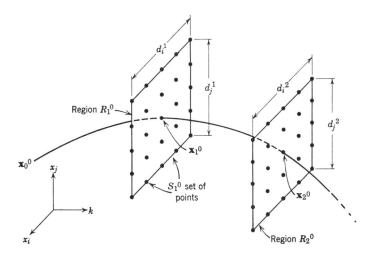

Figure 7-12. Illustration of an initial set of limited regions.

trajectory in state space. Corresponding to the initial trajectory in state space is an initial trajectory in policy space $\mathbf{m}_0^0, \mathbf{m}_1^0, \ldots, \mathbf{m}_{K-1}^0$; specification of the initial state-space trajectory generally defines the initial policy-space trajectory, and vice versa, because of the relationships given in 7-102.

The initial region R_k^0 for $\mathbf{x}_k$ could be assigned any number of shapes, but for convenience, a hyper-rectangular region is considered here. The rectangles shown in Figure 7-12 are suggestive of the n-dimensional case. The x_i, x_j coordinates in the figure represent n-dimensional state space, while the k coordinate is indicative of time. Note that the initial trajectory in state space passes through the centers of the limited rectangular regions.

If ξ^n rapid-access storage locations are available for storage of $\mathscr{F}_k(\mathbf{x}_k)$, each component of $\mathbf{x}_k$ may assume only ξ values in its limited region assuming each component of $\mathbf{x}_k$ is to be treated equally. By spacing these values evenly —they could be spaced unevenly—throughout the region R_k^0, an array of points S_k^0 is obtained, as in Figure 7-12. This completes the first strategy phase.

We may also use region-limiting in policy space; Figure 7-12 is suggestive of this if the $\mathbf{x}$'s and x's in the figure are replaced by $\mathbf{m}$'s and m's, respectively. Caution must be exercised, however, because the optimal policy trajectory may be quite discontinuous, as is evidenced in the example of Section 7-13.

The first dynamic programming phase is conducted in the standard way; either a forward or a backward dynamic programming algorithm may be employed. If a forward algorithm is applicable, with $\mathbf{x}_0$ fixed, $\mathscr{F}_1(\mathbf{x}_1)$ is evaluated first:

$$\mathscr{F}_1(\mathbf{x}_1) = \underset{\mathbf{m}_0 \in U_0}{\text{minimum}} \{f_1(\mathbf{x}_1, \mathbf{m}_0)\} \tag{7-124}$$

where $\mathbf{x}_1 \in S_1^0$ and where $\mathbf{m}_0$ must satisfy

$$\mathbf{x}_1 = \mathbf{q}(\mathbf{x}_0, \mathbf{m}_0, 0) \tag{7-125}$$

which is typically satisfied by only one value of $\mathbf{m}_0$ for given $\mathbf{x}_1$. The function $\mathscr{F}_1(\mathbf{x}_1)$ is computed at its grid of points S_1^0 and is stored.

Computation of $\mathscr{F}_2(\mathbf{x}_2)$ is next:

$$\mathscr{F}_2(\mathbf{x}_2) = \underset{\mathbf{m}_1 \in U_1}{\text{minimum}} \{f_2(\mathbf{x}_2, \mathbf{m}_1) + \mathscr{F}_1[\mathbf{v}(\mathbf{x}_2, \mathbf{m}_1, 1)]\} \tag{7-126}$$

where $\mathbf{x}_2 \in S_2^0$ and where, in general,

$$\mathbf{x}_k = \mathbf{v}(\mathbf{x}_{k+1}, \mathbf{m}_k, k), \quad k = 0, 1, \ldots, K-1 \tag{7-127}$$

Because $\mathscr{F}_1(\mathbf{x}_1)$ is stored at only those values of $\mathbf{x}_1$ which are contained in S_1^0, n-dimensional interpolation must be employed to evaluate

$$\mathscr{F}_1[\mathbf{v}(\mathbf{x}_2, \mathbf{m}_1, 1)]$$

Any value of m_1 for which the corresponding value of $v(x_2,m_1,1)$ is outside the region $R_1{}^0$ may be ignored in the evaluation of $\mathscr{F}_2(x_2)$. With $\mathscr{F}_2(x_2)$ computed at its grid of points $S_2{}^0$, $\mathscr{F}_3(x_3)$ may be computed, and so forth.

After $\mathscr{F}_K(x_K)$ is computed, the control sequence $m_0{}^1, m_1{}^1, \ldots, m_{K-1}^1$ (and the corresponding state sequence $x_0, x_1{}^1, x_2{}^1, \ldots, x_K{}^1$) is determined on the basis that, within the limited framework of the first dynamic programming solution, it resulted in a minimum of the performance measure P. The steps in this determination are as follows: first, the value $x_K{}^1$, an element of $S_K{}^0$, is determined on the basis that

$$\mathscr{F}_K(x_K{}^1) \le \mathscr{F}_K(x_K) \tag{7-128}$$

for any $x_K \in S_K{}^0$; second, the value m_{K-1}^1 of m_{K-1} is determined on the basis that

$$\{f_K(x_K{}^1, m_{K-1}^1) + \mathscr{F}_{K-1}[v(x_K{}^1, m_{K-1}^1, K-1)]\}, \qquad m_{K-1}^1 \in U_{K-1}$$

is less than

$$\{f_K(x_K{}^1, m_{K-1}) + \mathscr{F}_{K-1}[v(x_K{}^1, m_{K-1}, K-1)]\}, \qquad m_{K-1} \in U_{K-1}$$

for any m_{K-1} which results in satisfaction of

$$v(x_K{}^1, m_{K-1}, K-1) \in R_{K-1}^0$$

third, the value m_{K-1}^1 of m_{K-1} is used to obtain $x_{K-1}^1 = v(x_K{}^1, m_{K-1}^1, K-1)$; and in succession, the values $m_{K-2}^1, x_{K-2}^1, m_{K-3}^1, x_{K-3}^1, \ldots, m_1{}^1, x_1{}^1$, and $m_0{}^1$ are obtained similarly. This concludes the first dynamic programming phase.

For the second strategy phase, the values $x_1{}^1, x_2{}^1, \ldots, x_K{}^1$ are assigned the roles of centers of new arrays of points $S_1{}^1, S_2{}^1, \ldots, S_K{}^1$, respectively. If any of the points $x_1{}^1, x_2{}^1, \ldots, x_K{}^1$ lie on the respective boundaries of the old arrays $S_1{}^0, S_2{}^0, \ldots, S_K{}^0$, the distances $d_j{}^k$ (see Figure 7-12) are maintained at their previous values; if not, the values of the $d_j{}^k$'s are reduced by a multiplicative factor. In either case, grid points are evenly distributed in the new regions $R_1{}^1, R_2{}^1, \ldots, R_K{}^1$.

The second dynamic programming phase is like the first, and the process is continued in a cyclic manner. The fact that each new set of grid points contains the best trajectory obtained up to that point in the procedure insures that the value of P is improved, at least it does not increase, after each dynamic programming phase. After a specified number of phases or after all $d_j{}^k$'s are decreased below a predetermined value, the solution at that point is a winner, if perhaps not the optimum.

Solution schemes of the preceding type have been categorized in the literature [7.37] under the heading of *approximation in state space*. In like manner, the phrase *approximation in policy space* has been used in conjunction with region-limiting of control variables, but this phrase has also been used in conjunction with a quite different solution procedure which is described in Section 7-11.

7-11. APPROXIMATIONS IN FUNCTION AND POLICY SPACE

7-11a. A Control Problem

Consider the control problem of Section 7-10a with the following restrictions:[7] (1) the column matrix $\mathbf{q} = \mathbf{q}(\mathbf{x}_k, \mathbf{m}_k)$ in Equation 7-102 is formally independent of k; (2) the constraint set $U_k = U$ of $\mathbf{m}_k$ is independent of k; (3) the function $f_k(\mathbf{x}_k, \mathbf{m}_{k-1})$ in Equation 7-106 is formally independent of k, i.e., $f(\mathbf{x}_k, \mathbf{m}_{k-1}) = f_k(\mathbf{x}_k, \mathbf{m}_{k-1}) = f_i(\mathbf{x}_k, \mathbf{m}_{k-1})$ for any i; (4) the duration K of the process approaches infinity; (5) the state $\mathbf{x}_k$ can be determined at "time" k; and (6) the optimal control action $\mathbf{m}_k^* = \hat{\mathbf{m}}_k(\mathbf{x}_k^*)$ is to be generated at time k on the basis of the then known value $\mathbf{x}_k^*$ of the optimal response of the system.

This type of problem is typically called a *regulator problem* (Figure 7-13). In order for a solution to exist to the problem, it is necessary (but not sufficient) that control actions $\mathbf{m}_k \in U$ exist which drive the values of $f(\mathbf{x}_k, \mathbf{m}_{k-1})$ to zero as k approaches infinity. An optimal controller produces the absolute minimum of the sum of the $f(\mathbf{x}_k, \mathbf{m}_{k-1})$'s.

It is clear that the principle of optimality applies to this problem: whatever the initial state and the initial decisions are, the remaining decisions must be optimal with respect to the state resulting from the initial decision in order for

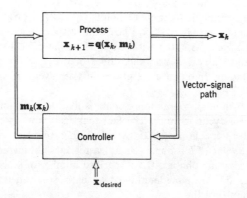

Figure 7-13. Feedback regulator.

---

[7] Restrictions 1 and 2 are equivalent to the assumption that the system is time-invariant; and restriction 3 is equivalent to the assumption that the performance measure is time-invariant.

Sect. 7-11 FUNCTION AND POLICY SPACE

the performance measure to be minimized. But observe that the problem which remains after the first stage is completed is identical to the original problem, except for a change in initial state, this occurring because an infinite number of stages remains and because of the time-invariance of the system. Thus it must be that the policy function $\hat{m}_k(x_k)$ is formally independent of k, i.e., $\hat{m}(x_k) \triangleq \hat{m}_k(x_k) = \hat{m}_i(x_k)$ for any i.

The backward dynamic programming algorithm 7-107 applies here in the following simplified form:

$$\mathscr{B}_{K-k}(x_k) = \underset{m_k \in U}{\text{minimum}} \{f[q(x_k,m_k),m_k] + \mathscr{B}_{K-k-1}[q(x_k,m_k)]\} \quad (7\text{-}129)$$

with $\mathscr{B}_0(x_K) = 0$. As the number of stages K approaches infinity, the return function $\mathscr{B}_{K-k}$ becomes independent of any finite k; thus,

$$\mathscr{B}(c) \triangleq \lim_{K \to \infty} \mathscr{B}_{K-k}(c) = \lim_{K \to \infty} \mathscr{B}_{K-k-1}(c) \quad (7\text{-}130)$$

where c is any $n \times 1$ column matrix of constants. Equation 7-129 is thereby simplified to

$$\mathscr{B}(x_k) = \underset{m_k \in U}{\text{minimum}} \{f[q(x_k,m_k),m_k] + \mathscr{B}[q(x_k,m_k)]\} \quad (7\text{-}131)$$

If the system is in state x_k at time k, the policy function $\hat{m}(x_k)$ yields the desired minimum in Equation 7-131:

$$\mathscr{B}(x_k) = f\{q[x_k,\hat{m}(x_k)],\hat{m}(x_k)\} + \mathscr{B}\{q[x_k,\hat{m}(x_k)]\} \quad (7\text{-}132)$$

Note that only two functions are unknown at the outset of the solution to this problem: the minimum-cost function $\mathscr{B}$ and the policy function $\hat{m}$. The optimal controller of Figure 7-13 is to generate a value of $\hat{m}$ as a function of the state at any time k. The following two subsections are devoted to this problem.

7-11b. An Approximation in Function Space

The type of approximation in function space considered here has its origins in classical analysis. A more descriptive but less succinct title for the solution mechanism involved in this "approximation in function space" is: "an iterative algorithm for generation of a minimum-cost function based on an initial guess of the function as a function of the state variables."

Assuming the form of the function can be guessed judiciously, the initial guess $\mathscr{B}^0(x_{k+1}) = \mathscr{B}^0[q(x_k,m_k)]$ is substituted into the right-hand member of Equation 7-131, and the minimized result is denoted by $\mathscr{B}^1(x_k)$:

$$\mathscr{B}^1(x_k) = \underset{m_k \in U}{\text{minimum}} \{f[q(x_k,m_k),m_k] + \mathscr{B}^0[q(x_k,m_k)]\} \quad (7\text{-}133)$$

In turn, $\mathscr{B}^1(\mathbf{x}_{k+1}) = \mathscr{B}^1[\mathbf{q}(\mathbf{x}_k,\mathbf{m}_k)]$ is substituted into the right-hand member of Equation 7-131 to obtain $\mathscr{B}^2(\mathbf{x}_k)$; in general,

$$\mathscr{B}^{j+1}(\mathbf{x}_k) = \underset{\mathbf{m}_k \in U}{\text{minimum}} \{f[\mathbf{q}(\mathbf{x}_k,\mathbf{m}_k),\mathbf{m}_k] + \mathscr{B}^{j}[\mathbf{q}(\mathbf{x}_k,\mathbf{m}_k)]\} \quad (7\text{-}134)$$

for $j = 0, 1, \ldots$. The conditions under which $\mathscr{B}^j(\mathbf{x}_k)$ approaches $\mathscr{B}(\mathbf{x}_k)$ are not at all clear in the general case. The rate of this convergence, assuming it does converge, is likewise vague. Note, however, that if the initial guess $\mathscr{B}^0(\mathbf{x}_k)$ equals identically zero, recurrence relations 7-129 and 7-134 are strikingly similar: they are identical for all practical purposes if $K - k - 1$ is identified with j. We would expect, therefore, that the iterated solution of 7-134 would converge to $\mathscr{B}(\mathbf{x}_k)$ when the initial guess $\mathscr{B}^0(\mathbf{x}_k)$ is set at zero—the main reason for it not to converge being that round-off and interpolation errors may accumulate. The rate of covergence when $\mathscr{B}^0(\mathbf{x}_k) = 0$ may be poor, however, and a more realistic choice of $\mathscr{B}^0(\mathbf{x}_k)$ is recommended. Once $\mathscr{B}(\mathbf{x}_k)$ is found, $\hat{\mathbf{m}}(\mathbf{x}_k)$ is generated on the basis that it minimizes the right-hand member of Equation 7-131.

7-11c. An Approximation in Policy Space

Bellman's approximation in policy space [7.5, 7.12, 7.22] is presented in this subsection. A more descriptive but less succinct title for the solution mechanism involved in this "approximation in policy space" is: "an iterative algorithm for generation of a policy function based on an initial guess of the policy function as a function of the state variables." The procedure is related to the approximation in function space of the preceding subsection; both procedures pertain to the solution of the problem of Section 7-11a.

Basically, we use an initial approximation $\mathbf{m}^0$ of the policy function $\hat{\mathbf{m}}$ to obtain a better policy function $\mathbf{m}^1$ which, in turn, is used to obtain $\mathbf{m}^2$, and so on. This approach is quite appealing in that many processes are controlled by known, but nonoptimal, policy functions upon which we would like to improve.

Concisely stated, the procedure is: First, evaluate the effect of using the policy $\mathbf{m}^0$ at each stage of the decision process for representative values of $\mathbf{x}_0$ (the resulting values of P are denoted here by $\mathscr{B}^0(\mathbf{x}_0)$). Second, evaluate the policy $\mathbf{m}^1$ which results in the minimum of the infinite-stage process when policy $\mathbf{m}^1$ is used at the first stage of the process, where the state is $\mathbf{x}_0$, and policy $\mathbf{m}^0$ is used thereafter; thus, $\mathbf{m}^1$ is the $\mathbf{m} \in U$ that yields the minimum of

$$f[\mathbf{q}(\mathbf{x}_0,\mathbf{m}),\mathbf{m}] + \mathscr{B}^0[\mathbf{q}(\mathbf{x}_0,\mathbf{m})]$$

Next, $\mathscr{B}^1(\mathbf{x}_0)$ is defined here, in contrast to the definition of the preceding subsection, to be the value of P, Equation 7-106, under the condition that policy $\mathbf{m}^1$ is used at every stage of the process with $\mathbf{x}_0$ being the initial state. And in cyclic fashion, the policy $\mathbf{m}^{j+1} = \mathbf{m}^{j+1}(\mathbf{x}_0)$ is obtained on the basis that it yields a minimum of P under the condition that policy $\mathbf{m}^{j+1}$ is used at the first stage, where the state is $\mathbf{x}_0$, and policy $\mathbf{m}^j$ is used thereafter; and $\mathcal{B}^{j+1}(\mathbf{x}_0)$ is formed from P by applying policy $\mathbf{m}^{j+1}$ at every stage with $\mathbf{x}_0$ being the initial state.

Analytically, the above steps may be cumbersome to put into effect—the iterative relationships are as follows:

$$\mathcal{B}^j(\mathbf{x}_0) = \sum_{i=1}^{\infty} f[\mathbf{x}_i^j, \mathbf{m}^j(\mathbf{x}_{i-1}^j)], \qquad j = 0, 1, \ldots \qquad (7\text{-}135)$$

where

$$\begin{aligned} \mathbf{x}_1^j &= \mathbf{q}[\mathbf{x}_0, \mathbf{m}^j(\mathbf{x}_0)] \\ &\vdots \\ \mathbf{x}_i^j &= \mathbf{q}[\mathbf{x}_{i-1}^j, \mathbf{m}^j(\mathbf{x}_{i-1}^j)] \\ &\vdots \end{aligned} \qquad (7\text{-}136)$$

The superscript j of $\mathbf{x}_i^j$ is introduced to emphasize the dependence of $\mathbf{x}_i^j$ on $\mathbf{m}^j$; given $\mathbf{x}_0$ and the policy function $\mathbf{m}^j$, $\mathcal{B}^j(\mathbf{x}_0)$ of 7-135 can be completely determined in terms of $\mathbf{x}_0$. Each new cost function $\mathcal{B}^j$ is used to find a new policy function, $\mathbf{m}^{j+1}(\mathbf{c}) \in U$ for any $n \times 1$ constant column matrix $\mathbf{c}$, from the relationship that

$$f[\mathbf{q}(\mathbf{c},\mathbf{m}^{j+1}(\mathbf{c})), \mathbf{m}^{j+1}(\mathbf{c})] + \mathcal{B}^j[\mathbf{q}(\mathbf{c},\mathbf{m}^{j+1}(\mathbf{c}))] \le f[\mathbf{q}(\mathbf{c},\mathbf{m}),\mathbf{m}] + \mathcal{B}^j[\mathbf{q}(\mathbf{c},\mathbf{m})] \qquad (7\text{-}137)$$

for any $\mathbf{m} \in U$. After each sequential iteration of 7-135 and 7-137, the new value of P obtained with a specific $\mathbf{x}_0 = \mathbf{c}$ is less than (or, at worst, equal to) the previous value of P with the same $\mathbf{x}_0 = \mathbf{c}$. This fact is substantiated by observing that policy $\mathbf{m}^j$ is one of the policies which $\mathbf{m}$ may assume in inequality 7-137; in that case, the left-hand member of 7-137 is the new value of P, and the right-hand member is the old value of P. As for convergence to the optimum or for rate of convergence, very little can be said in the general case. The reader may gain additional insight on the methods of this and the preceding subsection by solving Problem 7.19.

The procedure given in this subsection can be modified and combined in many ways with that of the preceding subsection. For example, an initial policy function $\mathbf{m}^0$ could be used to determine an initial cost function $\mathcal{B}^0$, as by using Equation 7-135, and this $\mathcal{B}^0$ could then be used as the initial cost function for the approximation-in-function-space technique of Section 7-11b.

7-11d. Nonoriented Minimal Chain Problems

Consider a chain network (Section 7-5) which has N nodes with the possibility of direct chain links (branches) between each and every node. Such a network is called a nonoriented network in contrast to the oriented ones of Section 7-5. Let S_T denote a nonempty set of terminal nodes and let S_0 denote a nonempty set of origin nodes. The object is to find a minimal path (minimum-length chain) of those that start on a node of S_0 and terminate on a node of S_T. Obviously, nodes from S_0 do not belong to S_T, and vice versa; the intersection of sets S_0 and S_T is empty. In a given case, S_0 and S_T could consist each of a single node.

Let $\mathscr{B}_j$ denote the length of the minimal chain that connects node j to S_T, and let $\mathscr{B}_{0T}$ denote the length of the minimal chain between S_0 and S_T, that is,

$$\mathscr{B}_{0T} = \min_{j \in S_0} \mathscr{B}_j \tag{7-138}$$

Also, let c_{ij} denote the length of the direct chain link between node i and node j. If no such links are given for particular i,j pairs, the corresponding c_{ij}'s are assigned large penalty values. In terms of this notation, it readily follows that

$$\mathscr{B}_i = \min_j (c_{ij} + \mathscr{B}_j), \qquad j \neq i \tag{7-139}$$

is a valid recurrence relation for which the boundary condition $\mathscr{B}_j = 0$ applies for any $j \in S_T$. But for other j's, note that $\mathscr{B}_j$ is unknown at the start, so that an approximation in function space is required for solution of 7-139.

In place of 7-139, consider the recurrence relation

$$\mathscr{B}_i^{k+1} = \min_j (c_{ij} + \mathscr{B}_j^k), \qquad j \neq i \tag{7-140}$$

for $k = 0, 1, 2, \ldots$, where $\mathscr{B}_j^0$ denotes the initial approximation of $\mathscr{B}_j$ as a function of j. Various ways of specifying initial approximations are considered in the literature. An effective one is

$$\mathscr{B}_j^0 = \underset{r \in S_T}{\text{minimum }} c_{jr}, \qquad j = 1, 2, \ldots \tag{7-141}$$

With this, the initial approximation $\mathscr{B}_{0T}^0$ corresponding to 7-138 is

$$\mathscr{B}_{0T}^0 = \min_{j \in S_0} (\min_{r \in S_T} c_{jr}) \tag{7-142}$$

And in particular, if S_0 contains only node number 1 and S_T contains only node number N, then $\mathscr{B}_{0T}^0$ reduces to c_{1N}.

A valuable feature of initial approximations 7-141 is that their use results in convergence of $\mathscr{B}_j^k$ to $\mathscr{B}_j$ in a finite number of iterations. To see this, observe that the chain corresponding to a particular $\mathscr{B}_j{}^0$ of 7-141 is the best of those *which are restricted to have but one link between node j and S_T*. Similarly, of all chains which start at node j and terminate in S_T, the chain corresponding to $\mathscr{B}_j^{k-1}$ is the best of those *which are restricted to have at most k links*. Now, an intrinsic property of a minimum-length chain is that it passes through any given node once at most, assuming positive lengths are assigned to each link. But because there are N nodes, at least one of which is in S_0 and at least one of which is in S_T, an upper bound on the number of direct links in the overall minimal chain is $N - 1$, and $\mathscr{B}_j^{N-2}$ necessarily equals the optimum $\mathscr{B}_j$. Actually, for any given chain network, it is highly probable that every $\mathscr{B}_j$ will have associated with it many fewer than $N - 1$ separate links. The form of 7-140 dictates that the optimal solution is obtained when $\mathscr{B}_j^K$ equals $\mathscr{B}_j^{K-1}$ for all j and for some $K \leq N - 2$.

Consider the case of a unique terminal node, node number N, and a unique origin node, node number 1. The evaluation of each $\mathscr{B}_i^{k+1}$ in 7-140 requires the comparison of at most $N - 1$ values. Because integer values of i from 1 through $N - 1$ are of interest in the evaluation of $\mathscr{B}_i^k$'s, there is a total of $(N - 1)^2$ comparisons at most for each value of k. Finally, the $\mathscr{B}_i^k$'s for k from 1 through $N - 2$ at most are to be evaluated, so the total number of comparisons for solution is at most $(N - 2)(N - 1)^2$; but this figure is very conservative. For example, if the number of links in any minimal chain is restricted to be no more than K and if each node has a direct link, say, to only 10 other designated nodes, then the total number of comparisons required is no more than $10(K - 1)(N - 1)$.

Generalizations of problems of the preceding type have been investigated in the literature (e.g., [7.16, 7.28, 7.77]). The examination of a particular generalization is left as an exercise (Problem 7.21).

7-12. CONTINUOUS DECISION PROCESSES: DISCRETE APPROXIMATIONS WITH n STATE VARIABLES

7-12a. A General Control Problem

The general problem considered in this section and in Section 7-13 is an extension of the control problem which is considered in Section 3-9. In both sections, the system under consideration is assumed to be characterized by a vector state equation of the form

$$\dot{\mathbf{x}} = \mathbf{q}(\mathbf{x},\mathbf{m},t) \qquad (7\text{-}143)$$

where $\mathbf{x} = \mathbf{x}(t) = [x_1(t)\ x_2(t)\ \cdots\ x_n(t)]'$ is the n-dimensional state vector of the system, $\dot{\mathbf{x}} = [(dx_1/dt)\ (dx_2/dt)\ \cdots\ (dx_n/dt)]'$ is the time derivative of the state vector, $\mathbf{m} = \mathbf{m}(t) = [m_1(t) \ m_2(t) \ \cdots \ m_r(t)]'$ is the r-dimensional control vector of the system, and

$$\mathbf{q} = \mathbf{q}(\mathbf{x},\mathbf{m},t) = [q_1(\mathbf{x},\mathbf{m},t) \ q_2(\mathbf{x},\mathbf{m},t) \ \cdots \ q_n(\mathbf{x},\mathbf{m},t)]'$$

is a given vector function of its arguments. The state equation 7-143 is assumed to be satisfied over some prescribed interval of time.

Unlike the results of Section 3-9, the major results of this section and the next hold for the case in which the control vector $\mathbf{m}$ is constrained, that is, $\mathbf{m}(t)$ is assumed to be restricted to lie in a closed and bounded set U of an r-dimensional space, and it is possible for $U = U(t)$ to be a function of time. For example, U could be a set of discrete, finite values each of which may be a function of time; or U could be defined by

$$m_{k,\min} \leq m_k \leq m_{k,\max}, \qquad k = 1, 2, \ldots, r \qquad (7\text{-}144)$$

where $m_{k,\max}$ and $m_{k,\min}$ are either finite constants or explicit finite functions of time for each integer k, $1 \leq k \leq r$. If strict inequality were required in constraint 7-144, the constraint set U would not be "closed" and the theory that follows would not be directly applicable.

As in Section 3-9, the problem is to specify the control vector $\mathbf{m}(t)$—here $\mathbf{m}(t) \in U(t)$—over the time interval defined by $t_a \leq t \leq t_b$ so as to minimize the functional J:

$$J = \int_{t_a}^{t_b} f(\mathbf{x},\mathbf{m},t) \, dt \qquad (7\text{-}145)$$

where $f = f(\mathbf{x},\mathbf{m},t)$ is a given real-valued function of its arguments and where t_a and t_b are values of time between which the state equation 7-143 is satisfied. It may be that t_a, t_b, $\mathbf{x}(t_a)$, and $\mathbf{x}(t_b)$ are completely specified, in which case the problem is said to have fixed end conditions; or it may be that the end conditions are partially defined by prescribed equations.

In Section 3-9, it is assumed that the functions f and $\mathbf{q}$ of Equations 7-145 and 7-143, respectively, are of class C^1 with respect to all of their arguments. In obtaining the major results of this section and those of the next, the same statement applies with respect to t and to the x_i's; that is, the first partial derivatives of f and $\mathbf{q}$ with respect to t and to the x_i's are assumed to be continuous in t and in the x_i's. Here, however, f and $\mathbf{q}$ need not be of class C^1 with respect to the m_i's. Only the variation of f and $\mathbf{q}$ with respect to the m_i's need be continuous (i.e., f and $\mathbf{q}$ are of class C^0 with respect to each m_i, $1 \leq i \leq r$).

7-12b. Recurrence Relations with Prespecified Time Increments

The optimal control problem of Section 7-12a may be solved by using discrete approximations in which the time interval $[t_a,t_b]$ is divided into K

time increments. The time increments are characterized by a strictly monotonic sequence of time instants: $t_a \triangleq t_0, t_1, t_2, \ldots, t_K \triangleq t_b$. In this subsection, the just-noted time instants are assumed to be specified in advance, that is, before the dynamic programming solution is initiated. A convenient and commonly used approach is that in which $t_{k+1} - t_k$ equals the same, fixed increment for any integer k, but certain other approaches, e.g., approaches based on Gaussian quadrature [7.60, 7.63], afford greater accuracy.

To simplify notation, let $\mathbf{x}_k \triangleq \mathbf{x}(t_k)$, $\mathbf{m}_k \triangleq \mathbf{m}(t_k)$, and $f(\mathbf{x}_k, \mathbf{m}_k, t_k) \triangleq f_k(\mathbf{x}_k, \mathbf{m}_k)$. If the difference $t_{k+1} - t_k$ is sufficiently small, the derivative $\dot{\mathbf{x}}(t)$ at $t = t_k$ can be equated, approximately, to a first-difference approximation, as follows:

$$\dot{\mathbf{x}}(t)|_{t=t_k} \simeq \frac{\mathbf{x}_{k+1} - \mathbf{x}_k}{t_{k+1} - t_k} \tag{7-146}$$

With this, Equation 7-143 can be replaced, approximately, by

$$\mathbf{x}_{k+1} = \mathbf{x}_k + (t_{k+1} - t_k)\mathbf{q}(\mathbf{x}_k, \mathbf{m}_k, t_k) \tag{7-147}$$

or, in more compact form,

$$\mathbf{x}_{k+1} = \mathbf{w}(\mathbf{x}_k, \mathbf{m}_k, k) \tag{7-148}$$

where

$$\mathbf{w}(\mathbf{x}_k, \mathbf{m}_k, k) \triangleq \mathbf{x}_k + (t_{k+1} - t_k)\mathbf{q}(\mathbf{x}_k, \mathbf{m}_k, t_k)$$

If the time increments $t_{k+1} - t_k$ are sufficiently small for all integer values of k and if the control action $\mathbf{m}(t)$ is relatively constant during the half-closed time intervals defined by $[t_k, t_{k+1})$ for all integer values of k, the performance measure of Equation 7-145 is approximately equivalent to

$$P = \sum_{k=0}^{K-1} (t_{k+1} - t_k) f_k(\mathbf{x}_k, \mathbf{m}_k) \tag{7-149}$$

Because of the assumption preceding 7-149, jump discontinuities of $\mathbf{m}(t)$ can occur only at the t_k instants of time in this approximate solution.

Equations 7-148 and 7-149 are basically the same as Equations 7-102 and 7-106, respectively, of the corresponding discrete decision process in Section 7-10—with minor changes both in index and notation, the reader may verify (Problem 7.22) the following backward recurrence relations which apply here and which correspond to Equation 7-107 of Section 7-10.

$$\mathscr{B}_1(\mathbf{x}_{K-1}) = \min_{\mathbf{m}_{K-1} \in U_{K-1}} (t_K - t_{K-1}) f_{K-1}(\mathbf{x}_{K-1}, \mathbf{m}_{K-1}) \tag{7-150a}$$

in which the constraint $\mathbf{x}_K = \mathbf{w}(\mathbf{x}_{K-1}, \mathbf{m}_{K-1}, K-1)$ must be taken into account in the minimization process if x_K is not completely free to be selected. In general,

$$\mathscr{B}_{K-k}(\mathbf{x}_k) = \underset{\mathbf{m}_k \in U_k}{\text{minimum}} \{(t_{k+1} - t_k)f_k(\mathbf{x}_k, \mathbf{m}_k) + \mathscr{B}_{K-k-1}[\mathbf{w}(\mathbf{x}_k, \mathbf{m}_k, k)]\} \quad (7\text{-}150\text{b})$$

for $k = K - 2, K - 3, \ldots, 1, 0$.

Forward recurrence relations which apply to the problem of this section can be obtained, provided that Equation 7-148 can be rearranged to obtain $\mathbf{x}_k$ as a single-valued function of $\mathbf{x}_{k+1}$, $\mathbf{m}_k$, and k, i.e.,

$$\mathbf{x}_k = \mathbf{v}(\mathbf{x}_{k+1}, \mathbf{m}_k, k) \quad (7\text{-}151)$$

Again, the reader may verify that

$$\mathscr{F}_1(\mathbf{x}_1) = \underset{\mathbf{m}_0 \in U_0}{\text{minimum}} (t_1 - t_0)f_0[\mathbf{v}(\mathbf{x}_1, \mathbf{m}_0, 0), \mathbf{m}_0] \quad (7\text{-}152\text{a})$$

in which the constraint $\mathbf{x}_0 = \mathbf{v}(\mathbf{x}_1, \mathbf{m}_0, 0)$ must be taken into account in the minimization process if $\mathbf{x}_0$ is not completely free to be selected. In general,

$$\mathscr{F}_{k+1}(\mathbf{x}_{k+1}) = \underset{\mathbf{m}_k \in U_k}{\text{minimum}} \{(t_{k+1} - t_k)f_k[\mathbf{v}(\mathbf{x}_{k+1}, \mathbf{m}_k, k), \mathbf{m}_k]$$
$$+ \mathscr{F}_k[\mathbf{v}(\mathbf{x}_{k+1}, \mathbf{m}_k, k)]\} \quad (7\text{-}152\text{b})$$

for $k = 1, 2, \ldots, K - 1$.

On the basis of the preceding development, it is clear that continuous decision processes—including problems of classical calculus of variations—can be approximated by discrete decision processes which are amenable to numerical solution via dynamic programming. All of the tools previously introduced for solution of discrete problems are at the disposal of those who wish to solve a continuous decision problem. This of itself is not too encouraging, however, when we have a problem with many state variables. The material of the following two subsections is of substantial help in this regard, but much remains to be done.

7-12c. A Continuous Recurrence Relation

From the dynamic programming point of view, it is expedient to imbed the control problem of Section 7-12a in *a set of infinitely many similar problems*. Thus, consider minimization of $J[\mathbf{x}(t), t, \mathbf{m}]$:

$$J[\mathbf{x}(t), t, \mathbf{m}] = \int_t^{t_b} f[\mathbf{x}(\tau), \mathbf{m}(\tau), \tau] \, d\tau \quad (7\text{-}153)$$

where, *for a given problem of the set, t assumes a particular value* in the range

Sect. 7-12 CONTINUOUS DECISION PROCESSES 427

$t_a \le t \le t_b$, and $\mathbf{x}(t)$ assumes a particular value at time $\tau = t$.[8] When $t = t_a$ in Equation 7-153, it is to be observed that the performance measure $J[\mathbf{x}(t_a), t_a, \mathbf{m}]$ is identical to the performance measure J, Equation 7-145, of the original problem—assuming, of course, that the same value of the initial state $\mathbf{x}(t_a)$ is associated with both performance measures. Also, when $t = t_b$, the value of $J[\mathbf{x}(t_b), t_b, \mathbf{m}]$ is obviously zero for any bounded $\mathbf{m}$.

To conform with the notation of the preceding paragraph, state equation 7-143 is rewritten as follows:

$$\frac{d\mathbf{x}(\tau)}{d\tau} = \mathbf{q}[\mathbf{x}(\tau), \mathbf{m}(\tau), \tau], \qquad t \le \tau \le t_b \qquad (7\text{-}154)$$

Different problems of the set are characterized by different values of t and $\mathbf{x}(t)$. As before, the control vector is assumed to be constrained by $\mathbf{m}(\tau) \in U(\tau)$.

Let the minimum of $J[\mathbf{x}(t), t, \mathbf{m}]$ with respect to $\mathbf{m} \in U$ be denoted by $\mathscr{B}[\mathbf{x}(t), t]$:

$$\mathscr{B}[\mathbf{x}(t), t] = \underset{\substack{\mathbf{m}(\tau) \in U(\tau) \\ t \le \tau \le t_b}}{\text{minimum}} \int_t^{t_b} f[\mathbf{x}(\tau), \mathbf{m}(\tau), \tau] \, d\tau \qquad (7\text{-}155)$$

As the reader may rightly surmise, the script $\mathscr{B}$ notation is suggestive of the backward nature of the recurrence relation that is developed in the following paragraph. In accord with prior convention, $\mathscr{B} = \mathscr{B}[\mathbf{x}(t), t]$ is called a minimum-cost function.

Question: Can the function $\mathscr{B}[\mathbf{x}(t), t]$ be related to the solution of a simpler problem? Well, suppose the integration in 7-155 is arranged as follows:

$$\mathscr{B}[\mathbf{x}(t), t] = \underset{\substack{\mathbf{m}(\tau) \in U(\tau) \\ t \le \tau \le t_b}}{\text{minimum}} \left\{ \int_t^{t+\epsilon} f[\mathbf{x}(\tau), \mathbf{m}(\tau), \tau] \, d\tau + \int_{t+\epsilon}^{t_b} f[\mathbf{x}(\tau), \mathbf{m}(\tau), \tau] \, d\tau \right\}$$

$$(7\text{-}156)$$

where ϵ is greater than zero. If ϵ is made sufficiently small and if the components of $\mathbf{m}(\tau)$ are both continuous and relatively constant over the half-open time interval defined by $t \le \tau < t + \epsilon$, the first integral in 7-156 can be approximated by $\epsilon f[\mathbf{x}(t), \mathbf{m}(t), t]$ with the result that

$$\mathscr{B}[\mathbf{x}(t), t] = \underset{\substack{\mathbf{m}(\tau) \in U(\tau) \\ t \le \tau \le t_b}}{\text{minimum}} \left\{ \epsilon f[\mathbf{x}(t), \mathbf{m}(t), t] + \int_{t+\epsilon}^{t_b} f[\mathbf{x}(\tau), \mathbf{m}(\tau), \tau] \, d\tau \right\} + \epsilon o_1(\epsilon)$$

$$(7\text{-}157)[9]$$

---

[8] The reader may well be puzzled by the convention adopted at this point with respect to the "variable" t. The utility of this convention, which is commonplace in the literature, is made apparent in this subsection.

[9] $o_i(\epsilon)$ is of the order ϵ, meaning that $\lim_{\epsilon \to 0} o_i(\epsilon) = 0$. Different integer values of the subscript i are used to differentiate between different functions, each of which is of order ϵ.

Now, since both $\mathbf{x}(t)$ and t are "fixed" at some values, the function

$$f[\mathbf{x}(t),\mathbf{m}(t),t]$$

is a function of $\mathbf{m}(\tau)$ at $\tau = t$ only. Hence, the minimization in 7-157 can be ordered as follows:

$$\mathcal{B}[\mathbf{x}(t),t] = \underset{\mathbf{m}(t) \in U(t)}{\text{minimum}} \left\{ \epsilon f[\mathbf{x}(t),\mathbf{m}(t),t] + \underset{\substack{\mathbf{m}(\tau) \in U(\tau) \\ t+\epsilon \leq \tau \leq t_b}}{\text{minimum}} \int_{t+\epsilon}^{t_b} f[\mathbf{x}(\tau),\mathbf{m}(\tau),\tau] \, d\tau \right\}$$
$$+ \epsilon o_1(\epsilon) \qquad (7\text{-}158)$$

But from the definition of $\mathcal{B}$, this amounts to

$$\mathcal{B}[\mathbf{x}(t),t] = \underset{\mathbf{m}(t) \in U(t)}{\text{minimum}} \{\epsilon f[\mathbf{x}(t),\mathbf{m}(t),t] + \mathcal{B}[\mathbf{x}(t + \epsilon), t + \epsilon]\} + \epsilon o_1(\epsilon) \quad (7\text{-}159)$$

Because of the assumed continuity of the m_k's in the interval defined by $t \leq \tau < t + \epsilon$ and because of the assumed continuity properties of $\mathbf{q}$ in 7-154, it must be that

$$\mathbf{x}(t + \epsilon) = \mathbf{x}(t) + \epsilon \dot{\mathbf{x}}(t) + \epsilon o_2(\epsilon) = \mathbf{x}(t) + \epsilon \mathbf{q}[\mathbf{x}(t),\mathbf{m}(t),t] + \epsilon o_2(\epsilon) \quad (7\text{-}160)$$

This is incorporated appropriately in 7-159 to obtain

$$\mathcal{B}[\mathbf{x}(t),t] = \underset{\mathbf{m}(t) \in U(t)}{\text{minimum}} \{\epsilon f[\mathbf{x}(t),\mathbf{m}(t),t] + \mathcal{B}[\mathbf{x}(t) + \epsilon \mathbf{q}(\mathbf{x}(t),\mathbf{m}(t),t)$$
$$+ \epsilon o_2(\epsilon), t + \epsilon]\} + \epsilon o_1(\epsilon) \quad (7\text{-}161)$$

and for sufficiently small but positive ϵ, $\epsilon o_1(\epsilon)$ and $\epsilon o_2(\epsilon)$ are omitted:

$$\mathcal{B}[\mathbf{x}(t),t] = \underset{\mathbf{m}(t) \in U(t)}{\text{minimum}} \{\epsilon f[\mathbf{x}(t),\mathbf{m}(t),t] + \mathcal{B}[\mathbf{x}(t) + \epsilon \mathbf{q}(\mathbf{x}(t),\mathbf{m}(t),t), t + \epsilon]\}$$
$$(7\text{-}162)$$

It is left to the reader to show the relationships between this continuous recurrence relation and the discrete approximation, Equation 7-150, previously derived.

7-12d. Recurrence Relations with Controlled Time Increments

Although continuous decision processes may be equated posthaste to discrete decision processes, as in Section 7-12b, such an approach masks a key feature of continuous decision processes; indeed, a degree of freedom is lost by a too hasty conversion to the discrete form. In this subsection, a version of the main attribute of Larson's *state-increment dynamic programming* [7.65, 7.66] is presented. The particulars, and there are many, of state-increment dynamic programming can be gleaned from the pertinent literature. The advantage of state-increment dynamic programming is that rapid-access storage space is conserved to a degree not generally obtainable with the direct dynamic programming approach.

Sect. 7-12   CONTINUOUS DECISION PROCESSES   429

To set the stage for the method, it is assumed that each state variable x_i, $i = 1, 2, \ldots, n$, must satisfy

$$\eta_i^- \leq x_i \leq \eta_i^+ \tag{7-163}$$

where η_i^- and η_i^+ may be functions of time and are limits between which the optimal solution is known to lie (or is required to lie). Of course, the time interval of interest is the closed interval $[t_a, t_b]$. In the $(n + 1)$-dimensional Euclidean space of the x_i's and time, the preceding statements define a closed and bounded region R of interest. Throughout R, we can imagine an array S of distributed points of which a representative sample is displayed in Figure 7-14. The x_i, x_j coordinates in the figure are suggestive of the n components of $\mathbf{x}$. A fixed spacing of magnitude Δt is shown between points in time, and a fixed spacing of magnitude Δx_i is shown between points in the x_i direction.

To apply the method, the Δx_i's and Δt must be selected sufficiently small so that interpolation may be used to find good estimates of $\mathscr{B}(\mathbf{x},t)$ between points in the array S. Accurate interpolation is possible if $\mathscr{B}(\mathbf{x},t)$ is continuous in all of its arguments. An additional restriction on the selection of Δt is that, for any $\epsilon \leq \Delta t$, the order-of-ϵ terms in Equation 7-161 must be negligible.

To evaluate $\mathscr{B}(\mathbf{x},t)$ at each point in the array S, we could let ϵ equal Δt in Equation 7-162 and proceed as in Section 7-12b, but this would necessitate having a great many values of $\mathscr{B}$ corresponding to time $t + \Delta t$ available in rapid-access storage to generate $\mathscr{B}$ at time t for even one value of $\mathbf{x}$. With the approach of this subsection, each new value of $\mathscr{B}$ is determined on the basis of a relatively small number, at most 2×3^n, of previously computed values of $\mathscr{B}$. For illustrative purposes, the black dots in Figure 7-15 represent

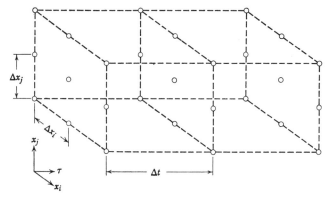

Figure 7-14. A representative sample of an array S of points.

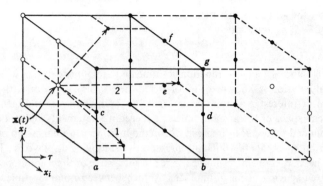

Figure 7-15. Indications of interpolations required to obtain the minimum-cost function at "point" $\mathbf{x}(t)$.

2×3^n points, ones with previously computed values of $\mathscr{B}$ which can be used in the evaluation of $\mathscr{B}$ at the general point labeled $\mathbf{x}(t)$.[10]

The evaluation of $\mathscr{B}[\mathbf{x}(t),t]$ proceeds as follows. In the first place, it is assumed that allowed values of $\mathbf{m}(t)$ can be represented by a set of values $U(t) = \{\mathbf{m}^1, \mathbf{m}^2, \ldots, \mathbf{m}^\nu\}$. For each $\mathbf{m}^k \in U(t)$, a positive value ϵ^k of ϵ for use in Equation 7-162 can be determined on the basis that the point corresponding to

$$\begin{bmatrix} \mathbf{x}(t) + \epsilon^k \mathbf{q}(\mathbf{x},\mathbf{m}^k,t) \\ t + \epsilon^k \end{bmatrix}$$

is on the surface of the solid "cube" in Figure 7-15. That is,

$$\begin{bmatrix} \mathbf{x}(t) + \epsilon^k |\mathbf{q}(\mathbf{x},\mathbf{m}^k,t)| \\ t + \epsilon^k \end{bmatrix} \leq \begin{bmatrix} \mathbf{x}(t) + \Delta \mathbf{x} \\ t + \Delta t \end{bmatrix} \qquad (7\text{-}164)$$

in which equality holds for at least one component, i.e.,

$$\epsilon^k = \min\left\{\frac{\Delta x_1}{|q_1(\mathbf{x},\mathbf{m}^k,t)|}, \frac{\Delta x_2}{|q_2(\mathbf{x},\mathbf{m}^k,t)|}, \ldots, \Delta t\right\} \qquad (7\text{-}165)$$

This equation is used to generate a value ϵ^k of ϵ for each value $\mathbf{m}^k$ of $\mathbf{m}$.

In order to perform the minimization in Equation 7-162, values of

$$\mathscr{B}[\mathbf{x} + \epsilon^k \mathbf{q}(\mathbf{x},\mathbf{m}^k,t), t + \epsilon^k]$$

must be found by using interpolation. For example, suppose that α is a specific value of k with

$$\epsilon^\alpha = \frac{\Delta x_i}{|q_i(\mathbf{x},\mathbf{m}^\alpha,t)|} \qquad (7\text{-}166)$$

[10] Points on the boundary of R require slightly different treatment which the reader may establish in due course.

for some integer i and, in addition, suppose that $q_i(\mathbf{x},\mathbf{m}^\alpha,t)$ is positive (e.g., see line number 1 in Figure 7-15). In this case, an n-dimensional interpolation scheme is used with the ith component of the state vector held fixed at $x_i + \Delta x_i$ (in Figure 7-15, points labeled a, b, c, and d can be used in an interpolation to estimate $\mathscr{B}[\mathbf{x} + \epsilon^\alpha \mathbf{q}(\mathbf{x},\mathbf{m}^\alpha,t), t + \epsilon^\alpha]$).

On the other hand, for a different value γ of k, suppose that Equation 7-165 yields $\epsilon^\gamma = \Delta t$. In that case (e.g., see line number 2 in Figure 7-15), an n-dimensional interpolation scheme is used with time held fixed at $t + \Delta t$ (in Figure 7-15, points labeled d, e, f, and g can be used in an interpolation scheme to estimate $\mathscr{B}[\mathbf{x} + \epsilon^\gamma \mathbf{q}(\mathbf{x},\mathbf{m}^\gamma,t), t + \epsilon^\gamma]$).

With this well-defined algorithm for determination of $\mathscr{B}$'s corresponding to ϵ^k's, it is conceptually straightforward to obtain $\mathscr{B}[\mathbf{x}(t),t]$:

$$\mathscr{B}[\mathbf{x}(t),t] = \min_k \{\epsilon^k f[\mathbf{x},\mathbf{m}^k,t] + \mathscr{B}[\mathbf{x} + \epsilon^k \mathbf{q}(\mathbf{x},\mathbf{m}^k,t), t + \epsilon^k]\} \qquad (7\text{-}167)$$

The advantage of the foregoing approach is that relatively few values of $\mathscr{B}$ need be held in rapid-access storage during the sometimes lengthy computations for the minimum in Equation 7-167. Actually, additional values of $\mathscr{B}$ may be stored in rapid-access storage to enable computation of a "block" of new values of $\mathscr{B}$—while the block of values is computed, no referral to slow-access storage need be made. The particular way in which this approach is implemented depends to a great extent on the available computational facilities. It is quite likely that special computers for dynamic programming will be developed for ultra-efficient solution of recurrence relations.[11] Even with the development of such computers, however, it is evident that there will always be an upper bound on the number of state variables and the number of control variables that can be accommodated in numerically detailed solutions of complex problems.

7-13. CONTINUOUS DECISION PROBLEMS: CALCULUS OF VARIATIONS AND EXTENSIONS [7.35, 7.83]

7-13a. The Problem and Its Forward Recurrence Relation

In this section, dynamic programming is used to derive key results of the calculus of variations and of Pontryagin's maximum principle [7.29, 7.80]. The problem considered is the same as that given in Section 7-12a, with qualifications. In the first place, questions involving end conditions are ignored throughout this section; the results (necessary conditions) obtained here are independent of the type of end conditions imposed. For variable-end-condition problems, transversality conditions—such as those obtained in

---

[11] Dr. R. C. Minnick formed some preliminary designs of such computers during the spring of 1966.

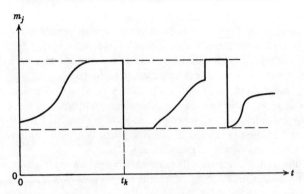

Figure 7-16. A control function $m_j = m_j(t)$ with jump discontinuities.

Chapters 3 and 8—are important additions to the necessary conditions obtained in this section.

A second qualification is made in regard to jump discontinuities of the control variables. It is assumed that the optimal control variables exhibit no more than a finite number of jump discontinuities, all of which occur at isolated instants of time. At a jump discontinuity—say, that of $m_j(t)$ at time t_k in Figure 7-16—the convention adopted in this section for the value of $m_j(t_k)$ is

$$m_j(t_k) = \lim_{\epsilon \to 0+} m_j(t_k - \epsilon) \tag{7-168}$$

Thus, at each isolated instant t_k of time, there is a finite neighborhood to the left of t_k over which all m_j's are continuous. This property is required to insure the validity of certain Taylor's series expansions that are made in the following development.

As in Section 7-12c, the first step in developing a continuous recurrence relation is to imbed the problem in a set of infinitely many similar problems. The minimum-cost function $\mathscr{F}[\mathbf{x}(t), t]$ is given by

$$\mathscr{F}[\mathbf{x}(t),t] \triangleq \min_{\substack{\mathbf{m}(\tau) \in U(\tau) \\ t_a \leq \tau \leq t}} \int_{t_a}^{t} f[\mathbf{x}(\tau),\mathbf{m}(\tau),\tau]\,d\tau \tag{7-169}$$

where, for a given problem of the set, t assumes a particular value of time in the range $t_a \leq t \leq t_b$, and $\mathbf{x}(t)$ assumes a particular value at time $\tau = t$. When $t = t_b$ in Equation 7-169, it is to be observed that $\mathscr{F}[\mathbf{x}(t_b),t_b]$ is the desired minimum for the problem.

By paralleling the development between Equations 7-155 and 7-159, the reader may substantiate the following forward recurrence relation:

$$\mathscr{F}[\mathbf{x}(t),t] = \min_{\mathbf{m}(t) \in U(t)} \{\epsilon f[\mathbf{x}(t),\mathbf{m}(t),t] + \mathscr{F}[\mathbf{x}(t-\epsilon),\,t-\epsilon]\} + \epsilon o_1(\epsilon) \tag{7-170}$$

Sect. 7-13   CONTINUOUS DECISION PROBLEMS   433

The previously adopted assumption in regard to jump discontinuities of $\mathbf{m}(t)$ enables us to relate $\mathbf{x}(t - \epsilon)$ to $\mathbf{x}(t)$ at any specific instant of time t:

$$\mathbf{x}(t - \epsilon) = \mathbf{x}(t) - \epsilon\dot{\mathbf{x}}(t) + \epsilon\mathbf{o}_2(\epsilon) \qquad (7\text{-}171)$$

which is valid for sufficiently small ϵ. This property is incorporated in Equation 7-170 with the result

$$\mathscr{F}[\mathbf{x}(t),t] = \underset{\mathbf{m}(t)\in U(t)}{\text{minimum}} \{\epsilon f[\mathbf{x}(t),\mathbf{m}(t),t]$$
$$+ \mathscr{F}[\mathbf{x}(t) - \epsilon\dot{\mathbf{x}}(t) + \epsilon\mathbf{o}_2(\epsilon), t - \epsilon]\} + \epsilon o_1(\epsilon) \quad (7\text{-}172)$$

Thus, for *fixed* $\mathbf{x}(t)$ and t, an optimal $\mathbf{m}(t)$, $\mathbf{m}^*(t)$, may be determined with the aid of 7-172.

7-13b. Hamilton-Jacobi Equations

At this point, a Taylor's series expansion of the $\mathscr{F}$ within the braces of 7-172 is inviting. In writing the expansion, it is supposed that the function $\mathscr{F}$ and its partial derivatives $\partial\mathscr{F}/\partial\mathbf{x}$ and $\partial\mathscr{F}/\partial t$ are continuous between $\mathbf{x}(t)$ and $\mathbf{x}(t - \epsilon)$ and between t and $t - \epsilon$ for each specific t in the range defined by $t_a \leq t \leq t_b$. Unfortunately, these continuity restrictions are not strictly satisfied in some of the cases for which Pontryagin's maximum principle is known to be valid (e.g., see Berkovitz [8.10, 8.11]). Thus, this dynamic programming approach to the maximum principle appears to lack complete generality.

Assuming $\mathscr{F}$ has the required continuity properties it follows that

$$\mathscr{F}[\mathbf{x}(t) - \epsilon\dot{\mathbf{x}}(t) + \epsilon\mathbf{o}_2(\epsilon), t - \epsilon] = \mathscr{F}[\mathbf{x}(t),t] - \epsilon\left[\frac{\partial\mathscr{F}[\mathbf{x}(t),t]}{\partial\mathbf{x}}\right]'\dot{\mathbf{x}}(t)$$
$$- \epsilon\frac{\partial\mathscr{F}[\mathbf{x}(t),t]}{\partial t} + \epsilon o_3(\epsilon) \quad (7\text{-}173)$$

And this is used in 7-172 with the result that

$$\mathscr{F}[\mathbf{x}(t),t] = \underset{\mathbf{m}(t)\in U(t)}{\text{minimum}} \left\{ f[\mathbf{x}(t),\mathbf{m}(t),t]\epsilon + \mathscr{F}[\mathbf{x}(t),t]\right.$$
$$\left. - \epsilon\left[\frac{\partial\mathscr{F}[\mathbf{x}(t),t]}{\partial\mathbf{x}}\right]'\dot{\mathbf{x}}(t) - \epsilon\frac{\partial\mathscr{F}[\mathbf{x}(t),t]}{\partial t}\right\} + \epsilon o_4(\epsilon) \quad (7\text{-}174)$$

Note that because $\mathbf{x}(t)$ and t are fixed for any given problem in the original set of problems, both $\mathscr{F}[\mathbf{x}(t),t]$ and $\partial\mathscr{F}[\mathbf{x}(t),t]/\partial t$ can be moved outside the braces in the right-hand member of 7-174; furthermore, the $\mathscr{F}[\mathbf{x}(t),t]$ on the right is exactly balanced by the $\mathscr{F}[\mathbf{x}(t),t]$ which is the left-hand member of 7-174, and both are removed to obtain

$$\epsilon \frac{\partial \mathscr{F}[\mathbf{x}(t),t]}{\partial t} = \underset{\mathbf{m}(t) \in U(t)}{\text{minimum}} \left\{ \epsilon f[\mathbf{x}(t),\mathbf{m}(t),t] - \epsilon \left[\frac{\partial \mathscr{F}[\mathbf{x}(t),t]}{\partial \mathbf{x}}\right]' \dot{\mathbf{x}}(t) \right\} + \epsilon o_4(\epsilon)$$

Division by $\epsilon \neq 0$ in the above equation and replacement of $\dot{\mathbf{x}}(t)$ by $\mathbf{q}[\mathbf{x}(t),\mathbf{m}(t),t]$ yields

$$\frac{\partial \mathscr{F}[\mathbf{x}(t),t]}{\partial t} = \underset{\mathbf{m}(t) \in U(t)}{\text{minimum}} \left\{ f[\mathbf{x}(t),\mathbf{m}(t),t] - \left[\frac{\partial \mathscr{F}[\mathbf{x}(t),t]}{\partial \mathbf{x}}\right]' \mathbf{q}[\mathbf{x}(t),\mathbf{m}(t),t] \right\} + o_4(\epsilon)$$

And finally, by allowing ϵ to be arbitrarily small, but not identically zero, it necessarily follows that

$$\frac{\partial \mathscr{F}[\mathbf{x}(t),t]}{\partial t} = \underset{\mathbf{m}(t) \in U(t)}{\text{minimum}} \left\{ f[\mathbf{x}(t),\mathbf{m}(t),t] - \left[\frac{\partial \mathscr{F}[\mathbf{x}(t),t]}{\partial \mathbf{x}}\right]' \mathbf{q}[\mathbf{x}(t),\mathbf{m}(t),t] \right\}$$
(7-175)

Let $\mathbf{m}^* = \mathbf{m}^*(\mathbf{x},t)$ denote the optimal control corresponding to any given $\mathbf{x}$ and t. Equation 7-175 must be satisfied by $\mathbf{m}^*$, as follows:

$$\frac{\partial \mathscr{F}(\mathbf{x},t)}{\partial t} = f(\mathbf{x},\mathbf{m}^*,t) - \left[\frac{\partial \mathscr{F}(\mathbf{x},t)}{\partial \mathbf{x}}\right]' \mathbf{q}(\mathbf{x},\mathbf{m}^*,t) \qquad (7\text{-}176)$$

Equation 7-176 is known as a Hamilton-Jacobi partial differential equation. The more classical form of this equation corresponds to the special case that $\dot{\mathbf{x}} = \mathbf{q}[\mathbf{x},\mathbf{m},t] = \mathbf{m}$; in this case, Equation 7-176 assumes the special form

$$\frac{\partial \mathscr{F}(\mathbf{x},t)}{\partial t} = f(\mathbf{x},\dot{\mathbf{x}}^*,t) - \left[\frac{\partial \mathscr{F}(\mathbf{x},t)}{\partial \mathbf{x}}\right]' \dot{\mathbf{x}}^* \qquad (7\text{-}177)$$

Although techniques exist for solving such partial differential equations, in conjunction with state equations and with other necessary conditions, it is not common practice to do so (but see [7.85]). The more usual approach involves the costate variables and the costate equations of the next subsection.

7-13c. Costate Equations

In this subsection, a set of first-order ordinary differential equations is derived to effectively take the place of the Hamilton-Jacobi partial differential equation of the preceding subsection. As the first step in the derivation, the derivative of Equation 7-176 is taken with respect to **x,** as follows:[12]

$$\frac{\partial^2 \mathcal{F}(\mathbf{x},t)}{\partial t \partial \mathbf{x}} = \frac{\partial f(\mathbf{x},\mathbf{m}^*,t)}{\partial \mathbf{x}} - \left[\frac{\partial^2 \mathcal{F}(\mathbf{x},t)}{\partial \mathbf{x}^2}\right]' \mathbf{q}(\mathbf{x},\mathbf{m}^*,t)$$

$$- \left[\frac{\partial \mathbf{q}(\mathbf{x},\mathbf{m}^*,t)}{\partial \mathbf{x}}\right]' \frac{\partial \mathcal{F}(\mathbf{x},t)}{\partial \mathbf{x}} + \left(\frac{\partial \mathbf{m}^*}{\partial \mathbf{x}}\right)' \frac{\partial}{\partial \mathbf{m}^*} [f(\mathbf{x},\mathbf{m}^*,t)$$

$$- \left(\frac{\partial \mathcal{F}(\mathbf{x},t)}{\partial \mathbf{x}}\right)' \mathbf{q}(\mathbf{x},\mathbf{m}^*,t)] \qquad (7\text{-}178)$$

In the case that **m*** is a stationary point for the minimum in Equation 7-175, the last term in Equation 7-178 is zero. Alternatively, the term $\partial \mathbf{m}^*/\partial \mathbf{x}$ is zero if **m*** is independent of **x** on the boundary of **U**. Assuming that one of these two conditions holds, Equation 7-178 reduces to

$$\frac{d}{dt}\left[\frac{\partial \mathcal{F}[\mathbf{x}^*(t),t]}{\partial \mathbf{x}}\right] = \frac{\partial f[\mathbf{x}^*(t),\mathbf{m}^*(t),t]}{\partial \mathbf{x}} - \left[\frac{\partial \mathbf{q}[\mathbf{x}^*(t),\mathbf{m}^*(t),t]}{\partial \mathbf{x}}\right]' \left[\frac{\partial \mathcal{F}[\mathbf{x}^*(t),t]}{\partial \mathbf{x}}\right]$$
(7-179)

in which the left-hand member is equal to the sum of the left-hand member in 7-178 and the second term on the right in 7-178, evaluated along an optimal path denoted by $\mathbf{x}^*(t)$. The result 7-179 is very interesting: the $n \times 1$ vector $\partial f/\partial \mathbf{x}$ and the $n \times n$ matrix $\partial \mathbf{q}/\partial \mathbf{x}$ are known functions of **x, m,** and t; if $\mathbf{x}^*(t)$ and $\mathbf{m}^*(t)$ were known for all t in the interval $t_a \leq t \leq t_b$, and if initial conditions were given, Equation 7-179 could be solved for $\partial \mathcal{F}[\mathbf{x}^*(t),t]/\partial \mathbf{x}$. Consider this in a different light. Consider the equation

$$\dot{\boldsymbol{\lambda}}^*(t) = \frac{\partial f[\mathbf{x}^*(t),\mathbf{m}^*(t),t]}{\partial \mathbf{x}} - \left[\frac{\partial \mathbf{q}[\mathbf{x}^*(t),\mathbf{m}^*(t),t]}{\partial \mathbf{x}}\right]' \boldsymbol{\lambda}^*(t) \qquad (7\text{-}180)$$

where $\boldsymbol{\lambda}^*(t) = [\lambda_1^*(t)\, \lambda_2^*(t) \cdots \lambda_n^*(t)]'$ is an $n \times 1$ matrix which is called the costate vector. Equations 7-179 and 7-180 are identical in form; $\boldsymbol{\lambda}^*(t)$ and $\partial \mathcal{F}[\mathbf{x}^*(t),t]/\partial \mathbf{x}$ are identical vector-valued functions, and the statements which follow Equation 7-179 also apply to $\boldsymbol{\lambda}^*(t)$. Corresponding to appropriate initial conditions on $\boldsymbol{\lambda}^*$ and corresponding to $\mathbf{x}^*(t)$ and $\mathbf{m}^*(t)$ over $t_a \leq t \leq t_b$, solutions for $\boldsymbol{\lambda}^*(t)$ exist.

The costate variables are also referred to as adjoint variables in the literature, but this designation is to be avoided, especially when nonlinear equations are involved, because it conflicts with other technical meanings of the word "adjoint." It is of interest to note that the costate variables are

---

[12] Again, this step can be taken provided that the appropriate second partial derivatives of $\mathcal{F}$ are well defined.

identical to the Lagrange multipliers of Chapter 3 in cases where the theory of Chapter 3 applies. This point is pursued in Section 7-13f.

7-13d. Hamiltonian Functions

Let the scalar function $\mathscr{H} = \mathscr{H}[\mathbf{x}(t),\boldsymbol{\lambda}(t),\mathbf{m}(t),t]$ be defined as follows:

$$\mathscr{H} \triangleq \boldsymbol{\lambda}(t)'\mathbf{q}[\mathbf{x}(t),\mathbf{m}(t),t] - f[\mathbf{x}(t),\mathbf{m}(t),t] \qquad (7\text{-}181)$$

where $\mathbf{x}$, $\mathbf{m}$, $\boldsymbol{\lambda}$, $\mathbf{q}$, and f are as previously defined. The utility of this definition is that it enables Equation 7-175 to be expressed in the following compact form:

$$\frac{\partial \mathscr{F}[\mathbf{x}^*(t),t]}{\partial t} = \underset{\mathbf{m}(t) \in U(t)}{\text{minimum}} \{-\mathscr{H}[\mathbf{x}^*(t),\boldsymbol{\lambda}^*(t),\mathbf{m}(t),t]\} \qquad (7\text{-}182\text{a})$$

or equivalently,

$$-\frac{\partial \mathscr{F}[\mathbf{x}^*(t),t]}{\partial t} = \underset{\mathbf{m}(t) \in U(t)}{\text{maximum}} \mathscr{H}[\mathbf{x}^*(t),\boldsymbol{\lambda}^*(t),\mathbf{m}(t),t] \qquad (7\text{-}182\text{b})$$

Again, it is emphasized that, for a given value of t and given $\mathbf{x}^*(t)$ and $\boldsymbol{\lambda}^*(t)$, an optimal $\mathbf{m}(t)$, $\mathbf{m}^*(t)$, may be obtained from 7-182. In other words, t, $\mathbf{x}^*(t)$, and $\boldsymbol{\lambda}^*(t)$ are treated as constants when the maximization is effected in 7-182b.

The function $\mathscr{H}$ is called a *Hamiltonian function* because of its origin in classical mechanics. A classical form of $\mathscr{H}$ corresponds to the special case that $\dot{\mathbf{x}} = \mathbf{q}(\mathbf{x},\mathbf{m},t) = \mathbf{m}$. In this case,

$$\mathscr{H} = \boldsymbol{\lambda}(t)'\dot{\mathbf{x}}(t) - f[\mathbf{x}(t),\dot{\mathbf{x}}(t),t] \qquad (7\text{-}183)$$

where $\boldsymbol{\lambda}(t)$ is called the momentum vector, and the Hamiltonian $\mathscr{H}$ corresponds to the sum of kinetic and potential energy in the system if the function f is of an appropriate form.

7-13e. Necessary Conditions: A Maximum Principle

In 1956, Pontryagin [7.29, 7.80] postulated a celebrated maximum principle which has been extended and "proved" in various ways by many authors. The meat of the material on Pontryagin's maximum principle is considered in Chapter 8, but the heart of this maximum principle has been obtained in the preceding subsections. In essence, Pontryagin's maximum principle consists of a set of necessary conditions which an optimal solution must satisfy. The central-most necessary condition, and the one from which the name "maximum principle" is derived, is obtained from Equation 7-182 and assumes the form

$$\mathscr{H}[\mathbf{x}^*(t),\boldsymbol{\lambda}^*(t),\mathbf{m}^*(t),t]_{\mathbf{m}^*(t) \in U(t)} \geq \mathscr{H}[\mathbf{x}^*(t),\boldsymbol{\lambda}^*(t),\mathbf{m}\dagger(t),t]_{\mathbf{m}\dagger(t) \in U(t)} \qquad (7\text{-}184)$$

Sect. 7-13   CONTINUOUS DECISION PROBLEMS   437

which holds for each t in the closed interval $[t_a,t_b]$. The notation $\mathbf{m}\dagger(t)$ is used here to denote any allowed value of $\mathbf{m}(t)$. The inequality in 7-184 is interpreted as follows: for each value of $t \in [t_a,t_b]$, the optimal value $\mathbf{m}^*(t)$ of $\mathbf{m}(t)$ is the one which maximizes the Hamiltonian $\mathscr{H}$, provided that the optimal state vector $\mathbf{x}^*(t)$ and the optimal costate vector $\boldsymbol{\lambda}^*(t)$ are also arguments of $\mathscr{H}$. It must be emphasized that Equation 7-184 does *not* give that $\mathscr{H}[\mathbf{x}(t),\boldsymbol{\lambda}(t),\mathbf{m}^*(t),t]$ is greater than or equal to $\mathscr{H}[\mathbf{x}(t),\boldsymbol{\lambda}(t),\mathbf{m}(t),t]$ for arbitrary $\mathbf{x}(t)$ and $\boldsymbol{\lambda}(t)$.

Other necessary conditions for the optimum include the original dynamics, which can be expressed in terms of $\mathscr{H}$, namely,

$$\dot{\mathbf{x}}^*(t) = \frac{\partial \mathscr{H}[\mathbf{x}^*(t),\boldsymbol{\lambda}(t),\mathbf{m}^*(t),t]}{\partial \boldsymbol{\lambda}} = \mathbf{q}[\mathbf{x}^*(t),\mathbf{m}^*(t),t] \qquad (7\text{-}185)$$

and the costate dynamics, Equation 7-180, which can also be expressed compactly in terms of $\mathscr{H}$:

$$\dot{\boldsymbol{\lambda}}^*(t) = -\frac{\partial \mathscr{H}[\mathbf{x}^*(t),\boldsymbol{\lambda}^*(t),\mathbf{m}^*(t),t]}{\partial \mathbf{x}} \qquad (7\text{-}186)$$

The paired sets of Equations 7-185 and 7-186 are appropriately called the *Hamiltonian canonical equations*.

In summary, simultaneous solutions are desired over the time interval $[t_a,t_b]$ for the following equations:

$$\dot{\mathbf{x}} = \frac{\partial \mathscr{H}(\mathbf{x},\boldsymbol{\lambda},\mathbf{m},t)}{\partial \boldsymbol{\lambda}} \qquad (7\text{-}187)$$

$$\dot{\boldsymbol{\lambda}} = \frac{-\partial \mathscr{H}(\mathbf{x},\boldsymbol{\lambda},\mathbf{m},t)}{\partial \mathbf{x}} \qquad (7\text{-}188)$$

and

$$\mathscr{H}(\mathbf{x},\boldsymbol{\lambda},\mathbf{m},t)|_{\mathbf{m}\in U} \geq \mathscr{H}(\mathbf{x},\boldsymbol{\lambda},\mathbf{m}\dagger,t), \quad \text{for any } \mathbf{m}\dagger \in U \qquad (7\text{-}189)$$

where terminal conditions on $\mathbf{x}$ and t may be partially or completely specified.[13] Functions $\mathbf{x}(t)$, $\mathbf{m}(t)$, and $\boldsymbol{\lambda}(t)$ which satisfy equations of the type 7-187, 7-188, and 7-189 are called *principal functions* (or principal solutions or principal curves) in this book.[14] Note that the "principal" of the phrase

---

[13] Transversality conditions and other necessary conditions on $\mathscr{H}^*$ are examined in Chapter 8. For example, if f of 7-145 and $\mathbf{q}$ of 7-143 do not have t as an explicit argument, it is shown that $\mathscr{H}^* = \mathscr{H}[\mathbf{x}^*(t),\boldsymbol{\lambda}^*(t),\mathbf{m}^*(t)]$ equals a constant throughout the interval $[t_a,t_b]$.

[14] The author is breaking with tradition here in order to give a more descriptive title to solutions of Equations 7-187, 7-188, and 7-189. Just as in Chapter 3, solutions to necessary conditions of variational-type problems are often called *extremal* solutions in the literature.

principal functions can be associated with "maximum *principle*" and with "*principle* of optimality." From the proof of this maximum principle, it is known that $\mathbf{x}^*(t)$, $\mathbf{m}^*(t)$, and $\boldsymbol{\lambda}^*(t)$ satisfy Equations 7-187, 7-188, and 7-189; but the proof does not rule out the possibility that other solutions to 7-187, 7-188, and 7-189 exist, and in general they do. In fact, the reader is cautioned to observe that

$$\mathscr{H}(\mathbf{x},\boldsymbol{\lambda},\mathbf{m},t)|_{\mathbf{m}\in U} \geq \mathscr{H}(\mathbf{x},\boldsymbol{\lambda},\mathbf{m}^*,t)|_{\mathbf{m}^*\in U} \qquad (7\text{-}190)$$

for any nonoptimal principal functions $\mathbf{x}$, $\mathbf{m}$, and $\boldsymbol{\lambda}$.

An attractive feature of condition 7-189 is that it often enables us to discover significant characteristics of the optimal control policy without actually having to solve two-point boundary-value problems associated with Equations 7-187, 7-188, and 7-189. The following example illustrates this fact. More detailed studies are given in Chapter 8.

Example 7-4. Consider the case in which the vector state equation 7-143 assumes the form

$$\dot{\mathbf{x}} = \mathbf{a}(\mathbf{x},t) + \mathbf{B}(\mathbf{x},t)\mathbf{m} \qquad (7\text{-}191)$$

which is linear in the $r \times 1$ control vector $\mathbf{m}$. The matrix $\mathbf{a}(\mathbf{x},t) = \mathbf{a} = [a_1 \; a_2 \; \cdots \; a_n]'$ is an $n \times 1$ column matrix with components which can be real-valued functions (of class C^1) of the state variables and of time. The matrix $\mathbf{B}(\mathbf{x},t)$ is of dimension $n \times r$:

$$\mathbf{B}(\mathbf{x},t) = \begin{bmatrix} b_{11} & b_{12} & \cdots & b_{1r} \\ b_{21} & b_{22} & & \cdot \\ \vdots & & \ddots & \vdots \\ b_{n1} & \cdot & \cdots & b_{nr} \end{bmatrix}$$

where each b_{ij} can be a real-valued function of $\mathbf{x}$ and t and is of class C^1.

Assume that the performance measure J is of the form

$$J = \int_{t_a}^{t_b} f(\mathbf{x},t)\, dt \qquad (7\text{-}192)$$

where the integrand $f(\mathbf{x},t)$ is of class C^1 and does not contain any components of the control vector $\mathbf{m}$. The components of the control vector are assumed to be constrained by the inequalities of 7-144.

For this example, the reader may verify that the Hamiltonian of Equation 7-181 assumes the form

$$\mathscr{H} = \boldsymbol{\lambda}'[\mathbf{a}(\mathbf{x},t) + \mathbf{B}(\mathbf{x},t)\mathbf{m}] - f(\mathbf{x},t) \qquad (7\text{-}193)$$

In order for $\mathscr{H}$ to be maximized with respect to $\mathbf{m} \in U$ when $\mathbf{x} = \mathbf{x}^*$ and

Sect. 7-13   CONTINUOUS DECISION PROBLEMS   439

$\lambda = \lambda^*$ at any time t in the interval $[t_a, t_b]$, it is necessary that

$$\lambda^{*\prime} \mathbf{B}(\mathbf{x}^*, t) \mathbf{m}^* \geq \lambda^{*\prime} \mathbf{B}(\mathbf{x}^*, t) \mathbf{m} \qquad (7\text{-}194)$$

for any $\mathbf{m} \in U$. This is equivalent to

$$\max_{\mathbf{m} \in U} \sum_{i=1}^{r} \alpha_i^*(t) m_i = \sum_{i=1}^{r} \alpha_i^*(t) m_i^* \qquad (7\text{-}195)$$

where

$$\alpha_i^*(t) = \sum_{k=1}^{n} \lambda_k^* b_{ki}(\mathbf{x}^*, t) \qquad (7\text{-}196)$$

In view of the assumed form of constraint set U, the $m_i^*(t)$'s of Equation 7-195 must be of the form

$$m_i^*(t) = \begin{cases} m_{i,\max} \text{ if } \alpha_i^*(t) > 0 \\ m_{i,\min} \text{ if } \alpha_i^*(t) < 0 \\ m_{i,\min} \leq m_i^* \leq m_{i,\max} \text{ if } \alpha_i^* = 0 \end{cases} \qquad (7\text{-}197)$$

The function $\alpha_i^*(t)$ is appropriately called a *switching function* because it dictates the instants at which the optimal control signal $m_i^*(t)$ switches between extreme values of the constraint set; the optimal control is of the so-called *bang-bang* variety, except in *singular cases* where one (or more) $\alpha_i^*(t)$ equals identically zero for some interval (intervals) of time contained in $[t_a, t_b]$. Singular cases require special treatment (see Chapter 8) for solution; fortunately, the singular case is the exception rather than the rule. Even in the nonsingular case, explicit evaluation of $\mathbf{m}^*(t)$ is generally a difficult task, as is to be observed in Chapter 8.

•

7-13f. Necessary Conditions: Classical Calculus of Variations

If the constraint set U is allowed to consist of all r-dimensional space, the control problem may still have a well-defined solution if the performance measure automatically penalizes all large values of $|m_i|$'s. This condition is assumed in Section 3-9 and it is also assumed in this subsection. In terms of the Hamiltonian, Equation 7-182, the optimal form $\mathbf{m}^*(t)$ of $\mathbf{m}(t)$ will correspond to the absolute maximum of $\mathscr{H}$ with respect to $\mathbf{m}(t)$. If the second partial derivatives of $\mathscr{H}$ with respect to the m_i's are continuous, the absolute maximum will correspond to a stationary point of $\mathscr{H}$ with respect to $\mathbf{m}(t)$. A necessary condition that must be satisfied by the m_i^*'s in this case is, therefore,

$$\left. \frac{\partial \mathscr{H}[\mathbf{x}^*(t), \lambda^*(t), \mathbf{m}(t), t]}{\partial m_i} \right|_{\mathbf{m} = \mathbf{m}^*} = 0 \qquad (7\text{-}198)$$

for $i = 1, 2, \ldots, r$ and for each t in the closed interval $[t_a, t_b]$.

Using the identity for $\mathscr{H}$ of 7-181, it follows from 7-198 that

$$\left\{ \boldsymbol{\lambda}^*(t)' \frac{\partial \mathbf{q}[\mathbf{x}^*(t), \mathbf{m}(t), t]}{\partial m_i} - \frac{\partial f[\mathbf{x}^*(t), \mathbf{m}(t), t]}{\partial m_i} \right\}_{\mathbf{m} = \mathbf{m}^*} = 0 \quad (7\text{-}199)$$

for $i = 1, 2, \ldots, r$. This equation and the Hamiltonian canonical equations 7-185 and 7-186 are to be identified with the Euler-Lagrange equations 3-141, 3-146, and 3-147 of Section 3-9. Thus, first-variational functions which satisfy Euler-Lagrange equations are a special class of principal functions.

7-14. QUADRATIC MINIMUM-COST FUNCTION AND CLOSED-LOOP CONTROL

7-14a. A General Case

The general approach of the preceding section can be used in an unlimited number of ways to obtain special results for special problem forms. A particularly interesting use is given in this section. The results obtained here are similar to those of Merriam [7.73, 7.74]. In essence, given both a linear, possibly time-variant state equation and an integral performance measure with a quadratic integrand, optimal controller parameters and optimal forcing functions are obtained, without the need for solution of two-point boundary-value problems.

Consider the state equation

$$\dot{\mathbf{x}} = \mathbf{A}\mathbf{x} + \mathbf{B}\mathbf{m} \quad (7\text{-}200)$$

where $\mathbf{A}$ is an $n \times n$ matrix with real-valued entries typified by $a_{ij} = a_{ij}(t)$, $\mathbf{B}$ is an $n \times r$ matrix with real-valued entries typified by $b_{ij} = b_{ij}(t)$, $\mathbf{m} = \mathbf{m}(t)$ is an $r \times 1$ control vector, and $\mathbf{x} = \mathbf{x}(t)$ is the $n \times 1$ state vector. The functions $a_{ij}(t)$ and $b_{ij}(t)$ are known bounded functions of time and are assumed to be piecewise continuous with a finite number of jump discontinuities between t_a and t_b. The special case where a_{ij} and b_{ij} are constants is obviously included.

The performance measure to be minimized here is

$$J = \int_{t_a}^{t_b} (\mathbf{e}'\mathbf{W}\mathbf{e} + \mathbf{m}'\mathbf{V}\mathbf{m}) \, dt \quad (7\text{-}201)$$

with interpretation as follows:

1. t_b is fixed.
2. $\mathbf{e}$ is a vector error and equals the difference $\mathbf{x} - \mathbf{x}_d$ between the state trajectory $\mathbf{x}(t)$ and a desired state trajectory $\mathbf{x}_d(t)$. The desired trajectory

$\mathbf{x}_d(t)$ is assumed to be continuous, and the derivative $d\mathbf{x}_d(t)/dt$ is assumed to be defined almost everywhere, if not everywhere in the closed interval $[t_a, t_b]$.

3. $\mathbf{V}$ is an $r \times r$ matrix with real-valued entries typified by $v_{ij}(t)$. Without loss of generality, the $v_{ij}(t)$'s can be selected such that $v_{ij}(t) = v_{ji}(t)$. Thus, $\mathbf{V} = \mathbf{V}'$. Furthermore, $\mathbf{V}$ is assumed to be positive-definite[15] for each t in $[t_a, t_b]$.

4. $\mathbf{W}$ is an $n \times n$ weighting matrix with real-valued entries typified by $w_{ij}(t)$. Here also, the w_{ij}'s can be selected such that $\mathbf{W} = \mathbf{W}'$, and $\mathbf{W}$ is assumed to be positive-semidefinite.

The continuity properties of the $v_{ij}(t)$'s and the $w_{ij}(t)$'s are the same as those assumed for the $a_{ij}(t)$'s.

The control vector $\mathbf{m}(t)$ is to be generated by a controller which has access to the error vector $\mathbf{e}(t)$ and the desired state trajectory $\mathbf{x}_d(t)$ (see Figure 7-17).

The assumption that $\mathbf{m}$ is an element of a closed and bounded set U *is not* made in this section. It will be seen, however, that the assumed positive-definite property of the $\mathbf{V}$ matrix insures the existence of a well-defined solution.

A key assumption which is made in the solution of the problem is that the minimum-cost function $\mathscr{B}[\mathbf{x}(t), t]$, Equation 7-162, is a quadratic function of $\mathbf{x}(t)$:

$$\mathscr{B}[\mathbf{x}(t), t] = (\mathbf{x} - \mathbf{x}_d)' \mathbf{H} (\mathbf{x} - \mathbf{x}_d) + \mathbf{k}'(\mathbf{x} - \mathbf{x}_d) + k_0 = \mathbf{e}' \mathbf{H} \mathbf{e} + \mathbf{k}' \mathbf{e} + k_0 \quad (7\text{-}202)$$

where $\mathbf{H}$ is an $n \times n$ matrix with entries to be determined which are typified by $h_{ij}(t)$, $\mathbf{k}$ is an $n \times 1$ matrix with entries to be determined which are typified by $k_j(t)$, and $k_0 = k_0(t)$ is a scalar function of t. Because of the quadratic nature of the integrand of performance measure 7-201 and because of the linear nature of state equation 7-200, the fact that $\mathscr{B}[\mathbf{x}(t), t]$ is indeed a quadratic function is well supported in the literature. In cases where the

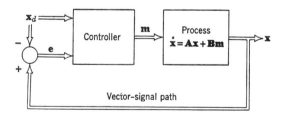

Figure 7-17. Closed-loop control.

[15] The phrase "positive-definite matrix" is defined in Section 2-4.

system is nonlinear and/or where the integrand of the performance measure is nonquadratic, the method given here may still be employed if the error **e** between the actual trajectory $\mathbf{x}(t)$ and the desired trajectory $\mathbf{x}_d(t)$ is sufficiently small to enable us to neglect higher-order terms (cubic terms and higher) in a Taylor's series expansion of $\mathscr{B}[\mathbf{x}(t),t]$ about $\mathbf{x}_d(t)$.

As a starting point, the reader is invited to develop the following necessary condition for the optimum:

$$\frac{-\partial \mathscr{B}}{\partial t} = \min_{\mathbf{m}} \left[f(\mathbf{x},\mathbf{m},t) + \left(\frac{\partial \mathscr{B}}{\partial \mathbf{x}}\right)' \mathbf{q}(\mathbf{x},\mathbf{m},t) \right] \qquad (7\text{-}203)$$

which is the backward recurrence relation counterpart to the forward recurrence relation of 7-175 established in Section 7-13b. Here, of course, $f(\mathbf{x},\mathbf{m},t) = \mathbf{e}'\mathbf{W}\mathbf{e} + \mathbf{m}'\mathbf{V}\mathbf{m}$, and $\mathbf{q}(\mathbf{x},\mathbf{m},t) = \mathbf{A}\mathbf{x} + \mathbf{B}\mathbf{m}$. These identities are used in 7-203 to obtain

$$\frac{-\partial \mathscr{B}}{\partial t} = \min_{\mathbf{m}} \left[\mathbf{e}'\mathbf{W}\mathbf{e} + \mathbf{m}'\mathbf{V}\mathbf{m} + \left(\frac{\partial \mathscr{B}}{\partial \mathbf{x}}\right)'(\mathbf{A}\mathbf{x} + \mathbf{B}\mathbf{m}) \right] \qquad (7\text{-}204)$$

Also required in the solution are the partial derivatives of $\mathscr{B}$, Equation 7-202. By straightforward vector differentiation (Appendix A), the reader may establish that

$$\frac{\partial \mathscr{B}}{\partial \mathbf{x}} = (\mathbf{H} + \mathbf{H}')\mathbf{e} + \mathbf{k} \qquad (7\text{-}205)$$

and

$$\frac{\partial \mathscr{B}}{\partial t} = \mathbf{e}'\dot{\mathbf{H}}\mathbf{e} + \mathbf{e}'[\dot{\mathbf{k}} - (\mathbf{H} + \mathbf{H}')\dot{\mathbf{x}}_d] - \mathbf{k}'\dot{\mathbf{x}}_d + \dot{k}_0 \qquad (7\text{-}206)$$

Because of the assumed conditions on **m** and **V**, the minimum of the right-hand member of 7-204 must correspond to the stationary point **m*** of **m**; thus,

$$\frac{\partial}{\partial \mathbf{m}} \left[\mathbf{e}'\mathbf{W}\mathbf{e} + \mathbf{m}'\mathbf{V}\mathbf{m} + \left(\frac{\partial \mathscr{B}}{\partial \mathbf{x}}\right)'(\mathbf{A}\mathbf{x} + \mathbf{B}\mathbf{m}) \right]_{\mathbf{m}=\mathbf{m}^*} = 0 \qquad (7\text{-}207)$$

or

$$(\mathbf{V} + \mathbf{V}')\mathbf{m}^* + \mathbf{B}'\left(\frac{\partial \mathscr{B}}{\partial \mathbf{x}}\right) = 0 \qquad (7\text{-}208)$$

and because of the assumed conditions on **V**,

$$\mathbf{m}^* = -\mathbf{V}^{-1}\mathbf{B}'(\partial \mathscr{B}/\partial \mathbf{x})/2 = -\mathbf{V}^{-1}\mathbf{B}'[(\mathbf{H} + \mathbf{H}')\mathbf{e} + \mathbf{k}]/2 \qquad (7\text{-}209)$$

In this equation, $\mathbf{V}^{-1}$ is the inverse of a specified weighting matrix, **B** is an assumed known matrix from the state equation, and **e** is the available error

QUADRATIC MINIMUM-COST FUNCTION

vector (see Figure 7-17). The $n \times n$ matrix $\mathbf{H} + \mathbf{H}' = \mathbf{H}(t) + \mathbf{H}(t)'$ and the $n \times 1$ matrix $\mathbf{k} = \mathbf{k}(t)$ must be available to the controller in order that $\mathbf{m}^*(t)$ of 7-209 can also be generated. The remainder of this section is devoted to the determination of $\mathbf{H} + \mathbf{H}'$ and $\mathbf{k}$. It is shown that $\mathbf{H} + \mathbf{H}'$ and $\mathbf{k}$ are solutions to sets of first-order differential equations with prescribed terminal conditions.

With $\mathbf{m}$ equal to $\mathbf{m}^*$, Equation 7-204 must be satisfied, as follows:

$$-\frac{\partial \mathscr{B}}{\partial t} = \mathbf{e}'\mathbf{W}\mathbf{e} + \mathbf{m}^{*\prime}\mathbf{V}\mathbf{m}^* + \left(\frac{\partial \mathscr{B}}{\partial \mathbf{x}}\right)'(\mathbf{A}\mathbf{x} + \mathbf{B}\mathbf{m}^*) \qquad (7\text{-}210)$$

The approach now is to substitute the identities of 7-205, 7-206, and 7-209 into Equation 7-210 for $\partial \mathscr{B}/\partial \mathbf{x}$, $\partial \mathscr{B}/\partial t$, and $\mathbf{m}^*$, respectively. By regrouping the terms in the resulting equation—a tedious but straightforward task—the following equation form is obtained:

$$\mathbf{e}'\mathbf{Z}_a\mathbf{e} + \mathbf{e}'\mathbf{z}_b + z_c = 0 \qquad (7\text{-}211)$$

where $\mathbf{Z}_a$, $\mathbf{z}_b$, and z_c are introduced to simplify notation and are defined by

$$\mathbf{Z}_a \triangleq \dot{\mathbf{H}} + \mathbf{W} + (\mathbf{H} + \mathbf{H}')\mathbf{A} - \tfrac{1}{4}(\mathbf{H} + \mathbf{H}')\mathbf{B}\mathbf{V}^{-1}\mathbf{B}'(\mathbf{H} + \mathbf{H}') \qquad (7\text{-}212)$$

$$\mathbf{z}_b \triangleq \dot{\mathbf{k}} + [\mathbf{A}' - \tfrac{1}{2}(\mathbf{H} + \mathbf{H}')\mathbf{B}\mathbf{V}^{-1}\mathbf{B}']\mathbf{k} + (\mathbf{H} + \mathbf{H}')(\mathbf{A}\mathbf{x}_d - \dot{\mathbf{x}}_d) \qquad (7\text{-}213)$$

and

$$z_c \triangleq \dot{k}_0 + \mathbf{k}'[\mathbf{A}\mathbf{x}_d - \dot{\mathbf{x}}_d - \tfrac{1}{4}\mathbf{B}\mathbf{V}^{-1}\mathbf{B}'\mathbf{k}] \qquad (7\text{-}214)$$

The matrix $\mathbf{Z}_a$ is $n \times n$, $\mathbf{z}_b$ is $n \times 1$, and z_c is a real scalar.

Observe that both $(\mathbf{H} + \mathbf{H}')$ and $\mathbf{B}\mathbf{V}^{-1}\mathbf{B}'$ appear at several points in 7-212, 7-213, and 7-214. It is notationally expedient, therefore, to introduce the following expressions:

$$\mathbf{\Psi} \triangleq \mathbf{H} + \mathbf{H}' \qquad (7\text{-}215)$$

and

$$\mathbf{\Gamma} \triangleq \mathbf{B}\mathbf{V}^{-1}\mathbf{B}' \qquad (7\text{-}216)$$

Note that $\mathbf{\Gamma}' = \mathbf{\Gamma}$ and $\mathbf{\Psi}' = \mathbf{\Psi}$. Equation 7-211 must be satisfied by any finite $\mathbf{e}(t)$; that is,

$$\sum_i \sum_j z_{aij} e_i e_j + \sum_i z_{bi} e_i + z_c \equiv 0 \qquad (7\text{-}217)$$

for any e_i, e_j combination. If all e_i's equal zero in 7-217, equality must still hold so that z_c must be zero; if any e_k for some k is the only nonzero e_i, the quadratic in e_k resulting from 7-217 will be identically zero only if both z_{akk} and z_{bk} equal zero; and with the preceding statements in mind, if e_k and e_r are the only nonzero e_i's, identical equality will hold in 7-217 if, and only if,

$z_{akr} + z_{ark} = 0$. In short, the entries of $\mathbf{Z}_a + \mathbf{Z}_a'$, $\mathbf{z}_b$, and z_c must all be identically zero.[16] It follows from 7-212, 7-213, 7-214, 7-215, and 7-216 that

$$\dot{\boldsymbol{\Psi}} = -2\mathbf{W} - \mathbf{A}'\boldsymbol{\Psi} - (\mathbf{A}'\boldsymbol{\Psi})' + \tfrac{1}{2}\boldsymbol{\Psi}\boldsymbol{\Gamma}\boldsymbol{\Psi} \tag{7-218}$$

$$\dot{\mathbf{k}} = \tfrac{1}{2}(\boldsymbol{\Psi}\boldsymbol{\Gamma} - 2\mathbf{A}')\mathbf{k} + \boldsymbol{\Psi}(-\mathbf{A}\mathbf{x}_d + \dot{\mathbf{x}}_d) \tag{7-219}$$

and

$$\dot{k}_0 = \mathbf{k}'[\tfrac{1}{4}\boldsymbol{\Gamma}\mathbf{k} + \dot{\mathbf{x}}_d - \mathbf{A}\mathbf{x}_d] \tag{7-220}$$

Note that $\boldsymbol{\Psi}$ is a diagonally symmetric matrix (actually $\mathbf{H}$ could have been so designated); and only $n(n+1)/2$ $\psi_{ij}(t)$'s, entries of $\boldsymbol{\Psi}$, need be generated.

Equations 7-218 and 7-219 are key differential equations; they must be solved if $\mathbf{m}^*(t)$ is to be generated, where

$$\mathbf{m}^*(t) = -\tfrac{1}{2}\mathbf{V}^{-1}\mathbf{B}'(\boldsymbol{\Psi}\mathbf{e} + \mathbf{k}) \tag{7-221}$$

which follows from 7-209 and 7-215. As previously defined, $\mathbf{e} = \mathbf{x} - \mathbf{x}_d$. To solve Equations 7-218 and 7-219, boundary conditions must be identified. Note that at the terminal time t_b, $\mathscr{B}[\mathbf{x}(t_b),t_b]$ of 7-202 must be zero, regardless of the bounded value of $\mathbf{e}(t_b)$; but this can be true only if all entries of $\boldsymbol{\Psi}(t_b)$, $\mathbf{k}(t_b)$, and $k_0(t_b)$ are identically zero. Thus, the required boundary conditions are

$$\boldsymbol{\Psi}(t_b) = \mathbf{0} \tag{7-222}$$

and

$$\mathbf{k}(t_b) = \mathbf{0} \tag{7-223}$$

Matrix equation 7-218 is a nonlinear equation of a special form known as a generalized Riccati equation (after Jacopo Francesco, Count Riccati, 1676–1754, who identified a scalar version of 7-218 in 1724). The nonlinear terms include products of the form $\psi_{ij}\psi_{kl}$. The fact that 7–218 contains nonlinear terms should not be disconcerting, however, for the boundary conditions are all specified at one point in time; in addition to straightforward numerical solution schemes, various analytical techniques are available for solution of Riccati equations [7.24, 7.33, 7.54]. Unlike numerical solutions of Euler-Lagrange equations, numerical solutions of Riccati equations tend to be numerically stable.

There is no conceptual difficulty associated with the solution of a set of differential equations backward in time, but most numerical solution schemes apply directly to forward solutions. On this basis, we may wish to introduce a dummy time variable, $\tau \triangleq t_b - t$, and to solve altered equations forward in τ. This approach is illustrated in Example 7-5.

---

[16] $\mathbf{Z}_a$ is a *skew symmetric matrix* because $\mathbf{Z}_a + \mathbf{Z}_a' = \mathbf{0}$.

Sect. 7-14　QUADRATIC MINIMUM-COST FUNCTION　445

Example 7-5. Consider a system governed by the following second-order differential equation:

$$\ddot{x} - a_{22}\dot{x} - a_{21}x = m, \qquad t_a \leq t \leq t_b \tag{7-224}$$

Let $x = x_1$ and $\dot{x} = x_2$; Equation 7-224 is thereby replaced by two first-order differential equations:

$$\dot{x}_1 = x_2 \tag{7-225}$$

and

$$\dot{x}_2 = a_{21}x_1 + a_{22}x_2 + m \tag{7-226}$$

or in matrix notation,

$$\dot{\mathbf{x}} = \mathbf{A}\mathbf{x} + \mathbf{b}m \tag{7-227}$$

where $a_{11} = 0$, $a_{12} = 1$, a_{21} and a_{22} are the entries of the 2 × 2 matrix $\mathbf{A}$, and $\mathbf{b} = [0 \ 1]'$.

The following conditions are assumed in this example: desired values of x_1 and x_2 are 1 and 0, respectively; the $\mathbf{W}$ matrix in the integrand of the performance measure 7-201 is a 2 × 2 identity matrix $\mathbf{I}$; and the $\mathbf{V}$ matrix of 7-201 is, in this case, a positive constant h.

The Riccati equations which must be solved are special cases of Equations 7-218 and 7-219; thus,

$$\dot{\boldsymbol{\Psi}} = -\begin{bmatrix} 0 & a_{21} \\ 1 & a_{22} \end{bmatrix}\boldsymbol{\Psi} - \boldsymbol{\Psi}\begin{bmatrix} 0 & 1 \\ a_{21} & a_{22} \end{bmatrix} + \tfrac{1}{2}\boldsymbol{\Psi}\boldsymbol{\Gamma}\boldsymbol{\Psi} - 2\mathbf{I} \tag{7-228}$$

where $\boldsymbol{\Psi}(t_b) = \mathbf{0}$ which necessitates solution of Equation 7-228 backward in time, and where

$$\boldsymbol{\Gamma} = \mathbf{B}\mathbf{V}^{-1}\mathbf{B}'$$

$$= [0 \ 1]' \frac{1}{h}[0 \ 1] = \frac{1}{h}\begin{bmatrix} 0 & 0 \\ 0 & 1 \end{bmatrix} \tag{7-229}$$

In terms of the individual entries in the 2 × 2 matrix $\boldsymbol{\Psi}$, Equation 7-228 yields

$$\dot{\psi}_{11} = -2a_{21}\psi_{12} + (1/2h)\psi_{12}^2 - 2, \qquad \psi_{11}(t_b) = 0 \tag{7-230}$$

$$\dot{\psi}_{12} = -\psi_{11} - a_{22}\psi_{12} - a_{21}\psi_{22} + (1/2h)\psi_{12}\psi_{22}, \qquad \psi_{12}(t_b) = 0 \tag{7-231}$$

and

$$\dot{\psi}_{22} = -2\psi_{12} - 2a_{22}\psi_{22} + (1/2h)\psi_{22}^2 - 2, \qquad \psi_{22}(t_b) = 0 \tag{7-232}$$

Similarly, corresponding to Equation 7-219,

$$\dot{\mathbf{k}} = -\begin{bmatrix} 0 & a_{21} \\ 1 & a_{22} \end{bmatrix}\mathbf{k} + \tfrac{1}{2}\Psi\begin{bmatrix} 0 & 0 \\ 0 & \frac{1}{h} \end{bmatrix}\mathbf{k} - \Psi\begin{bmatrix} 0 & 1 \\ a_{21} & a_{22} \end{bmatrix}\begin{bmatrix} 1 \\ 0 \end{bmatrix} \quad (7\text{-}233)$$

where $\mathbf{k}(t_b) = \mathbf{0}$ which necessitates solution of 7-233 backward in time. In terms of the individual entries in the 2×1 matrix $\mathbf{k}$, Equation 7-233 yields

$$\dot{k}_1 = -a_{21}k_2 + (1/2h)\psi_{12}k_2 - a_{21}\psi_{12}, \qquad k_1(t_b) = 0 \quad (7\text{-}234)$$

and

$$\dot{k}_2 = -k_1 - a_{22}k_2 + (1/2h)\psi_{22}k_2 - a_{21}\psi_{22}, \qquad k_2(t_b) = 0 \quad (7\text{-}235)$$

Once solutions for Ψ and $\mathbf{k}$ are obtained, they are used to generate the optimal $m(t)$, $m^*(t)$, by using Equation 7-221 which here reduces to

$$m^*(t) = [0 \quad -1/2h](\Psi \mathbf{e} + \mathbf{k}) = (-1/2h)(\psi_{21}e_1 + \psi_{22}e_2 + k_2) \quad (7\text{-}236)$$

where $e_1 = x_1 - 1$ and $e_2 = x_2$.

A particularly important case of this example is that in which $a_{21} = 0$. Observe that $\dot{k}_1$ and $\dot{k}_2$ of Equations 7-234 and 7-235 can *never* differ from zero when $a_{21} = 0$ because the equations then form a homogeneous pair with specified terminal conditions of zero. Thus, $k_1(t)$ and $k_2(t)$ are zero (if $a_{21} = 0$), and Equations 7-230, 7-231, 7-232, and 7-236 reduce to

$$\dot{\psi}_{11} = (1/2h)\psi_{12}^2 - 2, \qquad \psi_{11}(t_b) = 0 \quad (7\text{-}237)$$

$$\dot{\psi}_{12} = -\psi_{11} - a_{22}\psi_{12} + (1/2h)\psi_{12}\psi_{22}, \qquad \psi_{12}(t_b) = 0 \quad (7\text{-}238)$$

$$\dot{\psi}_{22} = -2\psi_{12} - 2a_{22}\psi_{22} + (1/2h)\psi_{22}^2 - 2, \qquad \psi_{22}(t_b) = 0 \quad (7\text{-}239)$$

and

$$m^*(t) = (-1/2h)[\psi_{21}(t)e_1(t) + \psi_{22}(t)e_2(t)] \quad (7\text{-}240)$$

To solve Equations 7-237, 7-238, and 7-239, it is convenient to let $t = t_b - \tau$ and $\psi_{ij}(t_b - \tau) \triangleq \phi_{ij}(\tau)$, with the result that

$$\dot{\phi}_{11}(\tau) = -(1/2h)\phi_{12}^2 + 2, \qquad \phi_{11}(0) = 0 \quad (7\text{-}241)$$

$$\dot{\phi}_{12}(\tau) = \phi_{11} + a_{22}\phi_{12} - (1/2h)\phi_{12}\phi_{22}, \qquad \phi_{12}(0) = 0 \quad (7\text{-}242)$$

and

$$\dot{\phi}_{22}(\tau) = 2\phi_{12} + 2a_{22}\phi_{22} - (1/2h)\phi_{22}^2 + 2, \qquad \phi_{22}(0) = 0 \quad (7\text{-}243)$$

which equations are solved forward in the dummy time variable τ, and the solutions are incorporated in Equation 7-240 by using the relationship $\psi_{ij}(t) = \phi_{ij}(t_b - t)$.

It is important to observe that the $\psi_{ij}(t)$'s are completely independent of the initial conditions on $\mathbf{x}(t)$ at $t = t_a$.

•

7-14b. Steady-State Riccati Equations

In addition to the conditions assumed in the preceding subsection, consider those cases where all entries in the $\mathbf{A}$, $\mathbf{B}$, $\mathbf{W}$, and $\mathbf{V}$ matrices are constants and where t_b approaches infinity. The performance measure J is then

$$J = \int_{t_a}^{\infty} (\mathbf{e}'\mathbf{W}\mathbf{e} + \mathbf{m}'\mathbf{V}\mathbf{m}) \, dt \tag{7-244}$$

Also, suppose that $\mathbf{x}_d$ satisfies the relationship

$$\dot{\mathbf{x}}_d = \mathbf{A}\mathbf{x}_d, \qquad t \geq t_a \tag{7-245}$$

Subtraction of terms of 7-245 from corresponding ones of 7-200 results in

$$\dot{\mathbf{e}} = \mathbf{A}\mathbf{e} + \mathbf{B}\mathbf{m}, \qquad t \geq t_a \tag{7-246}$$

Observe that zero values for $\mathbf{m}$ and $\mathbf{e}$ result in zero values for the components of $\dot{\mathbf{e}}$, which fact provides for the *maintainability* of $\mathbf{e}$ at zero in the steady state.

It is assumed that the error $\mathbf{e}(t)$ of 7-246 is *completely regulable*, which is defined here as follows: $\mathbf{e}(t)$ of 7-246 is completely regulable if there exists an $\mathbf{m}(t)$ for every bounded $\mathbf{e}(t_a)$ such that J of 7-244 can be maintained finite. For example, $\mathbf{e}(t)$ is completely regulable in the case that all eigenvalues of the matrix $\mathbf{A}$ have negative real parts, even if all entries of $\mathbf{B}$ are identically zero. Regulability of $\mathbf{e}(t)$ is insured if $\mathbf{e}(t)$ is *completely controllable* [7.55]; $\mathbf{e}(t)$ of 7-246 is completely controllable if there exists a bounded $\mathbf{m}(t)$ for every bounded $\mathbf{e}(t_a)$ such that $\mathbf{e}(t)$ can be forced to zero *in finite time*. Note that controllability implies regulability, but the counter statement is not true; for example, in the extreme case that the entries of $\mathbf{B}$ are identically zero and all eigenvalues of $\mathbf{A}$ have negative real parts, $\mathbf{e}(t)$ is completely regulable but is not completely controllable because $\mathbf{e}(t)$ approaches zero exponentially and does not reach zero identically in finite time.

Kalman [7.54, 7.55] introduced concepts of controllability and derived controllability conditions which apply to Equation 7-246. In brief, Kalman has shown that the system characterized by 7-246 is completely controllable if and only if the $n \times (nr)$ matrix

$$[\mathbf{B} \quad \mathbf{A}\mathbf{B} \quad \mathbf{A}^2\mathbf{B} \quad \cdots \quad \mathbf{A}^{n-1}\mathbf{B}]$$

has rank n. In the special case that $\mathbf{m}$ equals a scalar control action and $\mathbf{B} = \mathbf{b}$ is $n \times 1$, the above-noted matrix has rank n if and only if the column vectors $\mathbf{b}, \mathbf{A}\mathbf{b}, \ldots, \mathbf{A}^{n-1}\mathbf{b}$ are linearly independent.

The minimum of J of 7-244 is to be obtained. Because of the infinite upper limit of integration and because of the assumed time-invariance of $\mathbf{A}$, $\mathbf{B}$, $\mathbf{W}$, and $\mathbf{V}$, the minimum of J depends on $\mathbf{e}(t_a)$ *but not explicitly on t_a itself*. Shifts in the time origin of the problem do not affect the minimum obtained. It follows that the minimum-cost function $\mathscr{B}$ is a function $\mathscr{B}[\mathbf{e}(t)]$ of the error at time t, but is not an explicit function of t. Thus,

$$\mathscr{B}[\mathbf{e}(t)] = \mathbf{e}(t)'\mathbf{H}\mathbf{e}(t) + \mathbf{k}'\mathbf{e}(t) + k_0 \qquad (7\text{-}247)$$

which is of the same quadratic form as in 7-202; but in this case, the entries of $\mathbf{H}$, $\mathbf{k}$, and k_0 must be constants independent of t. The entries of $\dot{\mathbf{H}}$, $\dot{\mathbf{k}}$, and $\dot{k}_0$ are therefore equal to zero. With this and the assumption that $\dot{\mathbf{x}}_d = \mathbf{A}\mathbf{x}_d$, Equations 7-218, 7-219, and 7-220 can be reduced to the counterparts which follow:

$$0 = -2\mathbf{W} - \mathbf{A}'\mathbf{\Psi} - (\mathbf{A}'\mathbf{\Psi})' + \tfrac{1}{2}\mathbf{\Psi}\mathbf{\Gamma}\mathbf{\Psi} \qquad (7\text{-}248)$$

$$0 = \tfrac{1}{2}(\mathbf{\Psi}\mathbf{\Gamma} - 2\mathbf{A}')\mathbf{k} \qquad (7\text{-}249)$$

and

$$0 = \tfrac{1}{4}\mathbf{k}'\mathbf{\Gamma}\mathbf{k} \qquad (7\text{-}250)$$

where the latter two equalities are satisfied by $\mathbf{k} = \mathbf{0}$. Also, the optimal control action of 7-221 reduces here to

$$\mathbf{m}^* = -\tfrac{1}{2}\mathbf{V}^{-1}\mathbf{B}'\mathbf{\Psi}\mathbf{e} \qquad (7\text{-}251)$$

Equation 7-251 can be implemented as a feedback control law, provided that $\mathbf{e}(t)$, or a good approximation thereof, can be generated as required. In this regard, the ability to "observe" a system is a limiting factor. A system is said to be *observable* [7.55] if any initial state can be computed in finite time on the basis of measured data. For the system characterized by 7-246, assume that the measured output is a $p \times 1$ vector $\mathbf{y}$:

$$\mathbf{y} = \mathbf{C}\mathbf{e} \qquad (7\text{-}252)$$

where $\mathbf{C}$ is a $p \times n$ matrix of given real constants. It can be shown [7.55] that the system is observable if and only if the following $n \times np$ matrix has rank n:

$$[\mathbf{C}' \quad \mathbf{A}'\mathbf{C}' \quad \cdots \quad (\mathbf{A}')^{n-1}\mathbf{C}']$$

Lee and Markus [7.70] consider observability conditions for classes of nonlinear systems. Ogata [2.20] derives observability conditions for discrete and continuous linear systems.

The matrix $\mathbf{\Psi}$ remains to be determined. It is a solution of 7-248 which is an algebraic equation known as a steady-state Riccati equation. Because of the quadratic form of 7-248, multiple solutions exist. The optimum solution

Sect. 7-14   QUADRATIC MINIMUM-COST FUNCTION   449

is the one which yields stability of the state equation

$$\dot{\mathbf{e}} = (\mathbf{A} - \tfrac{1}{2}\mathbf{B}\mathbf{V}^{-1}\mathbf{B}'\mathbf{\Psi})\mathbf{e}$$
$$= (\mathbf{A} - \tfrac{1}{2}\mathbf{\Gamma}\mathbf{\Psi})\mathbf{e} \quad (7\text{-}253)$$

For multivariable systems (n large), Newton's method or other search techniques (Chapter 6) may be used to solve 7-248 for the optimal $\mathbf{\Psi}$. In some cases, it is easier to obtain a numerical solution of the corresponding differential equations 7-218 and to extract the steady-state solution which corresponds to the desired solution of 7-248.

Example 7-6. Reconsider Example 7-5, but with $t_b \to \infty$ and with $a_{21} = 0$. The desired $\mathbf{x}$ is characterized by $\mathbf{x}_d = [1 \quad 0]'$ and $\dot{\mathbf{x}}_d = [0 \quad 0]'$ which satisfy

$$\dot{\mathbf{x}}_d = \mathbf{A}\mathbf{x}_d = \begin{bmatrix} 0 & 1 \\ 0 & a_{22} \end{bmatrix}\begin{bmatrix} 1 \\ 0 \end{bmatrix} = \begin{bmatrix} 0 \\ 0 \end{bmatrix} \quad (7\text{-}254)$$

Matrix Equation 7-248 reduces here to Equations 7-237, 7-238, and 7-239 with $\dot{\psi}_{11} = \dot{\psi}_{12} = \dot{\psi}_{22} = 0$, as follows:

$$0 = \psi_{12}^2 - 4h \quad (7\text{-}255)$$
$$0 = -\psi_{11} - a_{22}\psi_{12} + (1/2h)\psi_{12}\psi_{22} \quad (7\text{-}256)$$

and

$$0 = -2\psi_{12} - 2a_{22}\psi_{22} + (1/2h)\psi_{22}^2 - 2 \quad (7\text{-}257)$$

The solution of 7-255 is

$$\psi_{12} = \pm 2h^{1/2} \quad (7\text{-}258a)$$

With this result, the quadratic equation 7-257 is solved,

$$\psi_{22} = 2ha_{22} \;\textcircled{\pm}\; 2[h^2 a_{22}^2 + h(1 \pm 2h^{1/2})]^{1/2} \quad (7\text{-}258b)$$

in which the circle around the first $\pm$ sign is used to distinguish it from the second such sign. Next, the results of 7-258 are substituted appropriately into 7-256 to obtain

$$\psi_{11} = \pm \;\textcircled{\pm}\; 2(ha_{22}^2 + 1 \pm 2h^{1/2})^{1/2} \quad (7\text{-}259)$$

To determine the correct terms of the $\pm$ and $\textcircled{\pm}$ signs, the optimal state equation 7-253 must be checked for stability. Thus, the matrix $\mathbf{A} - \tfrac{1}{2}\mathbf{\Gamma}\mathbf{\Psi}$ is of interest:

$$\mathbf{A} - \tfrac{1}{2}\mathbf{\Gamma}\mathbf{\Psi} = \begin{bmatrix} 0 & 1 \\ -\dfrac{\psi_{12}}{2h} & a_{22} - \dfrac{\psi_{22}}{2h} \end{bmatrix}$$

$$= \begin{bmatrix} 0 & 1 \\ -(\pm h^{-1/2}) & -\textcircled{\pm}(a_{22}^2 + h^{-1} \pm 2h^{-1/2})^{1/2} \end{bmatrix} \quad (7\text{-}260)$$

Using 7-260 in 7-253 gives

$$\dot{e}_1 = e_2 \tag{7-261}$$

and

$$\dot{e}_2 = -(\pm h^{-1/2})e_1 - \boxed{\pm}(a_{22}^2 + h^{-1} \pm 2h^{-1/2})^{1/2}e_2 \tag{7-262}$$

Cursory examination of these equations shows that, for stability, the + of the $\pm$ sign must be used, as also the + of the $\boxed{\pm}$ sign.

•

7-15. A STOCHASTIC CONTROL PROBLEM

The literature (e.g., [7.12, 7.62]) contains many results pertaining to the control of systems that are subjected to stochastic disturbances. Dynamic programming is useful in the development of optimal control policies for many such systems. The following problem and its dynamic programming solution illustrate pertinent concepts that are involved.

Consider a system that is characterized by a scalar state equation,

$$x_{k+1} = q(x_k, m_k, w_k, k), \quad k = 0, 1, \ldots, K - 1 \tag{7-263}$$

where, as in Section 7-6, x_k represents the state of the system at "time" k (actually, at the instant of time corresponding to the integer k), m_k is the decision variable at time k, and $q(x_k, m_k, w_k, k)$ is a known real-valued function of its arguments. The variable w_k at time k is probabilistic in character: for illustrative purposes in this section, the probability distribution associated with w_k is assumed to be discrete; w_k is assumed to take on one of two values, $w_k = 1$ with probability $p_{1k}(x_k) \equiv p_{1k}$, and $w_k = -1$ with probability $p_{-1k}(x_k) \equiv p_{-1k}$. Of course, the sum $p_{1k} + p_{-1k}$ equals 1 for any k and x_k—but also note that these probabilities may be functions of both k and x_k. Other forms of probability distributions for w_k can be readily treated (see Problem 7.31).

As in Section 7-6, m_k is assumed to be constrained to be an element of a compact (closed and bounded) set U_k. The performance measure of interest here is that of 7-41:

$$P = \sum_{i=1}^{K} f_i(x_i, m_{i-1}) \tag{7-41}$$

where a minimum of P is desired. However, the w_i's are known initially only in terms of probability distributions: at time k, the value of x_k is assumed to be measurable, but w_k and all w_i's for $i \geq k$ are known only in terms of probability distributions. Thus, the best that can be done in regard to the

Sect. 7-15 A STOCHASTIC CONTROL PROBLEM

selection of m_k at time k is that it be selected to minimize the expected value of that part of P which remains to be determined. Initially, the minimum $\mathcal{B}_K(x_0)$ of the expected value $E\{P\}$ of P is desired:

$$\mathcal{B}_K(x_0) = \underset{m_k\text{'s } \in U_k\text{'s}}{\text{minimum}} \underset{w_k\text{'s}}{E} \left\{ \sum_{k=1}^{K} f_k(x_k, m_{k-1}) \right\} \quad (7\text{-}264)$$

where the subscript K of $\mathcal{B}_K(x_0)$ denotes that K stages of the decision process remain, and the argument x_0 of $\mathcal{B}_K(x_0)$ is the state of the system at the beginning of this K-stage process. Note that $f_1(x_1, m_0)$ is independent of w_k's and m_k's for all k greater than zero; thus, Equation 7-264 can be rewritten in the form

$$\mathcal{B}_K(x_0) = \underset{m_0 \in U_0}{\text{minimum}} \underset{w_0}{E} \left\{ f_1(x_1, m_0) + \underset{\substack{m_k\text{'s} \in U_k\text{'s} \\ k \neq 0}}{\text{minimum}} \underset{\substack{w_k\text{'s} \\ k \neq 0}}{E} \sum_{k=2}^{K} f_k(x_k, m_{k-1}) \right\}$$

$$= \underset{m_0 \in U_0}{\text{minimum}} \underset{w_0}{E} \{ f_1(x_1, m_0) + \mathcal{B}_{K-1}(x_1) \} \quad (7\text{-}265)$$

And because $x_1 = q(x_0, m_0, w_0, 0)$,

$$\mathcal{B}_K(x_0) = \underset{m_0 \in U_0}{\text{minimum}} \underset{w_0}{E} \{ f_1[q(x_0, m_0, w_0, 0), m_0] + \mathcal{B}_{K-1}[q(x_0, m_0, w_0, 0)] \} \quad (7\text{-}266)$$

By following steps similar to those of the preceding paragraph (or simply by invoking the principle of optimality at this point), the following recurrence relation is established.

$$\mathcal{B}_{K-k}(x_k) = \underset{m_k \in U_k}{\text{minimum}} \underset{w_k}{E} \{ f_{k+1}[q(x_k, m_k, w_k, k), m_k] + \mathcal{B}_{K-k-1}[q(x_k, m_k, w_k, k)] \} \quad (7\text{-}267)$$

which applies for $k = 0, 1, \ldots, K - 2$. Because of the stochastic nature of the problem, the terminal state x_K cannot be assigned a specific value; x_K is assumed to be selected in an optimal fashion; and therefore,

$$\mathcal{B}_1(x_{K-1}) = \underset{m_{K-1} \in U_{K-1}}{\text{minimum}} \underset{w_{K-1}}{E} \{ f_K[q(x_{K-1}, m_{K-1}, w_{K-1}, K-1), m_{K-1}] \} \quad (7\text{-}268)$$

For notational simplification, let

$$\Omega_k \equiv \Omega_k(x_k, m_k, w_k)$$
$$\triangleq f_{k+1}[q(x_k, m_k, w_k, k), m_k] + \mathcal{B}_{K-k-1}[q(x_k, m_k, w_k, k)] \quad (7\text{-}269)$$

for $k = 0, 1, \ldots, K - 1$—with the understanding that $\mathcal{B}_0(x_K) \equiv 0$. Identity 7-269 is used in 7-267 and 7-268 to obtain

$$\mathcal{B}_{K-k}(x_k) = \underset{m_k \in U_k}{\text{minimum}} \underset{w_k}{E} \Omega_k \quad (7\text{-}270)$$

for $k = K - 1, K - 2, \ldots, 1, 0$.

For the particular probability distribution given in the second paragraph of this section, Equation 7-270 assumes the form

$$\mathscr{B}_{K-k}(x_k) = \minimum_{m_k \in U_k} [p_{1k}\Omega_k(x_k,m_k,1) + p_{-1k}\Omega_k(x_k,m_k,-1)] \quad (7\text{-}271)$$

As in the deterministic case (Section 7-6), the solution can be obtained recursively in tabular fashion.

7-16. ESTIMATION OF STATE VARIABLES IN THE PRESENCE OF NOISE

When random variables are included in the characterization of a physical system, probability theory (e.g., [7.78]) should be applied to analyze the system and to construct meaningful measures of system performance. In the preceding section, a control performance measure is based on expected values of future states of a system. In Section 7-16a, best estimates of the values of present and past state variables are desired, and the performance measure is a conditional probability density function which is maximized by use of the desired estimates. In Section 7-16b, the results of Section 7-16a are examined under special linear conditions: The optimum in this case corresponds both to the maximum of an appropriate probability density function and to the minimum of the expected value of a sum of squared errors.

Section 7-16a is based on work of Larson and Peschon [7.67, 7.68]. Related dynamic programming approaches are considered by Lcc [7.71] and Cox [7.31]. The approach to the discrete Kalman-Bucy filter of Section 7-16b is based on works of Ho and Lee [7.48], Greensite [7.42], and Larson and Peschon [7.67, 7.68].

7-16a. Modal Trajectory Estimation [7.67, 7.68]

Consider the vector state equation

$$\mathbf{x}_{k+1} = \mathbf{q}(\mathbf{x}_k, \mathbf{w}_k, k), \quad k = 0, 1, \ldots, K \quad (7\text{-}272)$$

where $\mathbf{x}_k$ is an $n \times 1$ state vector, $\mathbf{w}_k$ is a noise vector (with properties enumerated in the following), and $\mathbf{q}(\mathbf{x}_k,\mathbf{w}_k,k)$ is a known vector function of its arguments. Measurements on the system 7-272 are assumed to yield

$$\mathbf{z}_{k+1} = \mathbf{g}(\mathbf{x}_{k+1}, \mathbf{v}_{k+1}, k+1), \quad k = 0, 1, \ldots, K \quad (7\text{-}273)$$

where $\mathbf{z}_{k+1}$ is called a *measurement vector* (usually of dimension less than or equal to that of $\mathbf{x}_{k+1}$), $\mathbf{v}_{k+1}$ is a noise vector (with properties enumerated in

Sect. 7-16 ESTIMATION OF STATE VARIABLES 453

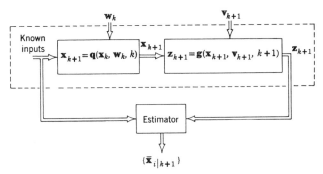

Figure 7-18. The set $\{\bar{\mathbf{x}}_{i|k+1}\} \equiv \{\bar{\mathbf{x}}_{0|k+1}, \bar{\mathbf{x}}_{1|k+1}, \ldots, \bar{\mathbf{x}}_{k+1|k+1}\}$ yields the maximum of the conditional probability density function $p(\{\mathbf{x}_i\}_{k+1} | \{\mathbf{z}_i\}_{k+1})$.

the following), and $\mathbf{g}(\mathbf{x}_{k+1}, \mathbf{v}_{k+1}, k+1)$ is a known vector function of its arguments. Figure 7-18 depicts the assumed conditions.

The problem is to determine a "best" estimate of the set $\{\mathbf{x}_0, \mathbf{x}_1, \ldots, \mathbf{x}_{K+1}\}$, denoted by $\{\mathbf{x}_i\}_{K+1}$, on the basis of known measurements $\{\mathbf{z}_1, \mathbf{z}_2, \ldots, \mathbf{z}_{K+1}\} \equiv \{\mathbf{z}_i\}_{K+1}$ and of known probability density functions for $\mathbf{x}_0$ and the pertinent noise vectors.

The noise vectors are assumed to be independent and white (uncorrelated in time): vector values of $\mathbf{w}_k$ and $\mathbf{v}_i$ depend solely on associated probability density functions $p(\mathbf{w}_k)$ and $p(\mathbf{v}_i)$, respectively.[17] The probability density functions $p(\mathbf{w}_k)$, $k = 0, 1, \ldots, K$, are assumed known, as are probability density functions $p(\mathbf{v}_i)$, $i = 1, 2, \ldots, K+1$, and a probability density function $p(\mathbf{x}_0)$ of the initial state $\mathbf{x}_0$.

Note that the explicit dependence of $\mathbf{q}$ in 7-272 on k enables us to account for forcing functions and/or control actions that are known to be applied over the interval between $k = 0$ and $k = K$. Also, if a given parameter of the system is initially unknown, it can be designated as a state variable, say $x_{j,k}$ is so designated with state equation $x_{j,k+1} = x_{j,k}$, and an estimate of $x_{j,k}$ may be obtained.

The *modal trajectory* $\{\bar{\mathbf{x}}_{0|K+1}, \bar{\mathbf{x}}_{1|K+1}, \ldots, \bar{\mathbf{x}}_{K+1|K+1}\} \equiv \{\bar{\mathbf{x}}_{i|K+1}\}$ is the estimate of $\{\mathbf{x}_i\}_{K+1}$ that yields a maximum of the conditional probability

---

[17] If an independent random variable that influences the system is correlated in time, it can be characterized, in general, by introducing additional state variables, the state equations for which contain only white noise forcing functions. For example, in a given case the nth state variable $x_{n,k}$ could be introduced and characterized simply by $x_{n,k+1} = 0.5 x_{n,k} + w_{1,k}$ where $w_{1,k}$ is a white noise forcing function, and $x_{n,k}$ is a correlated random variable which serves as a forcing function in certain of the other state equations. The resulting characterization satisfies the conditions assumed in this section.

density function $p(\{\mathbf{x}_i\}_{K+1}|\{\mathbf{z}_i\}_{K+1})$. Thus,

$$\begin{aligned}
p(\{\hat{\mathbf{x}}_{i|K+1}\}|\{\mathbf{z}_i\}_{K+1}) &= \max_{\{\mathbf{x}_i\}_{K+1}} p(\{\mathbf{x}_i\}_{K+1}|\{\mathbf{z}_i\}_{K+1}) \\
&= \max_{\mathbf{x}_{K+1}} \max_{\{\mathbf{x}_i\}_K} p(\{\mathbf{x}_i\}_{K+1}|\{\mathbf{z}_i\}_{K+1}) \\
&= \max_{\mathbf{x}_{K+1}} \mathscr{I}(\mathbf{x}_{K+1}, K+1) \quad (7\text{-}274)
\end{aligned}$$

where $\mathscr{I}(\mathbf{x}_{K+1}, K+1)$ is a maximum-return function,

$$\mathscr{I}(\mathbf{x}_{K+1}, K+1) \triangleq \max_{\{\mathbf{x}_i\}_K} p(\{\mathbf{x}_i\}_{K+1}|\{\mathbf{z}_i\}_{K+1}) \quad (7\text{-}275)$$

A recurrence relation between $\mathscr{I}(\mathbf{x}_{K+1}, K+1)$ and $\mathscr{I}(\mathbf{x}_K, K)$ is obtained by using a form of Bayes' theorem:

$$p(\{\mathbf{x}_i\}_{K+1}|\{\mathbf{z}_i\}_{K+1}) = \frac{p(\mathbf{z}_{K+1}|\{\mathbf{x}_i\}_{K+1},\{\mathbf{z}_i\}_K) p(\{\mathbf{x}_i\}_{K+1}|\{\mathbf{z}_i\}_K)}{p(\mathbf{z}_{K+1}|\{\mathbf{z}_i\}_K)} \quad (7\text{-}276)$$

in which, by the Markov nature of the process characterized by 7-272 and 7-273, the term $p(\mathbf{z}_{K+1}|\{\mathbf{x}_i\}_{K+1},\{\mathbf{z}_i\}_K)$ is equivalent to $p(\mathbf{z}_{K+1}|\mathbf{x}_{K+1})$, and the term $p(\{\mathbf{x}_i\}_{K+1}|\{\mathbf{z}_i\}_K)$ is equivalent to $p(\mathbf{x}_{K+1}|\mathbf{x}_K) p(\{\mathbf{x}_i\}_K|\{\mathbf{z}_i\}_K)$. It follows that

$$p(\{\mathbf{x}_i\}_{K+1}|\{\mathbf{z}_i\}_{K+1}) = \frac{p(\mathbf{z}_{K+1}|\mathbf{x}_{K+1}) p(\mathbf{x}_{K+1}|\mathbf{x}_K) p(\{\mathbf{x}_i\}_K|\{\mathbf{z}_i\}_K)}{p(\mathbf{z}_{K+1}|\{\mathbf{z}_i\}_K)} \quad (7\text{-}277)$$

The right-hand member of 7-277 is substituted for $p(\{\mathbf{x}_i\}_{K+1}|\{\mathbf{z}_i\}_{K+1})$ in 7-275 to obtain the desired recurrence relation:

$$\begin{aligned}
\mathscr{I}(\mathbf{x}_{K+1}, K+1) &= \max_{\{\mathbf{x}_i\}_K} \left[\frac{p(\mathbf{z}_{K+1}|\mathbf{x}_{K+1}) p(\mathbf{x}_{K+1}|\mathbf{x}_K) p(\{\mathbf{x}_i\}_K|\{\mathbf{z}_i\}_K)}{p(\mathbf{z}_{K+1}|\{\mathbf{z}_i\}_K)} \right] \\
&= \max_{\mathbf{x}_K} \left[\frac{p(\mathbf{z}_{K+1}|\mathbf{x}_{K+1}) p(\mathbf{x}_{K+1}|\mathbf{x}_K)}{p(\mathbf{z}_{K+1}|\{\mathbf{z}_i\}_K)} \mathscr{I}(\mathbf{x}_K, K) \right] \quad (7\text{-}278)
\end{aligned}$$

where standard techniques (e.g., Papoulis [7.78], p. 125) may be applied to determine $p(\mathbf{z}_{K+1}|\mathbf{x}_{K+1})$ on the basis of 7-273 and the known $p(\mathbf{v}_{K+1})$, and to determine $p(\mathbf{x}_{K+1}|\mathbf{x}_K)$ on the basis of 7-272 and the known $p(\mathbf{w}_K)$. At the instant corresponding to $k = K+1$, $\{\mathbf{z}_i\}_{K+1}$ is a known set of vector values, and therefore $p(\mathbf{z}_{K+1}|\{\mathbf{z}_i\}_K)$ is simply a scale factor in 7-278. It follows that the vector policy function $\hat{\mathbf{x}}_K(\mathbf{x}_{K+1}, K+1)$ for $\mathbf{x}_K$ that yields the maximum in 7-278 also yields the maximum in

$$\mathscr{F}(\mathbf{x}_{K+1}, K+1) = \max_{\mathbf{x}_K} [p(\mathbf{z}_{K+1}|\mathbf{x}_{K+1}) p(\mathbf{x}_{K+1}|\mathbf{x}_K) \mathscr{F}(\mathbf{x}_K, K)] \quad (7\text{-}279)$$

where $\mathscr{F}(\mathbf{x}_{K+1}, K+1)$ is a scaled maximum-return function. In general,

$$\mathscr{F}(\mathbf{x}_{k+1}, k+1) = \max_{\mathbf{x}_k} [p(\mathbf{z}_{k+1}|\mathbf{x}_{k+1}) p(\mathbf{x}_{k+1}|\mathbf{x}_k) \mathscr{F}(\mathbf{x}_k, k)] \quad (7\text{-}280a)$$

Sect. 7-16 ESTIMATION OF STATE VARIABLES 455

for $k = 1, 2, \ldots, K$, where

$$\mathscr{F}(\mathbf{x}_1, 1) = \max_{\mathbf{x}_0} [p(\mathbf{z}_1|\mathbf{x}_1)p(\mathbf{x}_1|\mathbf{x}_0)p(\mathbf{x}_0)] \quad (7\text{-}280\text{b})$$

After the return functions are generated, the modal value $\bar{\mathbf{x}}_{K+1|K+1}$ of $\mathbf{x}_{K+1}$ is obtained on the basis that

$$\mathscr{F}(\bar{\mathbf{x}}_{K+1|K+1}, K+1) = \max_{\mathbf{x}_{K+1}} \mathscr{F}(\mathbf{x}_{K+1}, K+1) \quad (7\text{-}281)$$

which is justified by 7-274, 7-278, and 7-279. In turn, $\bar{\mathbf{x}}_{K|K+1}$ is obtained from

$$\mathscr{F}(\bar{\mathbf{x}}_{K+1|K+1}, K+1) = \max_{\mathbf{x}_K} [p(\mathbf{z}_{K+1}|\bar{\mathbf{x}}_{K+1|K+1})p(\bar{\mathbf{x}}_{K+1|K+1}|\mathbf{x}_K)\mathscr{F}(\mathbf{x}_K, K)]$$

and because $p(\mathbf{z}_{K+1}|\bar{\mathbf{x}}_{K+1|K+1})$ is known,

$$p(\bar{\mathbf{x}}_{K+1|K+1}|\bar{\mathbf{x}}_{K|K+1})\mathscr{F}(\bar{\mathbf{x}}_{K|K+1}, K) \geq p(\bar{\mathbf{x}}_{K+1|K+1}|\mathbf{x}_K)\mathscr{F}(\mathbf{x}_K, K) \quad (7\text{-}282)$$

In cyclic fashion,

$$p(\bar{\mathbf{x}}_{k+1|K+1}|\bar{\mathbf{x}}_{k|K+1})\mathscr{F}(\bar{\mathbf{x}}_{k|K+1}, k) \geq p(\bar{\mathbf{x}}_{k+1|K+1}|\mathbf{x}_k)\mathscr{F}(\mathbf{x}_k, k) \quad (7\text{-}283)$$

for $k = K - 1, K - 2, \ldots, 1$.

As with other n-state-variable problems, the generation and storage of return functions may be accomplished with the aid of the techniques in Section 7-10.

The existence of a maximum at each stage of the decision process is assumed in the preceding development. This existence condition is not strictly satisfied in many problems of interest, as a limiting case of the following example illustrates; yet there is a straightforward way to obtain an optimal solution.

Example 7-7. Larson and Peschon [7.67, 7.68] give examples in which the known probability density functions are discrete; in effecting required maximizations, they obtain maximum-weight components of discrete density functions. The known probability density functions of this example are continuous, and the maxima obtained are peak values of continuous density functions.

Given:

$$\begin{bmatrix} x_{1,k+1} \\ x_{2,k+1} \end{bmatrix} = \begin{bmatrix} 0 & 1 \\ -0.4 & 1.3 \end{bmatrix} \begin{bmatrix} x_{1,k} \\ x_{2,k} \end{bmatrix} + \begin{bmatrix} \epsilon & 0 \\ 0 & 0.1 \end{bmatrix} \begin{bmatrix} w_{1,k} \\ w_{2,k} \end{bmatrix} + \begin{bmatrix} 0 \\ (0.5)^k \end{bmatrix} \quad (7\text{-}284)$$

where the real number ϵ is bounded by $0 < \epsilon \ll 0.1$. The known density functions for $w_{1,k}$ and $w_{2,k}$ are Cauchy density functions:

$$p(w_{1,k}) = \frac{1}{\pi} \frac{\beta_k}{\beta_k^2 + w_{1,k}^2} \quad (7\text{-}285)$$

where, for this example, all β_k's $= 1$; similarly,

$$p(w_{2,k}) = \frac{1}{\pi} \frac{1}{1 + w_{2,k}^2} \tag{7-286}$$

and it is assumed that $w_{1,k}$ is independent of $w_{2,k}$.

The initial-condition probability density function is also assumed to be Cauchy:

$$p(\mathbf{x}_0) = \frac{1}{\pi^2} \left(\frac{10}{1 + 100x_{1,0}^2} \right) \left(\frac{10}{1 + 100x_{2,0}^2} \right) \tag{7-287}$$

The measurement system gives

$$z_{k+1} = x_{1,k+1} + v_{k+1} \tag{7-288}$$

where

$$p(v_{k+1}) = \frac{1}{\pi} \frac{2}{1 + 4v_{k+1}^2} \tag{7-289}$$

Equations 7-284 and 7-285 are used to obtain

$$p(x_{1,k+1}|x_{1,k},x_{2,k}) = \left[\frac{p(w_{1,k})}{\frac{\partial}{\partial w_{1,k}}(x_{2,k} + \epsilon w_{1,k})} \right]_{w_{1,k} = (x_{1,k+1} - x_{2,k})/\epsilon}$$

$$= \frac{1}{\pi} \frac{\epsilon}{\epsilon^2 + (x_{1,k+1} - x_{2,k})^2} \tag{7-290}$$

and, similarly,

$$p(x_{2,k+1}|x_{1,k},x_{2,k}) = \frac{1}{\pi} \frac{0.1}{0.01 + [x_{2,k+1} + 0.4x_{1,k} - 1.3x_{2,k} - (0.5)^k]^2} \tag{7-291}$$

Because of the assumed independence of $w_{1,k}$ and $w_{2,k}$,

$$p(\mathbf{x}_{k+1}|\mathbf{x}_k) = \frac{1}{\pi^2} \left(\frac{\epsilon}{\epsilon^2 + (x_{1,k+1} - x_{2,k})^2} \right)$$

$$\times \left(\frac{0.1}{0.01 + [x_{2,k+1} + 0.4x_{1,k} - 1.3x_{2,k} - (0.5)^k]^2} \right) \tag{7-292}$$

The measurement system 7-288 and $p(v_{k+1})$ of 7-289 give

$$p(z_{k+1}|\mathbf{x}_{k+1}) = \frac{1}{\pi} \frac{2}{1 + 4(z_{k+1} - x_{1,k+1})^2} \tag{7-293}$$

Because $0 < \epsilon \ll 0.1$, the density function of 7-290 approximates a Dirac delta function centered at $x_{2,k} = x_{1,k+1}$ and it is apparent that the peak value of $p(\mathbf{x}_{k+1}|\mathbf{x}_k)$ is dominated by a $(1/\epsilon)$ factor when $\hat{x}_{2,k} = x_{1,k+1}$; this fact

Sect. 7-16   ESTIMATION OF STATE VARIABLES

and Equations 7-280, 7-292, and 7-293 are used to obtain

$$\mathscr{F}(\mathbf{x}_{k+1}, k+1) = \max_{x_{1,k}} \left[\frac{0.2}{\pi^3 \epsilon} \frac{1}{1 + 4(z_{k+1} - x_{1,k+1})^2} \right.$$
$$\times \frac{1}{0.01 + [x_{2,k+1} + 0.4x_{1,k} - 1.3x_{1,k+1} - (0.5)^k]^2}$$
$$\left. \times \mathscr{F}(x_{1,k}, x_{1,k+1}, k) \right]$$
(7-294)

where

$$\mathscr{F}(\mathbf{x}_1, 1) = \max_{x_{1,0}} \left[\frac{0.2}{\pi^3 \epsilon} \frac{1}{1 + 4(z_1 - x_{1,1})^2} \frac{1}{0.01 + [x_{2,1} + 0.4x_{1,0} - 1.3x_{1,1} - 1]^2} \right.$$
$$\left. \times \frac{100}{\pi^2} \frac{1}{1 + 100x_{1,0}^2} \frac{1}{1 + 100x_{1,1}^2} \right]$$
(7-295)

Because the factor $(0.2/\pi^3 \epsilon)$ is a common multiplicative scale factor in 7-294 and 7-295, it can be deleted in the generation of a new set of scaled maximum-return functions.

The approach to solution from this point forward is clear. Given a measured value z_1, the maximum in 7-295 is obtained for each pair of values of $x_{1,1}$ and $x_{2,1}$ over a representative range of values of each; since both $x_{1,0}$ and $x_{2,0}$ are likely to be in the neighborhood of zero (see Equation 7-287), the range of values for $x_{2,1}$ should be centered around $x_{2,1} = 1$, and the range of values for $x_{1,1}$ should be centered around $x_{1,1} = 0$. The resulting maximum-return function $(\pi^3 \epsilon/0.2)\mathscr{F}(\mathbf{x}_1, 1)$ is used in 7-294, once z_2 is known, to obtain a scaled $\mathscr{F}(\mathbf{x}_2, 2)$. The process is continued until a scaled $\mathscr{F}(\mathbf{x}_{K+1}, K+1)$ is obtained, at which point the modal trajectory is generated by using conditions 7-281, 7-282, and 7-283.

●

7-16b. Discrete Kalman-Bucy Filter

Consider the state equation

$$\mathbf{x}_{k+1} = \mathbf{A}_k \mathbf{x}_k + \mathbf{\Gamma}_k \mathbf{w}_k + \mathbf{b}_k(\mathbf{m}_k), \quad k = 0, 1, \ldots \quad (7\text{-}296)$$

which is a special case of state equation 7-272. $\mathbf{A}_k$ is a given $n \times n$ matrix, the entries of which may depend on k, as may the entries of $\mathbf{\Gamma}_k$, a given rectangular matrix of appropriate dimension. Both $\mathbf{A}_k$ and $\mathbf{\Gamma}_k$ are assumed independent of $\mathbf{x}_k$'s and $\mathbf{w}_k$'s. The vector $\mathbf{w}_k$ is a white, Gaussian, independent noise vector with zero mean value and with a specified covariance matrix[18]

$$\mathbf{S}_k \triangleq \text{Cov}(\mathbf{w}_k) = E(\mathbf{w}_k \mathbf{w}_k') \qquad (7\text{-}297)$$

The $\mathbf{b}_k(\mathbf{m}_k)$ part of 7-296 is a given $n \times 1$ vector function, perhaps a nonlinear function of the components of an $r \times 1$ control vector $\mathbf{m}_k$. At the kth sampling instant, the numerical vector value of $\mathbf{b}_k(\mathbf{m}_k)$ is assumed known. The initial state $\mathbf{x}_0$ is assumed to be characterized by a Gaussian probability density function with mean value $\bar{\mathbf{x}}_0$ and covariance matrix $\mathbf{P}_0$.

Measurements on the system 7-296 yield

$$\mathbf{z}_{k+1} = \mathbf{H}_{k+1} \mathbf{x}_{k+1} + \mathbf{v}_{k+1}, \qquad k = 0, 1, \ldots \qquad (7\text{-}298)$$

which is a special, linear case of the measurement system 7-273. $\mathbf{H}_{k+1}$ is a given rectangular matrix of appropriate dimension, and $\mathbf{v}_{k+1}$ is a white, Gaussian, independent noise vector with zero mean value and with a given covariance matrix:

$$\mathbf{R}_{k+1} = \text{Cov}(\mathbf{v}_{k+1}) = E(\mathbf{v}_{k+1} \mathbf{v}_{k+1}') \qquad (7\text{-}299)$$

The independent assumptions previously noted give

$$E(\mathbf{v}_i \mathbf{w}_j') = \mathbf{0} \text{ all } i, j \qquad (7\text{-}300\text{a})$$

$$E(\mathbf{x}_0 \mathbf{w}_j') = \mathbf{0} \text{ all } j \qquad (7\text{-}300\text{b})$$

$$E(\mathbf{x}_0 \mathbf{v}_j') = \mathbf{0} \text{ all } j \qquad (7\text{-}300\text{c})$$

$$E(\mathbf{w}_j \mathbf{w}_k') = \mathbf{S}_k \, \delta_{j,k} \text{ all } j, k \qquad (7\text{-}300\text{d})$$

and

$$E(\mathbf{v}_j \mathbf{v}_k') = \mathbf{R}_k \, \delta_{j,k} \text{ all } j, k \qquad (7\text{-}300\text{e})$$

where $\delta_{j,k}$ is the Kronecker delta function.

For systems of the preceding type, Kalman [7.53], Kalman and Bucy [7.56], and others (see, for example, the literature noted in references [7.30] and [7.86]) obtain a best estimate $\bar{\mathbf{x}}_{k+1|k+1}$ of the state vector at the $(k+1)$st sampling instant based on knowledge of the estimate $\bar{\mathbf{x}}_{k|k}$ and the measurement vector $\mathbf{z}_{k+1}$. Required computations can be effected with a digital computer to obtain the estimates in "real" time, and the estimation process is appropriately called the Kalman-Bucy filter. The Kalman-Bucy filter is

---

[18] The covariance matrix for $\mathbf{w}_k$ is a square matrix which equals the expected value of $(\mathbf{w}_k - \bar{\mathbf{w}}_k)(\mathbf{w}_k - \bar{\mathbf{w}}_k)'$, where the mean value $\bar{\mathbf{w}}_k$ equals zero in 7-297. For a Gaussian random variable, the mean and the covariance are used to completely characterize the probability density function.

Sect. 7-16   ESTIMATION OF STATE VARIABLES   459

best in the sense that a minimum of the expected value of

$$(\bar{\mathbf{x}}_{k+1|k+1} - \mathbf{x}_{k+1})'(\bar{\mathbf{x}}_{k+1|k+1} - \mathbf{x}_{k+1})$$

is obtained. In addition, the filter yields the maximum of the conditional probability density function 7-277, a fact which is shown by Ho and Lee [7.48], Greensite [7.42], and Larson and Peschon [7.67, 7.68]. The derivation that follows is based on these sources.

Under the assumed conditions of this subsection, it is well known [7.78] that the conditional probability density functions in 7-280 are Gaussian, in which case these functions are determined by appropriate expected values and covariances. Thus, $p(\mathbf{z}_{k+1}|\mathbf{x}_{k+1})$ is characterized by

$$\begin{aligned} E(\mathbf{z}_{k+1}|\mathbf{x}_{k+1}) &= E(\mathbf{H}_{k+1}\mathbf{x}_{k+1} + \mathbf{v}_{k+1}|\mathbf{x}_{k+1}) \\ &= \mathbf{H}_{k+1}\mathbf{x}_{k+1} \end{aligned} \quad (7\text{-}301)$$

and

$$\text{Cov}(\mathbf{z}_{k+1}|\mathbf{x}_{k+1}) = \mathbf{R}_{k+1} \quad (7\text{-}302)$$

which give

$$p(\mathbf{z}_{k+1}|\mathbf{x}_{k+1}) = c_1 \exp\left[-\tfrac{1}{2}(\mathbf{z}_{k+1} - \mathbf{H}_{k+1}\mathbf{x}_{k+1})'\mathbf{R}_{k+1}^{-1}(\mathbf{z}_{k+1} - \mathbf{H}_{k+1}\mathbf{x}_{k+1})\right] \quad (7\text{-}303)$$

where c_1 is a constant.

Similarly,

$$E(\mathbf{x}_{k+1}|\mathbf{x}_k) = \mathbf{A}_k\mathbf{x}_k + \mathbf{b}_k(\mathbf{m}_k) \quad (7\text{-}304)[19]$$

and

$$\text{Cov}(\mathbf{x}_{k+1}|\mathbf{x}_k) = \mathbf{\Gamma}_k\mathbf{S}_k\mathbf{\Gamma}_k' \quad (7\text{-}305)$$

which give

$$p(\mathbf{x}_{k+1}|\mathbf{x}_k) = c_2 \exp\left[-\tfrac{1}{2}(\mathbf{x}_{k+1} - \mathbf{A}_k\mathbf{x}_k - \mathbf{b}_k(\mathbf{m}_k))' \right. \\ \left. \times (\mathbf{\Gamma}_k\mathbf{S}_k\mathbf{\Gamma}_k')^{-1}(\mathbf{x}_{k+1} - \mathbf{A}_k\mathbf{x}_k - \mathbf{b}_k(\mathbf{m}_k))\right] \quad (7\text{-}306)$$

where c_2 is a constant.

Because the maximum-return function $\mathscr{F}(\mathbf{x}_k,k)$ is obtained recursively by using 7-280, it too is Gaussian and can be expressed by

$$\begin{aligned} \mathscr{F}(\mathbf{x}_k,k) &= \max_{(\mathbf{x}_i)_{k-1}} c_3 \exp\left[-\tfrac{1}{2}\sum_{i=0}^{k}(\mathbf{x}_i - \bar{\mathbf{x}}_{i|k})'\mathbf{P}_{i|k}^{-1}(\mathbf{x}_i - \bar{\mathbf{x}}_{i|k})\right] \\ &= c_4 \exp\left[-\tfrac{1}{2}(\mathbf{x}_k - \bar{\mathbf{x}}_{k|k})'\mathbf{P}_{k|k}^{-1}(\mathbf{x}_k - \bar{\mathbf{x}}_{k|k})\right] \end{aligned} \quad (7\text{-}307)$$

---

[19] Note that $\mathbf{b}_k(\mathbf{m}_k)$ is a given vector value at the kth instant.

where c_3 and c_4 are constants, and both the covariance matrix $\mathbf{P}_{k|k}$ and the estimate $\bar{\mathbf{x}}_{k|k}$ are generated by the Kalman-Bucy filter. Known initial values are

$$\bar{\mathbf{x}}_{0|0} = \bar{\mathbf{x}}_0 \qquad (7\text{-}308a)$$

and

$$\mathbf{P}_{0|0} = \mathbf{P}_0 \qquad (7\text{-}308b)$$

To obtain the Kalman-Bucy filter, the right-hand members of 7-303, 7-306, and 7-307 are substituted appropriately into 7-280:

$$\mathscr{F}(\mathbf{x}_{k+1}, k+1) = \max_{\mathbf{x}_k} c_5 \exp\{-\tfrac{1}{2}[(\mathbf{z}_{k+1} - \mathbf{H}_{k+1}\mathbf{x}_{k+1})'\mathbf{R}_{k+1}^{-1}(\mathbf{z}_{k+1} - \mathbf{H}_{k+1}\mathbf{x}_{k+1})$$
$$+ (\mathbf{x}_{k+1} - \mathbf{A}_k\mathbf{x}_k - \mathbf{b}_k(\mathbf{m}_k))'(\mathbf{\Gamma}_k\mathbf{S}_k\mathbf{\Gamma}_k')^{-1}$$
$$\times (\mathbf{x}_{k+1} - \mathbf{A}_k\mathbf{x}_k - \mathbf{b}_k(\mathbf{m}_k))$$
$$+ (\mathbf{x}_k - \bar{\mathbf{x}}_{k|k})'\mathbf{P}_{k|k}^{-1}(\mathbf{x}_k - \bar{\mathbf{x}}_{k|k})]\}$$
$$(7\text{-}309)$$

The policy $\hat{\mathbf{x}}_k(\mathbf{x}_{k+1})$ that minimizes the term within the brackets of 7-309 yields the desired maximum. The reader may verify that

$$\hat{\mathbf{x}}_k(\mathbf{x}_{k+1}) = [\mathbf{P}_{k|k}^{-1} + \mathbf{A}_k'(\mathbf{\Gamma}_k\mathbf{S}_k\mathbf{\Gamma}_k')^{-1}\mathbf{A}_k]^{-1}$$
$$\times [\mathbf{A}_k'(\mathbf{\Gamma}_k\mathbf{S}_k\mathbf{\Gamma}_k')^{-1}(\mathbf{x}_{k+1} - \mathbf{b}_k(\mathbf{m}_k)) + \mathbf{P}_{k|k}^{-1}\bar{\mathbf{x}}_{k|k}] \quad (7\text{-}310)$$

When the right-hand member of 7-310 is substituted for $\mathbf{x}_k$ in 7-309, the resulting function must be identical to 7-307 with k replaced by $k+1$. Thus, like coefficients of $\mathbf{x}_{k+1}$ in the exponents are equated to obtain the desired recurrence relations:[20]

$$\mathbf{P}_{k+1|k+1}^{-1} = \mathbf{H}_{k+1}'\mathbf{R}_{k+1}^{-1}\mathbf{H}_{k+1} + [\mathbf{A}_k\mathbf{P}_{k|k}\mathbf{A}_k' + \mathbf{\Gamma}_k\mathbf{S}_k\mathbf{\Gamma}_k']^{-1} \qquad (7\text{-}311)$$

and

$$\bar{\mathbf{x}}_{k+1|k+1} = \mathbf{A}_k\bar{\mathbf{x}}_{k|k} + \mathbf{b}_k(\mathbf{m}_k)$$
$$+ \mathbf{P}_{k+1|k+1}\mathbf{H}_{k+1}'\mathbf{R}_{k+1}^{-1}[\mathbf{z}_{k+1} - \mathbf{H}_{k+1}(\mathbf{A}_k\bar{\mathbf{x}}_{k|k} + \mathbf{b}_k(\mathbf{m}_k))] \quad (7\text{-}312)$$

A flow diagram for these discrete Kalman-Bucy filter equations is given in Figure 7-19.

Example 7-8. Consider the system of Example 7-7, but replace all probability density functions with Gaussian density functions. The Kalman-Bucy

---

[20] This tedious step requires the use of the matrix identity

$$\Theta^{-1} - \Theta^{-1}\mathbf{Y}[\Phi^{-1} + \Omega\Theta^{-1}\mathbf{Y}]^{-1}\Omega\Theta^{-1} = [\Theta + \mathbf{Y}\Phi\Omega]^{-1}$$

where Θ is any $n \times n$ nonsingular matrix, $\mathbf{Y}$ is any $n \times m$ matrix, Φ is any $m \times m$ nonsingular matrix, and Ω is any $m \times n$ matrix (see reference [7.27]). Here: $\Theta = \mathbf{\Gamma}_k\mathbf{S}_k\mathbf{\Gamma}_k'$, $\mathbf{Y} = \mathbf{A}_k$, $\Phi = \mathbf{P}_{k|k}$, and $\Omega = \mathbf{A}_k'$. Note that, in general, $[\Theta + \mathbf{Y}\Phi\Omega]^{-1} \neq \Theta^{-1} + (\mathbf{Y}\Phi\Omega)^{-1}$.

Sect. 7-16 ESTIMATION OF STATE VARIABLES 461

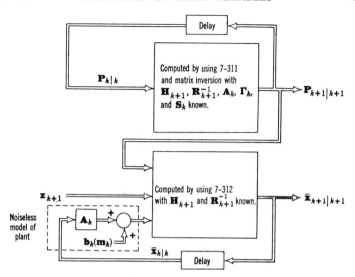

Figure 7-19. A Kalman-Bucy filter.

filter equations 7-311 and 7-312 are applicable; here,

$$\mathbf{A}_k = \mathbf{A} = \begin{bmatrix} 0 & 1 \\ -0.4 & 1.3 \end{bmatrix} \tag{7-313a}$$

$$\mathbf{\Gamma}_k = \mathbf{\Gamma} = \begin{bmatrix} \epsilon & 0 \\ 0 & 0.1 \end{bmatrix} \tag{7-313b}$$

$$\mathbf{b}_k(\mathbf{m}_k) = \begin{bmatrix} 0 \\ (0.5)^k \end{bmatrix} \tag{7-313c}$$

and

$$\mathbf{H}_{k+1} = \mathbf{h} = [1 \quad 0] \tag{7-313d}$$

The covariance matrix $\mathbf{S}_k$ for the system noise is assumed to be independent of k in this example,

$$\mathbf{S}_k = \mathbf{S} = \begin{bmatrix} \dfrac{\pi}{2} & 0 \\ 0 & \dfrac{\pi}{2} \end{bmatrix} \tag{7-314}$$

Similarly, the assumed covariance matrix for the measurement noise is

$$\mathbf{R}_k = r_{11} = \frac{\pi}{8} \tag{7-315}$$

Equation 7-311 assumes the form

$$\mathbf{P}_{k+1|k+1}^{-1} = \begin{bmatrix} \frac{8}{\pi} & 0 \\ 0 & 0 \end{bmatrix}$$

$$+ \begin{bmatrix} p_{22,k} + \frac{\epsilon^2 \pi}{2} & -0.4p_{12,k} + 1.3p_{22,k} \\ -0.4p_{12,k} + 1.3p_{22,k} & \left(0.16p_{11,k} - 1.04p_{12,k} \right. \\ & \left. + 1.69p_{22,k} + \frac{\pi}{200}\right) \end{bmatrix}^{-1} \quad (7\text{-}316)$$

where $p_{12,k} = p_{21,k}$ because of symmetry, and Equation 7-312 assumes the form

$$\bar{\mathbf{x}}_{k+1|k+1} = \begin{bmatrix} \bar{x}_{2,k} \\ -0.4\bar{x}_{1,k} + 1.3\bar{x}_{2,k} + (0.5)^k \end{bmatrix} + \mathbf{P}_{k+1|k+1} \begin{bmatrix} \frac{8}{\pi}(z_{k+1} - \bar{x}_{2,k}) \\ 0 \end{bmatrix} \quad (7\text{-}317)$$

Given a set of real z_{k+1}'s, Equations 7-316 and 7-317 can be evaluated iteratively (see Problem 7.36).

•

7-17. CONCLUSION

The approach to solution via dynamic programming is keyed to a *principle of optimality*. In the initial sections of this chapter (Sections 7-2 through 7-6), seemingly diverse optimization problems are solved with the aid of dynamic programming. In Section 7-8, the common bond, Bellman's principle of optimality, is brought to the fore. When confronted with a new optimization problem, we should check to see if this principle of optimality can be applied to obtain a solution.

A superlative quality of dynamic programming is that it can be used to obtain both numerical results through iterative operations on recurrence relations and analytical results, e.g., those which, as in Section 7-13, are associated with the calculus of variations and the maximum principle of Pontryagin. When numerical results are generated recursively, the problem of accumulated error should be considered (Section 7-7 and references [7.43, 7.44]), and the sensitivity of the numerical solution with respect to changes (or uncertainties) in assumed conditions may be of concern [7.84]. As for analytical results, those of Section 7-13 are only a sample of those which have been obtained with dynamic programming in the literature ([7.5, 7.6, 7.36, 7.45], and others).

For multivariable optimization problems of a general nature, the straightforward dynamic programming approach to numerical solution is hampered; both the number of computations and the amount of rapid-access computer storage required are exponential-type functions of the number of state variables. A variety of multivariable problems can be treated, however, through modifications and extensions of the straightforward dynamic programming approach; e.g., series approximations of functions (Section 7-10b), Lagrange multipliers (Section 7-10c), region-limiting strategies and iterated dynamic programming (Section 7-10d), approximations in function space and policy space (Section 7-11), and recurrence relations with controlled time increments (Section 7-12d). In addition, substantial computational and computer-storage savings can be gained if the performance measure and the constraint equations exhibit either convex or concave properties (the defining properties of a convex function are given in Section 6-2); Hadley [7.46] considers such cases.

When carefully applied, dynamic programming yields the absolute optimal solutions to a wide class of problems; nonoptimal but relative extrema are avoided, provided that absolute optima are obtained at each stage of a given solution. To obtain absolute optima at each stage of a dynamic programming solution, appropriate search techniques from Chapter 6 may be employed.

The problems treated in this chapter illustrate various areas of application. A wide variety of other applications and results obtained by use of dynamic programming are to be found in the literature. A sample of these is as follows: assorted control problems [7.1, 7.2, 7.59, 7.61, 7.64, 7.69, 7.88, 7.89, 7.90, 7.91, 7.95]; adaptive system considerations [7.3, 7.12, 7.20, 7.38]; thermo and hydro-thermo power coordination [7.23, 7.32, 7.39, 7.58, 7.72]; problems of statistical communication theory [7.18, 7.19]; problems of logic design [7.17, 7.40, 7.93]; general industrial engineering problems [7.16, 7.46, 7.49, 7.57]; discounted dynamic programming [7.25, 7.26]; research-and-development budgeting and project selection [7.47]; production-line inspection [7.81]; approximation of curves by line segments [7.11] and diode-squarer design [7.34]; pattern recognition [7.92]; decomposition of linear programs [7.76]; and fault detection in complex systems [7.41].

REFERENCES

[7.1] Adorno, D. S. "Optimum Control of Certain Linear Systems with Quadratic Loss, I." *Information and Control*, 5, 1–12, March 1962.

[7.2] Aoki, M. "On Optimal and Suboptimal Policies in the Choice of Control Forces for Final-Value Systems." *IRE International Convention Record*, 8, part 4, 15–22, 1960.

[7.3] Aoki, M. "On Performance Losses in Some Adaptive Control Systems, I." *Preprint Volume*, Joint Automatic Control Conference, 29–33, June 1964.

[7.4] Bellman, R. "Dynamic Programming and Lagrange Multipliers." *Proceedings of the National Academy of Sciences*, **42**, 767–769, Oct. 1956.

[7.5] Bellman, R. *Dynamic Programming*. Princeton Univ. Press, Princeton, N.J., 1957.

[7.6] Bellman, R. "Dynamic Programming and the Numerical Solution of Variational Problems." *Operations Research*, **5**, 277–288, April 1957.

[7.7] Bellman, R. "On the Application of the Theory of Dynamic Programming to the Study of Control Process." *Proceedings*, 1956 Symposium on Nonlinear Circuit Analysis, Polytechnic Institute of Brooklyn, Brooklyn, N.Y., pp. 199–213, 1957.

[7.8] Bellman, R. "On a Routing Problem." *Quarterly of Applied Mathematics*, **16**, 87–90, April 1958.

[7.9] Bellman, R. "Functional Equations and Successive Approximations in Linear and Nonlinear Programming." *Naval Research Logistics Quarterly*, **7**, 63–83, March 1960.

[7.10] Bellman, R. *Dynamic Programming and Linear Prediction Theory*. Paper P-2308, The RAND Corporation, Santa Monica, Calif., May 1961.

[7.11] Bellman, R. "On the Approximation of Curves by Line Segments Using Dynamic Programming." *Communications of the Association for Computing Machinery*, **4**, 284, June 1961.

[7.12] Bellman, R. *Adaptive Control Processes: A Guided Tour*. Princeton University Press, Princeton, N.J., 1961.

[7.13] Bellman, R. "Dynamic Programming, Generalized States, and Switching Systems." *Journal of Mathematical Analysis and Applications*, **12**, 360–363, Oct. 1965.

[7.14] Bellman, R., and S. E. Dreyfus. "Dynamic Programming and the Reliability of Multicomponent Devices." *Operations Research*, **6**, 200–206, March–April 1958.

[7.15] Bellman, R., and S. E. Dreyfus. "Functional Approximation and Dynamic Programming." *Mathematical Tables and Other Aids to Computation*, **13**, 247–251, Oct. 1959.

[7.16] Bellman, R., and S. E. Dreyfus. *Applied Dynamic Programming*. Princeton Univ. Press, Princeton, N.J., 1962.

[7.17] Bellman, R., J. Holland, and R. Kalaba. "On an Application of Dynamic Programming to the Synthesis of Logical Systems." *Journal of the Association for Computing Machinery*, **6**, 486–493, Oct. 1959.

[7.18] Bellman, R., and R. Kalaba. "On the Role of Dynamic Programming in Statistical Communication Theory." *IRE Transactions on Information Theory*, **IT-3**, 197–203, Sept. 1957.

[7.19] Bellman, R., and R. Kalaba. "On Communication Processes Involving Learning and Random Duration." *IRE National Convention Record*, **6**, part 4, 16–20, 1958.

[7.20] Bellman, R., and R. Kalaba. "Dynamic Programming and Adaptive Processes: Mathematical Foundation." *IRE Transactions on Automatic Control*, **AC-5**, 5–10, Jan. 1960.

[7.21] Bellman, R., and R. Kalaba. "Dynamic Programming, Invariant Imbedding and Quasilinearization: Comparisons and Interconnections." Pages 135–145 of *Computing Methods in Optimization Problems*, A. V. Balakrishnan and L. W. Neustadt (Editors). Academic Press, New York, 1964.

[7.22] Bellman, R., and R. Kalaba. *Dynamic Programming and Modern Control Theory*. Academic Press, New York, 1965.

[7.23] Bernholtz, B., and L. J. Graham. "Hydro-Thermo Economic Scheduling Part II, Extension of the Basic Theory." *AIEE Transactions*, **80**, part III, 1089–1096, Feb. 1961.

[7.24] Birkhoff, G., and Gian-Carlo Rota. *Ordinary Differential Equations*. Ginn, New York, 1962.

[7.25] Blackwell, D. "Discrete Dynamic Programming." *The Annals of Mathematical Statistics*, **33**, 719–726, June 1962.

[7.26] Blackwell, D. "Discounted Dynamic Programming." *The Annals of Mathematical Statistics*, **36**, 226–235, Feb. 1965.

[7.27] Bodewig, E. *Matrix Calculus*. Interscience, New York, 1959 (2nd edition).

[7.28] Boehm, B. W., and R. L. Mobley. *Adaptive Routing Techniques for Distributed Communications Systems*. Memorandum RM-4781-PR, The RAND Corporation, Santa Monica, Calif., Feb. 1966, 78 pp.

[7.29] Boltyanskii, V. G., R. V. Gamkrelidze, and L. S. Pontryagin. "On the Theory of Optimal Processes." *Doklady Akad. Nauk S.S.S.R.*, **110**, 7–10, 1956.

[7.30] Bucy, R. S., and P. D. Joseph. *Filtering for Stochastic Processes with Applications to Guidance*. Interscience Tracts in Pure and Applied Mathematics, Vol. 23, Wiley, New York, 1968.

[7.31] Cox, H. "On the Estimation of State Variables and Parameters for Noisy Dynamic Systems." *IEEE Transactions on Automatic Control*, AC-9, 5–12, Jan. 1964.

[7.32] Dahlin, E. B., and D. W. C. Shen. "Application of Dynamic Programming to Optimization of Hydroelectric/Steam Power-System Operation." *Proceedings, Institution of Electrical Engineers*, London, **112**, 2255–2260, Dec. 1965.

[7.33] Davis, H. T. *Introduction to Nonlinear Differential and Integral Equations*. United States Atomic Energy Commission, U.S. Government Printing Office, Washington, D.C., 1960.

[7.34] Deiters, R. M. "The Optimum Design of a Diode Squarer by Applying the Criterion of Square Root of the Integral of Per Cent Error Squared." *IEEE Transactions on Electronic Computers*, EC-14, 456–463, June 1965.

[7.35] Desoer, C. A. "Pontryagin's Maximum Principle and the Principle of Optimality." *Journal of the Franklin Institute*, **271**, 361–367, May 1961.

[7.36] Dreyfus, S. E. *Dynamic Programming and the Calculus of Variations*. Academic Press, New York, 1965.

[7.37] Durling, A. E. *Computational Aspects of Dynamic Programming in Higher Dimensions*. Ph.D. Thesis, Syracuse Univ., Syracuse, N.Y., 1964.

[7.38] Freimer, M. "A Dynamic Programming Approach to Adaptive Control Processes." *IRE National Convention Record*, **7**, part 4, 12–17, 1959.

[7.39] Fukao, T., T. Yamazaki, and S. Kimura. "An Application of Dynamic Programming to the Problem of Economic Operation of a Power System." *Electrotechnical Journal of Japan*, Tokyo, **5**, No. 2, 64–68, 1959.

[7.40] Glass, H. *The Application of Dynamic Programming to the Solution of the Zero-One Integer Linear Programming Problem.* Ph.D. Thesis, Washington Univ., St. Louis, Mo., 1966.

[7.41] Gluss, B. "An Optimum Policy for Detecting a Fault in a Complex System." *Operations Research*, 7, 468–477, July–Aug. 1959.

[7.42] Greensite, A. L. "Dynamic Programming and the Kalman Filter Theory." *Proceedings of the National Electronics Conference*, 20, 601–605, 1964.

[7.43] Guignabodet, J. *Analysis of Some Process Control Aspects of Dynamic Programming.* Ph.D. Thesis, Washington, Univ., St. Louis, Mo., 1961.

[7.44] Guignabodet, J. "Dynamic Programming: Cumulative Errors in the Evaluation of an Optimal Control Sequence." *Transactions of the ASME*, series D (*Journal of Basic Engineering*), 85, 151–156, June 1963.

[7.45] Gumowski, I. "Sur une Relation Entre le Calcul des Variations et la Programmation Dynamique." *Comptes Rendus Hebdomadaires Des Séances de L'Académie des Sciences*, 260, 4912–4915, May 1965.

[7.46] Hadley, G. *Nonlinear and Dynamic Programming.* Addison-Wesley, Reading, Mass., 1964.

[7.47] Hess, S. W. "A Dynamic Programming Approach to R and D Budgeting and Project Selection." *IRE Transactions on Engineering Management*, **EM-9**, 170–179, Dec. 1962.

[7.48] Ho, Y. C., and R. C. K. Lee. "A Bayesian Approach to Problems in Stochastic Estimation and Control." *IEEE Transactions on Automatic Control*, **AC-9**, 333–339, Oct. 1964.

[7.49] Howard, R. A. *Dynamic Programming and Markov Processes.* The M.I.T. Press, Cambridge, Mass., 1960.

[7.50] Junnarkar, N. V., and S. M. Roberts. "Dynamic Programming and Lagrangian Multipliers." *Industrial and Engineering Chemistry Fundamentals*, 4, 488–490, Nov. 1965.

[7.51] Kahne, S. J. "On Mobility in Constrained Dynamical Systems." *IEEE Transactions on Automatic Control*, **AC-9**, 318–319, July 1964.

[7.52] Kahne, S. J. *Feasible Control Computations Using Dynamic Programming.* Physical and Mathematical Sciences Research Papers, No. 94, Air Force Cambridge Research Laboratories, L. G. Hanscom Field, Bedford, Mass., April 1965.

[7.53] Kalman, R. E. "A New Approach to Linear Filtering and Prediction Problems." *Transactions of the ASME*, series D (*Journal of Basic Engineering*), 82, 35–44, March 1960.

[7.54] Kalman, R. E. "Contributions to the Theory of Optimal Control." *Boletin de la Sociedad Matematica Mexicana*, 102–119, 1960.

[7.55] Kalman, R. E. "On the General Theory of Control Systems." Pages 26–37 of *The Theory of Optimal Control* (reprinted from "Automatic and Remote Control," *Proceedings of the First International Congress of the International Federation of Automatic Control*, 1960), Butterworths, London, 1963.

[7.56] Kalman, R. E., and R. S. Bucy. "New Results in Linear Filtering and Prediction Theory." *Transactions of the ASME*, series D (*Journal of Basic Engineering*), 83, 95–108, March 1961.

[7.57] Kaufmann, A., and R. Cruon. *Dynamic Programming: Sequential, Scientific Management.* Academic Press, New York, 1967.

[7.58] Kirchmayer, L. K., and R. J. Ringlee. "Optimal Control of Thermal-Hydro System Operation." *Preprint Volume*, Joint Automatic Control Conference, 121–123, June 1963.

[7.59] Kolosov, G. E., and R. L. Stratonovich. "A Problem in the Synthesis of an Optimal Control Solved by the Method of Dynamic Programming." *Automation and Remote Control*, **24**, 1061–1067, 1963.

[7.60] Kronrod, A. D. *Nodes and Weights of Quadrature Formulas.* Consultants Bureau Enterprises, New York, 1965.

[7.61] Kushner, H. J. "Some Problems and Some Recent Results in Stochastic Control." *IEEE International Convention Record*, **13**, part 6, 108–116, 1965.

[7.62] Kushner, H. J. *Stochastic Stability and Control.* Academic Press, New York, 1967.

[7.63] Lanczos, C. *Applied Analysis.* Prentice-Hall, Englewood Cliffs, N.J., 1956.

[7.64] Lapidus, L., E. Shapiro, S. Shapiro, and R. E. Stillman. "Optimization of Process Performance." *AICHE Journal*, American Institute of Chemical Engineers, **7**, 288–294, June 1961.

[7.65] Larson, R. E. "An Approach to Reducing the High-Speed Memory Requirement of Dynamic Programming." *Journal of Mathematical Analysis and Applications*, **11**, 519–537, July 1965.

[7.66] Larson, R. E. "Dynamic Programming with Reduced Computational Requirements." *IEEE Transactions on Automatic Control*, **AC-10**, 135–143, April 1965.

[7.67] Larson, R. E., and J. Peschon. "Recursive Estimation of the Modal Trajectory." *Proceedings*, Third Annual Allerton Conference on Circuit and System Theory, Univ. of Illinois, Urbana, Ill., pp. 472–481, Oct. 1965.

[7.68] Larson, R. E., and J. Peschon. "A Dynamic Programming Approach to Trajectory Estimation." *IEEE Transactions on Automatic Control*, **AC-11**, 537–540, July 1966.

[7.69] Lee, E. B. "Design of Optimum Multivariable Control Systems." *Transactions of the ASME*, series D (*Journal of Basic Engineering*), **83**, 85–90, March 1961.

[7.70] Lee, E. B., and L. Markus. *Foundations of Optimal Control Theory.* Wiley, New York, 1967.

[7.71] Lee, R. C. K. *Optimal Estimation, Identification, and Control.* The M.I.T. Press, Cambridge, Mass., 1964.

[7.72] Lowery, P. G. "Generating Unit Commitment by Dynamic Programming." *IEEE Transactions on Power Apparatus and Systems*, **PAS-85**, 422–426, May 1966.

[7.73] Merriam, C. W., III. "An Optimization Theory for Feedback Control System Design." *Information and Control*, **3**, 32–59, March 1960.

[7.74] Merriam, C. W., III. *Optimization Theory and the Design of Feedback Control Systems.* McGraw-Hill, New York, 1964.

[7.75] Mond, B., and O. Shisha. "On the Approximation of Functions of Several Variables." *Journal of Research of the National Bureau of Standards*, Section B,

Mathematics and Mathematical Physics, **70B**, 211–218, July–Sept. 1966.

[7.76] Nemhauser, G. L. "Decomposition of Linear Programs by Dynamic Programming." *Naval Research Logistics Quarterly*, **11**, 191–196, June–Sept. 1964.

[7.77] Onaga, K. "Optimum Flows in General Communication Networks." *Journal of the Franklin Institute*, **283**, 308–327, April 1967.

[7.78] Papoulis, A. *Probability, Random Variables, and Stochastic Processes.* McGraw-Hill, New York, 1965.

[7.79] Pierre, D. A. "A Rational Fraction Approximation Formula for $\exp[-(sD)^{1/2}]$ with Applications." *Journal of the Society for Industrial and Applied Mathematics*, **12**, 93–104, March 1964.

[7.80] Pontryagin, L. S., V. G. Boltyanskii, R. V. Gamkrelidze, and E. F. Mishchenko. *The Mathematical Theory of Optimal Processes.* (Authorized Translation from the Russian; K. N. Trirogoff, Translator; L. W. Neustadt, Editor). Wiley, New York, 1962.

[7.81] Pruzan, P. M., and J. T. R. Jackson. "A Dynamic Programming Application in Production Line Inspection." *Technometrics*, **9**, 73–81, Feb. 1967.

[7.82] Roberts, S. M. "Dynamic Programming and Lagrangian Multipliers." *Industrial and Chemical Engineering Fundamentals*, **2**, 224–228, Aug. 1963.

[7.83] Rozonoer, L. I. "L. S. Pontryagin's Maximum Principle in the Theory of Optimum Systems." *Automation and Remote Control*, **20**, 1288–1302, 1405–1421, and 1517–1532 (three parts), 1959.

[7.84] Scherz, C. J. *An Accuracy Analysis of Some Problems in Optimal Control.* Ph.D. Thesis, Washington Univ., St. Louis, Mo., 1965.

[7.85] Snow, D. R. "Caratheodory-Hamilton-Jacobi Theory in Optimal Control." *Journal of Mathematical Analysis and Applications*, **17**, 99–118, Jan. 1967.

[7.86] Sorenson, H. W. "Kalman Filtering Techniques." Pages 219–292 of *Advances in Control Systems*, Vol. 3, C. T. Leondes (Editor). Academic Press, New York, 1966.

[7.87] Stagg, G. W., and M. Watson. "Dynamic Programming Transmission Line Design." *IEEE International Convention Record*, **12**, part 3, 55–61, 1964.

[7.88] Stratonovich, R. L. "Most Recent Development of Dynamic Programming Techniques and Their Application to Optimal Systems Design." *Automatic and Remote Control*, pp. 352–357, Theory Volume, Proceedings of the 2nd Congress of the International Federation on Automatic Control, 1963, Butterworths Inc., Washington, D.C., 1964.

[7.89] Tchamran, A. "On Bellman's Functional Equation and a Class of Time-Optimal Control Systems." *Journal of the Franklin Institute*, **280**, 493–505, Dec. 1965.

[7.90] Tou, J. T., and P. D. Joseph. "Modern Synthesis of Computer Control Systems." *IEEE Transactions on Applications and Industry*, **82**, 61–65, May 1963.

[7.91] Tung, F. "An Optimal Discrete Control Strategy for Interplanetary Guidance." *IEEE Transactions on Automatic Control*, **AC-10**, 328–335, July 1965.

[7.92] Udagawa, K., T. Fukumura, and K. Abe. "A Composing Method of Recognition Parameter Set Using Dynamic Programming." *Electronics and Communications in Japan*, **48**, 90–99, May 1965.

[7.93] Udagawa, K., and Y. Inagaki. "Method of Reliable Sequential Circuits by Means of Dynamic Programming." *Electronics and Communications in Japan*, **47**, 47–57, June 1964.
[7.94] Wall, H. S. *Continued Fractions*. Van Nostrand, Princeton, N.J., 1948.
[7.95] Wang, P. K. C., and F. Tung. "Optimum Control of Distributed Parameter Systems." *Preprint Volume*, Joint Automatic Control Conference, 16–32, 1963.

PROBLEMS

7.1 Verify the results of Tables 7-3 and 7-4.

7.2 A given function $f(x,\beta)$ is known to be concave with respect to its arguments.
a. Show that $\mathscr{F}(\beta)$ is also concave with respect to β, where $\mathscr{F}(\beta) = \max_{ax \leq \beta} f(x,\beta)$
(*Hint*: Make use of the fact that $(1 - \alpha)f(\hat{x}_a, \beta_a) + \alpha f(\hat{x}_b, \beta_b)$ is less than or equal to $f[(1 - \alpha)\hat{x}_a + \alpha \hat{x}_b, (1 - \alpha)\beta_a + \alpha \beta_b]$ with $\alpha \in [0, 1]$.)
b. Suppose that the f_i's, $i = 1, 2, \ldots, K$, of Section 7-2 are known to be concave, and that the $g_i(x_i)$'s are of the form 7-14. Are the $\mathscr{F}_i$'s also concave?
c. In addition to the conditions of Part b, assume that the x_i's are restricted to integer values. Are the $\mathscr{F}_i$'s concave?

7.3 Three work crews are to be scheduled to perform overtime work. Based on past experience under similar circumstances, the ith crew's performance on overtime is given by

$$c_i(1 - e^{-\alpha_i t_i}) \text{ (units of work)}$$

where c_i and α_i are constants, and t_i is the overtime (hours) worked by the ith crew. Overtime pay to the ith crew is given by $a_i t_i$ (dollars). The total cost of the overtime work is to be less than or equal to \$2,000. Each t_i is to be selected from the set of non-negative integers. The total number of units of work is to be maximized.
a. Are the maximum-return functions concave?
b. Find t_1^*, t_2^*, and t_3^* under the following conditions:
$a_1 = 100$, $a_2 = 200$, $a_3 = 300$, $c_1 = 5$, $c_2 = 9$, $c_3 = 13$, and $\alpha_1 = \alpha_2 = \alpha_3 = 0.4$.

7.4 A characteristic of a given system is represented by a function $g(\alpha_1, \alpha_2, \ldots, \alpha_K)$ where the α_i's denote parameters and g is of class C^2. Design-center values of the α_i's are μ_i's; the actual values of the α_i's are assumed to be sufficiently close to the corresponding μ_i's so that a truncated Taylor's series expansion of $g(\alpha_1, \alpha_2, \ldots, \alpha_K)$ gives

$$g(\alpha_1, \alpha_2, \ldots, \alpha_K) \cong \bar{g} + \sum_{i=1}^{K} (\alpha_i - \mu_i)a_i^{1/2}$$

where

$$\bar{g} \triangleq g(\mu_1, \mu_2, \ldots, \mu_K) \quad \text{and} \quad a_i \triangleq (\partial g / \partial \alpha_i)^2_{\alpha_i\text{'s} = \mu_i\text{'s}}$$

If each α_i is an independent random variable with normal density function characterized by mean value μ_i and standard deviation σ_i (variance σ_i^2), show that the variance of the characteristic g is given by

$$E[(g - \bar{g})^2] \cong \sum_{i=1}^{K} a_i \sigma_i^2$$

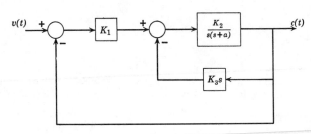

Figure 7-P5.

7.5 Consider the block diagram of Figure 7-P5. The percent overshoot of $c(t)$ that results from application of a step input can be shown to equal

$$100 \exp[-\xi\pi/(1 - \xi^2)^{1/2}]$$

where the damping factor ξ equals $(a + K_2 K_3)/2(K_1 K_2)^{1/2}$. Design-center values are $K_1^\circ = 16$, $K_2^\circ = 1$, $K_3^\circ = 3$, and $a^\circ = 1$; these values result in $\xi^\circ = 0.5$. Let scaled parameters be introduced in the following way: $z_1 = K_1/16$, $z_2 = K_2$, $z_3 = K_3/3$, and $z_4 = a$. The value of ξ is then

$$\xi = (z_4 + 3z_2 z_3)/8(z_1 z_2)^{1/2}$$

It is given that each of the z_i's is selected from a normal distribution, and the standard deviation of each density function can be specified at one of three values: 0.1, 0.04, or 0.02. The standard deviation σ_ξ of the density function of the damping factor ξ is to be less than or equal to 0.025, and of course the mean value of ξ is 0.5. Added system costs associated with the various standard deviations are listed in the following table:

	Added Cost			
σ	z_1	z_2	z_3	z_4
0.1	$ 0	$ 0	$ 0	$ 0
0.04	20	40	30	50
0.02	30	80	50	100

Use the approach of Section 7-2 to find the minimum added cost which results in $\sigma_\xi \leq 0.025$.

PROBLEMS

7.6 Write a digital computer program for dynamic programming solution of one-state-variable integer allocation problems.

7.7 Find the maximum of P in Section 7-4 under the following conditions: $b = 12$, $K = 3$, $p_1 = 0.5$, $p_2 = 0.4$, $p_3 = 0.3$, and

$$g_j(x_j) = \begin{cases} 0, & x_j = 0 \\ 1 + \alpha_j x_j, & x_j = 1, 2, \ldots \end{cases}$$

where $\alpha_1 = 0.9$, $\alpha_2 = 1.0$, and $\alpha_3 = 1.1$.

7.8 For the chain network of Figure 7-1, find the following:
a. The *maximum* oriented chain length from node (1,0) to node (1,6).
b. The maximum oriented chain length between $x_2 = 0$ and $x_2 = 6$.
c. The shortest oriented product chain between $x_2 = 0$ and $x_2 = 6$. (The "length" of a product chain is the product of the link lengths.)
d. The shortest oriented product chain from node (3,0) to node (1,6).

7.9 Consider the chain network of Figure 7-P9. Find the minimal length of connected chain between the upper left-hand corner and the lower right-hand corner and indicate the corresponding path (or paths).
a. Use a forward solution procedure.
b. Use a backward solution procedure.

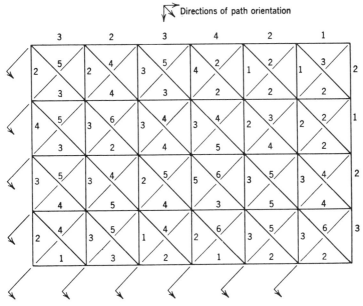

Figure 7-P9.

7.10 Use dynamic programming to solve Problem 2.27a.

7.11 Use dynamic programming to solve Problem 2.27a, given that x_k^d's $= 0$ and that $x_{j+3} = 0$.

7.12 Use dynamic programming to solve Problem 2.27a, given that $a = b = h = 1$, x_k^d's $= 1$, and $|m_k\text{'s}| \leq 1$. Values of m_j^* are to be determined for x_j's in the interval $[-1, 3]$.

7.13 Use dynamic programming to solve Problem 2.27a, given that $a = b = 1$, $h = 0$, x_k^d's $= 1$, and $|m_k\text{'s}| \leq 1$. Values of x_{j+3} are to be in the closed interval $[0.8, 1.2]$. From what range of values of x_j can the state variable be driven to $x_{j+3} \in [0.8, 1.2]$?

7.14 Derive Equations 7-58, 7-59, 7-60, and 7-61.

7.15 Show that the minimum-cost function associated with Figure 7-P15 has a jump discontinuity. Except for the nonmonotonic character of $v(\beta_1, m_0, 0)$ the conditions assumed here are the same as those given with Theorem 7-3.

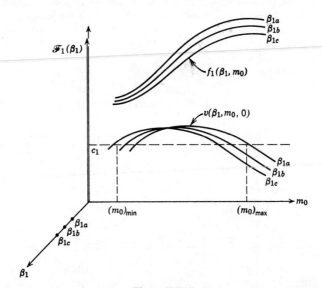

Figure 7-P15.

7.16 A function $\mathscr{F}(x_1, x_2)$ has been generated numerically over an evenly spaced grid of points: Δx_1 and Δx_2 are fixed increments and the values $\mathscr{F}(i \Delta x_1, j \Delta x_2)$ are available for $i, j = 0, 1, 2, \ldots$. Two-dimensional interpolation is to be used in the approximate evaluation of $\mathscr{F}$ at points other than grid points. Such interpolation is often required in obtaining noninteger dynamic programming solutions. Let $\mathbf{x}_s$ denote a specific point at which an approximation $\mathscr{F}_a(\mathbf{x}_s)$ of $\mathscr{F}(\mathbf{x}_s)$ is to be computed. Three grid values of $\mathscr{F}$ are to be used to obtain a linear approximation $\mathscr{F}_a(\mathbf{x}) = a_0 + a_1 x_1 + a_2 x_2$. The first grid point $\mathbf{x}^0$ is selected on the basis that it is the closest grid point to $\mathbf{x}_s$; the remaining grid points $\mathbf{x}^1$ and $\mathbf{x}^2$ are selected on the basis that the point $\mathbf{x}_s$ is contained in the triangle defined by $\mathbf{x}^0$, $\mathbf{x}^1$, and $\mathbf{x}^2$. Obtain expressions for a_0, a_1, and a_2 which result in satisfaction of $\mathscr{F}_a(\mathbf{x}^0) = \mathscr{F}(\mathbf{x}^0) \equiv \mathscr{F}^0$, $\mathscr{F}_a(\mathbf{x}^1) = \mathscr{F}(\mathbf{x}^1) \equiv \mathscr{F}^1$, and $\mathscr{F}_a(\mathbf{x}^2) = \mathscr{F}(\mathbf{x}^2) \equiv \mathscr{F}^2$.

7.17 Two machines are to be used to produce 3 products in a given production period of 80 hours. The number of lots of product j to be produced is x_j, and profit from production of x_j lots is $f_j(x_j) = c_j(1 - e^{-\alpha_j x_j})$ (dollars) where $c_1 = 800$, $c_2 = 600$, $c_3 = 400$, $\alpha_1 = (1/80)$, $\alpha_2 = (1/40)$, and $\alpha_3 = (1/20)$. Machine i requires $g_{ij}(x_j)$ hours to produce x_j lots of product j:

$$g_{ij}(x_j) = a_{ij}u(x_j - \epsilon) + d_{ij}x_j$$

where a_{ij} represents a setup time; $u(x_j - \epsilon) = 0$ if $x_j = 0$, but $u(x_j - \epsilon) = 1$ if $x_j > 0$. Specific values are $a_{11} = a_{12} = 4$, $a_{13} = 8$, $a_{21} = a_{22} = 4$, $a_{23} = 8$, $d_{11} = 0.3$, $d_{12} = 0.5$, $d_{13} = 0.7$, $d_{21} = 0.4$, $d_{22} = 0.5$, and $d_{23} = 0.6$. Machine 2 is to be operated the full 80 hours whereas machine 1 may operate less than 80 hours. Constraints are

$$\sum_{j=1}^{3} g_{1j}(x_j) \leq 80$$

and

$$\sum_{j=1}^{3} g_{2j}(x_j) = 80$$

a. Find the optimal x_j's. Use increments of 4 hours in the state variable and incorporate the second machine constraint by using a Lagrange multiplier approach. Use linear interpolation of return functions to account for the fact that fractions of hours are allowed.

b. Repeat Part a, but use the following expression for $f_j(x_j)$:

$$f_j(x_j) = \begin{cases} c_j\alpha_j x_j, & 0 \leq x_j \leq (1/\alpha_j) \\ (1 + \alpha_j x_j)(0.5)c_j, & (1/\alpha_j) \leq x_j \end{cases}$$

7.18 A given system is controlled by use of a digital computer and is characterized by two first-order difference equations:

$$x_1(k+1) = x_1(k) + x_2(k) + 0.3m(k)$$

and

$$x_2(k+1) = x_2(k) + 0.3m(k)$$

The control action $m(k)$ and the state variables $x_1(k)$ and $x_2(k)$ are constrained to be less than or equal to one in magnitude. A minimum of P is desired:

$$P = \sum_{k=1}^{5} k|x_1(k)|$$

where the minimum of P is a function of $x_1(0)$ and $x_2(0)$.

a. List the appropriate backward recurrence relations for dynamic programming solution of this problem.

b. Use a grid spacing of $\Delta x_1 = \Delta x_2 = 0.1$, $|x_1| \le 1$ and $|x_2| \le 1$, to obtain an approximate dynamic programming solution to the problem. Use the interpolation procedure suggested in Problem 7.16. (*Suggestion:* Write a computer program to effect the solution.)

c. With starting values of $x_1(0) = 1$ and $x_2(0) = 0$, use the solution obtained in Part b as the starting solution, $x_1(k)^0$ and $x_2(k)^0$, for an iterated dynamic programming solution in which values of Δx_1 and Δx_2 are 0.02 with bounds on $x_1(k)$ and $x_2(k)$ given by $|x_1(k) - x_1(k)^0| \le 0.1$ and $|x_2(k) - x_2(k)^0| \le 0.1$

7.19 Given P:

$$P = \sum_{k=1}^{\infty} f(x_k, m_{k-1}) = \sum_{k=1}^{\infty} [(x_k)^2 + (m_{k-1})^2]$$

where U is the set of real numbers and where

$$x_{k+1} = q(x_k, m_k) = \tfrac{1}{2} x_k + m_k$$

a. Use approximation in function space to find the return function $\mathscr{B}(x_k)$ and the policy function $\hat{m}_k(x_k)$. Use the initial guess of $\mathscr{B}^0(x_k) = 0$. (*Hint:* Your *initial* results will be $\mathscr{B}^1(x_k) = \min_{m_k} [(\tfrac{1}{2}x_k + m_k)^2 + m_k^2] = \tfrac{1}{8} x_k^2$ which corresponds to $m_k = -0.25 x_k$; and

$$\mathscr{B}^2(x_k) = \min_{m_k} \{(\tfrac{1}{2}x_k + m_k)^2 + m_k^2 + \mathscr{B}^1[q(x_k, m_k)]\}$$
$$= \min_{m_k} [(9/8)(\tfrac{1}{2}x_k + m_k)^2 + m_k^2]$$

for which the minimizing value of m_k equals $-(9/34)x_k$.)

b. Use approximation in policy space to find the return function $\mathscr{B}(x_k)$ and the policy function $\hat{m}_k(x_k)$. Use $m^0(x_k) = -0.7 x_k$ as the initial policy function.

7.20 Use approximation in function space to find the minimal nonoriented chain that connects the upper left-hand node of Figure 7-P9 to the upper right-hand one.

7.21 Consider the chain network of Figure 7-P9 and suppose that the center node is to be included in a nonoriented chain network between the upper left-hand node and the upper right-hand node. Modify the procedure of Section 7-11d and find the minimal chain which connects the three nodes in question.

7.22 Verify Equations 7-150a and 7-150b and show the "equivalence" of these equations to the continuous recurrence relation 7-162.

7.23 Derive Equation 7-170.

7.24 Consider the scalar state equation $x_{k+1} = a_k x_k + b_k m_k$, $k = 0, 1, \ldots, K-1$ where the a_k's and b_k's are given real values. The m_k's are to be selected to minimize P,

$$P = \sum_{j=1}^{K} [w_j(x_j - x_j^d)^2 + v_j m_{j-1}^2], \qquad w_j\text{'s and } v_j\text{'s} > 0$$

where the w_j's, x_j^d's, and v_j's are given real values. Show that the minimum of P is of the form $h_1 x_0^2 + h_2 x_0 + h_3$ where h_1, h_2, and h_3 are independent of the x_k's. (*Hint:* Use convolution summation to express x_j in terms of x_0 and m_k's.)

7.25 As a special case of the problem in Section 7-11a, consider

$$\mathbf{q}(\mathbf{x}_k, m_k) = \mathbf{A}\mathbf{x}_k + \mathbf{b} m_k$$

and

$$f(\mathbf{x}_k, m_{k-1}) = \mathbf{x}_k' \mathbf{W} \mathbf{x}_k + v m_{k-1}^2$$

where $\mathbf{A}$, $\mathbf{b}$, $\mathbf{W}$, and v are independent of k; $\mathbf{A}$ is an $n \times n$ matrix of real values, $\mathbf{b}$ is an $n \times 1$ matrix of real values, $\mathbf{W}$ is an $n \times n$ positive-semidefinite matrix of real values, and v is a positive real value. It is assumed that the constraint set U on the m_k's is unbounded and that the system is completely regulable. Under these conditions, it can be shown that the minimum-cost function $\mathscr{B}(\mathbf{x}_k)$ is a quadratic of the form

$$\mathscr{B}(\mathbf{x}_k) = \mathbf{x}_k' \mathbf{H} \mathbf{x}_k$$

where the real constants in the $n \times n$ matrix $\mathbf{H}$ are to be determined.
 a. Use Equation 7-131 to find the optimal m_k, m_k^*, in terms of $\mathbf{x}_k$, $\mathbf{A}$, $\mathbf{b}$, $\mathbf{W}$, $\mathbf{H}$, and v.
 b. Give a procedure for the evaluation of $\mathbf{H}$ by paralleling the approach of Section 7-14. The solution involves a steady-state, discrete Riccati equation.

7.26 Consider Problem 3.23, but with $x(T)$ not specified. Obtain the appropriate Riccati equation and solve for the optimal feedback control law.

7.27 If x_{1d} is the desired value of x_1 in Example 7-6 and is changed from the value 1 to some other value, what change is required in the solution?

7.28 Consider the linear oscillator equation, $\ddot{x} + x = m$, where x and $\dot{x}$ can be measured for purposes of feedback control. A minimum of J is desired:

$$J = \int_0^T (x^2 + \dot{x}^2 + m^2) \, dt$$

where T is a given positive value.
 a. Express the special cases of Equations 7-218, 7-219, and 7-221 that apply to this problem.
 b. Find the specific feedback control law when T is allowed to approach infinity.

7.29 Suppose that the term $\mathbf{m}'\mathbf{V}\mathbf{m}$ in Equation 7-201 is replaced by $(\mathbf{m} - \mathbf{m}_d)' \mathbf{V} (\mathbf{m} - \mathbf{m}_d)$ where $\mathbf{m}_d$ is a given, desired control vector as a function of time.
 a. Show the effect that this change has on the results of Section 7-14a.
 b. Suppose that all entries in the $\mathbf{A}$, $\mathbf{B}$, $\mathbf{W}$, $\mathbf{V}$, and $\mathbf{m}_d$ matrices are constants and that $t_b \to \infty$. Also suppose that $\mathbf{m}(t) \to \mathbf{m}_d$ as $t \to \infty$. What condition must be satisfied to insure the maintainability of $\mathbf{e}(t)$ at zero in the steady state?

7.30 Consider the block diagram of Figure 7-P30. The state variables x_1, x_2, and x_3 have initial conditions which are not shown in the diagram. Desired

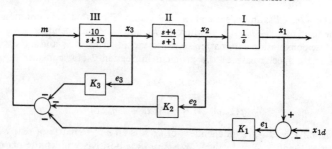

Figure 7-P30.

values of x_2 and x_3 are zero whereas the desired value of x_1 is a constant x_{1d}. The control action $m(t)$ is to be generated to obtain the minimum of J:

$$J = \int_0^\infty [(x_1 - x_{1d})^2 + x_2^2 + m^2] \, dt$$

a. Express the state equations of the system in the standard form $\dot{\mathbf{e}} = \mathbf{Ae} + \mathbf{b}m$.
b. The solution approach of Section 7-14b is to be used. Find $\mathbf{W}$ and $\mathbf{V}$ and express the optimal values of K_1, K_2, and K_3 in terms of ψ_{ij}'s.
c. Express the particular case of Equation 7-248 that applies here.
d. The Newton-Raphson method (Sections 6-3a and 6-6d) is suggested as a way to solve the equations of Part c. Give a detailed description of the method as it applies to this solution.
e. What difficulties would be encountered if blocks III and II were interchanged in Figure 7-P30?

7.31 Modify the results of Section 7-15 to account for the case where probabilities associated with w_k are as follows:

w_k	−1	−0.5	0	0.5	1
probability	0.1	0.25	0.3	0.25	0.1

7.32 A given discrete state equation is

$$x_{k+1} = x_k + (1 + 0.1 w_k) m_k, \quad |m_k| \leq 1$$

where $w_k = +1$ with probability 0.4 and -1 with probability 0.6. A minimum of the expected value of P is desired, where

$$P = \sum_{k=1}^{3} k(x_k - 1)^2$$

The initial state x_0 is in the interval $[-1, 3]$. Find $m_0^*(x_0)$.

7.33 In the general allocation problem of Section 7-2, suppose that each f_i is a function of an independent random variable w_i in addition to the deterministic variable x_i: $f_i = f_i(x_i, w_i)$. A discrete probability distribution is known for each w_i.

PROBLEMS

a. Outline a dynamic programming approach for the maximization of the expected value of P.
b. In Problem 7.3, suppose that each c_i is an independent random variable. Find the maximum of the expected number of work units that can be completed when the following discrete probability distributions are associated with the c_i's.

c_1	probability	c_2	probability	c_3	probability
4	0.1	8	0.3	10	0.4
5	0.5	9	0.4	13	0.5
7	0.4	10	0.3	15	0.1

7.34 Consider Example 7-7 with a new Γ matrix (see 7-296):

$$\Gamma = \begin{bmatrix} \epsilon_1 & 0 \\ 0 & \epsilon_2 \end{bmatrix}$$

To what does Equation 7-294 reduce if both ϵ_1 and ϵ_2 are taken as arbitrarily small, but positive values?

7.35 Supply the intermediate steps required to obtain Equations 7-309 through 7-312.

7.36 a. Write a computer flow diagram for the Kalman-Bucy filter (Equations 7-311 and 7-312).
b. Use a computer to iteratively evaluate 7-316 and 7-317 of Example 7-8. Given are $\epsilon = 0$, $\mathbf{P}_0 = 0.01\mathbf{S}$, $\bar{\mathbf{x}}_0 = \mathbf{0}$, and z_{k+1}'s:

$k+1$	1	2	3	4	5	6	7	8	9	10
z_{k+1}	0.2	1.1	1.5	2.6	2.0	2.0	1.4	1.3	1.2	1.0

7.37 Given:

$$x_{k+1} = x_k \quad \text{and} \quad z_{k+1} = x_{k+1} + v_{k+1}$$

for $k = 0, 1, \ldots, K$, where v_{k+1} satisfies the conditions assumed in Section 7-16b. The variance of v_{k+1} is r_{k+1}.
a. Express the particular versions of 7-311 and 7-312 that apply.
b. Write a computer program to solve the equations of Part a. Input data is to include $\bar{x}_0$, P_0, z_{k+1}'s, r_{k+1}'s, and K.
c. Obtain the Kalman-Bucy filter output under the following conditions:

$P_0 = 9$, $\bar{x}_0 = 8$, r_{k+1}'s $= 4$, $z_1 = 9$, $z_2 = 10$, $z_3 = 11$, $z_4 = 8$, $z_5 = 9$, and $z_6 = 10$.

7.38 Recast a problem from Chapter 3 into a form which can be solved by the use of straightforward dynamic programming and develop the appropriate recurrence relations.

A MAXIMUM PRINCIPLE

8

8-1. INTRODUCTION

Many systems can be characterized in terms of a set of first-order, ordinary differential equations in which control or policy variables are to be selected over time to obtain some desirable objectives in an optimal manner. In Chapter 3, certain problems of this type are considered, but the theory is not well suited for use on problems in which inequality constraints are imposed on the control variables. The maximum principle of this chapter is essentially that known as Pontryagin's maximum principle; not only inequality constraints, but constraints restricting the control actions to be elements of closed and bounded sets can be treated in the normal course of problem solution.

Pontryagin's maximum principle consists of a set of necessary conditions that must be satisfied by optimal solutions. All of these necessary conditions have origins in classical calculus of variations. Thus, the state and costate equations and the transversality conditions herein are essentially those of Section 3-9, but are formed here in a more systematic way by use of a Hamiltonian function. Also, as noted by many authors (e.g., [8.11, 8.62]), the maximizing condition of Pontryagin's maximum principle is in a direct sense an expression of the Weierstrass condition of Section 3-10g. Historically, Pontryagin's original work was accomplished in the mid-1950's and was preceded by much related work, especially the coupled works of Bliss [8.12] in 1930 and McShane [8.77] in 1939, the work of Valentine [8.107] in 1937, and that of Hestenes [8.46] in 1950. It was Pontryagin's work, however, that illuminated the full potential of the theory at hand. Since 1956 this maximum principle has been related, not only to calculus of variations [8.11, 8.55, 8.66, 8.67], but to dynamic programming [7.83, 8.29, 8.97] and to methods of gradient search [8.60]. Work such as that by Roxin [8.96] has added to

Sect. 8-2   PRELIMINARY CONCEPTS   479

geometrical interpretations of Pontryagin's maximum principle. Some of the basic papers in this area have been compiled for ready reference by Oldenburger [8.84].

Preliminary concepts and notation are given in Section 8-2. In Section 8-3, a canonical problem form is related to equivalent problem forms. The necessary conditions of the applicable maximum principle are then given (Section 8-4) for the normal and nonsingular general case. The necessity of these conditions for optimality is shown (Sections 7-13, 8-5, and 8-6), and special cases are examined to fully impart the proper interpretation of the necessary conditions. Alternative developments to that of Section 7-13 may be of interest to the reader—Athans and Falb [8.5] give a heuristic proof of the maximum principle in forty pages of text.

The necessary conditions of Pontryagin's maximum principle are the basic ingredients with which optimal solutions may be obtained in a variety of ways. For example, in regard to time optimal control of linear systems with bounded control variables (Section 8-7), there exist several ways in which the necessary conditions can be used to obtain feedback control laws. The approach to be used to find solutions that satisfy the necessary conditions is generally not obvious. Invariably a two-point boundary-value problem must be solved; the search procedures of Section 8-8 are representative of those advanced in the literature for digital computer solution of pertinent two-point boundary-value problems. Analog computer solutions have also been effected (e.g., [8.16, 8.27]).

The necessary conditions of Section 8-4 are incomplete for certain problems of the form described in Section 8-3. Both non-normal cases (Section 8-9) and singular cases (Section 8-10) exist. In non-normal cases (which are somewhat rare), the form of the Hamiltonian must be modified to insure the existence of appropriate costate variables. In singular cases (which are more common), the maximizing condition of Pontryagin must be augmented by additional necessary conditions. Finally, the canonical problem form that is given in Section 8-3 is not necessarily the most convenient form for some problems of the general type considered. With different canonical problem forms, different but related necessary conditions for optimality apply; such conditions are derived in Section 8-11.

8-2. PRELIMINARY CONCEPTS

The primary variables in this chapter are time t, state variables $\mathbf{x} = \{x_1, x_2, \ldots, x_n\}$, costate variables $\boldsymbol{\lambda} = \{\lambda_1, \lambda_2, \ldots, \lambda_n\}$, and policy or control variables $\mathbf{m} = \{m_1, m_2, \ldots, m_r\}$. Here, $\mathbf{x}$, $\boldsymbol{\lambda}$, and $\mathbf{m}$ are real-valued functions of t. As in previous chapters, it is operationally convenient to treat $\mathbf{x}$, $\boldsymbol{\lambda}$, and **m** as column matrices or vectors. Both **x** and **λ** can be viewed as vectors in n-dimensional Euclidean spaces, and **m** as a vector in an r-dimensional Euclidean space. These spaces are called X space, Λ space, and M space, respectively.

A closed and bounded subset U of M space is of interest, as is a time interval $[t_a, t_b]$. A given function $f(\mathbf{x},\mathbf{m})$ is said to be of class C^0 on $X \times U$ if it is continuous with respect to the x_i's and m_i's on the subset $X \times U$ of the $(n + r)$-dimensional product space $X \times M$. Similarly, $f(\mathbf{x},\mathbf{m})$ is said to be of class C^1 with respect to **x** on $X \times U$ if all $\partial f/\partial x_i$'s exist and are continuous at any finite point in $X \times U$. These continuity properties are assumed to hold for the functions of interest in this chapter.

The reason that continuity restrictions are imposed in optimization problems is generally to validate use of truncated Taylor's series for small displacements. If certain continuity restrictions are not satisfied by a given function, a satisfactory engineering approach is often that of approximation; the given function may be approximated by one which has the required degree of continuity. For example, the function $f_1(x)$,

$$f_1(x) \triangleq |x| \tag{8-1}$$

is of class C^0 but not of class C^1 because of the jump discontinuity in $df_1(x)/dx$ at $x = 0$. In place of $f_1(x)$, consider the use of $f_2(x)$:

$$f_2(x) \triangleq \begin{cases} |x|, & |x| \geq \epsilon \\ (1.5/\epsilon)x^2 - (0.5/\epsilon^3)x^4, & |x| \leq \epsilon \end{cases} \tag{8-2}$$

which, as the reader may verify, is of class C^1 for any positive ϵ. The function $f_2(x)$ is a good approximation to $f_1(x)$ in the sense that the norm

$$\int_{t_a}^{t_b} |f_1[x(t)] - f_2[x(t)]| \, dt$$

can be made arbitrarily small by making ϵ, $\epsilon > 0$, arbitrarily small. It is of interest that, although 8-1 does not satisfy continuity conditions specified in Section 8-3, solutions involving this $f_1(x)$ in the integrands of functionals have been obtained by using Pontryagin's maximum principle; in some cases [8.36], the resulting solutions yield an infinite number of vacillations in control variables as the terminal time t_b is approached.

For the problems considered, necessary conditions for optimality consist of state equations, costate equations, a maximizing condition, and transversality conditions. The first three of these conditions are considered in Section 7-13 with which the reader is assumed to be familiar. Assuming the existence of a solution which satisfies these necessary conditions, the solution would be unique if:

Sect. 8-3   CANONICAL FORM AND EQUIVALENT PROBLEMS   481

1. The state and costate variables were given at one specific instant of time.
2. At each $t \in [t_a, t_b]$, the maximizing condition was satisfied by a unique value $\mathbf{m}^*(t)$ of $\mathbf{m}$.
3. Singular solutions (Section 8-10) could not exist for the given problem.

The trouble with this list is that statement 1 *never* applies, while statements 2 and 3 are sometimes applicable. Various relations can be imposed on the end conditions $\mathbf{x}(t_a)$, $\mathbf{x}(t_b)$, t_a, and t_b. For example, $\mathbf{x}(t_b)$ and t_b could be required to lie on a particular manifold (subset) of an $(n + 1)$-dimensional space. It is possible that the end conditions imposed on the state variables of a given problem will not be satisfied by any choice of $\mathbf{m}(t)$ on $U \times [t_a, t_b]$, in which case no solution will satisfy the necessary conditions for optimality; it is possible that one unique solution will satisfy the necessary conditions for optimality; or it is possible that a multitude of solutions will satisfy the necessary conditions. Conditions for existence and uniqueness are considered in the literature (e.g., [8.5, 8.91, 8.98]).

8-3. A CANONICAL PROBLEM FORM AND EQUIVALENT PROBLEMS

As in Chapter 3, attention is centered on systems for which the measure of performance can be expressed in terms of an integral, which here assumes the form

$$J(\mathbf{m}) = \int_{t_a}^{t_b} f(\mathbf{x}, \mathbf{m}) \, dt \tag{8-3}$$

where $\mathbf{x} = \mathbf{x}(t)$ is the $n \times 1$ state vector of the system, $\mathbf{m} = \mathbf{m}(t)$ is the $r \times 1$ control vector of the system, and $\mathbf{x}$ and $\mathbf{m}$ are related by the vector state equation

$$\dot{\mathbf{x}} = \mathbf{q}(\mathbf{x}, \mathbf{m}) \tag{8-4}$$

The functions $q_1, q_2, \ldots, q_n$, and f are assumed to be of class C^1 with respect to the x_i's, and of class C^0 with respect to the m_i's. All functions are assumed to be real-valued with respect to $\mathbf{x}$'s and $\mathbf{m}$'s of interest. Various conditions may be imposed upon t_a, t_b, $\mathbf{x}(t_a)$, and $\mathbf{x}(t_b)$—if all of these are not specified numerically, transversality conditions (Section 8-6) are of consequence in the solution.

The functional $J(\mathbf{m})$ is to be minimized by appropriate selection of $\mathbf{m}(t)$ which is assumed to be constrained to belong to some known closed and bounded set U in r-dimensional space. Examples of possible forms of U are noted in Section 7-12a. For realistic solutions to some problems, e.g, on-line control problems, the final solution should be of the feedback variety, namely, at each instant $t \in [t_a, t_b]$, the optimal value $\mathbf{m}^*$ of $\mathbf{m}$ that is applied is that which is expected to lead to the following minimum:

$$\mathscr{B}[\mathbf{x}(t),t] = \underset{\mathbf{m} \in U}{\text{minimum}} \int_t^{t_b} f[\mathbf{x}(\tau),\mathbf{m}(\tau)]\, d\tau \qquad (8\text{-}5)$$

where the actual or best-estimate value of $\mathbf{x}(t)$ is used in the generation of $\mathbf{m}^*$ at time t. In this case, $\mathbf{m}^* = \mathbf{m}^*[\mathbf{x}(t),t]$ is a function of the actual or best-estimate state of the system at each $t \in [t_a, t_b]$. The problem of generating the function $\mathbf{m}^*[\mathbf{x}(t),t]$ is known as the control system *synthesis problem*. Practical solutions to this synthesis problem should include considerations of stability and sensitivity.

That the preceding statements serve as a canonical form for a large class of problems is shown in the remainder of this section. However, it is emphasized that, for any given problem of a class of equivalent problems, a different canonical form and corresponding necessary conditions for optimality might well be more *convenient*, both conceptually and computationally. Thus, sets of related conditions, e.g., those given in Section 8-11, are of value.

Case A. (Time-variant considerations.)

In the preceding canonical form, time t does not appear as an explicit argument in any of the f or q_i functions which are therefore of the time-invariant form. Suppose, however, that when we are formulating a particular problem, time t does appear explicitly in the pertinent functions—the question is: "What should be done to reduce the problem to the previously given canonical form?" To answer this question, suppose that the original system dynamics are characterized by $n - 1$, rather than n state variables. If an nth state variable x_n is introduced and defined by

$$\dot{x}_n \triangleq 1, \qquad x_n(t_a) = t_a, \quad \text{and} \quad x_n(t_b) = t_b \qquad (8\text{-}6)$$

then this state variable x_n can be used in place of t everywhere it appears as an explicit term in f or in a q_i function. Thus, a form of time-invariance is gained at the expense of the introduction of an additional, but artificial state variable.

Case B. (Isoperimetric constraints.)

Isoperimetric constraints, such as those considered in Sections 3-4 and 3-8a, can be imbedded in two related ways into the canonical form of this chapter. It suffices to consider one isoperimetric constraint,

$$\int_{t_a}^{t_b} f_j(\mathbf{x},\mathbf{m})\, dt = K_j \qquad \text{(a constant)} \qquad (8\text{-}7)$$

where $f_j = f_j(\mathbf{x},\mathbf{m})$ is a given function with continuity properties equivalent to those assumed for f.

Sect. 8-3   CANONICAL FORM AND EQUIVALENT PROBLEMS   483

One approach is that of replacing 8-7 with an equivalent state equation; thus, in identifying and labeling state variables of a given problem, we may choose to define $\dot{x}_j$ by

$$\dot{x}_j = f_j(\mathbf{x},\mathbf{m}) \triangleq q_j(\mathbf{x},\mathbf{m}), \qquad x_j(t_a) = 0 \quad \text{and} \quad x_j(t_b) = K_j \qquad (8\text{-}8)$$

the satisfaction of which guarantees satisfaction of 8-7. But note that, in such a case, x_j does not appear as an argument of the f and q_i functions. Because of this, the corresponding costate equation (see Equation 7-180) is $\dot{\lambda}_j = 0$, and therefore λ_j is a constant. This fact directly justifies the second and conventional approach which follows.

If the isoperimetric constraint 8-7 is to be satisfied by the optimal solution of a problem in canonical form (here $\dot{x}_j$ is not assumed to be defined by 8-8), an augmented functional $J(\mathbf{m},h_j)$ can be formed as follows:

$$J(\mathbf{m},h_j) = \int_{t_a}^{t_b} [f(\mathbf{x},\mathbf{m}) + h_j f_j(\mathbf{x},\mathbf{m})] \, dt \qquad (8\text{-}9)$$

where h_j is a time-invariant Lagrange multiplier. If a value of h_j exists such that the solution which yields the minimum of J in 8-9 also results in satisfaction of 8-7, the solution is the optimal one. The solution procedure must therefore include a systematic search for an appropriate h_j.

Case C. (Extremization of end-point values. See also Section 8-11.)

Suppose that a given function $f_p[\mathbf{x}(t_b),t_b]$ is to be minimized where $\mathbf{x}(t)$ satisfies the state equation 8-4 and where $\mathbf{x}(t_a)$ and t_a are specified. The function $f_p = f_p[\mathbf{x}(t),t]$ is assumed to be of class C^2, and therefore $f_p[\mathbf{x}(t_b),t_b]$ can be expressed in the form

$$f_p[\mathbf{x}(t_b),t_b] = \int_{t_a}^{t_b} (df_p/dt) \, dt + f_p[\mathbf{x}(t_a),t_a]$$

$$= \int_{t_a}^{t_b} [(d\mathbf{x}/dt)'(\partial f_p/\partial \mathbf{x}) + (\partial f_p/\partial t)] \, dt + f_p[\mathbf{x}(t_a),t_a]$$

$$= \int_{t_a}^{t_b} [\mathbf{q}'(\partial f_p/\partial \mathbf{x}) + (\partial f_p/\partial t)] \, dt + f_p[\mathbf{x}(t_a),t_a] \qquad (8\text{-}10)$$

With $\mathbf{x}(t_a)$ and t_a specified, minimization of $f_p[\mathbf{x}(t_b),t_b]$ in 8-10 is equivalent to minimization of J:

$$J \triangleq f_p[\mathbf{x}(t_b),t_b] - f_p[\mathbf{x}(t_a),t_a]$$

$$= \int_{t_a}^{t_b} [\mathbf{q}'(\partial f_p/\partial \mathbf{x}) + (\partial f_p/\partial t)] \, dt \qquad (8\text{-}11)$$

and the integrand of 8-11 can be identified with f of 8-3. The possible time dependence of the integrand in 8-11 can be circumvented by using the approach outlined under Case B of this section.

A commonly used form of $f_p[\mathbf{x}(t_b),t_b]$ is the linear form

$$f_p = \left[\sum_{i=1}^{n} c_i x_i(t_b)\right] + c_{n+1} t_b \qquad (8\text{-}12)$$

where the c_i's are weighting factors matched to problems of interest. The equivalent J of 8-11 is

$$J = \int_{t_a}^{t_b} \left[c_{n+1} + \sum_{i=1}^{n} c_i q_i(\mathbf{x},\mathbf{m})\right] dt \qquad (8\text{-}13)$$

As an additional case, consider that in which $c_1 = 1$ and all other c_i's are zero; under these conditions,

$$J = \int_{t_a}^{t_b} q_1(\mathbf{x},\mathbf{m})\, dt \qquad (8\text{-}14)$$

in which case $q_1(\mathbf{x},\mathbf{m})$ can be identified with f of Equation 8-3.

The motivation for 8-12 is that we can view the minimization problem geometrically as an effort to drive the system characterized by 8-4 from an initial state to a terminal state which is maximized in the

$$-\mathbf{c} = -[c_1 \quad c_2 \quad \cdots \quad c_{n+1}]'$$

vector direction in the $(n + 1)$-dimensional space of x_i's and t. In summary, any one of 8-3, 8-11, or 8-12 could serve as the performance measure in a canonical problem form of the type under consideration. All that is required is that we define state variables of a given problem of the class in an appropriate manner to fit a particular canonical form.

Example 8-1. If t_a, $\mathbf{x}(t_a)$, and $\mathbf{x}(t_b)$ are fixed and t_b is to be minimized, the weighting factor c_{n+1} in 8-12 is set to 1 with all other c_i's set at zero, with the result that J of 8-13 is reduced to

$$J = \int_{t_a}^{t_b} dt \qquad (8\text{-}15)$$

●

Example 8-2. Suppose that for a given problem it is not possible to drive from a given $\mathbf{x}(t_a)$ to a desired $\mathbf{x}^d$ of $\mathbf{x}(t_b)$ in a given time $t_b - t_a$. In such a case, we might choose to minimize a weighted sum of terminal errors squared:

$$f_p = \sum_{i=1}^{n} c_i [x_i(t_b) - x_i^d]^2 \qquad (8\text{-}16)$$

where x_i^d is the desired value of $x_i(t_b)$. The corresponding J of 8-11 is

$$J = \int_{t_a}^{t_b} \sum_{i=1}^{n} 2c_i[x_i(t) - x_i^d]q_i(\mathbf{x},\mathbf{m})\, dt \qquad (8\text{-}17)$$

●

8-4. A MAXIMUM PRINCIPLE

The maximum principle of this chapter is essentially that which is known as Pontryagin's maximum principle and consists of a set of necessary conditions that must be satisfied by the optimal solution of the canonical problem posed in Section 8-3. The conditions derived in Section 7-13 constitute one part of this maximum principle; other parts are linked with conditions on the Hamiltonian $\mathscr{H}$ of 8-18 and with transversality conditions. The major results are summarized in this section; the next two sections in conjunction with Section 7-13 provide justification.

Corresponding to the canonical problem of Section 8-3, a Hamiltonian is defined:[1]

$$\mathscr{H} = \mathscr{H}(\mathbf{x},\boldsymbol{\lambda},\mathbf{m})$$

$$\triangleq -f(\mathbf{x},\mathbf{m}) + \sum_{i=1}^{n} \lambda_i q_i(\mathbf{x},\mathbf{m}) \qquad (8\text{-}18)$$

The state equations for the problem are given by 8-4 which is equivalent to

$$\dot{\mathbf{x}} = \frac{\partial \mathscr{H}}{\partial \boldsymbol{\lambda}} \qquad (8\text{-}19)$$

The costate vector $\boldsymbol{\lambda}$ must satisfy the vector costate equation

$$\dot{\boldsymbol{\lambda}} = -\frac{\partial \mathscr{H}}{\partial \mathbf{x}} \qquad (8\text{-}20)$$

If $\mathbf{x}^*(t)$ and $\boldsymbol{\lambda}^*(t)$ denote optimal state and costate vectors at any given t in $[t_a, t_b]$, then the control value $\mathbf{m}^*(t)$ which leads to the *minimum* of J in 8-3 is the value which *maximizes* $\mathscr{H}$; that is,

$$\mathscr{H}^* \triangleq \mathscr{H}(\mathbf{x}^*,\boldsymbol{\lambda}^*,\mathbf{m}^*)|_{\mathbf{m}^* \in U} \geq \mathscr{H}(\mathbf{x}^*,\boldsymbol{\lambda}^*,\mathbf{m})|_{\mathbf{m} \in U} \qquad (8\text{-}21)$$

which holds for any piecewise continuous $\mathbf{m}(t)$ with values in the compact

---

[1] The form of $\mathscr{H}$ given at this point is for the *normal case*. In the non-normal case, which is rare, the $-f(\mathbf{x},\mathbf{m})$ part of $\mathscr{H}$ must be omitted. The nature of the non-normal case is examined further in Section 8-9.

set U. Equations 8-19, 8-20, and 8-21 are obtained directly from the results listed in Section 7-13e. Precautionary notes are sounded in Section 7-13e in regard to the interpretation of 8-21. These are equally important here and should be reviewed as necessary. In singular cases (Section 8-10), the maximizing condition must be augmented by additional necessary conditions.

As in Chapter 7, state and costate trajectories which satisfy 8-19, 8-20, and 8-21 are called *principal trajectories* (extremal trajectories) and are candidates for an optimal trajectory.

The general transversality condition of Pontryagin's maximum principle is

$$\left[-\mathscr{H}\,\delta t + \sum_{i=1}^{n} \lambda_i\,\delta x_i\right]_{t_a}^{t_b} = 0 \tag{8-22}$$

which must be satisfied when $\mathbf{x} = \mathbf{x}^*$, $\boldsymbol{\lambda} = \boldsymbol{\lambda}^*$, and $\mathbf{m} = \mathbf{m}^*$; and where the interpretations given to the δt and δx_i variations are exactly those given in the classical calculus of variations. A derivation of 8-22 and its interpretation in special cases are examined in Section 8-6. In any given case, specified endpoint conditions in conjunction with 8-22 supply $2n + 2$ boundary conditions that must be satisfied by t_a, t_b, $\mathbf{x}^*$, and $\boldsymbol{\lambda}^*$.

An additional, and quite remarkable feature is that $\mathscr{H}^*$ must equal a constant for all t in $[t_a,t_b]$,[2] that is,

$$\mathscr{H}^* = \mathscr{H}[\mathbf{x}^*(t),\boldsymbol{\lambda}^*(t),\mathbf{m}^*(t)] = c, \qquad t \in [t_a,t_b] \tag{8-23}$$

where c is a constant which is dependent upon other conditions of the problem. For example, if t_a is fixed and t_b is free to be selected in an optimal manner, the transversality condition 8-22 gives

$$\mathscr{H}[\mathbf{x}^*(t_b),\boldsymbol{\lambda}^*(t_b),\mathbf{m}^*(t_b)]\,\delta t_b = 0$$

which can be true for an arbitrary δt_b variation only if

$$\mathscr{H}[\mathbf{x}^*(t_b),\boldsymbol{\lambda}^*(t_b),\mathbf{m}^*(t_b)] = 0 \tag{8-24}$$

but since $\mathscr{H}^*$ must equal a constant for all t in $[t_a,t_b]$, the constant c must be zero in any free-terminal-time case.

8-5. THE CONSTANCY OF $\mathscr{H}^*$

If $\mathbf{x}^*(t)$, $\boldsymbol{\lambda}^*(t)$, and $\mathbf{m}^*(t)$ are known functions of t, $\mathscr{H}^*$ reduces to a function of time $\mathscr{H}^*(t)$. But in the preceding section, it is stated that $\mathscr{H}^*(t)$ is a

---

[2] This statement is the only one of the given necessary conditions which is not valid if $\mathscr{H}$ depends upon t explicitly, as in Equation 7-181. In that case, the corresponding necessary condition generalizes to $d\mathscr{H}(\mathbf{x}^*,\mathbf{m}^*,\boldsymbol{\lambda}^*,t)/dt = \partial\mathscr{H}(\mathbf{x}^*,\mathbf{m}^*,\boldsymbol{\lambda}^*,t)/\partial t$ at the continuity points of $\mathbf{m}^*(t)$, and $\mathscr{H}(\mathbf{x}^*,\mathbf{m}^*,\boldsymbol{\lambda}^*,t)$ is continuous and piecewise-differentiable on the interval (t_a,t_b) [7.83].

constant, a trivial function. Proof of this assertion is developed in two parts: First, it is shown that $\mathcal{H}^*(t)$ is a continuous function of t for $t \in [t_a, t_b]$; and second, it is shown that the derivative of $\mathcal{H}^*(t)$ is zero on any interval of time in which $\mathbf{m}^*(t)$ is continuous. The fact that $\mathcal{H}^*(t)$ equals a constant follows immediately.

To show that $\mathcal{H}^*(t)$ is continuous, consider any two instants of time t_1 and t_2 in $[t_a, t_b]$ and observe that, by Equation 8-21,

$$\mathcal{H}^*(t_2) \geq \mathcal{H}[\mathbf{x}^*(t_2), \boldsymbol{\lambda}^*(t_2), \mathbf{m}^*(t_1)] \tag{8-25}$$

and

$$\mathcal{H}^*(t_1) \geq \mathcal{H}[\mathbf{x}^*(t_1), \boldsymbol{\lambda}^*(t_1), \mathbf{m}^*(t_2)] \tag{8-26}$$

It follows from 8-25 and 8-26 that the difference $\mathcal{H}^*(t_2) - \mathcal{H}^*(t_1)$ is bounded between two values; that is,

$$\mathcal{H}^*(t_2) - \mathcal{H}^*(t_1) \geq \mathcal{H}[\mathbf{x}^*(t_2), \boldsymbol{\lambda}^*(t_2), \mathbf{m}^*(t_1)] - \mathcal{H}^*(t_1) \tag{8-27}$$

and

$$\mathcal{H}^*(t_2) - \mathcal{H}^*(t_1) \leq \mathcal{H}^*(t_2) - \mathcal{H}[\mathbf{x}^*(t_1), \boldsymbol{\lambda}^*(t_1), \mathbf{m}^*(t_2)] \tag{8-28}$$

Assuming that solutions $\mathbf{x}^*$ and $\boldsymbol{\lambda}^*$ exist,[3] they are continuous in time because $\mathbf{m}$ in 8-19 and 8-20 is bounded for all $t \in [t_a, t_b]$. Also, $\mathcal{H}(\mathbf{x}, \boldsymbol{\lambda}, \mathbf{m})$ of 8-18 is necessarily continuous in both $\mathbf{x}$ and $\boldsymbol{\lambda}$. The question to be resolved is the question of continuity in time of $\mathcal{H}^*(t)$ with respect to $\mathbf{m}^*(t)$ where $\mathbf{m}^*(t)$ may exhibit points of discontinuity. This question is answered by observing that the right-hand term of inequality 8-27 contains $\mathbf{m}^*(t_1)$ only, whereas the right-hand term of 8-28 contains $\mathbf{m}^*(t_2)$ only. Thus, if t_2 is allowed to approach the arbitrary $t_1 \in [t_a, t_b]$, both right-hand terms approach zero; therefore $\mathcal{H}^*(t_2) - \mathcal{H}^*(t_1)$ must also approach zero. This is the essence of continuity.

To show that $d\mathcal{H}^*(t)/dt = 0$ over any interval of time on which $\mathbf{m}^*(t)$ is continuous, consider inequality 8-27 modified as follows:

$$\frac{\mathcal{H}^*(t_2) - \mathcal{H}^*(t_1)}{t_2 - t_1} \geq \frac{\mathcal{H}[\mathbf{x}^*(t_2), \boldsymbol{\lambda}^*(t_2), \mathbf{m}^*(t_1)] - \mathcal{H}^*(t_1)}{t_2 - t_1} \tag{8-29}$$

In this proof, it is assumed that $\mathbf{m}^*(t)$ is continuous in the closed interval $[t_1, t_2]$, where t_2 is assumed greater than t_1 without loss of generality. In the

---

[3] The continuity conditions assumed in this chapter do not guarantee the existence of solutions. For example, if $\dot{x} = e^x$ and $x(0) = 1$, $x(t) = 1 + \ln[1/(1 - te)]$ when $t \in (0, e^{-1})$, but $x(t)$ does not exist when $t \geq e^{-1}$.

limit as t_2 approaches t_1, inequality 8-29 gives

$$\frac{d\mathcal{H}^*(t_1)}{dt} \geq \left[\frac{\partial\mathcal{H}[\mathbf{x}^*(t_1),\boldsymbol{\lambda}^*(t_1),\mathbf{m}^*(t_1)]}{\partial \mathbf{x}}\right]' \frac{d\mathbf{x}^*(t_1)}{dt}$$
$$+ \left[\frac{\partial\mathcal{H}[\mathbf{x}^*(t_1),\boldsymbol{\lambda}^*(t_1),\mathbf{m}^*(t_1)]}{\partial \boldsymbol{\lambda}}\right]' \frac{d\boldsymbol{\lambda}^*(t_1)}{dt} \quad (8\text{-}30)$$

But in view of 8-19 and 8-20, inequality 8-30 reduces to

$$\frac{d\mathcal{H}^*(t_1)}{dt} \geq 0 \quad (8\text{-}31)$$

In like manner, inequality 8-28 is equivalent to

$$\frac{\mathcal{H}^*(t_2) - \mathcal{H}^*(t_1)}{t_2 - t_1} \leq \frac{\mathcal{H}^*(t_2) - \mathcal{H}[\mathbf{x}^*(t_1),\boldsymbol{\lambda}^*(t_1),\mathbf{m}^*(t_2)]}{t_2 - t_1} \quad (8\text{-}32)$$

And in the limit as $t_2 \to t_1$,

$$\frac{d\mathcal{H}^*(t_1)}{dt} \leq [(\partial\mathcal{H}^*/\partial\mathbf{x})'(d\mathbf{x}^*/dt) + (\partial\mathcal{H}^*/\partial\boldsymbol{\lambda})'(d\boldsymbol{\lambda}^*/dt)]_{t=t_1} = 0 \quad (8\text{-}33)$$

But 8-31 and 8-33 can both be satisfied only if $d\mathcal{H}^*(t_1)/dt = 0$. Hence, the proof is complete.

Example 8-3. The utility of knowing that $d\mathcal{H}^*/dt \equiv 0$ is illustrated by the scalar case considered here. Given

$$\dot{x} + ax = bm, \quad t \in [t_a, t_b] \quad \text{and} \quad |m| \leq m_{\max} \quad (8\text{-}34)$$

where b, a, and $m_{\max}$ are specified positive values.[4] Corresponding to any initial time t_a and initial state $x(t_a)$, a minimum of J is desired.

$$J = \int_{t_a}^{t_b} (h_1|m| + h_2 x^2 + h_3 m^2 + h_4)\, dt \quad (8\text{-}35)$$

in which h_1 through h_4 are non-negative constant weighting factors. The specified terminal state is $x(t_b) = 0$, but t_b is not specified. An optimal feedback controller is desired as the final solution to the problem.

The three variables x^*, λ^*, and m^* are necessarily related by four conditions: the state equation 8-19, the costate equation 8-20, the maximizing condition 8-21, and the constancy of $\mathcal{H}^*$ condition 8-24. To obtain the feedback form of the control, *only the last two conditions are required in this scalar case*; λ^* can be eliminated between these two conditions, and a

---
[4] Actually, negative values of "a" could be taken into account; but in that case, only the initial states satisfying $|x| < (b/-a)m_{\max}$ would be controllable.

Sect. 8-5　　THE CONSTANCY OF $\mathscr{H}^*$　　489

feedback form of m^* as a function of x^* results. For completeness of the example, however, all conditions and their interrelationships are examined.

The Hamiltonian for this example is obtained on the basis of Equations 8-18, 8-34, and 8-35:

$$\mathscr{H} = -(h_1|m| + h_2 x^2 + h_3 m^2 + h_4) + (-ax + bm)\lambda \tag{8-36}$$

Thus,

$$\dot{x} = \partial \mathscr{H}/\partial \lambda = -ax + bm \tag{8-37}$$

and

$$\dot{\lambda} = -\partial \mathscr{H}/\partial x = 2h_2 x + a\lambda \tag{8-38}$$

At each point in time along the optimal trajectories $x^*(t)$ and $\lambda^*(t)$, m^* is the value of m, $|m| \leq m_{\max}$, which maximizes $\mathscr{H}(x^*,\lambda^*,m)$. If the stationary value m_s of m corresponding to $\mathscr{H}$ does not lie in the interval $[-m_{\max}, m_{\max}]$, the maximum of $\mathscr{H}(x^*,\lambda^*,m)$ is obtained at either $m_{\max}$ or $-m_{\max}$. The stationary point m_s of $\mathscr{H}(x^*,\lambda^*,m)$ is obtained on the basis that

$$\left.\frac{\partial \mathscr{H}(x^*,\lambda^*,m)}{\partial m}\right|_{m=m_s} = -h_1 \operatorname{sgn}(m_s) - 2h_3 m_s + b\lambda^* = 0 \tag{8-39}$$

where

$$\operatorname{sgn}(m_s) \triangleq \begin{cases} 1, & m_s > 0 \\ -1, & m_s < 0 \end{cases} \tag{8-40}$$

Although $\operatorname{sgn}(0) \triangleq 0$ is a common definition, it is not used here; the condition $m_s = 0$ is considered separately in the following. From 8-39,

$$m_s = \frac{b\lambda^* - h_1 \operatorname{sgn}(m_s)}{2h_3} \tag{8-41}$$

And corresponding to the two possibilities for $\operatorname{sgn}(m_s)$, m^* is

$$m^* = \begin{cases} m_{\max} & \text{if } (b\lambda^* - h_1)/2h_3 \geq m_{\max} \\ (b\lambda^* - h_1)/2h_3 & \text{if } 0 \leq (b\lambda^* - h_1)/2h_3 < m_{\max} \\ (b\lambda^* + h_1)/2h_3 & \text{if } -m_{\max} \leq (b\lambda^* + h_1)/2h_3 < 0 \\ -m_{\max} & \text{if } (b\lambda^* + h_1)/2h_3 < -m_{\max} \end{cases} \tag{8-42}$$

Equation 8-42 does not define m^* when λ^* is between $-h_1/b$ and h_1/b. Direct examination of 8-36 reveals that the part of $\mathscr{H}$ which depends on m, $-h_1|m| + b\lambda^* m - h_3 m^2$, is nonpositive for $\lambda^* \in [-h_1/b, h_1/b]$, and the maximum is obtained when $m = 0$. Figure 8-1 contains a graphical display of this result and that of Equation 8-42. Note that the results of Figure 8-1

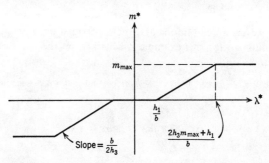

Figure 8-1. The optimal control as a function of the optimal costate variable.

do not depend on h_2, but that both the costate equation and the Hamiltonian are dependent on h_2.

The reader is invited to investigate several special cases of Figure 8-1: case A, a minimum-time problem corresponds to the case that h_1, h_2, and h_3 equal zero; case B, an integral-square-value problem corresponds to h_1 and h_4 equaling zero and $m_{max} \to \infty$; and case C, a minimum-fuel problem corresponds to h_2, h_3, and h_4 equaling zero.

It is apparent from the preceding considerations that the optimal feedback controller is nonlinear in general. With t_b free, Equations 8-22, 8-23, and 8-36 give

$$-(h_1|m^*| + h_2 x^{*2} + h_3 m^{*2} + h_4) + (-ax^* + bm^*)\lambda^* = 0 \quad (8\text{-}43)$$

And from Figure 8-1, the following relationships are observed:

$$|\lambda^*| \leq h_1/b \quad \text{if} \quad m^* = 0 \quad (8\text{-}44\text{a})$$

$$\lambda^* = (2h_3 m^* + h_1 \operatorname{sgn} m^*)/b \quad \text{if} \quad 0 < |m^*| < m_{max} \quad (8\text{-}44\text{b})$$

and

$$\lambda^* \begin{cases} \geq (2h_3 m_{max} + h_1)/b & \text{if} \quad m^* = m_{max} \\ \leq -(2h_3 m_{max} + h_1)/b & \text{if} \quad m^* = -m_{max} \end{cases} \quad (8\text{-}44\text{c})$$

The objective at this point is to use the information contained in 8-44 to eliminate λ^* from 8-43. The development is less involved if $h_2 = 0$, and $h_2 = 0$ is assumed in the remainder of this example. Cases analogous to that of h_2 being nonzero are treated in Sections 8-8 and 8-10 and are also suggested as exercises (Problem 8.1).

With $h_2 = 0$, relations 8-44 remain unchanged, but Equation 8-43 reduces to

$$-(h_1|m^*| + h_3 m^{*2} + h_4) + (-ax^* + bm^*)\lambda^* = 0 \quad (8\text{-}45)$$

Sect. 8-5   THE CONSTANCY OF $\mathcal{H}^*$ 491

If $m^* = 0$, relation 8-44a gives

$$\lambda^* = \alpha h_1/b, \qquad \alpha \in [-1, 1] \tag{8-46}$$

and Equation 8-45 reduces to

$$-h_4 - (a\alpha h_1/b)x^* = 0$$

which is equivalent to

$$x^* = (-bh_4/a\alpha h_1), \qquad \alpha \in [-1, 1] \text{ but } \alpha \not\equiv 0 \tag{8-47}$$

Thus, the range defined by $|x^*| \geq (bh_4/ah_1) \triangleq x_{c1}$ is a *coasting range* in which the optimal value m^* of m is zero.

Again with $h_2 = 0$, if $|m^*|$ is in the open interval $(0, m_{\max})$, Equation 8-44b can be used to eliminate λ^* from 8-45 as follows:

$$0 = -(h_1 m^* \operatorname{sgn} m^* + h_3 m^{*2} + h_4) + (-ax^* + bm^*)(2h_3 m^* + h_1 \operatorname{sgn} m^*)/b$$

$$= h_3 m^{*2} - (2ah_3 x^*/b)m^* - h_4 - (ah_1 x^*/b) \operatorname{sgn} m^* \tag{8-48}$$

When $m^* \neq 0$, it is clear from the dynamics of the system 8-34 and the form of the performance measure 8-35 that $\operatorname{sgn} m^* = -\operatorname{sgn} x^*$; so Equation 8-48 can be solved for m^*:

$$m^* = (a/b)x^* \pm [(ax^*/b)^2 - (ah_1/bh_3)|x^*| + (h_4/h_3)]^{1/2} \tag{8-49}$$

which is valid provided h_3 is not zero. Note that m^* differs in sign from x^* only if the $-$ of the $\pm$ in 8-49 is used when $x^* > 0$, and the $+$ of the $\pm$ is used when $x^* < 0$. Thus,

$$m^* = (a/b)x^* - (\operatorname{sgn} x^*)[(ax^*/b)^2 - (ah_1/bh_3)|x^*| + (h_4/h_3)]^{1/2} \tag{8-50}$$

which is applicable over a specific region of the x^* domain. The previously determined boundary, $x_{c1} = bh_4/ah_1$, of the coasting range is one boundary that applies.

To determine the range of applicability of 8-50, let $x^* = \alpha x_{c1} = \alpha bh_4/ah_1$, for real $\alpha \in [-1, 1]$. Equation 8-50 reduces to

$$m^* = \frac{h_4}{h_1}\alpha - (\operatorname{sgn} \alpha)\left[\left(\frac{h_4}{h_1}\right)^2 \alpha^2 - \frac{h_4}{h_3}|\alpha| + \frac{h_4}{h_3}\right]^{1/2}$$

$$= \frac{h_4}{h_1}\alpha - (\operatorname{sgn} \alpha)\left[\left(\frac{h_4}{h_1}\right)^2 \alpha^2 + \frac{h_4}{h_3}(1 - |\alpha|)\right]^{1/2} \tag{8-51}$$

At $\alpha = \pm 1$, that is, at $|x^*| = x_{c1}$, Equation 8-51 reduces to $m^* = 0$. Furthermore, when $\alpha \in (-1, 0)$, m^* of 8-51 is a monotonically increasing function of α, a fact which the reader may verify. It follows that if no upper bound

$m_{\max}$ of m^* were specified, the maximum value of m^* that would be applied would be $(h_4/h_3)^{1/2}$, which corresponds to $\alpha = 0-$ (and therefore $x^* = 0-$) in Equation 8-51. On the other hand, in order for an upper bound $m_{\max}$ on $|m^*|$ to influence the optimal solution, the value of $m_{\max}$ would have to be less than $(h_4/h_3)^{1/2}$. Another way to substantiate the latter remark is to solve Equation 8-48 for the critical value x_{c2} of x^* corresponding to $m^* = -m_{\max}$. The result is

$$x_{c2} = \frac{h_4 - h_3 m_{\max}^2}{(a/b)(h_1 + 2h_3 m_{\max})} \tag{8-52}$$

which is greater than zero only if $h_4 > h_3 m_{\max}^2$.

Additional insight on the solution is gained by examination of the costate equation 8-38. With $h_2 = 0$, the costate response is

$$\lambda(t) = \lambda(t_a) e^{a(t - t_a)}, \qquad t \geq t_a \tag{8-53}$$

which is a monotonically increasing function of t if $\lambda(t_a) > 0$, and is a monotonically decreasing function of t if $\lambda(t_a) < 0$. Suppose $\lambda(t_a)$ is in the open interval $(0, h_1/b)$. Figure 8-1 indicates three contiguous regions of control: $m^*(t) = 0$ for $t \in [t_a, t_1]$ where t_1 is determined on the basis that $\lambda(t_a) e^{a(t_1 - t_a)} =$

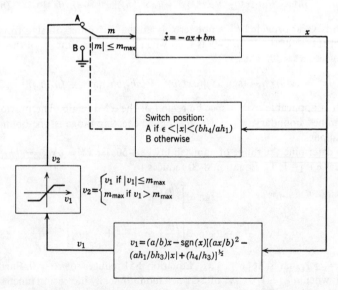

Figure 8-2. An optimal, nonlinear controller for a linear plant.

h_1/b; followed by

$$m^*(t) = \frac{h_1}{b}\frac{b}{2h_3}[e^{a(t-t_1)} - 1] = \frac{h_1}{2h_3}[e^{a(t-t_1)} - 1] \quad \text{for } t \in [t_1, t_2]$$

where t_2 is determined on the basis that $(h_1/2h_3)[e^{a(t_2-t_1)} - 1] = m_{\max}$; followed by $m^*(t) = m_{\max}$ for $t \in [t_2, t_b]$, where t_b is determined on the basis that $x^*(t_b) = 0$. As previously noted, the latter mode of control is not applicable if $m_{\max} > (h_4/h_3)^{1/2}$, in which case the definitive criterion for t_2 is $t_2 = t_b$, $x^*(t_b) = 0$.

When $x^* = 0$, the optimal control which holds it at zero is $m^* = 0$, a fact which is clearly evident on the basis of the state equation 8-37 and which is not obtained through the use of Pontryagin's maximum principle. The fact that the correct $\lambda(t_a)$, one corresponding to a given $x(t_a)$, is unknown at the onset, but is available in the solution, is characteristic of problems solved by using Pontryagin's maximum principle. The block diagram of Figure 8-2 depicts a controller for the process. The controller shown differs in one way from the optimal controller: a zone of zero control (a *dead zone*) has been included in the immediate neighborhood of $x = 0$. In a physical implementation of the "optimal" system, *chattering* of $m(x)$ would occur in the vicinity of $x = 0$. The dead zone is included as a simple way to prevent chattering.

•

8-6. THE GENERAL TRANSVERSALITY CONDITION

It is to be shown that

$$\left[-\mathcal{H}\, \delta t + \sum_{i=1}^{n} \lambda_i(t)\, \delta x_i \right]_{t=t_a}^{t=t_b} = 0 \tag{8-22}$$

is the general transversality condition that applies to the problem of Section 8-3. The Hamiltonian $\mathcal{H}$ is defined by 8-18, and the costate variables $\lambda_i(t)$'s are solutions to first-order differential equations 8-20. Equation 8-22 is shown in this section to be a special case of 3-100 in Chapter 3; and the δt and δx_i's are minute variations as interpreted in Chapter 3. Interpretations of 8-22 in special cases are given at the end of this section.

The encumbering feature of the problem in Section 8-3 is that the control vector **m** is constrained to belong to a closed and bounded (compact) set U in r-dimensional space, i.e., $\mathbf{m}(t) \in U$ at each instant $t \in [t_a, t_b]$. Actually, Equation 8-22 also applies if the boundaries of U are explicit continuous functions of time and if the f and q_i functions are explicit functions of time; the validation of 8-22 is made with the inclusion of these time-varying conditions in this section. From the development in Section 7-13, note that

Equations 8-19, 8-20, and 8-21 are also appropriate in time-varying situations; however, the same cannot be said of the constancy of $\mathcal{H}$, Equation 8-23.

Defining relationships for $U(t)$ could be of the following general forms:

$$g_i(\mathbf{m},t) \leq 0, \qquad i = 1, 2, \ldots, k_1 \qquad (8\text{-}54)$$

and

$$g_i(\mathbf{m},t) \in \{c_{i0}(t), c_{i1}(t), \ldots, c_{in}(t)\}, \qquad i = k_1 + 1, k_1 + 2, \ldots, k_2 \qquad (8\text{-}55)$$

where the g_i's are continuous functions of their arguments, and the $c_{ij}(t)$'s are continuous functions of the argument t. For example, a constraint of the form 8-54 could be simply

$$|m_k| - 1 \leq 0$$

and an elementary example of 8-55 is

$$m_j \in \{-1, 0, 1\}$$

which means that, at any time t, the jth control variable must assume one of three values in the set $\{-1, 0, 1\}$.

To derive 8-22 from 3-100, the general constraints 8-54 and 8-55 must be reduced to a form for which the theory of Chapter 3 applies. Consider the equality constraints

$$z_i \triangleq g_i(\mathbf{m},t) + m_{r+i}^2 = 0, \qquad i = 1, 2, \ldots, k_1 \qquad (8\text{-}56)$$

which are equivalent to 8-54 for real m_{r+i}'s (artificial control variables). Similarly, constraints 8-55 can be placed in equality form through the use of the product equivalent constraints that follow:

$$z_i \triangleq \prod_{j=1}^{n} [g_i(\mathbf{m},t) - c_{ij}(t)] = 0, \qquad i = k_1 + 1, k_1 + 2, \ldots, k_2 \qquad (8\text{-}57)$$

Suppose that $\dot{y}_i$'s, $i = 1, 2, \ldots, r + k_1$, are introduced and defined to be equal to corresponding m_i's. Let $\dot{\mathbf{y}}$ denote the vector set of these $\dot{y}_i$'s which are used in place of m_i's in all equations. In the classical calculus of variations, the derivative vectors, $\dot{\mathbf{x}}$ and $\dot{\mathbf{y}}$, are continuous between corner points of first-variational curves, but one or more components of $\dot{\mathbf{x}}$ and $\dot{\mathbf{y}}$ exhibit ordinary jump discontinuities at corner points.

On the basis of the preceding two paragraphs, the classical variational approach can be applied. An augmented functional J_a is formed,

$$J_a = \int_{t_a}^{t_b} \left\{ f(\mathbf{x}, \dot{\mathbf{y}}, t) + \boldsymbol{\lambda}'[\dot{\mathbf{x}} - \mathbf{q}(\mathbf{x}, \dot{\mathbf{y}}, t)] + \sum_{i=1}^{k_2} \lambda_{ai} z_i(\dot{\mathbf{y}}, t) \right\} dt$$

$$\equiv \int_{t_a}^{t_b} f_a \, dt \qquad (8\text{-}58)$$

Sect. 8-6　THE GENERAL TRANSVERSALITY CONDITION

where the λ_{ai}'s are additional Lagrange multipliers. Condition 3-100 applies to the integrand f_a of 8-58, in the modified form that follows:

$$\{[f_a\,\delta t - (\partial f_a/\partial\dot{\mathbf{x}})'\dot{\mathbf{x}}\,\delta t + (\partial f_a/\partial\dot{\mathbf{x}})'\,\delta\mathbf{x}] + [-(\partial f_a/\partial\dot{\mathbf{y}})'\dot{\mathbf{y}}\,\delta t + (\partial f_a/\partial\dot{\mathbf{y}})'\,\delta\mathbf{y}]\}_{t_a}^{t_b} = 0 \tag{8-59}$$

Terms to be identified in 8-59 are $\dot{\mathbf{x}}$, $\partial f_a/\partial\dot{\mathbf{x}}$, and $\partial f_a/\partial\dot{\mathbf{y}}$. From 8-4, $\dot{\mathbf{x}} = \mathbf{q}$; and from 8-58,

$$\frac{\partial f_a}{\partial\dot{\mathbf{x}}} = \boldsymbol{\lambda} \tag{8-60}$$

The identification of $\partial f_a/\partial\dot{\mathbf{y}}$ is somewhat more subtle. The vector Euler-Lagrange equation corresponding to $\mathbf{y}$ is

$$\frac{\partial f_a}{\partial\mathbf{y}} - \frac{d}{dt}\frac{\partial f_a}{\partial\dot{\mathbf{y}}} = \mathbf{0} \tag{8-61}$$

But $\mathbf{y}$ does not appear as an argument of f_a, so 8-61 reduces to

$$\frac{\partial f_a}{\partial\dot{\mathbf{y}}} = \mathbf{c} \tag{8-62}$$

where $\mathbf{c} = [c_1 \quad c_2 \quad \cdots \quad c_{r+k_1}]'$ is a column vector of constants. The results of 8-60 and 8-62 are used to reduce 8-59 to

$$\{[-\mathscr{H}\,\delta t + \boldsymbol{\lambda}'\,\delta\mathbf{x}] + [-\mathbf{c}'\dot{\mathbf{y}}\,\delta t + \mathbf{c}'\,\delta\mathbf{y}]\}_{t_a}^{t_b} = 0 \tag{8-63}$$

Observe that end conditions on $\mathbf{y}$ are not given; yet $\mathbf{y}(t_a)$ can be specified at zero without loss of generality because f_a is an explicit function of $\dot{\mathbf{y}}$ and not of $\mathbf{y}$. With $\mathbf{y}(t_a)$ so specified, the vector variation $\delta\mathbf{y}(t_a)$ is constrained to equal the zero vector, but $\delta\mathbf{y}(t_b)$ is an arbitrary vector variation, and therefore $\mathbf{c}$ must be a zero vector in 8-63 to insure equality. Hence, Equation 8-63 reduces to the stated result 8-22.

Case A. (t_b free.)

With t_b free to be selected in an optimal manner, the variation δt_b is independent of other variations, so that $\mathscr{H}^*(t_b) = \mathscr{H}(\mathbf{x}^*,\boldsymbol{\lambda}^*,\mathbf{m}^*,t)_{t=t_b}$ must equal zero in order for 8-22 to be satisfied by the optimal solution. As noted in Section 8-4, if t_b is free and $\mathscr{H}$ is formally independent of t, $\mathscr{H}(\mathbf{x}^*,\boldsymbol{\lambda}^*,\mathbf{m}^*) = 0$ for all $t \in [t_a,t_b]$.

Case B. ($\mathbf{x}(t_b)$ free.)

With $\mathbf{x}(t_b)$ free to be selected in an optimal manner, each $\delta x_i(t_b)$ variation is independent of other variations so that each $\lambda_i^*(t_b)$ must be zero in order for 8-22 to be satisfied by optimal $\lambda_i(t_b)$'s.

Case C. ($\mathbf{x}(t_b)$ required to satisfy a given equation $\phi_1(\mathbf{x}) = 0$, and other end-point conditions independent of $\mathbf{x}(t_b)$.)

Let S_1 denote the set of real x's which satisfy

$$\phi_1(\mathbf{x}) = 0 \tag{8-64}$$

where $\phi_1 = \phi_1(\mathbf{x})$ is a given real-valued function of class C^2. The set S_1 can be viewed as a hypersurface in an n-dimensional Euclidean space. It is assumed that the terminal value $\mathbf{x}^*(t_b)$ of an optimal trajectory is required to be an element of S_1 and that this requirement is consistent with other conditions of the problem.

At each point on S_1, the gradient $\partial \phi_1 / \partial \mathbf{x}$ defines a normal vector to the S_1 hypersurface, provided that $\partial \phi_1 / \partial \mathbf{x}$ is not the zero vector at any point on S_1. It is assumed that this latter property is satisfied, in which case S_1 is called a *smooth hypersurface*.

Consider a minute variation $\delta \mathbf{x} = \mathbf{x} - \mathbf{x}^*(t_b)$ from $\mathbf{x}^*(t_b)$. A Taylor's series expansion of $\phi_1(\mathbf{x})$ about the point $\mathbf{x}^*(t_b)$ is

$$\phi_1(\mathbf{x}) = \phi_1[\mathbf{x}^*(t_b)] + \left[\frac{\partial \phi_1}{\partial \mathbf{x}}\right]'_{\mathbf{x}=\mathbf{x}^*(t_b)} \delta \mathbf{x} + Rm \tag{8-65}$$

where Rm is of order $\delta \mathbf{x}' \delta \mathbf{x}$. Note that by assumption $\phi_1[\mathbf{x}^*(t_b)] = 0$; and in order for $\delta \mathbf{x}$ to be an allowed terminal variation, it must yield $\phi_1(\mathbf{x}) = 0$ in 8-65. But, therefore, a necessary condition on $\delta \mathbf{x}$ is

$$\left[\frac{\partial \phi_1}{\partial \mathbf{x}}\right]'_{\mathbf{x}=\mathbf{x}^*(t_b)} \delta \mathbf{x} = 0 \tag{8-66}$$

which defines a hyperplane that is tangent to S_1 at the point $\mathbf{x}^*(t_b)$. The gradient vector $[\partial \phi_1 / \partial \mathbf{x}]_{\mathbf{x}=\mathbf{x}^*(t_b)}$ is the normal to this hyperplane.

Now, consider

$$\boldsymbol{\lambda}(t_b)' \delta \mathbf{x} = 0 \tag{8-67}$$

which is that part of the transversality condition 8-22 of interest here and which must be satisfied when $\boldsymbol{\lambda}(t_b) = \boldsymbol{\lambda}^*(t_b)$. This result means that $\boldsymbol{\lambda}^*(t_b)$ must also be normal to the tangent hyperplane defined by 8-66. In geometric terms, therefore, *the transversality condition is satisfied if the costate vector $\boldsymbol{\lambda}(t_b)$ is normal to the tangent plane of the hypersurface S_1 at the terminal point $\mathbf{x}(t_b) \in S_1$*.

For purposes of computation, the δx_i's must be eliminated from the necessary conditions associated with this transversality condition. Because S_1 is assumed smooth, some $\partial \phi_1 / \partial x_j$ is nonzero at $\mathbf{x} = \mathbf{x}^*$ on S_1; Equation 8-66 can therefore be arranged to obtain δx_j as dependent on the remaining δx_i's:

$$\delta x_j = -\sum_{\substack{i \\ i \neq j}} \frac{\partial \phi_1 / \partial x_i}{\partial \phi_1 / \partial x_j} \delta x_i \tag{8-68}$$

Sect. 8-6   THE GENERAL TRANSVERSALITY CONDITION

If the right-hand member of 8-68 is used in place of δx_j in 8-67, the result is

$$\sum_{\substack{i \\ i \neq j}} \left[\lambda_i(t_b) - \frac{\partial \phi_1/\partial x_i}{\partial \phi_1/\partial x_j} \lambda_j(t_b) \right] \delta x_i = 0 \qquad (8\text{-}69)$$

And because the δx_i's remaining in 8-69 are independent, it follows that their coefficients must be zero, with the result that

$$\frac{\lambda_i(t_b)}{\lambda_j(t_b)} = \frac{\partial \phi_1/\partial x_i}{\partial \phi_1/\partial x_j}\bigg|_{t=t_b}, \qquad i \neq j \qquad (8\text{-}70)$$

which is equivalent to

$$\boldsymbol{\lambda}(t_b) = \alpha_1 \frac{\partial \phi_1}{\partial \mathbf{x}}\bigg|_{t=t_b} \qquad (8\text{-}71)$$

for a nonzero α_1. Equations 8-70 and 8-64 constitute a set of n conditions of the $2n + 2$ end-point conditions required to obtain principal solutions. The remaining $n + 2$ end-point conditions must be found in terms of t_a, t_b, and $\mathbf{x}(t_a)$.

Case D. ($\mathbf{x}(t_b)$ required to satisfy a set of equations $\phi_i(\mathbf{x}) = 0$, $i = 1, 2, \ldots$, $\nu < n$.)

Given is a set of equations

$$\phi_i(\mathbf{x}) = 0, \qquad i = 1, 2, \ldots, \nu < n \qquad (8\text{-}72)$$

where each real-valued $\phi_i(\mathbf{x})$ is assumed to be of class C^2. Let S_i denote the set of real $\mathbf{x}$'s which satisfy the ith equation in 8-72. The intersection of all S_i sets defines a manifold S; all $\partial \phi_i/\partial \mathbf{x}$'s are assumed to be linearly independent at each point on S.

The requirement on $\mathbf{x}^*(t_b)$ is, in this case, that it belongs to S. Proceeding as under Case C, each $\phi_i(\mathbf{x})$ can be expanded in a Taylor's series about an optimal terminal point $\mathbf{x}^*(t_b)$:

$$\phi_i(\mathbf{x}) = \phi_i[\mathbf{x}^*(t_b)] + \left[\frac{\partial \phi_i}{\partial \mathbf{x}}\right]'_{\mathbf{x}=\mathbf{x}^*(t_b)} \delta \mathbf{x} + Rm \qquad (8\text{-}73)$$

where allowed terminal variations $\delta \mathbf{x}$ are those for which both $\phi_i(\mathbf{x})$ and $\phi_i[\mathbf{x}^*(t_b)]$ are zero; this fact and the fact that the remainder Rm is negligible for minute $\delta \mathbf{x}$ lead to the following necessary conditions that must be satisfied by minute variations on S:

$$\left[\frac{\partial \phi_i}{\partial \mathbf{x}}\right]'_{\mathbf{x} \in S} \delta \mathbf{x} = 0, \qquad i = 1, 2, \ldots, \nu \qquad (8\text{-}74)$$

This set of equations defines a hyperplane which is tangent to the S manifold at any given point $\mathbf{x} \in S$.

As in Case C, the applicable part of 8-22 for this problem is

$$\lambda(t_b)' \, \delta\mathbf{x} = 0 \tag{8-75}$$

Comparison of 8-74 and 8-75 shows that the transversality condition is satisfied if $\lambda(t_b)$ is normal to the tangent plane of the manifold S at the terminal point $\mathbf{x}(t_b)$. For numerical purposes, the preceding statement is equivalent to the requirement that $\lambda(t_b)$ be a linear combination of the $\partial \phi_i / \partial \mathbf{x}$'s, that is,

$$\lambda(t_b) = \sum_{i=1}^{\nu} \alpha_i \left. \frac{\partial \phi_i}{\partial \mathbf{x}} \right|_{t=t_b} \tag{8-76}$$

where the α_i's are real values, at least one of which is nonzero. By evaluating the α_i's of 8-76 in terms of $\lambda_i(t_b)$'s and $\partial \phi_i[\mathbf{x}(t_b)]/\partial x_j$'s, the n equations in 8-76 can be reduced to $n - \nu$ conditions that are independent of the α_i's. The resulting $n - \nu$ conditions plus the ν conditions of 8-72 form a set of n end-point conditions. Additional end-point conditions in terms of t_a, t_b, and $\mathbf{x}(t_a)$ are applicable in any given case.

Example 8-4. Consider the *RLC* circuit diagram in Figure 8-3. The input voltage $m(t)$ is constrained by

$$0 \leq m(t) \leq E$$

The voltage across the fixed capacitor C is denoted by $x_1(t)$, and the current through the fixed inductor L is denoted by $x_2(t)$. A specified time interval $[0, T]$ is of interest; a maximum of $x_1(T)$ is to be obtained by proper selection of $m(t)$ over the time interval $[0, T]$. The values of $x_1(0)$ and $x_2(0)$ are assumed known, but $x_1(T)$ and $x_2(T)$ are free to be selected in an optimal manner.

Straightforward circuit analysis gives

$$\dot{\mathbf{x}} = \mathbf{A}\mathbf{x} + \mathbf{b}m \tag{8-77}$$

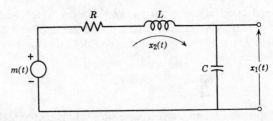

Figure 8-3. A series *RLC* circuit with a maximum of $x_1(T)$ desired.

Sect. 8-6   THE GENERAL TRANSVERSALITY CONDITION

in which

$$\mathbf{A} = \begin{bmatrix} 0 & 1/C \\ -1/L & -R/L \end{bmatrix} \quad \text{and} \quad \mathbf{b} = \begin{bmatrix} 0 \\ 1/L \end{bmatrix}$$

On the basis of 8-77, it is clear that maximization of $x_1(T)$ is equivalent to minimization of J:

$$J = \int_0^T -x_2(t)\, dt \tag{8-78}$$

Based on Equations 8-18, 8-77, and 8-78, the applicable Hamiltonian is

$$\mathscr{H} = x_2 + (x_2/C)\lambda_1 + (-x_1 - Rx_2 + m)\lambda_2/L \tag{8-79}$$

The maximizing condition on $\mathscr{H}$ is satisfied when

$$m^* = \begin{cases} E, & \lambda_2^* > 0 \\ 0, & \lambda_2^* < 0 \end{cases} \tag{8-80}$$

The vector costate equation is formed:

$$\begin{aligned}\dot{\boldsymbol{\lambda}} &= -\partial\mathscr{H}/\partial\mathbf{x} \\ &= -\mathbf{A}'\boldsymbol{\lambda} + \begin{bmatrix} 0 \\ -1 \end{bmatrix}\end{aligned} \tag{8-81}$$

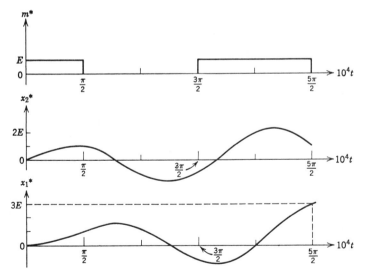

Figure 8-4. A particular case of the solution of Example 8-4: $L = 0.1$ millihenry, $C = 100$ microfarads, and $R = 0$.

As previously noted, the variations $\delta x_1(T)$ and $\delta x_2(T)$ are free, in which case the transversality condition 8-22 is satisfied only if

$$\lambda_1(T) = \lambda_2(T) = 0 \tag{8-82}$$

With the terminal values 8-82 known, 8-81 can be solved backward in time to obtain a unique solution $\lambda_2^*(t)$ of $\lambda_2(t)$. The reader may verify that

$$\lambda_2^*(t) = \frac{e^{-\xi\omega_n(T-t)}}{\omega_n(1-\xi^2)^{1/2}} \sin\left[\omega_n(1-\xi^2)^{1/2}(T-t)\right] \tag{8-83}$$

where $\omega_n = (1/LC)^{1/2}$ and $\xi = R/2L\omega_n$; it is assumed that ξ is less than 1. The sign-determining factor of 8-83 is the sinusoidal term, and therefore $m^*(t)$ of 8-80 can be expressed as

$$m^*(t) = \begin{cases} E, & \sin\left[\omega_n(1-\xi^2)^{1/2}(T-t)\right] > 0 \\ 0, & \sin\left[\omega_n(1-\xi^2)^{1/2}(T-t)\right] < 0 \end{cases} \tag{8-84}$$

A particular optimal solution is displayed in Figure 8-4.

The solution 8-84 is easily obtained because terminal conditions on the costate variables are known and because the costate equations 8-81 are independent of the state variables and $m(t)$. Unfortunately, these simplifying features apply only to a small minority of the problems that are of interest.

●

8-7. TIME OPTIMAL CONTROL

8-7a. Comments

One of the most pursued control objectives is that of forcing a system to change from an initial state to a desired terminal state in minimum time, subject to satisfying limitations imposed on the control and the state variables. Paiewonsky [8.85] has given a detailed account of the literature prior to 1961 on time optimal control of linear systems; only the barest outline of the history and the approaches to solution are considered in this subsection.

In a 1943 patent [8.33], Doll outlined the problem of controlling a linear, second-order, lightly damped system in minimum time when the control action is constrained in magnitude. In 1950, McDonald [8.75] considered practical approaches to control of various second-order linear systems in a minimum-time sense. The first rigorous solution of Doll's problem was obtained by Bushaw [8.18, 8.19] in a 1952 Ph.D. thesis, and the 1950 work [8.46] of Hestenes is striking in its equivalence to Pontryagin's maximum principle; but it can be said that the necessary conditions associated with time optimal control were not well known until after the 1956 work of Pontryagin.

TIME OPTIMAL CONTROL

For time optimal control of linear systems with control inputs constrained by magnitude limitations only, it is known (Example 7-4, Section 7-13e) that time optimal control laws are of the bang-bang type, except in singular cases (Section 8-10). The term bang-bang is used to emphasize the fact that the control actions switch between extremes of allowed values. When initial and terminal conditions on the state variables are known, as is usually assumed in time optimal control, the transversality condition 8-22 gives only that $\mathcal{H}^*(t_b) = 0$—appropriate end-point values on the costate variables are not given. Search procedures [8.35, 8.69, 8.81] have been advanced in the literature for determining time optimal trajectories of linear time-variant systems, but these procedures currently lack the ability to cope with feedback control of practical systems.

For linear time-invariant systems, a fruitful approach [8.5, 8.37, 8.91] has been that of determining switching boundaries or hypersurfaces that isolate regions of state space in which particular modes of bang-bang control are optimal. The switching surfaces can be obtained by integrating the state and costate equations in reverse time, where $\mathcal{H}^*(t_b) = 0$ gives one relationship that the $\lambda_i(t_b)$'s must satisfy and where a given vector value of $\boldsymbol{\lambda}(t_b)$ results in a set of switch *points* defined by state-space coordinates. By repeating the reverse-time integrations with numerous distinct $\boldsymbol{\lambda}(t_b)$'s, each of which satisfies $\mathcal{H}^*(t_b) = 0$, the switch points aggregate on hypersurfaces in state space, and the switching hypersurfaces can be determined from the aggregates. Athanassiades[5] and Smith [8.4] detail the approach for nth-order time-invariant systems with real eigenvalues and one control variable, in which case there is one time-optimal $(n - 1)$-dimensional switching surface—it consists of the locus of all trajectories which terminate at the origin (the desired state) and which result from $n - 2$ or fewer changes in sign of the bang-bang control. For complex eigenvalue systems, the switching surfaces assume intricate geometrical shapes that are difficult to store for use in feedback control. Even for real eigenvalue systems of third or higher order, the use of switching surface approximations is advocated [8.38, 8.40, 8.103].

For second-order systems the switching surfaces reduce to switch curves in the state plane. The second-order system of Section 8-7b is one which, because of its practical nature, has received much attention. To supplement the presentation of Section 8-7b, the reader may refer to numerous works (e.g., [8.5, 8.22, 8.74, 8.83, and 8.104]).

For high-order, linear, time-invariant systems with real eigenvalues, an alternative to the switching surface approach is that in which optimal switch times are iteratively computed in "fast" time for use in controlling a relatively slower process. One such method, developed by Lee [8.72] and Smith [8.102], is presented in Section 8-7c.

---

[5] Athanassiades subsequently shortened his name to Athans.

8-7b. A Second-Order System

Consider the block diagram of Figure 8-5. The process to be controlled, the plant, of the system is characterized by a second-order Laplace transform transfer function

$$\mathscr{L}[c(t)] = \frac{1}{s(s+a)} \mathscr{L}[m(t)] \tag{8-85}$$

where the constant a is assumed greater than zero—a variety of related second-order cases are considered in the literature (e.g., [8.22]). The controller is to be designed to reduce the error $e(t)$ and its derivative $\dot{e}(t)$ to zero in the shortest possible time. Although not shown in the block diagram, arbitrary initial conditions on both $c(t)$ and $\dot{c}(t)$ exist. The input $v(t)$ is a reference input. The control action $m(t)$ is constrained by $|m| \leq m_{\max}$.

The differential equation for the plant is

$$\ddot{e} + a\dot{e} = \ddot{v} + a\dot{v} - m, \qquad |m| \leq m_{\max} \tag{8-86}$$

However, the reference input v is assumed to be piecewise constant over relatively long periods of time so that $\dot{v}$ and $\ddot{v}$ are zero over the response time of the system.

To form the state equations, let

$$x_1 \triangleq -e \tag{8-87}$$

and

$$\dot{x}_1 \triangleq x_2 \tag{8-88a}$$

Equations 8-87 and 8-88a are used in 8-86 to obtain

$$\dot{x}_2 = -ax_2 + m, \qquad |m| \leq m_{\max} \tag{8-88b}$$

At any time t_a, $x_1(t_a)$ and $x_2(t_a)$ have some values. The controller is to force x_1 and x_2 to zero at an unspecified time t_b, where

$$J = \int_{t_a}^{t_b} dt \tag{8-89}$$

is minimized in the process.

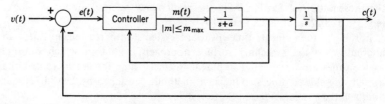

Figure 8-5. A second-order plant.

TIME OPTIMAL CONTROL

The Hamiltonian for the problem is formed on the basis of 8-88 and 8-89:

$$\mathcal{H} = -1 + \lambda_1 x_2 + \lambda_2(-ax_2 + m) \quad (8\text{-}90)$$

The costate equations are formed:

$$\dot{\lambda}_1 = -\frac{\partial \mathcal{H}}{\partial x_1} = 0 \quad (8\text{-}91\text{a})$$

and

$$\dot{\lambda}_2 = -\frac{\partial \mathcal{H}}{\partial x_2} = -\lambda_1 + a\lambda_2 \quad (8\text{-}91\text{b})$$

Because values of x_1 and x_2 are known at both t_a and t_b, the transversality condition 8-22 gives no end-condition information on the costate variables. Also, because t_b is free to be selected in an optimal manner, condition 8-22 reveals that $\mathcal{H}^*(t_b) = 0$, but no direct use of this fact is made here. What is to be observed is that the value m^* of m that maximizes $\mathcal{H}$ along the optimal state and costate trajectories is

$$m^* = m_{\max} \operatorname{sgn} \lambda_2^* \quad (8\text{-}92)$$

where the sgn function is defined in 8-40; and λ_2^*, being a solution of 8-91, *can have at most one change in sign* (the reader may verify this statement). Thus, it is known that the optimal control equals either $+m_{\max}$ or $-m_{\max}$ over $[t_1, t_b]$ and equals $-m_{\max}$ or $+m_{\max}$ over $[t_a, t_1]$ where, depending on the values of $x_1(t_a)$ and $x_2(t_a)$, it is possible for the switch time t_1 to equal t_a.

The fact that $m^* = \pm m_{\max}$ along the final trajectory can be used to calculate a switch curve in the state plane; when the state trajectory intersects the switch curve, control action is reversed. The final part of the state trajectory, the terminal trajectory, which leads to the origin of the state plane, is along the switch curve. Let x_{1f} and x_{2f} denote the set of state values along terminal state trajectories, and let m equal a constant b in 8-88. The ratio of corresponding members of 8-88b and 8-88a is

$$\frac{dx_{2f}}{dx_{1f}} = -a + \frac{b}{x_{2f}}$$

which is rearranged to obtain

$$\frac{x_{2f}\, dx_{2f}}{b - ax_{2f}} = dx_{1f} \quad (8\text{-}93)$$

Along the terminal trajectory that corresponds to $b = m_{\max}$, x_{2f} is negative and is being forced to zero by application of $m = m_{\max}$ in Equation 8-88b.

In this case, $m_{max} - ax_{2f}$ is greater than zero, so that integration of 8-93 gives

$$\frac{1}{a^2}[m_{max} - ax_{2f} - m_{max}\ln(m_{max} - ax_{2f})] = x_{1f} + c_1 \qquad (8\text{-}94)$$

where the constant c_1 is determined on the basis that $x_{2f} = 0$ when $x_{1f} = 0$. In short, Equation 8-94 reduces to

$$a^2 x_{1f} = -ax_{2f} - m_{max}\ln\left(1 - \frac{ax_{2f}}{m_{max}}\right) \qquad (8\text{-}95)$$

which applies for $x_{2f} < 0$.

For $x_{2f} > 0$, $b = -m_{max}$ is the applied value of m in 8-88b to force x_{2f} to zero, and the integration in 8-93 gives

$$\frac{1}{a^2}[-m_{max} - ax_{2f} + m_{max}\ln(ax_{2f} + m_{max})] = x_{1f} + c_2 \qquad (8\text{-}96)$$

where c_2 must be such that $x_{2f} = 0$ when $x_{1f} = 0$. It follows that

$$a^2 x_{1f} = -ax_{2f} + m_{max}\ln\left(1 + \frac{ax_{2f}}{m_{max}}\right) \qquad (8\text{-}97)$$

which applies for $x_{2f} > 0$.

Equations 8-95 and 8-97 are equivalent to one equation:

$$a^2 x_{1f} = -ax_{2f} + (\text{sgn } x_{2f})\, m_{max}\ln\left(1 + \frac{a|x_{2f}|}{m_{max}}\right) \qquad (8\text{-}98)$$

An example of this curve is given in Figure 8-6. For system states which are to the right of the switch curve shown in Figure 8-6, the appropriate control is $m^* = -m_{max}$—this fact is assured because the maximum principle indicates at most one switch point between the initial and the terminal times. For system states to the left of the switch curve, $m^* = +m_{max}$ is the optimal control. What is required is a way of determining whether the state of the system is to the left or to the right of the switch curve. A *switching function* is synthesized for this purpose.

Consider the function $g \equiv g(x_1, x_2, m_{max}, a)$,[6]

$$g = -a^2 x_1 - ax_2 + (\text{sgn } x_2)\, m_{max}\ln\left(1 + \frac{a|x_2|}{m_{max}}\right) \qquad (8\text{-}99)$$

On the basis of the switch-curve equation 8-98, it is clear that $g = 0$ when $x_1 = x_{1f}$ and $x_2 = x_{2f}$. For any fixed value of x_2, only one value of x_1 exists

---

[6] In many applications, a polynomial approximation of g is sufficient. See reference [8.79], for example.

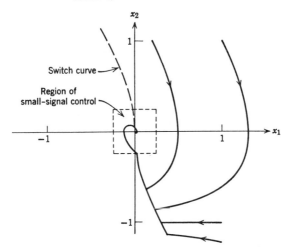

Figure 8-6. Switch curve and ideal responses with $a = m_{\max} = 1$. A region of small-signal control is included in the design.

for which $g = 0$; that is, the switch curve is the only set of points for which $g = 0$. Thus, it can be shown that g is less than zero if and only if the state is to the right of the switch curve, and g is greater than zero if and only if the state is to the left of the switch curve. For states off the switch curve, the optimal control is

$$m^* = m_{\max} \operatorname{sgn} [g(x_1,x_2,m_{\max},a)] \qquad (8\text{-}100)$$

Equation 8-100 does not apply for points on the switch curve because the maximizing condition on $\mathscr{H}$ (8-90) yields no information about m^* when $\lambda_2^* = 0$. There are many ways to account for points on the switch curve; for example, a small amount of time delay could be present in the switching so that the trajectory "parallels" the switch curve after switching occurs; or the switching function could be modified so that the trajectory branches slightly away from the switch curve. The latter approach is a natural one from the standpoint of sensitivity of response with respect to $m_{\max}$ and a. Consider the case where $m_{\max}$ and a are known to lie in the open intervals $(m_{\max,lb}, m_{\max,ub})$ and (a_{lb}, a_{ub}), respectively, where lb stands for lower bound, and ub stands for upper bound. If the $m_{\max,ub}$ and a_{ub} values are used to generate g, it can be shown that state trajectories branch away from the switch curve after switching occurs, as illustrated in Figure 8-7. On the other hand, if other values of a and $m_{\max}$ are used to generate the switching function, and a_{ub} and $m_{\max,ub}$ are the actual system values, the trajectories tend

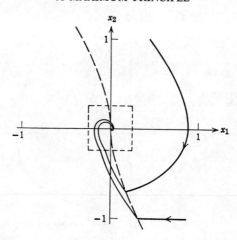

Figure 8-7. Typical responses when system parameters are $a = m_{\max} = 0.7$, whereas the switch curve is designed for $a = m_{\max} = 1$.

to recross the switch curve after having switched once, and the control action *chatters* back and forth, causing the trajectory to zigzag along the switch curve. The chattering action would slow the response of the system; and if mechanical relays are used to generate m, the chattering would cause the operational lifetime of the system to be reduced.

When $m_{\max, ub}$ and a_{ub} are used to generate the switching function g, the trajectories do not tend exactly to the origin in the state plane; and in the vicinity of the origin, chattering of m results if steps are not taken to avoid it. To avoid this chattering, an appropriate dead zone (see Example 8-3) could be used, but often more appropriate is a small-signal, linear mode of control. The use of one strategy for large-signal control and another for small-signal control is called *dual-mode control*.

As a particular example, consider an analog simulation[7] of the system with nominal values of $m_{\max}$ and a equal to one. When the switch-curve values of $m_{\max}$ and a are equal to corresponding values of the plant of the system, trajectories such as those of Figure 8-6 are as anticipated—a rather

---

[7] This simulation was done by J. R. Albers, B. A. Laws, D. V. Nafus, and O. J. Olson in a senior laboratory project entitled "Bang-Bang Control," May 1967. The switch curve was imbedded as a copper strip on a printed circuit board; an x-y plotter with appropriate electrical contacts was used to determine the position of the state of the analog model in relation to the switch curve, and a second plotter was used to plot responses. The instant at which the small-signal mode of control was to start was sensed by use of a separate rectangular copper strip about the origin of the state plane.

Sect. 8-7 TIME OPTIMAL CONTROL 507

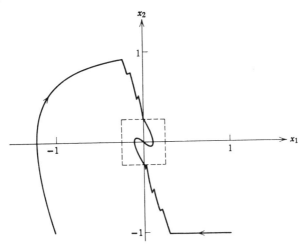

Figure 8-8. Typical responses when system parameters are $a = m_{\max} = 1.5$, whereas the switch curve is designed for $a = m_{\max} = 1$.

large region of linear-mode control is indicated about the state-plane origin. But when the switch curve is designed on the basis that $a = m_{\max} = 1$, and the plant has values of $m_{\max} = a = 1.5$, the responses shown in Figure 8-8 are typical if a small amount of delay is introduced between the time at which the switch curve is crossed and the time at which the control signal is switched. The smaller the amount of this delay, the greater the number of switchings along the switch curve. However, when $m_{\max} = a = 0.7$ in the plant, whereas the switch curve is computed on the basis that $m_{\max} = a = 1$, the responses of Figure 8-7 are obtained.

For the implemented system either a digital controller (with a sufficiently high sampling rate) or a continuously acting controller could be used to generate the switching function of Equation 8-99. The block diagram of a particular, but not necessarily the best, form of dual-mode controller is displayed in Figure 8-9a. The logarithmic factor in the diagram could be generated by use of appropriate diode circuits. The logic element in Figure 8-9a is shown connected to the amplifier with a dashed path and is included to obtain a small-signal mode of control about the origin in the state plane. When either $|e| > \epsilon_1$ or $|\dot{c}| > \epsilon_2$, or both, bang-bang control is in effect, and the output of the OR circuit is in the high position. But when both $|e| < \epsilon_1$ and $|\dot{c}| < \epsilon_2$, the output of the OR circuit is low, and the gated elements are in the low position for small-signal control.

As previously noted, the small-signal region shown in Figures 8-6, 8-7,

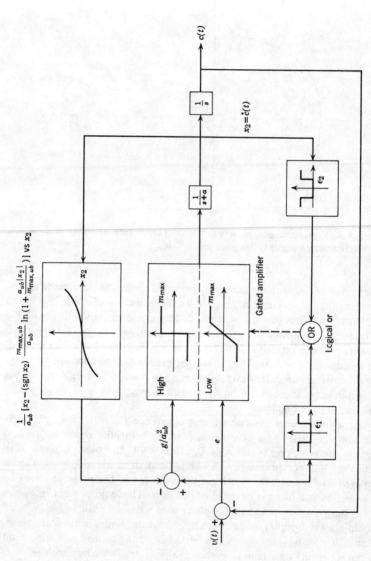

Figure 8-9a. An implementation of dual-mode control of a second-order system.

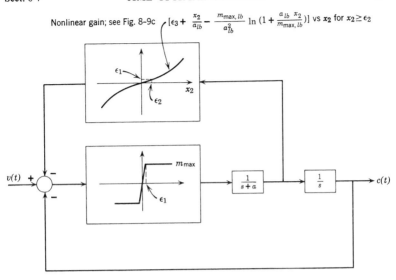

Figure 8-9b. Another form of dual-mode control of a second-order system.

and 8-8 is rather large. With more stringent requirements, this region could be made smaller, and the lower (upper) half of the switch curve could be shifted a small amount to the right (left) (see Problem 8.12). With a smaller region of small-signal control, however, it is often difficult to solidly capture the state trajectory in the small-signal mode; the trajectory may tend to pass through the region several times before holding therein. In such cases, a smoother response can be obtained if the boundaries of the small-signal region of Figure 8-9a are provided with hysteresis; when the state of the system is outside the small-signal region, the boundaries are defined by $|x_1| = \epsilon_{1a}$ and $|x_2| = \epsilon_{2a}$; but after the state of the system enters the preceding boundaries, new boundaries are set at $|x_1| = \epsilon_{1b} > \epsilon_{1a}$ and $|x_2| = \epsilon_{2b} > \epsilon_{2a}$.

Rauch and Howe [8.93] consider a different type of dual-mode control, one which yields system responses that are quite insensitive to parameter variations. A form of this dual-mode control is depicted in Figure 8-9b. The unsaturated mode of control for the configuration of Figure 8-9b is associated with state-plane points in the vicinity of the "switch curve" of Figure 8-9c; no logic operations are required to implement the control in this case. As with other control configurations, system sensitivity with respect to parameter changes should be considered in a detailed design (see Problem 8.13).

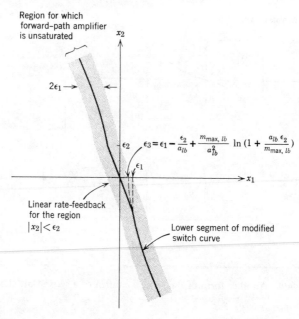

Figure 8-9c. This modified switch curve is especially shaped for use with the saturating-amplifier, dual-mode controller of Figure 8-9b. The lower segment of the switch curve is given by $0 = a_{lb}^2(x_1 - \epsilon_3) + a_{lb}x_2 + m_{\max,lb} \ln\left(1 + \dfrac{a_{lb}|x_2|}{m_{\max,lb}}\right)$.

8-7c. Optimal Switch-Time Evaluation

Consider the linear time-invariant system characterized by

$$\dot{\mathbf{x}} = \mathbf{A}\mathbf{x} + \mathbf{b}m, \qquad |m| \leq 1 \qquad (8\text{-}101)$$

where $\mathbf{A}$ is an $n \times n$ matrix and $\mathbf{b}$ is an $n \times 1$ matrix. The entries of $\mathbf{A}$ and $\mathbf{b}$ are real constants, and the system is assumed to be controllable (see Section 7-14b). The state $\mathbf{x}$ is to be driven from any initial point $\mathbf{x}(0)$ at time zero to a given terminal state $\mathbf{x}^*(T)$ in minimum time T.

The minimum-time controller for the system 8-101 is of the bang-bang type because the Hamiltonian (Equation 8-18) is, in this case,

$$\mathscr{H} = -1 + \boldsymbol{\lambda}'\mathbf{A}\mathbf{x} + \boldsymbol{\lambda}'\mathbf{b}m \qquad (8\text{-}102)$$

on the basis of which

$$\dot{\boldsymbol{\lambda}} = -\mathbf{A}'\boldsymbol{\lambda} \qquad (8\text{-}103)$$

and
$$m^* = \text{sgn}(\mathbf{b}'\boldsymbol{\lambda}^*) \tag{8-104}$$

The problem in solving 8-103 is that the end-point conditions on $\boldsymbol{\lambda}$ are unknown; for any given $\mathbf{x}(0)$, there must exist some $\boldsymbol{\lambda}(0)$ for which $\mathbf{x}^*(T)$ is the desired terminal state and T is minimum.

In this subsection, the approach to solution is based on the work of Lee [8.72] and Smith [8.102]. Attention is restricted to those systems of the form 8-101 in which the n eigenvalues, α_i's, associated with the $\mathbf{A}$ matrix are real and distinct. The eigenvalues of $-\mathbf{A}'$ are the negative of those of $\mathbf{A}$, and the λ_i^*'s that satisfy 8-103 are weighted sums of $e^{-\alpha_i t}$'s. It follows that $\mathbf{b}'\boldsymbol{\lambda}^*$ also is such a weighted sum,

$$\mathbf{b}'\boldsymbol{\lambda}^* = \sum_{i=1}^{n} c_i e^{-\alpha_i t} \tag{8-105}$$

where the c_i's are real and time-invariant but depend on $\mathbf{x}(0)$ and $\mathbf{x}^*(T)$. This weighted sum of real-valued exponentials can have at most $n - 1$ zeros which correspond to the switch times of m^* in Equation 8-104 [8.91, page 122].

To simplify the computational procedure, consider the particular *similarity transformation* (Appendix A) $\mathbf{SAS}^{-1}$ that converts $\mathbf{A}$ into the diagonal matrix $\mathbf{D}$:

$$\mathbf{D} = \mathbf{SAS}^{-1} = \begin{bmatrix} \alpha_1 & 0 & \cdots & 0 \\ 0 & \alpha_2 & & \vdots \\ \vdots & & \ddots & 0 \\ 0 & \cdots & 0 & \alpha_n \end{bmatrix} \tag{8-106}$$

Let $\mathbf{y} = \mathbf{Sx}$. Equation 8-101 then reduces to

$$\begin{aligned}\dot{\mathbf{y}} = \mathbf{S}\dot{\mathbf{x}} &= \mathbf{SAS}^{-1}\mathbf{Sx} + \mathbf{Sb}m \\ &= \mathbf{Dy} + \mathbf{w}m\end{aligned} \tag{8-107}$$

where $\mathbf{w}$ is defined to be the $n \times 1$ matrix $\mathbf{Sb}$. The solution of Equation 8-107 can be expressed as

$$\mathbf{y}(t) = e^{\mathbf{D}t}\mathbf{y}(0) + \int_0^t e^{\mathbf{D}(t-\tau)}\mathbf{w}m(\tau)\,d\tau, \qquad t \geq 0 \tag{8-108}$$

where, because of the diagonal form of $\mathbf{D}$,

$$e^{\mathbf{D}t} = \begin{bmatrix} e^{\alpha_1 t} & 0 & \cdots & 0 \\ 0 & e^{\alpha_2 t} & & \vdots \\ \vdots & & \ddots & 0 \\ 0 & \cdots & 0 & e^{\alpha_n t} \end{bmatrix}$$

In order for $y(T)$ to equal $Sx^*(T) = y^*(T)$, 8-108 gives

$$e^{-DT}y^*(T) - y(0) = \int_0^T e^{-D\tau}wm^*(\tau)\,d\tau \qquad (8\text{-}109)$$

It is known that $m^*(t)$ alternates in sign between plus and minus one at most $n-1$ times in the interval $[0,T)$. The initial value $m^*(0)$ may be plus or minus unity, depending on $x(0)$. To determine the correct initial value of m^*, both $+1$ and -1 are tested, and the one that yields T minimum is applied. Thus, consider an $m(t)$ of the form

$$m(t) = \begin{cases} \pm 1, & t_i \le t < t_{i+1}, \quad i = 0, 2, 4, \ldots \\ \mp 1, & t_j \le t < t_{j+1}, \quad j = 1, 3, \ldots \end{cases} \qquad (8\text{-}110)$$

where $t_0 = 0$, the reference initial time, and where the remaining t_i's are to be determined on the basis that $y(T) = y^*(T)$, minimum $T \le t_n$. If $x(0)$ is in the region of state space that requires $n-1$ switchings to reach the given $x^*(T)$ in minimum time, then $T = t_n$, but if $x(0)$ is in any other region, fewer than $n-1$ switchings are required.

Consider the case in which $n-1$ switchings are required in $[0,T)$. The combined use of 8-109 and 8-110 gives

$$e^{-DT}y^*(T) - y(0) = \pm \sum_{i=0}^{n-1} (-1)^i \int_{t_i}^{t_{i+1}} e^{-D\tau}w\,d\tau$$

$$= \pm \sum_{i=0}^{n-1} (-1)^i \begin{bmatrix} \dfrac{e^{-\alpha_1 t_{i+1}} - e^{-\alpha_1 t_i}}{-\alpha_1} & 0 & \cdots & 0 \\ 0 & & & \vdots \\ \vdots & & & \\ 0 & \cdots & & \dfrac{e^{-\alpha_n t_{i+1}} - e^{-\alpha_n t_i}}{-\alpha_n} \end{bmatrix} w$$

which reduces to

$$e^{-\alpha_k T}y_k^*(T) - y_k(0) = \pm \frac{w_k}{\alpha_k}[1 - 2e^{-\alpha_k t_1} + 2e^{-\alpha_k t_2} + \cdots + (-1)^n e^{-\alpha_k T}] \quad (8\text{-}111)$$

for $k = 1, 2, \ldots, n$. The unknowns in 8-111 are the t_i's and $T = t_n$. It is known [8.14] that the $n-1$ switch times to reach the desired state are unique, so 8-111 has a unique solution $t_1 < t_2 < \cdots < T$ corresponding to given $y(0)$ and $y^*(T)$. The equations can be solved by using an iterative linearization technique; the initial value t_i^0 of each t_i is guessed, and the $(j+1)$st iterates t_i^{j+1}'s of the t_i's are obtained from the jth iterates by replacing each $e^{-\alpha_k t_i}$ of 8-111 with the right-hand member of 8-112:

$$e^{-\alpha_k t_i^{j+1}} = e^{-\alpha_k t_i^j} e^{-\alpha_k(t_i^{j+1} - t_i^j)}$$
$$\cong e^{-\alpha_k t_i^j}[1 - \alpha_k(t_i^{j+1} - t_i^j)], \qquad j = 0, 1, \ldots \qquad (8\text{-}112)$$

which, as with Newton's procedure (Section 6-6d), is obtained on the basis of a truncated Taylor's series. The resulting approximations of 8-111 are linear algebraic equations in t_i^{j+1}. Solution procedures for the linear equations are well known. Conditions for stopping the search can be similar to those given in Chapter 6. Smith [8.102] reports that the only problem encountered in this approach is that the original ordering of the t_i^0's is not always preserved in later t_i^j's, especially when the t_i^0's are much smaller than the optimal t_i's.

In those cases where less than $n - 1$ switchings are required, Smith [8.102] has applied the same procedure as given in the preceding paragraph. He notes, however, that small limit cycles in the response of the system are usually generated as a result. Whether or not the procedure is feasible for on-line control depends on the computer facilities available and on the urgency of controlling the given process in a minimum-time sense. Smith considers special problems encountered in on-line control and gives illustrative examples in which the on-line minimum-time control is used.

8-8. SEARCH TECHNIQUES FOR SOLUTION OF BOUNDARY-VALUE PROBLEMS

8-8a. Comments

As noted in Chapter 6, there exists a multiplicity of search techniques, all with relative advantages and disadvantages, for use in the extremization of functions. The same can be said of search techniques that are applicable to the extremization of functionals. In fact, with the classical direct methods of the calculus of variations (Section 3-11), problems of extremization of functionals are approximated by problems of extremization of classical functions, in which case, techniques of Chapter 6 are directly applicable. It is apparent, however, that the information obtained from necessary conditions for optimality is essentially ignored with this "direct" approach. Such is not the case with the search techniques of this section.

On the basis of Section 8-4, the necessary conditions for optimality of the general problem of Section 8-3 include:

1. the state equations,
2. the costate equations,
3. the maximizing condition on $\mathcal{H}$, and
4. the transversality conditions.

While all of these conditions can be taken into account, the exact way in which they are used varies from method to method.

The methods that are most apparent are those in which search is conducted over unknown initial conditions on differential equations. The unknown initial conditions are implicit functions of certain known terminal conditions. With a first guess of unknown initial conditions, necessary conditions 1, 2, and 3 of the preceding paragraph lead to state and costate trajectories that generally miss required terminal values, but the information obtained may be used to adjust the initial conditions so as to decrease the degree of error obtained in subsequent trials. Kahne [3.14] outlines one such approach which is reported to work well for classical variational problems. But, as noted by Bellman [8.8], such methods invariably fail if the time span between initial and terminal times is relatively large. The reason for failure is that small computational errors are exponentially distorted in time when, as is normally the case, costate and/or state equations display unstable modes of response (see Problem 6.35).

The iterative procedure considered in Section 8-8b is one in which the maximizing condition on $\mathscr{H}$ is emphasized. The procedure is gradient-based; depending on the way in which a particular matrix is specified the method is equivalent to certain gradient methods of Bryson and Denham [8.17] or to a second-variation method of Kelley *et al.* [8.56]. The method is relatively simple, an advantage, and computational stability is typical, but rate of convergence often leaves much to be desired if the classical gradient procedure is used.

The method of Section 8-8c is called a generalized Newton-Raphson technique. Hestenes [8.45] conceived the basic idea of the method in 1949 when he called it the method of "differential variations." Bellman and Kalaba [8.9] have pursued the method and related methods under the title of *quasilinearization*. At the start of each iteration in this method, all pertinent necessary conditions are linearized about a current solution via a Newton-Raphson approach (Sections 6-3a and 6-6d), and superposition properties of linear differential equations are exploited to obtain a solution to the linearized equations; the solution is forced to satisfy required end conditions and can be used as the starting solution for the next iteration. Kenneth and Taylor [8.59] and Kenneth and McGill [8.58] have successfully extended the method to account for inequality constraints by using the slack-variable approach of Valentine [8.107]. Kopp and McGill [8.63] give a detailed comparison of the method with those of the preceding paragraph. While the complexity of the generalized Newton-Raphson search is greater, the rate of convergence is considerably better when the linearized state and costate trajectories are near optimum. Problems of computational stability can be encountered with the method however.

The methods of Sections 8-8d and 8-8e are iterative methods which entail repetitive solution of Riccati equations. With the approach of Noton *et al.* [8.82], the state equations are linearized via the Newton-Raphson scheme, and an approximate performance measure is formed by retaining all terms less than third-order in a Taylor's series expansion of the integrand of the given performance measure. The expansions of a given iteration are formed about the solution obtained as a result of the immediately preceding iteration. The approximate problem of each iteration is solved by use of Riccati equations that are similar to those obtained in Section 7-14. With an approach of Schley and Lee [8.100], linearization of the costate equations is used in place of series expansion in the performance measure, and an explicit Riccati transformation is found which relates the linearized state and costate solutions. In either case, the price of improved computational stability associated with the Riccati equations must be paid for by greater computer storage requirements.

The methods of this section make use of truncated Taylor's series and therefore might appear to be inapplicable when inequality constraints are to be taken into account. As noted previously, however, the slack variable approach of Valentine [8.107] (see also Section 3-8d) has been used with success in the generalized Newton-Raphson approach. Also, penalty functions can be used to account for all types of constraints in ways similar to those considered in Section 6-11. Pertinent work includes that of Fuller [8.42], Rothenberger and Lapidus [8.95], and Lasdon *et al.* [8.71].

The methods considered in this section are representative of those that have been and are being developed for determining optimal system response. They are generally unsuitable for closed-loop control because of their iterative nature, but they can be useful in determining plausible structure for suboptimal feedback control laws.

8-8b. Utilization of $\mathscr{H}$ in a Search Solution

Consider the canonical problem of Section 8-3, but with f and $\mathbf{q}$ assumed to be of class C^1 with respect to $\mathbf{m}$ and with the constraint $\mathbf{m} \in U$ incorporated in the integrand f of 8-3 through the use of differentiable penalty functions (see Section 6-11). The approach has a better chance of success if U is a convex region in r-dimensional Euclidean space. It is assumed here that the terminal time t_b is fixed but that the terminal state-variable values are free; it follows from the transversality condition 8-22 that the costate variables assume the value of zero at $t = t_b$. The Hamiltonian $\mathscr{H}$ and the costate equations are generated as before. A search approach that may lead to minimization of the performance measure under the preceding conditions is the following:

1. Judiciously, or otherwise, pick an initial control action $\mathbf{m}^0(t)$.
2. Let the index $k = 0$.
3. Using the known $\mathbf{m}^k$, integrate the state equations 8-19 forward in time and store the resulting $\mathbf{x}^k(t)$ state solution.

4. Using $\lambda_i(t_b) = 0$, $i = 1, 2, \ldots, n$, and having both $\mathbf{x}^k(t)$ and $\mathbf{m}^k(t)$ available, integrate the costate equations 8-20 backward in time.
5. At each step of the reverse-time costate integration, the estimate of $\mathbf{m}^*(t)$ is to be improved according to the rule

$$\mathbf{m}^{k+1}(t) = \mathbf{m}^k(t) + \alpha \mathbf{H} \frac{\partial \mathscr{H}}{\partial \mathbf{m}}\bigg|_{\mathbf{m}^k, \mathbf{x}^k, \lambda^k} \quad (8\text{-}113)$$

where α is a real number at each instant t, and $\mathbf{H}$ is an $r \times r$ direction-weighting matrix. As in general gradient search (Section 6-6a), both α and $\mathbf{H}$ can be selected in many ways. With classical gradient search, an $r \times r$ identity matrix is used for $\mathbf{H}$, but this is seldom the most efficient approach.
6. Test the degree of constancy of $\mathscr{H}^k(t)$ as a function of time, where

$$\mathscr{H}^k(t) = \mathscr{H}[\mathbf{x}^k(t), \lambda^k(t), \mathbf{m}^k(t)]$$

Check the maximum value of $[(\partial \mathscr{H}^k/\partial \mathbf{m})'(\partial \mathscr{H}^k/\partial \mathbf{m})]^{1/2}$ over time, and check to determine if k is less than or equal to N, a fixed upper bound on the number of iterations allowed. If tests for stopping are not satisfied, update k, $k + 1 \to k$, and return to step 3.

The whole of the procedure is well suited for programming on a general-purpose digital computer. The numerical integration in step 3 is easily performed if the state equations are stable. Stability of the state equations generally results in the costate equations being stable in reverse time, so it is typical for the integration in step 4 to be computationally feasible. In step 5, the objective is to adjust $\mathbf{m}$ to increase the value of $\mathscr{H}$ along the specific state and costate trajectories knowing that $\mathscr{H}$ is maximized with respect to $\mathbf{m}$ along the optimal state and costate trajectories. In step 6, the difference between the maximum and minimum values of $\mathscr{H}^k(t)$ is compared to some predetermined error bound; and the magnitude of the gradient of $\mathscr{H}$ is similarly tested. To complete all of the steps, the entire costate trajectory $\lambda^k(t)$ need not be stored; at any instant $t_1 \in [t_a, t_b]$ in the reverse-time integration, $\lambda^k(t_1)$ is used in the formation of $\mathscr{H}^k(t_1)$, $\mathbf{m}^{k+1}(t_1)$, and $\partial \mathscr{H}^k/\partial \mathbf{m}$ at time t_1, and in the formation of $\lambda^k(t_1 - \epsilon)$ for one or more small values of $\epsilon > 0$, but after this use, it can be discarded.

The method of Newton (Section 6-6d) is a second-variation method that can be applied to the maximization of $\mathscr{H}$. With this method, $\mathbf{H}$ is specified as the inverse matrix of the $r \times r$ matrix of second partial derivatives $\partial^2 \mathscr{H}/\partial m_i \, \partial m_j$, and α is equated to -1. The matrix $\mathbf{H} = \mathbf{H}(t)$ should be negative-definite at each $t \in [t_a, t_b]$ for convergence to a maximum. If the initial choice of $\mathbf{m}^0(t)$ is poor, the method may fail. But if $\mathbf{m}^0(t)$ is "close" to the optimum, convergence is much better than that of classical gradient search.

Sect. 8-8 SOLUTION OF BOUNDARY-VALUE PROBLEMS 517

Certain deficiencies of this second-variation method can be overcome by using other quadratic-convergent search techniques of Chapter 6.

8-8c. A Newton-Raphson Algorithm for Linearization of Differential Equations and Solution of Two-Point Boundary-Value Problems

Consider the set of equations

$$\begin{bmatrix} \dot{\mathbf{x}} \\ \dot{\boldsymbol{\lambda}} \end{bmatrix} = \begin{bmatrix} \partial \mathcal{H}/\partial \boldsymbol{\lambda} \\ -\partial \mathcal{H}/\partial \mathbf{x} \end{bmatrix}, \qquad t \in [t_a, t_b] \tag{8-114}$$

and

$$0 = \frac{\partial \mathcal{H}}{\partial \mathbf{m}}, \qquad t \in [t_a, t_b] \tag{8-115}$$

These equations apply to the canonical problem of Sections 8-3 and 8-4 under the condition that $\mathcal{H}$ is of class C^1 with respect to $\mathbf{m}$ (see also Section 3-9) and that the constraint $\mathbf{m} \in U$ is replaced by a differentiable penalty term in $f(\mathbf{x},\mathbf{m})$. Actually, for purposes of the development that follows, $\mathcal{H}$ is assumed to be of class C^2 with respect to all of its arguments. Of the boundary values for 8-114, $2n$ are assumed known, some at t_a and others at t_b. Both t_a and t_b are assumed fixed in this introductory development.

Suppose that some $\mathbf{x}^k$, $\boldsymbol{\lambda}^k$, and $\mathbf{m}^k$ are known functions of t which *do not* satisfy 8-114 and 8-115, but which hopefully are close to the optimal values of $\mathbf{x}$, $\boldsymbol{\lambda}$, and $\mathbf{m}$, respectively. A Taylor's series of the right-hand members of 8-114 and 8-115 about the known values $\mathbf{x}^k$, $\boldsymbol{\lambda}^k$, and $\mathbf{m}^k$ is proposed:

$$\begin{bmatrix} \dot{\mathbf{x}} \\ \dot{\boldsymbol{\lambda}} \end{bmatrix} = \begin{bmatrix} \dfrac{\partial \mathcal{H}^k}{\partial \boldsymbol{\lambda}} \\ -\dfrac{\partial \mathcal{H}^k}{\partial \mathbf{x}} \end{bmatrix}$$

$$+ \begin{bmatrix} \dfrac{\partial^2 \mathcal{H}^k}{\partial \mathbf{x}\, \partial \boldsymbol{\lambda}}(\mathbf{x} - \mathbf{x}^k) + \dfrac{\partial^2 \mathcal{H}^k}{\partial \boldsymbol{\lambda}^2}(\boldsymbol{\lambda} - \boldsymbol{\lambda}^k) + \dfrac{\partial^2 \mathcal{H}^k}{\partial \mathbf{m}\, \partial \boldsymbol{\lambda}}(\mathbf{m} - \mathbf{m}^k) \\ -\dfrac{\partial^2 \mathcal{H}^k}{\partial \mathbf{x}^2}(\mathbf{x} - \mathbf{x}^k) - \dfrac{\partial^2 \mathcal{H}^k}{\partial \boldsymbol{\lambda}\, \partial \mathbf{x}}(\boldsymbol{\lambda} - \boldsymbol{\lambda}^k) - \dfrac{\partial^2 \mathcal{H}^k}{\partial \mathbf{m}\, \partial \mathbf{x}}(\mathbf{m} - \mathbf{m}^k) \end{bmatrix} + \cdots$$

$$(8\text{-}116)$$

and

$$0 = \frac{\partial \mathcal{H}^k}{\partial \mathbf{m}} + \frac{\partial^2 \mathcal{H}^k}{\partial \mathbf{x}\, \partial \mathbf{m}}(\mathbf{x} - \mathbf{x}^k) + \frac{\partial^2 \mathcal{H}^k}{\partial \boldsymbol{\lambda}\, \partial \mathbf{m}}(\boldsymbol{\lambda} - \boldsymbol{\lambda}^k) + \frac{\partial^2 \mathcal{H}^k}{\partial \mathbf{m}^2}(\mathbf{m} - \mathbf{m}^k) + \cdots \tag{8-117}$$

where

$$\frac{\partial \mathcal{H}^k}{\partial \lambda} \triangleq \frac{\partial \mathcal{H}}{\partial \lambda}\bigg|_{\mathbf{x}=\mathbf{x}^k, \lambda=\lambda^k, \mathbf{m}=\mathbf{m}^k} \tag{8-118}$$

and

$$\frac{\partial^2 \mathcal{H}^k}{\partial \mathbf{x}\, \partial \mathbf{m}} \triangleq \begin{bmatrix} \dfrac{\partial^2 \mathcal{H}}{\partial x_1\, \partial m_1} & \dfrac{\partial^2 \mathcal{H}}{\partial x_2\, \partial m_1} & \cdots & \dfrac{\partial^2 \mathcal{H}}{\partial x_n\, \partial m_1} \\ \vdots & & & \vdots \\ \dfrac{\partial^2 \mathcal{H}}{\partial x_1\, \partial m_r} & & \cdots & \dfrac{\partial^2 \mathcal{H}}{\partial x_n\, \partial m_r} \end{bmatrix}_{\mathbf{x}^k, \lambda^k, \mathbf{m}^k} \tag{8-119}$$

Other partial derivatives of $\mathcal{H}$ are analogously defined.

If Equations 8-116 and 8-117 are truncated after the linear terms in $\mathbf{x}$, λ, and $\mathbf{m}$, as is done in ordinary Newton-Raphson search for solutions of equations (Sections 6-3a and 6-6d), the resulting linear, but time-varying, differential equations define a set of solutions $\mathbf{x}^{k+1}$, λ^{k+1}, and $\mathbf{m}^{k+1}$:

$$\begin{bmatrix} \dot{\mathbf{x}}^{k+1} \\ \dot{\lambda}^{k+1} \end{bmatrix} = \begin{bmatrix} \dfrac{\partial \mathcal{H}^k}{\partial \lambda} \\ -\dfrac{\partial \mathcal{H}^k}{\partial \mathbf{x}} \end{bmatrix} + \begin{bmatrix} \dfrac{\partial^2 \mathcal{H}^k}{\partial \mathbf{x}\, \partial \lambda}(\mathbf{x}^{k+1}-\mathbf{x}^k) + \dfrac{\partial^2 \mathcal{H}^k}{\partial \mathbf{m}\, \partial \lambda}(\mathbf{m}^{k+1}-\mathbf{m}^k) \\ -\dfrac{\partial^2 \mathcal{H}^k}{\partial \mathbf{x}^2}(\mathbf{x}^{k+1}-\mathbf{x}^k) - \dfrac{\partial^2 \mathcal{H}^k}{\partial \lambda\, \partial \mathbf{x}}(\lambda^{k+1}-\lambda^k) - \dfrac{\partial^2 \mathcal{H}^k}{\partial \mathbf{m}\, \partial \mathbf{x}}(\mathbf{m}^{k+1}-\mathbf{m}^k) \end{bmatrix} \tag{8-120}$$

and

$$0 = \frac{\partial \mathcal{H}^k}{\partial \mathbf{m}} + \frac{\partial^2 \mathcal{H}^k}{\partial \mathbf{x}\, \partial \mathbf{m}}(\mathbf{x}^{k+1} - \mathbf{x}^k) + \frac{\partial^2 \mathcal{H}^k}{\partial \lambda\, \partial \mathbf{m}}(\lambda^{k+1} - \lambda^k) + \frac{\partial^2 \mathcal{H}^k}{\partial \mathbf{m}^2}(\mathbf{m}^{k+1} - \mathbf{m}^k) \tag{8-121}$$

In 8-120, the $\partial^2 \mathcal{H}^k/\partial \lambda^2$ term is omitted because $\mathcal{H}$ is linear in λ. The set of linear algebraic equations in 8-121 can generally be solved at any $t \in [t_a, t_b]$ for $\mathbf{m}^{k+1}$ in terms of $\mathbf{x}^{k+1}$, λ^{k+1}, and t; the solution obtained is used to eliminate $\mathbf{m}^{k+1}(t)$ from the linear differential equations in 8-120. In turn, the boundary conditions on the linear equations in 8-120 may be satisfied by use of superposition. For example, suppose the n state variables are assigned specific values at $t = t_a$. The total solution is to be formed by use

Sect. 8-8   SOLUTION OF BOUNDARY-VALUE PROBLEMS   519

of a weighted sum of solutions, each such solution having an initial-condition set which is orthogonal to the other initial-condition sets. Let $\mathbf{x}^{k+1,0}$, $\boldsymbol{\lambda}^{k+1,0}$ denote the solution of 8-120 that is obtained by setting the initial conditions on the costate variables to zero and using the given initial values of the state variables. Similarly, let $\mathbf{x}^{k+1,i}$, $\boldsymbol{\lambda}^{k+1,i}$ denote the solution of 8-120 that is obtained by setting the ith costate variable to unity at $t = t_a$ and by assigning the zero value to all other initial state and costate variables. Solutions $\mathbf{x}^{k+1,i}$, $\boldsymbol{\lambda}^{k+1,i}$ are generated for $i = 1, 2, \ldots, n$. The desired solution $\mathbf{x}^{k+1}$, $\boldsymbol{\lambda}^{k+1}$ is the one that satisfies all end conditions, and, because superposition applies in the linear equation case,

$$\begin{bmatrix} \mathbf{x}^{k+1} \\ \boldsymbol{\lambda}^{k+1} \end{bmatrix} = \begin{bmatrix} \mathbf{x}^{k+1,0} + \sum_{i=1}^{n} \alpha_i \mathbf{x}^{k+1,i} \\ \boldsymbol{\lambda}^{k+1,0} + \sum_{i=1}^{n} \alpha_i \boldsymbol{\lambda}^{k+1,i} \end{bmatrix} \qquad (8\text{-}122)$$

where the n α_i's are determined on the basis that n terminal conditions at time t_b are satisfied (e.g., if $\mathbf{x}(t_b)$ is fixed, the n α_i's are adjusted so that $\mathbf{x}^{k+1}(t_b)$ equals the given vector; but if $\mathbf{x}(t_b)$ is free, the transversality condition 8-22 shows that $\boldsymbol{\lambda}(t_b)$ is a zero vector, in which case the criterion for selection of the α_i's is that $\boldsymbol{\lambda}^{k+1}(t_b)$ equals a zero vector).

Although $n + 1$ integrations of 8-120 must be effected in this solution procedure, only the end-point values of the solutions need be stored in order to generate the α_i's. After the α_i's are computed, all proper initial conditions are known, and the whole of the solution of 8-120 can be effected once more to obtain the whole of $\mathbf{x}^{k+1}(t)$, $\boldsymbol{\lambda}^{k+1}(t)$. As noted by Kenneth and Taylor [8.59], this last solution affords a check on the numerical stability of the previous integrations. If the time interval $t_b - t_a$ is large, numerical instability of computer solutions is likely, and only under special conditions can convergence to the optimal solution be guaranteed [8.54, 8.76]. When $\mathbf{x}^k$, $\boldsymbol{\lambda}^k$, and $\mathbf{m}^k$ are close to optimal values, the convergence to the optimum tends to be good; for quadratic cases such as that of Section 7-14, initial zero vectors for $\mathbf{x}^0$, $\boldsymbol{\lambda}^0$, and $\mathbf{m}^0$ give rise to the optimal solution in one iteration, provided numerical stability is maintained.

8-8d. Iterative Solutions with Stabilization via Riccati Equations

Again consider the problem of Section 8-3, but with $f(\mathbf{x},\mathbf{m})$ assumed to be of class C^2 with respect to all arguments, and $\mathbf{q}(\mathbf{x},\mathbf{m})$ assumed to be of class C^1 with respect to all arguments. As in the preceding subsection, suppose that a Newton-Raphson (truncated Taylor's series) approach is used on the state equation:

$$\dot{\mathbf{x}}^{k+1} \triangleq \mathbf{q}^k + \frac{\partial \mathbf{q}^k}{\partial \mathbf{x}}(\mathbf{x}^{k+1} - \mathbf{x}^k) + \frac{\partial \mathbf{q}^k}{\partial \mathbf{m}}(\mathbf{m}^{k+1} - \mathbf{m}^k) \qquad (8\text{-}123)$$

in which

$$\mathbf{q}^k \triangleq \mathbf{q}(\mathbf{x}^k, \mathbf{m}^k) \qquad (8\text{-}124)$$

and

$$\frac{\partial \mathbf{q}^k}{\partial \mathbf{m}} = \begin{bmatrix} \frac{\partial q_1}{\partial m_1} & \frac{\partial q_1}{\partial m_2} & \cdots & \frac{\partial q_1}{\partial m_r} \\ \vdots & & & \vdots \\ \frac{\partial q_n}{\partial m_1} & \cdots & & \frac{\partial q_n}{\partial m_r} \end{bmatrix}_{\mathbf{x}=\mathbf{x}^k, \mathbf{m}=\mathbf{m}^k} \qquad (8\text{-}125)$$

where $\mathbf{x}^k$ and $\mathbf{m}^k$ are given (or previously computed) suboptimal solutions in the interval $[t_a, t_b]$.

Rather than use the costate equations at this point, let the integrand of the functional J (8-3) be expanded in a truncated Taylor's series, including second-order terms, about $\mathbf{x}^k$ and $\mathbf{m}^k$:

$$\begin{aligned} J^{k+1} \triangleq \int_{t_a}^{t_b} \bigg[& f^k + \left(\frac{\partial f^k}{\partial \mathbf{x}}\right)'(\mathbf{x}^{k+1} - \mathbf{x}^k) + \left(\frac{\partial f^k}{\partial \mathbf{m}}\right)'(\mathbf{m}^{k+1} - \mathbf{m}^k) \\ & + \tfrac{1}{2}(\mathbf{x}^{k+1} - \mathbf{x}^k)' \frac{\partial^2 f^k}{\partial \mathbf{x}^2}(\mathbf{x}^{k+1} - \mathbf{x}^k) \\ & + (\mathbf{x}^{k+1} - \mathbf{x}^k)' \frac{\partial^2 f^k}{\partial \mathbf{m}\, \partial \mathbf{x}}(\mathbf{m}^{k+1} - \mathbf{m}^k) \\ & + \tfrac{1}{2}(\mathbf{m}^{k+1} - \mathbf{m}^k)' \frac{\partial^2 f^k}{\partial \mathbf{m}^2}(\mathbf{m}^{k+1} - \mathbf{m}^k) \bigg] dt \quad (8\text{-}126) \end{aligned}$$

in which

$$\frac{\partial f^k}{\partial \mathbf{m}} \triangleq \frac{\partial f(\mathbf{x},\mathbf{m})}{\partial \mathbf{m}}\bigg|_{\mathbf{x}=\mathbf{x}^k, \mathbf{m}=\mathbf{m}^k} \qquad (8\text{-}127)$$

and

$$\frac{\partial^2 f^k}{\partial \mathbf{m}\, \partial \mathbf{x}} \triangleq \begin{bmatrix} \frac{\partial^2 f}{\partial m_1\, \partial x_1} & \cdots & \frac{\partial^2 f}{\partial m_r\, \partial x_1} \\ \vdots & & \vdots \\ \frac{\partial^2 f}{\partial m_1\, \partial x_n} & \cdots & \frac{\partial^2 f}{\partial m_r\, \partial x_n} \end{bmatrix}_{\mathbf{x}=\mathbf{x}^k, \mathbf{m}=\mathbf{m}^k} \qquad (8\text{-}128)$$

Sect. 8-8   SOLUTION OF BOUNDARY-VALUE PROBLEMS   521

State equation 8-123 contains n linear differential equations in $\mathbf{x}^{k+1}$ and $\mathbf{m}^{k+1}$. The integrand of the performance measure J^{k+1} (8-126) is quadratic in $\mathbf{x}^{k+1}$ and $\mathbf{m}^{k+1}$. Thus, if the terminal values of the state variables are not specified in advance, the Riccati equation approach of Section 7-14a may be applied to find the values of $\mathbf{m}^{k+1}$ and $\mathbf{x}^{k+1}$ which yield a minimum of J^{k+1} subject to satisfaction of 8-123. Actually, a slight extension (Problem 8.22) of the approach in Section 7-14a is required here because of additional terms contained in 8-123 and 8-126. It is assumed that the penalty functions on $\mathbf{m}$ are sufficient to guarantee the existence of a minimum of J^{k+1}. After the appropriate $\mathbf{m}^{k+1}$ and $\mathbf{x}^{k+1}$ are obtained, the value of k can be updated, $k + 1 \to k$; and the procedure can be repeated as necessary to obtain a solution, but perhaps not *the* solution to the original problem. The advantage of this approach over that of the preceding subsection is that solutions of Riccati equations tend to be numerically stable for the type of minimization problem encountered in problems of optimal control.

8-8e. A Riccati Transformation

A variation of the preceding approach is considered in this subsection. A more general approach is given by Schley and Lee [8.100], who include consideration of end-point functionals. Consider Equations 8-114 and 8-115, and suppose that Equation 8-115 can be solved for $\mathbf{m}$ as a function of $\mathbf{x}$, λ, and t. In actuality, we might settle for an approximate solution, for example, the approximate solution obtained by using 8-121. With $\mathbf{m}$ known as a function of $\mathbf{x}$, λ, and t, 8-114 can be reduced to the form

$$\dot{\mathbf{x}} = \hat{\mathbf{q}}(\mathbf{x},\lambda,t) \qquad (8\text{-}129\text{a})$$

and

$$\dot{\lambda} = -\xi(\mathbf{x},\lambda,t) \qquad (8\text{-}129\text{b})$$

where $\hat{\mathbf{q}}$ and ξ are formed from $\partial \mathscr{H}/\partial \lambda$ and $\partial \mathscr{H}/\partial \mathbf{x}$, respectively, in the obvious way. With some suboptimal functions $\mathbf{x}^k$ and λ^k known over $[t_a, t_b]$, Newton-Raphson linearization can be applied to 8-129 to obtain

$$\dot{\mathbf{x}}^{k+1} = \hat{\mathbf{q}}^k + \left(\frac{\partial \hat{\mathbf{q}}^k}{\partial \mathbf{x}}\right)(\mathbf{x}^{k+1} - \mathbf{x}^k) + \left(\frac{\partial \hat{\mathbf{q}}^k}{\partial \lambda}\right)(\lambda^{k+1} - \lambda^k) \qquad (8\text{-}130\text{a})$$

$$\dot{\lambda}^{k+1} = -\xi^k - \left(\frac{\partial \xi^k}{\partial \mathbf{x}}\right)(\mathbf{x}^{k+1} - \mathbf{x}^k) - \left(\frac{\partial \xi^k}{\partial \lambda}\right)(\lambda^{k+1} - \lambda^k) \qquad (8\text{-}130\text{b})$$

If $\mathbf{x}(t_a)$ is given but $\mathbf{x}(t_b)$ is free, $\lambda(t_b)$ is a zero vector. In such a case, we would like to solve 8-130a forward in time and 8-130b backward in time.

A procedure with which 8-130a and 8-130b can be decoupled is based on the following Riccati transformation. Assume

$$\lambda^{k+1} = \Psi^{k+1} \mathbf{x}^{k+1} + \mathbf{k}^{k+1} \quad (8\text{-}131)$$

where Ψ^{k+1} is an $n \times n$ matrix function of time to be determined and $\mathbf{k}^{k+1}$ is an $n \times 1$ matrix function of time to be determined. Because $\lambda^{k+1}(t_b)$ equals a zero vector and $\mathbf{x}^{k+1}(t_b)$ is free, both $\mathbf{k}^{k+1}(t_b)$ and $\Psi^{k+1}(t_b)$ are matrices of zeros.

Consider the result obtained by differentiating 8-131:

$$\dot{\lambda}^{k+1} = \dot{\Psi}^{k+1} \mathbf{x}^{k+1} + \Psi^{k+1} \dot{\mathbf{x}}^{k+1} + \dot{\mathbf{k}}^{k+1} \quad (8\text{-}132)$$

In 8-132, $\dot{\lambda}^{k+1}$ can be replaced by the right-hand member of 8-130b; $\dot{\mathbf{x}}^{k+1}$ can be replaced by the right-hand member of 8-130a; and λ^{k+1} can be replaced, everywhere it appears, by the right-hand member of 8-131. The net result of these operations is

$$\left[\dot{\Psi}^{k+1} + \Psi^{k+1} \frac{\partial \hat{\mathbf{q}}^k}{\partial \lambda} \Psi^{k+1} + \frac{\partial \boldsymbol{\xi}^k}{\partial \lambda} \Psi^{k+1} + \Psi^{k+1} \frac{\partial \hat{\mathbf{q}}^k}{\partial \mathbf{x}} + \frac{\partial \boldsymbol{\xi}^k}{\partial \mathbf{x}} \right] \mathbf{x}^{k+1}$$

$$+ \left[\dot{\mathbf{k}}^{k+1} + \Psi^{k+1} \frac{\partial \hat{\mathbf{q}}^k}{\partial \lambda} \mathbf{k}^{k+1} + \frac{\partial \boldsymbol{\xi}^k}{\partial \lambda} \mathbf{k}^{k+1} + \Psi^{k+1} \boldsymbol{\omega}^k + \boldsymbol{\upsilon}^k \right] = 0 \quad (8\text{-}133)$$

in which

$$\boldsymbol{\omega}^k \triangleq \hat{\mathbf{q}}^k - \frac{\partial \hat{\mathbf{q}}^k}{\partial \mathbf{x}} \mathbf{x}^k - \frac{\partial \hat{\mathbf{q}}^k}{\partial \lambda} \lambda^k \quad (8\text{-}134)$$

and

$$\boldsymbol{\upsilon}^k \triangleq \boldsymbol{\xi}^k - \frac{\partial \boldsymbol{\xi}^k}{\partial \mathbf{x}} \mathbf{x}^k - \frac{\partial \boldsymbol{\xi}^k}{\partial \lambda} \lambda^k \quad (8\text{-}135)$$

Equation 8-133 is to hold for any initial state $\mathbf{x}(t_a)$ and for any $t \in [t_a, t_b]$; but this can be true only if the terms within brackets sum to zero. Thus,

$$-\dot{\Psi}^{k+1} = \Psi^{k+1} \frac{\partial \hat{\mathbf{q}}^k}{\partial \lambda} \Psi^{k+1} + \frac{\partial \boldsymbol{\xi}^k}{\partial \lambda} \Psi^{k+1} + \Psi^{k+1} \frac{\partial \hat{\mathbf{q}}^k}{\partial \mathbf{x}} + \frac{\partial \boldsymbol{\xi}^k}{\partial \mathbf{x}} \quad (8\text{-}136)$$

with terminal end-point condition $\Psi^{k+1}(t_b)$ equal to an $n \times n$ matrix of zeros; and

$$-\dot{\mathbf{k}}^{k+1} = \Psi^{k+1} \frac{\partial \hat{\mathbf{q}}^k}{\partial \lambda} \mathbf{k}^{k+1} + \frac{\partial \boldsymbol{\xi}^k}{\partial \lambda} \mathbf{k}^{k+1} + \Psi^{k+1} \boldsymbol{\omega}^k + \boldsymbol{\upsilon}^k \quad (8\text{-}137)$$

with terminal end-point condition $\mathbf{k}^{k+1}(t_b)$ equal to an $n \times 1$ matrix of zeros.

The procedure then is to solve the generalized Riccati equation 8-136 in reverse time. The solution $\Psi^{k+1}(t)$ is used in 8-137 to enable solution for

$\mathbf{k}^{k+1}(t)$. Both $\mathbf{\Psi}^{k+1}(t)$ and $\mathbf{k}^{k+1}(t)$ are used in 8-131 to determine $\boldsymbol{\lambda}^{k+1}(t)$ as a function of $\mathbf{x}^{k+1}(t)$, which relationship is used in Equation 8-129a to obtain a differential equation in $\mathbf{x}^{k+1}(t)$ that can be solved as an initial-value problem. Knowing $\mathbf{x}^{k+1}(t)$ and $\boldsymbol{\lambda}^{k+1}(t)$, we can determine $\mathbf{m}^{k+1}(t)$ by using the relationship assumed at the start of this subsection. As in the preceding subsection, the procedure is applied repetitively a specified number of times or until some measure of improvement from one iteration to the next is less than a specified value. For example, the measure of improvement could be

$$\max_{\substack{i \\ t \in [t_a, t_b]}} |x_i^{k+1} - x_i^k| \leq \epsilon \quad \text{(a specified constant)} \quad (8\text{-}138)$$

which, when satisfied, determines the terminal value of $k + 1$. The stabilization of the numerical solution via the Riccati approach is obtained at the expense of an obvious increase in the computer storage space that is required.

8-9. NON-NORMAL SOLUTIONS[8]

In the canonical problem of Section 8-3, let the variable x_0 be introduced and defined by

$$\dot{x}_0 = f(\mathbf{x}, \mathbf{m}), \quad t_a \leq t \leq t_b \quad \text{and} \quad x_0(t_a) \triangleq 0 \quad (8\text{-}139)$$

where the objective of minimizing $x_0(t_b)$ is equivalent to that of minimizing J (8-3). With the original vector state equation 8-4 augmented by 8-139, the Hamiltonian $\mathscr{H}$ of 8-18 assumes the modified form

$$\mathscr{H} = (-1 + \lambda_0) f(\mathbf{x}, \mathbf{m}) + \sum_{i=1}^{n} \lambda_i q_i(\mathbf{x}, \mathbf{m}) \quad (8\text{-}140)$$

where

$$\dot{\lambda}_0 = -\frac{\partial \mathscr{H}}{\partial x_0} = 0 \quad (8\text{-}141)$$

in which case, λ_0 is a constant, and 8-140 reduces to

$$\mathscr{H} = h_0 f(\mathbf{x}, \mathbf{m}) + \sum_{i=1}^{n} \lambda_i q_i(\mathbf{x}, \mathbf{m}) \quad (8\text{-}142)$$

where h_0 is a constant. The fact that $h_0 = -1$ can normally be used in the solution of the canonical problem is assured by the dynamic programming development of Section 7-13. There exist non-normal cases, however, in which the optimal state trajectory has no neighboring trajectories that satisfy the

[8] In some literature (e.g., [8.69, 8.70]), the term non-normal has been used in a different context to describe singular solutions which are considered in Section 8-10.

constraints of the problem; the recurrence relations of Section 7-13 are developed under the tacit assumption that such neighboring solutions to the state equations exist.

The solution of the canonical problem of Section 8-3 is said to be *normal* if there are no sets of multipliers $(h_0, \lambda_1, \lambda_2, \ldots, \lambda_n)$ with h_0 equal to zero. If the solution is normal, the Hamiltonian can be scaled to have h_0 equal -1. Bliss [8.13] gives necessary and sufficient conditions for normality of classical variational problems, and Berkovitz [8.11] develops sufficient (but not necessary) conditions for normality of a general control problem; unfortunately, the conditions are seldom easy to apply.

Example 8-5. To illustrate non-normality, consider an elementary scalar case:

$$\dot{x}_1 = m_1, \qquad |m_1| \leq 1 + \epsilon \tag{8-143}$$

where ϵ is a real parameter. Assumed conditions are listed as follows:

$$t_a = 0, \quad \text{and} \quad x_1(0) = 0 \tag{8-144}$$

$$x_1(t_b) = \phi(t)|_{t=t_b} \triangleq \frac{t_b^2 + 1}{2} \tag{8-145}$$

and

$$J = \int_0^{t_b} dt = t_b \tag{8-146}$$

In this case, $\mathscr{H}$ of 8-142 assumes the form

$$\mathscr{H} = h_0 + \lambda_1 m_1 \tag{8-147}$$

The costate equation is

$$\dot{\lambda}_1 = 0 \tag{8-148}$$

and therefore,

$$\lambda_1^* = c_1 \quad \text{(a constant)} \tag{8-149}$$

From the maximizing condition on $\mathscr{H}$,

$$m_1^*(t) = (1 + \epsilon) \operatorname{sgn} \lambda_1^* \tag{8-150}$$

provided that $\lambda_1^* \neq 0$. On the basis of Equations 8-143 and 8-150,

$$\dot{x}_1^* = (1 + \epsilon) \operatorname{sgn} \lambda_1^*, \qquad x_1^*(0) = 0 \tag{8-151}$$

and therefore

$$x_1^* = (1 + \epsilon) t \operatorname{sgn} \lambda_1^* \tag{8-152}$$

In order for $x_1^*(t_b)$ to satisfy 8-145, it is clear that λ_1^* must be positive; thus,

$$x_1^* = (1 + \epsilon) t \tag{8-153}$$

Because $x_1(t)$ is required to terminate on the given curve $\phi(t)$, a minute terminal variation of x_1 is related to a corresponding terminal variation of t_b:

$$\delta x_1|_{t=t_b} = \frac{\partial \phi}{\partial t}\bigg|_{t=t_b} \delta t_b = t_b\, \delta t_b \qquad (8\text{-}154)$$

This result is used in the transversality condition:

$$0 = [-\mathscr{H}^* \,\delta t + \lambda_1^* \,\delta x_1]_{t=t_b}$$
$$= [-h_0^* - \lambda_1^* m_1^* + \lambda_1^* t_b]\,\delta t_b$$

which, on the basis of 8-149 and 8-150, reduces to

$$(t_b - 1 - \epsilon)c_1 = h_0^* \qquad (8\text{-}155)$$

At the terminal time t_b, Equations 8-145 and 8-153 give

$$(1 + \epsilon)t_b = \frac{t_b^2 + 1}{2}$$

which is solved for the minimum time t_b:

$$t_b = (1 + \epsilon) - (2\epsilon + \epsilon^2)^{1/2} \qquad (8\text{-}156)$$

Substitution of the right-hand member of 8-156 for t_b in 8-155 gives

$$-(2\epsilon + \epsilon^2)^{1/2} c_1 = h_0^* \qquad (8\text{-}157)$$

If $\epsilon > 0$, a normal case prevails, and h_0^* can be assigned the value of -1. If $-2 < \epsilon < 0$, Equation 8-157 yields imaginary values—no solution exists that satisfies the constraints imposed. Also, for $\epsilon < -1$, the magnitude constraint on m_1 cannot be satisfied. But if $\epsilon = 0$, the non-normal case prevails; h_0^* must be zero for satisfaction of 8-157 with $\epsilon = 0$, and it should be noted that the *only* function $m(t)$, $|m(t)| \leq 1$, that results in satisfaction of the end conditions imposed on $x(t)$ is $m(t) = m^*(t)$ which equals essentially 1 throughout the time interval $[0, t_b]$.

8-10. SINGULAR SOLUTIONS[9]

It can occur that state and costate trajectories which satisfy 8-19, 8-20, and 8-21 are such that, over a measurable period of time $[t_1, t_2]$ the Hamiltonian $\mathscr{H}$ is independent of some control action $m_i(t)$. In the context of Pontryagin's maximum principle, a *singular arc* is a valid state and costate trajectory along which $\mathscr{H}$ is independent of one or more of the components of **m**. A *singular solution* is a solution that satisfies the necessary conditions and that

---

[9] As noted in Section 8-9, the use of "non-normal solution" in the place of "singular solution" has persisted in the literature to a limited extent.

contains one or more singular arcs. Note that the maximizing condition on $\mathcal{H}$ cannot be applied in the usual way along a singular arc; this rightly suggests that additional conditions are required to determine singular solutions.

A typical but far from universal form of $\mathcal{H}$ which may result in singular arcs is the following:

$$\mathcal{H} = \mathcal{H}_0(\mathbf{x},\boldsymbol{\lambda}) + \phi(\mathbf{x},\boldsymbol{\lambda})\mathcal{H}_1(\mathbf{x},\boldsymbol{\lambda},\mathbf{m}) \tag{8-158}$$

where $\mathcal{H}_0$, ϕ, and $\mathcal{H}_1$ are scalar-valued functions. The function $\mathcal{H}_1$ is assumed to be dependent on $\mathbf{m}$ for all possible combinations of $\mathbf{x}$ and $\boldsymbol{\lambda}$ which satisfy 8-19 and 8-20. Suppose that a solution of the necessary conditions exists with the property that

$$\phi(\mathbf{x},\boldsymbol{\lambda}) \equiv 0, \qquad t \in [t_1, t_2] \tag{8-159a}$$

But in order for 8-159a to hold, it must also be true that

$$0 = \dot{\phi}(\mathbf{x},\boldsymbol{\lambda}) = \ddot{\phi}(\mathbf{x},\boldsymbol{\lambda}) = \cdots, \qquad t \in (t_1, t_2) \tag{8-159b}$$

In the $2n$-dimensional space of $X \times \Lambda$, the equation $\phi(\mathbf{x},\boldsymbol{\lambda}) = 0$ defines a hypersurface. Over some portions of this surface, it may be possible for 8-159 to be satisfied by singular arcs; in which case, those portions are called *singular surfaces*. Remaining portions of the $\phi(\mathbf{x},\boldsymbol{\lambda}) = 0$ hypersurface do not support singular arcs, and a subset of these may constitute *mode-change* surfaces. A mode-change surface is called a *switching surface* in certain cases, e.g., in the case that $\mathcal{H}_1$ is linear in the components of $\mathbf{m}$.

Actually, the existence of a solution that satisfies 8-159 is unknown in advance. What we would like to do is:

1. determine under what conditions singular surfaces can exist for a given problem;
2. if they exist, derive the form that the singular arcs assume in the space of $X \times \Lambda$ and in the X space alone if possible;
3. test the optimality of singular arcs, i.e., determine if the optimal trajectory between two fixed points on a singular surface is a singular arc;
4. find an optimal trajectory which satisfies given end conditions—the optimal trajectory *may* consist of contiguous segments of nonsingular and singular arcs; and
5. relate $\mathbf{m}$ to $\mathbf{x}$ and $\boldsymbol{\lambda}$ along singular arcs of interest (for feedback control, $\mathbf{m}$ should be related to $\mathbf{x}$ alone along singular arcs of interest).

The accomplishment of the preceding tasks is seldom easy. For the particular form of $\mathcal{H}$ in 8-158, we have available conditions 8-159, wherein total derivatives of $\mathbf{x}$ and $\boldsymbol{\lambda}$ with respect to t can be eliminated by using the state and costate equations 8-19 and 8-20. In addition, when $\mathcal{H}$ is formally

Sect. 8-10   SINGULAR SOLUTIONS   527

independent of t, as in 8-158, it is known that the optimal $\mathcal{H}$ equals a constant c for $t \in [t_a, t_b]$; it must therefore be true that, along a singular arc,

$$\mathcal{H}_0(\mathbf{x}, \boldsymbol{\lambda}) = c, \qquad t \in [t_1, t_2] \tag{8-160a}$$

and

$$0 = \dot{\mathcal{H}}_0(\mathbf{x}, \boldsymbol{\lambda}) = \ddot{\mathcal{H}}_0(\mathbf{x}, \boldsymbol{\lambda}) = \cdots, \qquad t \in (t_1, t_2) \tag{8-160b}$$

Redundant information is generally obtained from higher-order derivatives in 8-159 and 8-160. On the basis of 8-19, 8-20, 8-159, and 8-160, we may solve for $\mathbf{m}$ in terms of $\mathbf{x}$ and $\boldsymbol{\lambda}$; and because $\mathbf{m} \in U$ is assumed, the singular surfaces in the $X \times \Lambda$ space are partially defined by this constraint.

It is possible to avoid consideration of singular solutions by incorporating additional, but "insignificant," penalty terms in the Hamiltonian of any given problem. Unfortunately, such inclusions usually complicate the two-point boundary-value problem that must be solved. Examples 8-6 and 8-7 which follow show subtle points that are encountered when singular solutions prevail. Theoretical work on necessary and sufficient conditions associated with singular solutions is reported in recent literature [8.7, 8.34, 8.57, 8.64, 8.87]. Johnson's review [8.52] can be consulted for an account of singular solutions in problems of optimal control. Ross [8.94] considers general network, pulse-processing problems for which singular solutions are common.

Example 8-6. Consider the circuit diagram of Figure 8-10. Let $\alpha = R/L$, and $m = e/L$. The state equation is

$$\dot{x} = -\alpha x + m, \quad 0 \leq m \leq m_{\max}, \quad x(0) = 0 \tag{8-161}$$

A specified time interval $[0, T]$ is of interest. A maximum of $x(T)$ is desired, but input energy is constrained by

$$\int_0^T mx \, dt = K, \quad \text{a constant} \tag{8-162}$$

By using the standard isoperimetric approach, an augmented functional J_a to be minimized is formed:

$$J_a = \int_0^T [-\dot{x}(t) + hm(t)x(t)] \, dt = \int_0^T [\alpha x - m + hmx] \, dt \tag{8-163}$$

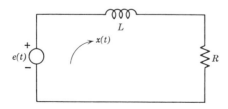

Figure 8-10. The voltage $e(t)$ is constrained in amplitude, $0 \leq e \leq e_{\max}$.

where h is a positive, time-invariant Lagrange multiplier. The Hamiltonian corresponding to 8-161 and 8-163 is

$$\mathcal{H} = -\alpha x + m - hmx + (-\alpha x + m)\lambda = (1 + \lambda - hx)m - (1 + \lambda)\alpha x \quad (8\text{-}164)$$

And the costate equation follows:

$$\dot{\lambda} = \alpha\lambda + \alpha + hm, \qquad \lambda(T) = 0, \qquad 0 \le m \le m_{\max} \quad (8\text{-}165)$$

where $\lambda(T) = 0$ is determined on the basis of the transversality condition 8-22.

The possibility of a singular solution exists because $\mathcal{H}$ of 8-164 is independent of m if $\lambda = \lambda_s$ and $x = x_s$, where

$$1 + \lambda_s - hx_s = 0 \quad (8\text{-}166)$$

In the x,λ plane, Equation 8-166 defines a line; if at some time t, x^* and λ^* are above the line, the maximizing condition on $\mathcal{H}$ is satisfied by $m^* = m_{\max}$; but if x^* and λ^* are below the line, the maximizing condition on $\mathcal{H}$ is satisfied by $m^* = 0$—see Figure 8-11.

To determine if a principal trajectory can exist on the switching line defined by 8-166, the relationship

$$\lambda_s = hx_s - 1 \quad (8\text{-}167)$$

is assumed to hold over some interval of time $[t_1,t_2]$ where $t_1 \ne t_2$, in which case the time derivative of 8-167 must be satisfied, as follows:

$$\dot{\lambda}_s = h\dot{x}_s, \qquad t \in (t_1,t_2) \quad (8\text{-}168)$$

and by using 8-161 and 8-165,

$$\alpha\lambda_s + \alpha + hm_s = h(-\alpha x_s + m_s)$$

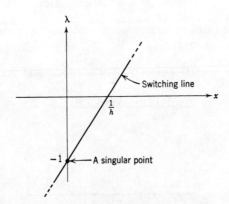

Figure 8-11. The state-costate plane for Example 8-6.

and therefrom,
$$\lambda_s = -hx_s - 1 \tag{8-169}$$

But observe that both 8-167 and 8-169 can be satisfied only if $x_s = 0$ and $\lambda_s = -1$. Thus, part of a solution could consist of $x_s = 0$ and $\lambda_s = -1$ over some subinterval $[t_1, t_2]$ of $[0, T]$, and on the basis of 8-161 and 8-165, m_s would have to equal zero over (t_1, t_2).

The solution would be straightforward if the initial condition on λ^* was known. On the basis of 8-165, it is evident that $\lambda^*(0)$ must be less than zero because, otherwise, $\lambda(T)$ could never be forced to zero. Similarly, $\lambda^*(0)$ must be greater than $-1 - (h/\alpha)m_{\max}$.

As a particular case of this example, suppose that
$$\alpha = m_{\max} = h = 1 \tag{8-170}$$

In this case, the reader may verify that values of $\lambda(0)$ less than -1 cannot satisfy both $\lambda(T) = 0$ and the maximizing condition on $\mathcal{H}$. For $\lambda(0) = -1 + \epsilon$ where $\epsilon \in (0, 1)$, the maximizing condition at $t = 0$ is satisfied by $m(0) = 1$. If $m(t) = 1$ for $t \in [0, t_c]$, Equation 8-165 gives
$$\lambda(t_c) = (-1 + \epsilon)e^{t_c} + \int_0^{t_c} 2e^{\tau} d\tau = (1 + \epsilon)e^{t_c} - 2 \tag{8-171}$$

and Equation 8-161 gives
$$x(t_c) = \int_0^{t_c} e^{-\tau} d\tau = 1 - e^{-t_c}$$

But then
$$e^{t_c} = \frac{1}{1 - x(t_c)}$$

which can be used to eliminate e^{t_c} from 8-171 with the result that
$$\lambda(t_c) = \frac{1 + \epsilon}{1 - x(t_c)} - 2 \tag{8-172}$$

Sets of $\lambda(t_c)$'s and $x(t_c)$'s which satisfy 8-172 are sketched in Figure 8-12, from which it is clear that $m(t) = m_{\max} = 1$ for all $t \in [0, T]$ if $\lambda(0) = -1 + \epsilon$.

Consider those cases for which $\lambda(0) = -1 + \epsilon, 0 \leq \epsilon < 1$. At the terminal time T, $\lambda(T) = 0$, and Equation 8-171 gives
$$T = \ln [2/(1 + \epsilon)] \tag{8-173}$$

If the given T is less than $\ln 2$, Equation 8-173 is satisfied by a unique $\epsilon \in (0, 1)$; but if the given T is greater than $\ln 2$, the singular solution, $x_s = m_s = 0$ and $\lambda_s = -1$, must apply for $t \in [0, T - \ln 2)$, and the

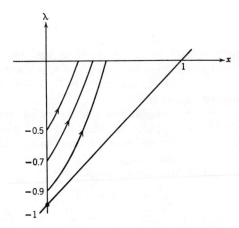

Figure 8-12. Optimal state-costate trajectories as a function of terminal time T: $\lambda(0) = -0.5$ gives $T = 0.288$, $\lambda(0) = -0.7$ gives $T = 0.432$, and $\lambda(0) = -0.9$ gives $T = 0.593$.

remainder of the optimal solution is obtained by using $m^*(t) = 1$ for $t \in [T - \ln 2, T]$.

Note that both x^* and m^* are dependent on h. With $h = 1$, as assumed in the two preceding paragraphs, a particular value of

$$\int_0^T x^* m^* \, dt$$

is obtained and is not necessarily equal to the given K of 8-162. In general, a search for an appropriate h must be made.

Considerations of singular solutions can be avoided in this example if an appropriate penalty function is added to the integrand of 8-163. One such penalty function is wm^2 where w is assigned an arbitrarily small but positive value. The corresponding two-point boundary-value problem is yet associated with the state Equation 8-161 and the costate Equation 8-165, but the maximizing condition on $\mathscr{H}$ gives

$$m^* = \begin{cases} m_{\max}, & m_0 > m_{\max} \\ m_0, & 0 \leq m_0 \leq m_{\max} \\ 0, & m_0 < 0 \end{cases} \quad (8\text{-}174)$$

where

$$m_0 \triangleq \frac{1 + \lambda^* - hx^*}{2w} \quad (8\text{-}175)$$

Example 8-7. Consider the state equation

$$\dot{\mathbf{x}} = \begin{bmatrix} 0 & 1 \\ a_{21} & a_{22} \end{bmatrix} \mathbf{x} + \begin{bmatrix} 0 \\ 1 \end{bmatrix} m, \qquad |m| \leq 1 \qquad (8\text{-}176)$$

The time interval $[0,T]$ is of interest, where T is free to be selected in the minimization of

$$J = \int_0^T \mathbf{x}'\mathbf{x}\, dt$$

$$= \int_0^T [x_1^2 + x_2^2]\, dt \qquad (8\text{-}177)$$

Initial conditions on $\mathbf{x}$ are given,

$$x_1(0) = x_{10}, \qquad x_2(0) = x_{20} \qquad (8\text{-}178a)$$

as also are terminal conditions

$$x_1(T) = 0, \qquad x_2(T) = 0 \qquad (8\text{-}178b)$$

The only restriction on $m(t)$ is that imposed in Equation 8-176.

In 1963, Wonham and Johnson [8.110] presented an n-state-variable generalization of this problem; and in a later work [8.53], they considered the effect of a penalty term of the form wm^2 in the integrand of 8-177. It is interesting to note that their results [8.53] suggest that several other authors, in earlier works, had overlooked a particular mode of control in the optimal control law.

The Hamiltonian for this problem is

$$\mathcal{H} = -x_1^2 - x_2^2 + x_2\lambda_1 + (a_{21}x_1 + a_{22}x_2 + m)\lambda_2 \qquad (8\text{-}179)$$

from which it follows that

$$\dot{\lambda}_1 = -a_{21}\lambda_2 + 2x_1 \qquad (8\text{-}180a)$$

and

$$\dot{\lambda}_2 = -\lambda_1 - a_{22}\lambda_2 + 2x_2 \qquad (8\text{-}180b)$$

The maximizing condition on $\mathcal{H}$ gives

$$m^* = \begin{cases} 1, & \lambda_2^* > 0 \\ -1, & \lambda_2^* < 0 \end{cases} \qquad (8\text{-}181)$$

But when $\lambda_2^* \equiv 0$, additional relations are required to determine m^*.

Because of the free terminal time T, it is known that $\mathcal{H}^* = \mathcal{H}(\mathbf{x}^*, \boldsymbol{\lambda}^*, m^*)$ must equal zero for all $t \in [0,T]$. Suppose that both $\mathcal{H}$ of 8-179 and $\lambda_2(t)$

equal identically zero over some subinterval $[t_1,t_2]$ of $[0,T]$; it follows from 8-179 that

$$\lambda_1 x_2 - x_1^2 - x_2^2 = 0, \qquad t \in [t_1,t_2] \tag{8-182}$$

Furthermore, it must also be true that $\dot\lambda_2 = 0$ for $t \in (t_1,t_2)$; by using 8-180b, therefore,

$$\lambda_1 = 2x_2, \qquad t \in (t_1,t_2) \tag{8-183}$$

And again, $\ddot\lambda_2 = 0$ for $t \in (t_1,t_2)$; therefore, from 8-180b,

$$\dot\lambda_1 = 2\dot x_2, \qquad t \in (t_1,t_2)$$

which is equivalent to

$$2x_1 = 2(a_{21}x_1 + a_{22}x_2 + m)$$

or

$$m = (1 - a_{21})x_1 - a_{22}x_2, \qquad t \in (t_1,t_2) \tag{8-184}$$

Higher-order derivatives of λ_2 for $t \in (t_1,t_2)$ yield no additional information in this case.

Equation 8-183 can be used to eliminate λ_1 from 8-182 with the result that

$$x_1 = \pm x_2, \qquad t \in (t_1,t_2) \tag{8-185}$$

This result indicates that state space (here the state plane) contains two lines, portions of which can support singular trajectories (see Figure 8-13). Corresponding to $x_1 = x_2$, Equation 8-184 gives

$$m = (1 - a_{21} - a_{22})x_2, \qquad t \in (t_1,t_2) \tag{8-186}$$

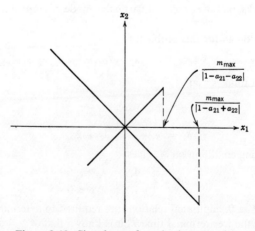

Figure 8-13. Singular surfaces in the state plane.

And Equation 8-176 gives
$$\dot{x}_1 = x_1$$
or
$$x_1 = ce^t, \quad t \in (t_1, t_2) \tag{8-187}$$

where c is a constant. Thus, along the singular surface $x_1 = x_2$, $|x_1| \leq m_{\max}/|1 - a_{21} - a_{22}|$, the values of x_1 and x_2 are forced away from the desired terminal values of zero, and any singular arcs on this singular surface are obviously to be excluded from optimal trajectories for the stated problem.

Corresponding to $x_1 = -x_2$, however, Equation 8-184 gives
$$m = (1 - a_{21} + a_{22})x_1, \quad t \in (t_1, t_2) \tag{8-188}$$
and Equation 8-176 gives
$$\dot{x}_1 = -x_1$$
or
$$x_1 = ce^{-t}, \quad t \in (t_1, t_2) \tag{8-189}$$

In this case, singular trajectories tend to the desired terminal values; but even so, it cannot be concluded at this point that some segment of the singular surface $x_1 = -x_2$, $|x_1| \leq m_{\max}/|1 - a_{21} + a_{22}|$, is part of the optimal trajectory which starts at a given point x_{10}, x_{20} in the state plane.

To keep the remainder of the example as lucid as possible, consider the *special case* in which both a_{21} and a_{22} are zero. *Question*: If the initial values of x_{10} and x_{20} are such that $x_{10} = -x_{20}$ and $|x_{10}| \leq 1$, does the optimal trajectory consist of that part of the singular arc from x_{10}, x_{20} to 0,0? As one means of answering this question—see Wonham and Johnson [8.110] for another—an examination is made in this paragraph and the next of the optimal solution that results if the restriction $|m| \leq 1$ is removed from the problem statement. In that case, Laplace transforms can be conveniently applied to the state and control variables, and minimization can be effected through use of Wiener-Hopf spectrum factorization (Chapter 4). In brief, with $a_{21} = a_{22} = 0$, the transforms corresponding to 8-176 are
$$sX_1(s) = X_2(s) + x_{10} \tag{8-190a}$$
and
$$sX_2(s) = M(s) + x_{20} \tag{8-190b}$$

where the capital letters denote, in this case, the one-sided Laplace transforms of the functions represented by corresponding small letters. By Parseval's theorem, Equation 8-177 is equivalent to
$$J = \frac{1}{2\pi j} \int_{-j\infty}^{j\infty} [X_1(s)X_1(-s) + X_2(s)X_2(-s)] \, ds \tag{8-191}$$

wherein both $X_1(s)$ and $X_2(s)$ can be replaced by appropriate functions of $M(s)$, as obtained by solving 8-190:

$$X_1(s) = \lim_{\epsilon \to 0+} [M(s) + x_{20} + sx_{10}]/(s + \epsilon)^2, \qquad \text{Re}(s) > -\epsilon \qquad (8\text{-}192\text{a})$$

and

$$X_2(s) = \lim_{\epsilon \to 0+} [M(s) + x_{20}]/(s + \epsilon), \qquad \text{Re}(s) > -\epsilon \qquad (8\text{-}192\text{b})$$

When the right-hand members of 8-192a and 8-192b are inserted appropriately in place of $X_1(s)$ and $X_2(s)$ in 8-191, the resulting integrand is of the general form 4-22, where $M(s)$—which assumes the role of $F(s)$ in 4-22—is to be selected for the minimum. The reader may verify that the optimal $M(s)$, $M^*(s)$, is

$$M^*(s) = \frac{-(x_{10} + x_{20})s - x_{20}}{s + 1}$$

$$= -(x_{10} + x_{20}) + \frac{x_{10}}{s + 1}, \qquad \text{Re}(s) > 0 \qquad (8\text{-}193)$$

which is obtained by using 4-47. The corresponding $m^*(t)$ is

$$m^*(t) = -(x_{10} + x_{20})\delta(t - 0+) + x_{10}e^{-t}, \qquad t \geq 0 \qquad (8\text{-}194)$$

where $\delta(t - 0+)$ is a Dirac delta function centered at $t = 0+$. Note that only on the singular surface $-x_{20} = x_{10}$ and $|x_{10}| \leq 1$ does the $m^*(t)$ of 8-194 comply with the requirement $|m| \leq 1$ of the original problem. But when it does comply, the solution exists and is unique, and must therefore be the optimum.

If a given trajectory intersects the particular singular surface $x_1 = -x_2$, $|x_1| \leq 1$, the results of the preceding paragraphs dictate that the remainder of the trajectory should lie on this singular arc for minimization of 8-177. The problem now is to find how optimal trajectories approach this singular arc. This is easily accomplished by using reverse-time integrations because values of state and costate variables are known at the instant t_1 when the singular arc is reached—$\lambda_2(t_1) = 0$ and $\lambda_1(t_1)$ is determined from 8-182. Thus, let $\mathbf{y}(\tau)$ be defined by

$$\mathbf{y}(\tau) = \begin{bmatrix} y_1(\tau) \\ y_2(\tau) \\ y_3(\tau) \\ y_4(\tau) \end{bmatrix} \triangleq \begin{bmatrix} x_1(t_1 - \tau) \\ x_2(t_1 - \tau) \\ \lambda_1(t_1 - \tau) \\ \lambda_2(t_1 - \tau) \end{bmatrix}, \qquad \tau \geq 0 \qquad (8\text{-}195)$$

SINGULAR SOLUTIONS

Use of this definition in 8-176 and 8-180, with $a_{21} = a_{22} = 0$, leads to

$$\frac{d\mathbf{y}(\tau)}{d\tau} = \begin{bmatrix} 0 & -1 & 0 & 0 \\ 0 & 0 & 0 & 0 \\ -2 & 0 & 0 & 0 \\ 0 & -2 & 1 & 0 \end{bmatrix} \mathbf{y}(\tau) + \begin{bmatrix} 0 \\ -1 \\ 0 \\ 0 \end{bmatrix} \operatorname{sgn} y_4(\tau) \quad (8\text{-}196)$$

Initial conditions that lead off the singular arc are $y_1(0+) \equiv y_{10} = x_1(t_1-)$, $y_{20} = -y_{10}$, $y_{30} = (y_{10}^2 + y_{20}^2)/y_{20} = -2y_{10}$, and $y_{40} = 0+$ or $0-$.

If $y_{40} = 0+$, sgn y_{40} starts at 1, and solutions to 8-196 sweep out a section of state space. Another section is swept out by use of $y_{40} = 0-$. But what is desired for purposes of feedback control is a switching function in terms of y_1 and y_2 (x_1 and x_2) rather than the known one in terms of y_4. To this end, the coupled equations in 8-196 are integrated forward in τ to the first value τ_s of τ at which y_4 changes sign.

The reader may verify that, for $\tau \in [0, \tau_s]$,

$$y_1 = (\operatorname{sgn} y_{40})\frac{\tau^2}{2} + (1 + \tau)y_{10} \quad (8\text{-}197a)$$

$$y_2 = -(\operatorname{sgn} y_{40})\tau - y_{10} \quad (8\text{-}197b)$$

$$y_3 = -2\left[y_{10} + y_{10}\tau + y_{10}\frac{\tau^2}{2} + (\operatorname{sgn} y_{40})\frac{\tau^3}{6}\right] \quad (8\text{-}197c)$$

and

$$y_4 = [(\operatorname{sgn} y_{40}) - y_{10}]\tau^2 - y_{10}\frac{\tau^3}{3} - (\operatorname{sgn} y_{40})\frac{\tau^4}{12} \quad (8\text{-}197d)$$

At $\tau = \tau_s$, $y_4(\tau_s) = 0$ and 8-197d can be used to obtain

$$\tau_s = -(\operatorname{sgn} y_{40})2y_{10} + [4y_{10}^2 + 12(1 - y_{10} \operatorname{sgn} y_{40})]^{1/2} \quad (8\text{-}198)$$

For a given sgn y_{40} and each y_{10}, $|y_{10}| \leq 1$, the pair $y_1(\tau_s), y_2(\tau_s)$ define a point on the switch curve in the x_1, x_2 plane. The complete switch curve is shown in Figure 8-14. The heavy segments of the switch curve are those obtained by using 8-197a, 8-197b, and 8-198. The dashed segments of the switch curve are obtained by starting on known segments, where initial conditions on the y_i's are known, and then by integrating 8-196 forward in τ to determine points in the state plane where $y_4(\tau) = 0$.

•

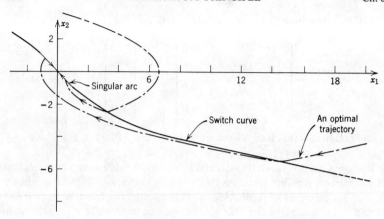

Figure 8-14. Typical trajectories associated with a dual-mode controller.

8-11. EQUIVALENT PRINCIPLES

The approach to this point in this chapter has been to use a *maximum* principle to obtain the *minimum* of a functional. Historical precedent is the only reason that this mixed approach has been taken. In Section 8-11a, an equivalent *minimum principle* is given for functional minimization. In Section 8-11b, a more convenient form of minimum principle is given for cases in which a function of end-point conditions is included in the performance functional.

8-11a. An Equivalent Minimum Principle

In the development of Section 7-13c, the costate vector λ is identified with the gradient $\partial \mathscr{F}(\mathbf{x},t)/\partial \mathbf{x}$ of the minimum-cost function $\mathscr{F}(\mathbf{x},t)$. As an alternative to that identity, suppose a costate vector λ_α is identified with $-\partial \mathscr{F}(\mathbf{x},t)/\partial \mathbf{x}$, i.e.,

$$\lambda_\alpha \triangleq -\lambda \tag{8-199}$$

The equivalent of Equation 7-180 is then

$$\dot{\lambda}_\alpha = -\frac{\partial f}{\partial \mathbf{x}} - \left(\frac{\partial \mathbf{q}}{\partial \mathbf{x}}\right)' \lambda_\alpha \tag{8-200}$$

and Equation 7-175 is appropriately modified to

$$\frac{\partial \mathscr{F}}{\partial t} = \underset{\mathbf{m}(t) \in U(t)}{\text{minimum}} (f + \lambda_\alpha' \mathbf{q}) \tag{8-201}$$

Sect. 8-11   EQUIVALENT PRINCIPLES   537

Let a Hamiltonian $\mathscr{H}_\alpha$ be defined equal to $-\mathscr{H}$ and therefore

$$\mathscr{H}_\alpha = f + \lambda_\alpha' \mathbf{q} \qquad (8\text{-}202)$$

which is only slightly changed in form from the $\mathscr{H}$ of 8-18. It follows from 8-200 and 8-202 that

$$\dot{\lambda}_\alpha = -\frac{\partial \mathscr{H}_\alpha}{\partial \mathbf{x}} \qquad (8\text{-}203)$$

It is also clear that

$$\dot{\mathbf{x}} = \frac{\partial \mathscr{H}_\alpha}{\partial \lambda_\alpha} \qquad (8\text{-}204)$$

But now, on the basis of 8-201 and 8-202, the optimal $\mathscr{H}_\alpha$ is minimized with respect to $\mathbf{m}$ at each time $t \in [t_a, t_b]$:

$$\mathscr{H}_\alpha[\mathbf{x}^*(t), \lambda_\alpha^*(t), \mathbf{m}^*(t)]_{\mathbf{m}^* \in U} \leq \mathscr{H}_\alpha[\mathbf{x}^*(t), \lambda_\alpha^*(t), \mathbf{m}(t)]_{\mathbf{m} \in U} \qquad (8\text{-}205)$$

for any $\mathbf{m} \in U$. The reader may verify that the general transversality condition 8-22 is equivalent to

$$[-\mathscr{H}_\alpha \, \delta t + \lambda_\alpha' \, \delta \mathbf{x}]_{t=t_a}^{t=t_b} = 0 \qquad (8\text{-}206)$$

which is unchanged in form. Similarly, the constancy condition 8-23 is equally applicable to $\mathscr{H}_\alpha^*$. *In short, with the defining relationship for the Hamiltonian of 8-202, the previously applied maximizing condition is converted to a minimizing one; all other necessary conditions of Section 8-4 are valid if $\mathscr{H}_\alpha$ is inserted in place of $\mathscr{H}$, and λ_α is inserted in place of λ.*

8-11b. Necessary Conditions for End-Point Functionals

Under Case C of Section 8-3, a direct approach is given to account for problems in which a function of end-point conditions is to be minimized. Although the approach is feasible as it stands, additional development is worthwhile: A transformation of costate variables results in transformed necessary conditions in which *only the transversality condition depends directly on the end-point functional.*

A minimum of J is desired, where here

$$J \triangleq \int_{t_a}^{t_b} \left[\left(\frac{\partial f_p(\mathbf{x})}{\partial \mathbf{x}} \right)' \frac{d\mathbf{x}}{dt} + f_\beta(\mathbf{x}, \mathbf{m}) \right] dt \qquad (8\text{-}207a)$$

or equivalently,

$$J = f_p(\mathbf{x})\big|_{t_a}^{t_b} + \int_{t_a}^{t_b} f_\beta(\mathbf{x}, \mathbf{m}) \, dt \qquad (8\text{-}207b)$$

in which $f_p = f_p(\mathbf{x})$ is a given real-valued function of class C^2, and the conditions associated with $f_\beta = f_\beta(\mathbf{x}, \mathbf{m})$ are the same as those assumed for

$f(\mathbf{x},\mathbf{m})$ in Section 8-3. The integrand of 8-207a is identified with $f(\mathbf{x},\mathbf{m})$ of Equation 8-3, and the Hamiltonian $\mathcal{H}_\alpha$ of 8-202 assumes the form

$$\mathcal{H}_\alpha = f_\beta + \left[\frac{\partial f_p}{\partial \mathbf{x}} + \boldsymbol{\lambda}_\alpha\right]' \mathbf{q} \tag{8-208}$$

Consider an augmented costate vector $\boldsymbol{\lambda}_\beta$:

$$\boldsymbol{\lambda}_\beta \triangleq \boldsymbol{\lambda}_\alpha + \frac{\partial f_p}{\partial \mathbf{x}} \tag{8-209}$$

With this, $\mathcal{H}_\beta = \mathcal{H}_\beta(\mathbf{x},\boldsymbol{\lambda}_\beta,\mathbf{m})$ is introduced and defined as

$$\mathcal{H}_\beta \triangleq \mathcal{H}_\alpha(\mathbf{x},\boldsymbol{\lambda}_\alpha,\mathbf{m})|_{\boldsymbol{\lambda}_\alpha + (\partial f_p/\partial \mathbf{x}) = \boldsymbol{\lambda}_\beta} \tag{8-210}$$

and therefore,

$$\mathcal{H}_\beta = f_\beta(\mathbf{x},\mathbf{m}) + \boldsymbol{\lambda}_\beta' \mathbf{q}(\mathbf{x},\mathbf{m}) \tag{8-211}$$

A differential equation in terms of $\boldsymbol{\lambda}_\beta$ is desired; from 8-209,

$$\dot{\boldsymbol{\lambda}}_\beta = \dot{\boldsymbol{\lambda}}_\alpha + \left(\frac{\partial^2 f_p}{\partial \mathbf{x}^2}\right)' \mathbf{q}$$

$$= -\frac{\partial \mathcal{H}_\alpha}{\partial \mathbf{x}} + \left(\frac{\partial^2 f_p}{\partial \mathbf{x}^2}\right)' \mathbf{q}$$

$$= -\left[\frac{\partial f_\beta}{\partial \mathbf{x}} + \left(\frac{\partial^2 f_p}{\partial \mathbf{x}^2}\right)' \mathbf{q} + \left(\frac{\partial \mathbf{q}}{\partial \mathbf{x}}\right)' \left(\boldsymbol{\lambda}_\alpha + \frac{\partial f_p}{\partial \mathbf{x}}\right)\right] + \left(\frac{\partial^2 f_p}{\partial \mathbf{x}^2}\right)' \mathbf{q}$$

$$= -\left[\frac{\partial f_\beta}{\partial \mathbf{x}} + \left(\frac{\partial \mathbf{q}}{\partial \mathbf{x}}\right)' \boldsymbol{\lambda}_\beta\right]$$

$$= -\frac{\partial \mathcal{H}_\beta}{\partial \mathbf{x}} \tag{8-212}$$

Also, it is evident that

$$\dot{\mathbf{x}} = \frac{\partial \mathcal{H}_\beta}{\partial \boldsymbol{\lambda}_\beta} \tag{8-213}$$

And the minimizing condition 8-205 assumes the form

$$\mathcal{H}_\beta(\mathbf{x}^*,\boldsymbol{\lambda}_\beta^*,\mathbf{m}^*)_{\mathbf{m}^* \in U} \leq \mathcal{H}_\beta(\mathbf{x}^*,\boldsymbol{\lambda}_\beta^*,\mathbf{m})_{\mathbf{m} \in U} \tag{8-214}$$

at any $t \in [t_a,t_b]$. Thus, Equations 8-211, 8-212, 8-213, and 8-214 *yield necessary conditions for optimality which are formally independent of the end-point functional* f_p.

To convert the transversality condition 8-206 to one in terms of $\boldsymbol{\lambda}_\beta$ and $\mathcal{H}_\beta$, Equation 8-209 is used to obtain

$$\left[-\mathcal{H}_\beta \, \delta t + \left(\boldsymbol{\lambda}_\beta - \frac{\partial f_p}{\partial \mathbf{x}}\right)' \delta \mathbf{x}\right]_{t=t_a}^{t=t_b} = 0 \tag{8-215}$$

It is this necessary condition, therefore, that accounts for the f_p part of J.

Example 8-8. Reconsider Example 8-2. The Hamiltonian corresponding to 8-18 is

$$\mathcal{H} = \sum_{i=1}^{n} [\lambda_i - 2c_i(x_i - x_i^d)]q_i(\mathbf{x},\mathbf{m}) \tag{8-216}$$

whereas the Hamiltonian corresponding to 8-211 is simply

$$\mathcal{H}_\beta = \sum_{i=1}^{n} \lambda_{\beta i} q_i(\mathbf{x},\mathbf{m}) \tag{8-217}$$

Necessary conditions for optimality in terms of $\mathcal{H}$ and λ are given by Equations 8-19, 8-20, 8-21, and 8-22, where 8-20 and 8-22 yield

$$\dot{\lambda}_j = -\frac{\partial \mathcal{H}}{\partial x_j}$$

$$= 2c_j q_j(\mathbf{x},\mathbf{m}) - \sum_{i=1}^{n} [\lambda_i - 2c_i(x_i - x_i^d)]\frac{\partial q_i}{\partial x_j} \tag{8-218}$$

and

$$\lambda_j(t_b) = 0 \tag{8-219}$$

for $j = 1, 2, \ldots, n$.

Necessary conditions in terms of $\mathcal{H}_\beta$ and λ_β are given by Equations 8-212, 8-213, 8-214, and 8-215, where 8-212 and 8-215 yield

$$\dot{\lambda}_{\beta j} = -\frac{\partial \mathcal{H}_\beta}{\partial x_j} = -\sum_{i=1}^{n} \lambda_{\beta i} \frac{\partial q_i}{\partial x_j} \tag{8-220}$$

and

$$\lambda_{\beta j}(t_b) = 2c_j[x_j(t_b) - x_j^d] \tag{8-221}$$

for $j = 1, 2, \ldots, n$. Observe that simplification is gained in the costate equations 8-220 at the expense of slightly more complicated terminal conditions 8-221.

●

8-12. CONCLUSION

In 1962, a bibliography [8.41] on optimal nonlinear control was published and contained 326 references dating back to the early 1930's with work such as that by Hazen [8.44]; however, 86 of the references were published in 1961. Since 1961 the number of technical papers per year on optimal control has easily exceeded 86. More recent literature on optimal control is cited in Paiewonsky's bibliography [8.86], 362 entries, and that of Athans [1.1], 237 entries. If we could list the number of papers devoted to all aspects of optimization, the total would be staggering. The maximum principle of this chapter is one which has stimulated much work and has yielded many practical results. In addition to the applications previously cited in this chapter, numerous others have been considered in the literature. Control applications are most numerous—applications are as varied as soft landing control of lunar vehicles [8.78] and the control of processes involving game-theoretic considerations [8.51, 8.61].[10] For communication systems, both optimal waveform design and maximization of signal-to-noise ratio have been considered [8.6, 8.32, 8.47, 8.106]. In regard to operation of industrial systems, Tracz's bibliography [8.105] contains 27 selected abstracts of works, dating from 1964 to 1967, dealing with the applications of optimal control theory to economic and business systems. Thus, the most economical operation of hydro and steam power systems may be based on maximum-principle conditions [8.25, 8.26, 8.43, 8.80]. A representative list of other actual and proposed applications is as follows: optimal pulse processing in electrical circuits [8.30, 8.94]; estimation of parameters and state variables in systems [8.31]; error analysis [8.48]; minimum-time frequency transitions in phase-locked loops [8.99]; and synthesis of nonuniform transmission lines [8.108].

The canonical problem of Section 8-3 is associated with systems characterized by a set of first-order, linear or nonlinear, ordinary differential equations. The general performance measure is an integral of a real-valued function of state and control variables; under conditions of convexity, simplifications in the necessary conditions can be shown to accrue [8.50]. Constraints on control variables are readily included in the original formulation of Pontryagin's maximum principle, but no allowance is made for inequality constraints on the state variables. It can be shown [8.1, 8.2, 8.10, 8.23, 8.65] that such constraints generally give rise to jump discontinuities in the costate responses. The extension of the theory to account for such constraints on state variables constitutes one generalization of the original principle. Another generalization is that of Butkovskii [8.20, 8.21] who was the first to formulate a maximum principle for distributed systems characterized by partial differential or multiple integral equations. Luenberger [8.73] developed a generalized maximum principle for systems that can be characterized by abstract-linear-space properties. And Kushner [8.68] has presented a stochastic maximum principle. But the most actively pursued generalizations, and the most controversial ones, are those grouped under the headings of discrete maximum principles [8.3, 8.24, 8.28, 8.49, 8.88, 8.89,

---

[10] The book by Isaacs [8.51] on differential games is based on work that was done prior to 1955 (prior to publication of the maximum principle by Pontryagin). Y. C. Ho shows the generality of the work by Isaacs in a book review published in the *IEEE Transactions on Automatic Control*, **AC-10**, 501–503, Oct. 1965.

8.92, 8.101]. For problems of time-optimal control, Pokoski [8.90] has compared the performance of systems with continuously acting controllers to comparable ones with sampled-data controllers.

REFERENCES

[8.1] Anorov, V. P. "Maximum Principle for Processes with Constraints of General Form. I." *Automation and Remote Control*, 357–367, March 1967.

[8.2] Anorov, V. P. "The Maximum Principle for Processes with General-Type Restrictions. II." *Automation and Remote Control*, 533–543, April 1967.

[8.3] Arimoto, S. "On a Multi-Stage Nonlinear Programming Problem." *Journal of Mathematical Analysis and Applications*, 17, 161–171, Jan. 1967.

[8.4] Athanassiades, M., and O. J. M. Smith. "Theory and Design of High-Order Bang-Bang Control Systems." *IRE Transactions on Automatic Control*, AC-6, 125–134, May 1961.

[8.5] Athans, M., and P. L. Falb. *Optimal Control*. McGraw-Hill, New York, 1966.

[8.6] Athans, M., and F. C. Schweppe. "Optimal Waveform Design via Control Theoretic Concepts." *Information and Control*, 10, 335–377, April 1967.

[8.7] Bass, R. W., and R. F. Webber. "Simplified Algebraic Characterization of Optimal Singular Control for Autonomous Linear Plants." *Preprint Volume*, Joint Automatic Control Conference, 465–469, June 1967.

[8.8] Bellman, R. "Functional Equations in the Theory of Dynamic Programming—XIII: Stability Considerations." *Journal of Mathematical Analysis and Applications*, 12, 537–540, Dec. 1965.

[8.9] Bellman, R., and R. Kalaba. *Quasilinearization and Nonlinear Boundary-Value Problems*. American Elsevier Publishing Co., New York, 1965.

[8.10] Berkovitz, L. D. "On Control Problems with Bounded State Variables." *Journal of Mathematical Analysis and Applications*, 5, 488–498, Dec. 1962.

[8.11] Berkovitz, L. D. "Variational Methods in Problems of Control and Programming." *Journal of Mathematical Analysis and Applications*, 3, 145–169, Aug. 1961.

[8.12] Bliss, G. A. "The Problem of Lagrange in the Calculus of Variations." *American Journal of Mathematics*, 52, 673–774, 1930.

[8.13] Bliss, G. A. *Lectures on the Calculus of Variations*. Univ. of Chicago Press, Chicago, 1946.

[8.14] Bogner, I., and L. F. Kazda. "An Investigation of the Switching Criteria for Higher-Order Servomechanisms." *Transactions of the AIEE*, 73, part II, 118–127, 1954.

[8.15] Boyadjieff, G., D. Eggleston, M. Jacques, H. Sutabutra, and Y. Takahashi. "Some Applications of the Maximum Principle to Second-Order Systems, Subject to Input Saturation, Minimizing Error, and Effort." *Transactions of the ASME*, series D (*Journal of Basic Engineering*), 86, 11–22, March 1964.

[8.16] Brennan, B. J., and A. P. Roberts. "Use of an Analogue Computer in the

Application of Pontryagin's Maximum Principle to the Design of Control Systems with Optimum Transient Response." *Journal of Electronics and Control*, 12, 345–352, April 1962.

[8.17] Bryson, A. E., and W. F. Denham. "A Steepest-Ascent Method for Solving Optimum Programming Problems." *Transactions of the ASME*, series E (*Journal of Applied Mechanics*), 29, 247–257, June 1962.

[8.18] Bushaw, D. W. "Optimal Discontinuous Forcing Terms." Pages 29–52 of *Contributions to the Theory of Nonlinear Oscillations*, Vol. 4, S. Lefschetz (Editor). Princeton Univ. Press, Princeton, N.J., 1958.

[8.19] Bushaw, D. W. *Differential Equations with a Discontinuous Forcing Term*. Report No. 469, Experimental Towing Tank, Stevens Institute of Technology, Hoboken, N.J., Jan. 1953.

[8.20] Butkovskii, A. G. "The Maximum Principle for Optimum Systems with Distributed Parameters." *Automation and Remote Control*, 22, 1156–1169, 1961.

[8.21] Butkovskii, A. G. "The Broadened Principle of the Maximum for Optimal Control Problems." *Automation and Remote Control*, 24, 292–304, 1963.

[8.22] Chang, S. S. L. *Synthesis of Optimum Control Systems*. McGraw-Hill, New York, 1961.

[8.23] Chang, S. S. L. "Optimal Control in Bounded Phase Space." *Automatica*, 1, 55–67, Jan.-March 1962.

[8.24] Chang, S. S. L., H. Halkin, and J. M. Holtzman. "On Convexity and the Maximum Principle for Discrete Systems." *IEEE Transactions on Automatic Control*, AC-12, 121–123, Feb. 1967.

[8.25] Dahlin, E. B. *Theoretical and Computational Aspects of Optimal Principles with Special Application to Power System Operation*. Ph.D. Thesis, Univ. of Pennsylvania, Philadelphia, 1964, 223 pp.

[8.26] Dahlin, E. B., and D. W. C. Shen. "Application of the Maximum Principle for Bounded State Space to the Hydro-Steam Dispatch Problems." *Preprint Volume*, Joint Automatic Control Conference, 651–659, June 1965.

[8.27] Darcy, V. J., and R. A. Hannen. "An Application of an Analog Computer to Solve the Two-Point Boundary-Value Problem for a Fourth-Order Optimal Control Problem." *IEEE Transactions on Automatic Control*, AC-12, 67–74, Feb. 1967.

[8.28] Denn, M. M. "On the Optimization of Continuous Complex Systems by the Maximum Principle." *International Journal of Control*, 1, 497, May 1965.

[8.29] Desoer, C. A. "Pontryagin's Maximum Principle and the Principle of Optimality." *Journal of the Franklin Institute*, 271, 361–367, May 1961.

[8.30] Desoer, C. A. "An Optimization Problem in Circuits." *IEEE Transactions on Circuit Theory*, CT-12, 28–31, March 1965.

[8.31] Detchmendy, D. M., and R. Sridhar. "Sequential Estimation of States and Parameters in Noisy Non-Linear Dynamical Systems." *Preprint Volume*, Joint Automatic Control Conference, 56–63, June 1965.

[8.32] Ditoro, D. M., and K. Steiglitz. "Application of the Maximum Principle to the Design of Minimum Bandwidth Pulses." *IEEE Transactions on Communication Technology*, COM-13, 433–438, Dec. 1965.

REFERENCES

[8.33] Doll, H. G. U.S. Patent No. 2,463,362, 1943.

[8.34] Dunn, J. C. "On the Classification of Singular and Nonsingular Extremals for the Pontryagin Maximum Principle." *Journal of Mathematical Analysis and Applications*, **17**, 1–36, Jan. 1967.

[8.35] Eaton, J. H. "An Iterative Solution to Time-Optimal Control." *Journal of Mathematical Analysis and Applications*, **5**, 329–344, Oct. 1962.

[8.36] Eggleston, D. M. "On the Application of the Pontryagin Maximum Principle Using Reverse Time Trajectories." *Transactions of the ASME*, series D (*Journal of Basic Engineering*), **85**, 478–480, Sept. 1963.

[8.37] Flügge-Lotz, I. *Discontinuous Automatic Control.* Princeton Univ. Press, Princeton, N.J., 1953.

[8.38] Flügge-Lotz, I., and H. A. Titus. "The Optimum Response of Full Third-Order Systems with Contactor Control." *Transactions of the ASME*, series D (*Journal of Basic Engineering*), **84**, 554–558, Dec. 1962.

[8.39] Flügge-Lotz, I., and H. D. Marbach. "On the Minimum Effort Regulation of Stationary Linear Systems." *Journal of the Franklin Institute*, **279**, 229–245, April 1965.

[8.40] Frederick, D. K., and G. F. Franklin. "Design of Piecewise-Linear Switching Functions for Relay Control Systems." *Preprint Volume*, Joint Automatic Control Conference, 594–604, Aug. 1966.

[8.41] Fuller, A. T. "Bibliography of Optimum Non-Linear Control of Determinate and Stochastic-Definite Systems." *Journal of Electronics and Control*, **13**, 589–612, Dec. 1962.

[8.42] Fuller, A. T. "The Replacement of Saturation Constraints by Energy Constraints in Control Optimization Theory." *International Journal of Control*, **6**, 201–227, Sept. 1967.

[8.43] Hano, I., Y. Tamura, and S. Narita. "An Application of the Maximum Principle to the Most Economical Operation of Power Systems." *IEEE Transactions on Power Apparatus and Systems*, **PAS-85**, 486–494, May 1966.

[8.44] Hazen, H. L. "Theory of Servomechanisms." *Journal of the Franklin Institute*, **218**, 279–330, Sept. 1934.

[8.45] Hestenes, M. R. *Numerical Methods of Obtaining Solutions of Fixed End Point Problems in the Calculus of Variations.* Report RM-102, RAND Corporation, Santa Monica, Calif., 1949.

[8.46] Hestenes, M. R. *A General Problem in the Calculus of Variations with Applications to Paths of Least Time.* Report RM-100, RAND Corporation, Santa Monica, Calif., 1950.

[8.47] Holtzman, J. M. "Signal-Noise Ratio Maximization Using the Pontryagin Maximum Principle." *The Bell System Technical Journal*, **45**, 473–489, March 1966.

[8.48] Howard, D. R., and Z. V. Rekasius. "Error Analysis with the Maximum Principle." *IEEE Transactions on Automatic Control*, **AC-9**, 223–229, July 1964.

[8.49] Hwang, C. L., and L. T. Fan. "A Discrete Version of Pontryagin's Maximum Principle." *Operations Research*, **15**, 139–146, Jan.-Feb. 1967.

[8.50] Isaacs, D., and C. T. Leondes. "Optimal Control System Synthesis for

Cost Functionals Involving Convex Single Valued Functions of State and Control Variables." *Information and Control*, **9**, 393–413, Aug. 1966.

[8.51] Isaacs, R. *Differential Games*. Wiley, New York, 1965.

[8.52] Johnson, C. D. "Singular Solutions in Problems of Optimal Control." Pages 209–267 of *Advances in Control Systems*, Vol. 2, C. T. Leondes (Editor). Academic Press, New York, 1965.

[8.53] Johnson, C. D., and W. M. Wonham. "On a Problem of Letov in Optimal Control." *Transactions of the ASME*, series D (*Journal of Basic Engineering*), **87**, 81–89, March 1965.

[8.54] Kalaba, R. "On Nonlinear Differential Equations, the Maximum Operation, and Monotone Convergence." *Journal of Mathematics and Mechanics*, **8**, 519–574, July 1959.

[8.55] Kalman, R. E. "The Theory of Optimal Control and the Calculus of Variations." Chapter 16 of *Mathematical Optimization Techniques*, R. Bellman (Editor). Univ. of California Press, Berkeley, Calif., 1963.

[8.56] Kelley, H. J., R. E. Kopp, and H. G. Moyer. "A Trajectory Optimization Technique Based upon the Theory of the Second Variation." AIAA Astrodynamics Conference, Yale Univ., New Haven, Conn., Aug. 1963.

[8.57] Kelley, H. J., R. E. Kopp, and H. G. Moyer. "Singular Extremals." Pages 63–101 of *Topics in Optimization*, G. Leitman (Editor). Academic Press, New York, 1967.

[8.58] Kenneth, P., and R. McGill. "Two-Point Boundary-Value-Problem Techniques." Pages 69–109 of *Advances in Control Systems*, Vol. 3, C. T. Leondes (Editor). Academic Press, New York, 1966.

[8.59] Kenneth, P., and G. E. Taylor. "Solution of Variational Problems with Bounded Control Variables by Means of the Generalized Newton-Raphson Method." Pages 471–487 of *Recent Advances in Optimization Techniques*, A. Lavi and T. P. Vogl (Editors). Wiley, New York, 1966.

[8.60] Knapp, C. H. "The Minimum Principle and the Method of Gradients." *IEEE Transactions on Automatic Control*, **AC-11**, 752–753, Oct. 1966.

[8.61] Kochetkov, Yu. A. "Pontryagin Maximum Principle Applied to Investigation of Minimal Problems in Control Processes." *Engineering Cybernetics*, 9–18, Sept.-Oct. 1965.

[8.62] Kopp, R. E. "On the Pontryagin Maximum Principle." Pages 395–402 of *International Symposium on Nonlinear Differential Equations and Nonlinear Mechanics*, J. P. LaSalle and S. Lefschetz (Editors). Academic Press, New York, 1963.

[8.63] Kopp, R. E., and R. McGill. "Several Trajectory Optimization Techniques, Part I: Discussion." Pages 65–89 of *Computing Methods in Optimization Problems*, A. V. Balakrishnan and L. W. Neustadt (Editors). Academic Press, New York, 1964.

[8.64] Kopp, R. E., and H. G. Moyer. "Necessary Conditions for Singular Extremals." *AIAA Journal*, **3**, 1439–1444, Aug. 1965.

[8.65] Koziorov, L. M., and Yu. I. Kupervasser. "Optimal Control for a Second-Order System with Constraints on the Phase Coordinates and Control." *Engineering Cybernetics*, 392–399, May-June 1966.

REFERENCES

[8.66] Krotov, V. F. "Methods of Solving Variational Problems on the Basis of the Sufficient Conditions for an Absolute Minimum. I." *Automation and Remote Control*, 23, 1473–1484, 1962.

[8.67] Krotov, V. F. "Methods of Solving Variational Problems. II. Sliding Regimes." *Automation and Remote Control*, 24, 539–553, 1963.

[8.68] Kushner, H.J. "On the Stochastic Maximum Principle: Fixed Time of Control." *Journal of Mathematical Analysis and Applications*, 11, 78–92, July 1965.

[8.69] LaSalle, J. P. "Time Optimal Control Systems." *Proceedings of the National Academy of Sciences*, 45, 573–577, April 1959.

[8.70] LaSalle, J. P. "The Time Optimal Control Problem." Pages 1–24 of *Contributions to the Theory of Nonlinear Oscillations*, Vol. 5, S. Lefschetz (Editor). Princeton Univ. Press, Princeton, N.J., 1960.

[8.71] Lasdon, L. S., R. K. Rice, and A. D. Waren. "An Interior Penalty Method for Inequality Constrained Optimal Control Problems." *Preprint Volume*, Joint Automatic Control Conference, 538–548, June 1967.

[8.72] Lee, E. B. "Mathematical Aspects of the Synthesis of Linear, Minimum Response-Time Controllers." *IRE Transactions on Automatic Control*, AC-5, 283–289, Sept. 1960.

[8.73] Luenberger, D. G. "A Generalized Maximum Principle." Pages 323–339 of *Recent Advances in Optimization Techniques*, A. Lavi and T. P. Vogl (Editors). Wiley, New York, 1966.

[8.74] McCausland, I. *Introduction to Optimal Control*. Wiley, New York, 1969.

[8.75] McDonald, D. "Nonlinear Techniques for Improving Servo Performance." *Proceedings of the National Electronics Conference*, 6, 400–421, 1950.

[8.76] McGill, R., and P. Kenneth. "A Convergence Theorem on the Iterative Solution of Nonlinear Two-Point Boundary Value Systems." XIVth IAF Congress, Paris, Sept. 1963.

[8.77] McShane, E. J. "On Multipliers for Lagrange Problems." *American Journal of Mathematics*, 61, 809–819, Oct. 1939.

[8.78] Meditch, J. S. "On the Problem of Optimal Thrust Programming for a Lunar Soft Landing." *IEEE Transactions on Automatic Control*, AC-9, 477–484, Oct. 1964.

[8.79] Meserve, W. E., and D. Jordan. "A Critical Study of Several Means of Stabilizing a Relay Control System." *ISA Transactions*, 4, 58–66, Jan. 1965.

[8.80] Narita, S. "The Application of the Maximum Principle to the Calculation of the Most Economical Operation of Power Systems." *Electrical Engineering in Japan*, 85, 23–33, Nov. 1965.

[8.81] Neustadt, L. W. "Synthesizing Time Optimal Control Systems." *Journal of Mathematical Analysis and Applications*, 1, 484–493, Dec. 1960.

[8.82] Noton, A. R. M., P. Dyer, and C. A. Markland. "Numerical Computation of Optimal Control." *IEEE Transactions on Automatic Control*, AC-12, 59–66, Feb. 1967.

[8.83] O'Donnell, J. J. "Bounds on Limit Cycles in Two-Dimensional Bang-Bang Control Systems with an Almost Time-Optimal Switching Curve." *IEEE Transactions on Automatic Control*, AC-9, 448–457, Oct. 1964.

[8.84] Oldenburger, R. (Editor). *Optimal and Self-Optimizing Control.* The M.I.T. Press, Cambridge, Mass., 1966.

[8.85] Paiewonsky, B. "Time Optimal Control of Linear Systems with Bounded Controls." Pages 333–365 of *International Symposium on Nonlinear Differential Equations and Nonlinear Mechanics,* J. P. LaSalle and S. Lefschetz (Editors). Academic Press, New York, 1963.

[8.86] Paiewonsky, B. "Optimal Control: A Review of Theory and Practice." *AIAA Journal,* 3, 1985–2006, Nov. 1965.

[8.87] Paraev, Y. I. "On Singular Control in Optimal Processes That Are Linear with Respect to the Control Inputs." *Automation and Remote Control,* 23, 1127–1134, 1962.

[8.88] Pearson, J. B., Jr., and R. Sridhar. "A Discrete Optimal Control Problem." *IEEE Transactions on Automatic Control,* AC-11, 171–174, April 1966.

[8.89] Pervozvanskiy, A. A. "Relationship Between the Basic Theorems of Mathematical Programming and the Maximum Principle." *Engineering Cybernetics,* 6–11, Jan.-Feb. 1967.

[8.90] Pokoski, J. L. *Time-Optimal and Suboptimal Control of Sampled-Data Systems.* Ph.D. Thesis, Montana State Univ., Bozeman, Mont., Aug. 1967, 129 pp.

[8.91] Pontryagin, L. S., V. G. Boltyanskii, R. V. Gamkrelidze, and E. F. Mischenko. *The Mathematical Theory of Optimal Processes.* Authorized Translation from the Russian, K. N. Trirogoff (Translator), L. W. Neustadt (Editor). Wiley, New York, 1962.

[8.92] Propoi, A. I. "On the Maximum Principle for Discrete Control Systems." *Automation and Remote Control,* 26, 1177–1187, 1965.

[8.93] Rauch, L. L., and R. M. Howe. "A Servo with Linear Operation in a Region about the Optimum Discontinuous Switching Curve." *Proceedings, 1956 Symposium on Nonlinear Circuit Analysis,* Polytechnic Institute of Brooklyn, Brooklyn, N.Y., pp. 215–223, 1957.

[8.94] Ross, M. E. *Network Theory for Optimal Pulse Design via Pontryagin's Maximum Principle.* Ph.D. Thesis, Montana State Univ., Bozeman, Mont., March 1968, 255 pp.

[8.95] Rothenberger, B. F., and L. Lapidus. "The Control of Nonlinear Systems: Part IV, Quasilinearization as a Numerical Method; Part V, Quasilinearization and State-Constrained Systems." *AICHE Journal,* American Institute of Chemical Engineers, 13, 973–988, Sept. 1967.

[8.96] Roxin, E. "A Geometrical Interpretation of Pontryagin's Maximum Principle." Pages 303–324 of *International Symposium on Nonlinear Differential Equations and Nonlinear Mechanics,* J. P. LaSalle and S. Lefschetz (Editors). Academic Press, New York, 1963.

[8.97] Rozonoer, L. I. "On the Variational Methods of Analysis of Automatic Control System Performance." Pages 477–480 of *Automatic and Remote Control,* 1, Butterworths, London, 1961.

[8.98] Rubio, J. E. "On the Uniqueness of the Solutions of Some Equations of Optimal Control Theory." *International Journal of Control,* 3, 69–78, Jan. 1966.

[8.99] Sanneman, R. W., and S. C. Gupta. "Optimum Strategies for Minimum

Time Frequency Transitions in Phase-Locked Loops." *IEEE Transactions on Aerospace and Electronic Systems*, AES-2, 570–581, Sept. 1966.

[8.100] Schley, C. H., Jr., and I. Lee. "Optimal Control Computation by the Newton-Raphson Method and the Riccati Transformation." *IEEE Transactions on Automatic Control*, AC-12, 139–143, April 1967.

[8.101] Shemer, J. E., and S. C. Gupta. "Applications of Butkovskii's Form of Discrete Maximum Principle." *ISA Transactions*, 5, 395–405, Oct. 1966.

[8.102] Smith, F. B., Jr. "Time-Optimal Control of Higher-Order Systems." *IRE Transactions on Automatic Control*, AC-6, 16–21, Feb. 1961.

[8.103] Smith, F. W. "Design of Quasi-Optimal Minimum-Time Controllers." *IEEE Transactions on Automatic Control*, AC-11, 71–77, Jan. 1966.

[8.104] Stout, T. M. "Switching Errors in an Optimum Relay Servomechanism." *Proceedings of the National Electronics Conference*, 9, 188–198, 1953.

[8.105] Tracz, G. S. "A Selected Bibliography on the Application of Optimal Control Theory to Economic and Business Systems, Management Science, and Operations Research." *Operations Research*, 16, 174–186, Jan.-Feb. 1968.

[8.106] Tufts, D. W., and D. A. Shnidman. "Optimum Waveforms Subject to Both Energy and Peak-Value Constraints." *Proceedings of the IEEE*, 52 1002–1007, Sept. 1964.

[8.107] Valentine, F. A. "The Problem of Lagrange with Differential Inequalities as Added Side Conditions." Pages 403–447 of *Contributions to the Calculus of Variations*. Univ. of Chicago Press, Chicago, 1937.

[8.108] Wohlers, M. R., R. E. Kopp, and H. G. Moyer. "Computational Techniques for the Synthesis of Optimum Nonuniform Transmission Lines Based on Variational Principles." *Proceedings of the National Electronics Conference*, 21, 135–140, 1965.

[8.109] Womack, B. F., and J. N. Dashiell. "A Weighted Time Performance Index for Optimal Control." *IEEE Transactions on Automatic Control*, AC-10, 201–202, April 1965.

[8.110] Wonham, W. M., and C. D. Johnson. "Optimal Bang-Bang Control with Quadratic Performance Index." *Preprint Volume*, Joint Automatic Control Conference, 101–112, June 1963.

PROBLEMS

8.1 A feedback control system is to be designed to obtain a minimum of J:

$$J = \int_{t_a}^{t_b} (x^2 + m^2)\, dt$$

where

$$\dot{x} = -x + m, \qquad |m| \leq 1$$

and where $x(t_b)$ is to equal zero and t_b is free. What optimal characteristic should be in the feedback block of Figure 8-P1?

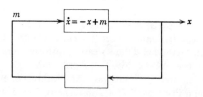

Figure 8-P1.

8.2 The terminal state of a given n-state-variable process is to lie on the surface defined by

$$\phi_1(\mathbf{x}) = \tfrac{1}{2}(\epsilon_1 x_1^2 + \epsilon_2 x_2^2 + \cdots + \epsilon_n x_n^2 - \epsilon_0) = 0$$

where the ϵ_i's are non-negative constants. Find appropriate transversality conditions.

8.3 At the unspecified terminal time $t = t_b$, each entry x_i of $\mathbf{x}$ is to equal a given differentiable function $\gamma_i(t)$. List the set of terminal conditions that apply to the associated minimization problem of Section 8-3.

8.4 At the unspecified terminal time $t = t_b$, a given function $\phi_1(\mathbf{x},t)$ is to equal zero. List the set of terminal conditions that apply to the associated minimization problem of Section 8-3.

8.5 Repeat Example 8-4 with the condition $0 \leq m \leq E$ replaced by $|m| \leq E$.

8.6 Repeat Example 8-4 with the performance measure of 8-78 replaced by

$$J = \int_0^T [-x_2 + hm^2]\,dt$$

where h is a positive constant.

8.7 Consider the circuit diagram of Figure 8-P7. Component values are as follows: $R_1 = 10^3$ ohms, $R_2 = 500$ ohms, $L_1 = 0.04$ henry, $L_2 = 0.01$ henry, $M = 0.015$ henry, and $C = 2 \times 10^{-7}$ farad. At $t = 0$, all currents are assumed zero. The source voltage $v_s(t)$ is constrained to lie between 0 and 10 volts.

  a. Find the state equations that correspond to state variables i_1, i_2, and v_0. Suggest scale changes as deemed appropriate.

  b. A *maximum* of $v_0(T)$ is desired, where $T = 2 \times 10^{-4}$ second. Find the costate equations and the optimal voltage waveforms.

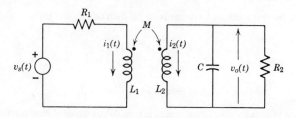

Figure 8-P7.

PROBLEMS

8.8 The block diagrams of Figure 8-P8 have equivalent input-output characteristics. State variables x_1 and x_2 are to be driven to zero in minimum time.

  a. Find an appropriate switch curve and switching function in terms of x_1 and x_2.

  b. If only c, $\dot{c}$, and m are measurable, what relationships can be used to compute x_1 and x_2?

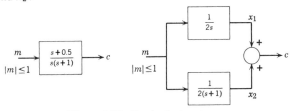

Figure 8-P8. Equivalent systems.

8.9 Consider the block diagram of Figure 8-P9. The rate of change $\dot{c}$ of c is limited by the inclusion of Block A. The controller is designed on the basis of the minimum-time control law of Section 8-7. Discuss the merit of this arrangement for minimum-time control subject to magnitude constraints on $\dot{c}$ and m.

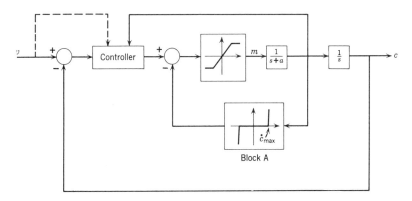

Figure 8-P9.

8.10 Repeat the development of Section 8-7b, but for the case that the constant $a = 0$. In place of the signal $\dot{c}$, which is fed back in Figure 8-9a, suppose that $x_2 = \dot{c} - \dot{v}$ is supplied as an input to the appropriate blocks. Can this system track a ramp input, of arbitrary finite slope, with zero steady-state error?

8.11 Suppose that the controller of Figure 8-5 and its output circuitry cause a relatively small time delay D in the signal applied to the plant. To compensate for this delay, it is proposed that a new switch curve be formed by rotating each point on the old switch curve by an amount which exactly compensates for the delay. Under the condition that $a = 0$, as in Problem 8.10, derive an equation that defines an appropriate rotated switch curve which is to be used outside the small-signal region of Figure 8-9a.

8.12 When $m_{\max,ub}$ and a_{ub} are used to form the switching function of Equation 8-99, a response overshoot invariably results in the system depicted by Figure 8-9a. Explain how the switch curve shifting proposed in the next-to-last paragraph of Section 8-7b might help to offset the overshoot.

8.13 Consider the configuration of Figure 8-9b with the switch curve designed on the basis that $a = m_{\max} = 1$. Given conditions are $\epsilon_1 = 0.1$ and $\epsilon_2 = 0.2$. Initial conditions are $x_1(0) = 1$ and $x_2(0) = 0$. Use a computer to generate state-plane trajectories under the following conditions: (1) actual plant values are $a = m_{\max} = 1$; (2) actual plant values are $a = m_{\max} = 1.5$; and (3) actual plant values are $a = m_{\max} = 0.7$. What design conclusions can you draw from your results? Should the lower-bound (*lb*) values of a and $m_{\max}$ be used in the design of the controller?

8.14 Given

$$\ddot{x} + x = m, \quad |m| \leq 1$$

where $x(0)$ and $\dot{x}(0)$ are given, and $x(T) = \dot{x}(T) = 0$. A minimum of T is desired. Find the appropriate switch curve in the $x,\dot{x}$ plane. (*Hint:* The switch curve consists of connected circular arcs. This is the classic example treated by Bushaw [8.18, 8.19]. Also see Pontryagin *et al.* [8.91].)

8.15 A given system is characterized by the following state equations:

$$\begin{bmatrix} \dot{x}_1 \\ \dot{x}_2 \\ \dot{x}_3 \end{bmatrix} = \begin{bmatrix} 0 & 1 & 0 \\ 0 & -1 & 1 \\ 0 & 0 & -2 \end{bmatrix} \begin{bmatrix} x_1 \\ x_2 \\ x_3 \end{bmatrix} + \begin{bmatrix} 0 \\ 0 \\ 1 \end{bmatrix} m$$

for $0 \leq t \leq T$, where $|m(t)| \leq 1$, $x_1(0) = 2$, $x_2(0) = 1$, $x_3(0) = 1$. The terminal time T is to be minimized, and the terminal values of the state variables are to equal zero.

a. Draw a block diagram that corresponds to the given state equations.

b. Find the eigenvalues and a set of appropriate eigenvectors (for a uniform answer, normalize the first nonzero entry in each eigenvector to 1). Use the eigenvectors to obtain an **S** matrix, as used in 8-106 and 8-107 (see Appendix A).

c. List the particular equations of the form 8-111 that apply. Solve these equations by using the iterative technique associated with 8-112. As initial guesses, use $t_1^0 = 2$, $t_2^0 = 3$, and $t_3^0 = 4$.

8.16 Consider minimization of

$$J = \int_{t_a}^{t_b} (w + |m|) \, dt$$

where w is a real constant and m is the control action associated with the following state equations:

$$\dot{x}_1 = x_2, \quad x_1(t_b) = 0$$
$$\dot{x}_2 = m, \quad x_2(t_b) = 0, \quad |m| \leq 1$$

in which t_b is free. Initial conditions are fixed.

PROBLEMS   551

  a. Find m^* as a function of λ_2^*.
  b. Give the costate equations.
  c. The results of Part a should suggest three modes of constant control action. Each such mode is appropriate over a specific region of the x_1,x_2 state plane. With $w = 1$, find these regions in terms of equations which define the boundaries of the regions. Make use of the fact that $\mathscr{H}^*(t) = 0$ (refer to pertinent statements in Section 8-7a).
  d. Repeat Part c with $w = 1/2$.

(More complicated problems of this type are treated by Flügge-Lotz and Marbach [8.39].)

8.17 Find the minimum of J:

$$J = \int_0^T (|x_1| + |m|)\, dt$$

where
$\dot{x}_1 = x_2$, $\dot{x}_2 = m$, $|m| \leq 1$, $x_1(0) = 2$, $x_2(0) = 0$, $x_1(T) = x_2(T) = 0$, and T is free to be selected in an optimal manner. (This is a special case of a problem treated by Boyadjieff et al. [8.15].)

8.18 Consider the system characterized by

$$\ddot{x} + x = m, \quad x(0) = 3, \quad \dot{x}(0) = 0$$

where a minimum of J is desired:

$$J = \int_0^5 \left(x^2 + \dot{x}^2 + \frac{w}{1-m^2}\right) dt$$

where w is a constant weighting factor. The $w/(1 - m^2)$ term is inserted in the integrand in lieu of the requirement that $|m| \leq 1$. An initial policy m^0 for m is $m^0 = -0.9 \sin t$, $0 \leq t \leq 5$. Use a computer in conjunction with the second-variation search technique of Section 8-8b (see the last paragraph of Section 8-8b) to obtain the second iterate m^2 for the case that $w = 0.2$. If at any instant t the use of 8-113 yields $|m^{k+1}(t)| > 0.99$, $m^{k+1}(t)$ is to be assigned the magnitude of 0.99.

8.19 A given temperature regulating system is characterized by

$$k_1 \dot{x}_1 = m + v_1(t) + k_2[v_2(t) - x_1 - x_d]$$

where x_1 is the temperature deviation from a desired constant temperature x_d, k_1 is a constant proportional to the mass that is being temperature-regulated, k_2 is a constant proportional to surface area and coefficient of heat conduction, $v_1(t)$ equals the number of BTU's per unit time that are generated by unregulated sources, $m = m(t)$ equals the number of BTU's per unit time that are generated by the controller, and $v_2(t)$ is the unregulated outside temperature. Both $v_1(t)$ and $v_2(t)$ are assumed to be known over a specified time period $[0, T]$ of interest. A minimum of J is desired:

$$J = \int_0^T (x_1^2 + w|m|)\, dt$$

where w is a constant weighting factor. The initial value $x_1(0)$ of x_1 is known, but $x_1(T)$ is free. At any instant t, the control action $m(t)$ is constrained to be one of three values: $m_{\max}$, 0, or $-m_{\max}$.

a. Give the Hamiltonian and the costate equation. Find a relationship between the optimal control action and the optimal costate variable. What is the terminal value of the costate variable?

b. In a particular case, the following conditions apply: $x_d = 70°$, $(m_{\max}/k_1) = 5°/\text{hr}$, $v_1/k_1 = 1°/\text{hr}$, $(k_2/k_1) = 1/4$, $w = (1/5k_1)$, $v_2(t) = 75 + 15 \sin [(2\pi/24)t]$. Problem time t is in hours: $t = 0$ corresponds to 9 a.m. on day one whereas $t = T = 24$ hours corresponds to the terminal time 9 a.m. on day two. Use a computer to apply a modified version of the search approach outlined in Section 8-8b. Step 5 of the search is simplified here because only 3 choices for the value of m exist at each instant of time. The constancy test in Step 6 is not applicable. Why?

8.20 Replace m as it appears in the state equation of the preceding problem with $k_1 x_2$ where x_2 is a second state variable governed by

$$\dot{x}_2 = k_3 \left(\frac{m}{k_1} - x_2 \right)$$

in which k_3 is a constant. Repeat Parts a and b of the preceding problem under this condition. Use $k_3 = 2$ in Part b, and base your initial guess of $m^*(t)$ on the result obtained in Part b of the preceding problem.

8.21 Consider the following nonlinear differential equation:

$$\ddot{x} + \dot{x} + 1.25x + x^3 = m$$

where, in mechanical systems, the cubic term may be associated with a hard spring. Initial conditions are $x(0) = 2$ and $\dot{x}(0) = 0$. Both $x(T)$ and $\dot{x}(T)$ are to equal zero. A minimum of J is desired:

$$J = \int_0^T m^4 \, dt$$

Use a computer in conjunction with the Newton-Raphson approach of Section 8-8c to generate optimal solutions for $T = 2$, $T = 4$, and $T = 8$.

8.22 Extend the Riccati approach of Section 7-14a to apply to the case considered in Section 8-8d.

8.23 Replace J of Problem 8.21 by

$$J = \int_0^T (x^2 + \dot{x}^2 + m^4) \, dt$$

and allow $x(T)$ and $\dot{x}(T)$ to be free. Other conditions of Problem 8.21 are unchanged.

a. The Riccati approach of Section 8-8e is to be used. Obtain the special cases of Equations 8-129, 8-130, 8-136, and 8-137 that apply.

b. With an initial guess $m^0(t) \equiv 0$ for all t, use a computer to solve the equations of Part a. Find the third iterate $m^3(t)$ corresponding to $T = 4$.

PROBLEMS

8.24 Consider the *maximization* of J:

$$J = \int_0^2 \frac{x}{2}\left(1 - 3\frac{m}{m_{max}}\right) dt$$

subject to

$$\dot{x} = -x + m, \quad x(0) = 0, \quad \text{and} \quad 0 \le m \le m_{max}$$

a. Give an appropriate Hamiltonian and a costate equation.
b. Solve for the optimal trajectories $\lambda^*(t)$, $m^*(t)$, and $x^*(t)$. (*Hint:* The problem is similar to Example 8-6, but with a different performance measure; a singular segment of the optimal trajectory is preceded and followed by nonsingular segments.)

8.25 Solve Equation 8-176 under the following conditions: $a_{21} = a_{22} = 0$, and $m(t)$ equals the $m^*(t)$ of Equation 8-194. Comment on the form of the solution and on the time required for $x_2(t)$ to equal $-x_1(t)$.

8.26 Derive Equations 8-196 through 8-198 of Example 8-7.

8.27 Given:

$$\dot{x}_1 = x_2 + m, \quad x_1(t_b) = 0, \quad |m| \le 1$$
$$\dot{x}_2 = -m, \quad x_2(t_b) = 0$$

where t_b is free. A feedback system is to be implemented to obtain a minimum of

$$J = \int_{t_a}^{t_b} (\tfrac{1}{2}x_1^2 + 1)\, dt$$

Determine an appropriate switch curve and check for optimality of singular solutions. (This is a special case of a problem considered by Womack and Dashiell [8.109].)

8.28 Given:

$$\dot{x} = -x + m, \quad x(0) = 5, \quad |m| \le 1, \quad 0 \le t \le T$$

where the terminal time T is not specified. A minimum of J is desired:

$$J = \tfrac{1}{2}x(T)^2 + \int_0^T (1 + \tfrac{1}{2}|m|)\, dt$$

a. List the pertinent necessary conditions that stem from the equivalent principle of Section 8-11b.
b. Solve the conditions of Part a to obtain the optimal $m(t)$.

8.29 Replace J of Problem 8.21 with

$$J = (x^2 + \dot{x}^2)_{t=T} + \int_0^T m^4\, dt$$

where $x(T)$ and $\dot{x}(T)$ are not specified, but other conditions of Problem 8.21 remain unchanged. List necessary conditions for optimality that stem from the equivalent principle of Section 8-11b.

MATRIX IDENTITIES AND OPERATIONS

appendix **A**

A listing of selected properties of matrices and matrix operations is given in this appendix. References [A.1, A.2, A.4, and A.6] can be consulted for comprehensive treatment of matrix theory.

Matrix

A matrix is an ordered, rectangular array of entries, usually numbers. Consider the matrix **A**:

$$\mathbf{A} = \begin{bmatrix} a_{11} & a_{12} & \cdots & a_{1n} \\ \vdots & & & \vdots \\ a_{m1} & & \cdots & a_{mn} \end{bmatrix} \quad \text{(A-1)}$$

which has m rows and n columns, i.e., **A** is an $m \times n$ matrix. A matrix which is $n \times 1$ is called a *column matrix* or a *column vector*; a matrix which is $1 \times n$ is called a *row matrix* or a *row vector*; and a matrix which is $n \times n$ is called a *square matrix*. For compactness, Equation A-1 is often expressed as $\mathbf{A} = (a_{ij})$. The entry a_{ij} of **A** is the entry from the ith row in the jth column. The entries of a vector are often called *components* or *elements*.

Scalar Multiplication

If α is a scalar,

$$\mathbf{A}\alpha = \alpha\mathbf{A} = (\alpha a_{ij}) = \begin{bmatrix} \alpha a_{11} & \alpha a_{12} & \cdots & \alpha a_{1n} \\ \alpha a_{21} & & & \cdot \\ \vdots & & & \vdots \\ \alpha a_{m1} & \cdot & \cdots & \alpha a_{mn} \end{bmatrix} \quad \text{(A-2)}$$

Addition

The matrix **A** can be added to the matrix **B** if and only if both **A** and **B** are of the same dimension:

$$\mathbf{A} + \mathbf{B} = \mathbf{B} + \mathbf{A} = (a_{ij} + b_{ij}) \tag{A-3}$$

Products of Matrices

The $m \times n$ matrix **A** can be multiplied on the right by any matrix **B** which is of dimension $n \times p$ for any positive integer p:

$$\mathbf{AB} = \left(\sum_{k=1}^{n} a_{ik} b_{kj}\right) \tag{A-4}$$

where the i,jth entry in the $m \times p$ product matrix **AB** is the sum indicated in the parentheses of A-4. *Note*: **AB** does not equal **BA**, in general; and the number of columns in **A** must equal the number of rows in **B** for A-4 to be well defined.

Associative Law

Given that **AB** and **BC** are well-defined matrix multiplications, it can be shown that

$$\mathbf{A(BC)} = \mathbf{(AB)C} \tag{A-5}$$

If **A** is an $n \times n$ matrix, the product of **A** times itself is denoted by $\mathbf{A}^2$. In general,

$$\mathbf{A}^k = \mathbf{A} \cdot \mathbf{A}^{k-1} \tag{A-6}$$

Distributive Law

$$\mathbf{A(B + C)} = \mathbf{AB} + \mathbf{AC} \tag{A-7}$$

and

$$\mathbf{(B + C)D} = \mathbf{BD} + \mathbf{CD} \tag{A-8}$$

Major Diagonal

The *major diagonal* of an $n \times n$ matrix is that which connects the upper left-hand corner to the lower right-hand corner.

Identity Matrix

An $n \times n$ *identity matrix*, denoted by **I**, is one which has 1's on the major diagonal and zeros elsewhere. For any $m \times n$ matrix **A**,

$$\mathbf{AI} = \mathbf{A} \tag{A-9}$$

Transpose

Given an $m \times n$ matrix $\mathbf{A}$ of real numbers, the transpose $\mathbf{A}'$ of $\mathbf{A}$ is an $n \times m$ matrix, the i,jth entry of which equals the j,ith entry of $\mathbf{A}$. It follows that

$$(\mathbf{A}')' = \mathbf{A} \tag{A-10}$$

and it can be shown that

$$(\mathbf{AB})' = \mathbf{B}'\mathbf{A}' \tag{A-11}$$

Symmetric Matrix

An $n \times n$ matrix $\mathbf{A}$ is said to be symmetric if $\mathbf{A} = \mathbf{A}'$.

Linear Combination

If $\mathbf{a}_1, \mathbf{a}_2, \ldots, \mathbf{a}_n$ are a set of vectors and $\alpha_1, \alpha_2, \ldots, \alpha_n$ are a set of scalars, the vector $\mathbf{b} = \alpha_1 \mathbf{a}_1 + \alpha_2 \mathbf{a}_2 + \cdots + \alpha_n \mathbf{a}_n$ is a linear combination of $\mathbf{a}_1, \mathbf{a}_2, \ldots, \mathbf{a}_n$.

Linear Independence

A set of row (column) vectors constitutes a linearly independent set if each of the vectors cannot be formed by a linear combination of the other vectors.

Basis

A set of n linearly independent $n \times 1$ vectors with real entries constitutes a *basis* for an n-dimensional space. Any vector in the n-dimensional space can be formed by a linear combination of the n linearly independent basis vectors.

Rank

Let $\mathbf{A}$ be an $m \times n$ matrix of real numbers. The *row rank* of $\mathbf{A}$ equals the number of linearly independent rows. The *column rank* of $\mathbf{A}$ equals the number of linearly independent columns. It can be shown that the row rank always equals the column rank, and the *rank* of $\mathbf{A}$ equals both the row and column ranks.

Augmented Matrices

Consider an $m \times n$ matrix $\mathbf{A}$ and an $m \times p$ matrix $\mathbf{B}$. If an $m \times (n + p)$ matrix is formed by placing the columns of $\mathbf{B}$ to the right of the columns of

A, the resulting matrix is called an augmented matrix and is denoted by $(\mathbf{A},\mathbf{B})$ or $[\mathbf{A}|\mathbf{B}]$.

Existence of Linear Equation Solutions

Consider the matrix equation
$$\mathbf{Ax} = \mathbf{b} \tag{A-12}$$
where $\mathbf{A}$ is $m \times n$, $\mathbf{x}$ is $n \times 1$, and $\mathbf{b}$ is $m \times 1$. The entries in $\mathbf{A}$ and $\mathbf{b}$ are assumed to be real numbers. The condition for solvability of A-12 is that the rank of $\mathbf{A}$ and the rank of the augmented matrix $(\mathbf{A},\mathbf{b})$ be the same.

If the rank of $\mathbf{A}$ does not equal the rank of $(\mathbf{A},\mathbf{b})$, it is sometimes useful to obtain an $\mathbf{x}$ which minimizes $(\mathbf{Ax} - \mathbf{b})'(\mathbf{Ax} - \mathbf{b})$. The reader may show that any vector value $\mathbf{x}^*$ of $\mathbf{x}$ which minimizes $(\mathbf{Ax} - \mathbf{b})'(\mathbf{Ax} - \mathbf{b})$ must satisfy
$$\mathbf{A'Ax}^* = \mathbf{A'b} \tag{A-13}$$

Partitioned Matrices

A partitioned matrix is one in which one or more of the entries is itself a matrix. The matrix entries of such a partitioned matrix are called *submatrices*.

Multiplication of Partitioned Matrices

Consider partitioned matrices of the form
$$\mathbf{A} = \begin{bmatrix} \mathbf{A}_{11} & \cdots & \mathbf{A}_{1p} \\ \vdots & & \vdots \\ \mathbf{A}_{q1} & \cdots & \mathbf{A}_{qp} \end{bmatrix} \tag{A-14}$$
and
$$\mathbf{B} = \begin{bmatrix} \mathbf{B}_{11} & \cdots & \mathbf{B}_{1r} \\ \vdots & & \vdots \\ \mathbf{B}_{p1} & \cdots & \mathbf{B}_{pr} \end{bmatrix} \tag{A-15}$$
where $\mathbf{A}$ and $\mathbf{B}$ are partitioned by horizontal and vertical lines.

If the submatrices $\mathbf{A}_{ij}$ and $\mathbf{B}_{jk}$ are such that products $\mathbf{A}_{ij}\mathbf{B}_{jk}$ are well defined for all i, j, and k, the product $\mathbf{AB}$ is
$$\mathbf{AB} = \begin{bmatrix} \mathbf{C}_{11} & \cdots & \mathbf{C}_{1r} \\ \vdots & & \vdots \\ \mathbf{C}_{q1} & \cdots & \mathbf{C}_{qr} \end{bmatrix} \tag{A-16}$$
where $\mathbf{C}_{ij} = \mathbf{A}_{i1}\mathbf{B}_{1j} + \mathbf{A}_{i2}\mathbf{B}_{2j} + \cdots + \mathbf{A}_{ip}\mathbf{B}_{pj}$.

MATRIX IDENTITIES AND OPERATIONS

Diagonal Matrix

An $n \times n$ matrix $\mathbf{A}$ is a diagonal matrix if all a_{ij} equal zero when $i \neq j$.

Generalized Diagonal Matrix

An $n \times n$ matrix $\mathbf{A}$ is a generalized diagonal matrix if its major diagonal entries are also major diagonal entries of partitioned square submatrices and if all entries not included in the square diagonal submatrices are zero. Such a matrix is denoted by

$$\mathbf{A} = \text{diag}(\mathbf{A}_1, \mathbf{A}_2, \ldots, \mathbf{A}_k) \tag{A-17}$$

where the submatrices $\mathbf{A}_1, \mathbf{A}_2, \ldots, \mathbf{A}_k$ are square matrices, not necessarily of equal dimension, which appear on the major diagonal.

Determinants

Consider an $n \times n$ matrix $\mathbf{A} = (a_{ij})$. The determinant of $\mathbf{A}$, denoted by $|\mathbf{A}|$, is a scalar which can be computed in many ways.

$$|\mathbf{A}| = \sum_{j=1}^{n} a_{ij} c_{ij}, \qquad i \in \{1, 2, \ldots, n\} \tag{A-18a}$$

or

$$|\mathbf{A}| = \sum_{i=1}^{n} a_{ij} c_{ij}, \qquad j \in \{1, 2, \ldots, n\} \tag{A-18b}$$

where

$$c_{ij} = (-1)^{i+j} |\mathbf{M}_{ij}|$$

in which $\mathbf{M}_{ij}$ is the $(n-1) \times (n-1)$ submatrix of $\mathbf{A}$ obtained by deleting the ith row and jth column. The determinant of $\mathbf{M}_{ij}$ is called the ijth *minor* of $\mathbf{A}$, and the scalar c_{ij} is called the *cofactor* of a_{ij}. Just as $|\mathbf{A}|$ is linearly related to the determinant of the $(n-1) \times (n-1)$ matrix $\mathbf{M}_{ij}$, so also is $|\mathbf{M}_{ij}|$ linearly related to the determinants of $(n-2) \times (n-2)$ matrices. In the limit, the determinants of 1×1 matrices, scalars, are required to compute $|\mathbf{A}|$; the determinant of a scalar equals the scalar itself.

Determinants of Matrix Products

Given two $n \times n$ matrices $\mathbf{A}$ and $\mathbf{B}$, $|\mathbf{AB}| = |\mathbf{A}| \cdot |\mathbf{B}|$.

Determinants of Generalized Diagonal Matrices

If $\mathbf{A} = \text{diag}(\mathbf{A}_1, \mathbf{A}_2, \ldots, \mathbf{A}_k)$ is a generalized diagonal matrix,

$$|\mathbf{A}| = |\mathbf{A}_1| \cdot |\mathbf{A}_2| \cdot |\mathbf{A}_3| \cdots |\mathbf{A}_k|$$

Singular Matrix

An $n \times n$ matrix $\mathbf{A}$ is singular if $|\mathbf{A}| = 0$, otherwise $\mathbf{A}$ is nonsingular. If rank $\mathbf{A}$ is n, $|\mathbf{A}| \neq 0$; but if rank $\mathbf{A}$ is less than n, $|\mathbf{A}| = 0$.

Inverse

If, and only if, $\mathbf{A}$ is nonsingular does there exist a unique inverse matrix denoted by $\mathbf{A}^{-1}$ with the property that $\mathbf{A}\mathbf{A}^{-1} = \mathbf{A}^{-1}\mathbf{A} = \mathbf{I}$, an $n \times n$ identity matrix.

Given an $n \times n$ matrix $\mathbf{A}$ and an $n \times 1$ matrix $\mathbf{b}$ with scalar entries, a unique $n \times 1$ solution $\mathbf{x} = \mathbf{A}^{-1}\mathbf{b}$ of $\mathbf{A}\mathbf{x} = \mathbf{b}$ exists if $\mathbf{A}^{-1}$ exists. Computational methods of evaluating $\mathbf{A}^{-1}$ are given in various works (e.g., [A.3 and A.5]).

Inverse Identities

If $\mathbf{A}^{-1}$ and $\mathbf{D}^{-1}$ exist, $(\mathbf{A}')^{-1} = (\mathbf{A}^{-1})'$ and $(\mathbf{A}\mathbf{D})^{-1} = \mathbf{D}^{-1}\mathbf{A}^{-1}$.

Frobenius's Relation for Inversion [A.2]

Consider the partitioned matrix

$$\mathbf{P} = \begin{bmatrix} \mathbf{A} & \mathbf{B} \\ (n \times n) & (n \times m) \\ \mathbf{C} & \mathbf{D} \\ (m \times n) & (m \times m) \end{bmatrix} \quad (A\text{-}19)$$

where $\mathbf{A}$ and $\mathbf{D}$ are assumed to be nonsingular. The inverse $\mathbf{P}^{-1}$ of $\mathbf{P}$ can be expressed as

$$\mathbf{P}^{-1} = \begin{bmatrix} \mathbf{A}^{-1} + \mathbf{A}^{-1}\mathbf{B}\mathbf{E}^{-1}\mathbf{C}\mathbf{A}^{-1} & -\mathbf{A}^{-1}\mathbf{B}\mathbf{E}^{-1} \\ -\mathbf{E}^{-1}\mathbf{C}\mathbf{A}^{-1} & \mathbf{E}^{-1} \end{bmatrix} \quad (A\text{-}20)$$

in which $\mathbf{E} = \mathbf{D} - \mathbf{C}\mathbf{A}^{-1}\mathbf{B}$.

Inverse of a Generalized Diagonal Matrix

The inverse $\mathbf{A}^{-1}$ of $\mathbf{A} = \text{diag}(\mathbf{A}_1, \mathbf{A}_2, \ldots, \mathbf{A}_k)$ is $\mathbf{A}^{-1} = \text{diag}(\mathbf{A}_1^{-1}, \mathbf{A}_2^{-1}, \ldots, \mathbf{A}_k^{-1})$.

Inner Product

If $\mathbf{x}$ and $\mathbf{y}$ are $n \times 1$ vectors, the product $\mathbf{x}'\mathbf{y}$ is a scalar and is called the inner product of $\mathbf{x}$ and $\mathbf{y}$.

MATRIX IDENTITIES AND OPERATIONS

Outer Product [Diadic Product]

If **x** and **y** are $n \times 1$ vectors, the product **xy**$'$ is an $n \times n$ matrix and is called an outer product.

Trace

Let **A** be an $n \times n$ matrix. The trace of **A** is the sum of the major-diagonal entries of **A**:

$$\text{trace } \mathbf{A} = \sum_{i=1}^{n} a_{ii} \qquad \text{(A-21)}$$

It can be shown that trace (**A**) equals the sum of the eigenvalues of **A**. Also, if **x** and **y** are $n \times 1$ vectors, $\mathbf{x}'\mathbf{y} = \text{trace }(\mathbf{xy}')$.

Nontrivial Vector

If **x** is a vector with at least one nonzero entry, **x** is a nontrivial vector.

Orthogonal Vectors

If **x** and **y** are nontrivial $n \times 1$ vectors and $\mathbf{x}'\mathbf{y} = 0$, **x** and **y** are mutually orthogonal.

Orthonormal Vectors

Two $n \times 1$ vectors **x** and **y** are orthonormal if they are orthogonal and

$$\mathbf{x}'\mathbf{x} = \mathbf{y}'\mathbf{y} = 1.$$

Projection of a Vector

Given two $n \times 1$ vectors **x** and **y** with real entries, the vector projection of **x** on **y** is $(\mathbf{x}'\mathbf{y}/\mathbf{y}'\mathbf{y})\mathbf{y}$.

Gradient Vector

Suppose a scalar function $f(\mathbf{x})$ of n variables $\mathbf{x} = \{x_1, x_2, \ldots, x_n\}$ is given and is of class C^1. The gradient vector $\partial f(\mathbf{x})/\partial \mathbf{x}$ can be expressed in terms of an $n \times 1$ column vector $[\partial f/\partial x_1 \quad \partial f/\partial x_2 \quad \cdots \quad \partial f/\partial x_n]'$.

Vector Derivative of a Row Matrix

Consider a row matrix $\mathbf{q} = [q_1(\mathbf{x}) \quad q_2(\mathbf{x}) \quad \cdots \quad q_k(\mathbf{x})]$ in which each entry is a scalar function of n variables and is of class C^1. The derivative of **q** with respect to $\mathbf{x}$ is an $n \times k$ matrix:

$$\frac{\partial \mathbf{q}}{\partial \mathbf{x}} = \begin{bmatrix} \frac{\partial q_1}{\partial \mathbf{x}} & \cdots & \frac{\partial q_k}{\partial \mathbf{x}} \end{bmatrix} \qquad (A\text{-}22)$$

Vector Derivative of a Column Matrix

Consider a column matrix $\mathbf{f} = [f_1(\mathbf{x}) \; f_2(\mathbf{x}) \; \cdots \; f_k(\mathbf{x})]'$ in which each entry is a scalar function of n variables and is of class C^1. By convention, the derivative $\partial \mathbf{f}/\partial \mathbf{x}$ of $\mathbf{f}$ with respect to $\mathbf{x}$ is a $k \times n$ matrix:

$$\frac{\partial \mathbf{f}}{\partial \mathbf{x}} = \begin{bmatrix} \frac{\partial f_1}{\partial x_1} & \frac{\partial f_1}{\partial x_2} & \cdots & \frac{\partial f_1}{\partial x_n} \\ \vdots & & & \vdots \\ \frac{\partial f_k}{\partial x_1} & & \cdots & \frac{\partial f_k}{\partial x_n} \end{bmatrix}, \quad k \neq 1 \qquad (A\text{-}23)$$

With this convention, $(\partial \mathbf{f}/\partial \mathbf{x})' = \partial(\mathbf{f}')/\partial \mathbf{x}$. The matrix of A-23 is sometimes called the *Jacobian matrix* of $\mathbf{f}$ with respect to $\mathbf{x}$.

Vector Derivative of an Inner Product

Consider two $k \times 1$ column vectors $\mathbf{f}$ and $\mathbf{g}$, the entries of which are of class C^1 with respect to the $n \times 1$ vector $\mathbf{x}$ of real variables. The gradient of $\mathbf{f}'\mathbf{g}$ with respect to $\mathbf{x}$ is given by

$$\frac{\partial}{\partial \mathbf{x}} \mathbf{f}'\mathbf{g} = \left(\frac{\partial}{\partial \mathbf{x}} \mathbf{f}' \right)\mathbf{g} + \left(\frac{\partial}{\partial \mathbf{x}} \mathbf{g}' \right)\mathbf{f} \qquad (A\text{-}24)$$

Common Matrix Derivatives

Let $\mathbf{x}$ be an $n \times 1$ vector of real variables, and let $\mathbf{A}$ be an $n \times n$ matrix which is independent of the entries of $\mathbf{x}$. Useful derivative identities are

$$\frac{\partial(\mathbf{x}'\mathbf{A})}{\partial \mathbf{x}} = \mathbf{A} \qquad (A\text{-}25a)$$

and

$$\frac{\partial(\mathbf{x}'\mathbf{A}\mathbf{x})}{\partial \mathbf{x}} = \mathbf{A}\mathbf{x} + \mathbf{A}'\mathbf{x} \qquad (A\text{-}25b)$$

MATRIX IDENTITIES AND OPERATIONS

Taylor's Series Expansion

If $f(\mathbf{x})$ is a real-valued function which is analytic at a point $\mathbf{x} = \mathbf{x}_a$ of a real n-dimensional Euclidean domain, $f(\mathbf{x})$ can be expanded in a Taylor's series about the point $\mathbf{x}_a$:

$$f(\mathbf{x}) = f(\mathbf{x}_a) + (\mathbf{x} - \mathbf{x}_a)' \frac{\partial f(\mathbf{x}_a)}{\partial \mathbf{x}} + \frac{1}{2!}(\mathbf{x} - \mathbf{x}_a)' \mathbf{A}(\mathbf{x} - \mathbf{x}_a) + \cdots \quad \text{(A-26)}$$

which is valid in some open neighborhood of the point $\mathbf{x}_a$. Operationally, both $\mathbf{x}$ and $\mathbf{x}_a$ are column vectors, and $\mathbf{A}$ is an $n \times n$ matrix:

$$\mathbf{A} = \begin{bmatrix} \frac{\partial^2 f}{\partial x_1^2} & \cdots & \frac{\partial^2 f}{\partial x_1 \partial x_n} \\ \vdots & & \vdots \\ \frac{\partial^2 f}{\partial x_n \partial x_1} & \cdots & \frac{\partial^2 f}{\partial x_n^2} \end{bmatrix}_{\mathbf{x}=\mathbf{x}_a} \quad \text{(A-27)}$$

which is called the *Hessian matrix* of f.

Quadratic Form

The function $f(\mathbf{x})$ is a quadratic form if

$$f(\mathbf{x}) = \mathbf{x}'\mathbf{A}\mathbf{x} \quad \text{(A-28)}$$

where $\mathbf{x}$ is an $n \times 1$ vector of real variables, and $\mathbf{A}$ is an $n \times n$ symmetric matrix of real numbers.

Positive-Definite and Positive-Semidefinite Quadratic Forms

A quadratic form $f(\mathbf{x})$ over the set of reals is positive-definite if $f(\mathbf{x})$ is greater than zero for all nontrivial $\mathbf{x}$ and $f(\mathbf{0}) = 0$; it is positive-semidefinite if it has no negative values.

A symmetric matrix $\mathbf{A}$ of reals is positive-definite (-semidefinite) if the corresponding $f(\mathbf{x})$ is positive-definite (-semidefinite).

Characteristic Roots or Eigenvalues

Given an $n \times n$ matrix $\mathbf{A}$ of scalars, the characteristic roots or *eigenvalues* of $\mathbf{A}$ are the roots of the nth-order *characteristic equation*

$$|\lambda \mathbf{I} - \mathbf{A}| = \lambda^n + c_{n-1}\lambda^{n-1} + \cdots + c_0 = 0 \quad \text{(A-29)}$$

where the c_i's are appropriate constants.

Characteristic Vector or Eigenvector

If and only if r is an eigenvalue of $\mathbf{A}$ does there exist a vector $\mathbf{z}$ such that

$$\mathbf{Az} = r\mathbf{z} \qquad (A\text{-}30)$$

The vector $\mathbf{z}$ is called a characteristic vector or an *eigenvector*.

Similarity

Two $n \times n$ matrices $\mathbf{A}$ and $\mathbf{B}$ are said to be *similar* if there exists a nonsingular matrix $\mathbf{P}$ such that

$$\mathbf{B} = \mathbf{P}^{-1}\mathbf{AP} \qquad (A\text{-}31)$$

Matrices Similar to Diagonal Matrices

If there exists a diagonal matrix $\mathbf{D}$ which is similar to a given matrix $\mathbf{A}$, the diagonal elements of $\mathbf{D}$ are the eigenvalues of $\mathbf{A}$. If the eigenvalues of $\mathbf{A}$ are distinct, $\mathbf{A} = \mathbf{PDP}^{-1}$ where the ith column of $\mathbf{P}$ is an eigenvector that corresponds to the eigenvalue in the ith column of $\mathbf{D}$. For any $n \times n$ symmetric matrix $\mathbf{A}$ of reals, there exists a set of n linearly independent eigenvectors which are mutually orthogonal; if these vectors are normalized to unity magnitude and are used as the columns of a transformation matrix $\mathbf{P}$, then $\mathbf{P}' = \mathbf{P}^{-1}$; and if the nonsingular transformation $\mathbf{x} = \mathbf{Py}$ is made, $\mathbf{x}'\mathbf{Ax} = \mathbf{y}'\mathbf{P}'\mathbf{APy} = \mathbf{y}'\mathbf{Dy}$ where $\mathbf{D}$ is a diagonal matrix of eigenvalues.

Diagonal Similarity Property for a Positive-Definite Matrix

A real, symmetric matrix $\mathbf{A}$ is positive-definite if and only if it is similar to a diagonal matrix with positive real diagonal entries.

Matrix Exponential

Given an $n \times n$ matrix $\mathbf{A}$ with real, constant entries, $e^{\mathbf{A}t}$ is used to denote an $n \times n$ matrix exponential:

$$e^{\mathbf{A}t} = \mathbf{I} + \mathbf{A}t + \frac{1}{2!}\mathbf{A}^2 t^2 + \cdots \qquad (A\text{-}32)$$

The matrix exponential is useful in the solution of sets of linear, first-order, differential equations. Consider the homogeneous set $\dot{\mathbf{x}} = \mathbf{A}\mathbf{x}$ where $\mathbf{x} = \mathbf{x}(t)$ is an $n \times 1$ matrix of real variables, and $\mathbf{x}(0)$ is specified. The solution $\mathbf{x}(t)$ for $t \geq 0$ equals $e^{\mathbf{A}t}\mathbf{x}(0)$.

A Linear Transformation

Consider the following set of first-order differential equations:

$$\dot{\mathbf{x}} = \mathbf{A}\mathbf{x} + \mathbf{B}\mathbf{m} \tag{A-33}$$

where $\mathbf{A}$ is an $n \times n$ matrix of real constants, $\mathbf{B}$ is an $n \times r$ matrix of real constants, $\mathbf{x} = \mathbf{x}(t)$ is an $n \times 1$ state vector, $\mathbf{m}$ is an $r \times 1$ control vector, and $\mathbf{x}(t_a)$ is specified. If there exists a diagonal matrix $\mathbf{D}$ which is similar to $\mathbf{A}$, the vector state equation A-33 may be solved more readily by introducing a new state vector $\mathbf{y}$:

$$\mathbf{y} = \mathbf{P}^{-1}\mathbf{x} \tag{A-34}$$

where the property $\mathbf{P}^{-1}\mathbf{A}\mathbf{P} = \mathbf{D}$ is used in A-33 to obtain

$$\begin{aligned}\dot{\mathbf{y}} &= \mathbf{P}^{-1}\dot{\mathbf{x}} \\ &= \mathbf{P}^{-1}\mathbf{A}(\mathbf{P}\mathbf{P}^{-1})\mathbf{x} + \mathbf{P}^{-1}\mathbf{B}\mathbf{m} \\ &= \mathbf{D}\mathbf{y} + \mathbf{P}^{-1}\mathbf{B}\mathbf{m}\end{aligned} \tag{A-35}$$

The solution for the $\mathbf{y}$ that satisfies a specified $\mathbf{y}(t_a)$ can be expressed in terms of a convolution integral:

$$\mathbf{y}(t) = e^{\mathbf{D}(t-t_a)}\mathbf{y}(t_a) + \int_{t_a}^{t} e^{\mathbf{D}(t-\tau)}\mathbf{P}^{-1}\mathbf{B}\mathbf{m}(\tau)\, d\tau \tag{A-36}$$

for $t \geq t_a$.

REFERENCES

[A.1] Bellman, R. *Introduction to Matrix Analysis*. McGraw-Hill, New York, 1960.

[A.2] Bodewig, E. *Matrix Calculus*. Interscience, New York, 1959 (2nd edition).

[A.3] Faddeeva, V. N. *Computational Methods of Linear Algebra*. Translated from the Russian by C. D. Benster. Dover Publications, New York, 1959.

[A.4] Frame, J. S. "Matrix Functions and Applications" (Five Parts: Part III coauthored by H. E. Koenig). *IEEE Spectrum*, **1**, March 1964 (Part I, 208–220); April 1964 (Part II, 102–108); May 1964 (Part III, 100–109); June 1964 (Part IV, 123–131); July 1964 (Part V, 103–109).

[A.5] Lanczos, C. *Applied Analysis*. Prentice Hall, Englewood Cliffs, N.J., 1961 (2nd printing).

[A.6] Perlis, S. *Theory of Matrices*. Addison-Wesley, Reading, Mass., 1958 (3rd printing).

TWO-SIDED LAPLACE TRANSFORM THEORY

appendix **B**

The Direct Transform

Let $f(t)$ be a real- and single-valued function, and assume that there is at least one finite σ for which the following integral exists:

$$\int_{-\infty}^{\infty} e^{-\sigma t} |f(t)| \, dt \tag{B-1}$$

Corresponding to such an $f(t)$, there exists a unique function $F(s) \equiv \mathscr{L}[f(t)]$:

$$F(s) \triangleq \int_{-\infty}^{\infty} f(t) e^{-st} \, dt \tag{B-2}$$

which is called the *two-sided Laplace transform* of $f(t)$.[1] For notational compactness in the following paragraphs, the phrase "Laplace transform" is used in place of the longer phrase "two-sided Laplace transform."

If $f(t)$ satisfies the above noted conditions, it can be shown that there exists a region $c_1 < \text{Re}(s) < c_2$ in the complex s plane called the *strip of convergence* for the pair $\{f(t), F(s)\}$. Any given s in the strip of convergence of $\{f(t), F(s)\}$ yields convergence of the integral B-2. For any s not in the strip of convergence, the integral B-2 does not exist. Of course, when the strip of convergence of a given $\{f(t), F(s)\}$ includes the $j\omega$ axis of the s plane, the pair $\{f(t), F(j\omega)\}$ is a *Fourier transform* pair.

---

[1] The two-sided Laplace transform is also called the bilateral Laplace transform.

TWO-SIDED LAPLACE TRANSFORM THEORY

For example, consider $f(t) = e^{-a|t|}$. The corresponding $F(s)$ is

$$F(s) = \int_{-\infty}^{\infty} e^{-a|t|} e^{-st} \, dt$$

$$= \int_{0}^{\infty} e^{-at} e^{-st} \, dt + \int_{-\infty}^{0} e^{at} e^{-st} \, dt$$

$$= \frac{1}{s+a} + \frac{1}{-s+a}$$

$$= \frac{2a}{(s+a)(-s+a)} \quad \text{for } -a < \text{Re}(s) < a \quad \text{(B-3)}$$

Consequently, the strip of convergence for the pair $\{e^{-a|t|}, 2a/(s+a)(-s+a)\}$ is specified by $-a < \text{Re}(s) < a$.

The Inverse Transform

It can be shown that the *inverse Laplace transform* $f(t)$ of a Laplace transform $F(s)$ is given by

$$f(t) = \frac{1}{2\pi j} \int_{c-j\infty}^{c+j\infty} F(s) e^{st} \, ds \quad \text{(B-4)}$$

where c is any real constant which lies within the strip of convergence of the pair $\{f(t), F(s)\}$. The inverse transformation is unique, at least within a set of functions which differ from each other by *zero measure*.

In many cases, it is convenient to evaluate the inverse transformation by using Cauchy's residue theorem from the theory of complex variables. Given that a function $Q(s)$ has an isolated pole[2] at $s = p$, the associated residue of $Q(s)$ can be determined on the basis of a Laurent expansion:

$$Q(s) = \sum_{k=-\infty}^{\infty} \frac{a_k}{(s-p)^k} \quad \text{(B-5)}$$

which, with appropriate scalar a_k's, is a valid representation of $Q(s)$ in an open neighborhood of $s = p$. By definition, the a_1 coefficient in B-5 is the *residue* associated with the pole of $Q(s)$ at $s = p$.

If $Q(s)$ has an nth-order pole at $s = p$, the residue a_1 equals:

$$a_1 = \frac{1}{(n-1)!} \lim_{s \to p} \left\{ \frac{d^{n-1}}{ds^{n-1}} [(s-p)^n Q(s)] \right\} \quad \text{(B-6)}$$

[2] A function $Q(s)$ has an isolated, nth-order pole at $s = p$ if: (1) the function $(s-p)^n Q(s)$ is analytic in an open neighborhood of $s = p$; and (2) the limit of $(s-p)^{n-1} Q(s)$ as s approaches p is unbounded.

For example, suppose that $Q(s) = e^{st}/(s + 1)^3$ for any finite value of t. This $Q(s)$ has a third-order pole at $s = -1$ with associated residue

$$a_1 = \frac{1}{2} \lim_{s \to -1} \frac{d^2}{ds^2} e^{st}$$

$$= \tfrac{1}{2} t^2 e^{-t} \tag{B-7}$$

Consider a Laplace transform $F(s)$ of the form

$$F(s) = [F_1(s) + F_2(s)]e^{sT} \tag{B-8}$$

where T is a real constant and where both $F_1(s)$ and $F_2(s)$ are ratios of polynomials in s with denominators of higher order than numerators; i.e., $F_1(s)$ and $F_2(s)$ are *proper rational fraction functions*. The functions $F_1(s)$ and $F_2(s)$ differ in that $F_1(s)$ is assumed to contain no poles to the right of the given strip of convergence of $\{f(t), F(s)\}$ whereas $F_2(s)$ is assumed to contain no poles to the left of the given strip of convergence. Obviously, we can in all cases expand a given proper rational fraction function into two functions $F_1(s)$ and $F_2(s)$ with the above assigned properties.

The inverse Laplace transform of the $F(s)$ in Equation B-8 is obtained by using the inversion formula B-4:

$$f(t) = \frac{1}{2\pi j} \int_{c-j\infty}^{c+j\infty} [F_1(s) + F_2(s)]e^{s(T+t)} \, ds \tag{B-9}$$

For $t \geq -T$, it can be shown that the integral of $[F_1(s) + F_2(s)]e^{s(t+T)}$ along the arc da_1b (see Figure B-1) approaches zero as R approaches infinity.

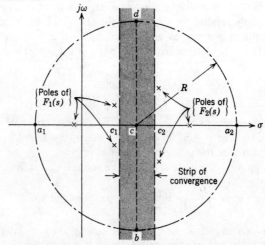

Figure B-1. s-plane plot ($s = \sigma + j\omega$) of typical poles of $[F_1(s) + F_2(s)]e^{s(t+T)}$.

Hence, for $t \geq -T$, the integral from $c - j\infty$ to $c + j\infty$ of $[F_1(s) + F_2(s)]e^{s(T+t)}$ must equal the integral of $[F_1(s) + F_2(s)]e^{s(T+t)}$ obtained by integrating around the closed contour a_1bcda_1 (as $R \to \infty$); but *Cauchy's residue theorem* specifies that this latter counterclockwise integral equals the sum of the residues of $F_1(s)e^{s(T+t)}$ at the poles of $F_1(s)$. Thus,

$$f(t) = \sum_{\substack{\text{over poles} \\ \text{of } F_1(s)}} \text{Residues of } [F_1(s)e^{s(T+t)}] \quad \text{for } t \geq -T$$

$$= \sum_{\substack{\text{over poles} \\ \text{of } F_1(s)}} \text{Residues of } [F_1(s) + F_2(s)]e^{s(T+t)}$$

$$= \sum_{\substack{\text{over poles} \\ \text{of } F_1(s)}} \text{Residues of } [F(s)e^{st}] \quad \text{for } t \geq -T \quad \text{(B-10)}$$

For $t < -T$, it can be shown that the integral of $[F_1(s) + F_2(s)]e^{s(t+T)}$ along the arc da_2b approaches zero as R approaches infinity. Hence, the integral from $c - j\infty$ to $c + j\infty$ of $[F_1(s) + F_2(s)]e^{s(t+T)}$ must equal the integral of $[F_1(s) + F_2(s)]e^{s(t+T)}$ obtained by integrating around the closed contour a_2bcda_2 (as $R \to \infty$); but again, as dictated by Cauchy's residue theorem, this latter clockwise integral equals the negative of the sum of the residues of $F_2(s)e^{s(T+t)}$ at the poles of $F_2(s)$. Thus,

$$f(t) = -\sum_{\substack{\text{over poles} \\ \text{of } F_2(s)}} \text{Residues of } [F_2(s)e^{s(T+t)}] \quad \text{for } t < -T$$

$$= -\sum_{\substack{\text{over poles} \\ \text{of } F_2(s)}} \text{Residues of } [F_1(s) + F_2(s)]e^{s(T+t)}$$

$$= -\sum_{\substack{\text{over poles} \\ \text{of } F_2(s)}} \text{Residues of } [F(s)e^{st}] \quad \text{for } t < -T \quad \text{(B-11)}$$

Combined use of Equations B-10 and B-11 gives the function $f(t)$ for all t. For a specific example of the above form of $F(s)$, consider the $F(s)$ given by

$$F(s) = \left[\frac{1}{s+a} + \frac{1}{-s+a}\right]e^{sT}, \quad -a < \text{Re}(s) < a \quad \text{(B-12)}$$

Equation B-10 yields

$$f(t) = e^{-a(t+T)} \quad \text{for } t \geq -T \quad \text{(B-13)}$$

and Equation B-11 yields

$$f(t) = e^{a(t+T)} \quad \text{for } t < -T \quad \text{(B-14)}$$

Thus,

$$f(t) = e^{-a|t+T|} \tag{B-15}$$

for all t.

In system analysis, we often have occasion to find the inverse Laplace transform of products of Laplace transformed functions. Given two Laplace transforms, $F(s)$ with strip of convergence $c_1 < \text{Re}(s) < c_2$, and $G(s)$ with strip of convergence $k_1 < \text{Re}(s) < k_2$, the product $F(s)G(s)$ defines a Laplace transform for which a unique inverse exists if and only if both $c_2 > k_1$ and $k_2 > c_1$, i.e., the intersection of the strips of convergence must be nonempty. The strip of convergence for $F(s)G(s)$ is the intersection of the strip of convergence of $F(s)$ with that of $G(s)$.

Parseval's Theorem

Consider two real-valued functions $x(t)$ and $y(t)$ which are assumed to be Laplace transformable. Also, assume that the product $x(t)y(t)$ is Laplace transformable and that the strip of convergence includes the $j\omega$ axis of the s plane. Under these conditions, Parseval's theorem dictates that

$$\int_{-\infty}^{\infty} x(t)y(t)\,dt = \frac{1}{2\pi j}\int_{-j\infty}^{j\infty} X(s)Y(-s)\,ds = \frac{1}{2\pi}\int_{-\infty}^{\infty} X(j\omega)Y(-j\omega)\,d\omega \tag{B-16}$$

A proof of this theorem can be found in most textbooks which develop Laplace transform theory.

Integral of the Squared Error

An important application of Parseval's theorem is that in which $x(t)$ equals $y(t)$ equals a dynamic system error $e(t)$ in B-16, perhaps one which is to be minimized with respect to parameters of a system. In such a case, the integral ISE of the squared error is given by

$$\text{ISE} = \frac{1}{2\pi j}\int_{-j\infty}^{j\infty} E(s)E(-s)\,ds = \frac{1}{2\pi}\int_{-\infty}^{\infty} |E(j\omega)|^2\,d\omega \tag{B-17}$$

and is often called the integral-square error. Booton *et al.* [B.1] have given a particularly convenient method for the evaluation of ISE when $E(s) = C(s)/D(s)$ is a proper rational fraction. The results of the method are given in Table B-1. As for the associated theory and an extended version of Table B-1, the reader may consult reference [B.2].

TWO-SIDED LAPLACE TRANSFORM THEORY

TABLE B-1.

$C(s) = c_{n-1}s^{n-1} + \cdots + c_1 s + c_0$, $D(s) = d_n s^n + \cdots + d_1 s + d_0$ where $D(s)$ has zeros in the left half-plane only.

n	$\dfrac{1}{2\pi j}\displaystyle\int_{-j\infty}^{j\infty} \dfrac{C(s)C(-s)}{D(s)D(-s)}\,ds$
1	$\dfrac{c_0^2}{2d_0 d_1}$
2	$\dfrac{c_1^2 d_0 + c_0^2 d_2}{2 d_0 d_1 d_2}$
3	$\dfrac{c_2^2 d_0 d_1 + (c_1^2 - 2c_0 c_2) d_0 d_3 + c_0^2 d_2 d_3}{2 d_0 d_3 (-d_0 d_3 + d_1 d_2)}$
4	$\dfrac{c_3^2(-d_0^2 d_3 + d_0 d_1 d_2) + (c_2^2 - 2c_1 c_3) d_0 d_1 d_4 + (c_1^2 - 2c_0 c_2) d_0 d_3 d_4 + c_0^2(-d_1 d_4^2 + d_2 d_3 d_4)}{2 d_0 d_4 (-d_0 d_3^2 - d_1^2 d_4 + d_1 d_2 d_3)}$
5	$\dfrac{1}{m_5}[c_4^2 m_0 + (c_3^2 - 2c_2 c_4) m_1 + (c_2^2 - 2c_1 c_3 + 2c_0 c_4) m_2 + (c_1^2 - 2c_0 c_2) m_3 + c_0^2 m_4]$ where $m_0 = (d_3 m_1 - d_1 m_2)/d_5 \qquad m_3 = (d_2 m_2 - d_4 m_1)/d_0$ $m_1 = -d_0 d_3 + d_1 d_2 \qquad\quad m_4 = (d_2 m_3 - d_4 m_2)/d_0$ $m_2 = -d_0 d_5 + d_1 d_4 \qquad\quad m_5 = 2 d_0 (d_1 m_4 - d_3 m_3 + d_5 m_2)$

For additional information on the two-sided Laplace transform and on linear systems theory in general, the reader may consult references [B.3, B.4, and B.5].

REFERENCES

[B.1] Booton, R. C., Jr., M. V. Mathews, and W. W. Seifert. *Nonlinear Servomechanisms with Random Inputs*. Report 70, Dynamic Analysis and Control Laboratory, Massachusetts Institute of Technology, Cambridge, Mass., 1953.

[B.2] Newton, G. C., Jr., L. A. Gould, and J. F. Kaiser. *Analytical Design of Linear Feedback Controls*. Wiley, New York, 1957.

[B.3] Pierre, D. A. "Transforms, Generalized Functions, and All That." *IEEE Transactions on Education*, E-9, 103–105, June 1966.

[B.4] Schwarz, R. J., and B. Friedland. *Linear Systems*. McGraw-Hill, New York, 1965.

[B.5] van der Pol, B., and H. Bremmer. *Operational Calculus Based on the Two-Sided Laplace Integral*. Cambridge Univ. Press, New York, 1950.

CORRELATION FUNCTIONS AND POWER-DENSITY SPECTRA

appendix C

Correlation Functions

Consider a function $y(t)$ which has the following properties:

$$\lim_{T \to \infty} \int_0^T |y(t)|\, dt \to \infty \tag{C-1}$$

$$\lim_{T \to \infty} \int_{-T}^0 |y(t)|\, dt \to \infty \tag{C-2}$$

and

$$|y(t)| \leq M \tag{C-3}$$

where M is an arbitrarily large but finite constant.

A function $y(t)$ with the above properties has an associated *autocorrelation function* $\phi_{yy}(\tau)$ which is defined by

$$\phi_{yy}(\tau) \triangleq \lim_{T \to \infty} \frac{1}{2T} \int_{-T}^T y(t)y(t+\tau)\, dt \tag{C-4}$$

For $\tau = 0$, $\phi_{yy}(\tau) = \phi_{yy}(0)$ is called the *mean-square value* of $y(t)$.

If two functions, $y(t)$ and $v(t)$, have properties identical to those assigned to $y(t)$ in the preceding paragraph, a *cross-correlation function* $\phi_{vy}(\tau)$ associated with $y(t)$ and $v(t)$ is defined by

$$\phi_{vy}(\tau) \triangleq \lim_{T \to \infty} \frac{1}{2T} \int_{-T}^T v(t)y(t+\tau)\, dt \tag{C-5}$$

CORRELATION FUNCTIONS AND POWER-DENSITY SPECTRA

A second cross-correlation function $\phi_{yv}(\tau)$ associated with $y(t)$ and $v(t)$ is given by the relation

$$\phi_{yv}(\tau) = \lim_{T \to \infty} \frac{1}{2T} \int_{-T}^{T} y(t)v(t + \tau)\, dt \qquad \text{(C-6)}$$

In the important case that $v(t)$ and $y(t)$ are real-valued functions of time, it is readily verified from Equations C-5 and C-6 that

$$\phi_{vy}(\tau) = \phi_{yv}(-\tau) \qquad \text{(C-7)}$$

In practice, we use a large value of T in Equations C-4, C-5, and C-6, rather than let T approach infinity, to evaluate correlation functions (e.g., [C.1, C.3]).

Random Signals

Autocorrelation functions and cross-correlation functions are introduced primarily as analytical tools to aid in the analysis and design of systems which are subjected to random signals. In contrast to a deterministic signal, a random signal has the property that its value cannot be determined precisely in advance. At best, we know the probability distributions associated with the signal. That is, at a given instant, we may know the probability that the value of the signal will be found to be between any two signal levels at any given time in the future; we may know the joint probability that the value of the signal will be found to lie both between two given signal levels at time t_1 and between two, possibly different, signal levels at time t_2 in the future; and so on, there being an infinite number of such joint probabilities.

A very important class of random signals are those which are called *stationary random signals*. A stationary random signal is a random signal for which the various probabilities mentioned above do not change when the time origin is shifted. To clarify this definition, consider as an example the thermo-electric noise voltage generated by a resistor which is placed in a fixed environment. An observer who samples the noise voltage across the resistor today will have the same likelihood of finding that the sampled voltage lies between two given voltage levels as an observer who samples the voltage tomorrow.

It is generally difficult to analyze and design a system directly in terms of the assorted probability distributions of random signals which act on the system, and no attempt at this is made in this appendix. On the other hand, the deterministic correlation functions of stationary random signals are relatively easy to work with, and significant results in regard to system analysis and design are obtainable therefrom. This fact is set forth in the following paragraphs and in Chapter 4.

Power-Density Spectra

To use correlation functions in the analysis and design of linear systems, we must first establish the relationships between correlation functions of input signals and correlation functions of output signals of a system. To this end, consider the stable, linear, time-invariant system depicted by the block-diagram of Figure C-1.

The real output signal $y(t)$ in Figure C-1 is obtained from the following convolution integral:

$$y(t) = \int_{-\infty}^{\infty} g(t_1) v(t - t_1) \, dt_1 \tag{C-8}$$

The definition given by Equation C-4 is employed to find $\phi_{yy}(\tau)$:

$$\phi_{yy}(\tau) = \lim_{T \to \infty} \frac{1}{2T} \int_{-T}^{T} \int_{-\infty}^{\infty} \int_{-\infty}^{\infty} g(t_1) v(t - t_1) g(t_2) v(t + \tau - t_2) \, dt_1 dt_2 dt$$

$$= \int_{-\infty}^{\infty} g(t_1) \left\{ \int_{-\infty}^{\infty} g(t_2) \left[\lim_{T \to \infty} \frac{1}{2T} \int_{-T}^{T} v(t - t_1) v(t + \tau - t_2) \, dt \right] dt_2 \right\} dt_1$$

Let t_3 denote $t - t_1$ in the bracketed term above, with the result that

$$\phi_{yy}(\tau) = \int_{-\infty}^{\infty} g(t_1) \left\{ \int_{-\infty}^{\infty} g(t_2) \left[\lim_{T \to \infty} \frac{1}{2T} \int_{-T-t_1}^{T-t_1} v(t_3) v(t_3 + \tau + t_1 - t_2) \, dt_3 \right] dt_2 \right\} dt_1$$

$$= \int_{-\infty}^{\infty} g(t_1) \left[\int_{-\infty}^{\infty} g(t_2) \phi_{vv}(\tau + t_1 - t_2) \, dt_2 \right] dt_1 \tag{C-9}$$

Quite often, the two-sided Laplace transform (Appendix B) of a correlation function exists. This is generally true, for example, in the case of a correlation function which is obtained from a stationary random signal with zero mean-value. Assuming $\phi_{yy}(\tau)$ is Laplace transformable the relationship of C-9 can be placed in a more compact form by use of

$$\Phi_{yy}(s) = \int_{-\infty}^{-\infty} \phi_{yy}(\tau) e^{-s\tau} \, d\tau \tag{C-10}$$

where $\Phi_{yy}(s)$ is called the *power-density spectrum*[1] of the signal $y(t)$.

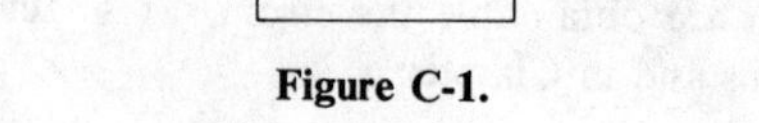

$v(t) \longrightarrow [g(t), G(s)] \longrightarrow y(t)$

Figure C-1.

CORRELATION FUNCTIONS AND POWER-DENSITY SPECTRA 575

When the right-hand member of C-9 is substituted for $\phi_{yy}(\tau)$ in Equation C-10, the result is

$$\Phi_{yy}(s) = \int_{-\infty}^{\infty} g(t_1) \int_{-\infty}^{\infty} g(t_2) \int_{-\infty}^{\infty} \phi_{vv}(\tau + t_1 - t_2) e^{-s\tau} \, d\tau \, dt_2 \, dt_1$$

$$= \int_{-\infty}^{\infty} g(t_1) \int_{-\infty}^{\infty} g(t_2) \int_{-\infty}^{\infty} \phi_{vv}(t_3) e^{-st_3} e^{-s(t_2 - t_1)} \, dt_3 \, dt_2 \, dt_1$$

$$= \int_{-\infty}^{\infty} g(t_1) e^{st_1} \, dt_1 \int_{-\infty}^{\infty} g(t_2) e^{-st_2} \, dt_2 \int_{-\infty}^{\infty} \phi_{vv}(t_3) e^{-st_3} \, dt_3$$

$$= G(s) G(-s) \Phi_{vv}(s) \tag{C-11}$$

Thus, Equation C-11 relates the power-density spectrum $\Phi_{yy}(s)$ of the output $y(t)$ in Figure C-1 to the power-density spectrum $\Phi_{vv}(s)$ of the input $v(t)$, and the autocorrelation function $\phi_{yy}(\tau)$ can be obtained through the use of the inversion formula

$$\phi_{yy}(\tau) = \frac{1}{2\pi j} \int_{-j\infty}^{j\infty} G(s) G(-s) \Phi_{vv}(s) e^{s\tau} \, ds \tag{C-12}$$

A second basic operation to be considered is that of summation of signals. Suppose that

$$e(t) = y_1(t) + [-y_2(t)] = y_1(t) - y_2(t) \tag{C-13}$$

where $\phi_{y_1 y_1}(\tau)$, $\phi_{y_2 y_1}(\tau)$, and $\phi_{y_2 y_2}(\tau)$ are known and where $y_1(t)$ and $y_2(t)$ are assumed to be real-valued functions of t. The autocorrelation function $\phi_{ee}(\tau)$ is desired:

$$\phi_{ee}(\tau) = \lim_{T \to \infty} \frac{1}{2T} \int_{-T}^{T} e(t) e(t + \tau) \, dt$$

$$= \lim_{T \to \infty} \frac{1}{2T} \int_{-T}^{T} [y_1(t) - y_2(t)][y_1(t + \tau) - y_2(t + \tau)] \, dt$$

$$= \lim_{T \to \infty} \frac{1}{2T} \left[\int_{-T}^{T} y_1(t) y_1(t + \tau) \, dt - \int_{-T}^{T} y_1(t) y_2(t + \tau) \, dt \right.$$

$$\left. - \int_{-T}^{T} y_2(t) y_1(t + \tau) \, dt + \int_{-T}^{T} y_2(t) y_2(t + \tau) \, dt \right]$$

$$= \phi_{y_1 y_1}(\tau) - \phi_{y_1 y_2}(\tau) - \phi_{y_2 y_1}(\tau) + \phi_{y_2 y_2}(\tau) \tag{C-14}$$

---

[1] In certain literature (for example, [C.5]), a multiplicative factor of $1/(2\pi)$ is included in the definition of the power-density spectrum $\Phi_{yy}(s)$. The dominant trend (for example, [C.1, C.2, C.4, and C.6]), however, is to define the power-density spectrum of $y(t)$ as the two-sided Laplace transform (and in particular, the corresponding Fourier

576   APPENDIX C

The corresponding power-density spectrum $\Phi_{ee}(s)$ is obviously given by

$$\Phi_{ee}(s) = \Phi_{y_1 y_1}(s) - \Phi_{y_1 y_2}(s) - \Phi_{y_2 y_1}(s) + \Phi_{y_2 y_2}(s) \qquad (C\text{-}15)$$

wherein the *cross-spectral density functions* $\Phi_{y_1 y_2}(s)$ and $\Phi_{y_2 y_1}(s)$ are related by

$$\Phi_{y_1 y_2}(s) = \Phi_{y_2 y_1}(-s) \qquad (C\text{-}16)$$

which is readily proved on the basis of the corresponding correlation function relationship C-7. It is equally true, of course, that

$$\Phi_{ee}(s) = \Phi_{ee}(-s) \qquad (C\text{-}17)$$

An additional relationship merits attention: the relationship between $\phi_{ee}(\tau)$ and the correlation functions of $v_1(t)$ and $v_2(t)$ in Figure C-2.
The results of the preceding paragraphs give

$$\begin{aligned}\Phi_{ee}(s) &= \Phi_{y_2 y_2}(s) - \Phi_{y_2 y_1}(s) - \Phi_{y_1 y_2}(s) + \Phi_{y_1 y_1}(s) \\ &= \Phi_{v_1 v_1}(s) G_1(s) G_1(-s) + \Phi_{v_2 v_2}(s) G_2(s) G_2(-s) - \Phi_{y_2 y_1}(s) - \Phi_{y_1 y_2}(s)\end{aligned}$$
$$(C\text{-}18)$$

What remains to be determined in Equation C-18 is a relationship for $\Phi_{y_1 y_2}(s) = \Phi_{y_2 y_1}(-s)$ in terms of the stationary random inputs $v_1(t)$ and $v_2(t)$. By definition,

$$\begin{aligned}\phi_{y_1 y_2}(\tau) &= \lim_{T \to \infty} \frac{1}{2T} \int_{-T}^{T} y_1(t) y_2(t + \tau)\, dt \\ &= \lim_{T \to \infty} \frac{1}{2T} \int_{-T}^{T} \int_{-\infty}^{\infty} \int_{-\infty}^{\infty} g_1(t_1) v_1(t - t_1) g_2(t_2) v_2(t + \tau - t_2)\, dt_1 dt_2 dt \\ &= \int_{-\infty}^{\infty} \int_{-\infty}^{\infty} g_1(t_1) g_2(t_2) \left[\lim_{T \to \infty} \frac{1}{2T} \int_{-T}^{T} v_1(t - t_1) v_2(t + \tau - t_2)\, dt \right] dt_1 dt_2\end{aligned}$$
$$(C\text{-}19)$$

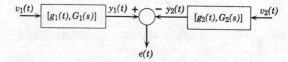

Figure C-2.

---

transform) of the autocorrelation function $\phi_{yy}(\tau)$. Essentially the same power-density spectrum concept is referred to by names such as "power spectrum" and "power spectral density." Also, most authors restrict the definition to correspond solely to the Fourier transform of $\phi_{yy}(\tau)$ with various notational conventions such as $\Phi_{yy}(j\omega)$, $\Phi_{yy}(\omega)$, $S_y(\omega)$, $S_{yy}(\omega)$, or $S_y(f)$ where $2\pi f = \omega$.

Let $t - t_1 = t_3$ in the bracketed term of equation C-19 to obtain

$$\phi_{y_1 y_2}(\tau) = \int_{-\infty}^{\infty} \int_{-\infty}^{\infty} g_1(t_1) g_2(t_2)$$

$$\times \left[\lim_{T \to \infty} \frac{1}{2T} \int_{-T-t_1}^{T-t_1} v_1(t_3) v_2(t_3 + t_1 + \tau - t_2) \, dt_3 \right] dt_1 dt_2$$

$$= \int_{-\infty}^{\infty} g_1(t_1) \left[\int_{-\infty}^{\infty} g_2(t_2) \phi_{v_1 v_2}(\tau + t_1 - t_2) \, dt_2 \right] dt_1 \quad \text{(C-20)}$$

Note that Equation C-20 is of the same form as Equation C-9. Thus, the same steps used in obtaining Equation C-11 are appropriate here, and the result is

$$\Phi_{y_1 y_2}(s) = G_1(-s) G_2(s) \Phi_{v_1 v_2}(s) \quad \text{(C-21)}$$

This identity is substituted appropriately into Equation C-18 to obtain the desired expression for $\Phi_{ee}(s)$,

$$\Phi_{ee}(s) = \Phi_{v_1 v_1}(s) G_1(s) G_1(-s) + \Phi_{v_2 v_2}(s) G_2(s) G_2(-s)$$
$$- \Phi_{v_1 v_2}(-s) G_1(s) G_2(-s) - \Phi_{v_1 v_2}(s) G_1(-s) G_2(s) \quad \text{(C-22)}$$

Example. A specific example based on the configuration of Figure C-2 serves to illustrate use of correlation functions and power-density spectra in system design. It is given that $v_1(t) = v_2(t)$ is a stationary random signal which is characterized by

$$\Phi_{v_1 v_1}(s) = \Phi_{v_2 v_2}(s) = \Phi_{v_2 v_1}(s) = \frac{2a}{(s+a)(-s+a)} \quad \text{(C-23)}$$

for $-a < \text{Re}(s) < a$. It is also given that

$$G_2(s) = K \quad \text{(C-24)}$$

and

$$G_1(s) = e^{sT} \quad \text{(C-25)}$$

where T is a positive constant, and K is a real parameter to be selected. Of course, the function $G_1(s) = e^{sT}$ is the generally unrealizable transfer function of a predictor. *The problem* is to make a device, with realizable transfer function $G_2(s) = K$, act as much like a predictor as possible. A useful measure of the predictive ability of $G_2(s)$ is the mean-square value $\phi_{ee}(0)$ of the difference between the output of a theoretical predictor and the output of $G_2(s)$. The smaller $\phi_{ee}(0)$ is, the better is the predictive action of $G_2(s)$. Thus, the problem is to minimize $\phi_{ee}(0)$ with respect to K.

The autocorrelation function $\phi_{v_1 v_1}(\tau)$ is obtained from the inversion formula:

$$\phi_{v_1v_1}(\tau) = \frac{1}{2\pi j}\int_{-j\infty}^{j\infty} \frac{2ae^{s\tau}}{(s+a)(-s+a)}\, ds$$

$$= \frac{1}{2\pi j}\int_{-j\infty}^{j\infty}\left[\frac{e^{s\tau}}{s+a} + \frac{e^{s\tau}}{-s+a}\right] ds$$

$$= e^{-a\tau}u(\tau) + e^{a\tau}u(-\tau)$$

$$= e^{-a|\tau|} \tag{C-26}$$

The identities given by Equations C-23, C-24, and C-25 are substituted appropriately into Equation C-22 to obtain

$$\Phi_{ee}(s) = \Phi_{v_1v_1}(s)(K^2 + 1 - Ke^{-sT} - Ke^{sT}) \tag{C-27}$$

The autocorrelation function $\phi_{ee}(\tau)$ is given by

$$\phi_{ee}(\tau) = \frac{1}{2\pi j}\int_{-j\infty}^{j\infty} \Phi_{ee}(s)e^{s\tau}\, ds$$

$$= \frac{1}{2\pi j}\int_{-j\infty}^{j\infty} \Phi_{v_1v_1}(s)(K^2 + 1 - Ke^{-sT} - Ke^{sT})e^{s\tau}\, ds \tag{C-28}$$

and the minimum of $\phi_{ee}(0)$ with respect to K is obtained in the standard way:

$$\frac{\partial}{\partial K}\phi_{ee}(\tau) = \frac{1}{2\pi j}\int_{-j\infty}^{j\infty} \Phi_{v_1v_1}(s)(2K - e^{-sT} - e^{sT})e^{s\tau}\, ds \tag{C-29}$$

which, on the basis of Equation C-26, is equivalent to

$$2Ke^{-a|\tau|} - e^{-a|\tau-T|} - e^{-a|\tau+T|} \tag{C-30}$$

At $\tau = 0$, C-30 yields

$$K = K^* = e^{-aT} \tag{C-31}$$

That this value K^* of K results in a minimum of $\phi_{ee}(0)$ is substantiated by the fact that the second partial derivative of $\phi_{ee}(0)$ with respect to K is always greater than zero, i.e.,

$$\frac{\partial^2}{\partial K^2}\phi_{ee}(0) = \frac{1}{2\pi j}\int_{-j\infty}^{j\infty} 2\Phi_{v_1v_1}(s)\, ds > 0 \tag{C-32}$$

REFERENCES

[C.1] Blackman, R. B., and J. W. Tukey. *The Measurement of Power Spectra*. Dover Publications, New York, 1958 (copyright by the American Telephone and Telegraph Company).

[C.2] Davenport, W. B., Jr., and W. L. Root. *An Introduction to the Theory of Random Signals and Noise*. McGraw-Hill, New York, 1958.

[C.3] Gauss, E. J. "Estimation of Power Spectral Density by Filters." *Journal of the Association for Computing Machinery*, **11**, 98–103, Jan. 1964.

[C.4] Mori, H., and E. P. Johnson, Jr. "Basic Statistical Theory." Chapter 10 of *Control Systems Engineering*, W. W. Seifert and C. W. Steeg, Jr. (Editors). McGraw-Hill, New York, 1960.

[C.5] Newton, G. C., Jr., L. A. Gould, and J. F. Kaiser. *Analytical Design of Linear Feedback Controls*. Wiley, New York, 1957.

[C.6] Papoulis, A. *Probability, Random Variables, and Stochastic Processes*. McGraw-Hill, New York, 1965.

INEQUALITIES AND ABSTRACT SPACES

appendix D

A selected set of inequality relations is given in this appendix. When applicable to a given extremization problem, an appropriate inequality may yield an elegant solution. Many inequalities can be viewed as special cases of general inequalities that are valid in a more abstract setting. Thus, the defining properties of certain abstract spaces are also listed.

Arithmetic and Geometric Means

Consider any set of real numbers $x_1, x_2, \ldots, x_n$ each of which is nonnegative. For such a set, the geometric mean value is less than or equal to the arithmetic mean:

$$(x_1 x_2 x_3 \cdots x_n)^{1/n} \leq \frac{1}{n}(x_1 + x_2 + \cdots + x_n) \tag{D-1}$$

Equality holds if and only if all x_i are equal.

Relationship between $(1 + x)^y$ and $1 + xy$

If y is a real number in the open interval (0, 1) and $x \geq -1$, then

$$(1 + x)^y \leq 1 + xy \tag{D-2}$$

But if y is outside [0, 1],

$$(1 + x)^y \geq 1 + xy \tag{D-3}$$

Equality holds in D-2 and D-3 if and only if $x = 0$.

INEQUALITIES AND ABSTRACT SPACES

Metric Space

A metric space consists of a nonempty set A and a distance measure or metric ρ which operates on the elements of A with the following results:

1. $\rho(a,b) \geq 0$ for all $a,b \in A$, and $\rho(a,b) = 0$ if and only if $a = b$.
2. $\rho(a,b) = \rho(b,a)$.
3. $\rho(a,b) \leq \rho(a,c) + \rho(c,b)$ for any $a, b, c \in A$.

Condition 3 is known as the *triangle inequality*.

Spaces of Real and Complex Numbers

An n-dimensional space consisting of real numbers is denoted by R_n. Elements of R_n are given in terms of n-tuples of real numbers, i.e., $\mathbf{x} = \{x_1, x_2, \ldots, x_n\}$ is an element of R_n if and only if each x_i is a real number.

An n-dimensional space consisting of complex numbers is denoted by C_n. Elements of C_n are given in terms of n-tuples of complex numbers, i.e., $\mathbf{x} = \{x_1, x_2, \ldots, x_n\}$ is an element of C_n if and only if each x_i is a complex number, where real numbers are special cases of complex numbers.

Example D-1. Consider the set R_n of reals. Let $\mathbf{x} = \{x_1, x_2, \ldots, x_n\}$ and $\mathbf{y} = \{y_1, y_2, \ldots, y_n\}$ be any two elements of R_n. The following distance measure $\rho(\mathbf{x},\mathbf{y})$ satisfies the conditions of a metric:

$$\rho(\mathbf{x},\mathbf{y}) = \left(\sum_{i=1}^{n} |x_i - y_i|^2\right)^{1/2} \tag{D-4}$$

•

Example D-2. Let the set $C(t_a, t_b)$ denote the set of all continuous real-valued functions over the interval $[t_a, t_b]$. Let $x = x(t)$ and $y = y(t)$ be any two elements of $C(t_a, t_b)$. The following distance measure $\rho(x,y)$ satisfies the conditions of a metric:

$$\rho(x,y) = \max_{t \in [t_a, t_b]} |x(t) - y(t)| \tag{D-5}$$

•

Cauchy Sequence

Consider a metric space A and a sequence $a_1, a_2, \ldots$ of elements from A. The sequence is called a *Cauchy sequence* if there exists an integer $k(\epsilon)$ for every $\epsilon > 0$ such that $\rho(a_i, a_j) < \epsilon$ for all $i, j > k(\epsilon)$.

Complete Metric Space

A metric space A is called a *complete* metric space if every Cauchy sequence in A converges to a limit which is also an element of A.

Linear Space

A set A of elements is called a linear space if its elements, which are called *vectors*,[1] satisfy:

1. $a + b = b + a$ for any $a, b \in A$.
2. $(a + b) + c = a + (b + c)$ for any $a, b, c \in A$.
3. $a + 0 = a$ where this 0 is the zero element of A, not the scalar 0 in general, and equality holds for any $a \in A$.
4. For any $a \in A$, an element $-a$ is also contained in A with the property that $a + (-a)$ equals the zero element of A.
5. $1a = a$ for any $a \in A$.
6. For any scalars α and β and for any elements a and b of A,

$$\alpha(\beta a) = (\alpha\beta)a,$$
$$(\alpha + \beta)a = \alpha a + \beta a,$$

and

$$\alpha(a + b) = \alpha a + \alpha b.$$

Example D-3. The set of all $n \times 1$ column matrices with real entries constitutes a linear space.

•

Normed Linear Space

A linear space A is called a normed linear space if there is defined an operation which maps each element $a \in A$ into a real number $\|a\|$ with the following properties:

1. $\|a\| > 0$ (scalar zero) if $a \neq 0$ (vector zero), and $\|a\| = 0$ if $a = 0$.
2. $\|a + b\| \leq \|a\| + \|b\|$ for any $a, b \in A$.
3. $\|\alpha a\| = |\alpha|(\|a\|)$ for any scalar α and any $a \in A$.

A Norm Inequality

If a and b are elements of a normed linear space,

$$|\,\|a\| - \|b\|\,| \leq \|a - b\| \tag{D-6}$$

---

[1] The column-vector (row-vector) concepts that are presented elsewhere in this book are special cases of a more general concept of a vector.

Banach Space

Consider a normed linear space in which a metric $\rho(a,b) = \|a - b\|$ is defined in terms of the norm. If the corresponding metric space is complete, the space is called a *Banach space*.

The $l_p(n)$ Spaces

A valid norm on either R_n or C_n is

$$\|\mathbf{x}\| = \left(\sum_{i=1}^{n} |x_i|^p\right)^{1/p}, \qquad 1 \leq p < \infty \tag{D-7}$$

where $\mathbf{x}$ is an element of either R_n or C_n, and the positive number p is bounded as indicated. The $l_p(n)$ spaces are Banach spaces. If $p = 2$, for example, the resulting space is an n-dimensional Euclidean space E^n.

It can be shown that D-7 defines a norm even if $n \to \infty$, provided that the components of $\mathbf{x}$ yield a bounded value of $\sum_{i=1}^{\infty} |x_i|^p$. The associated Banach space is denoted by l_p.

Similarly, if $p \to \infty$, it can be shown that the norm D-7 is yet valid and reduces to

$$\|\mathbf{x}\| = \max_{i} |x_i| \tag{D-8}$$

The associated Banach space is denoted by $l_\infty(n)$.

Minkowski's Inequality for Sums

Minkowski's inequality for sums is a statement of the fact that the norm D-7 satisfies the triangle inequality:

$$\left(\sum_{i=1}^{n} |x_i + y_i|^p\right)^{1/p} \leq \left(\sum_{i=1}^{n} |x_i|^p\right)^{1/p} + \left(\sum_{i=1}^{n} |y_i|^p\right)^{1/p} \tag{D-9}$$

where $\mathbf{x} = \{x_1, x_2, \ldots, x_n\}$ and $\mathbf{y} = \{y_1, y_2, \ldots, y_n\}$ are elements of either R_n or C_n.

The $L_p(t_a, t_b)$ Spaces

Consider the set of scalar-valued $x(t)$'s for which the following integration is well defined for a particular value of p:

$$\|x\| = \left(\int_{t_a}^{t_b} |x(t)|^p \, dt\right)^{1/p}, \qquad 1 \leq p < \infty \tag{D-10}$$

Each value of p in D-10 is associated with a Banach space denoted by $L_p(t_a,t_b)$.

If p is allowed to approach infinity in D-10 and if the $x(t)$'s are restricted to those functions which are piecewise continuous between simple jump discontinuities, the norm D-10 can be shown to be equivalent to

$$\|x\| = \max_{t\in[t_a,t_b]} |x(t)| \qquad (D\text{-}11)$$

Minkowski's Inequality for L_p Spaces

Minkowski's inequality for the L_p spaces is a statement of the fact that the norm D-10 satisfies the triangle inequality:

$$\left(\int_{t_a}^{t_b} |x(t)+y(t)|^p\,dt\right)^{1/p} \le \left(\int_{t_a}^{t_b} |x(t)|^p\,dt\right)^{1/p} + \left(\int_{t_a}^{t_b} |y(t)|^p\,dt\right)^{1/p} \qquad (D\text{-}12)$$

where $x(t)$ and $y(t)$ are elements of $L_p(t_a,t_b)$.

Conjugate Indexes

Consider real numbers p and q which are both greater than 1. The number p is said to be a conjugate index to q if

$$\frac{1}{p} + \frac{1}{q} = 1 \qquad (D\text{-}13a)$$

or, equivalently,

$$p = \frac{q}{q-1} \qquad (D\text{-}13b)$$

Hölder's Inequality for Sums

Let p and q be conjugate indexes, and let $\mathbf{x}$ be an arbitrary element of l_p and $\mathbf{y}$ be an arbitrary element of l_q. Hölder's inequality for sums is

$$\sum_i |x_i y_i| \le \left(\sum_i |x_i|^p\right)^{1/p} \left(\sum_i |y_i|^q\right)^{1/q} \qquad (D\text{-}14)$$

Equality holds in D-14 if and only if the $|x_i|^p$'s are related to the $|y_i|^q$'s by a common proportionality factor; that is, equality holds if and only if $|x_i|^p = c|y_i|^q$ for some $c > 0$ and for all i.

Hölder's Inequality for the L_p Spaces

Let p and q be conjugate indexes, and let $x = x(t)$ be an arbitrary element of $L_p(t_a,t_b)$ and $y = y(t)$ be an arbitrary element of $L_q(t_a,t_b)$. Hölder's inequality in this case is

INEQUALITIES AND ABSTRACT SPACES 585

$$\int_{t_a}^{t_b} |xy| \, dt \leq \left(\int_{t_a}^{t_b} |x|^p \, dt \right)^{1/p} \left(\int_{t_a}^{t_b} |y|^q \, dt \right)^{1/q} \quad \text{(D-15)}$$

Equality holds in D-15 if and only if the integral

$$\int_{t_a}^{t_b} (|x(t)|^p - c|y(t)|^q) \, dt = 0$$

for some real constant c. Because $(1/p) + (1/q) = 1$, equality holds in D-15 if $|x(t)| = c_1 |y(t)|^{q-1}$ where c_1 is a positive constant.

Absolute-Value Inequalities

Consider the following well-known inequality:

$$\left| \int_{t_a}^{t_b} x(t) y(t) \, dt \right| \leq \int_{t_a}^{t_b} |x(t) y(t)| \, dt \quad \text{(D-16)}$$

where $x(t)$ and $y(t)$ can be complex-valued functions. For reasons to be considered soon, $x(t) = |x(t)| e^{j\theta_1(t)}$ is assumed to be an element of $L_p(t_a, t_b)$, and $y(t) = |y(t)| e^{j\theta_2(t)}$ is assumed to be an element of $L_q(t_a, t_b)$. It can be shown that equality holds in D-16 if and only if $\theta_1(t) = -\theta_2(t) + \phi$, where ϕ is an arbitrary real constant over essentially all of $[t_a, t_b]$. If $x(t)$ and $y(t)$ are restricted to be real-valued functions, the condition for equality in D-16 reduces to $xye^{j\beta} \geq 0$ for essentially all $t \in [t_a, t_b]$ where β can be either zero or pi radians.

Equality holds in D-15 and D-16 simultaneously if and only if

$$x(t) = c|y(t)|^{q-1} e^{-j\theta_2(t) + j\phi}, \qquad t \in [t_a, t_b] \quad \text{(D-17)}$$

in which all symbols are as previously defined. If $x(t)$ and $y(t)$ are restricted to be real-valued functions, condition D-17 reduces to

$$x(t) = c|y(t)|^{q-1} (\text{sgn } y) e^{j\beta} \quad \text{(D-18)}$$

The reader may wish to develop the corresponding condition for simultaneous equality in D-14 and D-19:

$$\left| \sum_i x_i y_i \right| \leq \sum_i |x_i y_i| \quad \text{(D-19)}$$

Example D-4. The real-valued output $y(t)$ of a particular dynamic system is characterized at a specific time T by

$$y(T) = \int_0^T g(T, \tau) m(\tau) \, d\tau \quad \text{(D-20)}$$

where $g(T,\tau)$ is a known real-valued function, $y(T)$ is a specific value, and $m(t)$ is to be selected to obtain a minimum of

$$\|m\|_p \triangleq \left(\int_0^T |m(\tau)|^p \, d\tau\right)^{1/p}, \quad p \geq 1 \tag{D-21}$$

Hölder's inequality and the absolute-value inequality give

$$|y(T)| = \left|\int_0^T g(T,\tau)m(\tau) \, d\tau\right|$$

$$\leq \left(\int_0^T |g(T,\tau)|^q \, d\tau\right)^{1/q} \left(\int_0^T |m(\tau)|^p \, d\tau\right)^{1/p} \tag{D-22}$$

where p and q are restricted to satisfy $p = [q/(q-1)] \geq 1$. Because the left-hand member in D-22 is a fixed positive value, as also is the first term in the right-hand member, the minimum value $\|m^*\|_p$ of $\|m\|_p$ is obtained when equality holds in D-22. For equality to hold, the version of Equation D-18 that applies in this case is

$$m^*(t) = c|g(T,t)|^{q-1}[\text{sgn } g(T,t)]$$

where the real constant c is determined on the basis that

$$y(T) = \int_0^T c|g(T,\tau)|^q \, d\tau \tag{D-23}$$

and therefore,

$$c = \frac{y(T)}{\int_0^T |g(T,\tau)|^q \, d\tau} \tag{D-24}$$

For example, if p is assigned the value of 2, q must also equal 2, and $m^*(t)$ reduces to

$$m^*(t) = \frac{y(T)g(T,t)}{\int_0^T [g(T,\tau)]^2 \, d\tau}, \quad p = 2 \tag{D-25}$$

But if $p \to \infty$, and therefore $q \to 1$,

$$m^*(t) = \frac{y(T)[\text{sgn} \cdot g(T,t)]}{\int_0^T |g(T,\tau)| \, d\tau}, \quad p \to \infty \tag{D-26}$$

Example D-5. (A Matched Filter)
Reconsider Equation D-20, but assume here that $y(T)$ is to be maximized, $m(\tau)$ is a known real-valued function, and $g(T,\tau)$ is to be selected in an optimal manner, subject to the constraint that

$$\|g\|_p = \left(\int_0^T |g(T,\tau)|^p \, d\tau\right)^{1/p} = K \tag{D-27}$$

where K is a specified positive value. All other assumed conditions remain unchanged from the preceding example. If $p \to \infty$, Equation D-27 reduces to

$$\|g\|_\infty = \max_{\tau \in [0,T]} |g(T,\tau)| = K \tag{D-28}$$

Hölder's inequality and the absolute-value inequality give

$$|y(T)| = \left| \int_0^T g(T,\tau) m(\tau) \, d\tau \right|$$

$$\leq \|g\|_p \|m\|_q = K \|m\|_q \tag{D-29}$$

where $\|m\|_q$ can be evaluated on the basis of the known $m(t)$. From the standpoint of obtaining the maximum $y(T)$, the best that can be done is the attainment of equality in D-29. Equality holds in D-29 when the optimal form $g^*(T,t)$ of $g(T,t)$ is used:

$$g^*(T,t) = c|m(t)|^{q-1}[\text{sgn } m(t)] \tag{D-30}$$

where the positive constant c is determined on the basis that $\|g^*\|_p = K$:

$$K = c \left(\int_0^T |m(t)|^{(q-1)p} \, dt \right)^{1/p}$$

$$= c \left(\int_0^T |m(t)|^q \, dt \right)^{1/p} = c(\|m\|_q)^{q/p}$$

$$= c(\|m\|_q)^{q-1} \tag{D-31}$$

It follows that

$$c = K/(\|m\|_q)^{q-1} \tag{D-32}$$

and

$$g^*(T,t) = \frac{K|m(t)|^{q-1}[\text{sgn } m(t)]}{(\|m\|_q)^{q-1}} \tag{D-33}$$

In the special case that $q = p = 2$,

$$g^*(T,t) = \frac{Km(t)}{\|m\|_2} \tag{D-34}$$

And in the special case that $p \to \infty$ and $q \to 1$,

$$g^*(T,t) = K \text{ sgn } m(t) \tag{D-35}$$

●

Jensen's Inequality

For any real values of p and q which satisfy $0 < q < p$,

$$\left(\sum_i |x_i|^p\right)^{1/p} \leq \left(\sum_i |x_i|^q\right)^{1/q} \tag{D-36}$$

Inner Product Space

A linear space A is called an inner product space if there is an associated rule which assigns a scalar $\langle a,b \rangle$ to every pair of elements $a,b \in A$; the scalar $\langle a,b \rangle$ is called an inner product and must satisfy:

1. $\langle b,a \rangle = \overline{\langle a,b \rangle}$, where $\overline{\langle a,b \rangle}$ denotes the conjugate of $\langle a,b \rangle$,
2. $\langle a_1 + a_2, b \rangle = \langle a_1, b \rangle + \langle a_2, b \rangle$,
3. $\langle \alpha a, b \rangle = \alpha \langle a,b \rangle$ for any scalar α, and
4. $\langle a,a \rangle > 0$ unless $a \equiv 0$, in which case $\langle a,a \rangle = 0$.

Hilbert Space

If the norm of a given Banach space is appropriately derived from an inner product the Banach space is called a Hilbert Space. Let H denote a Hilbert space and let $x \in H$. The norm $\|x\|$ of x is defined to be

$$\|x\| \triangleq \langle x,x \rangle^{1/2} \tag{D-37}$$

Cauchy's Inequality

Given that x and y are elements of a Hilbert space H, Cauchy's inequality is applicable, as follows:

$$|\langle x,y \rangle| \leq \|x\| \, \|y\| \tag{D-38}$$

This inequality is also called the Cauchy-Schwarz inequality.

Example D-6. The space $l_2(n)$ is a Hilbert space with inner product

$$\langle \mathbf{x}, \mathbf{y} \rangle = \sum_{k=1}^{n} x_k \bar{y}_k \tag{D-39}$$

where $\mathbf{x}$ and $\mathbf{y}$ may be elements of C_n but can be restricted to be elements of R_n.

●

INEQUALITIES AND ABSTRACT SPACES 589

Example D-7. The space $L_2(t_a,t_b)$ is a Hilbert space with inner product

$$\langle x(t),y(t)\rangle = \int_{t_a}^{t_b} x(t)\bar{y}(t)\, dt \tag{D-40}$$

where $x(t)$ and $y(t)$ are elements of $L_2(t_a,t_b)$.

•

Parallelogram Law

Given an inner product space V and arbitrary elements x and y of V,

$$\|x + y\|^2 + \|x - y\|^2 = 2\|x\|^2 + 2\|y\|^2 \tag{D-41}$$

where $\|x\| = \langle x,x\rangle^{1/2}$.

Orthogonal Vectors

Two nontrivial vectors x and y of an inner product space V are said to be *orthogonal* if, and only if,

$$\langle x,y\rangle = 0 \tag{D-42}$$

Orthogonal Projection Theorem

Let U and V be Hilbert spaces, and let U be contained in V (let U be a subspace of V). Corresponding to any vector $x \in V$, there exists a vector $x_p \in U$ with the properties

$$\|x - x_p\| \leq \|x - y\| \quad \text{for any } y \in U \tag{D-43}$$

and

$$\|x - x_p\| = \min_{y \in U} \|x - y\| \tag{D-44}$$

The vector x_p is called the *orthogonal projection* of x onto U.

The *orthogonal projection theorem gives that D-43 and D-44 are equivalent to*

$$\langle x - x_p, y\rangle = 0 \tag{D-45}$$

for $x_p \in U$ and *all* $y \in U$. If orthogonal basis vectors for U are known, they can be used with D-45 to determine the $x_p \in U$ corresponding to a given $x \in V$. The following examples illustrate the approach.

Example D-8. Let V be E^3, and let U be the set of real vectors in the plane defined by orthogonal vectors $[1 \quad 0 \quad 1]'$ and $[1 \quad 1 \quad -1]'$. The projection $\mathbf{x}_p \in U$ of a particular $\mathbf{x} = [2 \quad 1 \quad 3]'$ is desired. Any basis vector of U can be identified with **y** in D-45 to obtain a relationship between the components x_{p1}, x_{p2}, and x_{p3} of $\mathbf{x}_p$; thus,

$$[1 \quad 0 \quad 1](\mathbf{x} - \mathbf{x}_p) = 2 - x_{p1} + 3 - x_{p3} = 0 \quad \text{(D-46)}$$

and

$$[1 \quad 1 \quad -1](\mathbf{x} - \mathbf{x}_p) = 2 - x_{p1} + 1 - x_{p2} - 3 + x_{p3} = 0 \quad \text{(D-47)}$$

Furthermore, for $\mathbf{x}_p$ to be in U,

$$\mathbf{x}_p = c_1 \begin{bmatrix} 1 \\ 0 \\ 1 \end{bmatrix} + c_2 \begin{bmatrix} 1 \\ 1 \\ -1 \end{bmatrix} \quad \text{(D-48)}$$

where c_1 and c_2 are real constants which are eliminated between the three equations in D-48 to obtain

$$x_{p3} + 2x_{p2} - x_{p1} = 0 \quad \text{(D-49)}$$

Equations D-46, D-47, and D-49 yield the solution $x_{p1} = 2.5$, $x_{p2} = 0$, and $x_{p3} = 2.5$.

●

Example D-9. Let V be $C(-\pi,\pi)$ with inner product defined by

$$\langle x(t), y(t) \rangle = \int_{-\pi}^{\pi} x(t)y(t)\, dt \quad \text{(D-50)}$$

where $x(t)$ and $y(t)$ are elements of $C(-\pi,\pi)$ (see Example D-2). The norm $\|x(t)\|$ is $\langle x(t), x(t)\rangle^{1/2}$. Let U be the space spanned by orthogonal basis vectors 1, $\cos t$, $\sin t$, $\cos 2t$, $\sin 2t$, ..., $\cos nt$, and $\sin nt$, where n is a finite integer. Given a continuous $x(t)$, its projection $x_p(t)$ onto U is a truncated Fourier series,

$$x_p(t) = a_0 + \sum_{i=1}^{n} (a_k \cos kt + b_k \sin kt) \quad \text{(D-51)}$$

in which the a_k's and b_k's are determined by using

$$\langle x - x_p, 1 \rangle = \langle x, 1 \rangle - \langle x_p, 1 \rangle = 0 \quad \text{(D-52)}$$

$$\langle x - x_p, \cos kt \rangle = \langle x, \cos kt \rangle - \pi a_k = 0 \quad \text{(D-53)}$$

and

$$\langle x - x_p, \sin kt \rangle = \langle x, \sin kt \rangle - \pi b_k = 0 \quad \text{(D-54)}$$

●

For detailed derivations and a guide to extensions of the theory abstracted herein, references [D.2, D.3, D.7, D.8, D.10, and D.12] may be consulted. Banach space generalizations of the nonlinear programming problem are considered in references [D.11 and D.15], and multivariable extensions of Examples D-4 and D-5 are given in reference [D.5]. A representative set of additional applications are to be found in the remaining references.

REFERENCES

[D.1] Aoki, M. "Minimum Norm Problems and Some Other Control System Optimization Techniques." Chapter 8, pp. 319–354 of *Modern Control Systems Theory*, C. T. Leondes (Editor). McGraw-Hill, New York, 1965.

[D.2] Beckenbach, E. F., and R. Bellman. *Inequalities*. Springer Verlag, New York, 1961.

[D.3] Hardy, G. H., J. E. Littlewood, and G. Pólya. *Inequalities*, Cambridge Univ. Press, New York, 1952.

[D.4] Hsieh, H. C., and R. A. Nesbit. "Functional Analysis and Its Applications to Mean Square Error Problems." Chapter 3, pp. 97–120 of *Modern Control Systems Theory*, C. T. Leondes (Editor). McGraw-Hill, New York, 1965.

[D.5] Kranc, D. M., and P. E. Sarachik. "An Application of Functional Analysis to the Optimal Control Problem." *Transactions of the ASME*, series D (*Journal of Basic Engineering*), **85**, 143–150, June 1963.

[D.6] Kuo, M. S. Y., and L. F. Kazda. "Minimum Energy Problems in Hilbert Function Space." *Journal of the Franklin Institute*, **283**, 38–54, Jan. 1967.

[D.7] Liusternik, L., and V. Sobolev. *Elements of Functional Analysis*. F. Ungar Publishing Co., New York, 1962.

[D.8] Luenberger, D. G. *Optimization by Vector Space Methods*. Wiley, New York, 1969.

[D.9] Papoulis, A. "Truncated Sampling Expansions." *Preprint Volume*, Joint Automatic Control Conference, 40–42, June 1967.

[D.10] Porter, W. A. *Modern Foundations of Systems Engineering*. The Macmillan Co., New York, 1966.

[D.11] Ritter, K. "Duality for Nonlinear Programming in Banach Space." *SIAM Journal on Applied Mathematics*, **15**, 294–302, March 1967.

[D.12] Shilov, G. E. *An Introduction to the Theory of Linear Spaces*. Translated from the Russian by R. A. Silverman. Prentice-Hall, Englewood Cliffs, N.J., 1961.

[D.13] Stubberud, A. R., and J. M. Swiger. "Minimum Energy Control of a Linear Plant with Magnitude Constraint on the Control Input Signals." *Preprint Volume*, Joint Automatic Control Conference, 398–406, June 1965.

[D.14] Swiger, J. M. "Application of the Theory of Minimum-Normed Operators to Optimum-Control-System Problems." Pages 151–218 of *Advances in Control Systems Theory and Applications*, Vol. 3, C. T. Leondes (Editor). Academic Press, New York, 1966.

[D.15] Varaiya, P. P. "Nonlinear Programming in Banach Space." *SIAM Journal on Applied Mathematics*, **15**, 284–293, March 1967.

AUTHOR INDEX

Aaron, M. R., 352
Abe, K., 468
Ablow, C. M., 352
Abramson, P., 258
Adorno, D. S., 463
Akhiezer, N. I., 126
Albers, J. R., 506n
Amara, R. C., 248, 255
Anderson, B. D., 185
Andreyev, N. I., 255, 352
Anorov, V. P., 541
Aoki, M., 463, 591
Archer, D. H., 358
Arimoto, S., 541
Arnoff, E. L., 255
Asher, D. T., 240, 255
Athanassiades, M., 501, 541
Athans, M., 2, 27, 479, 501n, 539, 541

Bakes, M. D., 255
Bass, R. W., 541
Beale, E. M. L., 255, 348
Beattie, L. A., 143, 184
Beckenbach, E. F., 591
Beckman, F. S., 314, 352
Bedrosian, E., 126
Bell, M. D., 356
Bellman, R., 367, 368, 420, 464, 465, 514, 541, 565, 591
Bereanu, B., 255
Bergsman, J., 255
Berkovitz, L. D., 126, 433, 524, 541
Bernholtz, B., 465
Bingulac, S. P., 126

Birkhoff, G., 465
Bishop, A. B., 353
Blackman, R. B., 579
Blackwell, D., 465
Bliss, G. A., 126, 478, 524, 541
Bode, H. W., 48, 52
Bodewig, E., 465, 565
Boehm, B. W., 465
Bogner, I., 541
Boltyanskii, V. G., 465, 468, 546
Bolza, O., 67, 126
Bongenaar, W., 52
Booth, A. D., 353
Booton, R. C., Jr., 184, 570, 571
Box, G. E. P., 272, 296, 353
Boyadjieff, G., 541, 551
Brand, L., 52
Bremmer, H., 571
Brennan, B. J., 541
Breuer, M. A., 255
Brigham, G., 352
Bromberg, N. S., 293, 329, 353
Brooks, R., 353
Brooks, S. H., 288, 329, 353
Brown, A., 249, 255
Brown, R. R., 353
Brown, W. M., 184
Brugler, J. S., 52, 54
Bryan, G. L., 28
Bryson, A. E., 353, 356, 514, 542
Bucy, R. S., 183, 458, 465, 466
Buehler, R. J., 353, 357
Burt, D. A., 358
Bushaw, D. W., 500, 542, 550
Butkovskii, A. G., 540, 542

593

AUTHOR INDEX

Carroll, C. W. 339, 340, 353
Cauchy, A., 265, 297, 331, 353
Centner, R. M., 353
Chalk, J. H. H., 171, 184
Chandler, C. B., 329–331, 354
Chang, C. S., 353
Chang, S. H., 52, 59
Chang, S. S. L., 152, 184, 542
Charnes, A., 217, 224, 255, 353
Cheney, L. K., 255
Chernoff, H., 353
Chien, R. T., 247, 258
Churchman, C. W., 5, 27
Clunies-Ross, C., 249, 256
Cochran, W. T., 184
Cook, R. A., 256
Cooper, W. W., 255
Cottle, R. W., 354
Courant, R., 353
Cox, H., 452, 465
Crockett, J. B., 308, 353
Cruon, R., 467
Cruz, J. B., Jr., 127
Curry, H. B., 353

Dahlin, E. B., 465, 542
Dantzig, G. B., 205, 218, 224, 229, 256, 354
Darcy, V. J., 542
Dashiell, J. N., 547, 553
Davenport, W. B., Jr., 579
Daves, H. T., 465
Davidon, W. C., 274, 320, 354
Davis, M. C., 184
Davis, R. S., 359
Davis, S. A., 52, 53
Deiters, R. M., 465
DeLand, E. C., 354
Denham, W. F., 353, 514, 542
Denn, M. M., 542
Dennis, J. B., 235, 256
DeRusso, P. M., 353
Desoer, C. A., 465, 542
Detchmendy, D. M., 542
de Trove, N. C., 52
Diamond, J., 184
Ditoro, D. M., 542
Doll, H. G., 500, 543
Dov, A. G. B., 240, 256
Downing, J. J., 184
Drake, F. D., 52, 54
Drake, J. H., 127

Dreyfus, S. E., 464, 465
Duda, R. O., 52, 56
Duffin, R. J., 354
Dunn, J. C., 543
Durling, A. E., 465
Dyer, P., 545

Eaton, J. H., 543
Eggleston, D. M., 541, 543
Eisenberg, E., 354
Elsgolc, L. E., 127
Euler, L., 121
Everett, H., 265, 342, 354

Faddeeva, V. N., 565
Falb, P. L., 479, 541
Fan, L. T., 543
Favin, D. L., 184
Fend, F. A., 329–331, 354
Fiacco, A. V., 334, 339, 340, 354
Finkel, R. W., 312, 354
Fletcher, R., 319–322, 348, 349, 354
Flood, M. M., 351, 354
Flügge-Lotz, I., 543, 551
Fomin, S. V., 127
Forsythe, G. E., 309, 312, 341, 354, 363
Fox, L., 127
Frame, J. S., 565
Francesco, J. (Count Riccati), 444
Frank, M., 354
Franklin, G. F., 543
Franklin, P., 52
Franks, L. E., 184
Frederick, D. K., 543
Freimer, M., 465
Friedland, B., 571
Frisch, I. T., 248, 256
Fromovitz, S., 356
Fukao, T., 465
Fukumura, T., 468
Fuller, A. T., 515, 543

Gale, D., 227, 256
Gal'perin, M. V., 354
Gamkrelidze, R. V., 465, 468, 546
Gartner, W. W., 52
Gass, S. I., 258
Gauss, E. J., 579
Gauss, K. F., 292
Gelfand, I. M., 127
Geoffrion, A., 353

AUTHOR INDEX

Gerst, I., 184
Gibrat, R., 240, 257
Glass, H., 466
Gluss, B., 466
Gnoyenskiy, L. S., 355
Goldstick, G. H., 249, 256
Gomory, R. E., 247, 256
Goodman, L. M., 52, 57, 61
Gould, L. A., 185, 571, 579
Graham, L. J., 465
Greber, H., 1, 27
Greene, J. C., 52, 54
Greensite, A. L., 452, 459, 466
Griffith, R. E., 355
Guignabodet, J., 395, 466
Gumowski, I., 466
Gupta, S. C., 184, 546, 547
Gurevich, A. M., 256

Hadamard, J., 352, 355
Hadley, G., 218, 256, 355, 463, 466
Halkin, H., 542
Hall, A. D., 5, 27
Hamming, R. W., 184
Hancock, H., 29, 52
Hancock, J. C., 184
Hannen, R. A., 542
Hano, I., 543
Hanson, G. E., 256
Hardy, G. H., 591
Hazen, H. L., 539, 543
Hedvig, T. I., 127
Helms, H. D., 184
Henderson, A., 255
Hess, S. W., 466
Hestenes, M., 314, 355, 478, 500, 514, 543
Higgins, T. J., xix
Ho, Y. C., 355, 452, 459, 466, 540n
Holland, J., 464
Holtzman, J. M., 542, 543
Hooke, R., 322, 355
Horowitz, I. M., 178, 183, 184
Hovanessian, S. A., 256
Howard, D. R., 543
Howard, R. A., 466
Howe, R. M., 509, 546
Hsieh, H. C., 184, 591
Hu, T. C., 247, 256
Huber, E. A., 357
Hurd, F. K., 127
Husson, S. S., 249, 256

Hutchinson, C. E., 184
Hwang, C. L., 543

Idelsohn, J. M., 353
Inagaki, Y., 469
Isaacs, D., 543
Isaacs, R., 540n, 544
Ivakhnenko, A. G., 355

Jackson, A. S., 238, 257, 355
Jackson, J. T. R., 468
Jackson, R., 185
Jacques, M., 541
Jaffe, R., 185
James, G., 355
James, R. C., 355
Jeeves, T. A., 322, 355
John, F., 334, 355
Johnson, C. D., 527, 531, 533, 544, 547
Johnson, E. P., Jr., 185, 579
Johnson, S. M., 281n, 355
Jordan, D., 545
Joseph, P. D., 465, 468
Josephs, H. C., 52, 54
Juncosa, M., 244–247, 257, 259
Junnarkar, N. V., 466
Jury, E. I., 185

Kaenel, R. A., 184
Kahne, S. J., xix, 126, 127, 401n, 466, 514
Kaiser, J. F., 185, 571, 579
Kalaba, R., 244–248, 257, 259, 368, 464, 465, 514, 541, 544
Kalman, R. E., 183, 447, 458, 466, 544
Kantorovich, L. V., 46, 52
Kasahara, Y., 52
Katzenelson, J., 185
Kaufmann, A., 467
Kazda, L. F., 541, 591
Kelley, H. J., 297, 355, 514, 544
Kempthorne, O., 353, 357
Kenneth, P., 127, 355, 514, 519, 544, 545
Kershner, R. B., 27
Kiefer, J., 281, 355
Kim, M., 257
Kimura, S., 465
King, R. P., 334, 355
Kirchmayer, L. K., 467
Kliger, I., 127, 131
Knapp, C. H., 185, 191, 544
Kochetkov, Yu. A., 544

AUTHOR INDEX

Koenig, H. E., 565
Kokotovic, P., 127, 356
Kolb, R. C., 185
Kolosov, G. E., 467
Kondo, B., 358
Kopp, R. E., 514, 544, 547
Korotkevich, G. I., 354
Koziorov, L. M., 544
Kranc, D. M., 591
Krasovskii, N. N., 240, 257
Krichmayer, L. K., 127
Kronrod, A. D., 467
Krotov, V. F., 545
Kruskal, J. B., Jr., 247, 257
Krylov, V. I., 46, 52
Kuehn, D. R., 243, 244, 257
Kuhn, H. W., 227, 256, 334, 356
Kundert, W. R., 52, 54
Kuo, M. S. Y., 591
Kupervasser, Yu. I., 544
Kushner, H. J., 356, 467, 540, 545
Kuznetsov, Yu. K., 356

Lagrange, J. P., 21, 62
Lanczos, C., 467, 565
Lang, W. W., 184
Lapidus, L., 299, 356, 467, 515, 546
Larson, R. E., 428, 452, 455, 459, 467
LaSalle, J. P., 545
Lasdon, L. S., 358, 515, 545
Lawler, E. L., 356
Laws, B. A., 506n
Lee, E. B., xix, 448, 467, 501, 511, 545
Lee, I., 515, 521, 547
Lee, R. C. K., 452, 459, 466, 467
Leggate, J. W., 325n
Lehman, R. S., 257
Lemke, C. E., 224, 257
Leon, A., 346-349, 351, 354, 356
Leondes, C. T., 184, 543
Levine, L., 356
Levinson, N., 257
Ling, S. T., 52
Littlewood, J. E., 591
Liusternik, L., 591
Llewellyn, R. W., 201, 257
Lootsma, F. A., 340, 356
Lorchirachoonkul, V., 53, 243, 257
Lowery, P. G., 467
Luenberger, D. G., 357, 540, 545, 591

McCausland, I., 545
McCormick, G. P., 340, 354
MacCracken, L. C., 185
McDonald, D., 500, 545
McGill, R., 514, 544, 545
Mackie, D. G., 249, 256
MacPhie, R. H., 185
McReynolds, S. R., 356
McShane, E. J., 478, 545
Madansky, A., 257
Magee, E. J., 356
Mangasarian, O. L., 257, 356
Marbach, H. D., 543, 551
Markland, C. A., 545
Markus, L., 448, 467
Masse, P., 240, 257
Mathews, M. V., 571
Maxwell, J. C., 29, 52
Mayall, R. B., 127
Meditch, J. S., 545
Meissinger, H. F., 356
Meer, S. A., 185
Mellon, B., 255
Merchav, S. J., 183, 185
Merriam, C. W., III, 440, 467
Meserve, W. E., 545
Miehle, W., 257
Miele, A., 115, 127
Mikhlin, S. G., 127
Minnick, R. C., 431n
Minsker, I. N., 354
Mishchenko, E. F., 468, 546
Mobley, R. L., 465
Mond, B., 356, 467
Mori, H., 579
Mostov, P. M., 127, 143, 185
Motzkin, T. S., 309, 312, 354
Movshovich, S. M., 355
Moyer, H. G., 544, 547
Mulligan, J. E., 8, 28
Murata, T., 357

Nafus, D. V., 506n
Narita, S., 543, 545
Nelson, D. E., 184
Nemhauser, G. L., 468
Nesbit, R. A., 591
Neuringer, J. L., 127, 185
Neustadt, L. W., 545
Newton, G. C., Jr., 185, 571, 579

Norris, R. C., 357
Noton, A. R. M., 514, 545

O'Donnell, J. J., 545
Ogata, K., 35n, 52, 448
Oldenburger, R., 479, 546
Oliver, L. T., 281n
Olson, O. J., 506n
Onaga, K., 468
Orden, A., 218, 229, 256

Paiewonsky, B., 500, 539, 546
Papoulis, A., 454, 468, 579, 591
Paraev, Y. I., 546
Paris, D. T., 127
Pearson, J. B., Jr., 546
Perkins, W. R., 127
Perlis, S., 565
Pervozvanskiy, A. A., 257, 546
Peschon, J., 452, 455, 459, 467
Peterson, E. L., 354
Pierre, D. A., 53, 56–59, 185, 257, 468, 571
Pinsker, I. Sh., 312, 357
Pokoski, J. L., 541, 546
Pollack, M., 247, 248, 257
Pólya, G., 591
Ponstein, J., 270, 356, 357
Pontryagin, L. S., 436, 465, 468, 478, 500, 546, 550
Porcelli, G., 257
Porter, J., 243, 244, 257
Porter, W. A., 591
Powell, M. J. D., 277, 280, 314, 320–322, 348–351, 354, 357
Propoi, A. I., 257, 546
Pruzan, P. M., 468
Pyne, I. B., 236, 258

Rapaport, H., 258
Rauch, L. L., 509, 546
Rechtin, E., 185
Reeves, C. M., 319, 320, 354
Rekasius, Z. V., 543
Rice, R. K., 358, 545
Riddle, A. C., 185
Rigney, D. S., 127, 185
Riley, V., 258
Ringlee, R. J., 467
Ritter, K., 591
Rivlin, L., 358
Roberts, A. P., 183, 186, 541

Roberts, S. M., 466, 468
Robins, H. M., 186
Rohrer, R. A., 127
Root, W. L., 579
Rosen, J. B., 357
Rosenbloom, P. C., 357
Rosenbrock, H. H., 348–350, 357
Ross, M. E., 53, 527, 546
Rota, Gian-Carlo, 465
Rothenberger, B. F., 515, 546
Roxin, E., 478, 546
Rozonoer, L. I., 468, 546
Rubio, J. E., 546
Rudberg, D. A., 325n
Rushforth, C. K., 173n
Rutman, R. S., 127
Rybashov, M. V., 357
Rybasov, V. I., 354

Saaty, T. L., 357
Sakawa, Y., 243, 258
Sakrison, D. J., 357
Sancho, N. G. F., 183, 186
Sanneman, R. W., 546
Sarachik, P. E., 591
Sard, E. W., 52, 54
Scanlan, J. O., 53
Scheibe, P. O., 357
Scherz, C. J., 468
Schley, C. H., Jr., 515, 521, 547
Schwarz, R. J., 571
Schwarzlander, H., 184
Schweppe, F. C., 541
Seifert, W. W., 571
Sengupta, S. S., 255
Shah, B. V., 312, 313, 348, 353, 357
Shapiro, E., 356, 467
Shapiro, S., xix, 356, 467
Shaw, F. S., 258
Shelly, M. W., II, 28
Shemer, J. E., 547
Shen, D. W. C., 465, 542
Shilov, G. E., 591
Shisha, O., 467
Shnidman, D. A., 547
Sidar, M., 127, 131
Sigalov, A. G., 127
Siljak, D. D., 356
Singleton, J. S., 53
Sinnott, J. F., Jr., 357
Smith, C. S., 349, 350, 357

Smith, F. B., Jr., 501, 511, 513, 547
Smith, F. W., 547
Smith, G. W., Jr., 53, 61
Smith, O. J. M., 501, 541
Snow, D. R., 468
Sobolev, V., 591
Sobral, M., Jr., 127
Sridhar, R., 542, 546
Sokkappa, B. G., 358
Solem, R. J., 184
Solymar, L., 127, 129
Sommer, R. C., 53
Sondak, N. E., 359
Sorenson, H. W., 468
Southwell, R. V., 292, 358
Sreedharan, V. P., 258
Srinivasan, A. V., 258
Stagg, G. W., 407, 468
Stakhovskii, R. I., 128
Steiglitz, K., 186, 542
Stewart, R. A., 355
Stiefel, E., 314, 355
Stillman, R. E., 356, 467
Stout, T. M., 256, 547
Stratonovich, R. L., 467, 468
Strutt, J. W. (Lord Rayleigh), 27, 28
Stubberud, A. R., 591
Sugai, I., 128
Sugie, N., 283, 358
Sutabutra, H., 541
Suzuki, T., 358
Swann, W. H., 358
Swiger, J. M., 591
Synge, J. L., 293, 358

Takahashi, Y., 541
Tamura, Y., 543
Taylor, G. E., 127, 355, 514, 519, 544
Tchamran, A., 468
Tezuka, Y., 52
Titus, H. A., 543
Tomovic, R., 128
Torng, H. C., 258
Totty, R. E., 184
Tou, J. T., 186, 468
Tozer, R. F., 186, 190
Tracz, G. S., 540, 547
Tretter, S. A., 186
Tricomi, F. G., 173, 186
Tseitlin, B. M., 312, 357
Tucker, A. W., 227, 256, 334, 356

Tufts, D. W., 186, 190, 547
Tukey, J. W., 579
Tung, F., 468, 469
Turin, G. L., 186
Tyndall, W. F., 258

Udagawa, K., 468, 469

Valentine, F. A., 96, 128, 478, 514, 515, 547
Van De Panne, C., 358
van der Pol, B., 571
Varaiya, P. P., 591
Volz, R. A., 322, 358
von Neumann, J., 227, 258

Wall, H. S., 469
Wang, P. K. C., 469
Waren, A. D., 358, 545
Watson, M., 407, 468
Webber, R. F., 541
Weisman, J., 324, 358
Weiss, E. A., 358
Whalen, B. H., 258
Whinston, A., 358
Wiener, N., 186
Wilde, D. J., 281n, 283, 358
Williams, A. C., 258
Wilson, K. B., 272, 296, 353
Wing, O., 247, 248, 258
Witte, B. F., 348
Wohlers, M. R., 547
Wolfe, P., 201, 256, 258, 334, 354, 358
Womack, B. F., 547, 553
Wonham, W. M., 531, 533, 544, 547
Wood, C. F., 358
Wood, H., 127

Yamazaki, T., 465
Yang, H. C., 249, 255
Youla, D. C., 186
Young, P. C., 358
Young, R. D., 258

Zadeh, L. A., 240, 258
Zangwill, W. I., 277, 349, 358
Zellnick, H. E., 299, 359
Zener, C., 354
Zotov, M. G., 186

SUBJECT INDEX

Absolute value, constraints (*see* Constraints)
criteria (*see* Performance measures)
Absolute-value inequalities, 585
Acceleration steps, 309–313
Adaptation, in systems, 4
to reduce sensitivity, 192
with dynamic programming, 463
with search techniques, 266, 326, 351
Admissible curve for classical variational calculus, 64
Admissible function, 8
Admissible parameter, 8
Admissible variation (*see* Variation)
Algebraic equation solution, by conjugate-gradient method, 314
by search, 331–332, 363
(*see also* Linear algebraic equations)
Algorithm, 5
Allocation problems, for reliability with redundancy, 378–380
for variance of a characteristic, 371–372, 469–470
investments, 370
power generation, 370
space allocation, 370
utilizing Lagrange multipliers, 412–414
with n state variables, 409–410
with one resource (state) variable, 369–380, 402
with random variables, 476–477
with 2 state variables, 412–414
Allowable variation (*see* Variation)
Analog simulation, of time optimal control, 506–507

Analog solutions, for linear programming, 235–239
for minimum-chain problems, an analog of strings, 234–235
Zener-diode analog, 235
for steepest ascent, 298–299
of two-point boundary-value problems, 479
Analogies, conducting sheet analogy for Laplace's equation, 251–253
for temperature distribution and steady-state fluid flow, 251–253
gravitational analogy of a stationary point, 31
in general, 234–235
in search, bunny-hop search, 325–328
density concept, 268, 329–331
mountain range, 267–268
Approximation, by a cubic, 277
by a quadratic, 275, 307, 360
discrete approximation of continuous systems, 425–426
for continuity improvement, 480
for dynamic programming solution, 411–412, 424–426
in function space, 419–423
in policy space, 417, 420–421
in state space, 417–420
of derivatives (*see* Derivative approximations)
of functions in state space, 402
with orthonormal functions, 60–61
with straight line, 273
(*see also* Curve fitting, Direct methods, *and* Interpolation)

SUBJECT INDEX

Area, of surface of revolution, 129
  under a curve, 129
Arithmetic mean, 580
Artificial variables, for transversality
    conditions, 494
  in linear programming, 217–218
Associative laws, of matrix theory, 556
  of spectrum factorization, 148–149
Augmented function (*see* Performance
    measures)
Augmented matrix, 557–558
Autocorrelation, 572–573

Backward solutions, comparison with
    forward, 385–386
  of a stochastic control problem, 450–452
  of continuous decision processes,
    425–431, 441–444, 448
  of discrete control problems, 387–390,
    411, 419–421
  of minimal chain problems, 383–386,
    422–423
Banach spaces, 583–587
Bandwidth, an integral-square measure
    of, 190
  for an "ideal" filter, 191
  in relation to sensitivity, 176–182
Bang-bang control, 501
Basic solutions of linear programming,
    degenerate, 200–201, 215–217
  feasible, 200, 205–206, 217–218
  nondegenerate, 200–201
Basis vectors, 557, 589–590
Bayes' theorem, 454
Bellman's principle of optimality, 21,
    368, 402–403
Block diagram (*see* Control system design)
Bode diagram, 179–182
Bolza, problem of, 66–68
Boundary, defined by an inequality, 272
  determined by a hyperplane, 236–237
  of a convex set, 201–202
  of a two-dimensional region, 43–44
  (*see also* Constraints)
Boundary conditions, for two-point
    boundary-value problems, 73–74
  (*see also* End-point conditions *and*
    Transversality conditions)
Boundary-value problems, search
    techniques for, 125, 513–523

Bounded set, 424
Branches, of a chain network, 380
  of a communications network, 244
  of moment-rosetta search, 331

Calculus of variations, fundamental
    lemma of, 70–71
  general scope of, 62–63
  some problem types, 17, 66–68, 100–102,
    124–125
  with dynamic programming, 431–440
  with Pontryagin's maximum principle,
    439–440, 478
Cancellation of poles with zeros, when to
    be avoided, 179–181, 190–191
Cartesian coordinates, 267
Cauchy density functions, 455–456
Cauchy sequences, 581
Cauchy's inequality, 588
Cauchy's residue theorem, 569
Center of mass, 329n
Centroid, 329–330
Chain link, 234–235
Chain networks, nonoriented ones, 234–235,
    422–423
  oriented minimal chain, 380–386,
    402–403, 471
Chain rule for derivatives, application of,
    36–37, 45
Characteristic equation, 563
Characteristic value, for an integral
    equation, 173
  (*see also* Eigenvalue)
Characteristic vector, 564
Chatter in control, 506–507
Circuit design, by allocation of parameter
    variance, 370–371, 469–470
  for a voltage-current characteristic,
    260–261, 463
  for steady-state energy transfer, 55–56
  for triggering level, 57
  of a clipping circuit, 58–59
  of a gating circuit, 57–58
  of a lead network, 34–35
  of an interstage coupling network,
    290–291
  of a peaking circuit, 364–365
  of frequency division circuit, 39–41
  with probability distributions given for
    parameters, 249
  with search techniques, 266

SUBJECT INDEX

Circuit design (*continued*)
  with worst-case design, 248–251
  (*see also* Filter design *and* Pulse-shape problems)
Class C^k (*see* Continuity)
Clipping circuit, 58–59
Closed interval, 32
Closed-loop system (*see* Feedback)
Closed set, 424
Coasting range of control, 491
Communications systems, a chain network problem, 380
  a linear communications network,
    optimal design of, 246–247
    optimal utilization of, 244–246
    prescheduling future system capacity, 247
  a minimum-time problem, 247
  dynamic programming approaches to, 463
  intersymbol and interchannel interference, 171
  link leasing problem, 247
  Pontryagin's maximum principle applied to, 540
  probabilistic communications networks, 248
  the maximum capacity route, 247
  with a pulse-code-modulated signal, 74–75
  (*see also* Pulse-shape problems)
Compensator, phase lead, 34–35
  (*see also* Controller)
Completely controllable system, 447
Completely regulable system, 447
Complete sequence of approximating functions, 119–120
Computer flow diagram, for an upper-bounding algorithm, 225
  for a steepest descent search, 302–303
  for bunny-hop search, 328
  for DFP search, 321
  for discrete Kalman-Bucy filter, 461
Computer storage space, conservation of
  in dynamic programming, 395–396, 411–418, 428–431
  in solution of two-point boundary-value problems, 515
Concave function, 269–270
Conducting sheet analog, 251–253
Congruent matrices, 316
Conjugate directions, 314–318
Conjugate indexes, 584

Conjugate point, 110–113
Constrained minima and maxima
  (*see* Constraints *and* Minima and maxima)
Constraint breakthrough, 238, 263, 339–340
Constraints, differential equations, 92–95, 100
  differential inequalities, 95–98
  energy, 77, 142, 170–171
  equality constraints on functions of parameters, 36–42, 334–335, 341, 344
  examples in a table, 10–16
  for linear programming, 194–197
  geometrical interpretations of, 200–204, 272
  hard versus soft, 233, 243
  inequality constraints on functions of parameters, 43–45, 334–335, 339–340, 342–344
  integral-square value, 137–138
  isoperimetric, 75–82, 91–92, 482–483
  isoperimetric inequality, 98
  linear inequality, 195–197, 236–252, 259–263
  lower bound, 194–196
  magnitude, 197
  mean-square value, 140–141
  on a linear integral, 115
  physical basis for, 7–8
  power, 140–141
  upper bound, 196–200, 224–226
  (*see also* Integer programming, State equations, *and particular problem types*)
Continuity, approximations for, 480
  of class C^0, 424
  of class C^1, 18, 32, 267
  of class C^k, 32–35
  of minimum-cost functions, 395–402
Continuous recurrence relations, 426–434
Contours of equimagnitude, 267
Control function, 22
Control law, 22
  (*see also* Policy function)
Control system design, a general continuous-system problem, 423–424
  an overshoot consideration, 470
  a peak-seeking regulator, 365–366
  a stochastic control problem, 450–452
  bibliographies on, 539
  for a distributed-parameter system, 243

Control system design (*continued*)
  for bandwidth reduction, 190
  for closed-loop control with quadratic functional, 440–450
  for curve-tracking with overcontrol elimination, 198–200
  for docking of space ships, 131–132
  for fuel problems, 132n, 242, 490
  for gear-ratio selection, 53
  of a temperature regulator, 551–552
  sampled-data systems and linear programming, 240–242
    closed-loop control and time-weighted error criteria, 242
    minimal action problems, 242
    minimal time problem, 242
    terminal error problems, 241
  scalar case of feedback control, 488–493
  scalar dynamic programming cases, 472
  sensitivity, 173–182
  singular control problems, 526–527, 531–536
  spectrum-factorization for distributed systems, 183
  via calculus of variations, 100–104
  via dynamic programming and the regulator problem, 418–421
  via Pontryagin's maximum principle, 481–486, 540
  with bang-bang control, 438–439
  with disturbance in the loop, 186, 190
  with dual-mode control, 506–510
  with intergral-square-error criteria, in general, 137–139, 150–153, 183
  with an energy constraint, 102–104
  with an integral-square-velocity constraint, 79–82
  with an unstable plant, 190–191
  with nonminimum-phase plant, 158–162, 180–182
  with time-delay in the plant, 189–190
  with ITAE criterion and saturation constraint, 365
  with Riccati equations (*see* Riccati equations)
  with sampled squared-errors criterion, 60
  with search techniques, 266
  with straightforward dynamic programming approach, 386–394, 403, 410–411

Control system design (*continued*)
  with time-variable feedback coefficients, 131, 444–447
  (*see also* Feedback *and* Time optimal control)
Control system design philosophy, 22, 102, 174–182, 481–482
Control vector, 100, 410, 424, 481
Controllability, 447
Controller, cascade, 160–162, 173–174
  dual mode, 506–510
  feedback, 173–174
  pattern-search controller, 324
  (*see also* Control system design)
Convex function, 269–270
convex set, 200–204
Convolution integral, 565
Corner conditions, scalar case, 72–73, 87–89
  vector case, 91
Corner frequency, 179, 182
Corner points, 72, 88–89
Cost, as a factor in performance measures, 6–7
Cost coefficients, in relation to Lagrange multipliers, 38
  for a routing problem, 246
Costate equation, 101–103, 434–437, 485
Costate variables, relation to Lagrange multipliers, 101, 435–436
Costate vector, 101–103, 434–437, 485
Covariance matrix, 458
Cross-correlation, 572–573
Cross-over frequency, zero db, 182
Cross-spectral density function, 576
Cubic convergence, 275
Curve fitting, with a polynomial, 60
  with linear line segments, 260
  (*see also* Approximation)

Dead zone of control, 493
Decision process, 367–368, 402–403
Degrees of control freedom, 174–178
Deleted neighborhood, 30–31
Demand coefficients, in a communications problem, 244
Density concept in search, 268, 329–331
Density function (*see* Power-density spectrum *and* Probability density functions)

SUBJECT INDEX

Derivative approximations, first, 122, 425
  partial, 252
  second, 293
Design center, 40n
Determinants, evaluation of, 559
  for sufficiency tests, 35–36, 112–113
Diagonal matrix, 316, 559–560
Difference equation, Fibonacci, 283
Difference equations (*see* Constraints *and* State equations)
Differential-difference equation for a continuum of state variables, 4
Digital computer consideration, for linear programming, 193
  for search techniques, 346–351
  (*see also* Computer flow diagrams *and* Computer storage space)
Dimensionality, limitations of straightforward dynamic programming, 408–411
  reduction of, 332–333, 363
Dirac delta function, 143, 456
  (*see also* Impulse)
Direct methods, concept of, 19–21
  Galerkin's method, 46–47
  of the calculus of variations, finite differences, 121–123
  Ritz-Galerkin methods, 119–121
  (*see also* Search techniques)
Direction vector, 286
Distributed-parameter systems
  (*see* Control system design *and* Conducting sheet analog)
Disturbance rejection
  (*see* Control system design *and* Filter design)
Dither in search, 292
Dot product, 238
  (*see also* inner product)
Dual-mode control, 506–510
Dual problems, in general, 21, 227, 351
  of linear programming, mixed dual, 229–232
    symmetric dual, 227–229
    unsymmetric dual, 229
Dual variables, 227–230
Dynamic programming, as an approach to solution, 17, 21
  assorted applications of, 463
  for continuous decision processes, 423–450

Dynamic programming (*continued*)
  for stochastic systems, 450–460, 476–477
  for time-invariant infinite-stage processes, 418–421
  framework for, 367–369
  iterated form of, 414–417
  straightforward procedures of, 368–369, 410–411
  with Lagrange multipliers, 412–414
  with series approximations, 411–412

Economics problems, allocation problems, 370–371
  an advertising and production-scheduling problem, 132–133
  interest rates considered in, 359
  manpower assignment, 259
  maximum-principle solutions, 540
  overtime scheduling, 469
  production scheduling, 197–198, 266, 361, 409–410, 473
  subjective criteria for, 5–7
  types of, 239–240
Eigenvalues, 563
Eigenvector, 564
Elastic stops, 97–98
Electrical circuits (*see* Circuit design)
End-point conditions, in dynamic programming, 381–386, 388–391
  (*see also* Transversality conditions)
End-point functionals, 66–68, 483–485, 537–539
End-point variations, 82–87, 90, 495–498
Energy, dissipation in circuits, 55, 77, 98
  (*see also particular problem types*)
Environmental factors, 7–8, 24
Equilibrium points, 31
Equimagnitude contours, 267
Erdmann-Weierstrass conditions (*see* Corner conditions)
Error criteria, 5–6, 63
  (*see also* Performance measures *and particular problem types*)
Estimation, of a parameter, 453
  of state variables, 452–462, 540
Euclidean distance, 128, 297
Euclidean space E^n, 30–31, 267–269, 583
Euler-Lagrange equations, for a general control problem, 101
  from Pontryagin's maximum principle, 440

SUBJECT INDEX

Euler-Lagrange equations (*continued*)
  scalar cases of, 73–75
  state-vector cases of, 90
  with constraints included, 91–99
Excess function, 117
Existence of solutions, effect of
    constraints on, 36, 56
  to a given differential equation, 487n
  to linear algebraic equations, 558
  to min-max problems, 31
Expected value, of a function, 17
  in unbiased cases, 271
  (*see also* Performance measures with
    probabilistic facets)
Extrema (*see* Minima and maxima)
Extremal curves, 72, 437n
  (*see also* First-variational curves *and*
    Principal curves)
Extreme point of a convex set, 203
Extremum (*see* Minima and maxima)
Extremum point, 31

Factorial search, 288–289
Family of solutions (*see* Fields of solutions)
Feasible design, 8
Feasible solution of linear programming,
    200, 205–206, 217–218
Feedback, with constant coefficients, 448
  for minimum-time transition, 501–513
  in a linear sampled-data case, 60
  philosophy of, 4, 20, 22, 102, 481–482
  with time-varying coefficients, 131,
    444, 446
  (*see also* Control system design)
Fibonaccian numbers, 283–284
Field problem, 251–253
Fields of first-variational curves, 109–112
Fields of solutions, central field, 108–109
  proper field, 107
  (*see also* Conjugate point)
Figure of merit (*see* Performance
    measures)
Filter design, inequalities used in, 586–587
  integral-square error, 137–139, 151–152
  Kalman-Bucy, a scalar case, 477
    discrete filter, 457–462
    relation to Wiener filter, 183
  Wiener mean-square-error filter, 140–141,
    153–158, 188–189
  with search techniques, 266, 364

Finite difference methods (*see* Direct
    methods)
First integral, 74
First-variational curves, as a class of
    principal curves, 440
  scalar case, 72
  state-vector case, 90
Forward solutions, of allocation
    problems, 410
  comparison with backward, 385–386
  for discrete control problems, 390–391,
    416–417
  in estimation and filter problems,
    454–460
  of a minimal chain problem, 381–383
  of continuous decision processes, 432–436
  of transmission-line problem, 406–408
Fourier series, 590
Fourier transform, 135, 566
Fractional change, of a function, 49–50
  of a functional, 123–124
Frequency-division circuit, 39–41
Friction, static, 53
  viscous, 79–81
Frobenius's relation, 560
Fuel criterion, 132n, 242, 490
Function, admissible, 8
  concave, 269–270
  convex, 269–270
  cubic, 277
  global properties of, 267–270
  local properties of, 266–267
  minimum-phase, 159
  monotonic, 398
  noninteracting or separable, 270
  of a function (*see* Functional)
  of class C^k (*see* Continuity)
  of order ϵ, 70
  orthogonal, 61, 589
  parametric form, 299
  proper rational fraction, 568
  quadratic, 274, 307
  regional properties of, 267–270
  unimodal, 269
Functional,
  augmented, for isoperimetric constraints,
    76, 483
    for general constraints, 91–96, 494
  concept of, 6, 62
  dependent on higher-order derivatives,
    91

SUBJECT INDEX

Functional (*continued*)
   end-point, 66–68, 483–485, 537–539
   integral-square error, 63, 138, 570
   mean-square error, 140, 572, 577
   quadratic, 73, 113, 118, 440
   strongly continuous, 65
   with integrand independent of t, 74
   with integrand independent of x, 74–75
   (*see also* Performance measures *and*
      Search techniques for functionals)

Galerkin's method, 46–47
   (*see also* Direct methods)
Game theory relation to optimal control, 540
Gating circuit, 57–58
Gaussian noise, 458–459
Gear-ratio selection, 53
Geometric mean, 580
Geometrical interpretations, of constraints, 200–204, 272
   of functions, 266–269
   of linear programming, 200
   of Pontryagin's maximum principle, 484
   of search, 19–20
      partan search, 312
      steepest ascent, 297
   of transversality conditions, 496–498
Geometric programming, 352
Golden-section ratio, 286
Gradient, as a direction vector, 236, 297
   in general, 48, 266–267, 561
   of a linear function, 236
Gradient projection method, 334
Graph problem, 380
Green's theorem, 114–115
Grid expansion, 401–402
Grid point, 288–289

Hamiltonian, constancy of, 486–493
   for end-point functionals, 538
   for maximum principle, 436, 485
   for minimum principle, 537
Hamiltonian canonical equations, 437
Hamilton-Jacobi equations, 433–434
Hard constraints, 233, 243
Hard spring, 552
Hessian matrix, 563
Hilbert space, 588–590
Hölder's inequalities, 584–585
Hyperplane, 200–201, 267, 496–497

Hyperpolyhedron, 20, 200
Hyper-rectangle, 330
Hypersurface, 267, 272, 496
Hysteresis for dual-mode control, 509

Identification, in relation to sensitivity, 24–25
   (*see also* Filter design)
Identity matrix, 299, 556
Ill-conditioned matrix, 307n
Imbedding concept, 367–369
Immobial states, 401n
Impulse, 78
   (*see also* Dirac delta function)
Impulse response, 164
Increment of a functional, 66
Indirect methods, 19
Industrial engineering problems, references to, 463, 540
   (*see also* Economics problems)
Inequalities, a norm inequality, 582
   of absolute values, 585
   of Hölder, 584–585
   of Jensen, 588
   of Minkowski, 583–584
   relating $(1 + x)^y$ to $1 + xy$, 580
   relating mean values, 580
Inequality constraints (*see* Constraints)
Inner product, 560
Inner product space, 588
Input admittance, 142
Integer n-tuple, 31
Integer programming, as a problem type, 17
   references to, 254, 352
   with dynamic programming, 370–380, 469–471, 476, 477
   with Lagrange multipliers, 342, 413
   (*see also* Factorial search)
Integral equation of the second kind, 173
Integral performance measures (*see* Functional *and* Performance measures)
Integral-square error, 63, 138, 570
   (*see also particular problem types*)
Interpolation, Davidon's equations, 277
   general considerations, 394–395
   for Lagrange multiplier, 414
   linear two-dimensional, 472–473
   projected linear interpolation, 393–394
   (*see also* Approximation)

SUBJECT INDEX

Intersymbol and interchannel interference, 171
Interval, closed, 32
 half-open, 112, 427
 of uncertainty, 280–283
 open, 110
Inverse, Laplace transform, 567–570
 matrix, 307, 560
ISE (*see* Integral-square error)
Isoline, 267
Isoperimetric constraint (*see* Constraint)
ITAE criterion, 63, 365
Iterated dynamic programming, 416–417

Jacobi condition, 109–112
Jacobian matrix, 562
Jump discontinuity, 432

Kalman-Bucy filter (*see* Filter design)
Kernal of an integral equation, 173
Key column of a simplex tableau, 211
Key row of a simplex tableau, 211

Lagrange, problem of, scalar case, 68–75
 state-vector case, 66–68, 90–91
Lagrange multipliers, as affecting search, 341, 363
 Everett's generalization of, 342–345
 for differential equations, 92–95
 for isoperimetric constraints, 75–76, 91–92, 483
 for m equality constraints, 42–43
 for one equality constraint, 36–41
 in comparison with cost coefficients, 38
 in comparison with weighting factors, 20–21
 in relation to costate variables, 101, 435–436
 with dynamic programming, 412–414
 with inequality constraints, 43–45, 95–96
Laplace transform, direct transform, 315, 566
 inverse transform, 567–570
Laplace's equation, 252
Laurent expansion, 567
Lead network, 34–35
Legendre condition, 105–106, 112
Linear algebraic equations, existence of solution for, 558
 uniqueness of solution of, 560
 (*see also* Linear programming)

Linear differential equations, solution via convolution, 565
 (*see also* State equations)
Linear integral, 115
Linear programming, applications of, 239–254
 a standard problem form, 194
 conversion to standard form, 195–200, 259–260
 repetitive linearization, 239, 249, 351
 (*see also* Simplex technique)
Linear space, 582
Linear transformation, 315–316, 362, 565
Links, of a chain network, 234–235, 380, 422
 of a communications network, 244–248
Local minima and maxima (*see* Minima and maxima)
Logic systems, 4–5, 254, 463
Loop transmission, 174
Low-pass system, 176

Maintainability of a system state, 447
Manifold, 272
Markovian process, 367–368
Matched filter, 164
Matrices, addition of, 556
 an inversion identity, 460n
 associative laws for, 556
 augmented matrix, 557–558
 column matrix, 555
 congruent matrices, 316
 covariance matrix, 458
 derivative of a column matrix, 562
 derivative of a row matrix, 561–562
 determinants of, 559
 diagonal matrix, 316, 559
 distributive laws for, 556
 entries of, 555
 exponential, 511, 564
 generalized diagonal matrix, 559
  determinant of, 559
  inverse of, 560
 Hessian matrix, 49, 563
 identity matrix, 299, 556
 ill-conditioned, 307n
 inverse, 307, 560
 inverse of products of, 560
 Jacobian matrix, 562
 major diagonal of, 556
 partitioned, 558

Matrices (*continued*)
  partitioned inverse of, 560
  positive-definite matrix, 35, 563–564
  positive-semidefinite matrix, 563
  products of, 556
  products of when partitioned, 558
  row and column rank of, 557
  row matrix, 555
  scalar multiplication of, 555
  sensitivity matrices, 49
  similar matrices, 511, 564
  singular matrix, 560
  skew symmetric matrix, 444n
  square matrix, 555
  symmetric, 557
  trace of, 561
  transpose of, 557
Maxima (*see* Minima and maxima)
Maximizing condition of Pontryagin's maximum principle, 436–439, 485–486
Maximum-return functions, 371
  concave cases, 463, 469
  for probabilistic cases, 454–460
  series approximation of, 411–412
  (*see also* Minimum-cost functions)
Mayer, problem of, 66–67
Mean-square error, 140
Mean-square value, 572
Measurement vector, 452, 458
Metric, 581
Metric space, 581
  complete metric space, 582
Minima and maxima, absolute, 31, 164, 269
  constrained (*see* Constraints)
  point of, 30–31
  strong local, 30–31, 267–268
  strong relative, 64–65
  weak local, 30–31
  weak relative, 64–65
  (*see also* Necessary conditions *and* Sufficient conditions)
Minimum-cost functions, 371n
  a discontinuous case, 472
  for probabilistic cases, 451–452
  quadratic cases, 441–450, 474–475
  (*see also* Maximum-return functions)
Minimum-phase function, 159
Minimum principle (*see* Pontryagin's maximum principle)

Minimum-time problem (*see* Time optimal control *and* Communications systems)
Minkowski's inequalities, 583–584
Min-max theory (*see* Minima and maxima)
Modal trajectory, 452–455
Mode-change surface, 526
Monostable flip-flop response, 57
Monotonic function, 398
Monte Carlo method, 288
Move penalties, 243
Multistage decision process, 367–368, 402–403

Necessary conditions, definition of, 18
  first necessary condition of the calculus of variations, 71–72
  for constrained local extrema of functions, 36–45, 334, 338, 340
  for constrained relative extrema of functionals, 90–99
  (*see also* Pontryagin's maximum principle)
  for corner points, 72–73, 91
  for local extrema of functions, 32–35
  for optimality of a functional of a Laplace transform, 145
  for optimal values of constrained end-point values (*see* Transversality conditions)
  for relative extrema of a mixed functional-function problem, 125–126
  in terms of the second variation, 106
Neighborhood, deleted, 30–31
Newton-Raphson search (*see* Search techniques)
Nodes, of a chain network, 234–235, 380, 422
  of a communications network, 244
  of an electrical network, 252–253
Noise, characterized by autocorrelation functions and power-density spectra, 153, 188
  considered in feedback design, 177–178
  effect on search, 271–272, 352
  stationary random, 573
  white, 453
  (*see also* Probability density functions)
Noise vector, 452–453

SUBJECT INDEX

Nonlinear programming problem, 333–334
  (*see also* Search techniques)
Nonlinear systems (*see particular system and problem types*)
Nonlinear voltage-current characteristics, 260–261, 463
Nonminimum-phase function, 159
Non-normal solutions, 523–525
Normed linear space, 582

Objective criteria (*see* Performance measures)
Observability, 448
Open-loop system, 140
Optimality, conditions for (*see* Necessary conditions *and* Sufficient conditions)
Optimization, in perspective, 1–3
  representative problem types, 8–18
Orthogonal initial condition sets, 519
Orthogonal projection theorem, 589–590
Orthogonal vectors, 589
Orthonormal functions, 61
Orthonormal vectors, 561
Outer product, 561
Overcontrol elimination, 198, 242
Overshoot, 470

Parallelogram law, 589
Parseval's theorem, 570
Partan, 312–314
Partitioned matrix, 558
Pattern recognition, references to, 254, 266, 463
Peaking circuits (*see* Circuit design *and* Pulse-shape problems)
Peak-seeking regulator, 365–366
Penalty coefficient, 217, 335–340
Penalty functions, compared to performance weighting, 20–21
  inside penalty function, 339–340, 363
  outside penalty function, 334–339
  with functionals, 97–98, 515, 551
Pencil point of a central field, 108
Performance measures, augmented, 37, 94, 101, 217, 341–342
  composite, 341
  convex or concave, 269–270
  discrete time-weighted absolute-value, 391
  examples in a table, 9–18
  integral, 45 (*see also* Functional)

Performance measures (*continued*)
  linear algebraic, 193
  noninteracting, 369
  objective and subjective criteria for, 5–7
  penalized, 21, 38, 335–340, 515
  with probabilistic facets, 17, 74–75, 140–141, 247–249, 370–371, 450–462, 469–470, 476–477
  (*see also particular problem types*)
Performance weighting, 20–21
Phase variables, 3n
Plant (fixed part) of a system, 7, 137
Point, in an n-dimensional space, 194n, 286
  trial or search point, 19–20, 23, 273
Point of maximum (minimum), 30–32
Pole of a function, 567
Policy function, for an allocation problem, 372–373
  for a control problem, 389
  a time-invariant one, 419–421
  (*see also* Control law)
Policy variable, for a control problem, 386
Pontryagin's maximum principle, an equivalent minimum principle, 536–537
  applications of, 540
  existence and uniqueness of solutions of, 481, 487n
  for a canonical problem form, 481–484
  generalizations of, 540
  modified for end-point functionals, 537–539
  relation to calculus of variations, 478
  relation to dynamic programming, 431–439, 478
  relation to gradient search, 478
Positive-definite, 35, 563–564
Power (*see* Constraints *and particular problem types*)
Power generation, allocation of, 370
Power-density spectra, 574–577
Predictor problems, a gain selection case, 577–578
  with ISE criteria, 139
  with mean-square-error criteria, 140–141
Primal problem, 227–232
Primal variables, 228–230
Principle of Optimality (*see* Bellman's principle of optimality)
Principal trajectories, 437–438, 486

SUBJECT INDEX

Probability density functions, Cauchy density function, 455–456
  Gaussian density functions, 458–459
  to obtain an expected value, 17
Production (*see* Economics problems)
Projection theorem, 589–590
Proper rational fraction, 147, 568
Pulse-shape problems, a solution containing delta functions, 134
  for an ion accelerator, 143
  for average output, 76–79, 144, 164
  for partial output energy, 170–173
  for peak output, 143, 163–169
  for peak output of RLC circuit, 498–500
  for time-weighted output, 141–142, 162–163
  inequalities used in solution of, 585–586
  singular solutions, 527–530
Pythagorean theorem, 297

Quadratic convergence, 24, 273–274, 307, 309, 320
Quadratic form, 563
Quadratic function, 274, 307
Quadratic functional (*see* Functional)
Quadratic minimum-cost functions, 441, 448, 474–475
Quadratic programming, 17, 334, 352
Quasilinearization, 514

Random search, 288, 329
Rank of a matrix, 557
Rational fraction function, 147, 568
Ravines, 267–268
Realizability, 8
Recurrence relations, backward versus forward, 385–386, 388–391, 425–426
  concept of, 368
  continuity considerations for, 395–402
  for allocation problems, 373–380, 409–410
  for chain problems, 382–385
  for continuous systems, 426–434
  for discrete control problems, 388–391, 411
  for discrete process and infinite number of stages, 419–421
  for discrete solution of continuous process, 425–431
  for estimating state variables, 454–460
  for transmission-line problem, 406–408

Recurrence relations (*continued*)
  with expected values, 451–452
  with iterated dynamic programming, 416–417
Region-limiting strategies, 414–417
Regulator problem, 418–419
Relaxation methods, 253, 292–293
Reliability, as a performance measure, 6–8
  with redundancy, 379–380
Residue, as an error measure, 46
  associated with a pole, 567–569
Residue theorem, 569
Resource variable, 371
Return function, 371n
Reverse-time integrations, of Riccati equations, 444, 446
  to obtain switching surfaces, 501, 534–535
  to stabilize costate solution, 516
Riccati equations, discrete version, 475
  for second-order system, 445–447, 449–450
  generalized, 444
  steady-state, 447–449, 475–476
  used with search techniques, 521–523
Riccati transformation, 522
Ridges, 267–268, 313
Ritz's method, 119
Root locus, 152, 266
Root-square locus, 152n

Saddle point, 31–33, 267–268, 295
Sampled-data systems, spectrum factorization for, 183
  (*see also* Control system design *and* Filter design)
Scaling, column scaling, 219–220
  effect on search, 270–271, 300, 310
  for linear programming, 218–222
  row scaling, 219–220
Scheduling problems (*see* Economics problems)
Search cycle, 349
Search iteration, 349
Search philosophy, 264–265, 345–351
Search sequence notation, 273n
Search techniques, acceleration-step, 309–312
  best-step steepest ascent, 299–300
  bunny-hop, 325–329
  comparison of, 345–350

Search techniques (*continued*)
  conjugate-direction methods, 314–319
  continued partan, 313–314
  continuous gradient, 297–299
  DFP (modified Davidon's method) 320–322
  discrete steepest descent, 299–306
  for functionals, 352, 365–366, 513–515
    generalized Newton-Raphson, 517–519
    gradient and second-variation methods, 515–517
    Riccati equation approaches, 519–523
  general gradient, 296
  geometrical interpretation of, 19–20, 266–272
  global versus local, 322
  iterated linear programming, 239, 249, 351
  Newton, 307–309
  nonsequential, 23, 287–288
    factorial, 288–289
    random (Monte Carlo), 288
  of Fletcher and Reeves, 319–320
  one-dimensional, cubic-convergent search, 274–277
    Fibonacci search, 280–284
    golden-section search, 284–286
    in n-dimensional space, 286–287
    Newton-Raphson search, 272–274
    quadratic-convergent search, 277–280
  partan, 312–314
  relaxation search, 292–296
  Southwell-Synge, 293
  SUMT, 340
  univariate, 292
  with Lagrange multipliers, 341–344, 363
  with penalty coefficients, 338–340
  without derivatives, centroid methods, 329–330
    creeping random search, 329
    direct search, 322
    directed array search, 324–325
    modified Powell's search, 277–280, 314
    moment-rosetta search, 330–331
    pattern search, 322–324
Search test functions, 346–350
Sensitivity, as a constraint, 7–8
  column matrix, 49–50
  considerations in time optimal control, 505–510
  effect of feedback on, 173–183

Sensitivity (*continued*)
  influenced by adaptive loop, 192
  information from dynamic programming, 386
  information from search, 351
  in linear programming, 232–234
  macroscopic versus microscopic, 24–26
  measure of Bode, 48
  of a functional, 123–124
  square matrix, 49–50
  worst case, 175, 363
Sensitivity range, 233–234
Sequence, complete, 119–120
Series approximation, 60–61, 119–121, 411–412, 590
Set, bounded, 424
  closed, 424
  constraint sets, 334, 342
    (*see also* Constraints)
  convex, 200–204
    (*see also* Interval)
sgn function, 489
Simplex, k-dimensional, 205–206
Simplex tableau, check column of, 212–213
  key column of, 211
  key row of, 211
  pivotal entry of, 212
Simplex technique, mechanics of, 209–226
  number of iterations of, 206
  scaling for, 218–222
  theory for, 204–209
  θ rule of, 207
Singular arc, 525–527, 533–534
Singular matrix, 560
Singular point (*see* Stationary point)
Singular solutions of Pontryagin's maximum principle, 525–535, 553
Singular surface, 526, 532–533
Slack variables, for calculus of variations, 96
  for linear programming, 196, 210, 217
  for ordinary min-max problems, 44
Slope function of a field, 107
Smooth hypersurface, 496
Space C_n, 581
Space $C(t_a, t_b)$, 581
Space E_n (*see* Euclidean space)
Space $L_p(t_a, t_b)$, 583–585, 589
Space $l_p(n)$, 583, 588
Space R_n of reals, 581
Spectral matrix, 183

SUBJECT INDEX 611

Spectrum factorization, notation for, 147–149
   process of, 144–150
Stability, in regard to systems, 26–27, 161
   of Riccati equation solution, 444
   (*see also* Two-point bountary-value problems)
Stages, of dynamic programming solutions, 21, 367
State equations, difference type, 240, 386, 391, 410
   differential, 100, 423–424, 481
   a differential-difference case, 4
   (*see also particular problem types*)
State space, 19–20
State variables, 3–5, 188, 367, 402–403, 408–411
State vector, 20, 67, 90, 100, 410, 481
State-increment dynamic programming, 428
Static friction, 53
Stationary point, 31, 35, 266
   of a cubic, 277
   of a quadratic, 274, 275, 278–280, 307–308
Stationary random signals, 573
Steepest ascent, in linear programming, 208, 236
   in nonlinear search, 297–306
Step function, 144
Stochastic linear programming, 254
Stochastic systems, 4
   (*see also* Performance measures *and problem types*)
Straight line, 286
Strip of convergence, 566–570
Structural variables, 228
Subjective criteria (*see* Performance measures)
Suboptimum, 2
Sufficient conditions, based on physical reasoning, 63, 78
   for absolute extrema in the Wiener-Hopf case, 150, 164
   for absolute extrema with dynamic programming, 352, 415, 463
   for local extrema of functions, 33–36
   for strong relative extrema of functionals, 115–118
   for weak relative extrema of functionals, 106, 112–113
   meaning of, 18

Sufficient conditions (*continued*)
   with linear programming, 204, 208
   with nonlinear programming. 334
Surface, planar, 75
   (*see also* Hypersurface)
Surface of revolution, 129
Switch curve, obtained by reverse-time integration, 534–535
   to form a control law, 503–510
Switching center, in a communications problem, 244–246
Switching function, as a function of states, 504–505
   as a function of t, 439
Switching surfaces, 501, 526
Switch-time evaluation, 510–513, 550
Sylvester's theorem, 35n
Symmetric matrix, 557
Synthesis, for control system design, 482
   as part of problem solution, 1
Systems, ability concepts for, 26–27
   a classification of types, 3–5
   (*see also particular system and problem types*)

Taylor's series, as generated by Maclaurin's series, 54
   in n variables, 48, 307, 563
   in one variable, 33, 174
   to linearize differential equations, 517–520
Temperature control, 551–552
Terminal conditions (*see* End-point conditions)
Terminal design task, 248–249
Time delay, effect on filtering, 153
   effect on time-optimal control, 549
   in a control problem, 189
Time invariance, in general, 418–419
   in linear systems, 135
   obtained artificially, 482
Time optimal control, in general, 500–502
   of second-order systems, 131, 502–510, 549–550
   of systems with real eigenvalues, 510–513, 550
   with integral constraint on the control variable, 132
   with sampled-data controllers, 541

SUBJECT INDEX

Tolerance considerations, 24, 40
  in circuit design, 248–251
  (*see also* Sensitivity)
Topographic map, 267–268
Tower-position pair, 405
Trace of a matrix, 561
Trajectory, in control space, 20
  in policy space, 416
  in state space, 20, 403, 416
Transfer function, 79, 137, 141, 502
Transmission-line tower-placement
    problem, 404–408
Transpose of a matrix, 557
Transversality conditions, for end-point
    functionals, 538–539
  for a general control problem, 101–103, 132
  of Pontryagin's maximum principle, 486, 493–498
  scalar cases, 82–87
  state-vector case, 90, 130
Triangle inequality, 581, 583–584
Tuned coupling circuit, 290–291
Two-phase method of linear programming, 218
Two-point boundary-value problems,
    general reference to, 22, 73, 125, 438, 440, 513–523
  stability of solution, 365–366, 514–515
  (*see also* Transversality conditions)

Unbounded solution, in control, 100–101, 534
  in linear programming, 214
  of a differential equation, 487n
Uncertainty interval, 280–283
Unimodal function, 269
Unit vectors, 211
Univariate search, 292
Upper-bounding, 224–226

Valentine's method, 96
Value coefficients, 194
Variance of a system function, 371
Variation, of end-point values (*see*
    Transversality conditions)
  of a function, 65
    admissible variation, 66
    allowable variation, 145
  of a functional, first, 66, 69

Variation (*continued*)
  obtained using Maclaurin's series, 66, 134
  second, 66, 105–106
Vector derivative, of a column matrix, 562
  of an inner product, 562
  of a row matrix, 561–562
Vectors, A-conjugate, 314–320, 361–362
  basis vectors, 557, 589–590
  components or elements of, 555
  direction, 286
  generalized notion of, 582
  gradient, 561
  (*see also* Gradient)
  inner product of, 560
  linear combination of, 557
  nontrivial vector, 561
  orthogonal, 561
  orthogonal component of, 238
  orthogonal (normal) to a hyperplane, 236, 267, 496
  orthonormal, 561
  outer product of, 561
  projection of, 561
  row and column, 555
  spanning, 317
  unit vectors, 211
Vertex, at the optimum, 204
  finding initial one, 205, 217
  of a polyhedron, 203
  of a simplex, 205
Viscous friction, 79–81
Volume, of surface of revolution, 129

Weierstrass condition, 115–118
Weierstrass function, 117–118
Weighting factor, as related to a Lagrange multiplier, 20
  in a pulse-shape problem, 142
  subjective selection of, 6–7
Weighting function, discrete, 241
White noise, 453
Wiener filters, 183
Wiener-Hopf equation, 146
Wiener-Hopf spectrum factorization, a
    general problem and solution, 144–150
Worst-case design, 233, 248–251
Worst-case sensitivity function, 175

Zero-memory system, 4–5

A CATALOG OF SELECTED
DOVER BOOKS
IN SCIENCE AND MATHEMATICS

A CATALOG OF SELECTED
DOVER BOOKS
IN SCIENCE AND MATHEMATICS

QUALITATIVE THEORY OF DIFFERENTIAL EQUATIONS, V.V. Nemytskii and V.V. Stepanov. Classic graduate-level text by two prominent Soviet mathematicians covers classical differential equations as well as topological dynamics and ergodic theory. Bibliographies. 523pp. 5⅜ × 8½. 65954-2 Pa. $10.95

MATRICES AND LINEAR ALGEBRA, Hans Schneider and George Phillip Barker. Basic textbook covers theory of matrices and its applications to systems of linear equations and related topics such as determinants, eigenvalues and differential equations. Numerous exercises. 432pp. 5⅜ × 8½. 66014-1 Pa. $9.95

QUANTUM THEORY, David Bohm. This advanced undergraduate-level text presents the quantum theory in terms of qualitative and imaginative concepts, followed by specific applications worked out in mathematical detail. Preface. Index. 655pp. 5⅜ × 8½. 65969-0 Pa. $13.95

ATOMIC PHYSICS (8th edition), Max Born. Nobel laureate's lucid treatment of kinetic theory of gases, elementary particles, nuclear atom, wave-corpuscles, atomic structure and spectral lines, much more. Over 40 appendices, bibliography. 495pp. 5⅜ × 8½. 65984-4 Pa. $12.95

ELECTRONIC STRUCTURE AND THE PROPERTIES OF SOLIDS: The Physics of the Chemical Bond, Walter A. Harrison. Innovative text offers basic understanding of the electronic structure of covalent and ionic solids, simple metals, transition metals and their compounds. Problems. 1980 edition. 582pp. 6⅛ × 9¼. 66021-4 Pa. $15.95

BOUNDARY VALUE PROBLEMS OF HEAT CONDUCTION, M. Necati Özisik. Systematic, comprehensive treatment of modern mathematical methods of solving problems in heat conduction and diffusion. Numerous examples and problems. Selected references. Appendices. 505pp. 5⅜ × 8½. 65990-9 Pa. $11.95

A SHORT HISTORY OF CHEMISTRY (3rd edition), J.R. Partington. Classic exposition explores origins of chemistry, alchemy, early medical chemistry, nature of atmosphere, theory of valency, laws and structure of atomic theory, much more. 428pp. 5⅜ × 8½. (Available in U.S. only) 65977-1 Pa. $10.95

A HISTORY OF ASTRONOMY, A. Pannekoek. Well-balanced, carefully reasoned study covers such topics as Ptolemaic theory, work of Copernicus, Kepler, Newton, Eddington's work on stars, much more. Illustrated. References. 521pp. 5⅜ × 8½. 65994-1 Pa. $12.95

PRINCIPLES OF METEOROLOGICAL ANALYSIS, Walter J. Saucier. Highly respected, abundantly illustrated classic reviews atmospheric variables, hydrostatics, static stability, various analyses (scalar, cross-section, isobaric, isentropic, more). For intermediate meteorology students. 454pp. 6⅛ × 9¼. 65979-8 Pa. $14.95

CATALOG OF DOVER BOOKS

ROTARY-WING AERODYNAMICS, W.Z. Stepniewski. Clear, concise text covers aerodynamic phenomena of the rotor and offers guidelines for helicopter performance evaluation. Originally prepared for NASA. 537 figures. 640pp. 6⅛ × 9¼. 64647-5 Pa. $15.95

DIFFERENTIAL GEOMETRY, Heinrich W. Guggenheimer. Local differential geometry as an application of advanced calculus and linear algebra. Curvature, transformation groups, surfaces, more. Exercises. 62 figures. 378pp. 5⅜ × 8½. 63433-7 Pa. $8.95

INTRODUCTION TO SPACE DYNAMICS, William Tyrrell Thomson. Comprehensive, classic introduction to space-flight engineering for advanced undergraduate and graduate students. Includes vector algebra, kinematics, transformation of coordinates. Bibliography. Index. 352pp. 5⅜ × 8½. 65113-4 Pa. $8.95

A SURVEY OF MINIMAL SURFACES, Robert Osserman. Up-to-date, in-depth discussion of the field for advanced students. Corrected and enlarged edition covers new developments. Includes numerous problems. 192pp. 5⅜ × 8½. 64998-9 Pa. $8.95

ANALYTICAL MECHANICS OF GEARS, Earle Buckingham. Indispensable reference for modern gear manufacture covers conjugate gear-tooth action, gear-tooth profiles of various gears, many other topics. 263 figures. 102 tables. 546pp. 5⅜ × 8½. 65712-4 Pa. $14.95

SET THEORY AND LOGIC, Robert R. Stoll. Lucid introduction to unified theory of mathematical concepts. Set theory and logic seen as tools for conceptual understanding of real number system. 496pp. 5⅜ × 8¼. 63829-4 Pa. $10.95

A HISTORY OF MECHANICS, René Dugas. Monumental study of mechanical principles from antiquity to quantum mechanics. Contributions of ancient Greeks, Galileo, Leonardo, Kepler, Lagrange, many others. 671pp. 5⅜ × 8½. 65632-2 Pa. $14.95

FAMOUS PROBLEMS OF GEOMETRY AND HOW TO SOLVE THEM, Benjamin Bold. Squaring the circle, trisecting the angle, duplicating the cube: learn their history, why they are impossible to solve, then solve them yourself. 128pp. 5⅜ × 8½. 24297-8 Pa. $4.95

MECHANICAL VIBRATIONS, J.P. Den Hartog. Classic textbook offers lucid explanations and illustrative models, applying theories of vibrations to a variety of practical industrial engineering problems. Numerous figures. 233 problems, solutions. Appendix. Index. Preface. 436pp. 5⅜ × 8½. 64785-4 Pa. $10.95

CURVATURE AND HOMOLOGY, Samuel I. Goldberg. Thorough treatment of specialized branch of differential geometry. Covers Riemannian manifolds, topology of differentiable manifolds, compact Lie groups, other topics. Exercises. 315pp. 5⅜ × 8½. 64314-X Pa. $8.95

HISTORY OF STRENGTH OF MATERIALS, Stephen P. Timoshenko. Excellent historical survey of the strength of materials with many references to the theories of elasticity and structure. 245 figures. 452pp. 5⅜ × 8½. 61187-6 Pa. $11.95

CATALOG OF DOVER BOOKS

TENSOR CALCULUS, J.L. Synge and A. Schild. Widely used introductory text covers spaces and tensors, basic operations in Riemannian space, non-Riemannian spaces, etc. 324pp. 5⅜ × 8¼. 63612-7 Pa. $8.95

A CONCISE HISTORY OF MATHEMATICS, Dirk J. Struik. The best brief history of mathematics. Stresses origins and covers every major figure from ancient Near East to 19th century. 41 illustrations. 195pp. 5⅜ × 8½. 60255-9 Pa. $7.95

A SHORT ACCOUNT OF THE HISTORY OF MATHEMATICS, W.W. Rouse Ball. One of clearest, most authoritative surveys from the Egyptians and Phoenicians through 19th-century figures such as Grassman, Galois, Riemann. Fourth edition. 522pp. 5⅜ × 8½. 20630-0 Pa. $10.95

HISTORY OF MATHEMATICS, David E. Smith. Nontechnical survey from ancient Greece and Orient to late 19th century; evolution of arithmetic, geometry, trigonometry, calculating devices, algebra, the calculus. 362 illustrations. 1,355pp. 5⅜ × 8½. 20429-4, 20430-8 Pa., Two-vol. set $23.90

THE GEOMETRY OF RENÉ DESCARTES, René Descartes. The great work founded analytical geometry. Original French text, Descartes' own diagrams, together with definitive Smith-Latham translation. 244pp. 5⅜ × 8½. 60068-8 Pa. $6.95

THE ORIGINS OF THE INFINITESIMAL CALCULUS, Margaret E. Baron. Only fully detailed and documented account of crucial discipline: origins; development by Galileo, Kepler, Cavalieri; contributions of Newton, Leibniz, more. 304pp. 5⅜ × 8½. (Available in U.S. and Canada only) 65371-4 Pa. $9.95

THE HISTORY OF THE CALCULUS AND ITS CONCEPTUAL DEVELOPMENT, Carl B. Boyer. Origins in antiquity, medieval contributions, work of Newton, Leibniz, rigorous formulation. Treatment is verbal. 346pp. 5⅜ × 8½. 60509-4 Pa. $8.95

THE THIRTEEN BOOKS OF EUCLID'S ELEMENTS, translated with introduction and commentary by Sir Thomas L. Heath. Definitive edition. Textual and linguistic notes, mathematical analysis. 2,500 years of critical commentary. Not abridged. 1,414pp. 5⅜ × 8½. 60088-2, 60089-0, 60090-4 Pa., Three-vol. set $29.85

GAMES AND DECISIONS: Introduction and Critical Survey, R. Duncan Luce and Howard Raiffa. Superb nontechnical introduction to game theory, primarily applied to social sciences. Utility theory, zero-sum games, n-person games, decision-making, much more. Bibliography. 509pp. 5⅜ × 8½. 65943-7 Pa. $12.95

THE HISTORICAL ROOTS OF ELEMENTARY MATHEMATICS, Lucas N.H. Bunt, Phillip S. Jones, and Jack D. Bedient. Fundamental underpinnings of modern arithmetic, algebra, geometry and number systems derived from ancient civilizations. 320pp. 5⅜ × 8½. 25563-8 Pa. $8.95

CALCULUS REFRESHER FOR TECHNICAL PEOPLE, A. Albert Klaf. Covers important aspects of integral and differential calculus via 756 questions. 566 problems, most answered. 431pp. 5⅜ × 8½. 20370-0 Pa. $8.95

CATALOG OF DOVER BOOKS

CHALLENGING MATHEMATICAL PROBLEMS WITH ELEMENTARY SOLUTIONS, A.M. Yaglom and I.M. Yaglom. Over 170 challenging problems on probability theory, combinatorial analysis, points and lines, topology, convex polygons, many other topics. Solutions. Total of 445pp. 5⅜ × 8½. Two-vol. set.
Vol. I 65536-9 Pa. $7.95
Vol. II 65537-7 Pa. $6.95

FIFTY CHALLENGING PROBLEMS IN PROBABILITY WITH SOLUTIONS, Frederick Mosteller. Remarkable puzzlers, graded in difficulty, illustrate elementary and advanced aspects of probability. Detailed solutions. 88pp. 5⅜ × 8½.
65355-2 Pa. $4.95

EXPERIMENTS IN TOPOLOGY, Stephen Barr. Classic, lively explanation of one of the byways of mathematics. Klein bottles, Moebius strips, projective planes, map coloring, problem of the Koenigsberg bridges, much more, described with clarity and wit. 43 figures. 210pp. 5⅜ × 8½. 25933-1 Pa. $5.95

RELATIVITY IN ILLUSTRATIONS, Jacob T. Schwartz. Clear nontechnical treatment makes relativity more accessible than ever before. Over 60 drawings illustrate concepts more clearly than text alone. Only high school geometry needed. Bibliography. 128pp. 6⅛ × 9¼. 25965-X Pa. $6.95

AN INTRODUCTION TO ORDINARY DIFFERENTIAL EQUATIONS, Earl A. Coddington. A thorough and systematic first course in elementary differential equations for undergraduates in mathematics and science, with many exercises and problems (with answers). Index. 304pp. 5⅜ × 8½. 65942-9 Pa. $8.95

FOURIER SERIES AND ORTHOGONAL FUNCTIONS, Harry F. Davis. An incisive text combining theory and practical example to introduce Fourier series, orthogonal functions and applications of the Fourier method to boundary-value problems. 570 exercises. Answers and notes. 416pp. 5⅜ × 8½. 65973-9 Pa. $9.95

THE THEORY OF BRANCHING PROCESSES, Theodore E. Harris. First systematic, comprehensive treatment of branching (i.e. multiplicative) processes and their applications. Galton-Watson model, Markov branching processes, electron-photon cascade, many other topics. Rigorous proofs. Bibliography. 240pp. 5⅜ × 8½. 65952-6 Pa. $6.95

AN INTRODUCTION TO ALGEBRAIC STRUCTURES, Joseph Landin. Superb self-contained text covers "abstract algebra": sets and numbers, theory of groups, theory of rings, much more. Numerous well-chosen examples, exercises. 247pp. 5⅜ × 8½. 65940-2 Pa. $7.95

Prices subject to change without notice.
Available at your book dealer or write for free Mathematics and Science Catalog to Dept. GI, Dover Publications, Inc., 31 East 2nd St., Mineola, N.Y. 11501. Dover publishes more than 175 books each year on science, elementary and advanced mathematics, biology, music, art, literature, history, social sciences and other areas.